Advanced Metallization Conference 2008 (AMC 2008)

The Advanced Metallization Conference gratefully acknowledges the continued support of Applied Materials for the printing of these conference proceedings.

think it. apply it.™
THIS IS NOT A CHAMBER.
IT'S THE MOST COMPREHENSIVE PORTFOLIO OF FAB RESOURCES IN THE WORLD.
It tackles the toughest challenges in the industry. It pushes the limits of physics in volume production. It extends the life of your tools with new process and hardware innovations. It's sub-30nm patterning and atomic-level deposition. It's the end of leakage. It's every advantage Applied Materials gives you, across technologies and across the fab.

Advanced Metallization Conference 2008 (AMC 2008)

Advanced Metallization Conference 2008 (AMC 2008)

Proceedings of the Conference held September 23–25, 2008, San Diego, California, U.S.A. and October 8–10, 2008, at The University of Tokyo, Tokyo, Japan. This combined Conference is affiliated with the Materials Research Society and sponsored by the Department of Continuing Education in Engineering, University Extension, University of California at Berkeley, California, U.S.A.

EDITORS:

Mehul Naik
Applied Materials Inc.
Santa Clara, California, U.S.A.

Roey Shaviv
Novellus Systems, Inc.
San Jose, California, U.S.A.

Takashi Yoda
Toshiba Corporation
Yokohama, Japan

Kazuyoshi Ueno
Shibaura Institute of Technology
Tokyo, Japan

MRS

Materials Research Society
Warrendale, Pennsylvania

Single article reprints from this publication are available through
University Microfilms Inc., 300 North Zeeb Road, Ann Arbor, Michigan 48106

CODEN: MRSPDH

Published by:

Materials Research Society
506 Keystone Drive
Warrendale, PA 15086
Telephone (724) 779-3003
Fax (724) 779-8313
Web site: http://www.mrs.org/

Manufactured in the United States of America

CONTENTS

METALLIZATION

MATERIALS AND CHARACTERIZATION

3D INTEGRATION AND PACKAGING

LOW-k MATERIALS

RELIABILITY AND YIELD

PREFACE

The 25th session of the Advanced Metallization Conference (AMC 2008) took place at the Del Mar Hilton Hotel in San Diego, California, on September 23–25, 2008. The conference Co-Chairs were Dr. Roey Shaviv (Novellus Systems Inc.) and Dr. Mehul Naik (Applied Materials Inc.). The Executive Committee for AMC 2008 was responsible for guiding the technical content of the conference and is itself comprised of 24 internationally distinguished researchers from the field of interconnect technology. In Japan, a parallel meeting, the Advanced Metallization Conference 18th Asian session (ADMETA 2008), was held at the Sanjo Conference Hall, The University of Tokyo in Tokyo, Japan on October 8–10, 2008. The Chairman for ADMETA 2008 was Dr. Takashi Yoda (Toshiba Corp.). The meeting was facilitated through the efforts of the conference Executive Committee and members. ADMETA 2008 was co-sponsored by the Japan Society of Applied Physics (JSAP), Silicon Technology Division, and the AMC is affiliated with the Materials Research Society.

AMC is a single-session workshop-atmosphere conference featuring the latest R&D and manufacturing results, as well as "real-world" integration and reliability results on the application of metallization and related technologies for advanced IC devices. AMC 2008 hosted scientists and engineers from all over the world, who came to listen and present state-of-the-art work in the field of ULSI interconnect technology including metallization, dielectrics, integration, packaging, design, and vertical integration. The conference included 14 invited talks, 31 regular oral presentations, and 51 poster submissions. The Executive Committee carefully reviewed all conference submissions and the presented papers were well received by conference participants. A highlight of the conference was the Keynote speech by Dr. Gary Patton, Vice President, Semiconductor Research & Development Center, IBM Systems and Technology Group, IBM Inc. Dr. Patton shared with the conference his perspectives on the merit of alliance-based R&D, stating that semiconductor process R&D innovation and collaboration has never been more critical than it is today. The development of leading-edge silicon technology requires long-term investments and collaboration in fundamental research to solve the significant challenges of the future technology nodes at an affordable cost. Thus, the benefits of a collaborative R&D model and the details of the future direction and challenges of semiconductor process R&D were presented and discussed.

The ADMETA meeting included two Keynote presentations including a business strategy and challenges for interconnect technology by Mr. Shozo Saito (Toshiba Corp.), and an overview of nanotechnology development in Albany given by Prof. Makoto Hirayama (SUNY, Albany). In addition to these talks, there were 6 invited talks, 29 regular oral presentations, and 28 poster submissions. Topical sessions at ADMETA included 3D Si and Cu deposition, CMP and characterization, low-k process, modeling/simulation, barrier/seed and contact, reliability, and integration.

Prof. Robert S. Blewer from the University of New Mexico at Albuquerque facilitated a short course titled "Micro and Nano-Scale Interconnect Techniques" that preceded AMC 2008. Immediately following the AMC conference in San Diego, on September 26, SEMATECH conducted a workshop on "Manufacturing and Reliability Challenges for 3D Interconnects using Through-Silicon Vias (TSVs)." Prior to ADMETA 2008, a full-day short course summarizing the current status of advanced ULSI metallization and interconnects was given.

The AMC 2008 marked the 25th anniversary of the Advanced Metallization Conference. The Executive Committee marked this landmark with a special award presented to Professor Robert S. Blewer (Sandia National Laboratories and the University of New Mexico at Albuquerque) and Dr. Rajiv V. Joshi (IBM Inc.) recognizing their 25 years of outstanding service as the founding fathers of the Advanced Metallization Conference.

AMC 2009 is scheduled to be held October 13–15, 2009 in Baltimore, Maryland and will be Co-Chaired by Dr. Paul Besser (AMD Inc.) and Stefan E. Schulz (Chemnitz University of Technology). ADMETA 2009 is scheduled to be held October 19–21, 2009 at the Sanjo Conference Hall, The University of Tokyo in Tokyo, Japan. For more information on AMC 2009 please contact Mrs. Melody Wells at the University of California by telephone at 510-642-4151; fax 510-642-6027; email amc@unex.berkeley.edu; or visit the URL http://www.unex.berkeley.edu/eng/metal/ For more information on ADMETA, please contact the conference secretary email at jimukyoku@admeta.org; or visit the URL http://www.admeta.org.

Mehul Naik
Roey Shaviv
Takashi Yoda
Kazuyoshi Ueno

February 2009

ACKNOWLEDGMENTS

The continued and growing success of both the Advanced Metallization Conference and ADMETA is due to the combined efforts of many individuals and organizations from the global semiconductor community. Of particular note are the contributions of keynote, invited, and oral speakers, poster presenters, short course instructors and panelists who bring the technology to the forefront at this conference. The editors would especially like to thank the active members of the Executive Committees of both the North American and Asian conferences, listed below, for their work in soliciting invited speakers, reviewing abstracts and full papers and serving as session chairs. AMC 2008 received generous support from Applied Materials Inc. The editors would also like to thank Melody Wells of the University of California at Berkeley, and Maki Yoshida, the ADMETA secretary, for their valuable help and advice throughout year.

2008 Advanced Metallization Conference Committee

Mehul Naik, Applied Materials Inc. (Co-Chair)
Roey Shaviv, Novellus Systems Inc. (Co-Chair)
Paul Besser, Advanced Micro Devices Inc.
Robert S. Blewer, Sandia National Laboratories
Timothy S. Cale, Process Evolution Ltd.
Reinhold Dauskardt, Stanford University
John Ekerdt, The University of Texas at Austin
Daniel Edelstein, IBM
Cindy Goldberg, Freescale France
Paul-Henri Haumesser, CEA-Leti
Chia-Hong Jan, Intel Corporation
Rajiv Joshi, IBM Research Division
Heinrich Körner, Infineon Technologies
Andrew McKerrow, Novellus Systems Inc.
Michael Mills, The Dow Chemical Company
Takayuki Ohba, The University of Tokyo
Peter Ramm, Fraunhofer Institute FhG-IZM
Steve Russell, Micron Technology Inc.
Stefan E. Schulz, Chemnitz University of Technology
Yosi Shacham-Diamand, Tel-Aviv University
Zsolt Tökei, IMEC
Joaquin Torres, STMicroelectronics
Shi-Qing Wang, On Wafer Technologies
James J. Watkins, University of Massachusetts
Melody Wells, UC Berkeley Extension

2008 ADMETA Executive Committee

Takashi Yoda, Toshiba (Chair)
Yukihiro Shimogaki, The University of Tokyo (Vice-Chair)
Takuya Fukuda, Future Advanced Technology Research Laboratory
Eiichi Kondoh, University of Yamanashi
Takao Kinoshita, Sharp
Shuji Kodaira, ULVAC
Noriyoshi Shimizu, Fujitsu Laboratories
Nobuhiro Shimoyama, NTT Microsystem Integration Laboratories
Kazuyoshi Ueno, Shibaura Institute of Technology

2008 Advisory Members
Yasuhiro Horiike, NIMS
Hiroshi Komiyama, The University of Tokyo
Kazuo Tsubouchi, Tohoku University
Masanori Murakami, Ritsumeikan
Nobuyoshi Kobayashi, ASM Japan
Yukio Yasuda, Kochi University of Technology
Takamaro Kikkawa, Hiroshima University
Takayuki Ohba, The University of Tokyo
Shigeaki Zaima, Nagoya University
Tomohiro Ohta, The University of Tokyo
Kazuya Masu, Tokyo Institute of Technology
Manabu Tsujimura, Ebara
Akihiko Ohsaki, Renesas Technology
Shoso Shingubara, Kansai University

2008 ADMETA Members
Kazuhide Abe, Oki Electric Industry
Koichi Abe, Hitachi Chemical
Nobuo Aoi, Panasonic
Kouyu Asai, Renesas Technology
Shunichi Hizume, Novellus Systems Japan
Akira Hoshino, Canon ANELVA
Fuminori Ito, NEC Electronics
Toru Imori, Nippon Mining & Metals
Takanobu Kaitsuka, Tokyo Electron
Yoshifumi Kawamoto, CASMAT
Hisao Kawasaki, Mitsubishi Heavy Industries
Junichi Koike, Tohoku University
Terukazu Kokubo, JSR
Seiichi Kondo, Selete
Hideaki Machida, The University of Tokyo
Osamu Nakatsuka, Nagoya University
Takeshi Nogami, IBM
Masashi Ohtsuka, Toshiba
Nobuo Owada, Hitachi Kokusai Electric
Shunji Sasabe, Dainippon Screen Mfg.
Tadashi Shimazu, Mitsubishi Heavy Industries
Hiroshi Shinriki, ASM Japan
Masakazu Sugiyama, The University of Tokyo
Shingo Takahashi, Sony
Mayumi B. Takeyama, Kitami Institute of Technology
Tsuyoshi Tamaru, Hitachi
Yasuaki Tsuchiya, NEC Electronics
Yasutake Uchida, Teikyo University
Yutaka Wada, Ebara
Daniel L. Diehl, Applied Materials Japan
Lih J. Chen, National Tsing Hua University
Young-Chang Joo, Seoul National University
Rakesh Kumar, A*STAR Institute of Microelectronics
Gil Heyun Choi, Samsung Electronics
Hyungjun Kim, Pohang University of Science and Technology
Winston S. Shue, Taiwan Semiconductor Manufacturing Company
Jin-Goo Park, Hanyang University
Chih-Chien Liu, United Microelectronics

MATERIALS RESEARCH SOCIETY CONFERENCE PROCEEDINGS

Tungsten and Other Refractory Metals for VLSI Applications, Robert S. Blewer, 1986; ISSN 0886-7860; ISBN 0-931837-32-4

Tungsten and Other Refractory Metals for VLSI Applications II, Eliot K. Broadbent, 1987; ISSN 0886-7860; ISBN 0-931837-66-9

Tungsten and Other Refractory Metals for VLSI Applications III, Victor A. Wells, 1988; ISSN 0886-7860; ISBN 0-931837-84-7

Tungsten and Other Refractory Metals for VLSI Applications IV, Robert S. Blewer, Carol M. McConica, 1989; ISSN 0886-7860; ISBN 0-931837-98-7

Tungsten and Other Advanced Metals for VLSI/ULSI Applications V, S. Simon Wong, Seijiro Furukawa, 1990; ISSN 1048-0854; ISBN 1-55899-068-2

Tungsten and Other Advanced Metals for ULSI Applications in 1990, Gregory C. Smith, Roc Blumenthal, 1991; ISSN 1048-0854; ISBN 1-55899-112-3

Advanced Metallization for ULSI Applications, Virendra V.S. Rana, Rajiv V. Joshi, Iwao Ohdomari, 1992; ISSN 1048-0854; ISBN 1-55899-152-2

Advanced Metallization for ULSI Applications 1992, Timothy S. Cale, Fabio S. Pintchovski, 1993; ISSN 1048-0854; ISBN 1-55899-192-1

Advanced Metallization for ULSI Applications in 1993, David P. Favreau, Yosi Shacham-Diamand, Yasuhiro Horiike, 1994; ISSN 1048-0854; ISBN 1-55899-235-9

Advanced Metallization for ULSI Applications in 1994, Roc Blumenthal, Guido Janssen, 1995; ISSN 1048-0854; ISBN 1-55899-279-0

Advanced Metallization and Interconnect Systems for ULSI Applications in 1995, Russell C. Ellwanger, Shi-Qing Wang 1996; ISSN 1048-0854; ISBN 1-55899-341-X

Advanced Metallization and Interconnect Systems for ULSI Applications in 1996, Robert Havemann, John Schmitz, Hiroshi Komiyama, Kazuo Tsubouchi, 1997; ISSN 1048-0854; ISBN 1-55899-385-1

Advanced Metallization and Interconnect Systems for ULSI Applications in 1997, Robin Cheung, Jeffrey Klein, Kazuo Tsubouchi, Masanori Murakami, Nobuyoshi Kobayashi, 1998; ISSN 1048-0854; ISBN 1-55899-412-2

Advanced Metallization Conference in 1998 (AMC 1998), Gurtej S. Sandhu, Heinrich Koerner, M. Murakami, Y. Yasuda, N. Kobayashi, 1999; ISSN 1048-0854; ISBN 1-55899-484-X

Advanced Metallization Conference 1999 (AMC 1999), Mihal E. Gross, Thomas Gessner, N. Kobayashi, Y. Yasuda, 1999; ISSN 1048-0854; ISBN 1-55899-539-0

Advanced Metallization Conference 2000 (AMC 2000), Dan Edelstein, Girish Dixit, Yukio Yasuda, Takayuki Ohba, 2001; ISSN 1048-0854; ISBN 1-55899-574-9

MATERIALS RESEARCH SOCIETY CONFERENCE PROCEEDINGS

Advanced Metallization Conference 2001 (AMC 2001), Andrew J. McKerrow, Yosi Shacham-Diamand, Shigeaki Zaima, Takayuki Ohba, 2002; ISSN 1048-0854; ISBN 1-55899-670-2

Advanced Metallization Conference 2002 (AMC 2002), Bradley M. Melnick, Timothy S. Cale, Shigeaki Zaima, Tomohiro Ohta, 2003; ISSN 1048-0854; ISBN 1-55899-719-9

Advanced Metallization Conference 2003 (AMC 2003), Gary W. Ray, Tom Smy, Tomohiro Ohta, Manabu Tsujimura, 2004; ISSN 1048-0854; ISBN 1-55899-757-1

Advanced Metallization Conference 2004 (AMC 2004), Darrell Erb, Peter Ramm, Kazuya Masu, Akihiko Osaki, 2005; ISSN 1048-0854; ISBN 1-55899-814-4

Advanced Metallization Conference 2005 (AMC 2005), Sywert H. Brongersma, Thomas C. Taylor, Manabu Tsujimura, Kazuya Masu, 2006; ISSN 1048-0854; ISBN 1-55899-865-9

Advanced Metallization Conference 2006 (AMC 2006), Stephen W. Russell, Michael E. Mills, Akihiko Osaki, Takashi Yoda, 2007; ISSN 1048-0854; ISBN 978-1-55899-947-3

Advanced Metallization Conference 2007 (AMC 2007), Andrew J. McKerrow, Yosi Shacham-Diamand, Shoso Shingubara, Yukihiro Shimogaki, 2008; ISSN 1048-0854; ISBN 978-1-55899-992-3

Advanced Metallization Conference 2008 (AMC 2008), Mehul Naik, Roey Shaviv, Takashi Yoda, Kazuyoshi Ueno, 2009; ISSN 1048-0854; ISBN 978-1-60511-125-4

Keynote Presentation

Conference Proceedings AMC XXIV © 2009 Materials Research Society

Leading-Edge Silicon Technology through Innovation & Collaboration

Gary L. Patton
IBM Systems and Technology Group, Semiconductor Research & Development Center, 2070 Route 52, 42F, Hopewell Junciton, NY, 12533

ABSTRACT

Never has semiconductor process R&D innovation and collaboration been more critical than it is today. The development of leading-edge silicon technology requires long-term investments and collaboration in fundamental research to solve the significant challenges of the future technology nodes at an affordable cost. This talk will describe the benefits of a collaborative R&D model and detail the future direction and challenges of semiconductor process R&D.

INTRODUCTION

Chip technology has had a profound impact on the world, entering almost all facets of modern life. The catalyst for this has been the exponential growth in computing power – roughly fifteen orders of magnitude over the past 100 years. The driving force for this change has been economics – smaller features results in better performance and cost/function, which in term leads to more applications and larger markets. Those economics and the requirements for technology scaling have both changed in a fundamental way over the past decade causing many companies to go "fabless" – exiting semiconductor process development and manufacturing, and others to enter into semiconductor process R&D collaborations with companies who they may have previously viewed as competitors – a kind of pre-competitive cooperation – to reduce their R&D costs. Never has semiconductor process R&D innovation been more critical than it is today. This requires long-term investments in fundamental research and collaboration to solve the significant challenges of the future technology nodes at an affordable cost.

DISCUSSION

Over the past fourty years, escalating process R&D costs have far outpaced industry revenue growth. This trend has accelerated over the past decade. The revenue growth rate has declined - approximately 4% CGR since 1995 compared to roughly 14% in the prior fifteen years, while the cost of development has been growing at an exponential rate driven by the challenges and cost of lithography and the end of traditional scaling.

For decades, developing the next technology was achieved by simply scaling all of the critical parameters (Voltage, $t_{ox,}$ W_g, L_g, etc.). The result was higher density and speed at a constant power density. At around the 90nm node, the gate oxide thickness reached a thickness of only a few atomic layers and traditional scaling could no longer continue (Fig. 1). Scaling is still important for driving down cost, but scaling no longer drove performance and resulted in increasing power density. Innovations in integrated devices, process technology and materials have become the key to achieving higher performance at acceptable power for future technology nodes (Fig. 2).

Collaboration brings together greater R&D resources and talent, often synergistic skills, results in shared learning between companies, creates opportunities to extend the R&D collaborations into the design enablement and design IP space (another area of growing R&D expense), and provides manufacturing sourcing flexibility. The semiconductor process innovation roadmap have three major thrusts: continuing to extend existing silicon technology, subsystem integration, and technologies beyond silicon FETs. Continuing existing silicon technology requires a balanced approach of innovations in device, interconnect, and patterning technology (Fig. 3). From the 90nm technology node to 22nm and beyond, FEOL innovations (Fig. 4) such as strained technology [1,2,3,4], UT-SOI/FinFET [5,6], metal gate high-k [7,8,9], and carton nanotube [10,11], interconnect innovations (Fig. 5) such as copper wiring, ultra-low k [12], air-gap [13], 3D stacking [14], and lithography innovations [15] (Fig. 6) are the driving force for Si technology scaling.

This talk will discuss the benefits of a collaborative R&D model and detail the future direction and challenges of semiconductor process R&D.

REFERENCES

[1] T. Ghani, et al., IEDM Tech. Dig., p. 978, 2003.

[2] H. S. Yang et al., IEDM Tech. Dig., p.1075, 2004.

[3] C. H. Chen et al., Symp. VLSI Tech Dig., p. 56, 2004.

[4] Z. Ren et al., Symp. VLSI Tech. Dig., p.172, 2008.

[5] A. Majumdar et al., IEEE Elec. Dev. Lett., Vol. 29, No. 5, May 2008.

[6] M. Guillorn, et al., Symp. VLSI Tech. Dig., p.12. 2008.

[7] V. Narayanan et al., Symp. VLSI Tech. Dig., p.224, 2006.

[8] M Chudzik et al., Symp. VLSI Tech. Dig., p.194, 2007.

[9] X. Chen et al., Symp. VLSI Tech. Dig., p.88, 2008.

[10] S. J. Tans et al., Nature (London), Vol. 393, p.49-52, 1998.

[11] Z. Chen et al., Science 24 Vol. 311. no. 5768, p. 1735, March 2006.

[12] S. Sankaran et al., IEDM Tech. Dig., p. 1, 2006.

[13] S. Nitta et al., Interconnect Technology Conference, p. 191, 2008.

[14] A. W. Topol et al., Proc. Elec. Components and Technology Conference, p. 931, 2004.

[15] A. Rosenbluth et al., Proc. SPIE, 6520, p. 65200P-1, 2007.

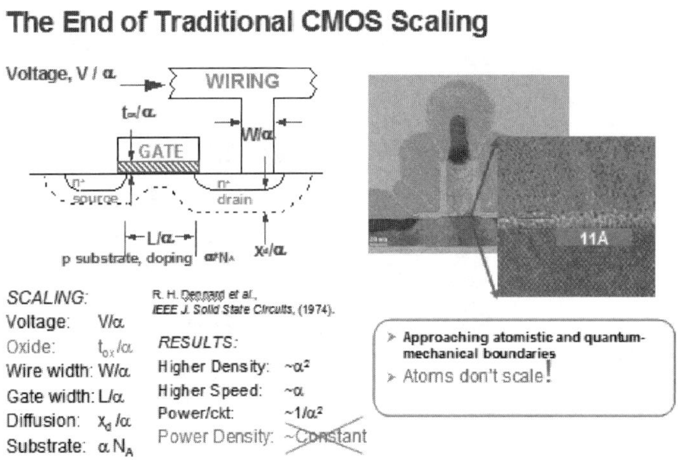

Fig. 1. End of Traditional CMOS Scaling.

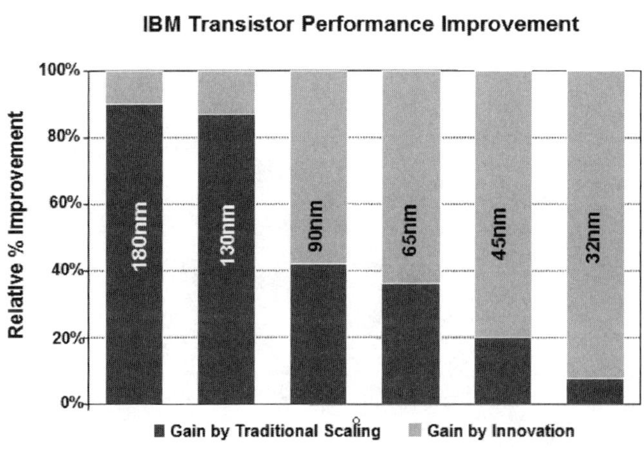

Fig. 2 Transistor performance improvement by traditional scaling and innovation.

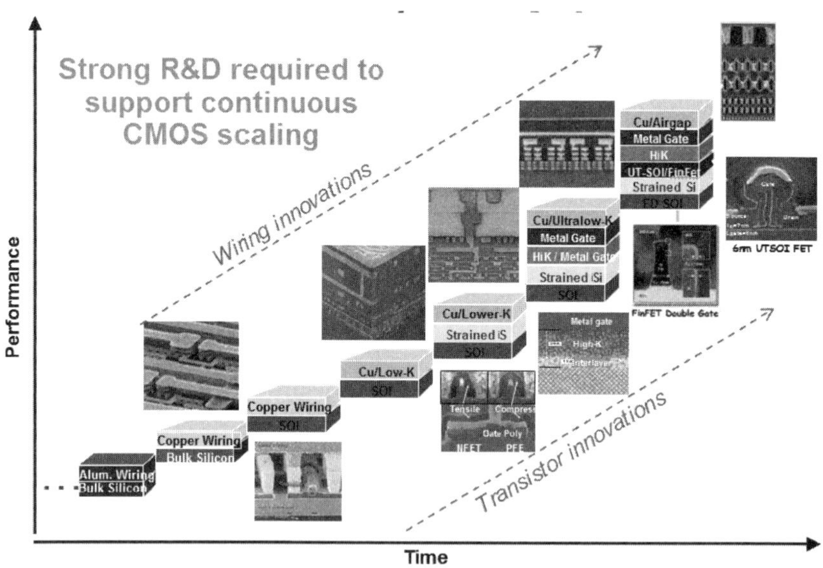

Fig. 3 Semiconductor Roadmap. Scaling by innovation.

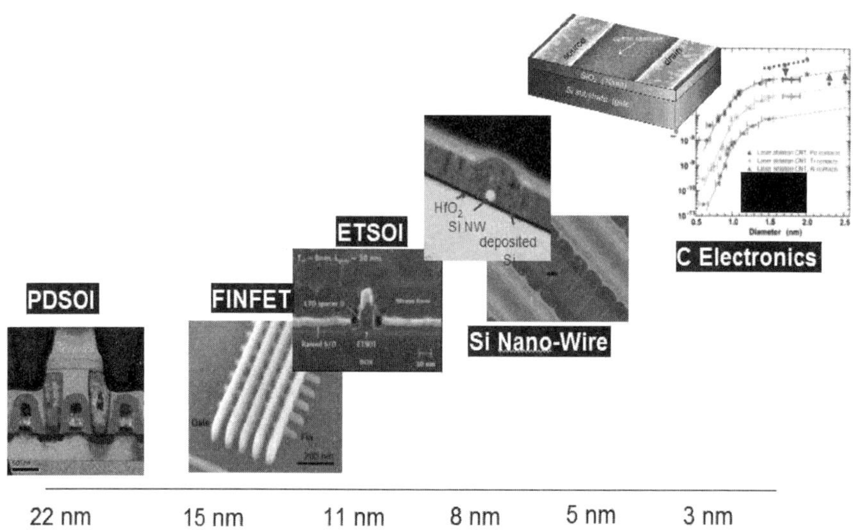

Fig. 4 FEOL roadmap at 22nm technology node and beyond.

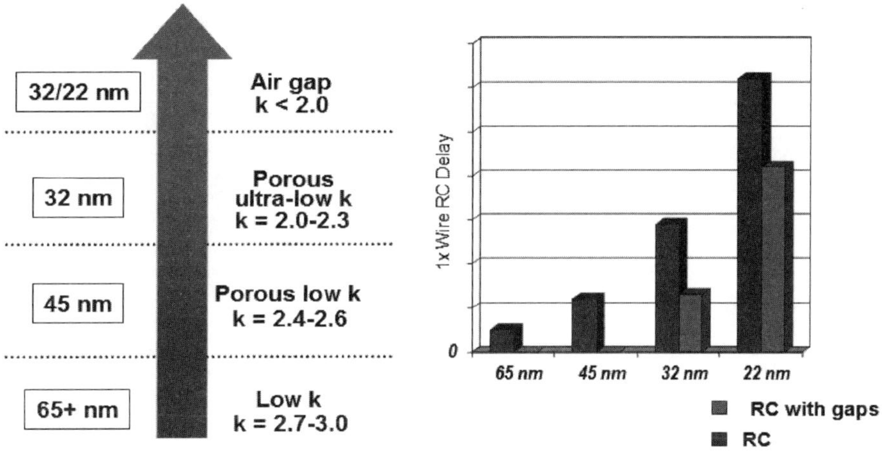

Fig. 5 BEOL roadmap

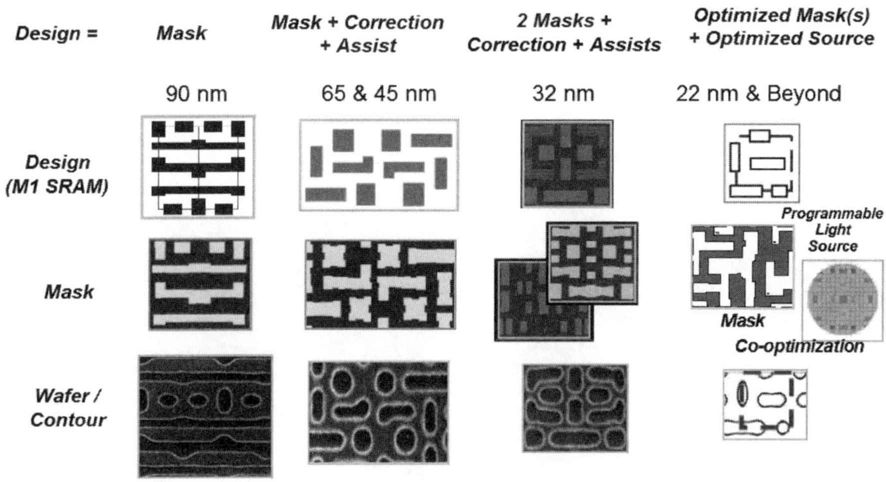

Fig. 6: 193 nm Optical Lithography Evolution: Lower k1 Factors

Integration

Conference Proceedings AMC XXIV © 2009 Materials Research Society

Metal Hard Mask Dual Damascene Architecture: an industrial solution to face scalability challenges

David Galpin[1], C. Monget[1], E. Richard[1], N. Posséme[2], Ph. Brun[2], M. Guillermet[2], C. Verove[1], A. Ostrovsky[1], Y. Trouiller[1], M. Mellier[1], S. Petitdidier[1], V. Caubet[1], M. Rivoire[1], O. Hinisinger[1], G. Imbert[1], Y. Le Friec[1], L. Favennec[1], I. Thomas[1], J. Guillan[1], K. Haxaire[1], P. Caubet[1], P. Vannier[1], Ph. Normandon[1], S. Chhun[1], V. Arnal[1], P. Ferreira[1], T. Berger[1], C. Besset[1], L. Arnaud[2], and D. Roy[1]

[1]STMicroelectronics, 850 rue Jean Monnet, Crolles, 38926, France

[2]CEA-LETI-Minatec, Grenoble, France

ABSTRACT

The choice of copper interconnect architecture is key for device performance, process manufacturability and future scalability. Trench First Hard Mask (TFHM) architecture is an alternative to Via First Trench Last (VFTL). Among the demonstrated TFHM benefits, the major advantage is eliminating all photo-resist contact to low-k materials through the use of a dense dielectric layer. Specific etch process developments are necessary to take full advantages of this architecture. With a reduced number of process steps and introduction of only a simple metal hard mask step, this approach is a highly attractive cost efficient manufacturing solution. Moving successively from 90nm to 65nm and onto 45nm nodes, it has been demonstrated that the TFHM architecture has facilitated progressively enhanced rates of yield learning, as well as minimizing the impact of introducing new lower k dielectrics. In anticipation of the new materials required for 32nm and beyond, an ultra low k ULK film with dielectric constant k=2.3 has been successfully integrated with minimum development.

In the challenging context of advanced backend integration, the TFHM architecture is foreseen as the cost efficient solution for volume manufacturing as well as the time effective solution for yield learning.

INTRODUCTION

Implementation of porous low-k materials is becoming mandatory in order to compensate the metal line-width resistance increase to preserve RC. These fragile low k materials are easily damaged by process steps, increasing the dielectric constant, and present incompatibilities with regular photolithography consumables: the resist poisoning effect [1]. ULK material introduction combined with device scaling is known to make backend integration more and more challenging [2, 3]. In this context, the choice of process architecture clearly plays a major role enabling both technology reuse, and increasing the rate of yield learning as well.

TFHM architecture is presented as an alternate industrial solution. Architecture improvement through 3 CMOS technology generations is presented, as well as the process and manufacturing benefits. 45nm technology node backend performances are reported and architecture scalability is fully discussed.

ARCHITECTURE EVOLUTION THROUGH TECHNOLOGY NODES

TFHM architecture is used for all STMicroelectronics' manufacturing over three successive CMOS generations. It was introduced into manufacturing at the 90nm node on both

200mm and 300mm [4]. This architecture has been progressively improved for over 8 years to successfully meet manufacturing and scalability requirements down to the 40nm node.

The fundamentals of the metal hard mask scheme as introduced at the 90nm node are presented on figure 1. A thin metallic hard mask layer is deposited on top of a thin dense dielectric layer, which serves to encapsulate the underlying low-k material. Only the metal hard mask layer is etched after trench photolithography, allowing the use of a thin single layer resist. The dense dielectric layer is used as an Etch Stop Layer (ESL) preventing any resist contact with the underlying low-k material. Via patterning is then performed on a very shallow hard mask topography, which is further planarized through use of an organic Bottom Anti Reflective Coating (BARC). Vias are etched down to a via ESL, which is commonly a silicon carbide based film (i.e. SiC, SiCN). Via resist plug fill and etch back steps are then performed, protecting the via bottom during line etch. Finally, line etching is performed using the metal hard mask, without any additional photolithography steps. After line etching, a last step ensures resist plug removal and finally via ESL opening. This scheme has demonstrated its manufacturing robustness at the 90nm node.

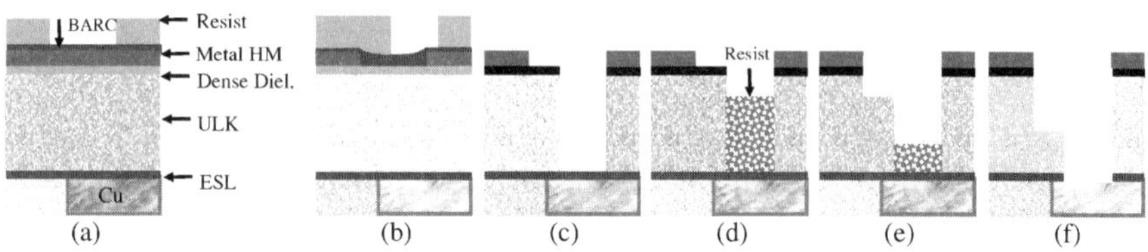

Figure 1: Original 90nm node TFHM architecture: a) line lithography with organic BARC, b) HM etch and via BARC/lithography, c) via etch and resist removal, d) via resist plug fill and recess, e) line etch, f) resist plug removal and ESL open.

Introduction of ULK materials requires optimization of the original scheme to gain further in process robustness. Both the via fill and final plug removal steps remain challenging with ULK materials. The i-line resist plug height control has a determinant impact on ULK moisture absorption and further can generate an etch-stop mechanism during line patterning [5]. In addition, final plug removal requires an oxygen based plasma which contributes to ULK dielectric constant degradation. In order to suppress both vulnerabilities with ULK materials, an enhanced integration scheme is used: the plug free TFHM as shown on figure 2. In this optimized plug free scheme, the subsequent via etch, line etch and via ESL open are processed in a single etch operation without air break, corresponding to steps c) and d) in figure 2. This completely eliminates the insertion, subsequent etch back and final removal of the via i-line resist plug. It completely minimizes ULK material exposure to atmospheric contamination and damage from plasma-stripping chemistries. This plug free integration scheme was introduced for 65nm ULK integration and was pursued at the 45nm node [6, 7].

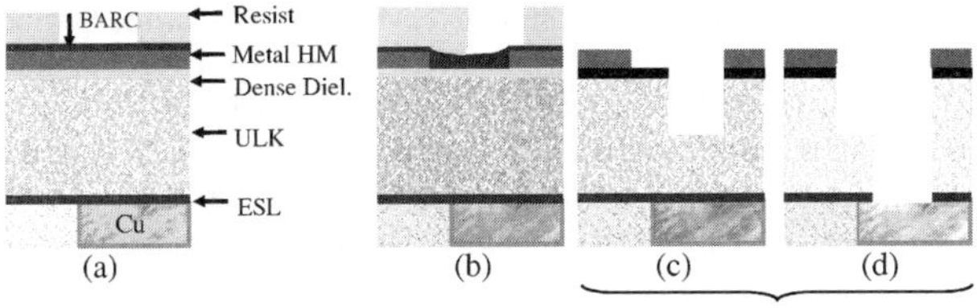

Figure 2: Plug free TFHM architecture sequence: a) line lithography with organic BARC, b) HM etch and via BARC/lithography, c) via etch and resist removal, d) line etch and ESL open.

METAL TFHM BENEFITS AS AN INDUSTRIAL SOLUTION

Line & Via Lithography on low topography

Line photolithography is performed on a flat metallic hard mask surface, allowing use of organic BARC. Subsequent via photolithography also encounters a very limited topography challenge. The shallow hard mask step is further reduced by BARC prior to via resist coating as shown in figure 3. This integration is very attractive in terms of lithography robustness and scalability for both line and via steps.

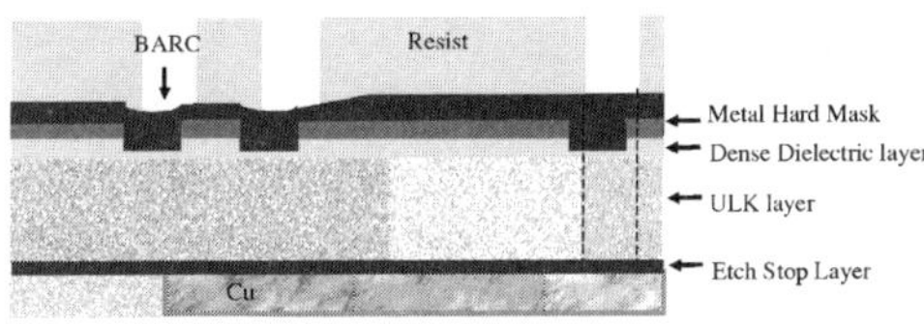

Figure 3: Schematic of via photolithography with minimized hard mask topography further compensated by BARC coating.

ULK integrity preservation

Because there is no direct contact between photo resist or BARC and any ULK or low-k material throughout the entire process integration, the much published via poisoning mechanism is completely avoided by construction. The dense dielectric TEOS layer encapsulating the low-k material definitively secures against resist poisoning. In addition, via and line lithography rework steps can be performed using standard toolsets and processes without possible damage to the underlying low dielectric constant films. This configuration also enables use of a very broad choice of materials and chemistries including nitrogen based layers or processes. Such nitrogen containing processes are generally undesirable in a VFTL approach because of resist poisoning concerns.

Yield learning curve acceleration

TFHM architecture dramatically simplifies the introduction of new dielectric materials, which translates directly into a faster yield learning curve. As illustration, the progression of the 65nm low-k and ULK defect density with respect to the 90nm node with low-k is shown in figure 4. A six metal layer inverter chain test vehicle with 76M vias (14mm^2) was used for yield and defectivity monitoring. The 300mm TFHM integration at the 90nm node required around 20 months to achieve a given and significantly low density level. For the same yield progression, the low-k 65nm node technology only requires almost half of this period, taking advantage of previous technology development by suppressing all systematic effects. This learning curve enhancement is even more pronounced with porous ULK material introduction at the 65nm node. Less than 8 months were necessary to achieve same back end defectivity performance.

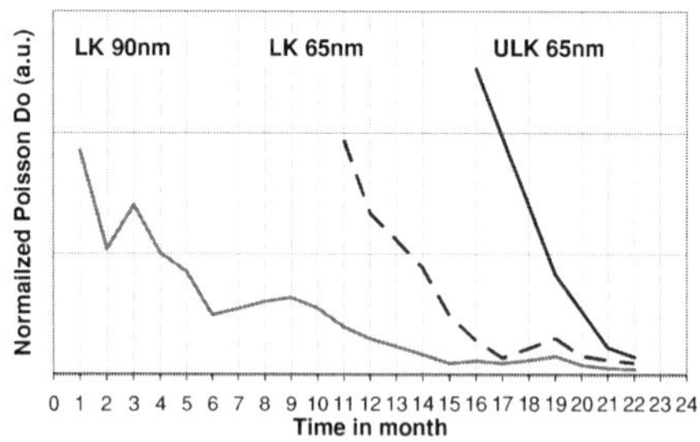

Figure 4: Yield learning trend for 3 TFHM generations illustrates tremendous yield learning acceleration on the ULK 65nm node.

Manufacturability benefits

TFHM architecture is based on a simple integration scheme that has scaled over 3 generations without introduction of any additional steps. Indeed, the plug free introduction at 65nm node further reduces the BEOL dual damascene cost by 50%. Cycle time also benefits from this flow simplification. Currently, a dual damascene brick only requires 13 process steps positioning this option as a very attractive industrial solution compared to VFTL.

KEY PROCESS STEPS FOR INTEGRATION ROBUSTNESS

Metal hard mask deposition

Metal hard mask deposition and polishing are straightforward processes. Metal hard mask deposition is performed with Physical Vapor Deposition (PVD) and processed in a regular chamber without any specific hardware upgrade. No toolset acquisition is necessary as this kind of chamber is commonly used for both aluminum based RIE BEOL processes, and also the aluminum pad layers commonly used before bonding.

Metal hard mask etching

Metal hard mask etch step needs to be optimized to ensure hard mask opening on all density structures [8]. At the same time, excessive over etch into the dense dielectric layer is prohibited to avoid ULK material exposure. The metal hard mask etch is characterized by a good selectivity to the metal hard mask and an optimized over etch step to respectively secure ULK encapsulation and line opens/shorts yield.

Misaligned via etching

Via patterning is performed just after hard mask etch. This implies potential via size reduction with via misalignment if the etch process is not correctly tuned. The via breakthrough step needs to be non selective to the metal hard mask to allow for misaligned vias as shown in figure 5. Tight via resistance dispersions, and solid via yield even with +25nm misalignment confirm the overall process robustness as shown in figures 6 & 7. The appropriate via etch process optimization is also beneficial for overlay and line-via alignment margin as well.

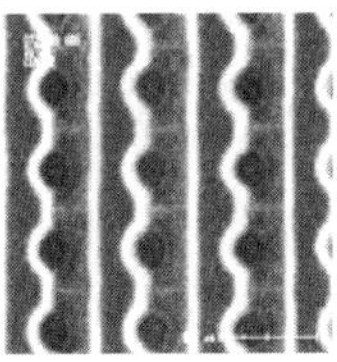

Figure 5: Misaligned via patterning through metal hard mask.

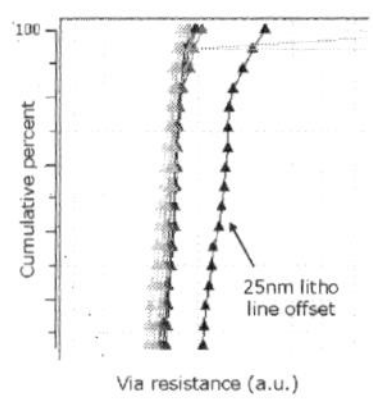

Figure 6: 25nm overlay misalignment does not affect via resistance yield of 25M high density chain.

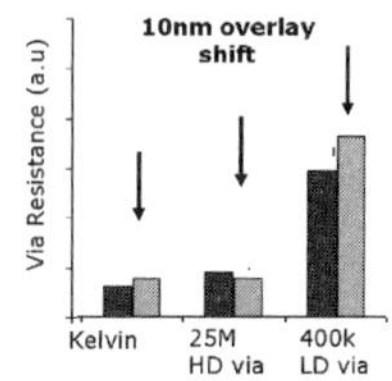

Figure 7: Via resistance is stable on 3 chain structures with 10nm overlay misalignment.

Integrated via-line etching

The plug free scheme combines via and line etches in a single step. As reported earlier, this approach presents major advantages but also leads to new etch challenges. Particularly, etch selectivity between ULK and via ESL remains a critical parameter to avoid underlying copper exposure before the ESL opening step. Introducing a partial via etching scheme with adequate etch chemistries successfully achieves industrial via yield performances.

Metal hard mask polishing

From a CMP point of view, the metal hard mask integration is virtually transparent. No specific slurry is required to polish this layer and its polish rate is about 50% higher than a regular TaN based barrier as reported in table 1.

Material	Normalized Polish rate (ref. ULK)
ULK	10
TEOS	30
Barrier	40
TiN	60

Table 1: Normalized CMP polish rates using a typical barrier slurry for materials used in a 45nm ULK BEOL.

INDUSTRIAL PERFORMANCES OF 45nm TECHNOLOGY

The 45nm node is the latest technology industrialized using the TFHM integration scheme. The TEM cross section in figure 8 shows a 45nm SRAM back-end with 7 metal levels.

At all the minimum pitch layers, ULK with k= 2.5 is integrated using 193nm hyper-NA > 1.1 immersion lithography.

The ULK 45nm backend yield performance is compared to the low-k 65nm BEOL using the same yield test vehicle: a 21mm^2 test chip using 121M vias connecting through 7 metal levels and including 77nm line feature for the 5 minimum pitch metal levels. The yield trend for this defectivity test vehicle is presented in figure 9. Again, it can be seen that yield learning cycle is accelerated by a factor of 2 between the 65nm and 45nm nodes, which includes also the major process change moving from a low k=3.0 to ULK k=2.5 dielectric as well. Only 10 months of yield learning are necessary to achieve an equivalent defectivity performance compared to the 65nm node. Further, the yield loss mechanisms are identified and attributable to classical BEOL defectivity mechanisms. There are no new defectivity mechanisms associated with the TFHM architecture.

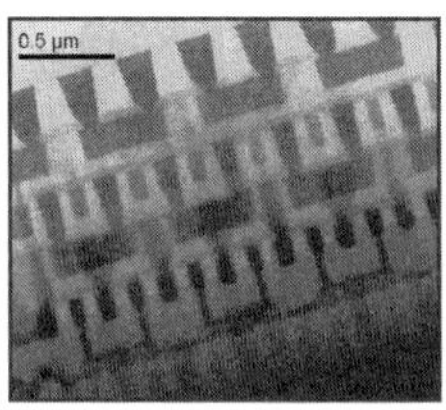

Figure 8: TEM image of 45nm ULK backend.

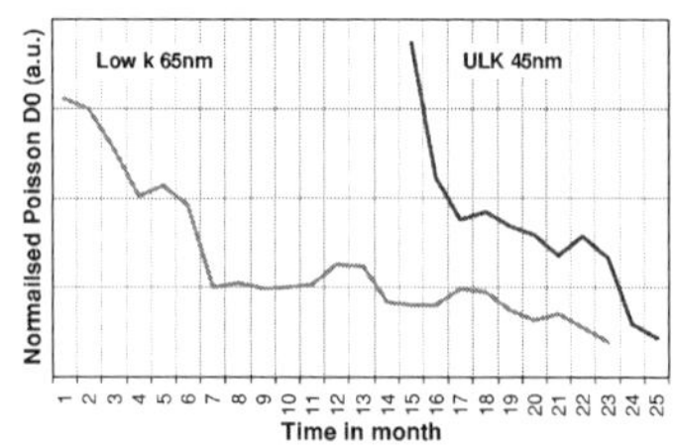

Figure 9: Yield learning trend comparison between low-k 65nm and ULK 45nm.

To ensure high-density device functionality, excellent parametric yield performance and process control are mandatory. As presented in figure 10, RC performance meets the generic model targets with a tight distribution for 45nm minimum pitch features. The C_p review of key parametric parameters such as line resistances, via resistances and lateral capacitances gives a good indicator of process capability and robustness. As shown in figure 11 the majority of C_p values are above 2.0 showing the 45nm backend integration demonstrates solid industrial process control.

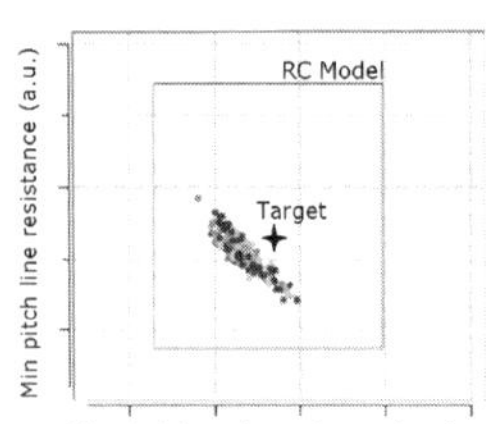

Figure 10: RC curve for ULK 45nm minimum pitch interconnects.

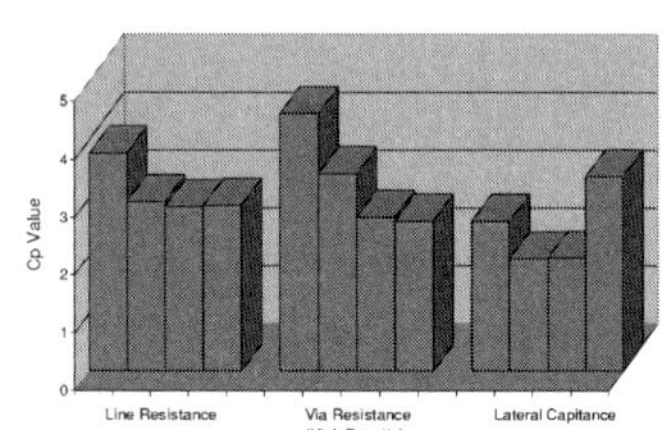

Figure 11: Process capability C_p values for key electrical parameters for the 4 minimum pitch layers.

Finally, to complete the overview of the 45nm TFHM integrated ULK solution, intrinsic backend reliability is reported for both Time Dependence Dielectric Breakdown (TDDB) and

Electro Migration (EM). The 45nm TFHM architecture demonstrates capability to successfully exceed all backend reliability criteria. Indeed, over the last 3 technology generations no new challenges specifically related to the TFHM scheme has been identified from a reliability standpoint.

TFHM ARCHITETURE SCALABILITY DISCUSSION

Both technology shrink and porous low k introduction [9, 10] represent challenges for advanced technology development. It is mandatory for the backend architecture to facilitate new materials introduction, feature scalability and also innovative reliability enhancers.

An Extreme Low-k (ELK) porous dielectric (k= 2.3) was integrated using the plug free TFHM architecture [11]. Four metal level build was performed using a 45nm design rule test chip. Zero process development was necessary to carry out this first demonstration. ELK dense via chains yields are compared to ULK yields from splits performed within the same batch of wafers as shown in figure 12. As theoretically expected, a 4.5% gain in RC performances confirms the integrated ELK material benefits as reported in figure 13. This confirms that the TFHM architecture is able to deliver fast electrical feedback for the next generation porous dielectric materials.

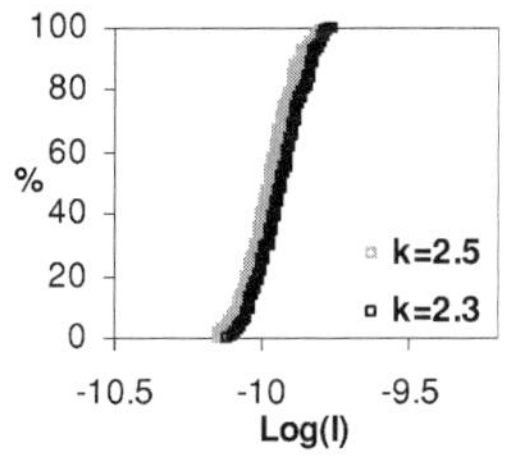

Figure 12: Normalized 3M dense via chain resistance.

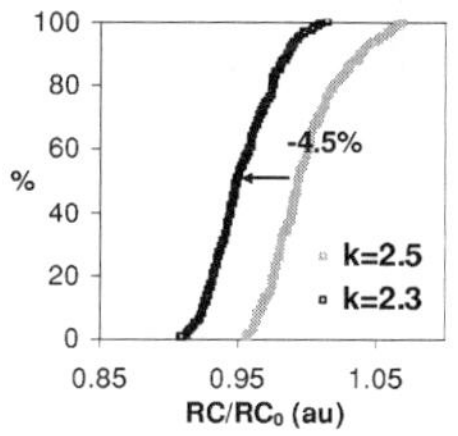

Figure 13: Normalized RC for 2 porous dielectrics.

With the scaling down of line features, the barrier/seed PVD process becomes more sensitive to line profile and specifically the line top opening which is a potential source of PVD pinch off. With the hard mask integration, the line geometry can be further modulated through process optimization to facilitate PVD deposition.

The last major concern beyond the 45nm node is EM reliability. Integration of Self Aligned Barrier (SAB) schemes is widely investigated to enhance copper adhesion properties and in the same way enhance reliability performances. Electro-less barriers such as CoWP were successfully integrated into the TFHM scheme using a k=2.5 ULK dielectric material [12]. Specifically the CoWP integration was successfully performed without any significant process development. The TFHM approach is fully compatible with the advanced SAB development required for advanced technology nodes.

CONCLUSION

The TFHM architecture is more than an optional scheme to the VFTL architecture. For over 8 years of manufacturing and continuous improvement since the 90nm node, through to today's most advanced process development, its performances and key benefits have been demonstrated time and time again. This approach is proposed as an alternative industrial solution reusing integration legacy for the new generations of porous dielectric materials. The very simple dielectric stack with a limited number of operations makes this option extremely attractive for manufacturing. Metal hard mask etching and integrated via and line etching remain challenging processes to fully benefit from the TFHM advantages. The accelerated yield learning curve and solid process control are confirmed on the 45nm technology node. Integration of advanced porous materials with the same competitive manufacturing scheme presents major interest and places TFMH architecture as the competitive solution for future technology nodes.

REFERENCES

1. A.K. Stamper et al., Advanced Metallization Conference 2006 (AMC 2006). Proceedings, 2007, p 53.
2. O. Hinsinger et al., 2004 International Electron Devices Meeting (IEEE Cat. No.04CH37602), 2005, p 317-20.
3. R.J.O.M. Hoofman et al., Proceedings of the IEEE 2005 International Interconnect Technology Conference (IEEE Cat. No. 05TH8780), 2005, p 85-7.
4. T. Devoivre et al., Proceedings of the 2002 IEEE International Workshop on Memory Technology, Design and Testing (MTDT2002), 2002, p 157-62.
5. N. Posseme et al., Proceedings of the IEEE 2006 International Interconnect Technology Conference (IEEE Cat. No. 06TH8862C), 2006, p 3 pp.
6. R. Fox et al., International Electron Devices Meeting 2005 (IEEE Cat. No.05CH37703C), 2005, p 4 pp.
7. E. Richard et al.., 2007 IEEE Symposium on VLSI Technology, 2007, p 178-9.
8. S.J. Kirk et al, 2007 IEEE/SEMI Advanced Semiconductor Manufacturing Conference, 2007, p 366-70.
9. C.C. Hsia., Microelectronic Engineering, v 83, n 11-12, Nov.-Dec. 2006, p 2055-8.
10. Y.N. Su et al, International Electron Devices Meeting 2005 (IEEE Cat. No.05CH37703C), 2005, p 4 pp.
11. K. Hamioud et al., Proceedings of Advanced Metallization Conference 2007 (MRS).
12. S. Chhun et al., Proceedings of the IEEE 2006 International Interconnect Technology Conference (IEEE Cat. No. 06TH8862C), 2006, p 3 pp.

Conference Proceedings AMC XXIV © 2009 Materials Research Society

Development of selective Co CVD capping process for reliability improvement of advanced Cu interconnect

Emiko Nakazawa[1], Koji Arita[1], Yasuaki Tsuchiya[1], Yumi Kakuhara[2], Shinji Yokogawa[2], Tetsuya Kurokawa[3], Nobuyuki Sasaki[4], Seshadri Ganguli[5], Hyoung-Chan Ha[5], Wei Ti Lee[5], Sang-Ho Yu[5], and Makoto Sekine[1]

[1]Process Technology Division, NEC Electronics Corporation, 1120 Shimokuzawa, Sagamihara Kanagawa, 229-1198, Japan

[2]Advanced Device Development Division, NEC Electronics Corporation, 1120 Shimokuzawa, Sagamihara Kanagawa, 229-1198, Japan

[3]Test and Analysis Engineering Division, NEC Electronics Corporation, 1120 Shimokuzawa, Sagamihara Kanagawa, 229-1198, Japan

[4]Applied Materials Japan, 3-20-20 kaigan, minato-ku Tokyo, 108-8444, Japan

[5]Applied Materials, 3050 Bowers Ave., Santa Clara, CA, 95054

ABSTRACT

This paper describes Selective Co CVD capping process for advanced Cu/Low-k interconnection with higher reliability. More than 100 times longer EM lifetime with no resistivity increase has been obtained by using this technology. Excellent selectivity has been confirmed from line to line leakage and TDDB lifetime measurements. The Selective Co CVD capping process is the candidate for 32nm node generation and beyond.

INTRODUCTION

For future devices, as interconnect dimensions decrease, wiring reliability improvement and resistivity reduction become challenging. Impurity doping technology using alloy seed [1] or self-aligned silicide [2], and metal capping technology [3] are proposed for reliability improvement. However, impurity doping technologies cannot simultaneously achieve high reliability and low resistivity because reliability performances heavily depend on doping amount [4]. On the contrary, metal cap process, such as CoWP electroless plating can achieve higher reliability without increase in resistivity. Unfortunately, line to line leakage has been major concern for electroless Co cap process. Pretreatment and chemical optimizations have been developed for selectivity improvement. But as long as the dielectric field exposes the plating solution containing metal ions during the deposition, metallic contamination anyway leads to selectivity loss. Recently, newly developed selective Co deposition utilizing CVD process has been developed by AMAT. We applied this process for Co capping process to improve reliability of advanced Cu/Low-k interconnects.

EXPERIMENT

Selective Co CVD capping process was performed by using common metal CVD chamber with Applied Materials' Endura2 platform. Pretreatment, Co deposition and post treatment were sequentially performed in the single chamber.

H2 anneal pretreatment was applied to remove surface contamination such as CuO and moisture, followed by selective Co deposition step. Post nitrogen containing plasma was applied to remove C and lower the resistivity.

Deposition rate on Cu and SiO2, film density and barrier property for Cu were investigated with blanket wafers.

Two-level dual damascene structure with NEC Electronics 55nm manufacturing process as shown in Fig.1 was used to demonstrate electrical characteristics and reliability performance (EM and TDDB).

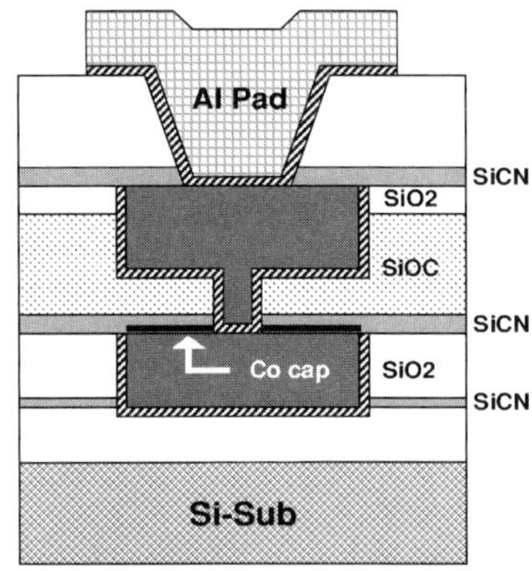

Fig.1 Schematic of the two-level interconnect structure with Co cap for E-test

RESULT AND DISCUSSION

Co film characteristics

Co deposition rate comparison between on Cu and on SiO2 is shown in Fig.2. Co deposition rate is obtained on Cu surface without incubation time. On the contrary, Co deposition on SiO2 has long incubation time. Theoretically, infinite selectivity of Co deposition could be obtained during incubation time on SiO2.

Figure 3 is the backside SIMS analysis result after 425°C 1hr annealing. This result shows Co film worked as Cu diffusion barrier. Co film density is 8.6g/cm3 and almost same as bulk density (8.9g/cm3). High Co density can potentially suppress Cu diffusion.

These results indicate that CVD-Co film could be potential capping metal for Cu, besides, might be able to eliminate the dielectric barrier in the future.

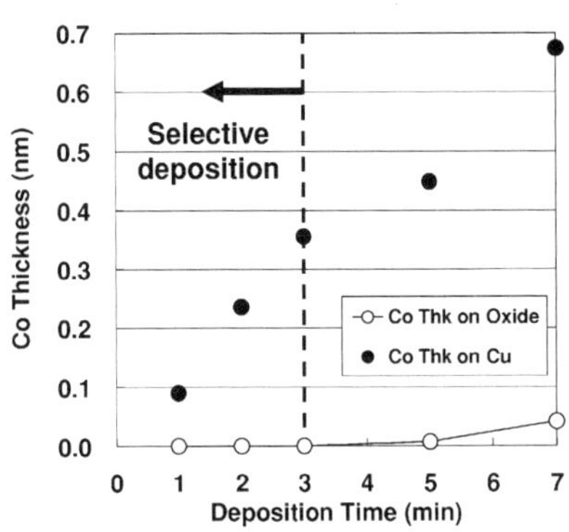

Fig.2 Co deposition rate comparison between on Cu and on SiO2

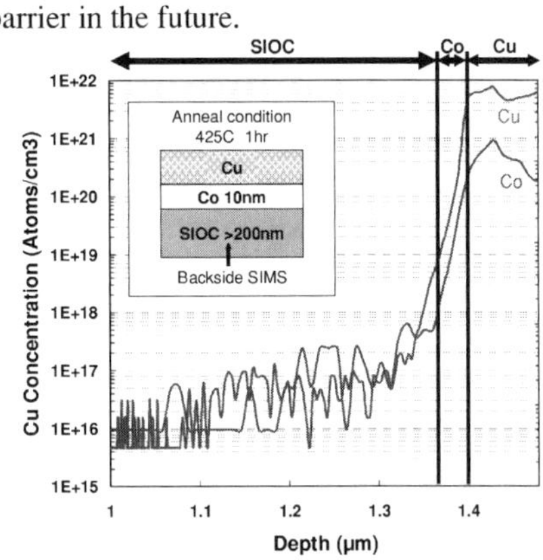

Fig.3 Backside SIMS analysis results after 425°C 1hr annealing with a sample structure shown in the inset

A cross-section TEM image of Co cap on Cu interconnects in two-level dual damascene structure is shown in Fig.4. Only Cu interconnects surface are successfully covered with less than 3nm thick Co layer.

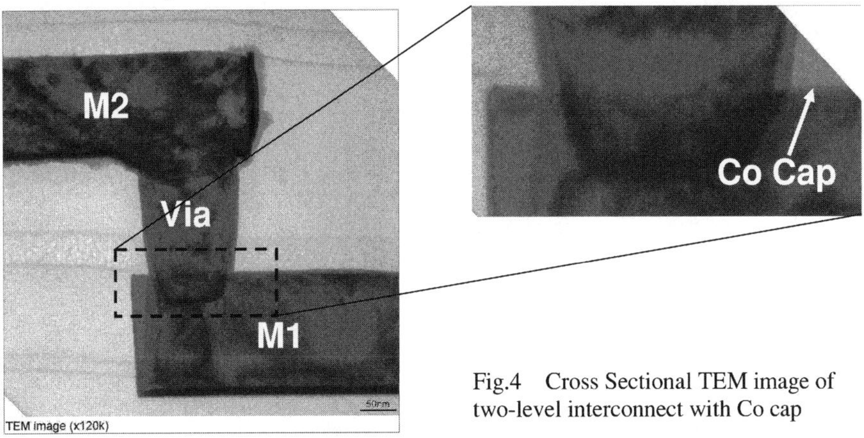

Fig.4 Cross Sectional TEM image of two-level interconnect with Co cap

Electrical characteristic and reliability performance

Figure 5 shows 4.2M via chain yield with 90nm vias. 100% via chain yield is obtained for the cases with and without Co cap. In addition, via resistance matches in both cases, because TEM image shows Co cap under via was removed by Dry etch process.

Figure 6 shows distribution of sheet resistance for 90nm wiring width. Unlike other capping process such as CuSiN and Cu-Alloy seed, no increase of line resistance or change of distribution is observed between with and without Co cap process. These results indicate that Cu-Co alloy or intermetallic compound is not formed between Cu and Co cap.

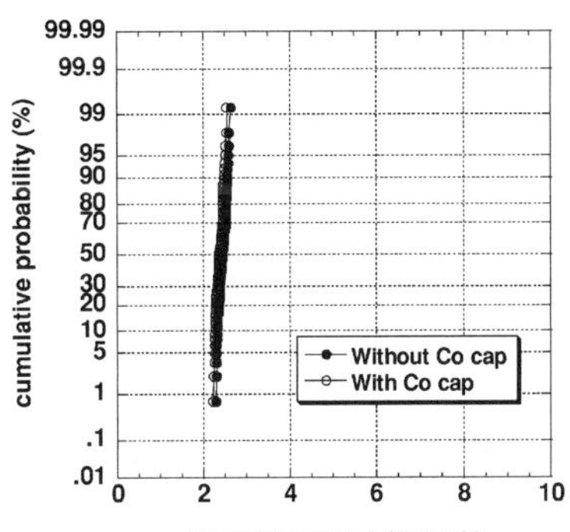

Fig.5 Distribution of via resistance with 4.2M via chain pattern

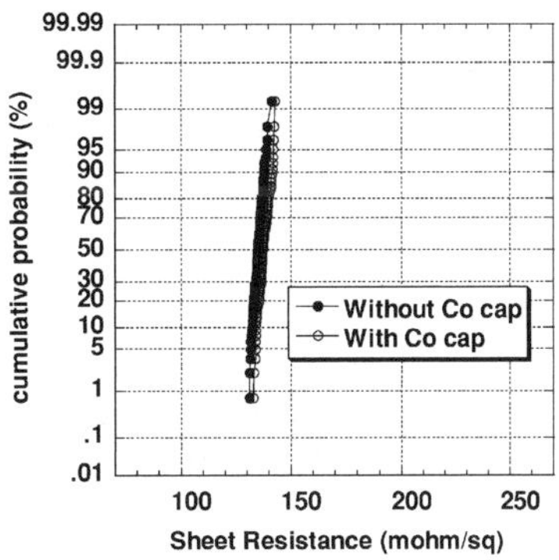

Fig.6 Distribution of sheet Resistance

Figure.7 shows schematic diagram of EM test structure. The test structure contains of 40 vias then the test structure could test 20 vias at the same time for downstream. The failure time was defined as the time at which the first via failed. The failure criteria was 0.5% abrupt resistance change, it corresponded to the first via failure from previous study [5]. The EM lifetime for each sample was plotted with the failure probability of a via using the cumulative hazard method.

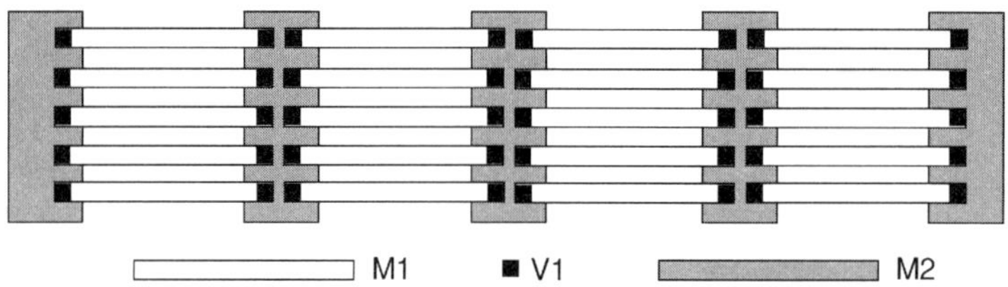

Fig.7 EM test structure for down stream evaluation.

Figure 8 shows EM lifetime for each samples. EM lifetime of Co cap sample is 100X longer and sigma is comparable with controlled sample.

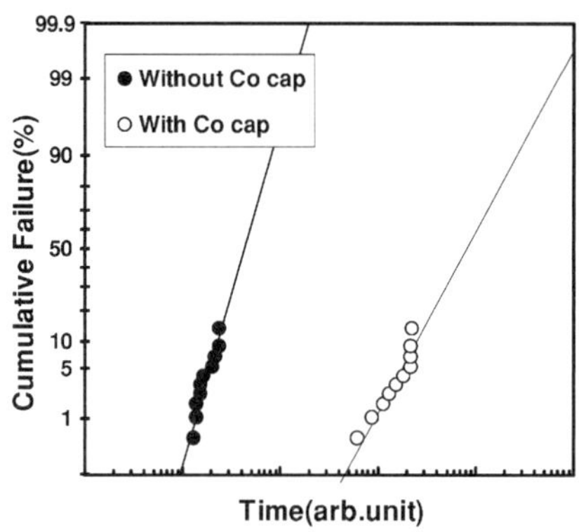

Fig.8 Via EM Lifetime distribution in 55nm node two-level dual damascene structure

Figure 9 shows the line to line leakage current with 90nm/90nm comb pattern. Co cap process has no impact on leakage current even for dense pattern. In Fig. 10, the comparison of TDDB lifetime between control sample and Co cap is shown. Excellent selectivity of Co CVD capping process is confirmed from no degradation of line to line leakage and TDDB.

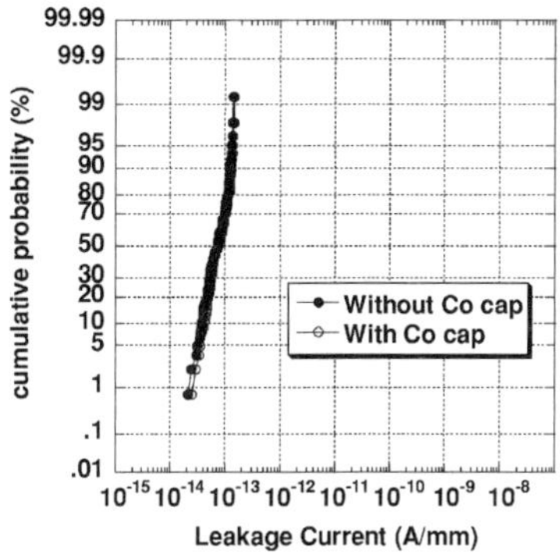

Fig.9 Distribution of line to line Leakage with 90nm/90nm comb pattern

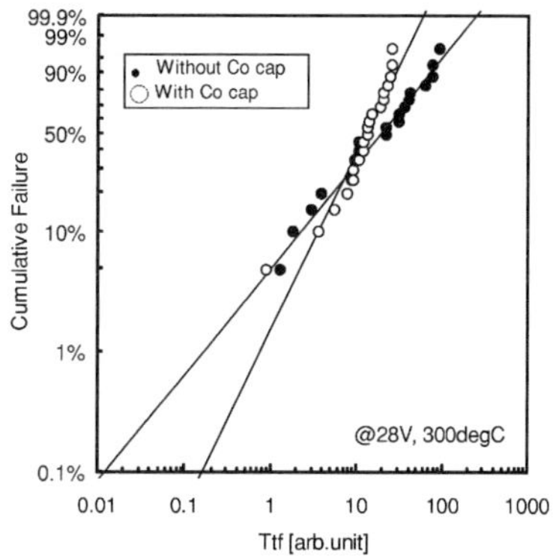

Fig.10 TDDB Lifetime measurement with 90nm/90nm comb pattern

CONCLUSIONS

Newly developed Selective Co CVD capping process has demonstrated improvement of Cu interconnects reliability such as EM lifetime. The reliability performance with Co CVD capping process was improved without increase in line resistance. Co CVD capping process had no impact on line to line leakage and TDDB. These results revealed excellent selectivity about this technique. In addition, this Co cap had good barrier property, thus Co cap may substitute for dielectric barrier such as SiCN. The selective Co CVD capping process is suitable for 32nm node and beyond.

ACKNOWLEDGMENTS

The authors would like to thank J. Lu, A. C-T Chan and S-E Phan for their technical supports and discussions.

REFERENCES

[1] T. Tonegawa et. al., Proc. of IITC2003, p216 (2003)
[2] T. Usami et. al., Proc. of IITC2006, p125 (2006)
[3] C. K. Hu et. al., Appl. Phys. Lett., Vol.84, p4986 (2004)
[4] S. Yokogawa et. al.,IEEE transaction on electron devices vol.55, p350
[5] H. Tsuchiya et al., Microelectron. Reliab., Vol.46, p1415 (2006)

Conference Proceedings AMC XXIV © 2009 Materials Research Society

Impact of Gas Silylation Process on Wire-to-wire Leakage Current for Low-k / Cu Dual-Damascene Interconnects

Yumi Hayashi[1], Akihiro Kojima[1], Naofumi Nakamura[1], Hideaki Masuda[2], Koichi Sato[2], Hisataka Hayashi[2], Noriaki Matsunaga[1], and Hideki Shibata[1]

[1]Center for Semiconductor Research & Development, Semiconductor Company, Toshiba Corporation, 8, Shinsugita-cho, Isogo-ku, Yokohama, 235-8522, Japan

[2]Process & Manufacturing Engineering Center, Semiconductor Company, Toshiba Corporation, 8, Shinsugita-cho, Isogo-ku, Yokohama, 235-8522, Japan

ABSTRACT

Damage silylation process is known to restore damaged low-k SiOC, and improve electrical properties and reliability of stress-induced voiding (SiV). Gas silylation, furthermore, improved wire-to-wire leakage current, with the result that gas silylation also has the effect of providing carbon inside SiOC, not only at damaged SiOC surface. The conduction mechanism of the improvement was clarified by leakage current and temperature-dependent I-V characteristics. That is, there are more defect sites in a sample w/o silylation, but the activation energy depth is almost the same in both w/ and w/o silylation samples in Poole-Frenkel (P-F) conduction region.

INTRODUCTION

For high performance of interconnects, damage silylation is promising in low-k inter-layer dielectric (ILD) structure. Silylation is known to improve electrical properties and SiV reliability [1-6]. But it is not clear that the silylation reacts not only at damaged surface of SiOC but also inside SiOC. In addition, J. Liu et al. reported that silylation has the effect of improving leakage current [7], but the conduction mechanism of the improvement is not known. In this study, we focused on wire-to-wire leakage current, and investigated mechanisms of leakage current between w/ and w/o silylation by comparing I-V characteristics and their temperature-dependence.

EXPERIMENTAL DETAILS

Double-level Cu dual-damascene (DD) structure was built up by using SiOC (k=2.5) as shown in Fig. 1. The pitch of the wiring is 140nm. Gas silylation was conducted after dual-damascene (DD) formation. Top-view of the wafer was checked by scanning electron microscope (SEM) before and after wet treatment (Fig. 2). Line resistance and inter-level capacitance were evaluated as electrical properties. I-V characteristics were examined in the second-level wiring (M2) at the pattern of line/space=70/70nm, whose total length is 8.8m. The testing of high-temperature short leakage was conducted in M2 layer, at the pattern of line/space=70/70nm and the total length of 0.9 m, in order to evaluate the effect of silylation.

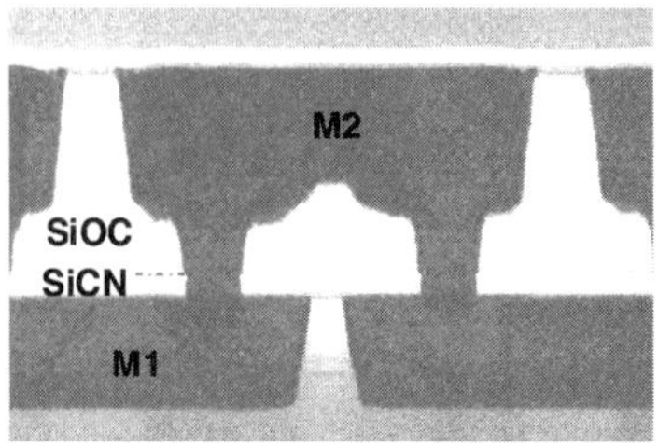

Figure 1 Cross-sectional SEM image of 2-level interconnects.

Figure 2 Flow of sample preparation.

RESULTS AND DISCUSSION

Restoring damage

Figure 3 shows top-view SEM images of before and after wet treatment. In w/o silylation sample, a layer was peeled off after wet treatment as shown in Fig. 3 (a) and (b). The layer peeled off is SiO_2 as shown in Fig. 4. SiO_2 was formed by plasma-enhanced chemical vapor deposition. The plasma may damage SiOC layer under SiO_2 and the damaged interfacial surface between SiOC and SiO_2 was peeled off during wet treatment. On the other hand, by conducting silylation the peeling off is suppressed, suggesting silylation process also restores not only the surface of SiOC, which is damaged by reactive ion etching (RIE), but also inside SiOC layer (Fig. 3 (c) and (d)).

Figure 5 shows RC-plot. It indicates that RC is lowered by silylation as reported previously [4, 5].

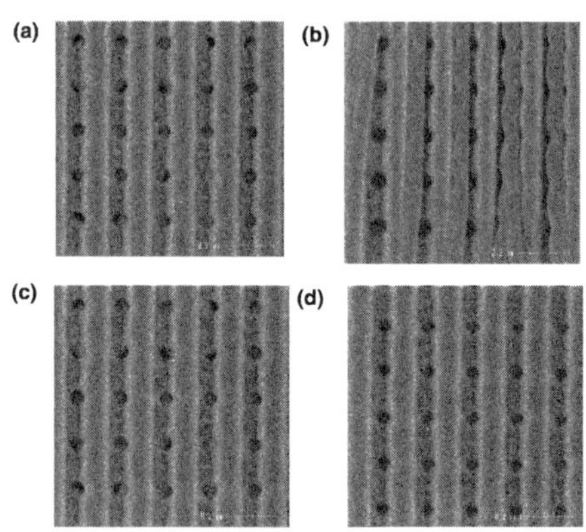

Figure 3 Top-view SEM images of (a) w/o silylation before wet treatment, (b) w/o silylation after wet treatment, (c) w/ silylation before wet treatment and (d) w/ silylation after wet treatment.

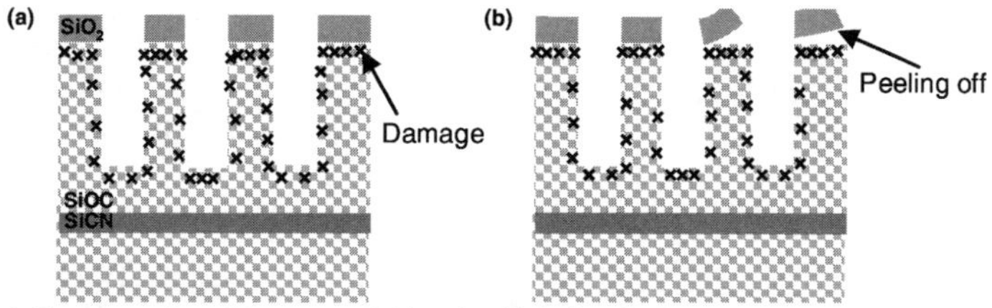

Figure 4 Cross-sectional images of (a) w/o silylation before wet treatment and (b) w/o silylation after wet treatment.

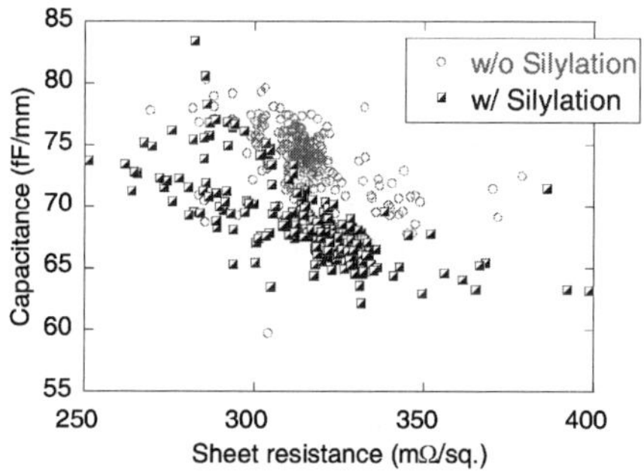

Figure 5 RC-plot of w/o and w/ silylation samples.

Analysis of wire-to-wire leakage current

I-V characteristics were evaluated in M2 layer at the pattern of line/space=70/70nm (Fig. 6 (a) and (b)). Some chips were broken down before the testing was finished in w/o silylation samples. Fig. 6 (c) and (d) shows $V^{0.5}$ vs. I/V plots of Fig. 6 (a) and (b). In the range of $V^{0.5}$ = 1.5 ~ 2.5 $(MV/cm)^{0.5}$, I/V are proportional to $V^{0.5}$, with the result that the conduction is P-F emission for both samples [8]. Furthermore, the range of $V^{0.5} < 1.2$ $(MV/cm)^{0.5}$ shows ohmic conduction, because I/V = constant.

Figure 7 shows leakage current at 1 and 4 MV/cm extracted from I-V characteristics as shown in Fig. 6. At 1 MV/cm, which is in the range of ohmic conduction, leakage current is almost the same for w/ and w/o silylation. On the other hand, in the range of P-F conduction, the sample of w/o silylation shows leakage current one order larger than that of w/ silylation sample. As a result, there are more defect sites in w/o silylation sample. Accordingly, it is clarified for the first time that silylation has the effect of decreasing the defect sites that mainly contribute to P-F emission.

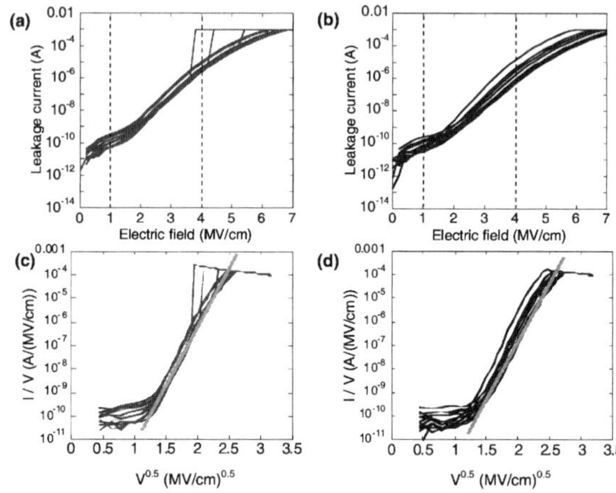

Figure 6 I-V characteristics of (a) w/o silylation and (b) w/ silylation. The plot of $V^{0.5}$ vs. I/V for (c) w/o silylation and (d) w/ silylation.

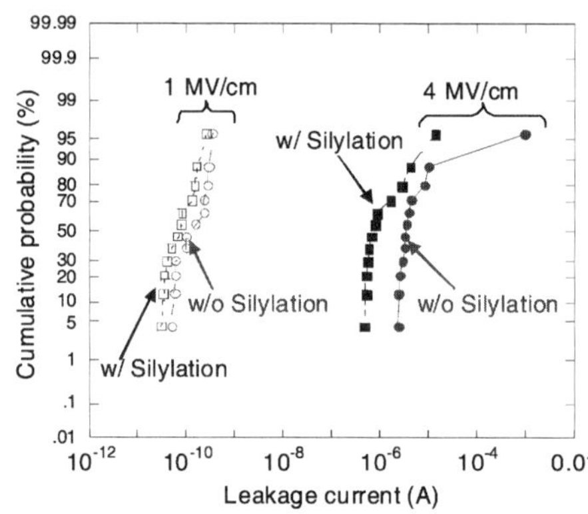

Figure 7 Cumulative distribution of leakage current at 1 and 4 MV/cm extracted from I-V characteristics of Fig. 6.

In order to study the mechanisms of the leakage current in detail, electron energy-loss spectroscopy (EELS) analysis was conducted. Figure 8 shows carbon signal maps of EELS. A SiCN/SiOC interface shows dark contrast, suggesting that the carbon is depleted. Accordingly, SiOC was damaged around the interface. The damage may be caused by chemical mechanical polishing or plasma of SiCN deposition. Carbon line profile was extracted from the line of A-B in Fig. 8 as shown in Fig. 9. W/o silylation sample shows carbon depletion in SiOC near the interface of SiCN/SiOC. On the other hand, the sample w/ silylation shows small depletion around the interface of SiCN/SiOC. Therefore, gas silylation process provides enough carbon inside SiOC and restores damaged sites, although silylation gas only showers a surface for which RIE is conducted directly.

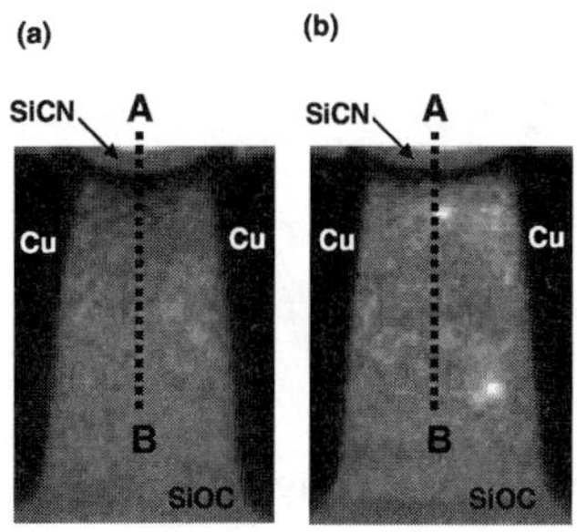

Figure 8 EELS map of carbon. (a) w/o silylation and (b) w/ silylation.

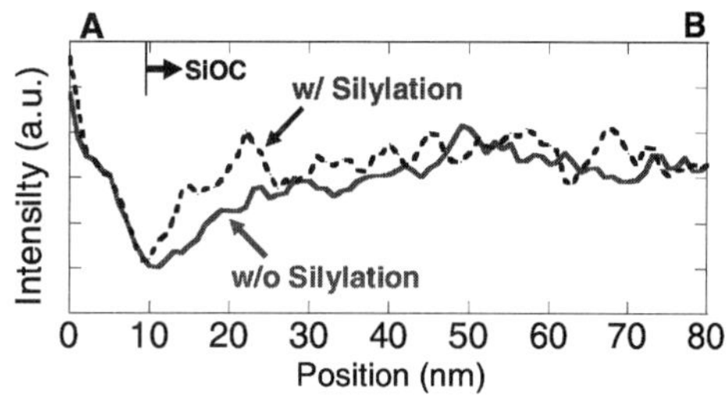

Figure 9 EELS carbon profile extracted from line A-B in Fig. 8.

Figure 10 shows temperature-dependent I-V characteristics at 30 ~ 175 deg C. From the characteristics, activation energy (E_a) of P-F condcution was caluculated as shown in Fig. 11. Because E_a is proportional to $V^{0.5}$, the conduction is P-F emission as Fig. 6 shows [9]. Comparing w/ and w/o silylation in Fig. 11, the value of E_a is almost the same. It indicates that the activation energy depth of defect sites is almost the same between w/ and w/o silylation samples, although the number of defect sites becomes smaller owing to silylation.

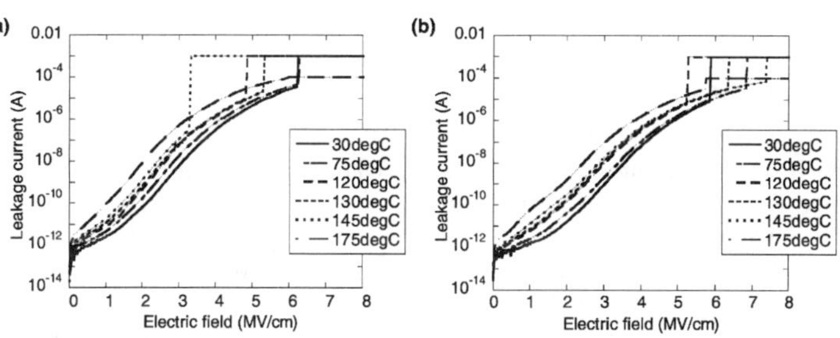

Figure 10 Temperature-dependent I-V characteristics of (a) w/o silylation and (b) w/ silylation.

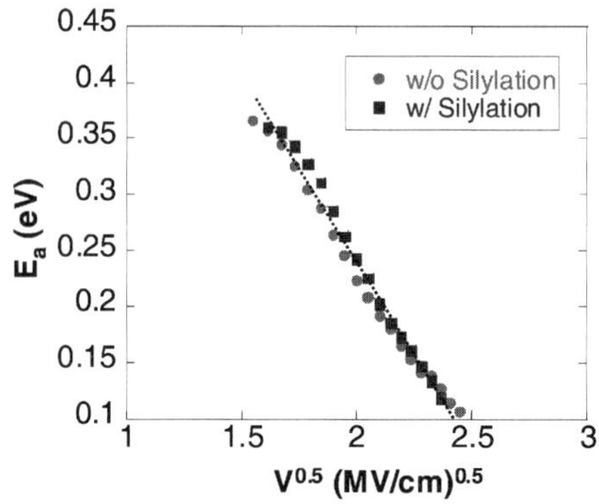

Figure 11 Field-dependent activation energy.

CONCLUSION

Gas silylation process was conducted after DD formation and wire-to-wire leakage current was evaluated to investigate the effects of silylation in detail. Top-view SEMs, which show peeling off of SiO_2, and I-V characteristics suggested that silylation can restore damage inside SiOC layer, not only at surface for which RIE is performed. Leakage current and E_a showed that the sample w/o silylation has more defect sites, although the activation energy depth of the defect sites is almost the same in w/ and w/o silylation samples. As a result, silylation works for the improvement of leakage current in the P-F conduction region, not in the ohmic conduction region.

REFERENCES

1. A. Bhanap, T. Ramos, A. Camarena, A. Naman, *Advanced Metallization Conference 2003*, (Mater. Res. Soc. Proc., 2003) pp.519-523
2. N. Nakamura, T. Yoshizawa, T. Watanabe, H. Miyajima, S. Nakao, N. Yamada, K. Fujita, N. Matsunaga and H. Shibata, *Proceedings of the IEEE 2004 International Interconnect Technology Conference*, (2004) pp.228-230
3. N. Matsunaga, N. Nakamura, K. Higashi, H. Yamaguchi, T. Watanabe, K. Akiyama, S. Nakao, K. Fujita, H. Miyajima, S. Omoto, A. Sakata, T. Katata, Y. Kagawa, H. Kawashima, Y. Enomoto, T. Hasegawa and H. Shibata, *Proceedings of the IEEE 2005 International Interconnect Technology Conference*, (2005) pp.6-8
4. N. Nakamura, N. Yamada, S. Nakao, K. Akiyama, H. Miyajima, N. Matsunaga, Y. Enomoto and H. Shibata, *Advanced Metallization Conference* 2005, (Mater. Res. Soc. Proc., 2005) pp.707-712

5. H. Chaabouni, LL. Chapelon, T. Chevolleau, O. Joubert and J. Torres, *Advanced Metallization Conference* 2007, (Mater. Res. Soc. Proc., 2007) pp293-299
6. A. Kojima, N. Nakamura, N. Matsunaga, H. Hayashi, K. Kubota, R. Asako, K. Maekawa, H. Shibata, T. Yoda and T. Ohiwa., *Advanced Metallization Conference* 2006, (Mater. Res. Soc. Proc., 2006) pp.301-305
7. J. Liu, W. Kim, J. Bao, H. Shi, W. Baek and P. S. Ho, *J. Vac. Sci. Technol. B*, **25(3)**, 906 (2007)
8. N. Suzumura, S. Yamamoto, D. Kodama, H. Miyazaki, M. Ogasawara, J. Komori and E. Murakami, *Proc. of IEEE 46th Annual International Reliability Physics Symposium*, (2008) pp.138-143
9. N. Suzumura, S. Yamamoto, D. Kodama, K. Makabe, J. Komori, E. Murakami, S. Maekawa and K. Kubota, *Proc. of IEEE 44th Annual International Reliability Physics Symposium*, (2006) pp.484-489

Conference Proceedings AMC XXIV © 2009 Materials Research Society

Highly-Reliable Seamless Low-k SiOCH Stacks (SEALS) with High Adhesion Strength Interface by Advanced Plasma Co-polymerization Process

Hironori Yamamoto, Fuminori Ito, Tsuneo Takeuchi, Naoya Furutake, and Yoshihiro Hayashi
LSI Fundamental Research Laboratory, NEC Electronics Corporation, 1120, Shimokuzawa, Sagamihara, Kanagawa, 229-1198, Japan

ABSTRACT

A seamless low-k SiOCH stacks (SEALS) by in-situ single-step sequential process has an advantage in interlayer adhesion property due to gradient change in the film properties within a thin transition layer. In this study, we investigated the adhesion strength as a function of the interface chemical-composition and the interface stress around the transition layer in the SEALS structure. The adhesion strength was increased by Si-oxidation, or essentially removal of carbon content and lowering the interface stress. The comprehensive control of the interface chemistry and the stress is a key factor for multi-layer Cu interconnects in 32nm-node and beyond.

INTRODUCTION

The interface adhesion strength of Cu interconnect is a big concern because of process compatibility with ULSIs fabrication processes such as CMP and packaging [1, 2]. Recently, we have reported new low-k /Cu BEOL concept such as seamless low-k SiOCH stacks (SEALS) by in-situ single-step sequential process based on plasma co-polymerization (PcP) technology [3]. By controlling the precursor gas mixing ratio of the hexagonal and linear types of organo-siloxene molecules, density-modulated SEALS was fabricated successfully, in which silica-amorphous carbon composite (SACC) film, a molecular pore stacked (MPS) film and a co-polymerized MPS (cMPS) film were stacked for HM/cap, IMD and ILD, respectively (Fig. 1) [4].

The in-situ single-step sequential process without air break has promise as low-cost, high throughput and high reliability. However, we found that the adhesion strength between the SACC-HM and the MPS-IMD was relatively weak and depended on the gas mixing sequence of two precursors.

In this paper, we investigated the effects of interfacial chemical and stress controls in the in-situ sequential deposition on the adhesion strength between the SACC-HM and MPS-IMD films, in order to improve the interface adhesion and clarify its mechanism.

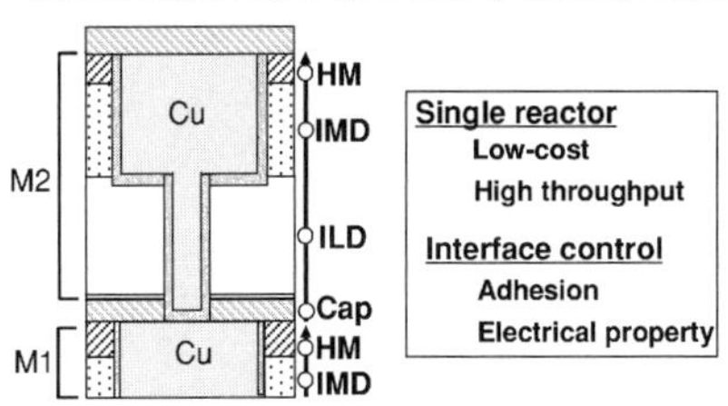

Fig. 1 Schematic drawing of in-situ sequential process for M1 and M2 structure. This process is carried out using only one reactor and has advantages of low cost, high throughput, and high reliability.

EXPERIMENT

We prepared two types of the organo-siloxane molecules as the liquid source for the plasma-polymerization [5]. A ring molecule for the MPS had a hexagonal Si-O structure with the side chains of vinyl and large hydrocarbon (R1). A chain-type molecule for SACC had the side chains of vinyl and hydrocarbons (R2, R3). As shown in Fig. 2, the source molecules were separately vaporized in a liquid injection-type vaporizer and were introduced into a PE-CVD reactor with He carrier gas. The in-situ sequential process was controlled by gas mixing ratio of the ring-type and the chian-type molecules. The k-value, chemical composition, and elastic modulus of the films deposited were measured by mercury–probe, XPS (X-ray photoelectron spectroscopy), and nano-indenter, respectively.

Table 1 Two kinds of molecules are prepared for this study. One is hexagonal Si-O ring structure, and the other is Si-O chain structure.

	Ring molecule	Chain molecule
Structure	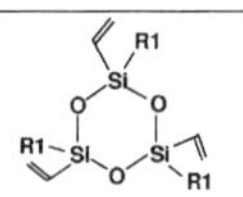	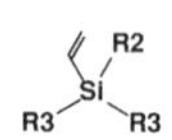
Side Chain	Vinyl, R1	Vinyl, R2, R3

R1,2,3=Hydrocarbon

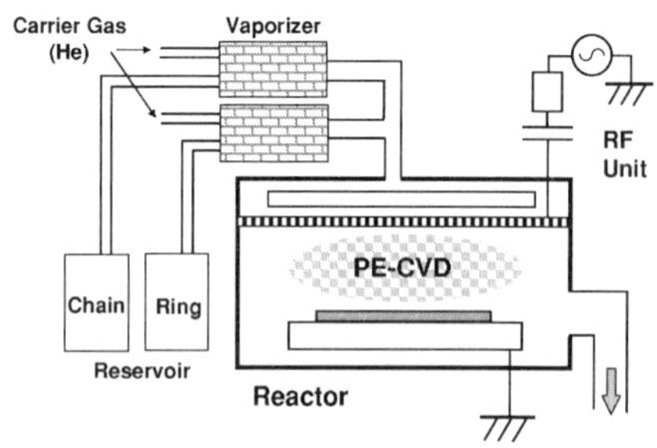

Fig. 2 Schematic drawing of the PE-CVD equipment. Ring and chain molecules were separately vaporized, and introduced into a reactor with He carrier gas.

RESULTS

Film properties

Figure 3 shows the k-value, chemical composition, and elastic modulus of the films as a function of the mixing ratio of the ring and the chain molecules. The mixing ratio of zero means the SiOCH film derived from 100% of the chain one, while the ratio of 1.0 means the SiOCH film from 100% of the ring molecule. The intermediates correspond to co-polymerized films from the ring and the chain molecules. The k-value of SiOCH film decreased with increasing the ring molecule ratio, or essentially increasing the carbon content. The elastic modulus conforms to tendency of the k-value. For the in-situ sequential process, molar ratio of 0, 0.3 and 1.0 were applied to SACC (Hard mask, Cap), cMPS (ILD), and MPS (IMD), respectively.

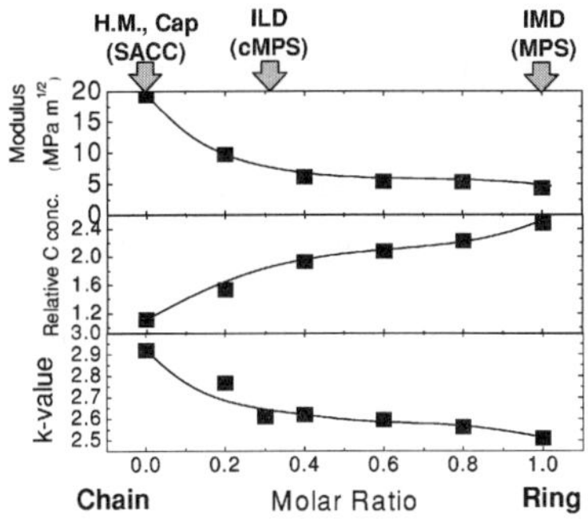

Fig. 3 Molar ratio of chain and ring molecules dependences of k-value, chemical composition and elastic modulus.

Evaluation of adhesion strength

The adhesion strength was evaluated by modified Edge Lift-off Test (m-ELT) which was carried out on blanket film stacks, as shown in Fig. 4. Here, two structures were evaluated for the m-ELT; (1) M1 structure of the stack of SACC-HM/MPS-IMD and (2) M2 structure of SACC-HM/MPS-IMD/cMPS-ILD/SACC-cap. The delamination occurred at the interface between SACC-HM and MPS-IMD irrespective of the M1 or M2 structures.

In order to improve the adhesion strength of SACC/MPS, four types of deposition processes were designed to modify the upper surface of MPS-IMD (Fig. 5). Process 1 is a conventional multi-step deposition, in which SACC-HM and MPS-IMD are deposited separately. Process 2 is an in-situ single-step sequential deposition in one plasma reactor, where the transition layer is controlled only by changing the source gas mixing ratio. The film deposition sequences for the in-situ single-step sequential process are shown in Fig. 6. In cases of Process 3-1 and 3-2, an in-situ adhesion layer (IAL) is deposited on the surface of MPS-IMD. The in-situ adhesion layers are introduced by increasing the He flow (Process 3-1) or increasing both of He flow-rate and RF-Power (Process 3-2). In process 4, the He plasma treatment was performed on the MPS-IMD surface.

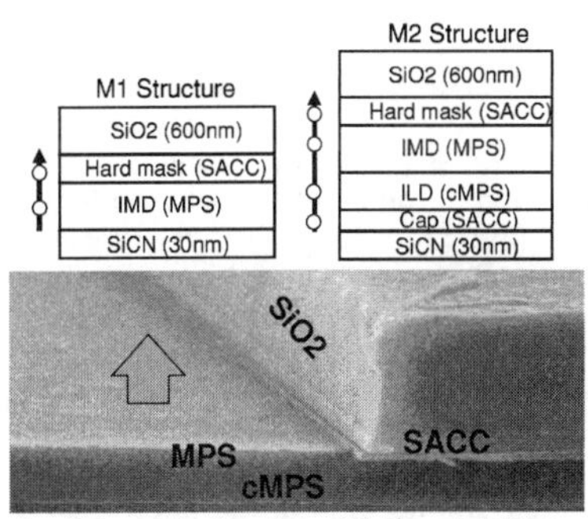

Fig. 4 Schematic drawing of M1 and M2 structure for m-ELT. Below is cross sectional SEM image after m-ELT. The delamination occurred at the interface between hard mask and IMD.

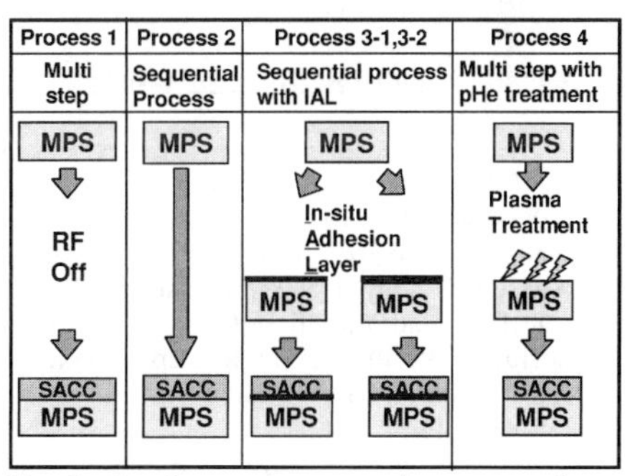

Fig. 5 Interface treatment on the upper surface of MPS (IMD) for delamination measures. In process 3, two types of adhesion layers were applied.

By the m-ELT evaluation (Fig. 7), the adhesion strength at the interface between SACC-HM and MPS-IMD were measured. The in-situ sequential depositions of Process 2, 3-1 and 3-2 had stronger adhesion than the multi-step deposition as Process 1. In case of Process 1, the adhesion strength was too weak to apply the m-ELT for both of M1 and M2 structures. Sequential process without IAL (Process 2) showed insufficient adhesion strength on M2 structure. Especially, the Process 3-2 exhibited high adhesion strength comparable to the Process 4 with He plasma treatment on the MPS-IMD. It is noted that the M2 structure had lower adhesion strength than the M1 structure in spite of the common delamination interface of SACC/MPS, which will be discussed later.

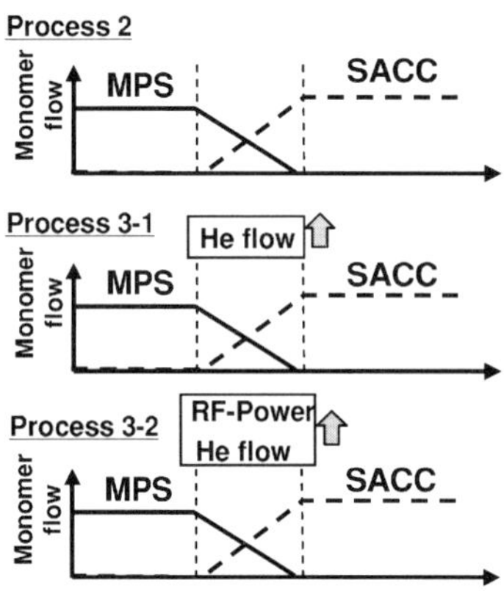

Fig. 6 Film deposition sequence for in-situ sequential process.

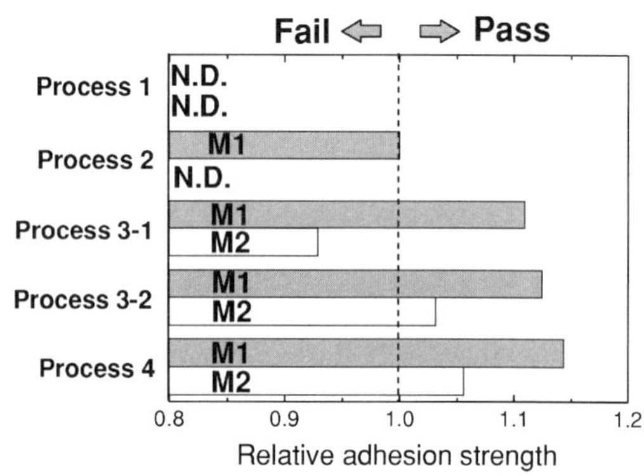

Fig. 7 Adhesion strength between hard mask and IMD measured by m-ELT. M1 structure has higher adhesion strength than M2 structure.

DISCUSSION

Chemical analysis at the interface

In order to clarify differences in the adhesion strength among interface treatments of the SACC/MPS, we analyzed the chemical composition of the interface between SACC-HM and MPS-IMD. Figure 8 shows depth profile of O/Si atomic ratio using XPS. By Process 1 with the weakest adhesion, transition layer is narrow and the O concentration decreased abruptly at the SACC/MPS interface. In case of Process 4 with the strongest adhesion of SACC/MPS, the interface had the highest oxygen concentration by direct He-plasma treatment on the MPS surface. The in-situ single-step sequential deposition processes with IAL (Process 3-1, and 3-2) had wider O-rich transition layer than the conventional multi-step deposition process (Process 1).

Figure 9 shows the carbon and the oxygen atomic ratio at the interfaces. Abundance ratio of the carbon and the oxygen depends on the adhesion strength between MPS-IMD and SACC-HM interface. The interface of in-situ sequential deposition process (Process 2, 3-1 and 3-

2) had higher oxygen and lower carbon concentrations than those of the multi-step deposition (Process 1). The oxygen concentration was higher in the order of Process 3-2 > 3-1 > 2. This result shows that the monomer dissociation would be advanced by increments of the He gas flow and the RF-power, resulting in the in-situ O-rich adhesion layer.

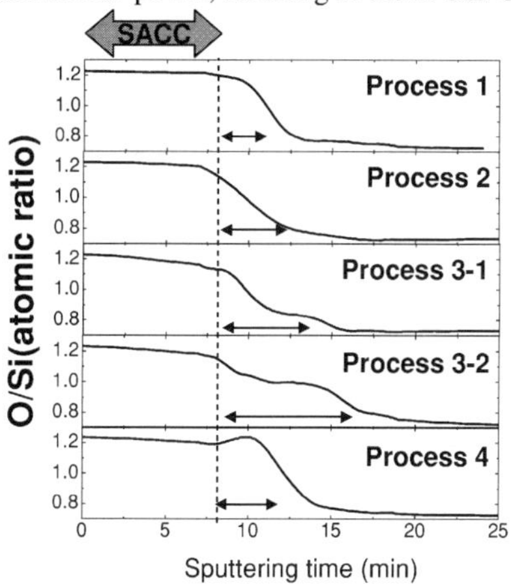

Fig. 8 XPS depth profiles of relative oxygen composition at transition area between hard mask and trench layers. Transition or altered layer is indicated by arrows.

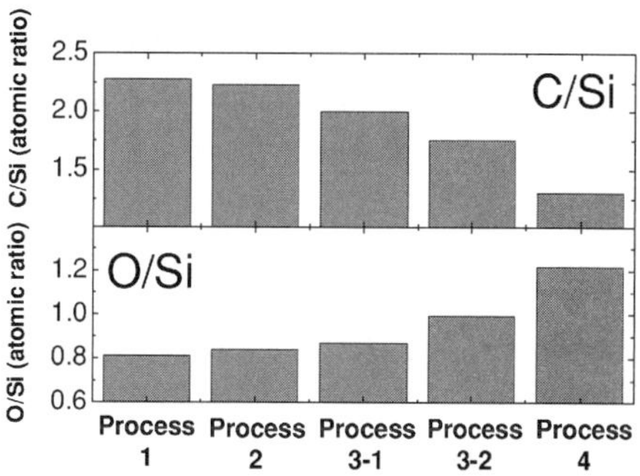

Fig. 9 Relative oxygen and Carbon composition at interface layers by XPS. As adhesion strength becomes higher, interface composition shows a drop in carbon concentration and an increase in oxygen concentration.

Relation between adhesion strength and film stress

Figure 10 shows warpage after SACC-HM deposition on the Si wafer, M1and M2 structures. The amount of warpage depends on both of the film thickness and the film stress. Most of the low-k films have tensile stress, and the wafer warped upward after the low-k film deposition. The SACC-HM, on contrast, had small compressive stress, and the wafer warped slightly downward. The difference of warpage between the M2 structure and the SACC-HM was larger than that between the M1 structure and the SACC-HM due to the film thickness difference. Therefore, the stronger adhesion strength of SACC-HM/MPS-IMD interface is required for the M2 structure, because the warpage difference between compressive SACC and tensile under-layer (MPS/cMPS) is larger than that of M1 structure.

We also evaluated the effects of interfacial chemical and stress controls in the in-situ sequential low-k deposition on the adhesion strength between SACC-HM and MPS-IMD. The film stress of SACC was changed toward the tensile by controlling the deposition condition. There is no difference in the chemical composition between the tensile SACC and the compressive one. The results of m-ELT evaluation are shown in Fig.11. The stress-controlled SACC-HM/MPS-IMD interface showed higher adhesion strength.

Figure 12 shows the adhesion strength as a function of wafer warpage and O/Si ratio at the interface layer. In case of the tensile SACC, while the oxygen concentration at the interface was low, the interface between SACC-HM and the underling layer had high adhesion strength. In addition to the interfacial chemical modification, the stress control toward smaller difference between HM and the underling layer is very effective to improve the adhesion strength.

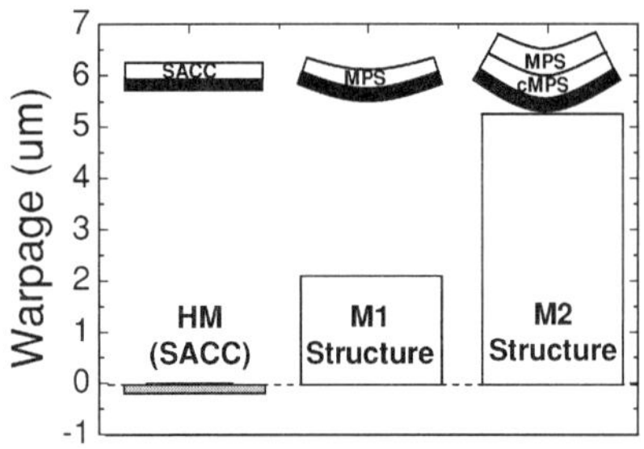

Fig.10 Wafer warpage of M1 and M2 structures. The SACC film has compressive stress and warps toward opposite direction against M1 and M2 structures.

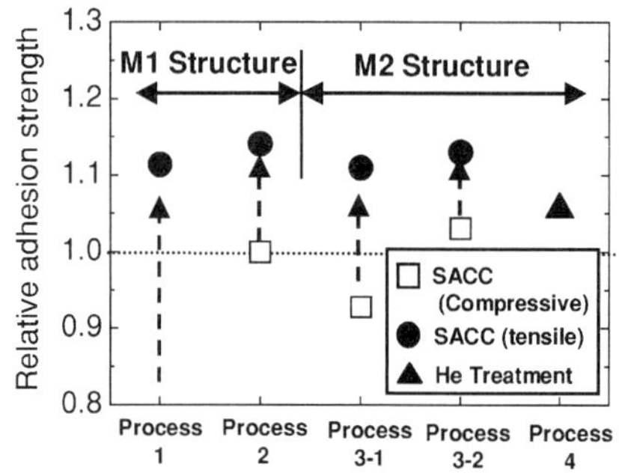

Fig.11 Adhesion strength of SACC-HM/MPS interface. The SACC film with tensile stress shows higher adhesion than that with compressive stress.

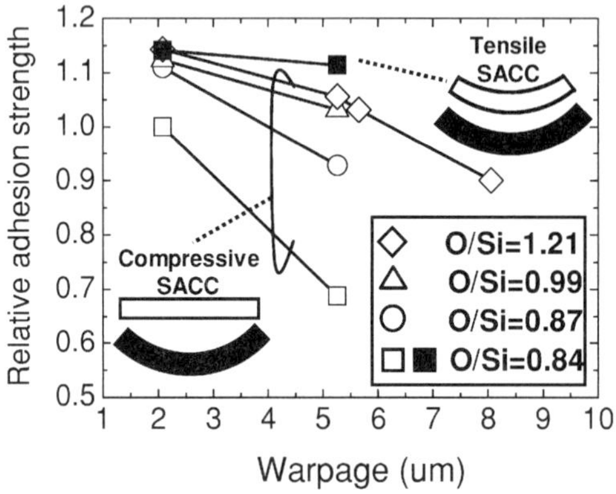

Fig.12 Relative adhesion strength as a function of wafer warpage and interface composition. Solid symbol (tensile-stress-controlled SACC) shows the highest adhesion strength.

CONCLUSIONS

We investigated the effects of interfacial chemical and stress controls in the in-situ sequential low-k deposition on the adhesion strength between HM and IMD. The adhesion strength was increased by the interfacial chemical controls such as Si oxidation or essentially the removal of carbon contents. The interfacial stress control is promising to enhance the mechanical properties of multi-level Cu interconnects for 32nm-node and beyond.

ACKNOWLEDGMENTS

The authors would like to thank Drs. H. Watanabe of NEC Electronics Corporation for research support. The authors would like to thank TOSOH FINECHEM CORPORATION and TOSOH CORPORATION for the precursor supplements of the organo-siloxanes.

REFERENCES

1. Y. H. Wang et al., "The mechanical properties of ultra-low-dielectric-constant films", *Thin Solid Films 462-463*, pp.227-230, 2007.
2. L. L. Mercado, S.-M. Kuo, C. Goldberg and D. Frear., "Impact of flip-chip packaging on copper/low-k structures", *IEEE Trans. Adv. Pack. Vol. 26 No. 4*, pp.433-440, Nov. 2003.
3. M. Tada et al., "Plasma co-polymerization technology with molecular-level structure tightening in in-situ SiOCH stacks for 32nm-node Cu interconnects", *Proceedings of IEEE IEDM 2006*, pp.351-354, Dec. 2006.
4. M. Ueki et al., "Cost-effective and high performance Cu interconnects (keff=2.75) with continuous SiOCH stacks incorporating a low-k barrier cap (k=3.1)", *Proceedings of IEEE IEDM 2007*, pp.973-976, Dec. 2007.
5. Y. Hayashi et al., "Novel molecular-structure design for PECVD porous SiOCH films toward 45nm-node, ASICs with k=2.3", *Proceeding of IEEE IITC 2004*, pp.225-227, Dec. 2004.

Conference Proceedings AMC XXIV © 2009 Materials Research Society

Airgap Structures by using Sacrificial Wet Etch: Fabrication, Thermal and Mechanical Behavior, Reliability

Knut Schulze[1,2], Stefan E. Schulz[1,2], Heinrich Körner[3], and Thomas Gessner[1,2]

[1]Center for Microtechnologies, Chemnitz University of Technology, Reichenhainer Strasse 70, Chemnitz, 09126, Germany

[2]Fraunhofer ENAS, Reichenhainer Strasse 88, Chemnitz, 09126, Germany

[3]Infineon Technologies AG, Otto-Hahn-Ring 6, Munich, 81739, Germany

ABSTRACT

There is an ongoing competition between ultra *low-k* (*ULK*) dielectric integration and airgap structures to achieve ultra low effective k-values and thus low parasitic capacitance in interconnect coupling. Since 2005 airgap structures are declared alternatives to *low-k* integration and became more and more promising since sufficient ultra *low-k* integration is delayed due to an immense number of integration problems. This paper discusses the impact of both, *low-k* integration and airgap structures, on the thermal behaviour of interconnect systems by *FEM* simulation in comparison to conventional Cu/SiO_2 interconnect system. Related to this a very advantageous behaviour of airgap structures is shown. Secondly the mechanical behaviour (one of the *low-k* integration issues) of two airgap technologies called "*mask*" and "*spacer*" is investigated by *FEM* simulation. The results are discussed in comparison to experimental data in terms of critical stress and adhesion strength. The mechanical loads do not exceed the critical yield stress of *SiC* films (3 GPa) and the *Cu/SiC* interfacial adhesion strength of 1.48 GPa. Finally, results of electromigration life time measurements are shown. A similar behaviour of airgap structures and conventional Cu/SiO_2 interconnects was verified.

INTRODUCTION

The formation of airgap structures is a promising alternative to achieve *ULK* performance for future technology nodes. It has been shown that k_{eff} - values of 2.2 are well achievable for different integration schemes and *MPU* generations [1],[2]. Beyond this, several airgap approaches largely share conventional processing of *Cu* and SiO_2. Two similar sacrificial layer airgap approaches ("*mask*" and "*spacer*") were developed at *TU Chemnitz* (schematics of processing are shown in Fig. 1 and Fig. 2). The feasibility of both technologies was demonstrated

Fig. 1: "*mask*" approach: *a)* after *CMP* and "wet-etch-mask" patterning; *b)* and *c)* partially wet etched by *buffered HF*; *d)* fully removed *USG* dielectrics, encapsulated airgap structures (using non-conformal *CVD*)

Fig. 2: "*spacer*" approach: *a)* after *CMP* and "wet-etch-mask" patterning; *b)* and *c)* partially wet etched by *buffered HF*; *d)* fully removed *USG* dielectrics, encapsulated airgap structures (using non-conformal *CVD*)

in [3]. Interconnect systems according to both technologies were characterized in terms of their electrical properties (reduction of inter-metal capacitances and k_{eff}) [1]. Both approaches use chemical selectivity. On the one hand the sacrificial material *PECVD* SiO_2 is well removable by *buffered HF* (hydrofluoric acid). On the other hand *SiC*-films, conductive diffusion barriers and copper lines are proper chemically stable against *HF* containing solutions [2],[4],[5]. Impact of *HF* based wet etch media as well as processing challenges of airgap formation of both technologies are discussed in [4].

The first part of this paper discusses the impact of airgap structures on the thermal behaviour of a complex interconnect system in comparison to the integration of *low-k* materials and conventional SiO_2/Cu interconnect architectures. The results of the thermal simulation shown here were discussed more in detail in [6]. Second part is dedicated to the mechanical behaviour of airgap structures of the "*mask*" and "*spacer*" approach during encapsulation, exposure to atmospheric pressure as well as *CMP* treatment. The impact on cantilever films and relevant interfaces of the interconnect system will be discussed in terms mechanical stress. Finally some results of life time measurements of airgap structures will be compared to the results of dense SiO_2 interconnects without sacrificial etch. These investigations have been performed at a single layer metal system (no via chains) of 1 μm interconnect width.

THEMAL BEHAVIOUR OF AIRGAP STRUCTURES

The global heat flux within *IC* metallization systems is perpendicular to the silicon substrate, which is usually the actively cooled heat sink of the whole system. This direction of global heat flow is mainly caused by the huge ratio of footprint area to the stack thickness of such architectures. The heat flow of complex metallization systems of 45 nm node geometries and up to 5 differently densely packed metal levels was investigated using *FEM* simulation. Furthermore the impact of the following conditions was checked:

- Dielectric material (SiO_2, porous *low-k* material, airgaps) and integration scheme (full or hybrid)
- Availability of vias and via density (interconnect length between two vias)
- Interconnect and via arrangement (matched and mismatched)
- Thickness of additional functional films (hard mask films, etch stop films, ...)

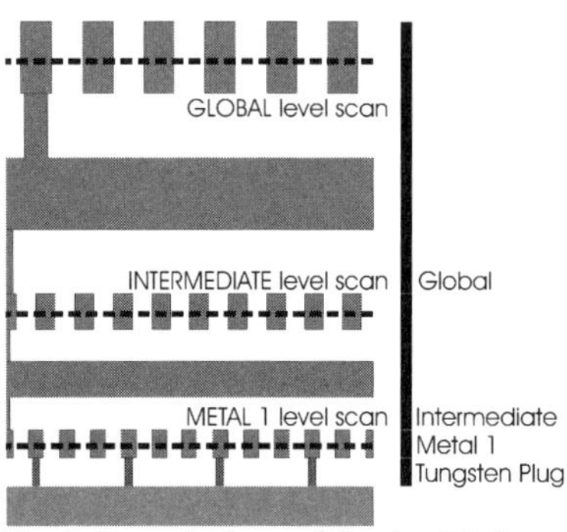

Fig. 3: General scheme of the simulated interconnect system of the 45 nm technology node and plotted scan paths (dotted lines) for temperature analysis

Simulation models

Fig. 3 shows a general schematic of interconnect system which was used for the thermal simulation. Five metal levels of dimensions according to the 45 nm technology node of the *ITRS* 2007 requirements were investigated. Additionally to the line adjustment the scan paths for temperature plots are shown in Fig. 3 (dotted lines).

They are placed across the periodic metal lines of *metal 1*, *intermediate*, and *global wiring*. The following boundary conditions were used during the *FEM* simulation. A passivation stack consisting of 300 nm *PECVD* SiO_2 = 1.05 W/mK and 500 nm *PECVD* Si_3N_4 = 0.7 W/mK was applied on top of all structures. Heat dissipation at the surface of the passivation stack was therefore set to 0 W/m². The substrate temperature was constant. Symmetrical conditions were concerned for the side surfaces. This means that no heat exchange through these areas is assumed to occur. The results of the simulations were the temperature increases in the interconnect systems compared to the substrate temperature. All simulated configurations/models are shown in Table 1 and Table 2.

Table 1: Summary of all simulated airgap and SiO_2 architectures

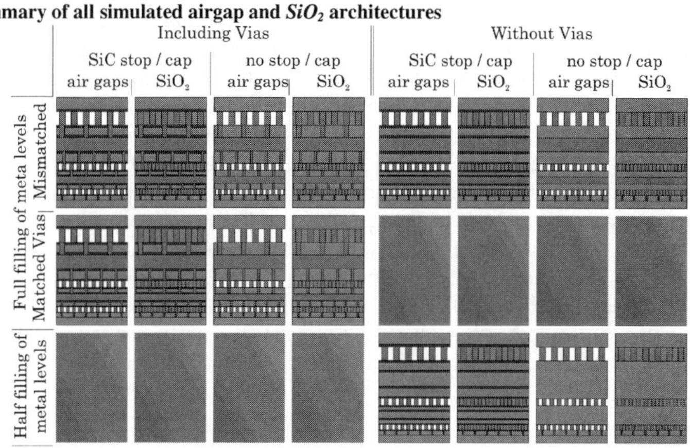

The sub-divisions include a variation of the dielectric material - airgaps (λ_{air} = 0.026 /mK) and *PECVD* SiO_2 (λ_{SiO2} = 1.05 W/mK) in Table 1, porous ultra *low-k* ($\lambda_{low\text{-}k}$ = 0.15 W/mK) dielectric in Table 2. Furthermore the integration scheme of the *low-k* case is varied in Table 2. In these

Table 2: Summary scheme of all simulated hybrid *low-k* and full *low-k* architectures

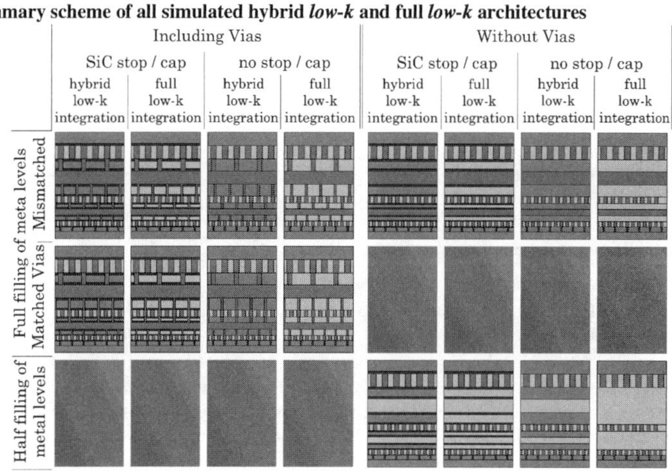

cases "*hybrid*" means a *low-k* material integration in the metal levels and *PECVD* SiO_2 in the via levels. "*Full low-k integration*" defines a full integration of the *low-k* material in both via and metal levels of the interconnect systems. All functional films (used for patterning, hard mask, etc.) were stipulated to be *PECVD SiC* with a thermal conductivity of λ_{SiC} = 0.3 W/mK and a thickness of 10 nm for *metal 1* level, 20 nm for *intermediate* dimensions, and 30 nm for *global wiring* (represented by the table rows identified with "*SiC stop / cap*" of Table 1 and Table 2). For the evaluation of the impact of the *SiC* films, cases of zero-thickness-*SiC* were introduced ("*no stop / cap*"). The schemes of the left side of each table use vias ("*Including vias*") and the schemes of the right half no vias.

Via integration is illustrated in two principle ways – "*mismatched*" and "*matched vias*". For "*matched vias*" the metal lines are contacted on top and at the bottom. On the other hand the "*mismatched vias*" have an alternation of top and bottom contacting. Both cases are similar, but there are differences in their thermal behaviour. Heat flux of "*mismatched vias*" through the metal areas (lines and vias) is interrupted in comparison to the "*matched vias*". This enforces a heat flux through the dielectrics and this increases the impact of different dielectrics/integration schemes. Independent of this, via-free architectures have been examined, too. These cases are shown on the right half of both tables, indicated by "*without vias*". Two types are shown. The "*full filling*" rows of both tables illustrate that all 5 metal levels of Fig. 3 contain dense metal lines. "*Half filling*" is characterized by alternating metal levels filled with dielectric material and metal levels filled with dense metal lines. They provide the most extensive dielectric areas of all simulation models. Joule heating caused by a constant current density in the interconnect lines was applied to all metal areas. These homogenous heat sources represent the loads of the simulations.

Simulation results

According to the models of Table 1 and Table 2 the absolute distributions of temperatures were simulated. These maximum temperature loads are summarized in Fig. 4 and Fig. 5. The most obvious impact is caused if vias are not available. The temperature rises of the models which include vias are negligibly low compared to the rises of the via free models. For that reason these results are not part of the discussion here (for more details see [6])

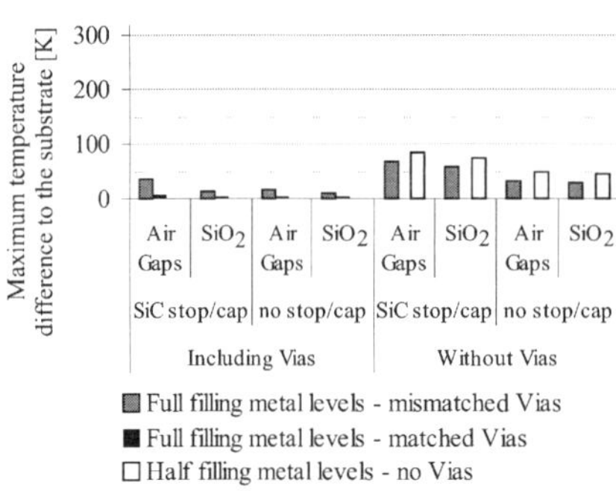

Fig. 4: Maximum occurring temperature rise of interconnect systems using airgaps or SiO_2 (see the cohort of models in Table 1) for dielectrics depending on cap and stop film materials and via constellation

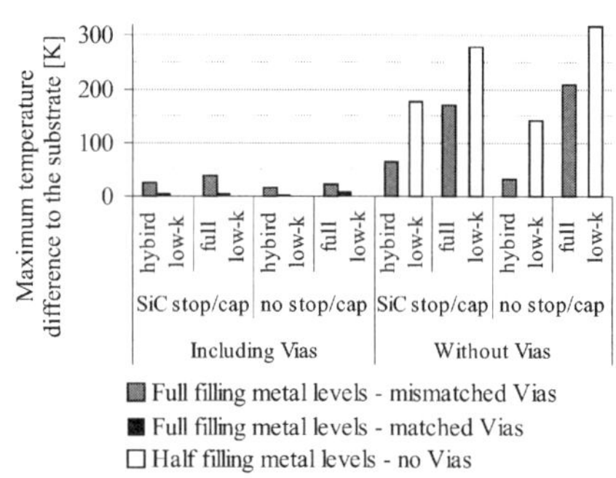

Fig. 5: Maximum occurring temperature rise of interconnect systems using *low-k* or full *low-k* integration scheme (see the cohort of models in Table 2) depending on cap and stop film materials and via constellation

"Without-Via" cases:

The worst case is the "*full low-k integration*" in combination with "*half filling*" of the metal lines. Large areas of very low thermal conductivity are responsible for an intensive heat accumulation in the upper levels of the interconnect system. The temperature scans showed discrete temperate levels for each metal level. This is caused by the absence of copper vias which take the main heat flux within each via level due to the low thermal conductivity of dielectric films ($\lambda_{dielectric\_film} \leq 1$ W/mK) compared to copper ($\lambda_{Cu} \approx 400$ W/mK). The main heat flux occurs perpendicular to the *Si* substrate / heat sink. The temperature distributions within each metal level show no significant fluctuation. The maximum temperature increase is directly coupled to the size of the dielectric area and dielectric material (value of thermal conductivity). According to this, *PECVD* SiO_2 filled structures offer the lowest min/max temperature distribution - only insignificantly better than the airgap variants. Because of the discrete temperature levels (no inter-metal temperature gradient) in all these cases, there is no necessity for an inter-metal heat flux (assumed that air is exclusively used for *IMD*). Taking this into account, the low thermal conductivity of air does not deteriorate heat dissipation remarkably. The performance of airgap technology is nearly as good as conventional SiO_2. This evidence is supported by the results of Shieh et al. [7].

Impact of SiC stop / cap / mask films

The impact of *SiC* film thickness depends on the integration scheme and the integrated material. Because of the lower thermal conductivity of *SiC* in comparison to SiO_2 furthermore thinned *SiC* films give an improvement of heat dissipation for the SiO_2, hybrid *low-k*, and airgap based technologies (Fig. 4). In case of full *low-k* integration *SiC* films of higher thickness are responsible for a reduction of the maximum temperature because of their higher thermal conductivity, λ_{SiC} = 0.30 W/mK, in comparison to values of porous *low-k* materials of about 0.15 W/mK.

MECHANICAL BEHAVIOR OF AIRGAP STRUCTURES

The definition of geometries and design rules for airgap based interconnect architectures requires an estimation regarding mechanical loads within the necessary encapsulation films. For that reason *FEM* simulations have been carried out using *Ansys*® simulation software.

Modelling and input data

Both approaches ("*mask*" and "*spacer*", Fig. 1 and Fig. 2) were included within these simulations, moreover symmetric and asymmetric configurations were considered for each approach. Schematics of the developed models are shown in Table 3.

Table 3: 2d-models for the *FEM* simulation of both approaches ("*mask*" and "*spacer*") including symmetric and asymmetric configurations

symmetric configurations		asymmetric configuration	
"*mask*" approach	"*spacer*" approach	"*mask*" approach	"*spacer*" approach
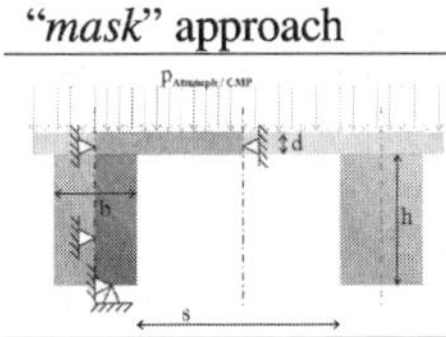	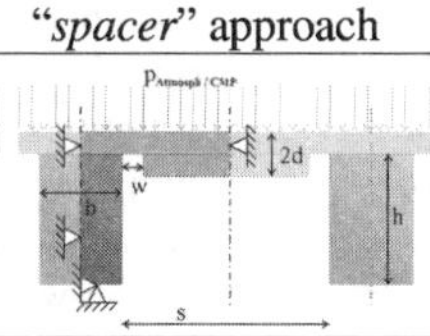	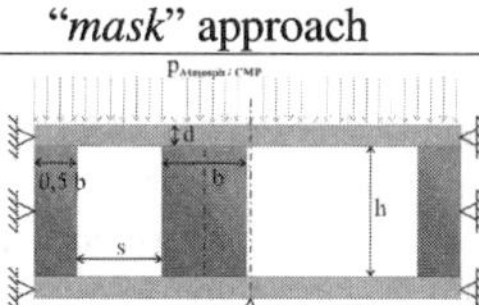	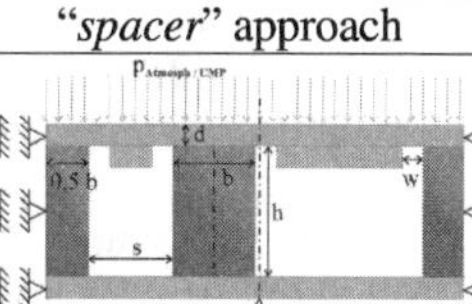

Using these models mechanical load and deformation of the cantilever-type structures as well as the stress at the *Cu/SiC*:H interface have been investigated. Geometry parameters and initial intrinsic stress within the *SiC:H* layer have been varied as given in Table 4. Necessary material parameters for simulations are given in Table 5. These parameters were taken from literature or determined by additional experiments. Elastic modulus was determined applying nanoindentation on *SiC:H* films of 300 nm and 500 nm thickness, respectively. Critical yield (flow) stress of 3.2 GPa was determined by "*cyclic partial loading-unloading nanoindentation*" and data analysis using *Elastica*® software. The critical yield stress of the *SiC:H* film is

Table 4: Variation range of *FEM* simulations for symmetric and asymmetric interconnect configurations ("*spacer*" and "*mask*" approach)

	Parameter	Symbol	Unit	Variation range
Symmetric configuration:	Interconnect width	b	nm	17, 38, 54, 75
	Interconnect height	h	nm	44, 62, 87, 119
	Interconnect spacing	s	nm	20 ... 150
	SiC:H thickness	$d_{SiC:H}$	nm	3 ... 30
	Spacer width	ω	nm	3 ... 9
	SiC:H intrinsic stress	$\sigma_{SiC:H}$	MPa	± 50
Asymmetric configuration:	Interconnect width	b	nm	17, 38, 54, 75
	Interconnect height	h	nm	44, 62, 87, 119
	Interconnect outer spacing	g	nm	5b ... 9b
	Interconnect inner spacing	S	nm	b ... $(g-b)/2$
	SiC:H thickness	$d_{SiC:H}$	nm	3 ... 30
	Spacer width	ω	nm	3 ... 9
	SiC:H intrinsic stress	$\sigma_{SiC:H}$	MPa	±50

the most essential parameter for the evaluation of the mechanical stability. A further important parameter is the adhesion strength of the *Cu/SiC:H* interface. Using results obtained from 4-point bending tests (measurement of adhesion energy) a critical interface stress of 1.48 GPa could be calculated by an approximation [8]. This is the weakest interface in comparison to the critical stresses σ_{db} of the interfaces of conductive and dielectric diffusion barriers (σ_{db} >>5 GPa) [9].

Table 5: Material parameters for Cu and SiC:H used within simulations

Material	E-modulus [GPa]	Poisson's ratio	Yield stress [GPa]
Cu	$E_{Cu} = 150$	$\nu_{Cu} = 0{,}32$	not necessary
SiC:H	$E_{SiC:H} = 32$	$\nu_{SiC:H} = 0{,}14$	$\sigma_{SiC:H\ critical} \geq 3{,}2$

Mechanical loads on airgap architectures can be classified in 3 states: The removal of sacrificial material leads to a deformation of the *SiC:H* films caused by intrinsic film stress. This deformation causes a mechanical load of the membrane-like structures and their fixing points, what is considered as ***load state 1***. During subsequent enclosure by low pressure deposition (*LPCVD*) a pressure typically below 1 mBar is encapsulated inside the gaps. That generates a mechanical load from the atmospheric pressure (app. 1 Bar) if the wafer is released from the deposition tool and should be considered as ***load state 2***. Within further processing, e.g. of multi level architectures, *CMP* steps become necessary, which introduce an additional mechanical load and are considered as ***load state 3***. Load state 2 and load state 3 can be simplified by a worst-case consideration. Therefore, the oxide thickness for the gap enclosure is considered to be zero. That means, the whole mechanical load acts to the cantilever *SiC:H* films only. Moreover, *CMP* pressure is assumed to be 1 Bar (typically that pressure does not exceed 0.6 Bar - app. 9 psi). So the maximum extrinsic mechanical load occurs in load state 3 with a total load of 2 Bar.

Results

Mechanical stress of cantilever SiC:H films

Fig. 6 shows the resulting maximum stresses within the *SiC:H* film of load state 2 (1 Bar) and load state 3 (2 Bar) in dependence on geometry parameters and initial intrinsic stress (see Table 4) for airgap structures based on the *"mask"* approach. The load state 1 is not included because no additional stress occurs in comparison to the intrinsic film stresses. Moreover, this state (load state 1) is statically instable in the case of compressive tensile stress. Because of this stress-free or tensile intrinsic stress is beneficial to improve reproducibility of airgap structures according to the *"mask"* approach. Load state 2 and 3 caused a negligibly bending of the cantilever films in the range of 0 nm up to 3 nm, directed into the cavities. With increasing spacing and decreasing *SiC:H* thickness the resulting film stress is rising. However, the maximum found stress did never exceed 300 MPa. This value is

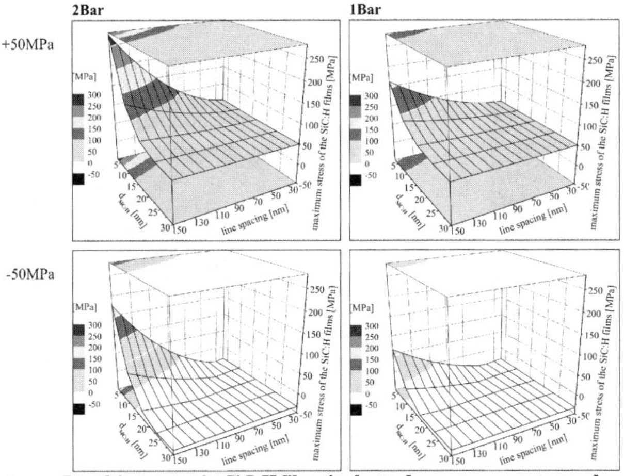

Fig. 6: Resulting stress in *SiC:H* films in dependence on geometry and intrinsic stress for airgap architectures based on mask approach and different external loads (load state 3 and 2); symmetric configuration

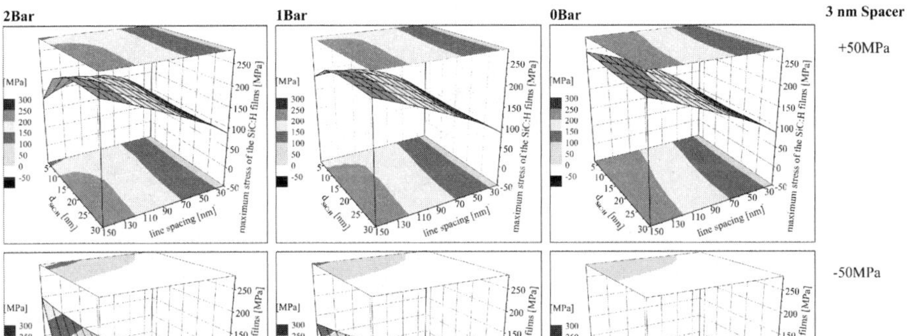

Fig. 7: Resulting stress in *SiC:H* films in dependence on geometry and intrinsic stress for airgap architectures based on spacer approach with 3 nm spacer width and different external loads (load state 3, 2, and 1); symmetric configuration

approximately one order of magnitude lower than the critical yield stress of these films. The results for the "*spacer*" approach with a spacer width of 3 nm is given in Fig. 7. Even there is no external load (load state 1) the calculated stresses show a dependency on the geometrical dimensions and differ tremendously from the intrinsic stress of the cantilever films. This behaviour is related to the *SiC:H* double layer as necessary for the "*spacer*" approach. Fig. 8 shows the basic motion of the cantilever *SiC:H* films in the case of tensile stress. The contrary behaviour is given by Fig. 9 for compressive intrinsic stress. The entirety of all calculations

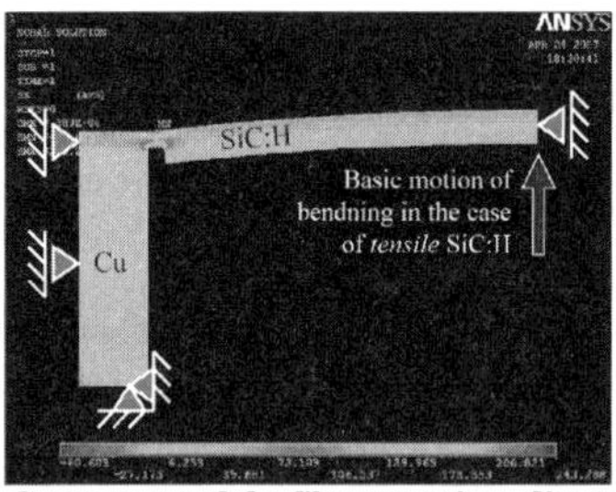

Fig. 8: Component of the film stress in x-direction and schematical bending of the cantilever *SiC:H* film for $\sigma_{SiC:H}$ = +50 MPa (tensile), load state 1, interconnect line spacing s = 150 nm, $d_{SiC:H}$ = 3 nm, spacer thickness w = 3 nm

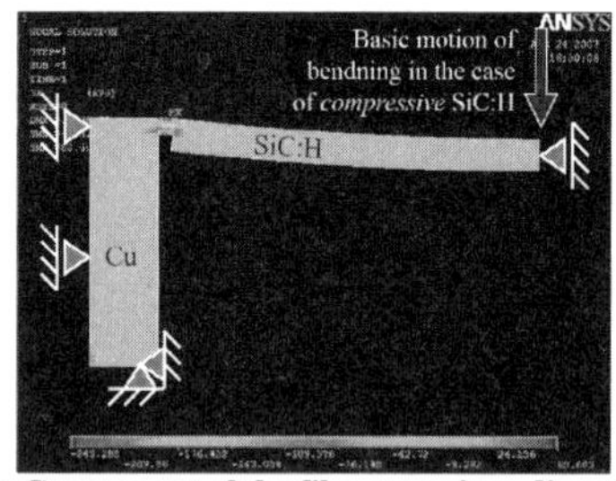

Fig. 9: Component of the film stress in x-direction and schematical bending of the cantilever *SiC:H* film for $\sigma_{SiC:H}$ = -50 MPa (compressive), load state 1, interconnect line spacing s = 150 nm, $d_{SiC:H}$ = 3 nm, spacer thickness w = 3 nm

("*spacer*" approach, compressive and tensile stresses, load state 1) show bending in a non-critical range of ± 0.5 nm. Due to the bending behaviour shown in Fig. 8 external loads reduce the stresses occurring within the cantilever films having intrinsic tensile stress. The maximum stresses are located in the range of the spacer slots next to the *Cu* interconnect lines (single layer *SiC:H* films, see Fig. 8 and Fig. 9). In the case of compressive stressed cantilever films the

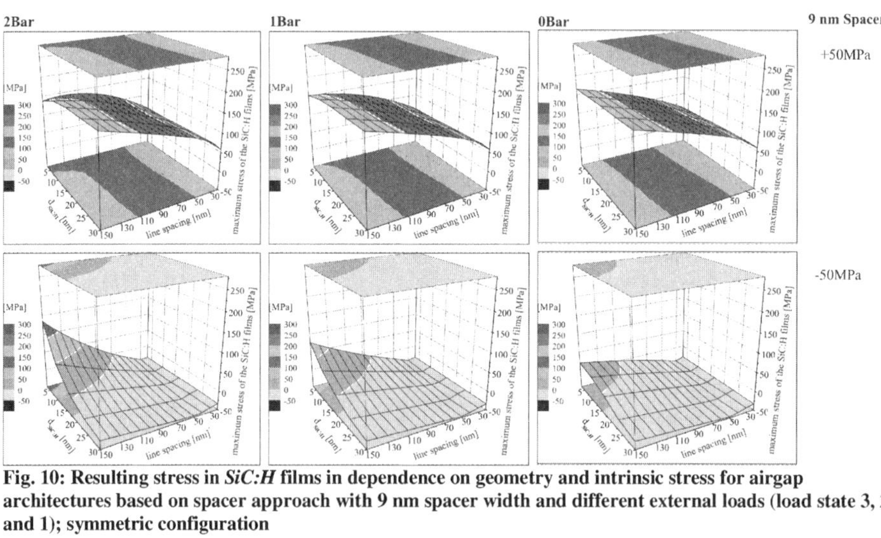

Fig. 10: Resulting stress in *SiC:H* films in dependence on geometry and intrinsic stress for airgap architectures based on spacer approach with 9 nm spacer width and different external loads (load state 3, 2, and 1); symmetric configuration

structures according to the "*mask*" and the "*spacer*" approach show a similar behaviour. The maximum stresses occur for the largest interconnect spacing (large membranes under external pressure loads), but interconnect structures of such dimensions are not intended to be carried out with airgaps. The increase of the spacer width from 3 nm (see Fig. 7) to 9 nm (Fig. 10) nm leads to a decrease of the resulting stress within the *SiC:H* film in all cases. Maximum stress here was found at 170 MPa, approximately 65 MPa lower than the value of the 3 nm spacer variant.

Even though the calculated stress values of all models show a strong dependency on geometrical dimensions as well as on external mechanical loads, these values appear non-critical since they are at least one order of magnitude lower than the critical yield stress of *SiC:H* films. Concerning the "*spacer*" approach compressive *SiC:H* films are favoured.

Mechanical stress of the SiC:H/Cu interface

Fig. 11 shows the results regarding stress and adhesion strength at the *Cu/SiC:H*-interface in dependence on geometry and initial intrinsic stress for the "*mask*" approach. The stress values

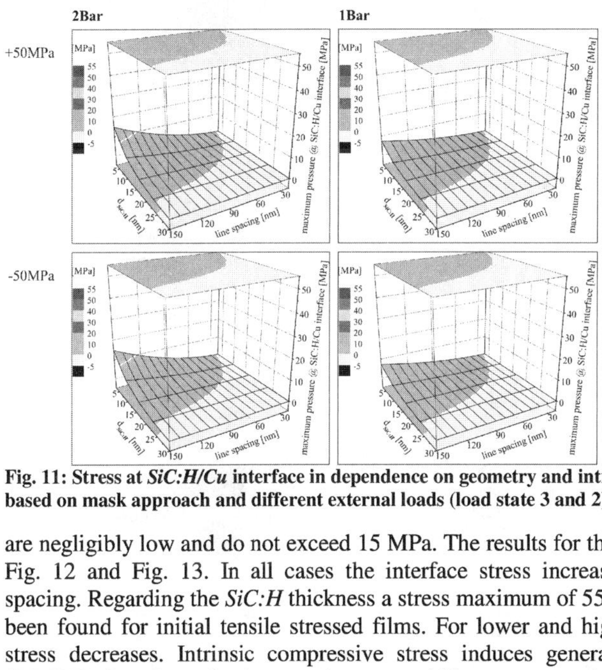

Fig. 11: Stress at *SiC:H/Cu* interface in dependence on geometry and intrinsic stress for airgap architectures based on mask approach and different external loads (load state 3 and 2); symmetric configuration

are negligibly low and do not exceed 15 MPa. The results for the "*spacer*" approach are given in Fig. 12 and Fig. 13. In all cases the interface stress increases with increasing interconnect spacing. Regarding the *SiC:H* thickness a stress maximum of 55 MPa at a thickness of 15 nm has been found for initial tensile stressed films. For lower and higher *SiC:H* thicknesses interface stress decreases. Intrinsic compressive stress induces generally no significant force to the *SiC:H/Cu* interface. An increased spacer width causes a reduction of the induced interfacial force. Compared to the critical stress σ_k of the *SiC:H/Cu* interface with > 1 GPa the stress values of Fig. 11, Fig. 12, and Fig. 13 can be neglected. There is no hint for a delamination at the *SiC:H/Cu* interface at all.

For asymmetric interconnect configurations an absolutely similar behaviour has been found and therefore needs not to be discussed in detail again.

Summarizing all mechanical simulations the following conclusions can be drawn:

- The maximum resulting stress within the *SiC:H* films (appr. 300 MPa) caused by intrinsic and extrinsic loads is far away from the critical yield stress (3.2 GPa) for that material.
- The maximum resulting interface stress (65 MPa) at the *Cu/SiC:H* interface is also far away from the critical value (1.48 GPa).
- With an ongoing shrinking of feature sizes the stress values will further decrease.
- In consideration of the feature sizes given by the *ITRS* roadmap intrinsic and extrinsic mechanical loads do not generate any limitations of the airgap design.

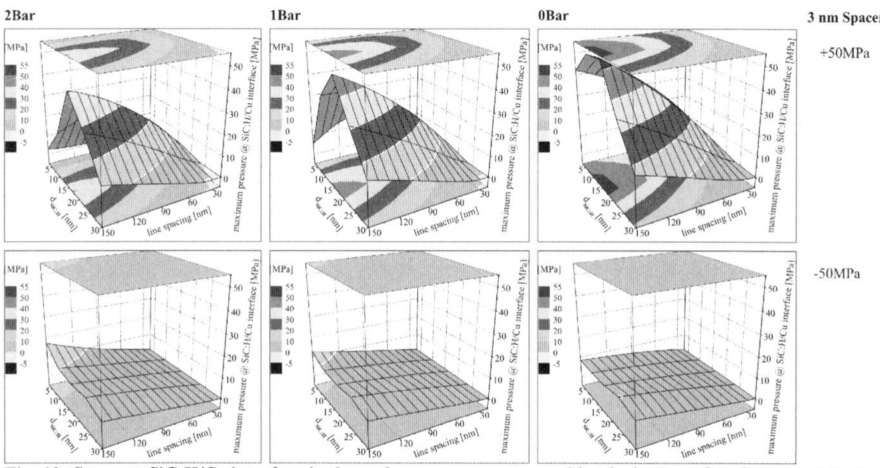

Fig. 12: Stress at *SiC:H/Cu* interface in dependence on geometry and intrinsic stress for airgap architectures based on spacer approach with 3 nm spacer width and different external loads (load state 3, 2, and 1); symmetric configuration

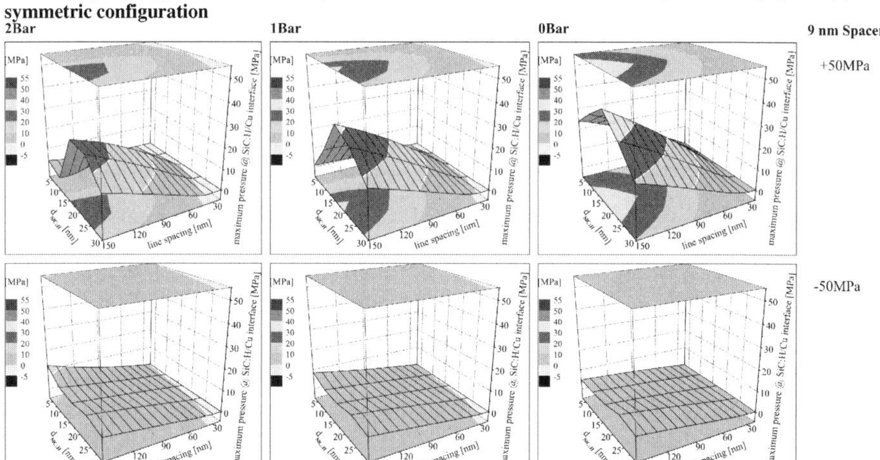

Fig. 13: Stress at *SiC:H/Cu* interface in dependence on geometry and intrinsic stress for airgap architectures based on spacer approach with 9 nm spacer width and different external loads (load state 3, 2, and 1); symmetric configuration

LIFE TIME MEASUREMENTS

Sample preparation

Test wafers were fabricated according to the "*spacer*" approach. On the one hand side the structures were processed without sacrificial wet etch (*SiO_2* reference, see example in Fig. 14). On the other hand local airgaps were produced by *buffered HF* treatment (2 min exposure time, see airgap structures of Fig. 15). The following geometries were used: interconnect length = 800 µm, line width = 1 µm and 2 µm, line height = 500 nm, airgap width ≈ 500 nm … 1 µm (due to the spherical etch profile). 50 % of the interconnect line surface was exposed to *buffered HF* during the wet etch treatment (ratio of hard mask opening, see the schematics of Fig. 2).

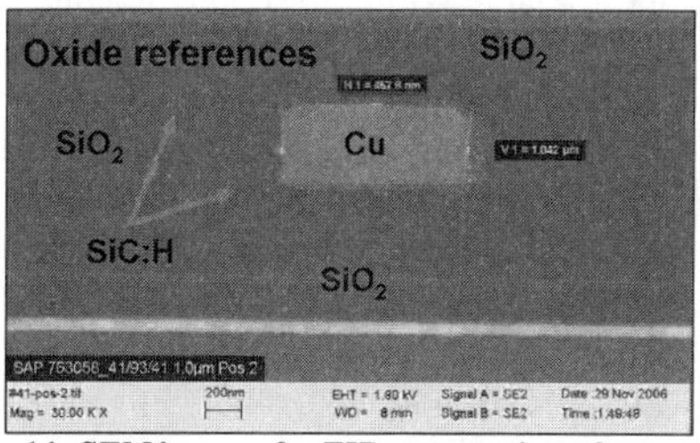

Fig. 14: *SEM* image of a *FIB* preparation of an interconnect line, embedded in dense *SiO_2* (reference sample)

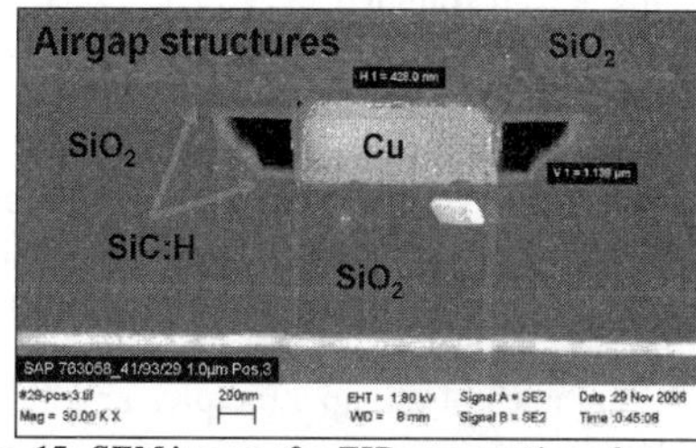

Fig. 15: *SEM* image of a *FIB* preparation of an interconnect line, surrounded by local airgaps

Measurements

177 test structures have been selected for the reliability tests. The average resistivity of the interconnect lines embedded in *SiO_2* was $\rho_{Cu(SiO2)}$ = 1.77 µΩ·cm (average ohmic resistance $R_{1\mu m}$ = 28.5 Ω, $R_{2\mu m}$ = 14.2 Ω). The according value for the lines surrounded by airgaps was $\rho_{Cu(airgaps)}$ = 1.77 µΩ·cm (average ohmic resistance $R_{1\mu m}$ = 28.3 Ω, $R_{2\mu m}$ = 14.1 Ω). Deviation of the resistivity and resistance values of *SiO_2* and airgap variants do not exceed the deviation due to geometrical tolerances of the preparation. Electromigration tests have been performed on packaged samples under accelerated test conditions (T = 250 °C, J = 35 mA/µm²). Failure times t(50%) (using a fail criteria dR/R0 ≥ 20 %) were determined to be approximately 1000 h for the dense *SiO_2* and approximately 1200 h for the airgap structures. Fig. 16 is the corresponding plot of cumulative failures versus time. It shows that both groups exhibit a lognormal failure distribution and a very comparable spread of failures versus time, independent from line width. Extrapolation of the experimental results to a potential set of

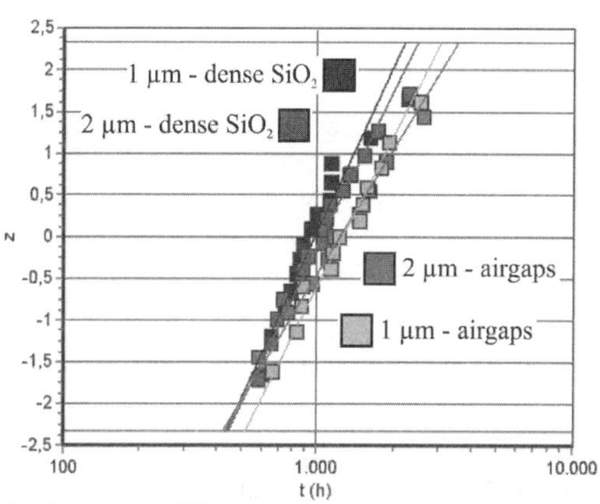

Fig. 16: Results of *EM*-measurements of dense *SiO_2* reference samples and airgap structures of 1 µm and 2 µm interconnect width (early *EM*-failure excluded, t < 500 h)

operation conditions (T_{OP} = 100 °C, t_{OP} = 11.4 a, 0.1 % failure rate) reveals that an aggressive design current density of 10 mA/µm² would be easily met (in excess of approximately 10x) for both the airgap samples and the dense SiO_2 references.

CONCLUSIONS

The impact of progressive dielectrics (*low-k* dielectrics and airgap structures) on the thermal behaviour of an interconnect system was investigated by *FEM* simulation compared to conventional dense SiO_2. The results have shown that selective airgap approaches have the potential to provide a thermal behavior comparable to SiO_2 based interconnect systems. Moreover the results show an extensive heat accumulation for full *low-k* integration schemes. This is well-founded since densely packed areas have a large portion of highly conductive metal materials, which take over the main heat conduction.

Investigations of the mechanical behaviour of airgap structures (using up-to-date node geometries) during the encapsulation of airgaps (*CVD* film deposition), exposure to air and *CMP* treatment have shown distinct loads to cantilever films and interfaces of the interconnect system. Nevertheless, compared to critical yield stress and interfacial adhesion strength, these loads are non-critical and give no hint for any interface delamination or collapse of cantilever films.

First reliability measurements using relaxed interconnect dimensions showed comparable *EM*-failure rates of airgap structures and Cu/SiO_2 integrations without the treatment of sacrificial wet etch.

REFERENCES

[1] K. Schulze, S.E. Schulz, M. Rennau, T. Gessner, Proceedings of the Materials for Advanced Metallization Conference (MAM) 2006, published in Microelectronic Engineering 83 (2006), pp. 2324-2328

[2] R. Gras *et al.*, Proceedings of the International Interconnect Technology Conference (IITC) 2008, p. 196-198

[3] K. Schulze, S.E. Schulz, M. Rennau, T. Gessner, Proceedings of the Advanced Metallization Conference (AMC) 2005, p. 309-316

[4] K. Schulze, S.E. Schulz, T. Gessner, Proceedings of the Advanced Metallization Conference (AMC) 2007, p. 301-308

[5] J. P. Gueneau de Mussy, O. Richard, G. Beyer, K. Maex, Electrochemical and Solid-State Letters 7 (11) 2004, p. G286-G289

[6] K. Schulze, S.E. Schulz, T. Gessner, Proceedings of the Advanced Metallization Conference (AMC) 2006, p. 445-451

[7] P. Shieh, L.C. Bassman, D.-K. Kim, K.C. Saraswat, M.D. Deal, J.P. McVittie, IEEE, IITC Conference 1998, pp. 125-127

[8] S.-Y. Chang, Y.-Sh. Lee, and Ch.-L. Lu, J. Electrochem. Soc. 154 (4) 2007, p. D241-248

[9] K. Schulze, PhD-thesis, 2008, http://archiv.tu-chemnitz.de/pub/2008/0066/index_en.html (german language)

ACKNOWLEDGEMENT

This research is supported by the European Commissions Information Society Technologies Programme, under contract No. IST-507587 (NanoCMOS) and IST-026828 (PULLNANO). The authors acknowledge Bernd Willems and Wolfgang Keller (Infineon Technology AG) for the *EM* measurements and preparations.

Conference Proceedings AMC XXIV © 2009 Materials Research Society

Air Gap Integration of 45nm Node Cu Interconnects and Evaluations of Electrical and Mechanical Characteristics

T. Harada, A. Ueki**, S. Kido*, K. Tomita*, Y. Kanda, T. Sasaki**, H. Tsuji**, T. Furuhashi*, T. Kabe, J. Shibata, A. Iwasaki**, J. Izumitani*, S. Matsumoto, Y. Kawano* and T. Ueda

ULSI Process Technology Development Center, Matsushita Electric Industrial Co., Ltd., Kyoto, Japan
* Process Technology Development Div., Renesas Technology Corp., Hyogo, Japan
** Engineering Department, Panasonic Semi-Conductor Engineering Co., Ltd., Kyoto, Japan

TEL : +81-75-662-8461, FAX : +81-75-662-8995, E-MAIL : harada.takeshi@jp.panasonic.com

ABSTRACT

Cu interconnects with air gaps were fabricated at 45nm node. Large air gaps were successfully formed between Cu lines. Air gap formation was properly controlled by AGE (Air Gap Exclusion). Keff as low as 2.3 was obtained, which surpassed the ITRS requirement for hp45 by far. Mechanical characteristics of Cu interconnects with air gaps were evaluated systematically for the first time. Young modulus and hardness changed little with air gap density. In addition, it was observed that air gap formation enhanced adhesion strength.

INTRODUCTION

The reduction of Keff has been strongly required because the impact of RC delay is becoming larger with the shrinkage of devices.

Air gap formation between lines is considered to be a promising method to reduce Keff [1-7]. There are several approaches to form air gaps. Among them, selective air gap formation based on non-conformal CVD has advantages as follows.

(1) High compatibility to existing process flows
(2) Wide design flexibility
(3) Extremely low Keff

Previously, we have successfully fabricated Cu interconnects with air gaps according to this approach at 130nm node [2] and 65nm node [4]. In this paper, Cu interconnects with air gaps are demonstrated at 45nm node and structural features and electrical characteristics are reported. In addition, mechanical characteristics such as Young modulus, hardness and adhesion strength are discussed.

PROCESS FLOW

The process flow to integrate Cu interconnects and air gaps is shown in Fig. 1. At first, Mx lines are formed in an ILD and a liner is deposited as shown in Fig. 1(a). In this work, a SiOC film (K≈3.0, E=22.3GPa) and a SiC based film were used for the ILD and the liner, respectively.

Secondly, a resist pattern is formed on the liner as shown in Fig. 1(b). The resist pattern is used to exclude air gaps from undesired regions such as peripheries of vias and wide spaces. This idea has been known as AGE (Air Gap Exclusion) [2].

Then, the liner and the ILD are etched to form deep gaps between the Mx lines as shown in Fig. 1(c). The deep gap is the key point to achieve extremely low Keff [3]. In general, high etching selectivity between the Mx lines and the ILD can be obtained so that the gaps are formed between the Mx lines in a self-aligned manner. From this reason, no fine lithography is needed and an inexpensive mask can be used for AGE.

After that, as shown in Fig. 1(d), an extremely thin liner and an ILD are deposited using non-conformal CVD. Air gaps are formed between the Mx lines by a pinch-off effect automatically. Usually, the pinch-off points are located above the top of the Mx lines. As a result, the coupling of electric field between the top of the Mx lines is decreased and extremely low Keff can be obtained. Finally, Vx vias and Mx+1 lines are formed by conventional dual damascene method as shown in Fig. 1(e).

In the process flow shown in Fig.1, "AGE Litho.", "Deep Gap Etching" and "Extremely Thin Liner Depo." are the steps necessary for the air gap formation. These steps can be added-on to any process flows very easily. That is, this approach has a strong advantage in compatibility.

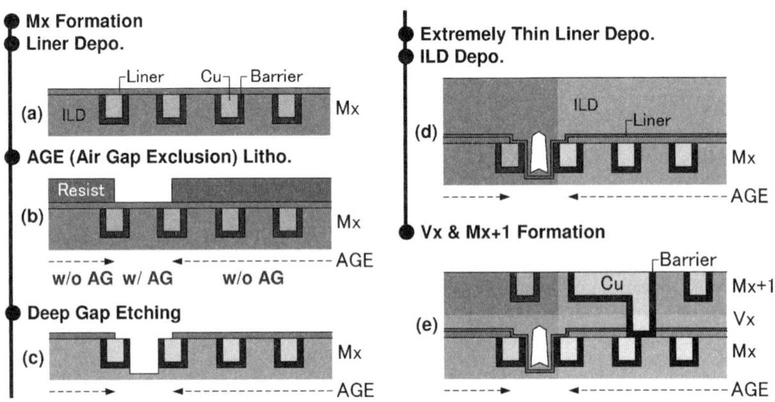

Fig. 1 Process Flow

EXPERIMENT

Cu interconnects having M1-V1-M2-V2-M3 were fabricated at 45nm node. Air gaps were formed at M2 based on the process flow shown in Fig. 1. A variety of elements with different air gap densities were created by ingenious design of the mask for AGE. Additionally, Cu interconnects were fabricated using a porous SiOC film (K≈2.4, E=7.8GPa) for reference. In this case, no air gaps were formed.

Physical analyses using TEM and SEM were carried out for doneness. Electrical characteristics such as capacitance, resistance, via chain yield and leakage current were evaluated especially focusing on M2.

The structures shown in Fig. 1(d) were used to evaluate mechanical characteristics. Evaluations were performed in two ways. One was "nano-indentation" and the other was "nano-scratch" as shown in Fig. 2. In "nano-indentation", CSM (Continuous Stiffness Measurement) mode was used. That is, a load with a small oscillating component was applied to a probe on a surface. By analyzing the response to the oscillating component, Young modulus and hardness were obtained as a function of displacement into surface. On the other hand, in "nano-scratch", a surface was scratched by a probe with an increasing load while recording penetration depth. The load to trigger delamination (critical load) was treated as the index of adhesion strength. The delamination was monitored both visually and numerically.

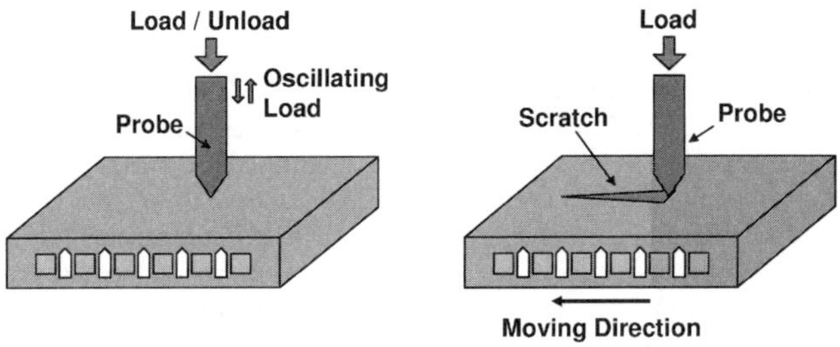

(a) Nano-indentation (b) Nano-scratch

Fig. 2 Mechanical characteristic evaluations

RESULTS AND DISCUSSIONS

Structural Analyses

Figure 3 shows the TEM photograph of the Cu interconnects with the air gaps. It was found that the bottom of the air gaps was very low thanks to the deep gap etching. In addition, the pinch off point was high enough to cut off the coupling of the top of the M2 lines. Therefore, low Keff was expected from this result. Some rounding of the M2 lines was observed. This was due to the damage by the gap etching.

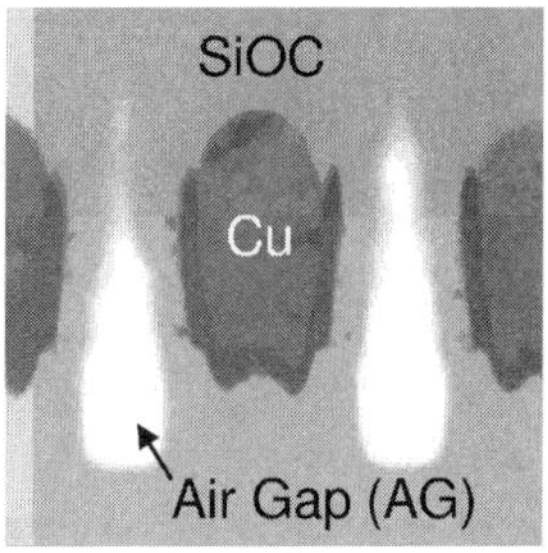

Fig. 3 Cu lines with air gaps

Cu interconnects having different air gap densities are shown in Fig. 4. In the case of "Air Gap (AG) 0%", all the spaces were masked by the resist pattern during the gap etching and no air gaps were formed. As for "AG 50%", two spaces were masked but neighboring two spaces were opened during the gap etching. As a result, two air gaps were formed in four spaces. With regard to "AG 100%", all the spaces were opened during the gap etching and air gaps were formed in every space. Thus, it was confirmed that AGE went well just as intended. This offers wide design flexibility for designers.

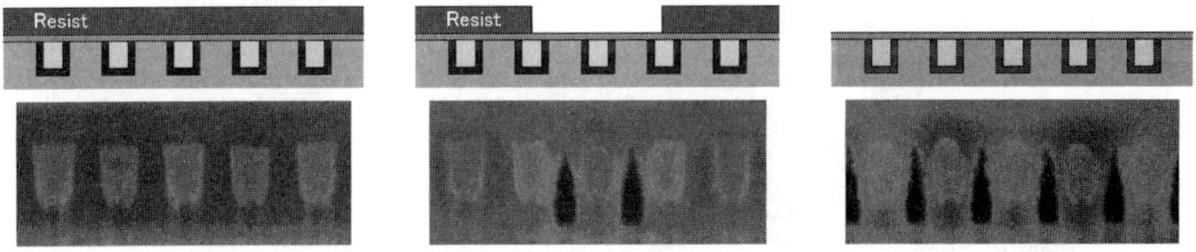

(a) Air Gap 0% (b) Air Gap 50% (c) Air Gap 100%

Fig. 4 Effect of AGE

Electrical Characteristics

Distributions of M2 capacitance are plotted in Fig. 5(a). "AG 100%" gave 35% lower capacitance than "AG 0%". This drastic reduction was as expected from the TEM photograph shown in Fig. 3. However, as shown in Fig. 5(b), slight increase of M2 resistance was observed by the air gap formation, which was attributed to the rounding of the top of the M2 lines. It was not straightforward to compare the performance of these interconnects, so RC products were calculated as a 1st order approximation. The results are shown in Fig. 5(c), indicating that the performance of "AG 100%" was improved by 29% due to the air gap formation. Projected Keff for "AG 100%" was 2.3 which surpassed the ITRS requirement for hp45 (2.6~2.9) by far.

Figure 6 shows V2 chain yield as a function of M2-V2 misalignment. Yield degradation caused by the contact of vias and air gaps was known as a show stopper of interconnects with air gaps. But high V2 chain yield was obtained independently of M2-V2 misalignment in this work. This result came from AGE around the V2 vias to avoid the the contact of the V2 vias and the air gaps. Leakage current between the M2 lines are summarized in Fig. 7. No dependence of leakage current on air gap density was observed.

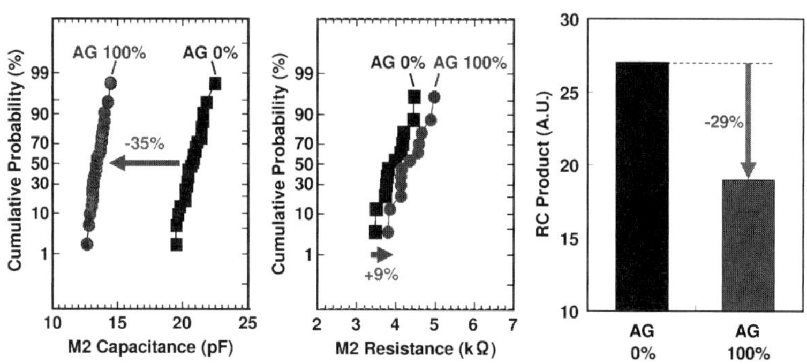

(a) Capacitance (b) Resistance (c) RC Product

Fig. 5 Capacitance and resistance

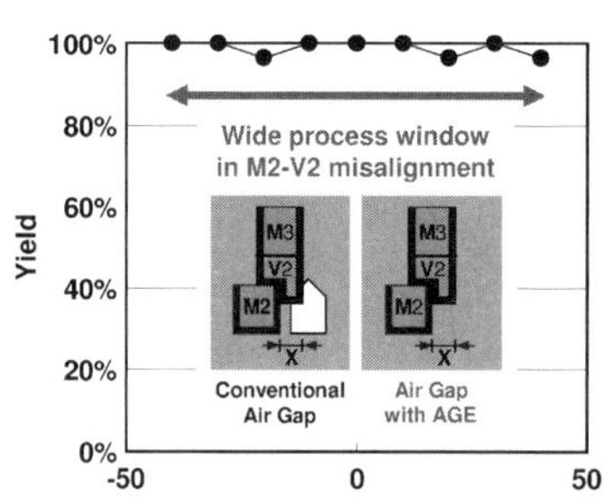

Fig. 6 Via chain yield

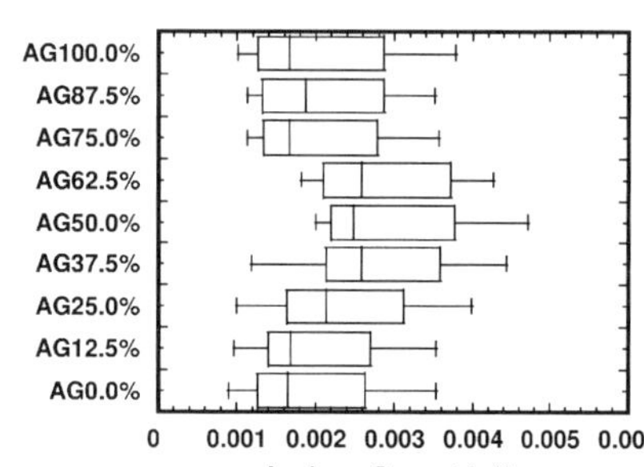

Fig. 7 Leakage Current

Mechanical Characteristics

The element specially designed for mechanical characteristics evaluations and the traces of "nano-indentation" are shown in Fig. 8. The element contained a large number of Cu interconnects with air gaps. The size of the element was 75um*60um, which was adequately large for "nano-indentation" and "nano-scratch". In "nano-indentation", Berkovich indenter was used, where the size of the projected area was the penetration depth multiplied by 7.5. The size was large enough compared to Cu interconnects. Five measurements were performed for each element to check the repeatability.

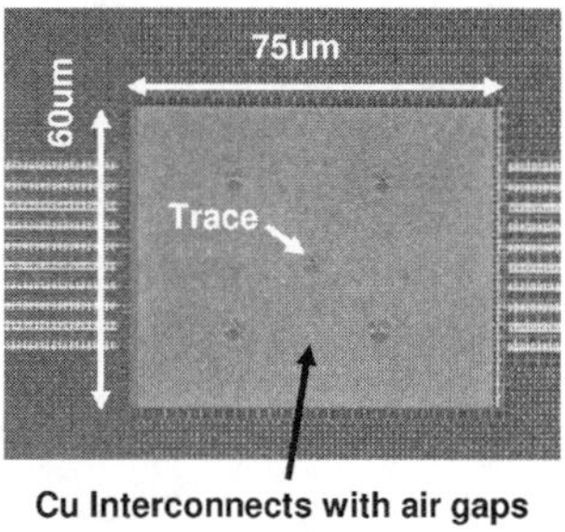

Fig. 8 Element and traces of measurement

Figure 9 shows Young modulus as a function of displacement into surface. The repeatability was fairy good showing that the measurements were performed successfully. Fig. 10 compares most probable Young modulus of the four structures. Similar values to the blanket films (22.3GPa for SiOC and 7.8GPa for porous SiOC) were obtained. The structure fabricated with the porous SiOC film showed the lowest Young modulus. It was interesting to note that Young modulus changed little with air gap density.

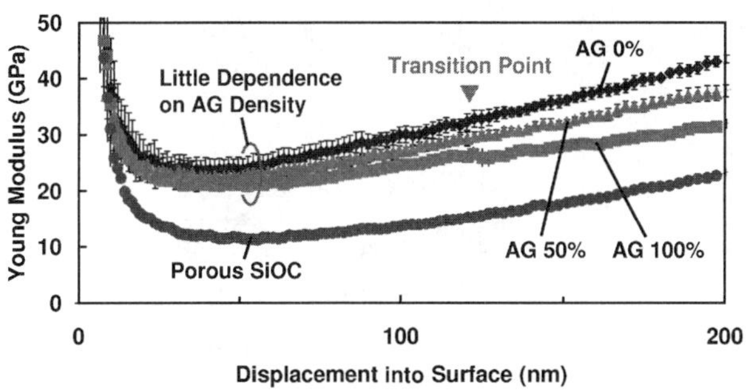

Fig. 9 Young modulus as function of displacement into surface

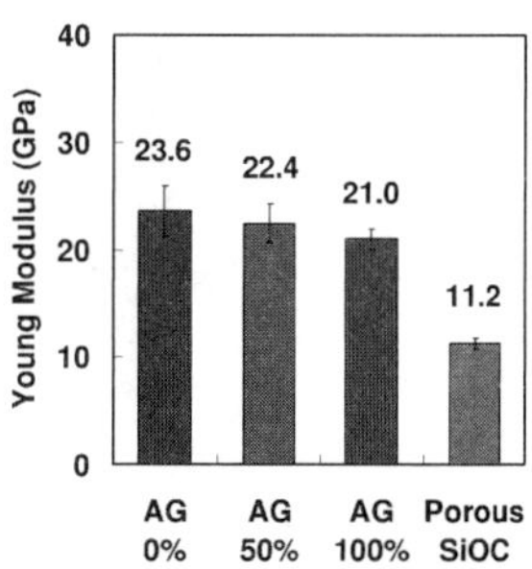

Fig. 10 Comparison of Young modulus

Hardness as a function of displacement into surface is shown in Fig. 11. In the case that displacement was below 130nm, which was important in actuality, there was little influence of air gap density on hardness. But in the extreme case that displacement exceeded 130nm, hardness decreased with air gap density. That is, there was a transition point of hardness in the case with air gaps. The structure fabricated with the porous SiOC film gave the lowest hardness throughout the entire range.

Structural analyses were carried out to see what had happened at the transition point. Figure 12 shows the results. Below the transition point, there was nothing peculiar. In this region, the Cu interconnects were considered to keep their original structures. But above the transition point, a mechanical breakdown of the Cu interconnects was observed. In other words, the dependence of hardness on air gap density appeared as a result of the mechanical breakdown of the Cu interconnects.

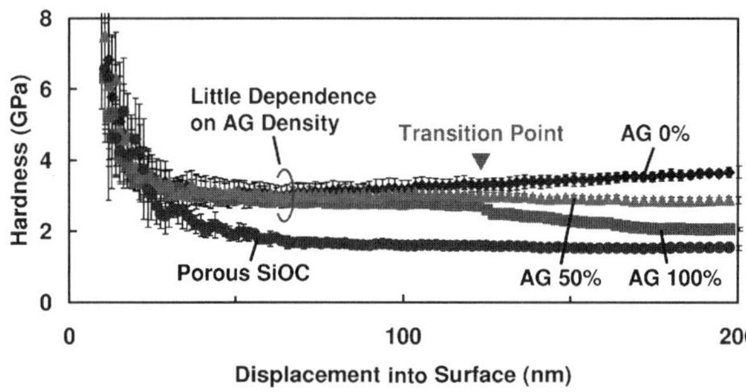

Fig. 11 Hardness as function of displacement into surface

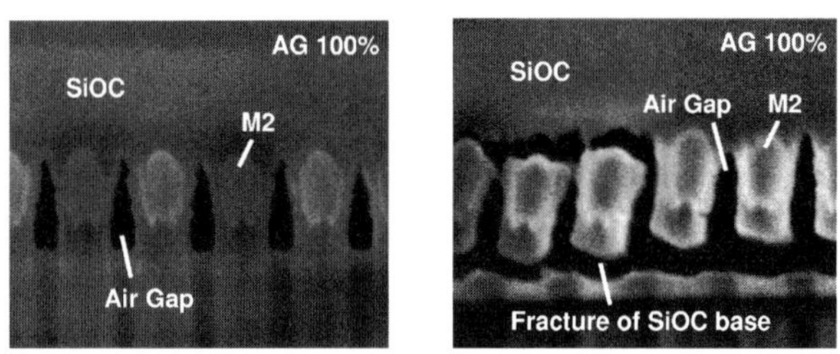

(a) Below transition point **(b) Above transition point**

Fig. 12 SEM photographs taken below and above transition point

Figure 13 exemplifies the result of "nano-scratch". Delamination was observed both visually and numerically. Typical scratch distance to cause the delamination was 7 ~ 10um, which was large enough compared to the size of the Cu interconnects. Three measurements were performed for each element.

Comparison of critical load is shown in Fig. 14. The structure fabricated with the porous SiOC film gave the lowest critical load. Surprisingly, the structure with the air gaps showed the highest critical load.

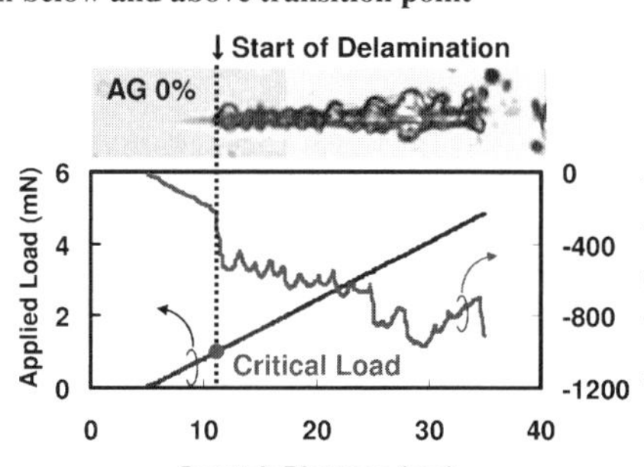

Fig. 13 Result of "nano-scratch" for "AG 0%"

The mechanism of the result seen in Fig. 14 was examined by structural analyses. As shown in Fig. 15, the delamination of SiOC/liner interface was observed in the structure without the air gaps. However, the delamination seemed to be suppressed in the structure with the air gaps. In the case with the air gaps, the SiOC film and the liner film were filled in the gaps between the M2 lines and it was reasonable to consider that the delamination was suppressed due to "anchor effect" of those films.

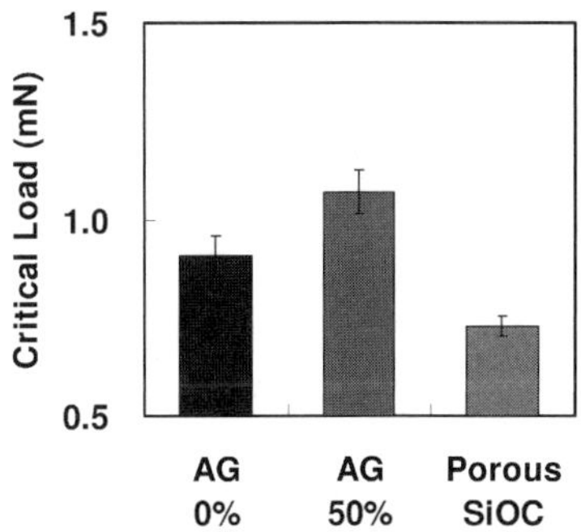

Fig. 14 Comparison of Critical Load

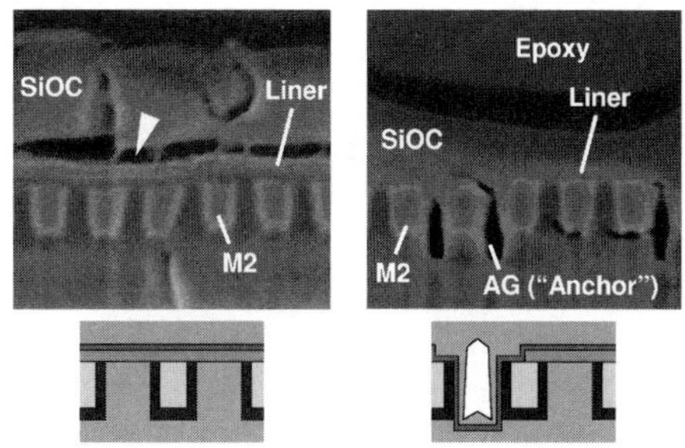

(a) AG 0% (b) AG 50%

Fig. 15 SEM photographs taken after "nano-scratch"

CONCLUSIONS

45nm node Cu interconnects with air gaps were fabricated. Good electrical characteristics were obtained. It was also confirmed that the air gap formation did not degrade mechanical characteristics. This technology is very promising for 32nm node and beyond because it can reduce Keff drastically and has a high compatibility to existing process flows.

ACKNOWLEDGMENTS

The authors would like to thank Kobelco Research Institute for technical supports and discussions.

REFERENCES

[1] T. Ueda et al., VLSI Symp., 1998.
[2] T. Harada et al., IITC, 2006.
[3] J. Noguchi et al., IITC, 2006.
[4] T. Harada et al., IITC, 2007.
[5] L. G. Gosset et al., IITC, 2007.
[6] R. Daamen et al., IITC, 2007.
[7] S. Nitta et al., AMC, 2007.

Conference Proceedings AMC XXIV © 2009 Materials Research Society

A novel materials solution to enable CoWP metal capping by reducing line-to-line leakage

Paul R. Besser[1], Axel Preusse[2], Chi-I Lang[3], Bob Fiordalice[3], Markus Nopper[2], Robert Seidel[2], Oliver Aubel[2], Daniel Berhe[3], Jinhong Tong[3], Nitin Kumar[3], and Tony Chiang[3]
[1]Technology Research Group, Advanced Micro Devices, Inc., Sunnyvale, CA, 94088
[2]AMD Fab36 LLC & Co. KG, Advanced Micro Devices Saxony LLC & Co. KG, Dresden, Germany
[3]Intermolecular, Inc., San Jose, CA, 95134

ABSTRACT

High-productivity combinatorial (HPC) methods and processing equipment have been used to create of a non-conductive, aqueous molecular masking layer (MML). The MML deposits selectively on low-K dielectrics and modifies the dielectric surface to be more hydrophilic. MML retards selectivity loss during CoWP deposition while not inhibiting CoWP electroless deposition onto Cu. Selective CoWP metal capping onto Cu lines was improved by MML deposition. MML was successfully integrated into a 45 nm technology node process technology with 300 mm wafers. The CoWP process with MML improves the line-to-line leakage (over the CoWP process alone) without significantly degrading the electrical reliability.

INTRODUCTION

In the microelectronics industry, leading-edge microprocessors contain several hundred million transistors. The interconnect complexity has increased with the number of transistors, and more than 11 layers of high-density interconnects are commonly used to connect the transistors into circuits. To increase the speed of the devices and reduce manufacturing costs, the critical dimensions are consistently reduced with technology node. Copper (Cu) interconnects in a dual inlaid architecture with low-K dielectrics are commonly employed in high-performance technologies and microprocessors. In the inlaid process, the interconnect lines are patterned and etched into the dielectric material, and the Cu is filled into the trenches. The Cu fill process involves the deposition of a barrier to contain the Cu and a seed layer to enable Cu fill with electrochemical deposition. Excess Cu and barrier in the field are removed with chemical-mechanical polishing (CMP) to electrically isolate the Cu interconnect lines. A dielectric capping layer is deposited after CMP to fully encapsulate the Cu interconnects.

As the dimensions of the interconnect lines shrink, however, Cu performance actually degrades. Specifically, the current density (or current-carrying capacity) is reduced by half with each generation of technology [1]. High current densities can lead to failure by means of electromigration (EM). EM is the migration of metal atoms due to momentum transfer from electrons [2] moving through the Cu wires. As the Cu atoms migrate, vacancies remain, then coalesce and migrate to form electromigration-induced voids. The fastest diffusion path for electrons/atoms is along the Cu/dielectric interface [3-4].

Selective electroless deposition of a ternary alloy (CoWP) with good barrier properties and reasonable conductivity (~60 $\mu\Omega$-cm) as a metal cap on the Cu lines has been demonstrated as one option for reducing diffusion along the interface [1,5], thereby improving the current density of the interconnect lines. The CoWP metal cap is deposited onto Cu interconnect lines as an additional process step in the integrated damascene Cu process flow. Selective metal capping has not yet been inserted into volume manufacturing due to the additional cost of the processing and the challenges of implementing CoWP, such as surface preparation, selectivity of CoWP to the intra-layer dielectric (ILD), selectivity of CoWP to Cu, and reliability degradation. While providing significant EM reliability improvement, the mode of reliability degradation has been shown to be electrical leakage between interconnect lines, as quantified by a voltage ramp breakdown (VRDB) or a time-dependent dielectric breakdown (TDDB) test [6].

In the current work, we seek to further enable selective metal capping by selectively depositing a film to modify the dielectric surface and retard selectivity loss while not inhibiting electroless CoWP deposition onto Cu. High-productivity combinatorial (HPC) methods and processing equipment were used for the materials discovery, unit process, and integration development of a non-conductive, aqueous layer, herein referred to as a molecular masking layer (MML). The MML deposits selectively on ILD with the goal of modifying the dielectric surface to be more hydrophilic without affecting the dielectric constant of the ILD. MML is a sacrificial protective layer that is dissolved during the CoWP-deposition or post-deposition cleans or decomposed during dielectric capping (pre-clean). We will also show that MML can be successfully inserted into 45 nm technology node process technology without significantly degrading the line-to-line leakage or the electrical reliability. The first results of promising TDDB performance in 45 nm technology were shown by Aubel *et al.* [7], providing confidence that process approaches like MML can achieve good overall reliability with CoWP in future technologies nodes.

EXPERIMENT

Blanket films of Cu and low-K ILD (K = 3.0) were employed for the MML development. Using HPC methods, 60 molecule types were screened by executing 7,635 experiments. The molecular weight, functionality, concentration, and enhancement reagents of the MML were specifically formulated so MML deposits on the dielectric while not depositing on, or inhibiting deposition of, CoWP onto Cu. Deposition onto ILDs was indirectly verified through contact angle measurements with spectroscopic ellipsometry. X-ray fluorescence (XRF) confirmed that CoWP did not deposit on ILD exposed to MML + CoWP and that CoWP did deposit onto Cu exposed to MML. The change in Cu resistance after MML and CoWP was determined electrically. The MML effect on the dielectric constant was quantified by mercury probing.

A 45 nm technology node integrated flow with K = 3.0 ILD was used as a test vehicle for evaluation of MML integration. The initial materials, unit process, and integration studies were performed on wafer pieces using a HPC workflow. Success metrics were selectivity, electrical resistance, CoWP thickness, and voltage ramp testing. We quantified selectivity by atomic force microscopy (AFM) scans and electrical measurement of leakage between Cu lines in a serpentine-comb leakage structure; we electrically quantified electrical resistance on a minimum pitch serpentine Cu line structure; we quantified CoWP thickness by optical inspection, AFM, and XRF; and, we used voltage ramp testing to electrically verify the leakage between lines on the ILD. In the voltage ramp test, the current between lines was measured as the voltage was ramped from 0 to 2 volts. Several MML formulations were evaluated in series to determine the optimal MML and integration.

The CoWP deposition between CMP and dielectric capping was performed with two different tools. For the process development on blanket films and on wafer pieces, we used a basic electroless CoWP deposition (ELD) process. For the full-flow, 300 mm wafers, we used a commercially available electroless deposition tool (including a pre- and post-deposition clean) to deposit CoWP using established chemistries [5].

The technology was scaled to 300 mm wafers. In the integrated flow, wafers were pulled from the 45 nm production line after metal2 Cu CMP. After the MML was qualified for introduction to the fabrication facility, and its compatibility with subsequent integration steps demonstrated, the MML was deposited onto the wafers. After the MML process, the CoWP metal cap was electrolessly deposited using a commercially available deposition tool (at the tool vendor) with a baseline, optimized process for our technology. XRF was used to demonstrate that CoWP did deposit on wafers with MML and without MML. The wafers were processed to end-of-line and electrical parameters were extracted at final test. Three structures of interest were tested for leakage: wide, narrow, and misaligned. We used VRBD testing to test the reliability of the nominal structure (0.072 μm width and 0.072 μm spacing with a via-to-via minimum distance of 0.093 μm) of two lengths (20 cm and 2 cm, with 1,000 and 66 inter-digitated fingers, respectively). The structures are in windmill format to cover all possible overlay in the process. The failure voltage was determined by a current increase to more than 1 μA (or a sudden jump in current for more then one decade).

RESULTS AND DISCUSSION

For the integration to be successful, MML must deposit on the dielectric while not depositing on or inhibiting deposition of CoWP onto Cu. Deposition of MML onto ILDs was indirectly verified through contact angle measurements with spectroscopic ellipsometry, as illustrated in Figure 1. Four different MML formulations are included in Figure 1. All four formulations reduce the contact angle from a range of 67-84° to 35-42°, effectively changing the dielectric from hydrophobic to hydrophilic. The hydrophilic surface is desirable for improved wetting during aqueous processing and to reduce penetration into the porous dielectric. XRF confirmed that CoWP did not deposit on ILD exposed to MML + CoWP (Figure 2) and confirmed that CoWP did deposit onto Cu exposed to MML (Figure 3). The change in Cu resistance after MML and CoWP was less than 2% (Figure 4), which is an acceptable increase in the process technology. The MML does not deleteriously affect the dielectric constant, as quantified by a mercury probe test.

A 45 nm technology node integrated flow with K = 3.0 ILD was used as a test vehicle for evaluation of MML integration on structure wafers. Some refinement of the MML was required when migrating to patterned wafers. Figure 5 includes the AFM images for a CoWP integrated process without (left) and with (right) MML. The AFM images reveal that the nodules are reduced with MML. Figure 6 is a plot of the leakage measured between lines during the voltage ramp test. Three samples were tested: as-received, ELD-only, and MML plus ELD. The leakage increases for the ELD-only sample, compared to the as-received (control) sample; however, the leakage of the ELD + MML sample is equivalent to that of the as-received (no MML or CoWP) sample.

With positive results on blanket films and wafer pieces, the technology was scaled to 300 mm wafers. In the integrated flow, wafers were pulled from the 45 nm production line after metal2 Cu CMP. MML was deposited onto the wafers at Intermolecular. After the MML process, the CoWP metal cap was electrolessly deposited. XRF was used to demonstrate that CoWP did deposit on wafers with MML and without MML (Figure 7). The wafers were processed to end-of-line and electrical parameters were extracted at final test. While all of the

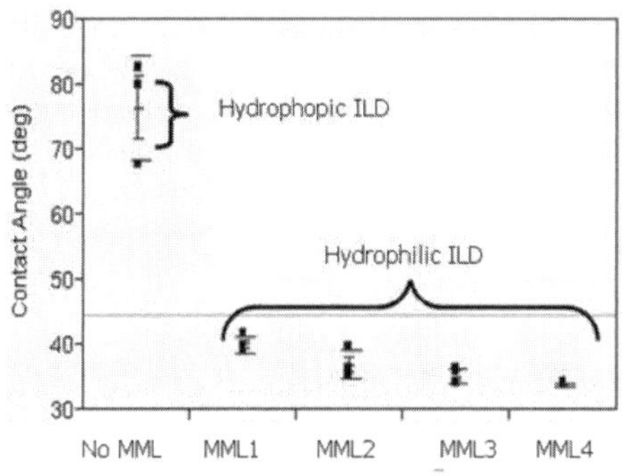

Figure 1. Effect of MML deposition on contact angle.

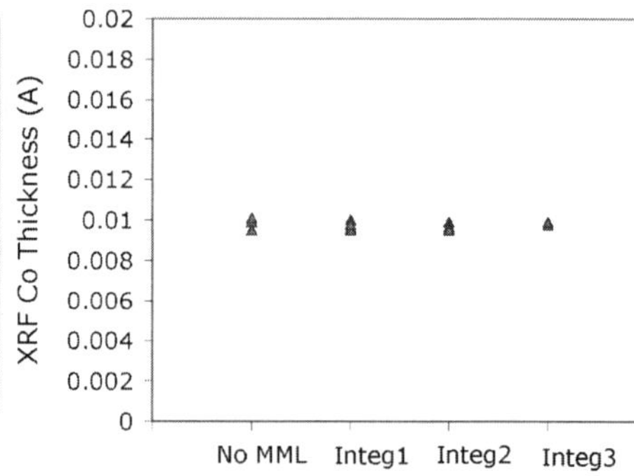

Figure 2. XRF Co Thickness on blanket ILD exposed to MML + CoWP processing for three integrations and one control.

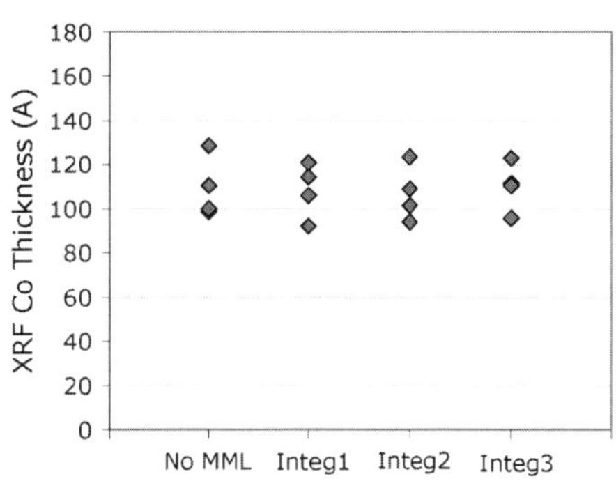

Figure 3. XRF Co thickness on blanket films of Cu exposed to MML and CoWP processing, for three integrations and one control.

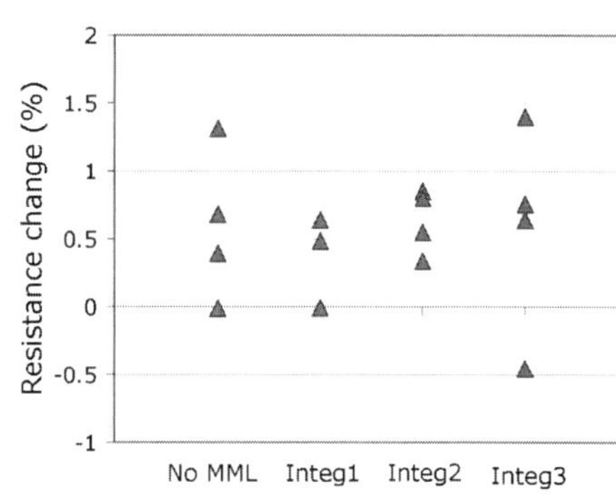

Figure 4. Cu resistance measured on blanket films of Cu exposed to MML and CoWP processing, for three integrations and control.

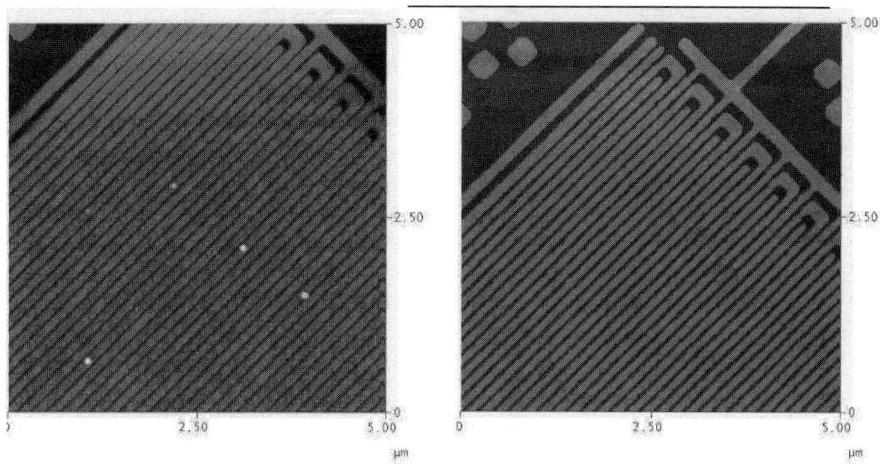

Figure 5. This figure illustrates the AFM images for a CoWP process without (left) and with (right) MML. The AFM images reveal that the nodules are reduced with MML.

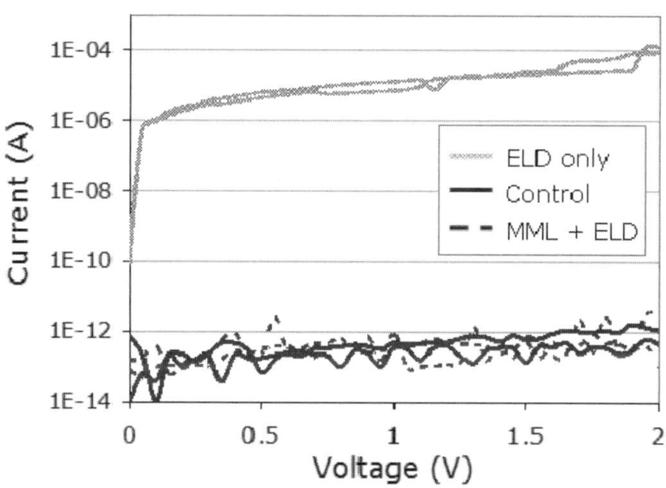

Figure 6. Leakage current measured on wafer pieces during voltage ramping from 0 to 2 V. Three samples were tested: as-received (control), ELD only, and MML plus ELD.

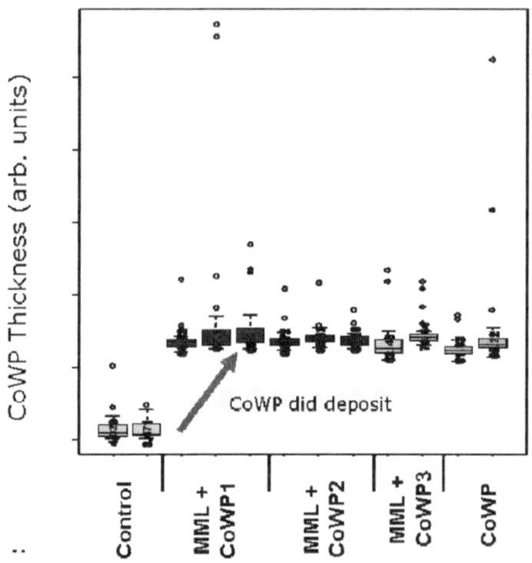

Figure 7. CoWP thickness vs. split.

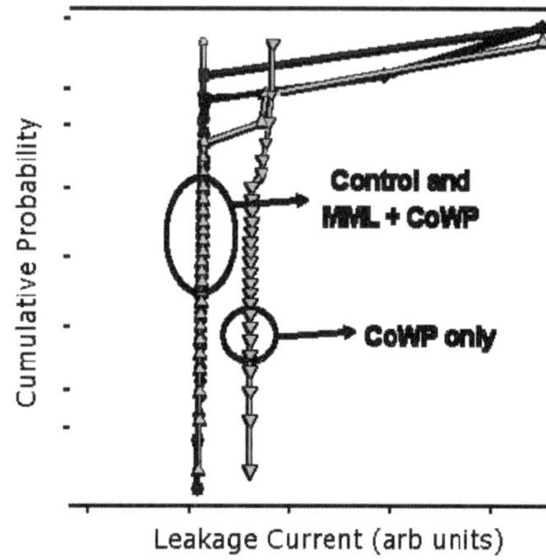

Figure 8. Cumulative probability plot of leakage vs. split for wide spacings.

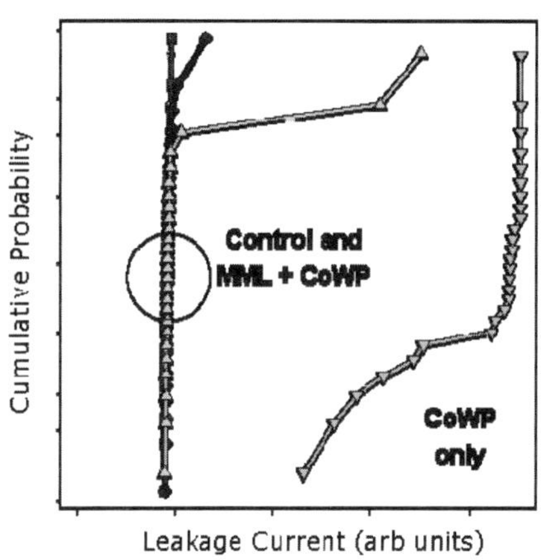

Figure 9. Cumulative probability plot of leakage vs. split for narrow spacing.

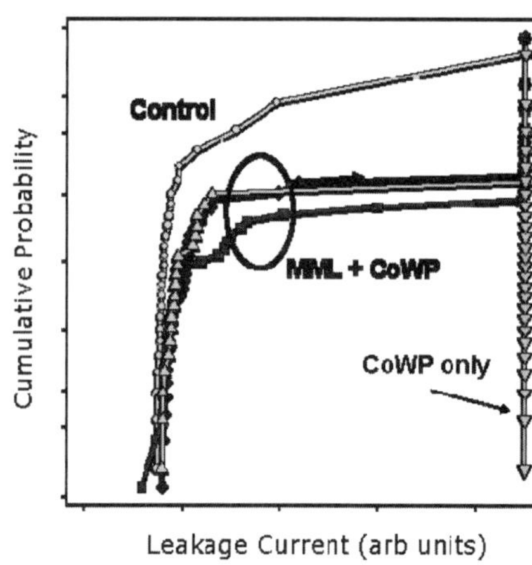

Figure 10. Cumulative probability plot of leakage vs. split for misaligned structures.

electrical parameters were collected and reviewed, we do not include all of them here due to a lack of space in the manuscript. Of particular relevance, however, is the leakage between lines. Three structures were tested for leakage: wide, narrow, and misaligned. The electrical results for these structures are included in Figures 8-10. For wide structures (Figure 8) with larger-than-minimum spacing between lines, the CoWP-capped wafers exhibited an order of magnitude higher leakage than either the uncapped control or the MML + CoWP-capped wafers. Narrow structures (minimum spacing) revealed that CoWP (without MML) led to high leakage between lines (Figure 9) and that MML reduced the leakage between lines. For the 20% misaligned structures (smaller than minimum spacing), wafers with MML had slightly higher leakage than control but performed orders of magnitude better than the CoWP baseline (Figure 10). The overall degraded VRDB performance of these CoWP samples compared to the standard CoWP baseline [7] in the fab (without transfer) could be due to the travel of the material. The CoWP in the current experiment was deposited at the tool vendor in the U.S.A. Nevertheless, the results still indicate a significant improvement using MML compared to CoWP (without MML).

The final critical test of a CoWP integration is reliability. In this experiment, VRBD was tested between interconnect lines (Figure 11). The CoWP baseline split (no MML) was not tested due to high leakage at end-of-line testing; however, the MML + CoWP and control (no MML) were tested. Figure 11 reveals that the MML + CoWP split was slightly degraded relative to uncapped control, but the difference was relatively small. In particular, the larger structure (20 cm) shows that no systematic defect issue (almost no extrinsic defect branch) was observed, suggesting that MML can help enable CoWP to be inserted into the technology.

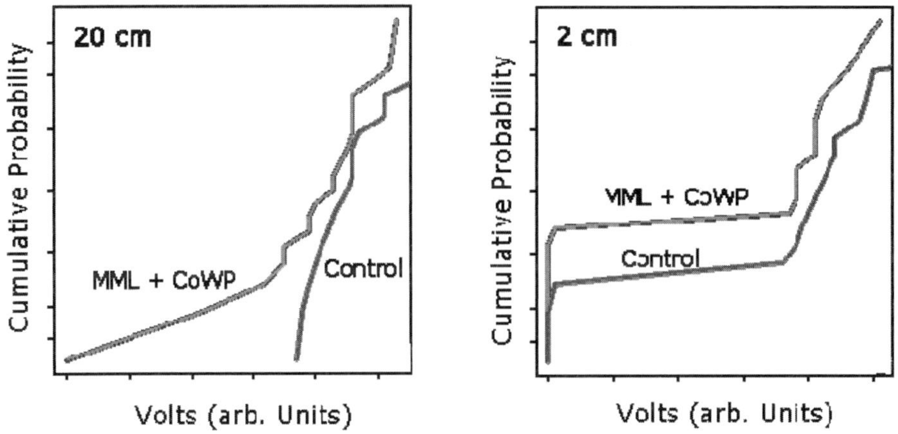

Figure 11. Cumulative failure distributions for VRDB tests on 20-cm and 2-cm structures, for the control sample and the MML+ CoWP sample.

CONCLUSIONS

High-productivity combinatorial methods and processing equipment were used for the materials, unit process, and integration development of a non-conductive, aqueous molecular masking layer (MML). It was shown that the MML deposits selectively on low-K ILDs and modifies the dielectric surface to be more hydrophilic. MML retards selectivity loss during CoWP deposition while not inhibiting CoWP deposition onto Cu. Selective metal capping of CoWP onto Cu lines was improved by MML deposition. MML was successfully integrated into a 45 nm technology node process technology. The CoWP process with MML improved the line-to-line leakage (over the CoWP process alone). The results strongly infer that MML can deliver very good overall reliability performance (assuming the expected EM performance boost is maintained) without reducing the TDDB performance significantly. This yields the conclusion that MML provides extendibility of the stable CoWP baseline on 45 nm to future technologies.

ACKNOWLEDGMENTS

Wafers were fabricated at AMD-Dresden Fab36. We thank the staff and technicians for processing, shipping, and electrically testing the wafers. Intermolecular performed the HPC characterization, the materials discovery, and wafer processing. CoWP was deposited at an equipment supplier, and we thank them for processing the wafers.

REFERENCES

1. C.-K. Hu *et al.*, Proc. of IRPS 2004, p. 222 (2004).
2. H.B. Huntington and A.R. Grone, J. Phys. Chem. Solids 20, 76 (1961).
3. C.K. Hu, Appl. Phys. Lett. 83, 869 (2003)
4. A.v. Glasow *et al.*, Proc. of IEEE IRPS 2003, p. 146 (2003)
5. A. Preusse *et al.*, Proc. of IEEE IITC 2008, 123 (2008)
6. F. Chen *et al.*, Proc. of IEEE IITC 2007, 120 (2007)
7. O. Aubel *et al.*, Proc of IEEE IRPS 2008 (2008).

Conference Proceedings AMC XXIV © 2009 Materials Research Society

Multi levels Air Gap Integration using Sacrificial Material Approach for Advanced Cu Interconnects Technologies

R. Gras[1, 3], F. Gaillard[2], D. Bouchu[2], P.H. Haumesser[2], G. Imbert[1], L. Vandroux[2], A. Farcy[1], T. Chevolleau[3], G. Passemard[1], J. Torres[1], P. Ancey[1]
[1]STMicroelectronics, 850 rue Jean Monnet 38926 Crolles France
[2]CEA-LETI-MINATEC, Grenoble France
[3]CNRS-LTM, Grenoble France
raphael.gras@st.com

ABSTRACT

Air gap integration is a very promising solution to achieve extreme low k values required by the 22 nm technological node and below. Proposed air gap integration scheme is based on the use of a sacrificial material removed by wet etching and allows a low cost introduction of localized air cavities. Processes were optimized to successfully integrate two air gap metal levels covered by one intermediate standard level using 65 nm design rules. Two alternative architectures were also investigated and their feasibility was demonstrated. These architectures attend to increase air gap reliability by using wells for wet etching chemistry diffusion, and to minimize air gap industrialization cost by integrating air gap from conventional dielectric stack.

INTRODUCTION

For the 22 nm technological node and beyond, very aggressive k values (< 2.0) are required to isolate interconnects [1]. It was shown that a limitation in term of maximal porosity introducible inside conventional porous Ultra Low K (ULK) materials has been reached [2], so porous materials will not be able to meet such low k values. Thus, new architectures are required and, until now, the introduction of air cavities between copper lines appears to be the only solution to meet ITRS requirement for the 22 nm node and below. From non conformal deposition approach [3-5] to sacrificial materials introduction [6-7], different air gap architectures and integration schemes have been proposed in the literature. Despite remarkable k values reductions obtained with conformal deposition technique, two major concerns remain with this approach: i) level by level air cavities integration is problematic for misaligned vias, which can lead to a breakthrough of cavities during via etching, and ii) the need for lithographic masks at critical dimensions on each level is too expensive for industrialization. Thus, air gap introduction after multi metal level integration in a sacrificial material removed during a specific technological step has been presented to solve these issues [8]. Proposed sacrificial materials are Thermally Degradable Polymer (TDP) [7] or conventional dielectrics such as SiO_2 removed by wet etching [9]. However, TDP approach does not allow a local integration of air cavities which can cause mechanical issue during packaging step for large air gap areas, whereas an additional non critical lithography can be used for air gap localization in case of sacrificial SiO_2 approach. Integration feasibility has been demonstrated in both cases with air gap introduced at line level and permanent material kept at via level. A new architecture, allowing localized air gaps integration formed at both line and via levels, has then been proposed to achieve the lowest k values [10]. In this paper, optimization steps of this architecture and alternative architectures description and feasibility are presented.

AIR GAP INTEGRATION SCHEME

Air gap integration scheme used to introduce air cavities is presented on figure 1. After multi metal level integration using sacrificial SiO_2 at both line and via levels, air gaps are created by isotropic HF etching of the sacrificial SiO_2 and localized using specific HF diffusion pathways defined in the SiCN capping. In that case, stand alone Self Aligned Barriers (SAB), such as Co alloys, are introduced to cap lower metal level lines in replacement of standard dielectric barriers. In fact, SABs are mandatory to allow HF chemistry diffusion needed for air gap formation at lower metal levels. After multi levels air gap integration, SiCN apertures are closed during upper metal level dielectric non conformal deposition and standard damascene processes made. A three metal level interconnect demonstrator is performed using 65 nm design rules on 300 mm wafers, integrating air cavities on two metal levels and covered by one standard intermediate metal level. Figure 2 illustrates the complete removal of the sacrificial SiO_2 at M1-M2 levels in a via chains structure above M3 level integrated with dense SiO_2. Electrical measurements associated with reliability tests on air gap structures have already been presented in [10].

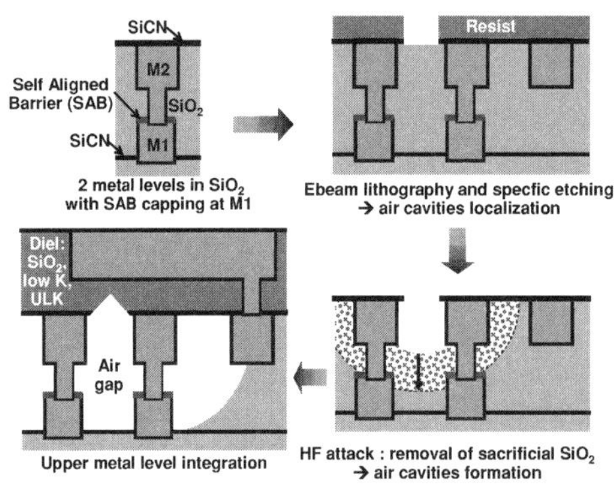

Fig. 1 : Schematic representation of air gap integration scheme

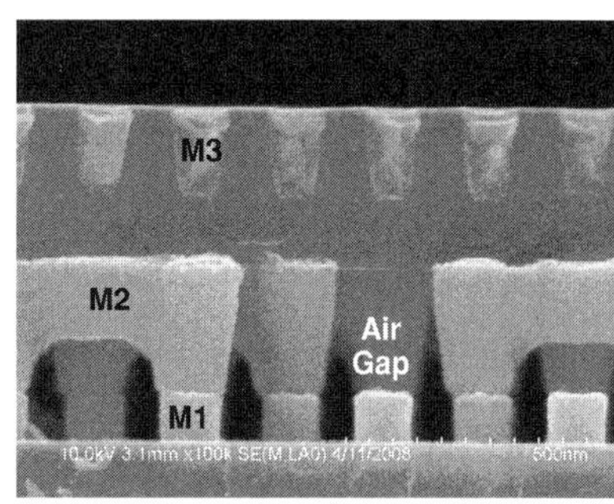

Fig. 2 : SEM cross section of air gap at M1 & M2 levels and dense dielectric at M3 level with minimum design rules

PROCESS AND ARCHITECTURE DEVELOPMENT

Non conformal deposition

After multi levels air gap introduction, non-conformal properties of PECVD deposition allow HF diffusion apertures closing. Closing aperture height has to be as low as possible to avoid any breakthrough of a cavity during upper level line etching (Fig. 1). Non conformal deposition process optimisation is presented to ensure low closing height and minimum deposition into air cavities at lower metal levels. The principle of this study is presented on figure 3: after trench etching, PECVD deposition is performed. Cavities width, depth, and closing height are measured for initial trenches of 100 nm for different tested materials (SiO_2, dense and porous SiOC) and process conditions. In particular, two precursors (SiH_4 and TEOS) are compared in case of SiO_2 depositions. Results are summarized in table 1. Best process leading to both low closing height and conformity is obtained for a sequential deposition of SiO_2 using SiH_4 then TEOS precursors. In addition, closing aperture height can be reduced by decreasing their dimensions and increasing underlying air-gap height [3].

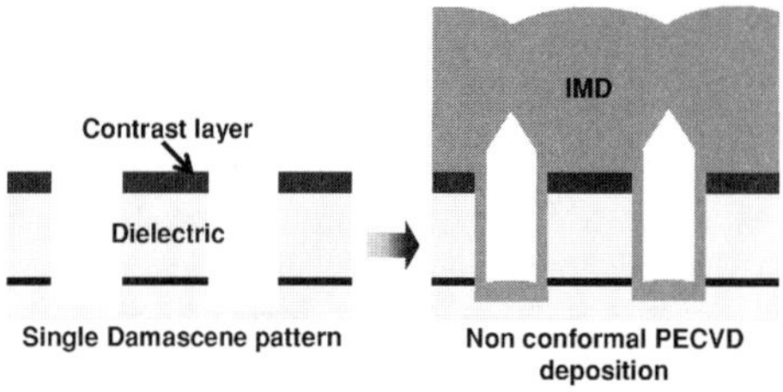

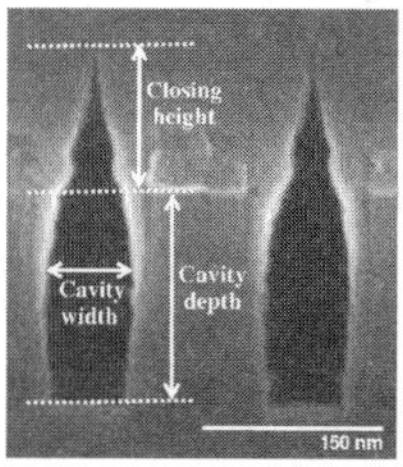

Fig. 3 : Non conformal PECVD deposition study (left) principle and (right) SEM picture example of SiO_2 silane deposition

	Closing height (nm)	Cavities width (nm)	Cavities depth (nm)
TEOS	80	50	165
SiH_4	120	75	185
SiH_4 + TEOS 01	110	75	185
SiH_4 + TEOS 02	90	75	185
Dense SiOC	90	70	190
Porous SiOC	140	70	180

Table 1 : Closing properties of PECVD depositions for initial trenches of 100 nm

Drastic copper degradation was also observed when SiO_2 deposition plasmas directly interact with metal lines (fig. 4 left). So thin SiCN liner was first deposited to act as a protection of lines during M3 Inter Metal Dielectric (IMD) deposition (fig. 4 right). Thus, developed optimised process allowed M3 integration using intermediate level dimensions (Fig. 2) showing the approach feasibility on aggressive dimensions.

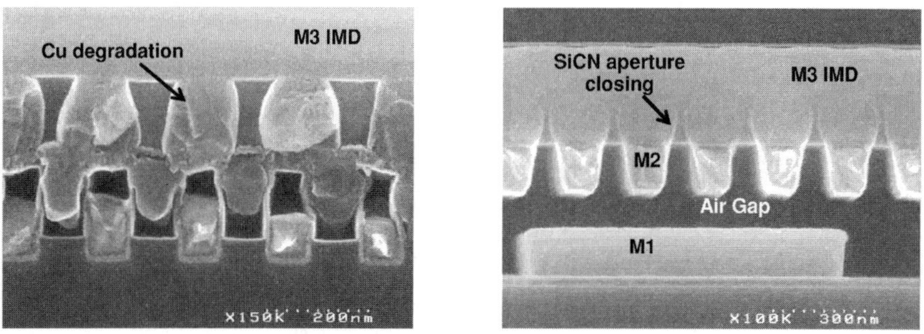

Fig. 4: SEM pictures after M3 IMD deposition : (left) bad process and (right) optimised process

Well approach

To limit previously discussed issues on closing height during upper level dielectric deposition and to enhance mechanical properties of the stack, another integration scheme is also proposed (fig. 5). In this approach, multi metal levels are formed integrating sacrificial material at lower levels and a permanent dielectric at upper level. Then, a specific lithography and an etching of upper level IMD on all its height in localized areas defines HF wells. To prevent M3 IMD removal during air cavities formation, an additional SiCN protection liner is further deposited on well sidewalls. Finally HF chemistry diffuses into the wells and removes sacrificial material introducing air gaps at lower metal levels and preserving upper level IMD. Integrating air gap using HF diffusion wells has two main advantages: i) the high aspect ratio of wells favours a low closing height of next level IMD non conformal deposition increasing approach reliability and ii) the presence of an additional dense level above air gap areas decreases delamination risks during sensitive steps such as copper CMP, thus increasing mechanical reliability of the stack.

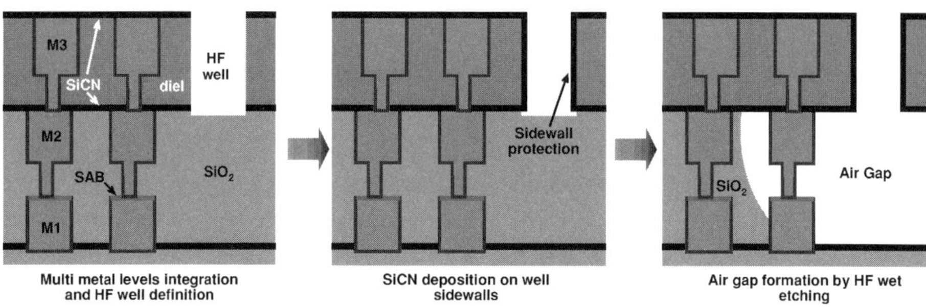

Fig. 5: Principle of well approach

Sidewalls protection feasibility using a thin SiCN layer was studied on one metal level demonstrator (fig. 6 left). After wells formation, a conformal SiCN liner was first deposited and then selectively etched at well bottom without modifying the sidewalls. Efficiency of such sidewall protection was demonstrated as shown on SEM picture of figure 6 right. In fact, SiO_2 in which wells were etched was not removed by HF chemistry during air gap formation step.

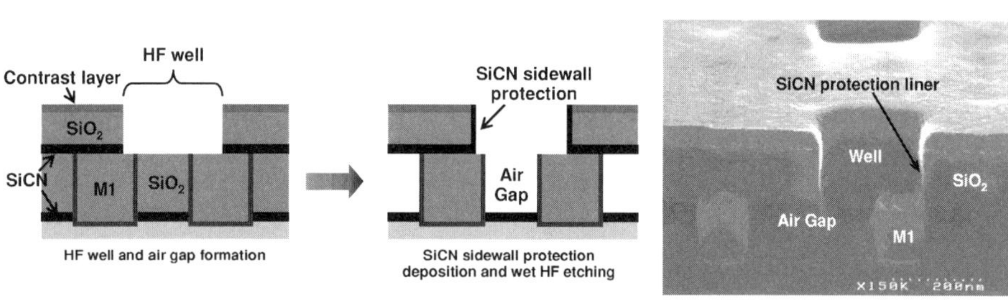

Fig. 6 : Well approach on one metal level demonstrator (left) principle of the study and (right) SEM picture showing feasibility

Sacrificial porous SiOC material

To minimize process developments and so make easier industrialization of air gap technology, there is also a great interest to integrate air gap keeping stack and materials currently used in the 45 nm technological node. In this stack, porous SiOC with k value of 2.5 is used as IMD and SiCN barriers cap metal lines. As no SABs are used in that case and air cavities are introduced after multi levels integration to avoid via misalignment issues, HF wells should be defined to etch all SiCN barriers capping levels where air gaps have to be introduced (fig. 7).

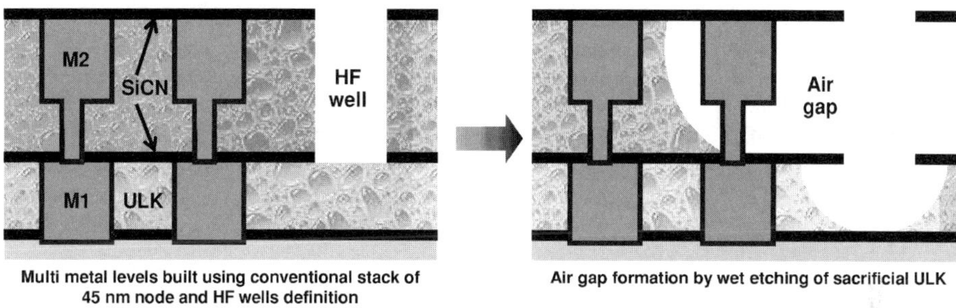

Fig. 7 : Alternative architecture using 45 nm technological node conventional stack combined with HF diffusion wells to integrate air gap

Thus, porous SiOC was also evaluated as sacrificial material considering the wet HF etching approach. One metal level demonstrator was performed where M1 level was fabricated using standard direct CMP integration scheme [11]. SEM pictures of figure 8 illustrate the behaviour of porous SiOC for different HF attack durations. Porous SiOC was shown to be etched by HF chemistry forming successfully air gap between copper lines. However, porous SiOC areas near copper lines sidewalls e.g. which are modified by single damascene etching plasma are removed faster than unmodified porous SiOC parts. This result emphasizes that this air gap integration scheme can be extended to aggressive technological nodes using porous SiOC material as sacrificial material.

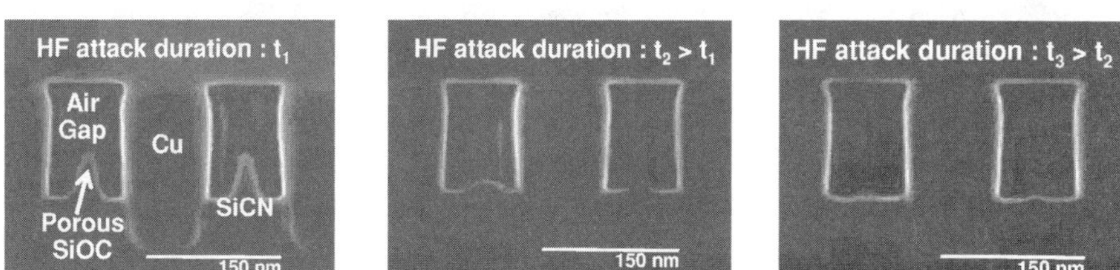

Fig. 8 : SEM pictures of air gap formed by wet etching of porous SiOC for different HF attack durations on one metal level demonstrator

CONCLUSIONS

Air cavities were integrated in two metal levels and covered by one standard intermediate level using 65 nm design rules on 300 mm wafers. In the proposed architecture, air gaps are integrated after multi metal levels fabrication by HF wet etching of a sacrificial dielectric, and localized by HF chemistry diffusion through apertures define in the SiCN top capping layer. SiCN apertures are closed during upper level IMD deposition due to non conformal properties of PECVD deposition. Deposition process was optimized to allow critical dimensions metal integration above air gap structures without degradation of underneath copper lines. Alternative architecture based on HF well diffusion was also proposed to enhance reliability of air gap approach and its feasibility was shown on one metal level demonstrator. Finally, it was demonstrated that patterned porous SiOC was etched by HF chemistry so this ULK material can be used as sacrificial material in this air gap approach. All these integration results make air gap a very promising solution to meet k values required by 22 nm technological node and beyond.

ACKNOWLEDGMENTS

The authors would like to acknowledge L.G. Gosset for his help. This work is supported by the European Commission's Information Society Technologies Programme (IST) under PULLNANO project Contract No. IST -026828.

REFERENCES

1. International Technology Roadmap for Semiconductors, 2006
2. V. Jousseaume et al., *MAM 2008*, to be published in Microelectronics Engineering
3. V. Arnal et al., *Proceedings of IITC 2001*, p. 298-300
4. T. Harada et al., *Proceedings of IITC 2005*, p. 15-17
5. S. Nitta et al., *Proceedings of IITC 2008*, p. 191-192
6. L.G. Gosset et al., *Proceedings of IITC 2007*, p. 58-60
7. R. Daamen et al., *Proceedings of IITC 2005*, p. 240-242
8. L.G. Gosset et al., *Microelectronic Engineering 82*, 321–332 (2005)
9. F. Gaillard et al., *Microelectronic Engineering 83*, p. 2309–2313 (2006)
10. R. Gras et al., *Proceedings of IITC 2008*, p.196-198
11. S. Kondo et al., *Proceedings of IITC 2006*, p.164-166

Conference Proceedings AMC XXIV © 2009 Materials Research Society

Novel Self-hydrophobication and Pre-reinforcement of Fine Pore Self-assembled Scalable Porous Silica film for quick Turn-Around-Time Low-k/Cu Interconnect Process

S. Chikaki[1], T. Kubota[1], H. Tanaka[2] K. Kohmura[2], M. Hirakawa [3], T. Nakayama[3], Y. Seino[4], N. Oda [1], and S. Saito [1]

[1] Research Dept.2, Selete, Inc., 16-1 Onogawa, Tsukuba, Ibaraki, 305-8569, Japan
[2] Performance Materials Business Sector, Mitsui Chemicals, Inc., Nagaura, Sodegaura, Chiba 299-0265, Japan
[3] Tsukuba Institute for Super Materials, ULVAC, Inc., 5-9-6, Tohkohdai, Tsukuba, Ibaraki 305-2635, Japan
[4] NIRC, AIST-Tsukuba-West, 16-1, Onogawa, Tsukuba, Ibaraki 305-8569, Japan
E-mail: chikaki.shinichi@selete.co.jp, Phone: +81-29-849-1652, Fax: +81-29-849-1452

ABSTRACT

Scalable porous-silica film is a promised film as low-k dielectric of advanced Cu damascene interconnect. Hydrophobication and reinforcement by a silylation treatment using TMCTS agent on the film formation achieve the high performance to the film. However, the silylation treatment takes long process time. Self-hydrophobication process has developed for simultaneous performing of silylation at curing step. The process holds high performance without the silylation step on film formation. Pre-reinforcement process developed for increase of mechanical strength at spin-on step. The process holds high performance with reduction of total process time of the film formation. Both of the new processes form 140 nm pitch porous – silica/Cu damascene successfully with reduction of RC time constant by 10 % in comparison with conventional (k=2.7)/Cu damascene.

INTRODUCTION

The self-assembled scalable porous silica (Po.SiO) has a distinguished property of high porosity holding high mechanical strength due to the non-agglomerated pour formation in higher strength SiO_2 skeletal. 1,3,5,7- tetramethylcyclotetrasiloxane (TMCTS) vapor treatment has been required for film formation of Po.SiO as a silylation process performing a linking reaction between silanol (Si-OH) on silica pore surface and Si-H on TMCTS [1,2,3]. The silylation process using TMCTS performs not only hydrophobication of silica pore surface but also reinforcing the skeletal due to covering a mechanical defect. However, application limit of the treatment such as performance for finer porous material was not known well, a handling method was not mature, for example, serious particles trouble occurred under high pressure condition to boost the performance. An obligation of the additional process is another demerit in a view of cost performance [4]. In this study, novel conceptual SOD chemical alternative to TMCTS treatment was evaluated.

CONCEPT

Conventional Po.SiO film process is formed by spin coating of SOD(C) chemical and following three treatments as shown in the left of Fig. 1. New self-hydrophobication process as shown in the center of Fig. 1 uses new SOD chemical named SOD(S). Silylation promoter was added to SOD(S). The target of the process is eliminating both treatment of reinforcement and hydrophobication after pore generation. Reinforcement and silylation perform simultaneously at pore generation step with UV & Thermal cure due to an effect of the additive. Another new pre-reinforcement process as shown in the right of Fig. 1 uses new SOD chemical named SOD(G). Gelation agent was added to SOD(G). The target of the process is reduction of total process time on film forming. Reinforcement starts with gelation additive at spin-on step due to gelation of siloxane oligomer. Reinforcement completes at UV & thermal step due to reduction of defects by gelation additive. HMDS agent is useful as hydrophobication alter UV and thermal step.

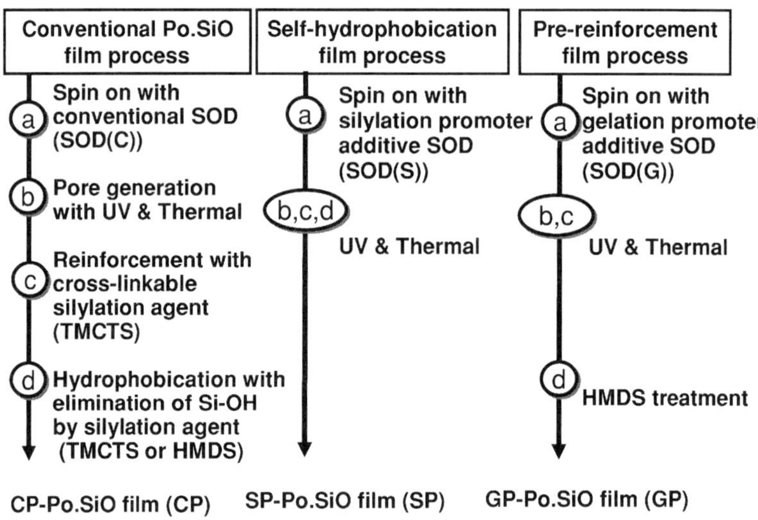

Fig. 1. Concepts of reducing time around time at film formation steps of porous silica.

EXPERIMENT

SOD chemicals of SOD(C), SOD(S) and SOD(G) formed porous silica film named CP, SP and GP, respectively as shown in Fig. 2. A surfactant which was a component of SOD chemical assembled a micelle. The micelle grew on drying procedure at spinner and following hot plate. Micelle evaporated leaving pore at 350°C UV irradiation. TMCTS and HMDS vapor treatment was performed at 350°C for CP and GP-Po.SiO, respectively. Nanoindentation was performed to measure mechanical strength of Po.SiO. Dielectric constant was calculated from capacitance measurement by Hg probing. Pore size distribution was measured by SAXS. Hydrophobicity was analyzed with IR absorbance under Brewster's angle condition. Stacked dielectric film of CVD-SiC(k=4.0), Po.SiO (k=2.0-2.5), CVD-SiOC (k=3.0) was fabricate for inter-metal dielectric, 140 nm pitch Cu damascene was formed by immersion-ArF lithography and CF_4 etching, ashing and wet cleaning, followed by silylation recovery and SiC (k =3.7) side wall formation as shown in Fig.3.

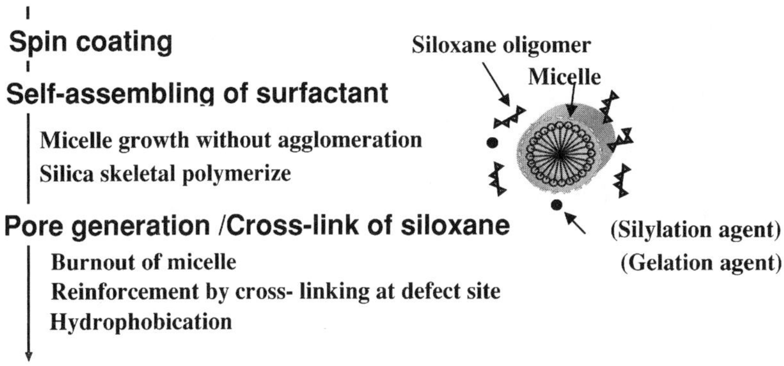

Fig. 2. Schematic flow of formation of self-assembled scalable porous silica film.

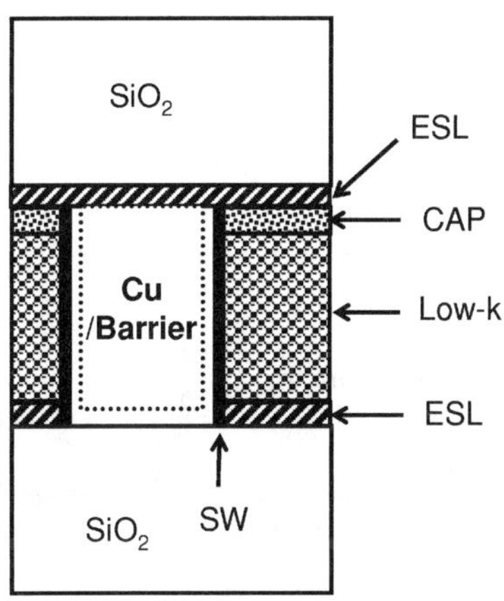

Fig. 3. Schematic cross section of fabricated 140nm pitch Po.SiO/Cu interconnect.

RESULTS AND DISCUSSION

Figure.4 shows pore size distribution and dielectric constant of porous silica film formed by self-hydrophobication process as a parameter of concentration ratio of surfactant in the SOD chemical. As decreasing of the concentration, pore size distribution shifted to smaller size, however, k-value was stable at 2.1. Fine porous of 1.73 nm was obtained. It was a half size of that of conventional Po.SiO (CP-Po.SiO). That indicates finer porous film without increase of k-value achieves by increase of concentration of surfactant.

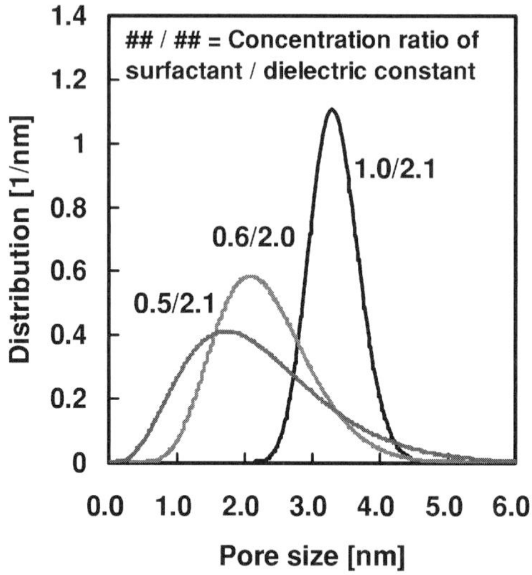

Fig. 4. Distribution of pore diameter of SP-Po.SiO (SP) as a parameter of relative template concentration.

The amount of remained Si-OH in the porous silica by self-hydrophobication process was smaller than the porous silica by conventional process as shown in Fig. 5. That indicates self-hydrophobication process performed on fine porous Po.SiO with self-hydrophobication .

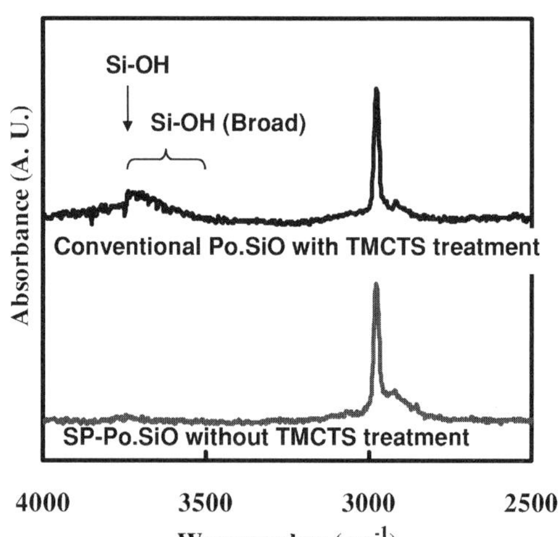

Fig. 5. FTIR absorbance of conventional Po.SiO (CP) with TMCTS treatment and SP-Po.SiO (SP) without TMCTS treatment.

All types of Po.SiO films had same strength of skeletal of Ew =34 GPa, greater than non self-assembled porous low-k film as shown in Fig. 6.

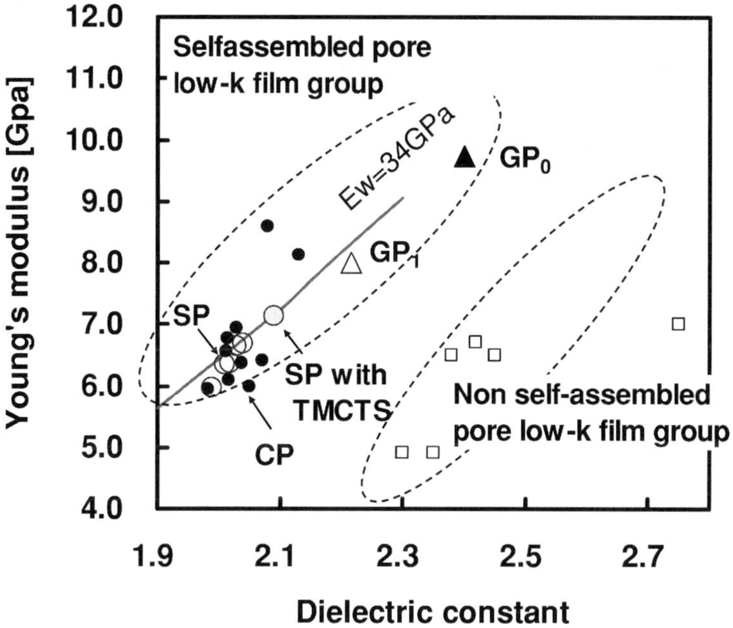

Fig. 6. Young's modulus vs. dielectric constant of CP, SP and GP self-assembled Po.SiO films and non self-assembled porous low-k films.

Figure 7 shows k-value and young's modulus of porous silica films as a function of storage term of SOD chemical. Fluctuation of the values shown on SOD(G). It found that the fluctuation was due to measurement of thickness of the film. Any significant shift of value was not observed. Both storage life of SOD(S) and SOD(G) were over 1 month.

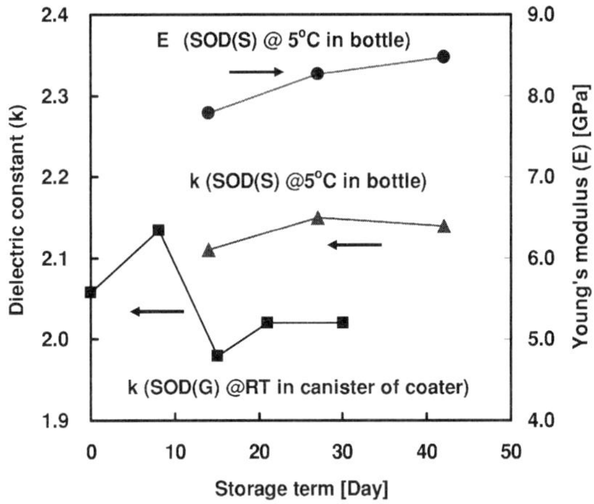

Fig. 7. Dielectric constant and young's modulus of SP and GP Po.SiO films as a function of storage term.

10% reduction of RC time constant was achieved by all SP, GP and CP-Po.SiO/Cu damascene in comparison with ULK(k=2.7)/Cu damascene as shown in Fig. 8.

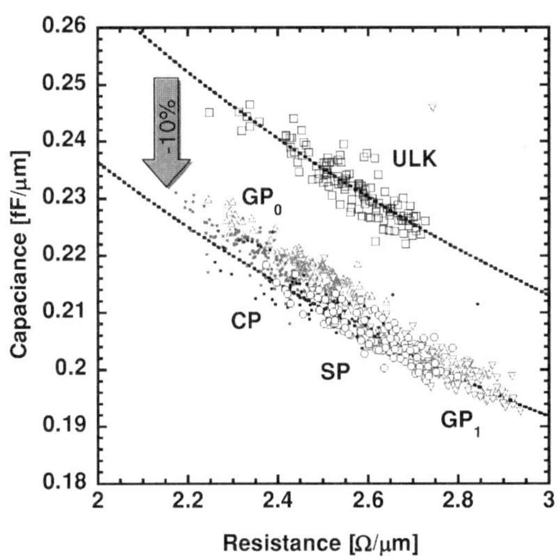

Fig. 8. Distribution of capacitance and resistance of low-k/Cu damascene as a parameter of porous low-k material.

This indicate reduction of turn around time at film formation was successfully achieved by using SOD(S) or SOD(G) without losing original property of self-assembled scalable porous silica as ultra low-k film.

.

CONCLUSIONS

Self-hydrophobication process has developed for simultaneous performing of silylation at curing step. The process holds high performance without the silylation step on film formation.

Pre-reinforcement process developed for increase of mechanical strength at spin-on step. The process holds high performance with reduction of total process time of the film formation.

Both of the new processes form 140 nm pitch porous –silica/Cu damascene successfully with reduction of RC time constant by 10 % in comparison with conventional (k=2.7)/Cu damascene.

ACKNOWLEDGEMENT

This work was supported by NEDO Japan

REFERENCES

1. K. Kohmura *et al,* Mat. Res Soc. Proc. 812, p85, 2004
2. T. Nakayama *et al*, Proc. MAM2006 , pp191-192
3. K. Kinoshita *et al.*, Proc. AMC2007, pp.344-345, 2007
4. S. Chikaki *et al.*, Tech. Digest IEDM2007, p969, 2007

Conference Proceedings AMC XXIV © 2009 Materials Research Society

Interconnect Performance and Scaling Strategy For the 22 nm Node and Beyond

James H.-C. Chen[1], Lijun Jiang[2], Alina Deutsch[2], Matthew S. Angyal[3], and Terry A. Spooner[1]
[1]IBM Research, International Business Machines, 255 Fuller Road, Albany, NY, 12203-3698
[2]IBM Research, International Business Machines, 1101 Kitchawan Road, Yorktown Heights, NY, 10598
[3]IBM Systems &Technology Group, International Business Machines, 2070 Route 52, Hopewell Junction, NY, 12533-6683

ABSTRACT

After four decades of continuous scaling on the CMOS technology, many innovations have been realized for improving interconnects RC performance since the emergence of copper. However, wire resistance has become a dominant factor on interconnect RC performance at the 22nm node. Optimization of metal aspect ratio and barrier thickness are important directions for mitigating the increase of wire resistance. In addition to reducing the dielectric constant of the ULK IMD, reducing the dielectric constant of the copper capping layer and the transitional layer of the ULK IMD are approaches for lowering the parasitic capacitance. In order to further boost interconnect performance, an air-gap integration is an option for reducing the capacitance without scaling metal thickness. Therefore, conventional scaling can be extended to the next generation.

INTRODUCTION

Determination of interconnect wire resistance and parasitic capacitance is not only a fundamental step in circuit design [1-4], but is also important for determining the BEOL (back-end-of-the-line) architecture [5] and optimizing the processes [6-8]. In order to address the roadmap directions for interconnect development, a novel methodology for creating contour plots of R and C as a function of geometry and material properties is presented. As a result, the direction for process optimization can be realized much easier from a two dimensional perspective. In addition, the relationship between RC performance, metal aspect ratio and dielectric constant of IMD (inter-metal dielectric) are presented within a single chart, which enables a new perspective on BEOL technology definition. Based on the simulation results, it is found that the non-scalable material layers, such as the diffusion barrier layer and the transitional layer in ULK (ultra-low k) starts to play important roles in interconnect RC performance. Here, an advanced on-chip wiring optimization tool, named AQUAIA [9], was utilized for the extraction of parasitic capacitance and wire resistance. Nevertheless, the tool allows the calculation of frequency-dependent resistive and inductive effects using accurate 2D or 3D field-solvers and complete transmission line parameters, R(f)L(f)C, are used in circuit simulation of interconnect performance. The automated framework allows multiple and conditional inputs for all design and process parameters and easy optimization target selection. Therefore, a contour plot of R or C can be derived, with an accuracy which could be better than 0.01% in less than 5 iterations.

SIMULATION ASSUMPTIONS

Two scaling scenarios were evaluated in this work to reflect the technology expectations and processes capability. In the first case, the best case scaling scenario assumes metal and via heights, the thickness of etch stop layer (ESL), and the thickness of the transitional layer, or adhesion layer, are reduced by the same ratio as the pitch. In addition, the sidewall barrier thickness is decreasing as far as 1 nm. As a result, this is a combination of the possible improvements of the industry. On the other hand, due to the reliability and processing capabilities constraints, some of the materials thickness may be invariable. Therefore, the worst case scaling scenario assumes the metal and via heights, the thickness of ESL, and the thickness of the transitional layer, as well as the barrier are the same as previous generation. The metal line width is assumed to be half of the pitch and the wire length is assumed to scale by the same ratio as the pitch for the calculation of scaled R and C.

RESULTS and DISCUSSIONS

The simulation begins with R and C as a function of the minimum pitch at intermediate wiring levels. In order to maintain the R×C performance, relationships of dielectric constant as a function of aspect ratio is derived by varying the metal line height. Based on a binary search algorithm, a constant C_{Total} curve can be plotted in a few iterations, by varying the dielectric constant and thickness of copper capping layer simultaneously. Using this technique, the constant C_{Total} and R×C can be plotted against film thickness and dielectric constant.

R and C against pitch scaling

In figure 1a, changes in wire resistance and parasitic capacitance are plotted by percentage against the pitch for the worst case scaling scenario. The technology expectations are shown by the κ of 2.2 and an 80% scaling on the thickness. However, the dramatic increase of wire resistance due to the thickness scaling, suggests the necessity for reducing the barrier thickness. On the other hand, for the best case scaling scenario, as shown in figure 1b, conventional thickness scaling also results in a significant performance drop, even when the dielectric constant of IMD is decreased from 2.4 to 2.2. Therefore, increasing of the metal aspect ratio is an option when wire resistance dominates the R×C performance, provided a reasonable coupling ratio is maintained.

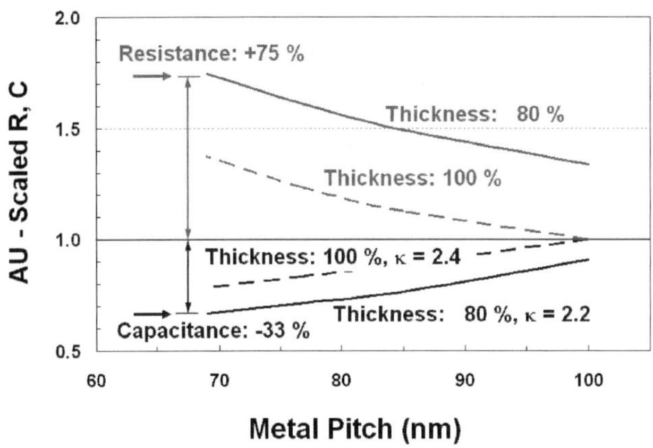

Figure 1a. Scaled R and C versus metal pitch for the worst case scaling scenario. The dramatic increase in wire resistance suggests the necessity for scaling of barrier thickness.

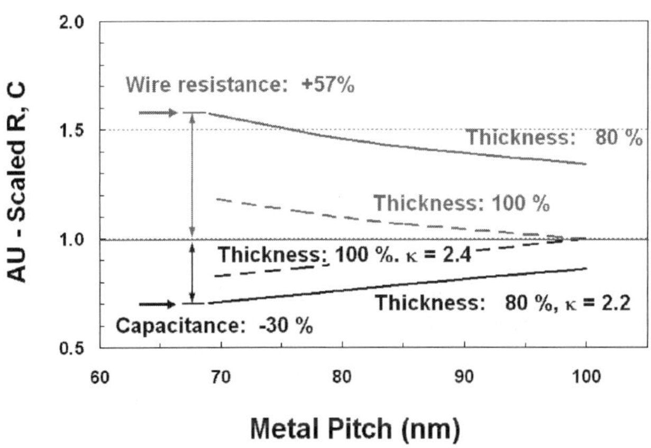

Figure 1b. Scaled R and C versus metal pitch for the best case scaling scenario. Thickness scaling results in a performance drop, even with an IMD of lower dielectric constant.

R×C performance and process capability

In figure 2, the IMD dielectric constant required for each scaling scenario in order to maintain the scaled R×C is plotted against the metal aspect ratio. The right Y-axis shows the performance with respect to an IMD of dielectric constant of 2.2. The plot indicates that reducing metal thickness requires an IMD with lower dielectric constant. Therefore, from the performance standpoint, it is not likely to keep the metal aspect ratio less than 2.7 for the conventional scaling scenario, since a dielectric constant of IMD lower than 1.9 is required. In order to keep the metal aspect ratio less than 2.5, an 80% pitch scaling scenario is recommended. In addition, an air-gap integration is a possible option for performance boost by reducing C without scaling metal thickness. With this approach, the conventional scaling can be extended to the next generation.

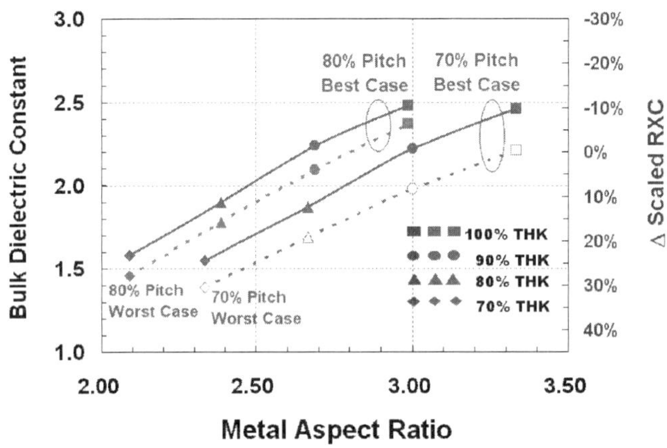

Figure 2. The IMD dielectric constant versus metal aspect ratio. Reducing the metal thickness such as from 100% to 70%, means that a IMD with lower dielectric constant is required.

Reduction of the parasitic capacitance

In figure 3a and figure 3b, contours of C_{Total} are plotted against the dielectric constant and thickness of the copper capping layer and the transitional layer of IMD, respectively. The plots indicate that for both of these dielectric layers, C_{Total} is a strong function of the dielectric constant rather than the thickness. This may suggest a direction for process improvement by reducing the interconnect capacitance.

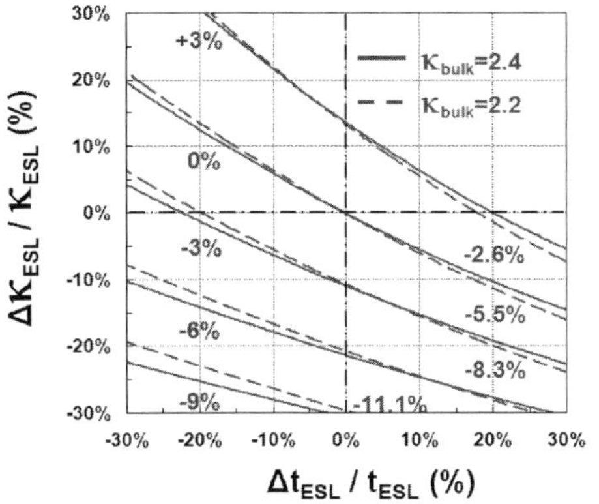

Figure 3a. Contour plots of constant C_{Total} versus dielectric constant and thickness of copper capping layer. The C_{Total} curves show a strong dependency on the dielectric constant than the thickness.

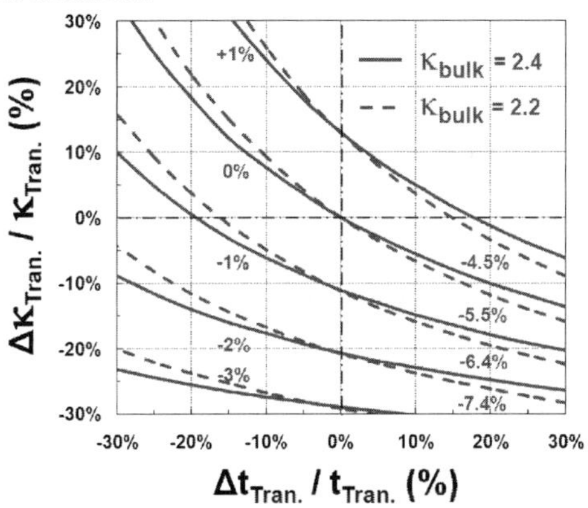

Figure 3b. Contour plots of constant C_{Total} versus dielectric constant and thickness of transitional layer of ULK (3b). The C_{Total} curves show a strong dependency on the dielectric constant than the thickness.

Reduction of barrier thickness

In figure 4, the performance loss in percentage is plotted against barrier thickness for various metal aspect ratios. A sidewall barrier thinner than 3 nm, is required to maintain R×C for a metal aspect ratio of 1.9 [10] or less [11]. This is a challenge for both copper diffusion and reliability. Moreover, metal with high aspect ratio will cause higher coupling. Thin barrier and air-gap are the options, if coupling is an issue.

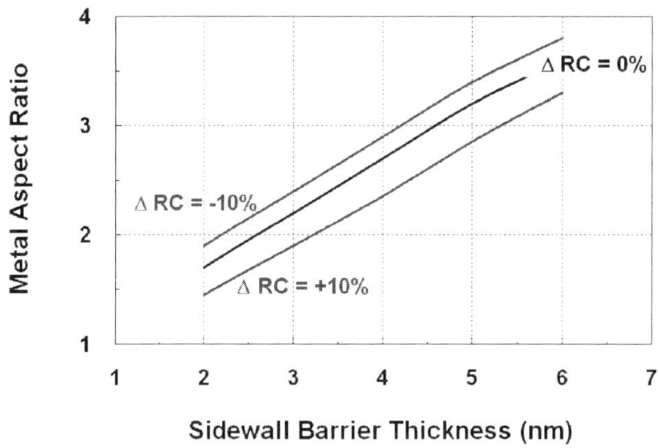

Figure 4. RC performance versus sidewall barrier thickness and metal aspect ratio. A barrier thickness thinner than 3nm is required to maintain performance for a metal aspect ratio less than 2.0.

Geometry effect

In figure 5, contours of C_{Total} are plotted against the metal trench ratio and width ratio. The trench ratio is the percentage of the metal height to the thickness of IMD, while the width ratio is the percentage of metal width to the pitch. Reduce of metal height or metal width decreases C_{Total}. On the other hand, contours of R×C are plotted in figure 6. Here, a metal width wider than half of the pitch decreases the R×C. This might indicate a different approach for optimizing the BEOL architecture, when wire resistance dominates the performance.

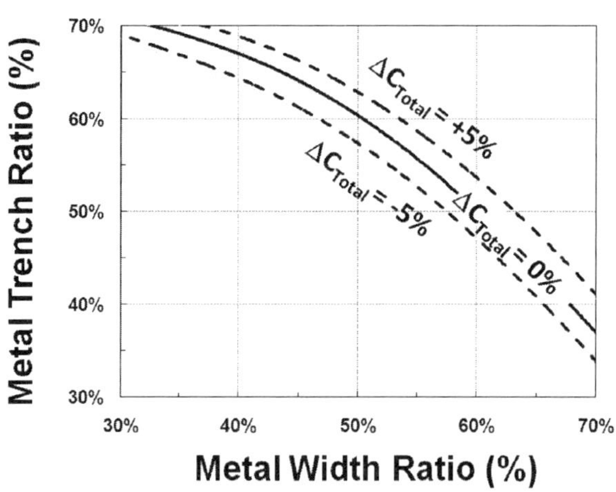

Figure 5. Constant C_{Total} curves versus metal trench ratio and metal width ratio.

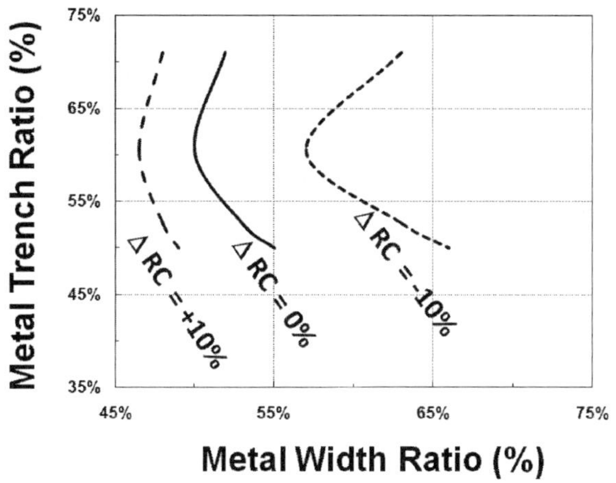

Figure 6. Constant RC curves versus metal trench ratio and metal width ratio. In this plot, a metal width wider than half of the pitch decreases the R×C.

CONCLUSIONS

In this paper, optimization of BEOL architecture for thin wire levels at 22nm node and beyond is presented. When the wire resistance dominates the RC performance, a conventional scaling scenario does not show benefits even with a dielectric constant of 2.2 material. As a result, an integration process for high aspect ratio metals is one of the directions for maintaining the RC performance. In addition, an air-gap integration is also another possible direction for boosting the RC performance and reducing the coupling ratio. From the material aspect, reducing the dielectric constant of the transitional layer and the etch stop layer is more important than reducing the thickness of these layers. On the other hand, reducing the barrier thickness is necessary for extending the conventional scaling scenario.

ACKNOWLEDGMENTS

The authors would like to thank Dr. Daniel Edelstein for his encouragement and guidance on the direction of this study.

REFERENCES

1. Albert E. Ruehli and Pierce A. Brennan, "Accurate Metallization Capacitances for Integrated Circuits and Packages", *IEEE J. of Solid-State Circuits*, pp. 289-290, August (1973).
2. T. Sakurai and K. Tamaru, "Simple formulas for two- and three-dimensional capacitances", *IEEE Trans. on Electron Devices*, Vol. ED-30, pp. 183-185, February (1983).

3. Ying Liu, S. R. Nassif et al., "Impact of interconnect variations on the clock skew of a gigahertz microprocessor", *Proc. Design Automation Conference*, pp. 168–171, (2000).
4. Ching-Chao Huang and Jue-Hsien Chern, "Accurate modeling of capacitive, resistive and inductive effects of interconnect", *WESCON Conference Record*, pp. 115-117, November (1995).
5. Mark Bohr, "Interconnect Scaling—The Real Limiter to High Performance ULSI", *Proc. International Electron Devices Meeting*, IEEE Press, New York, pp. 241-244, (1995).
6. Y. Tsukamoto, T. Kunikiyo et al., "Realistic scaling scenario for sub-100nm embedded SRAM based on 3-dimensional interconnect simulation", *Proc. of SISPAD*, pp. 63-66, September (2002).
7. Simon Lin, Changming Jin et al., "Low-k dielectrics characterization for damascene integration", *Proc. of the 4th IITC*, San Francisco, pp. 146-148, June (2001).
8. T. Kokubo, A. Das et al., "Cu/LKD-5109 damascene integration demonstration using FF-02 low-k spin-on hard-mask and embedded etch-stop", *Proc. of the 5th IITC*, San Francisco, pp. 51-53, June (2002).
9. I. M. Elfadel, M. B. Anand et al., "Aquaia: A CAD tool for on-chipinterconnect modeling, analysis and optimization", *Dig. Electr. Perf. Electronic Packaging*, Monterey CA, Vol. 11, pp. 337–340, October (2002).
10. International Technology Roadmap for Semiconductors, 2006 update version. (http://www.itrs.net)
11. A. Deutsch et al., "On-Chip Wiring Design Challenges for Gigahertz Operation", *IEEE Proceedings*, pp. 529-555, vol. 89, no. 4, April (2001).

Implementation of Ru based barriers in 50 nm half pitch single damascene Cu/SiCOH (k=2.5) structures

Laureen Carbonell[1], Henny Volders[1], Nancy Heylen[1], Kristof Kellens[1], Zsolt Tokei[1], Dirk Hendrickx[1], Herbert Struyf[1], Martine Claes[1], Janko Versluijs[1], Els Van Besien[1], Rudy Caluwaerts[1], Andrew Cockburn[2], Virginie Gravey[2], Kavita Shah[3], Amir Al-Bayati[3], X. Fu[3], D. Lubben[3], A. Sundarrajan[3], and Gerald P. Beyer[1]

[1]IMEC vzw, Kapeldreef 75, Leuven, B-3000, Belgium

[2]Applied Materials Europe, Kapeldreef 75, Leuven, B-3000, Belgium

[3]Applied Materials, 3050 Bowers Avenue, P.O. Box 58039, Santa Clara, CA, 95054

ABSTRACT

In this study, we present results on the implementation of Ru- based barriers in 50 nm half pitch single damascene Cu/SiCOH (k=2.5) structures obtained by double patterning. High electrical yields and low Cu defectivity could be obtained when keeping the metallization steps closely coupled in time. The electrical and reliability performance are very sensitive to the double patterning approach and to the metallization scheme involved. The highest electrical yields were obtained when using a Ru- metallization scheme involving the use of a dedicated Ru slurry limiting damage to the surface of the IMD low k material and reducing the corrosion of the Cu in the trenches. On the other hand, the specific resistivity of Cu measured in the 50 nm half pitch features was the lowest among the investigated metallization schemes, around 3.75 μOhm.cm. This could reflect different surface scattering effects and differences in the Cu microstructure (recrystallization, impurity incorporation) when using different barrier/liner materials.

INTRODUCTION

Efficient copper metallization and interconnect reliability are becoming real challenges for future manufacturing as the feature sizes decrease with each technology node. Various characteristics of the PVD Ta-based diffusion barrier and Cu seed layer, such as wettability, conformality and continuity may lead to subsequent gap fill failure if not optimal. Recently alternative liner metals such as Ru [1] and Ru alloys [2] have been introduced to act as adhesion enhancement layers and nucleation layers to enable Direct Plating. However the incorporation of such metals may also affect the Cu fill ability, electrical properties and reliability. In this paper we report on the implementation of Ru-based barriers in 50 nm half pitch Cu/SiCOH (k=2.5) single damascene trenches obtained by double patterning and show that it is a potential enabling technology for sub- 32 nm / SiCOH integration schemes.

EXPERIMENTAL DETAILS

Single damascene 50 nm half pitch trenches were formed in BDII™ (k=2.5) by double patterning using a 30 nm TiN metal hard mask (MHM) [3]. A first exposure step (M1A) is followed by a hard mask etch & ash (figure 1). The BARC opening step was done using a CF_4 based chemistry and the TiN opening was done using a Cl_2 based chemistry and was followed by an ash step using a N_2/H_2 strip compatible with the low k dielectric material. A careful fine-tuning of the BARC opening, MHM etch and strip steps are required in order to obtain the desired trench CD, MHM profile and CD uniformity. The second litho step (M1B) is directly aligned to M1A. After M1B hard mask opening and strip (similar processes as used for M1A patterning), the 50 nm half-pitch patterns are transferred into the underlying low k dielectric using a CF_4 based chemistry. After the M1B patterning, all wafers were subjected to an APM clean to remove residual polymers and the remaining TiN metal hard mask layer.

For the reference metallization scheme, a 3 nm PVD TaN/Ta barrier layer was deposited prior to a 20 nm PVD Cu seed layer to provide electrical contact for Cu fill. The filling of the features was done by electroplating (ECP) using a medium acid 3-component chemistry-followed by a post-plating anneal at 180ºC under N_2/H_2 to optimize the Cu microstructure. The excess of barrier and Cu on the field areas was finally removed by a 2-step CMP process.

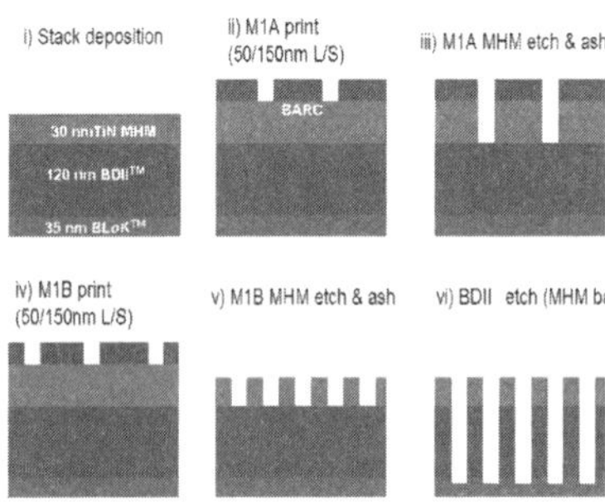

Figure 1: Double patterning scheme used to integrate 50 nm half pitch single damascene trenches in BDII™ (k=2.5)

	Barrier	Seed	ECP Cu/anneal	CMP
Features with h_{Cu}=84.5 nm	3 nm PVD TaN/Ta **(Ref.)**	20 nm PVD Cu	500 nm Cu + anneal @ 180°C under N_2/H_2	- Commercial selective Cu slurry - Experimental selective Ta- barrier slurry
	2.5 nm RuTa	20 nm advanced PVD Cu		- Commercial selective Cu slurry - Commercial Ru- barrier slurry
	4 nm RuTa			
Features with h_{Cu}=190 nm	6 nm PVD TaN/Ta **(Ref.)**	60 nm PVD Cu	800 nm Cu + anneal @ 180°C under N_2/H_2	- Commercial selective Cu slurry - Commercial non-selective Ta- barrier slurry
	2 nm RuTa	60 nm advanced PVD Cu		- Commercial selective Cu slurry - Commercial Ru- barrier slurry
	4 nm RuTa			
	2 nm TaN / 2 nm RuTa			

Table 1: Metallization splits used for the 100 nm pitch features and the larger features used for the ρ_{Cu} measurements

In this work, PVD RuTa alloy liners were used in combination with a PVD Ta- based diffusion barrier or alone, and an advanced 20 nm PVD Cu seed. Their impact on the Cu filling efficiency (SEM,TEM), electrical properties, reliability performance (breakdown) and specific resistivity of copper were evaluated. All metallization splits considered may be found in Table 1.

RESULTS AND DISCUSSION

Cu defectivity

Specific defects are introduced at each process step of the Cu metallization process, which can, in turn, influence the subsequent processing of the wafers and the final electrical performance of the devices. In the metallization schemes considered in this work, the Cu defectivity in 100 nm pitch features is mainly caused by a poor compatibility between the electroplating + post-plating anneal and the selective Cu slurry used during the CMP process. A delay of more than 24 hours between the plating + anneal steps and the CMP process results in the appearance of yield-killing Cu defects. High electrical yields could only be achieved when keeping the ECP, post-plating anneal and CMP steps closely coupled in time. Copper pitting during CMP is known to be directly related to the Cu microstructure and impurity incorporation during the electroplating step [4]. Based on the previous results and additional characterization measurements (not shown here), comparable final Cu grain sizes (~3 μm wide in bond pad areas) were observed whether a delay was introduced during processing or not. This may suggest that there is no direct relationship between the Cu grain size and the line defectivity. However an incomplete recrystallization and an uneven distribution of impurities in the copper after anneal, evidenced by the presence of small grains within the thickness of the layers, are likely responsible for the pronounced line defectivity observed when a delay is introduced between the plating + anneal and CMP steps.

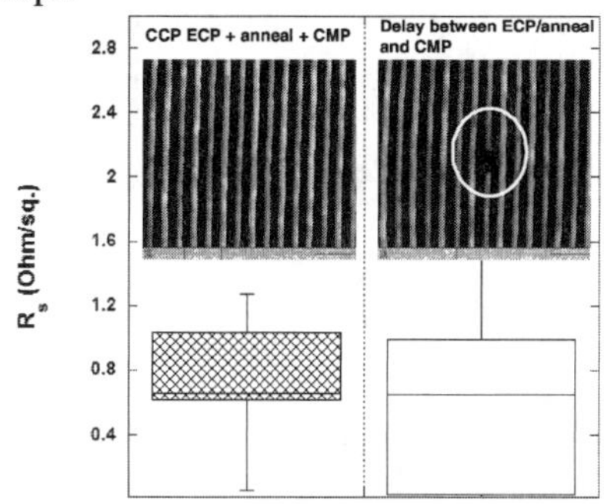

Figure 2: Effect of a delay introduced between ECP/post-plating anneal and CMP on Cu defectivity as observed by SEM with back-scattered electron imaging and electrical yield of 100 nm pitch meander-fork structures with a 3 nm TaN/Ta barrier + 20 nm Cu seed combination.

Electrical performance

When using double patterning, the layout of the test structures is split into two images. Structures such as the meander-fork (MF) structures have the meanders printed in one step and the forks printed in the second step [3]. Therefore, the electrical performance of the test structures relates to the design layout properties (figure 3). The MF structures with meanders printed in M1A and forks printed in M1B (referred to as M1A) have lower electrical yields than MF structures with meanders printed in M1B and forks printed in M1A (referred to as M1B). The latter exhibit lower Rs values, tighter distributions and highest yields suggesting slight differences between the two M1A and M1B patterning steps. Such differences relate to small overlay errors or to the final etch/strip step which is very sensitive to the resist profile control. The features formed during the 2nd litho step, M1B, are slightly wider and deeper than those formed at the M1A step (figure 5a & 5b), thus exhibiting lower Rs values.

Within a same group of MF structures, highest yields (up to ~95%) were obtained for all structures using a Ru- based barrier as compared to the reference TaN/Ta barrier. Comparable results were obtained for all Ru barrier splits. Also, whatever the barrier/seed combination, top-view SEM imaging of the M1A MF structures did not reveal any yield-killing defects but revealed the presence of sidewall defects due to poor adhesion of the barriers to the low k materials (figure 4). Such defects relate to remaining polymers that were not removed by the APM post-patterning clean and contribute to limiting the electrical performance of the MF structures. Though, the additional yield loss observed for the structures using the TaN/Ta metallization scheme is indicative of differences in the trench top surface quality as illustrated by the TEM cross-section pictures presented in figures 5a and 5b.

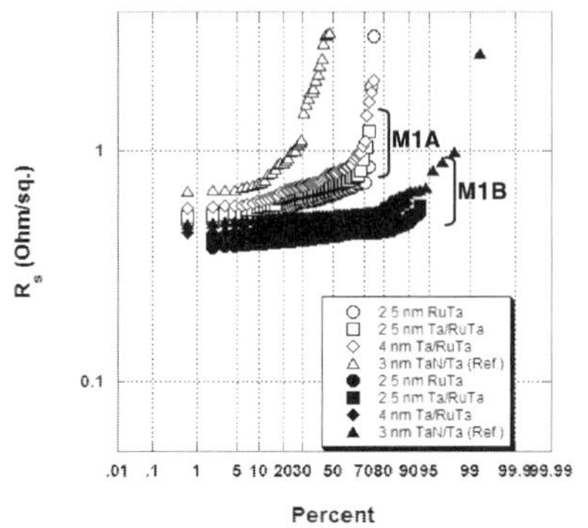

Figure 3: R_s distributions of 100 nm pitch M1A and M1B meander fork structures with different barrier/liner combinations

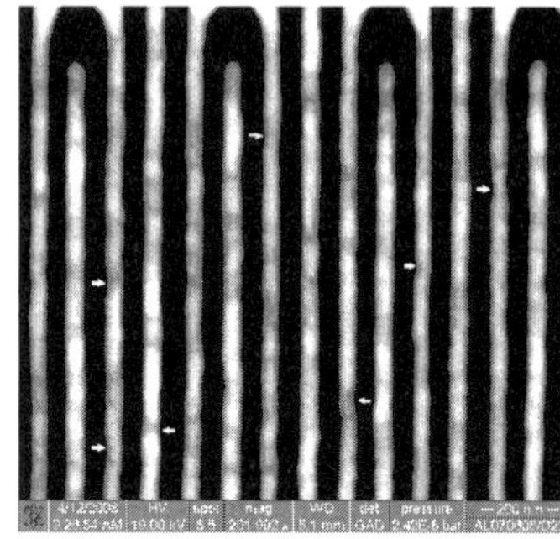

Figure 4: SEM back-scattered image of a 100 nm pitch M1A MF structure with a RuTa + 20 nm advanced Cu seed combination

Less Cu corrosion and less IMD surface damage are observed when using Ru-based barriers in combination with the Ru barrier slurry, thus resulting in an enhanced electrical performance.

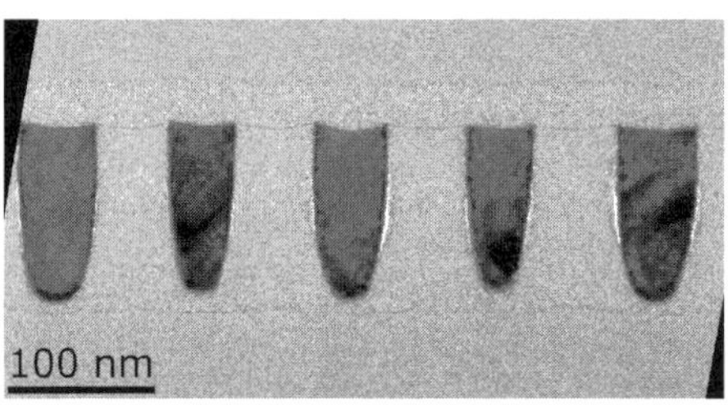

Figure 5a: TEM cross-section of a M1A MF structure using a 2.5 nm RuTa liner

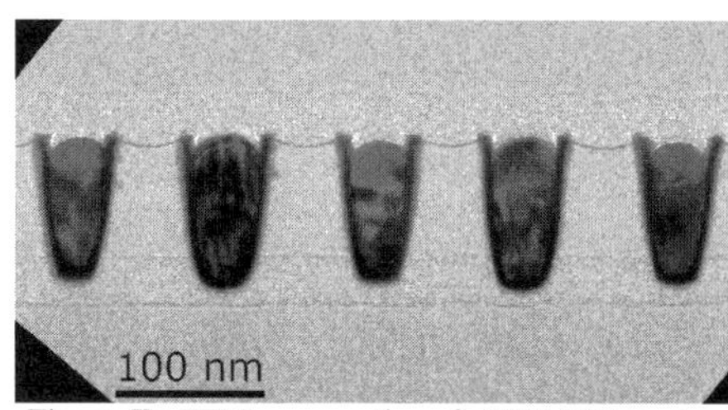

Figure 5b: TEM cross-section of a M1A MF structure using a 3 nm TaN/Ta barrier

Reliability performance

Low k dielectric properties are prone to be affected by various process steps during process integration which will modulate the reliability behavior of the interconnects [5]. Reliability testing of yielding 100 nm pitch MF structures was initiated with breakdown measurements carried out at 100°C with a ramp up to 50V in 3 seconds. These results show reasonable breakdown fields for the different barrier splits considered (figure 6). Moreover the breakdown field is dependent both on the type of structure and the type of barrier.

M1B MF structures exhibit larger breakdown fields than M1A MF structures. When scaling down the critical dimensions to 50 nm half pitch, the impact of IMD spacing variations introduced during double patterning and Line Edge Roughness (LER) becomes important and results in many localized high electrical fields, thus lowering the breakdown voltage [6]. Though, comparable LER (3σ=6 nm) and small spacing variations (~5 nm) of the IMD thickness may explain on the own the pronounced difference in breakdown voltage between the two types of structures, especially when involving Ru- based barriers. Differences in the I-V curves recorded for both types of structures suggest that also different breakdown mechanisms occur (figure 7a & 7b).

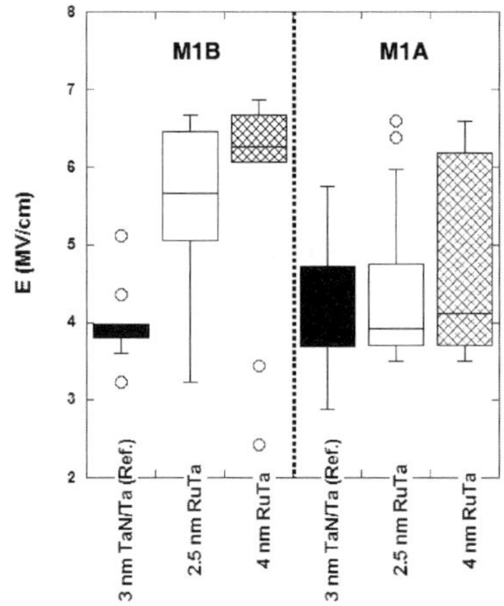

Figure 6: Distribution of breakdown fields of 1 cm long M1A & M1B MF structures measured @ 100°C and calculated using the median CD values extracted from the TEM cross-section analysis

In the case of the M1B MF structures, the high probability of "up-jumps" of the leakage current indicate that the dielectric breakdown dominates the breakdown mechanism. For the M1A MF structures, a high probability of "down-jumps" of the leakage current is an indication that the meander metal line fails before the dielectric. The narrower M1A meanders can not sustain high current densities and thus fail earlier than the M1B meanders. Different process-related effects can contribute to this early failure such as the presence of post-etch residues that may significantly reduce the barrier integrity and adhesion to the dielectric in the narrowest features.

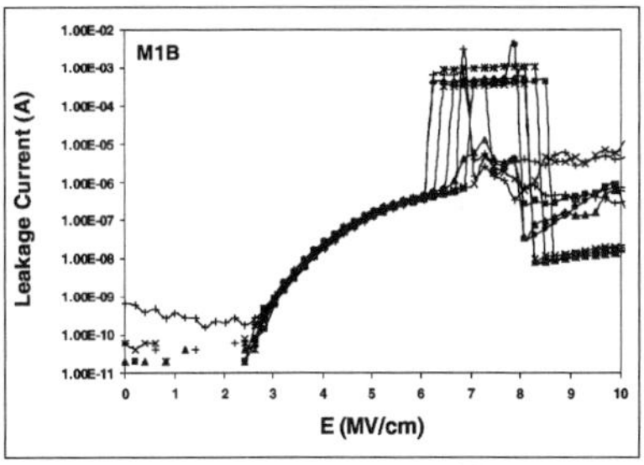

Figure 7a: I-V curves measured at 100°C on M1B MF structures with a 4 nm RuTa barrier

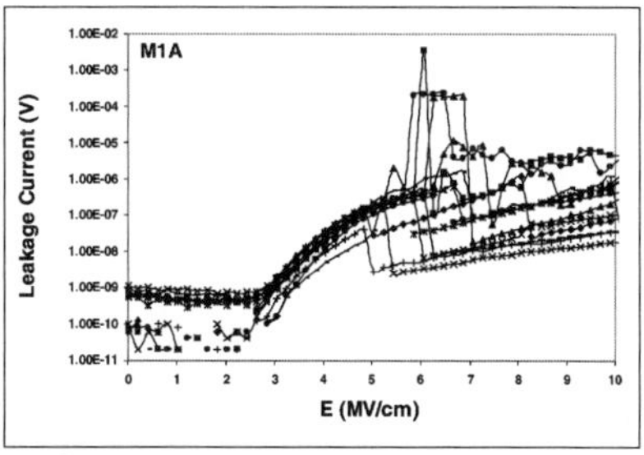

Figure 7b: I-V curves measured at 100°C on M1A MF structures with a 4 nm RuTa barrier

Within the same M1B MF structures, larger breakdown fields are obtained when using Ru-based barriers. As the double patterning process was common to all splits considered, the CMP interface and the diffusion/barrier dielectric interface are likely the other two parameters that significantly impact the breakdown fields.
Due to the characteristics of the CMP process and the fast diffusivity of Cu ions in the low k dielectric, the CMP interface is readily contaminated by copper when the low-k material is modified. Thus, it is important to limit the damage induced to the low k material during the barrier polish step and the subsequent plasma treatment prior to the dielectric cap deposition. The larger breakdown fields measured for the M1B structures indicate that the dedicated slurry used for the polishing of Ru- based barriers is more compatible with the inter-metal dielectric material (IMD) thus reducing the IMD damage. This is also illustrated in the TEM pictures presented in figures 5a & 5b. The breakdown fields measured on the M1B structures using a 4 nm Ru- based barrier are higher compared to those measured on structures using a 2.5 nm Ru- based barrier. The thinner RuTa layer might be discontinuous and permeable to Cu, thus resulting in lower breakdown fields.

Specific resistivity of copper

The specific resistivity of Cu (ρ_{Cu}) was extracted from line resistance vs. temperature measurements [7] and plotted as a function of line width after measuring the height of the Cu in the trenches by TEM analysis (figure 8). As the specific resistivity and cross-sectional area depend on the line resistance and its variation with temperature, any factor influencing these parameters (CD variation, line height, impurity incorporation, grain size, etc…) will be reflected in changes of the Cu specific resistivity in the lines.

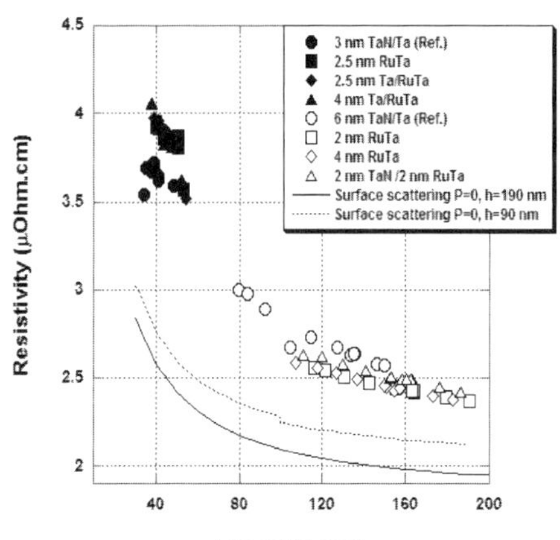

Figure 8: Impact of the novel barrier materials on the Cu resistivity of trenches with different heights (h_{Cu}=84.5 nm and h_{Cu}=190 nm)

The changes in resistivity are related to the feature sizes and various barriers used. However, the effect of the barriers on ρ_{Cu} is different depending on the feature size considered. In large structures (h_{Cu}=190 nm), compared to the reference structures with a TaN/Ta barrier, a pronounced decrease in ρ_{Cu} (~-2-3%) is observed when using Ru- based barrier/liners. In smaller features (h_{Cu}=84.5 nm), an opposite trend is observed as an increase in ρ_{Cu} (~+5%) is observed when using Ru- based barrier/liners. Such differences could be related to modified surface scattering effects (e.g. Ru/Cu intermixing) and differences in the Cu microstructure (recrystallization, impurity incorporation, etc...).

No significant differences in ρ_{Cu} could be noticed between the different Ru- barrier splits used in the smallest dimensions. However more differences were observed in the large features with the lowest ρ_{Cu} obtained when using a 2 nm RuTa liner without any Ta- based barrier.

CONCLUSIONS

In this work, we report on the successful integration of Ru- based barriers in 50 nm half pitch Cu/SiCOH (k=2.5) single damascene trenches obtained by double patterning. High electrical yields and low Cu defectivity could be obtained when keeping the metallization steps closely coupled in time. Therefore first Cu resistivity and reliability measurements on 100 nm pitch features could be successfully achieved.

ACKNOWLEDGEMENTS

The authors would like to thank W. Alaerts, A. Vandervorst, B. Deweerdt, O. Richard, C. Drijbooms, P. Van Marcke and H. Bender for technical support and Y. Li for fruitful discussions.

REFERENCES

[1] Xin-Ping Qu et al., APL vol. 88 p151912 (2006).
[2] Chun-Wei Chen et al., J. Electrochem. Soc. vol. 155 (6) p H438 (2008).
[3] Jan Van Olmen et al., Proceedings of the Advanced Metallization Conference 2007 (2007).
[4] Hsien-Ping Feng et al., J. Electrochem. Soc. vol. 155(1) p H21 (2008).
[5] Yunlong Li et al., Semicond. Sci. And Tech. Vol. 22 p320 (2007).
[6] Michele Stucchi et al. proceedings of the International Interconnect Technology Conference 2008 (2008).
[7] Hua Li et al., Proceedings of the Advanced Metallization Conference 1999 (1999).

Conference Proceedings AMC XXIV © 2009 Materials Research Society

A simple de-embedding method for characterization of on-chip four-port networks

Shuhei Amakawa[1], Hiroyuki Ito[2], Noboru Ishihara[1], and Kazuya Masu[1,2]
[1]Integrated Research Institute, Tokyo Institute of Technology, 4259-R2-17 Nagatsuta, Midori-ku, Yokohama, 226-8503, Japan
[2]Precision and Intelligence Laboratory, Tokyo Institute of Technology, 4259-R2-17 Nagatsuta, Midori-ku, Yokohama, 226-8503, Japan

ABSTRACT

A simple de-embedding method that requires only a THRU pattern is proposed for characterizing on-chip four-port networks such as differential transmission lines. It makes use of an orthogonal transformation of the scattering matrix for simplifying the problem, and the only requirement for it to work is that the THRU have a certain symmetry property. The proposed thru-only de-embedding method is applied to a pair of coupled transmission lines up to 50 GHz. The result is found to be approximately consistent with that from the standard open-short method, which requires two dummy patterns: OPEN and SHORT. SHORT is apt to deviate from the ideal at multi-GHz frequencies and the proposed method has the added advantage of being able to dispense with it.

INTRODUCTION

Demand for accurate high-frequency characterization of on-chip devices has been escalating along with the accelerated development of high-speed digital signaling and radio-frequency (RF) circuits. High-frequency characterization of an on-chip device is typically done by using a vector network analyzer (VNA). Since an on-chip device under test (DUT) is inevitably "embedded" in such intervening structures as probe pads (Fig. 1a) and they leave definite traces in the scattering parameters measured by a VNA, the characteristics of the DUT have to be "de-embedded [1]" in some way from the as-measured data. While there have been a number of de-embedding methods proposed for a two-port network, very few (e.g. [2]) have been proposed for a four-port network in spite of the fact that many important devices, such as differential transmission lines, are represented as four-port networks. In this paper, we propose a simple four-port de-embedding method requiring only a THRU pattern.

THEORY OF MODE TRANSFORMATION AND DE-EMBEDDING

A network matrix of a $2n$-port (Fig. 2) can be transformed into a different representation by changing the basis sets for voltage and current by the following transformations [3].

$$\boldsymbol{v} = \boldsymbol{K}_{\mathrm{V}}\tilde{\boldsymbol{v}}, \quad \boldsymbol{i} = \boldsymbol{K}_{\mathrm{I}}\tilde{\boldsymbol{i}}, \quad \boldsymbol{K}_{\mathrm{V}}^{\mathrm{T}}\boldsymbol{K}_{\mathrm{I}}^{*} = \mathbf{1}_{2n}. \tag{1}$$

Here ^{T} denotes transpose and * denotes complex conjugate. $\boldsymbol{v}$ and $\boldsymbol{i}$ are, respectively, the port voltage vector and the port current vector in the conductor domain. $\mathbf{1}_{2n}$ is a $2n \times 2n$ identity matrix. The

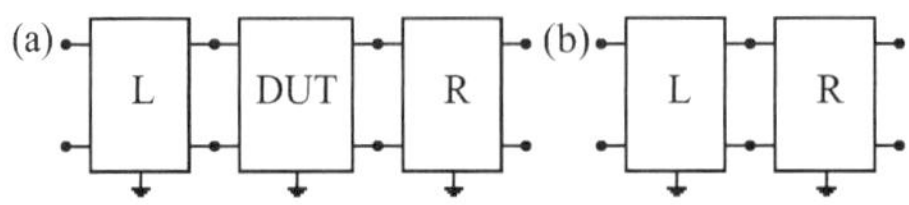

Figure 1: (a) A model of a four-port measured by a VNA. (b) A model of a THRU (Fig. 3a).

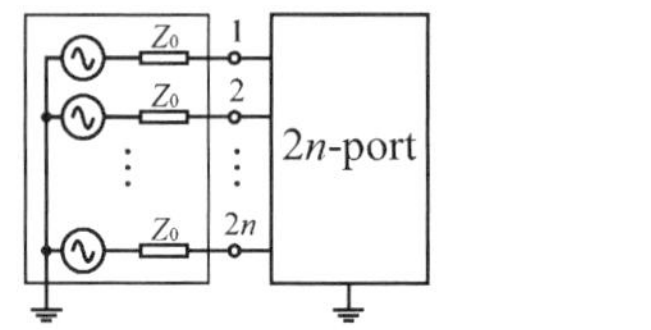

Figure 2: The right $2n$-port is the network under measurement. The left one is the measurement system including a VNA.

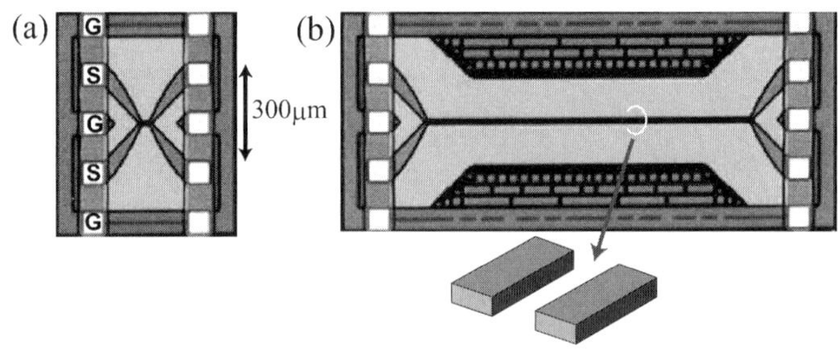

Figure 3: (a) A micrograph of the THRU pattern with GSGSG pads. (b) A pair of 1 mm-long transmission lines. The line width is 6 μm and the spacing between the lines is 4.6 μm. The nominal differential characteristic impedance is 100 Ω.

last equation in (1) ensures that the power flux remains invariant under the change of bases [3]; for example, $\boldsymbol{a}^{\mathrm{T}}\boldsymbol{a}^* = \tilde{\boldsymbol{a}}^{\mathrm{T}}\tilde{\boldsymbol{a}}^*$. (1) suggests that $\boldsymbol{K}_{\mathrm{V}}$ and $\boldsymbol{K}_{\mathrm{I}}$ can be expressed in terms of a unitary matrix $\boldsymbol{K}_{\mathrm{U}}$ ($\boldsymbol{K}_{\mathrm{U}}^{\dagger} = \boldsymbol{K}_{\mathrm{U}}^{-1}$) and a Hermitian matrix $\boldsymbol{K}_{\mathrm{P}}$ ($\boldsymbol{K}_{\mathrm{P}}^{\dagger} = \boldsymbol{K}_{\mathrm{P}}$) as $\boldsymbol{K}_{\mathrm{V}} = \boldsymbol{K}_{\mathrm{U}}\boldsymbol{K}_{\mathrm{P}}$ and $\boldsymbol{K}_{\mathrm{I}} = \boldsymbol{K}_{\mathrm{U}}\boldsymbol{K}_{\mathrm{P}}^{-1}$ by polar decomposition, where † denotes conjugate transpose. Since $\boldsymbol{v}$ and $\boldsymbol{i}$ are related by the impedance matrix as $\boldsymbol{v} = \boldsymbol{Z}\boldsymbol{i}$, impedance matrices undergo the following transformation.

$$\tilde{\boldsymbol{Z}} = \boldsymbol{K}_{\mathrm{V}}^{-1}\boldsymbol{Z}\boldsymbol{K}_{\mathrm{I}} = \boldsymbol{K}_{\mathrm{P}}^{-1}\boldsymbol{K}_{\mathrm{U}}^{\dagger}\boldsymbol{Z}\boldsymbol{K}_{\mathrm{U}}\boldsymbol{K}_{\mathrm{P}}^{-1}. \tag{2}$$

A generalized scattering matrix $\boldsymbol{S}$ of a $2n$-port (Fig. 2) relates the vector, $\boldsymbol{a}$, of power waves of a given frequency incident on the $2n$-port to the vector, $\boldsymbol{b}$, of outgoing waves [4].

$$\boldsymbol{b} = \begin{bmatrix} \boldsymbol{b}_1 \\ \boldsymbol{b}_2 \end{bmatrix} = \begin{bmatrix} \boldsymbol{S}_{11} & \boldsymbol{S}_{12} \\ \boldsymbol{S}_{21} & \boldsymbol{S}_{22} \end{bmatrix} \begin{bmatrix} \boldsymbol{a}_1 \\ \boldsymbol{a}_2 \end{bmatrix} = \boldsymbol{S}\boldsymbol{a}, \quad \boldsymbol{v} = \boldsymbol{v}^+ + \boldsymbol{v}^-, \quad \boldsymbol{i} = \boldsymbol{i}^+ + \boldsymbol{i}^-, \tag{3}$$

$$\boldsymbol{a} = \boldsymbol{R}_0^{-1/2}\boldsymbol{v}^+ = \frac{1}{2}\boldsymbol{R}_0^{-1/2}(\boldsymbol{v} + \boldsymbol{Z}_0\boldsymbol{i}), \quad \boldsymbol{b} = \boldsymbol{R}_0^{-1/2}\boldsymbol{v}^- = \frac{1}{2}\boldsymbol{R}_0^{-1/2}(\boldsymbol{v} - \boldsymbol{Z}_0^*\boldsymbol{i}). \tag{4}$$

$\boldsymbol{Z}_0$ is the reference impedance matrix used to define the scattering matrix and $\boldsymbol{R}_0 = \Re(\boldsymbol{Z}_0)$. It is the impedance matrix of the terminating $2n$-port shown in Fig. 2 on the left (with all the signal sources shunted) and is also transformed by (2). $\boldsymbol{Z}_0$ is usually set by a VNA to be a real scalar matrix $\boldsymbol{Z}_0 = \boldsymbol{R}_0 = R_0\boldsymbol{1}_{2n}$. If the reference impedance matrix is a scalar matrix, scattering matrices undergo a unitary transformation $\tilde{\boldsymbol{S}} = \boldsymbol{K}_{\mathrm{U}}^{\dagger}\boldsymbol{S}\boldsymbol{K}_{\mathrm{U}}$.

A four-port under measurement can be represented as a cascade of three four-ports (Fig. 1a), and its transfer matrix can be written as $\boldsymbol{T}_{\mathrm{meas}} = \boldsymbol{T}_{\mathrm{L}}\boldsymbol{T}_{\mathrm{DUT}}\boldsymbol{T}_{\mathrm{R}}$. Once the intervening structures L and R are somehow characterized, the properties of the DUT can be de-embedded by $\boldsymbol{T}_{\mathrm{DUT}} = \boldsymbol{T}_{\mathrm{L}}^{-1}\boldsymbol{T}_{\mathrm{meas}}\boldsymbol{T}_{\mathrm{R}}^{-1}$. The four-port de-embedding method proposed by Han et al. [2] follows this idea. Their method requires that the upper ports and the lower ports of both L and R consist of *uncoupled* two-ports as in Fig. 4b. This condition may be fulfilled by appropriately configured off-chip systems [2]. However, on-chip THRU patterns (e.g. Fig. 3a) can hardly meet this requirement. Note, however, that an on-chip THRU, typically having GSGSG probe pads, can often be made symmetrical about the horizontal line shown in Fig. 4a. In that case, the THRU can be decomposed into a pair of uncoupled two-ports by the transformation from the original conductor domain (Fig. 4a) to the even/odd domain (Fig. 4b). The change of bases is performed by the matrices $\boldsymbol{K}_{\mathrm{Ve/o}}$ and $\boldsymbol{K}_{\mathrm{Ie/o}}$ as follows:

$$\boldsymbol{v} = \begin{bmatrix} V_1 \\ V_2 \\ V_3 \\ V_4 \end{bmatrix} = \boldsymbol{K}_{\mathrm{Ve/o}}\boldsymbol{v}_{\mathrm{e/o}},\ \boldsymbol{i} = \begin{bmatrix} I_1 \\ I_2 \\ I_3 \\ I_4 \end{bmatrix} = \boldsymbol{K}_{\mathrm{Ie/o}}\boldsymbol{i}_{\mathrm{e/o}},\ \boldsymbol{K}_{\mathrm{Ve/o}} = \boldsymbol{K}_{\mathrm{Ie/o}} = \frac{1}{\sqrt{2}}\begin{bmatrix} 1 & 0 & 1 & 0 \\ 0 & 1 & 0 & 1 \\ 1 & 0 & -1 & 0 \\ 0 & 1 & 0 & -1 \end{bmatrix}, \tag{5}$$

$$\boldsymbol{v}_{\mathrm{e/o}} = \begin{bmatrix} V_{\mathrm{e1}} \\ V_{\mathrm{e2}} \\ V_{\mathrm{o1}} \\ V_{\mathrm{o2}} \end{bmatrix} = \frac{1}{\sqrt{2}}\begin{bmatrix} V_1 + V_3 \\ V_2 + V_4 \\ V_1 - V_3 \\ V_2 - V_4 \end{bmatrix},\quad \boldsymbol{i}_{\mathrm{e/o}} = \begin{bmatrix} I_{\mathrm{e1}} \\ I_{\mathrm{e2}} \\ I_{\mathrm{o1}} \\ I_{\mathrm{o2}} \end{bmatrix} = \frac{1}{\sqrt{2}}\begin{bmatrix} I_1 + I_3 \\ I_2 + I_4 \\ I_1 - I_3 \\ I_2 - I_4 \end{bmatrix}. \tag{6}$$

Note that $\boldsymbol{K}_{\mathrm{Ve/o}}$ is orthogonal and therefore $\boldsymbol{K}_{\mathrm{U}} = \boldsymbol{K}_{\mathrm{Ve/o}}$ and $\boldsymbol{K}_{\mathrm{P}} = \boldsymbol{1}_4$ in this case.

Let $\boldsymbol{S}$ be the conductor-domain 4×4 scattering matrix as measured by a VNA. The scattering matrix in the even/odd domain, $\boldsymbol{S}_{\mathrm{e/o}}$, is given by the following orthogonal transformation.

$$\boldsymbol{S}_{\mathrm{e/o}} = \begin{bmatrix} \boldsymbol{S}_{\mathrm{ee}} & \boldsymbol{S}_{\mathrm{eo}} \\ \boldsymbol{S}_{\mathrm{oe}} & \boldsymbol{S}_{\mathrm{oo}} \end{bmatrix} = \left[\begin{array}{cc|cc} S_{\mathrm{e1e1}} & S_{\mathrm{e1e2}} & S_{\mathrm{e1o1}} & S_{\mathrm{e1o2}} \\ S_{\mathrm{e2e1}} & S_{\mathrm{e2e2}} & S_{\mathrm{e2o1}} & S_{\mathrm{e2o2}} \\ \hline S_{\mathrm{o1e1}} & S_{\mathrm{o1e2}} & S_{\mathrm{o1o1}} & S_{\mathrm{o1o2}} \\ S_{\mathrm{o2e1}} & S_{\mathrm{o2e2}} & S_{\mathrm{o2o1}} & S_{\mathrm{o2o2}} \end{array}\right] = \boldsymbol{K}_{\mathrm{Ve/o}}^{-1}\boldsymbol{S}\boldsymbol{K}_{\mathrm{Ve/o}}. \tag{7}$$

The reference impedance matrix $\boldsymbol{Z}_{0\mathrm{e/o}}$ equals $\boldsymbol{Z}_0$ because $\boldsymbol{K}_{\mathrm{P}} = \boldsymbol{1}_4$, which is in contrast with the common/differential transformation [5, 6].

If the THRU has the said symmetry, the off-diagonal submatrices $\boldsymbol{S}_{\mathrm{eo}}$ and $\boldsymbol{S}_{\mathrm{oe}}$ are zero, meaning that the four-port consists of a pair of uncoupled two-ports as shown in Fig. 4b. Then, $\boldsymbol{T}_{\mathrm{L}}$ and $\boldsymbol{T}_{\mathrm{R}}$ can be determined fairly accurately by applying the π equivalent shown in Fig. 5 [7, 8] to each of the resultant two-ports described by $\boldsymbol{S}_{\mathrm{ee}}$ and $\boldsymbol{S}_{\mathrm{oo}}$, respectively. The conversion of $\boldsymbol{S}$ to/from $\boldsymbol{T}$ is done via $\boldsymbol{S}'$ given by

$$\boldsymbol{S}' = \begin{bmatrix} \boldsymbol{S}'_{11} & \boldsymbol{S}'_{12} \\ \boldsymbol{S}'_{21} & \boldsymbol{S}'_{22} \end{bmatrix} = \left[\begin{array}{cc|cc} S'_{11} & S'_{12} & S'_{13} & S'_{14} \\ S'_{21} & S'_{22} & S'_{23} & S'_{24} \\ \hline S'_{31} & S'_{32} & S'_{33} & S'_{34} \\ S'_{41} & S'_{42} & S'_{43} & S'_{44} \end{array}\right] = \left[\begin{array}{cc|cc} S_{11} & S_{13} & S_{12} & S_{14} \\ S_{31} & S_{33} & S_{32} & S_{34} \\ \hline S_{21} & S_{23} & S_{22} & S_{24} \\ S_{41} & S_{43} & S_{42} & S_{44} \end{array}\right], \tag{8}$$

$$\begin{bmatrix} \boldsymbol{T}_{11} & \boldsymbol{T}_{12} \\ \boldsymbol{T}_{21} & \boldsymbol{T}_{22} \end{bmatrix} = \begin{bmatrix} \boldsymbol{S}'^{-1}_{21} & -\boldsymbol{S}'^{-1}_{21}\boldsymbol{S}'_{22} \\ \boldsymbol{S}'_{11}\boldsymbol{S}'^{-1}_{21} & \boldsymbol{S}'_{12} - \boldsymbol{S}'_{11}\boldsymbol{S}'^{-1}_{21}\boldsymbol{S}'_{22} \end{bmatrix}, \tag{9}$$

$$\begin{bmatrix} \boldsymbol{S}'_{11} & \boldsymbol{S}'_{12} \\ \boldsymbol{S}'_{21} & \boldsymbol{S}'_{22} \end{bmatrix} = \begin{bmatrix} \boldsymbol{T}_{21}\boldsymbol{T}_{11}^{-1} & \boldsymbol{T}_{22} - \boldsymbol{T}_{21}\boldsymbol{T}_{11}^{-1}\boldsymbol{T}_{12} \\ \boldsymbol{T}_{11}^{-1} & -\boldsymbol{T}_{11}^{-1}\boldsymbol{T}_{12} \end{bmatrix}. \tag{10}$$

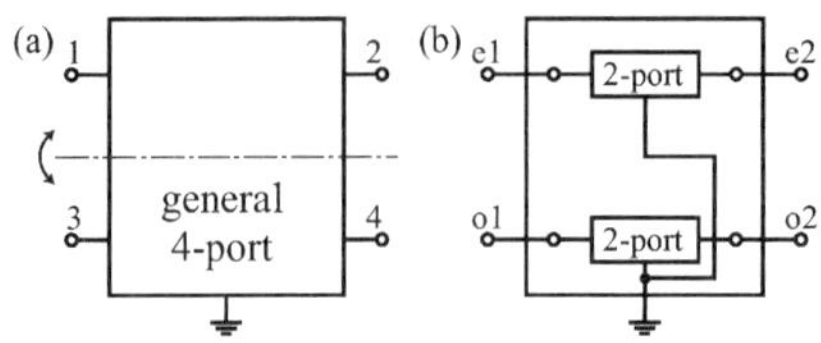

Figure 4: (a) A general four-port. (b) A four-port consisting of a pair of uncoupled two-ports.

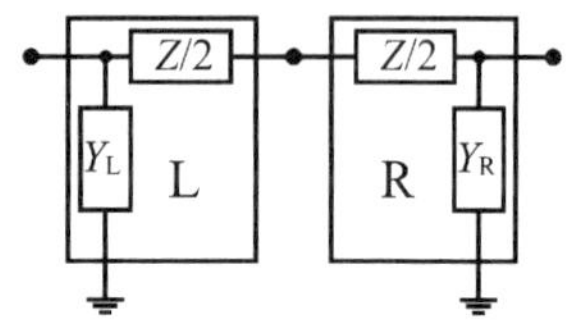

Figure 5: The π equivalent for characterizing the upper/lower half of the transformed THRU (Fig. 4b).

EXPERIMENTAL RESULTS

We applied the proposed de-embedding method to samples fabricated by a 0.18 μm CMOS process. The frequency ranged from 100 MHz to 50 GHz. The two-step open-short method [9] originally proposed for a two-port was also used for comparison. Fig. 6 shows the de-embedded characteristics of a symmetric THRU itself (Fig. 3a). The reflection coefficients obtained by the proposed method ($S_{11\mathrm{TH}}$ and $S_{33\mathrm{TH}}$) stay very close to the center of the Smith chart and the transmission coefficients ($S_{21\mathrm{TH}}$ and $S_{43\mathrm{TH}}$) at its right end as they should, thereby demonstrating the validity of the model shown in Fig. 5. Fig. 7 shows even- and odd-mode transmission coefficients for a pair of 1 mm-long transmission lines (Fig. 3b). A comparatively large difference is seen between the results from the two de-embedding methods for the even mode. One likely cause of the difference is the nonideal behavior of the SHORT [10]. The odd-mode results, on the other hand, agree very well. This suggests that the odd mode is, as expected, immune to the problem that plagued the even mode.

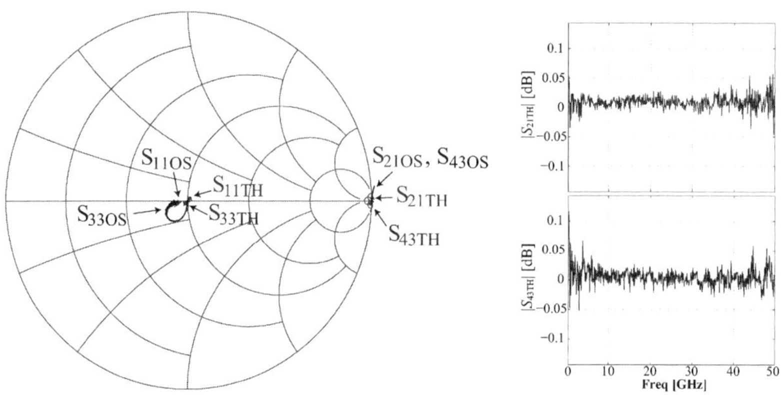

Figure 6: Characteristics of the THRU (Fig. 3a) after performing the thru-only ($S_{11\mathrm{TH}}$, $S_{33\mathrm{TH}}$, $S_{21\mathrm{TH}}$, $S_{43\mathrm{TH}}$) or open-short ($S_{11\mathrm{OS}}$, $S_{33\mathrm{OS}}$, $S_{21\mathrm{OS}}$, $S_{43\mathrm{OS}}$) de-embedding. Ideally, S_{11} and S_{33} should stay at the center of the Smith chart and S_{21} and S_{43} at its right end.

CONCLUSIONS

We have proposed a simple thru-only de-embedding method for characterizing on-chip four-port devices such as coupled transmission lines. It is applicable if the THRU is symmetrical about the

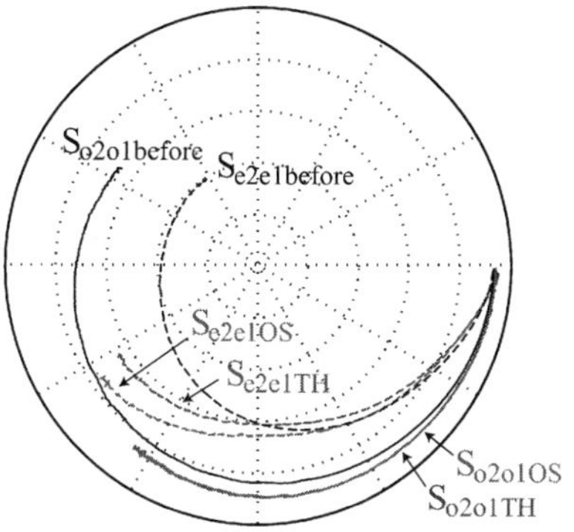

Figure 7: Even-mode (broken lines) and odd-mode (solid lines) transmission coefficients for a pair of transmission lines (Fig. 3b) before and after de-embedding (thru-only or open-short).

horizontal line shown in Fig. 4a. The proposed thru-only de-embedding method was applied to a pair of on-chip transmission lines (Fig. 3b) up to 50 GHz and the result was promising. Its applicability should further be assessed with more test cases and electromagnetic simulations.

ACKNOWLEDGMENTS

This work was partially supported by JSPS.KAKENHI, MEXT.KAKENHI, STARC, MIC.SCOPE, and VDEC in collaboration with Agilent Technologies Japan, Ltd., Cadence Design Systems, Inc., and Mentor Graphics, Inc.

REFERENCES

[1] R. F. Bauer and P. Penfield, Jr, IEEE Trans. Microw. Theory Tech., vol.22, no.3, pp. 282–288, March 1974.

[2] D.-H. Han, T. G. Ruttan, and L. A. Polka, 61st Automatic RF Techniques Group (ARFTG) Conf., pp.37–43, June 2003.

[3] C. R. Paul, *Analysis of Multiconductor Transmission Lines*, 2nd edition, Wiley-Interscience, 2008.

[4] K. Kurokawa, IEEE Trans. Microw. Theory Tech., vol.13, no.2, pp.194–202, March 1965.

[5] K. Yanagawa, K. Yamanaka, T. Furukawa, and A. Ishihara, Instrumentation and Measurement Technol. Conf., pp.866–869, 1994.

[6] D. E. Bockelman and W. R. Eisenstadt, IEEE Trans. Microw. Theory Tech., vol.43, no.7, pp.1530–1539, July 1995.

[7] D. C. Laney, *Modulation, Coding and RF Components for Ultra-Wideband Impulse Radio*, PhD thesis, University of California, San Diego, 2003.

[8] H. Ito and K. Masu, IEEE MTT-S International Microwave Symposium Digest, pp.383–386, June 2008.

[9] M. C. A. M. Koolen, J. A. M. Geelen, and M. P. J. G. Versleijen, Proc. Bipolar Circuits and Technology Meeting, pp.188–191, 1991.

[10] Y. Goto, Y. Natsukari, and M. Fujishima, Jpn. J. Appl. Phys., vol.47, no.4, pp.2812–2816, April 2008.

Conference Proceedings AMC XXIV © 2009 Materials Research Society

Enabling Metal Deposition Technology for High-k Metal Gate Integration

Naomi Yoshida[1], Rongjun Wang[2], Xianmin Tang[2], Dave Liu[2], Lara Hawrylchak[2], Osbert Chang[2], Khaled Ahmed[2], Tushar Mandrekar[2], Christopher Lazik[2], Giuseppina Conti[2], Steven Hung[2], Hao Chen[2], Chorng-Ping Chang[2], and Srinivas Gandikota[2]

[1]SSG, MTCG, Applied Materials, 3225 Oakmead Village Drive, M/S 1252 P.O.Box 58039, Santa Clara, CA, 95054

[2]Applied Materials, Santa Clara, CA, 95054

ABSTRACT

Uniform and low damage metal gate deposition was achieved using a new concept PVD based RF-PVD technology. MOSCAP data showed optimization of RF-PVD TiN process to achieve uniform flat band voltage across the wafer, as well as low interface state density. Successful effective work function tuning for NMOS was demonstrated by combining with LaOx capping layer also deposited by RF-PVD below TiN metal gate.

INTRODUCTION

In the past years, high-k gate dielectric has come to the real chip manufacturing world. Metal gate is required for hafnium based high-k dielectric integration. Gate first flow requires high thermal budget after metal gate process, which limits work function tuning by metal gate material change. One of the most promising approaches is the combination of one mid gap metal gate with two metal oxide capping layers to obtain the required work function for both PMOS and NMOS [1-4]. In this approach, metal gate process should cause no degradation to the gate dielectric and the interface, no effect on equivalent oxide thickness (EOT), and provide highly uniform and repeatable thin film deposition. On the other hand, PVD is known to be one of the most robust technologies for metallization. However, the conventional PVD chamber was designed to deposit relatively thick film with good step coverage which usually requires high plasma density. Hence, novel PVD based process is desired for metal gate disposition for high-k integration.

The purpose of this study is to develop low damage, production-worthy metal gate process using the novel RF-PVD technology, which has already successfully demonstrated the production worthy performance for metal oxide capping layer deposition for effective work function tuning [3]. In this study, titanium nitride was mainly investigated as mid-gap metal for gate first flow with metal oxide capping layer for effective band-edge work function.

EXPERIMENT

MOSCAP devices were fabricated with the process flow shown in Figure 1 on 300mm p-type Si wafers. First, gate dielectric film stack was grown with HfO_2 as the high-k film by atomic layer deposition. Next, 5-10nm metal gate was deposited in various conditions, followed by doped poly deposition. High temperature anneal by RTP was performed after poly to simulate high thermal budget of gate first process flow.

MOSCAP structures were formed by lithography and etch process after TiN film deposition by PVD as the top electrode. At last, samples were annealed in forming gas at 425°C.

HFCV and IV sweep tests were performed with MOSCAP samples to extract electrical parameters, such as EOT, flat band voltage (V_{FB}), gate leakage current.

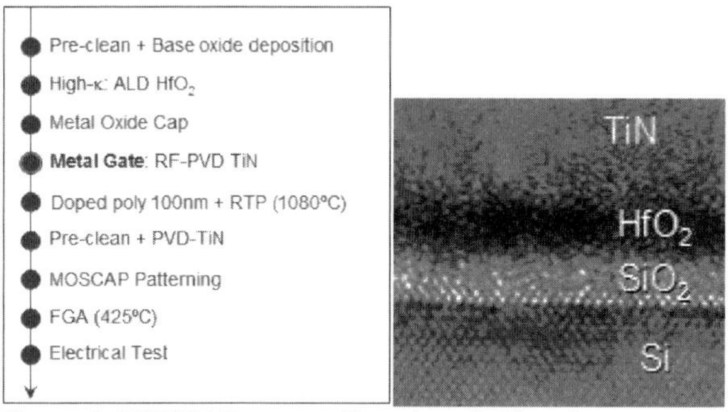

Figure 1. MOSCAP process flow chart (left) and a cross-sectional TEM image of high-k and metal gate (right)

DISCUSSION

Plasma induced damage and within wafer uniformity

Prior to MOSCAP test, plasma induced damage was investigated using SPIDER wafers [5]. It showed no degradation in gate leakage current after RF-PVD TiN deposition. This confirmed no significant plasma induced damage by RF-PVD.

After optimizing TiN blanket film thickness uniformity, the process was further optimized to improve electrical parameter uniformity using MOSCAP. Figure 2 shows the V_{FB} within-wafer uniformity (WiW) with different RF-PVD TiN process conditions on SiO_2 gate. Good WiW uniformity of V_{FB} was achieved by process condition "C", less than 2%, 1sigma, by optimizing process parameters such as gas flow rate and RF power during TiN deposition.

Impact on EOT and gate leakage current

The impact of metal gate on EOT was investigated with various TiN process conditions using HfO_2 gate stack. Figure 3 shows the effect of TiN composition on the EOT. It was found that high nitrogen conditions caused higher capacitance equivalent thickness (CET) than lower N concentration after high thermal budget. TiN composition also affected the flat band voltage. Higher nitrogen showed higher V_{FB}. RF-PVD metal gate process was optimized for minimal impact on the high-k dielectric and the interfaces as well as for good within wafer uniformity of electrical parameters. Figure 4 shows the leakage current performance with TiN metal gate deposited by three different technologies: 1) conventional PVD; 2) atomic layer deposition (ALD); 3) RF-PVD. RF-PVD showed the lowest gate leakage at thin CET compared to the other technologies. RF-PVD also showed low interface damage comparable to ALD-TiN (Table 1), which is lower than the conventional PVD process.

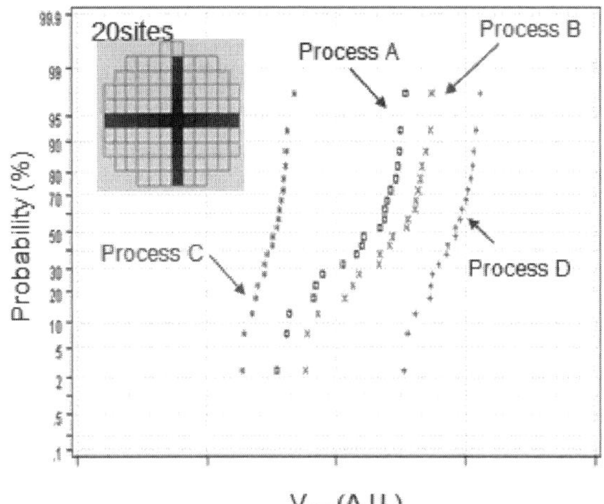

Figure 2. Probability plots of flat band voltage with TiN metal gate on SiO_2. 20sites were tested within a wafer as shown in the map. The uniformity was optimized by changing process parameters such as gas flow rate and RF power. Process "C" shows tight distribution of V_{FB} across the wafer.

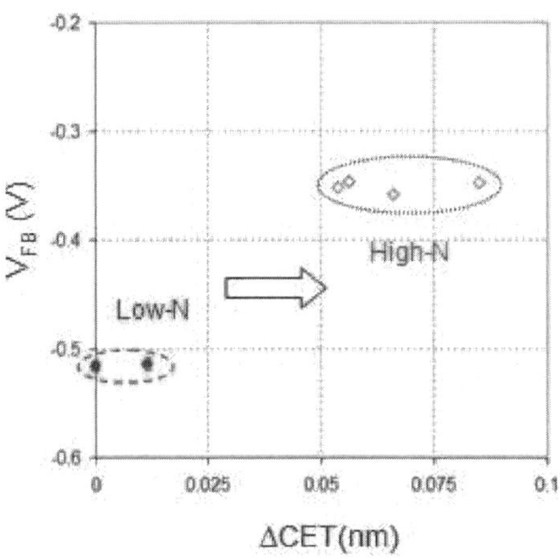

Figure 3. High N concentration in TiN shows higher CET in various RF-PVD process conditions. N-rich TiN also provided higher flat band voltage on SiO_2/HfO_2 gate stack.

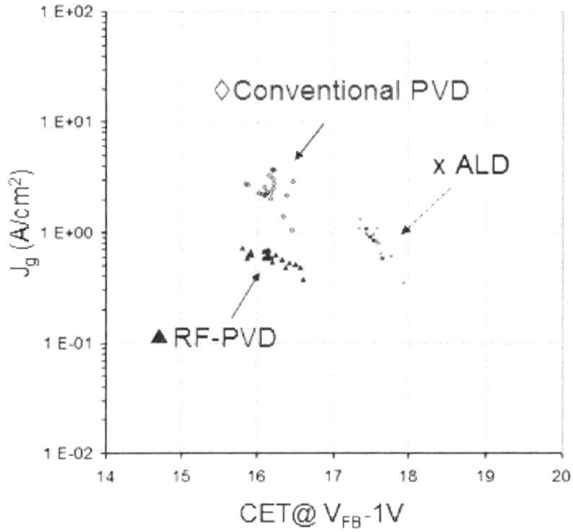

Figure 4. J_g-CET plots with TiN metal gate deposited with three different hardware; the conventional PVD, ALD and RF-PVD. RF-PVD TiN shows better scalability than conventional PVD or ALD.

Table I. Interface state density with various TiN by the conductance method

TiN Process	Gate Dielectric	D_{it} [$cm^{-2}eV^{-1}$]
Conventional PVD	SiO_2+HfO_2 (2nm)	1.3e+12
RF-PVD	SiO_2+HfO_2 (2nm)	2.2e+11
ALD	SiO_2+HfO_2 (2nm)	2.1e+11
Poly Ref.	SiO_2 only (2nm)	5.5e+10

Combination with capping layer for effective work function tuning with TiN metal gate

As shown above, RF-PVD TiN achieved low damage and uniform metal gate deposition which meet the high-k metal gate integration requirements for gate first flow.

For the gates first flow, metal oxide capping layer can be used to tune effective work function with mid gap metal gate. As sown in Figure 5, it was successfully demonstrated that TiN metal gate and LaO_x capping layer deposited by RF-PVD achieved flat band voltage shift close to NMOS target work function with no CET change. RF-PVD TiN also demonstrated superior uniformity (Fig. 6) and particle performance during the extended run.

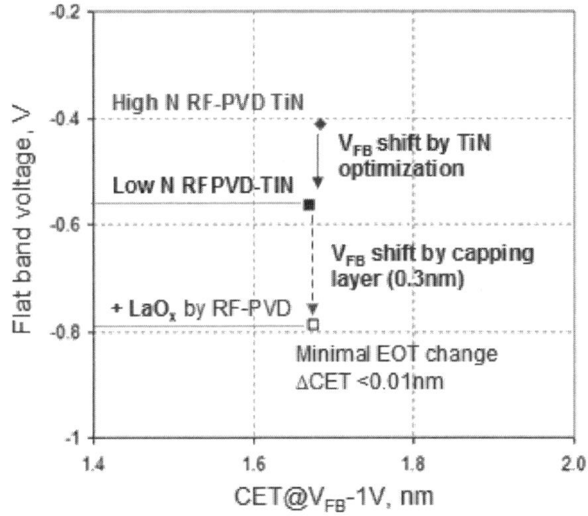

Figure 5. Effect of metal gate TiN condition and LaO_x capping layer addition on CET and flat band voltage

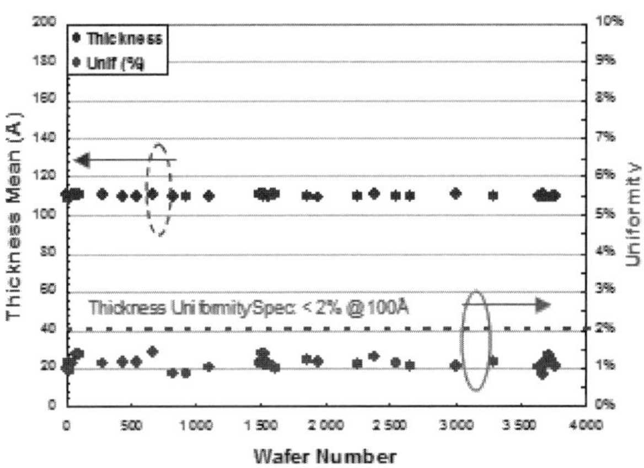

Figure 6. RF-PVD TiN extended run thickness uniformity.

CONCLUSIONS

Thus, a novel RF-PVD technology was established for metal gate deposition for high-k integration. It was confirmed as a production worthy technology of metal gate deposition which provides low damage and uniform film to meet the stringent requirements of high-k metal gate integration. In addition, RF-PVD can be applied for different material depositions by changing the target material. It was successfully demonstrated that combination of metal gate and metal oxide capping layer by RF-PVD provides a critical solution for gate-first MOSFET gate integration.

ACKNOWLEDGMENTS

The authors would like to thank B. Zheng, H. Gang, C. Jin, D. Dobo, I. Liu, P. Xu, C. Cai, M. Okazaki, M. Jin, B. McDougal and MTCG operations for their support in MOSCAP fabrication and analysis.

REFERENCES

1. P. Sivasubramani et al., VLSI, p.68, 2007
2. H. Harris et al., IRPS, p.661, 2006
3. N. Yoshida et al., ECST, Vol.13, Issue1, p.143, 2008
4. H.C. Wen et al., VLSI, p.160, 2007
5. P. Aum et al., T-ED, Vol.45, No.3, p.722, 1998

Conference Proceedings AMC XXIV © 2009 Materials Research Society

O2 Plasma Damage and Dielectric Recoveries to Patterned CDO Low-k Dielectrics

H. Huang[1], J. Bao[1], H. Shi[1], P. S. Ho[1], M L McSwiney[2], M D Goodner[2], M Moinpour[2], and G M Kloster[2]

[1]Laboratory for Interconnect and Packaging, Microelectronics Research Center, The University of Texas at Austin, Austin, TX, 78758

[2]Intel Corporation, Logic Technology Development, Hillsboro, OR, 97124

ABSTRACT

This paper investigated the mechanism of oxygen plasma damage to patterned low-κ structures and dielectric recovery by silylation. Plasma damage was induced to patterned structures using a remote hybrid plasma source with separate ions and atomic radicals. In addition, the oxygen plasma damage to blanket low-κ films and its angular dependence were studied. The damage was characterized by a combination of analytical techniques and the results indicated that damage by ions was anisotropic while damage by radicals was isotropic. The carbon depletion depth was found to be controlled by diffusion of radical species. Sidewall recovery by trimethylchlorosilane (TMCS) vapor silylation was performed and it was found to be more effective for recovery of the sidewall carbon loss induced by O_2 radicals compared to that by hybrid O_2.

INTRODUCTION

With continuing scaling beyond the 45nm technology node, implementation of ultra low-κ (ULK) dielectrics into Cu interconnects becomes necessary [1]. Plasma processing can degrade low-κ dielectrics by depleting the methyl groups leading to moisture uptake and increases in the effective dielectric constant and leakage current of the low k structure [2, 3]. For ULK dielectrics, the incorporation of porosity can enhance the plasma damage due to increase in the penetration of the plasma species. So far, few reports are available on plasma damage to patterned low-κ dielectrics due to the limitations of metrology techniques [4]. This study investigates the mechanism of O_2 plasma damage and dielectric recovery for patterned low-κ structures. Multiple analytical techniques were used to analyze the damage by oxygen plasma on low κ films at different incident angles (0° to 90°). The results enabled us to understand the angular dependence of damage by ions and radicals on surface chemistry, composition, and density of the low-κ film. This was supplemented by the use of electron energy loss spectroscopy (EELS) in transmission electron microscope to examine the oxygen plasma damage at the sidewall of patterned low-κ structures. Finally, dielectric recovery by trimethylchlorosilane (TMCS) vapor silylation was performed on the patterned samples and it was found to be more effective for recovery of the sidewall carbon loss induced by O_2 radicals than that by hybrid O_2.

EXPERIMENT DETAILS

The pristine CDO films were deposited in a standard PECVD chamber to yield a film with about 25% porosity, dielectric constant of 2.5, density of 1.25 g/cm3 and refractive index at 633 nm of 1.33. The patterns were formed by plasma etching and subsequent DHF clean. The samples were treated by remote plasmas using an atom/ion hybrid source which can generate hybrid ions and neutrals or atomic radical only. The operation of the plasma source has been described elsewhere [5]. The typical ion flux in hybrid beam was estimated to be of the order of 10^{15} atoms/ (cm^2•s) by a Langmuir probe.

XPS surface analyses were performed using a PHI 5700 XPS system. TEM/EELS analysis was performed using an FEI TECNAI G2 F20 Analytical TEM. Film thickness and refractive index were measured by a J. A. Woollam VASE Spectroscopic Ellipsometer. Chemical bond structures were investigated by a Magna 560 FTIR spectrometer. Film density was extracted from x-ray reflectivity measured by an X'Pert MRD system. Water contact angle was measured by a CA100 Ramé-Hart goniometer.

Dielectric recovery was performed in a vapor silylation chamber. TMCS was stored in the bubblers in the liquid phase at room temperature. They were carried into the silylation chamber by CO_2 gases. An optimized experimental condition was reached using a high CO_2 flow rate while maintaining the temperature of the chamber wall, tube, and substrate around 100°C. The damaged samples were treated by the silylation agents for identical 60 minutes at 250 torr. Detailed description of using vapor silylation agents for dielectric recovery can be found in ref. 6.

RESULT and DISCUSSION

Damage to patterned low-κ structures

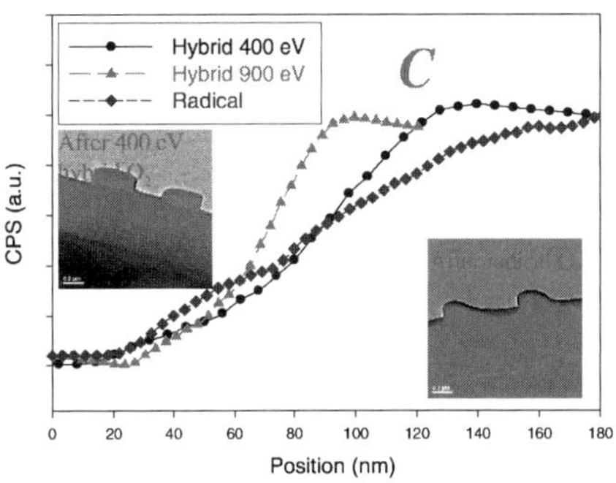

Figure. 1. EELS line scan of the carbon concentration on the sidewall after the treatment of 400 eV hybrid O_2, 900 eV hybrid O_2 and radical O_2. Insets are TEM cross section images of trenches after hybrid and radical O_2 damage.

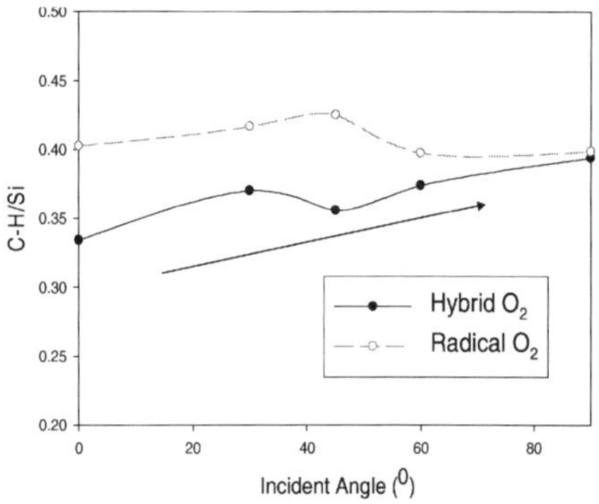

Figure. 2. Relation between C-H/Si and the beam incident angles after 400 eV hybrid O_2 and radical O_2.

The damage to patterned low-κ structures after hybrid beam treatment was different from that after radical beam treatment. TEM and EELS analyses were performed and the results are shown in Fig.1. To avoid contamination during TEM sample preparation, the damaged patterned low-κ films were coated with a protective layer of ~30 nm Cr and ~20 nm Au. The TEM images (insets in Fig.1) showed more distortion in the trench geometry by the radical beam. The EELS line scans across the middle of the trench sidewall (Fig. 1) indicated plasma treatment induced a gradient carbon depletion region from the surface into the bulk. Compared with hybrid beams, the radical beams are isotropic and thus interact more effectively with the carbon atoms on the sidewall. This led to a carbon depletion depth after radical O_2 treatment as large as 140 nm, as compared with a depletion depth of 100nm after hybrid beam treatment at 400 eV. Increasing the hybrid beam energy to 900 eV decreased the depletion depth to 70 nm, which was still less than that of the radical beam. These results indicate that hybrid beams are more directional and penetrate less and cause less carbon loss on the sidewall.

Angular dependence of damage to low-κ films

Previous studies performed in our laboratory showed that hybrid oxygen beam treatment of blanket low-κ films produced more CO_2 and H_2O uptake than radical oxygen beam treatment. XPS and FTIR analysis indicated that the energetic ions in the hybrid beam enhanced the formation of C=O and C-O bonds on the low-κ surface and released subsequently more CO_2 gases in the reaction byproducts [7].

In the patterned low-κ structure, the sidewall can be regarded as a surface that interacts with ions and neutrals in the plasma at different incident angles. To study the nature of the directional effect, we investigated plasma damage on low-κ films as a function of incidence angles. The blanket low-κ film was rotated from 0° to 90° corresponding to the beam normal to the sample surface at 0° and parallel at 90°.

To examine the surface chemistry change, we focused on the loss of surface carbon concentration due to methyl depletion. For this purpose, the XPS carbon 1s peak after hybrid beam treatment was deconvoluted into three sub-peaks of C-H (-285.2 eV), C-O (-286.7 eV) and C=O (-288.2 eV). Here the C-H bond represents the characteristic hydrophilic methyl groups. Fig.2 showed that the C-H/Si ratio after hybrid O_2 treatment increased with increasing incidence angles. In contrast, the C-H/Si ratio remained almost constant after radical O_2 treatment.

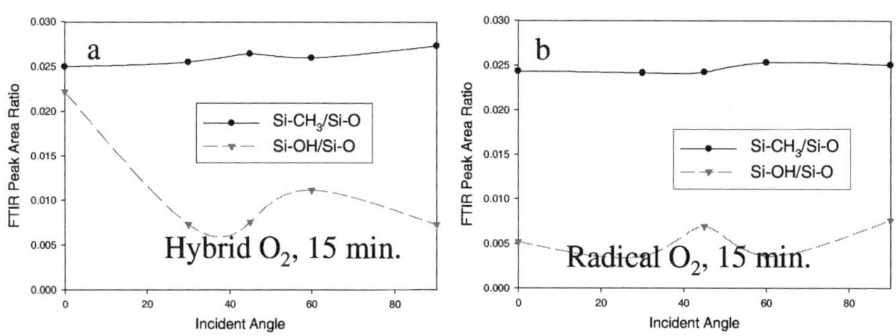

Figure. 3. Relation between FTIR peak area ratio of Si-CH3/Si-O and Si-OH/Si-O and the beam incident angles after (a) 400 eV hybrid O_2 and (b) radical O_2.

The FTIR peak ratios of $Si\text{-}CH_3$ (1274 cm^{-1}) and Si-O (1250~950 cm^{-1}), Si-OH (3600~3200 cm^{-1}) and Si-O are shown in Fig.3a and 3b. After hybrid O_2 treatment, the Si-CH_3/Si-O ratio increased while the Si-OH/Si-O decreased with increasing incidence angles.

Interestingly, the Si-CH_3/Si-O and Si-OH/Si-O ratios in the low k film after radical O_2 beam treatments at varying angles did not show notable differences.

Measurement of water contact angles and refractive indexes showed a similar trend. The water contact angle provides a measure of the intensity of the surface polar terms. In Fig. 4 (a), the contact angle gradually increased from ~68° to ~90° as the O_2 hybrid beam impinging from the vertical to the parallel direction. For radical O_2 beam, the contact angles remained at around 85°, independent of the beam incident angles where the small ±5° variation in the contact angle was probably due to process variations. To examine the extent of the dielectric loss, the refractive index n which can be correlated to the dielectric constant was measured as a function of incidence angle. In Fig. 4 (b), the refractive index at 633 nm was found to decrease with increasing incidence angles after hybrid O_2 beam treatment. In contrast, there was almost no angular dependence of the refractive index with the incidence angle for the radical beam.

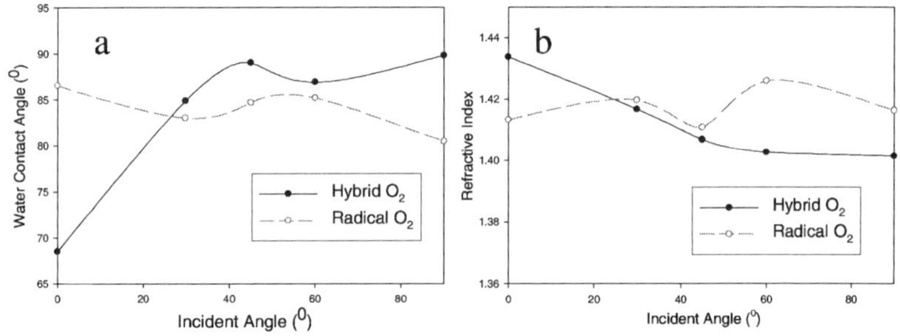

Figure. 4. Relation between (a) water contact angles, (b) refractive indexes and the beam incident angles after 400 eV hybrid O_2 and radical O_2.

Table 1. Density, thickness and roughness of the low-κ films after hybrid and radical O_2 treatment at 0^0 and 90^0.

Sample treatment	Density (g/cm^3)	Thickness (nm)	Roughness (nm)
Hybrid O_2, 0^0 (Vertical)	1.69 (top)	34.9	0.5
	1.34 (bottom)	57.1	~10
Hybrid O_2, 90^0 (Parallel)	1.67 (top)	13.6	0.6
	1.25 (bottom)	76.0	~20
Radical O_2, 0^0 & 90^0	1.52 (top)	47.3	0.6
	1.42 (bottom)	46.1	~15

The film density, thickness and roughness were extracted from x-ray reflectivity measurements. The results in Table.1 show a large difference in interface roughness between the top and the bottom layers, confirming the carbon density gradient observed by EELS. The highest density of the damaged layer after the hybrid beam treatment was found to be at 0°, which can be attributed to the directional ion bombardment of the hybrid beam. At 90°, with the sample surface parallel to the beam direction, the hybrid O_2 beam caused only a limited surface densification while the bulk remained largely unchanged. Compared with the 0° vertical hybrid O_2 treatment, the thickness of the densification layer was also reduced. The density gradient after radical O_2 treatment was similar at 0° and 90°. Compared with hybrid O_2, the radical O_2 induced a thicker densification layer although its density was lower. The density of the bottom layer after radical O_2 treatment was larger than that after hybrid beam treatment. This indicates that the neutral oxygen can diffuse deep into the bulk of the low-κ film to cause methyl loss while the surface densified layer after hybrid O_2 treatment blocked the penetration of ions and neutrals.

Dielectric recovery by silylation

Sidewall recovery was attempted with an optimized TMCS vapor silylation and the effect was examined by nano-beam EELS. Considering the bulk is much less damaged and is therefore regarded as a reference point, the signal ratio of EELS carbon K edge at -284 eV on the sidewall to that in the bulk was used to evaluate the carbon recovery on damaged patterned films(Fig. 5). The ratio was only 0.56 on the hybrid O_2 damaged trenches, and it rose to 1.52 on radical O_2 damaged trenches. These results showed only partial recovery of surface carbon on the sidewall damaged by hybrid O_2. However, the recovery was more complete on the sidewall damaged by radical O_2. The difference of damage recovery could be attributed to more surface densification by ions that inhibited the rate-limiting mass transport of vapor chemicals. As indicated by Table 1, the average densities of the damaged layer after hybrid O_2 and radical O_2 at 90° were 1.67 g/cm^3 and 1.52 g/cm^3 respectively. The more densified surface after hybrid O_2 damage might inhibit the mass transfer of silylation agent, and thus weakened the dielectric recovery.

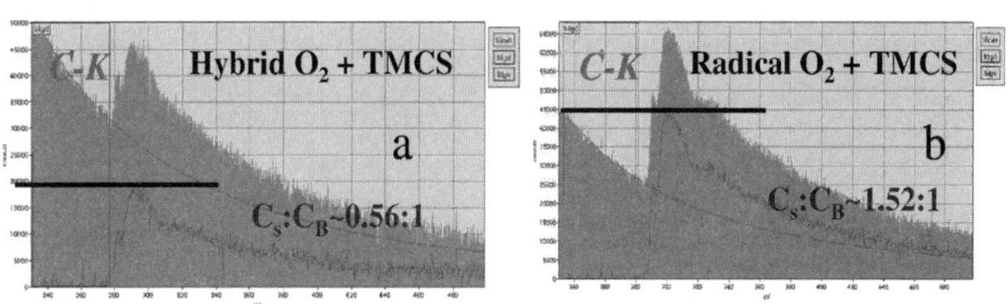

Fig. 5. Nano-beam EELS analysis of carbon recovery on the trench sidewall damaged by (a) hybrid and (b) radical O_2 plasma respectively. (Beam size=0.7 nm). C_S and C_B are carbon K edge height at sidewall surface and bulk.

CONCLUSIONS

Plasma damage to patterned low-κ dielectrics is a complicated phenomenon depending on the plasma species and the pattern structures. The energetic ions can remove methyl groups and induce surface densification but the effect is anisotropic and thus depends on the impingement angles to the low-κ surface. Radicals, with less kinetic energy, are chemically active but are isotropic, which induce relatively uniform damage independent of the impingement angle. The depth of carbon depletion is determined by the diffusion of neutral species into low-κ. Therefore, the radicals can cause deeper damage to the sidewall than ions, particularly for porous low-κ dielectrics. In comparison, dielectric recovery by TMCS vapor silylation was more complete on the sidewall after radical plasma damage because the surface densification by ions can block the mass transport of silylation agents.

ACKNOWLEDGMENTS

This work was performed in part at the Microelectronics Research Center at UT Austin of National Nanofabrication Infrastructure Network supported by National Science Foundation under award # 0335765.

REFERENCES

1. International Technology Roadmap for Semiconductors, Semiconductor Industry Association, 2007
2. D. Shamiryan, M. R. Baklanov, S. Vanhaelemeersch and K. Maex, *J. Vac. Sci. Technol. B* **20**, 1923 (2002)
3. H. Shi, J. Bao, J. Liu, H. Huang, S. Smith, Q. Zhao, P.S. Ho, M.D. Goodner, M. Moinpour, G.M. Kloster, *Proc. of AMC*, **64**, (2007)
4. T. Abell, J. Lee, M. Moinpour, *Mater. Res. Soc. Symp. Proc.* **914**, F04-02 (2006)
5. J. Bao, H. Shi, J. Liu, H. Huang, P. S. Ho, M. D. Goodner, M. Moinpour and G. M. Kloster, *J. Vac. Sci. Technol. B* **26**, 219 (2008).
6. H. Shi, J. Bao, H. Huang, J. Liu, S. Smith, W. Kim, Y. Sun, P. S. Ho, M. L. McSwiney, M. Moinpour and G. M. Kloster, *Mater. Res. Soc. Symp. Proc.* **1079**, N02-10, (2008)
7. J. Bao, H. Shi, H. Huang, P. S. Ho, M. L. McSwiney, M. D. Goodner, M. Moinpour and G. M. Kloster, *AVS 54th Symp.*, 2007

Conference Proceedings AMC XXIV © 2009 Materials Research Society

Capping thermal budget and IC line geometry impacts on post Si(C)N hillocks in copper lines

Anaël Timma[1], Pierre Caubet[1], Laurin Dumas[1], Bernard Chenevier[2], Katia Haxaire[1], and Philippe Normandon[1]
[1]STMICROELECTRONICS, 850, rue Jean Monnet, Crolles, 38926, France
[2]LMPG, Minatec 3, parvis Louis Neel, Grenoble, 38016, France

ABSTRACT

Post Si(C)N hillocks density on Cu interconnects with thicknesses of 0.5-4μm and varied line widths of 0.4-30μm were measured for standard Damascene process electroplated Cu lines. For 3μm thick lines, the influence of copper line width and local density on post Si(C)N hillocks formation was studied. As local copper line density increases, hillocks height proportionally increases for narrow lines, but remains constant for wider ones. Impact of seed layer nature on hillocks formation was also studied. Results shown that hillocks density is reduced by 75 % when using a CuAl(1 atomic-%) seed layer for thick copper lines .
In a second part, a Design Of Experiment (DOE) was drafted on annealed copper thin films after polishing. The aim was to simulate the thermal budget undergone by copper films during Si(C)N capping with varied thermalization temperature and duration and then characterize resulting hillocks. Activation energy of hillocks growth on copper films was determined about 0.7eV. Hillocks total volume is linearly dependent of thermal budget.

INTRODUCTION

Post (Si(C)N) capping copper hillocks were often reported as a major issue in copper interconnects used in CMOS manufacturing [1]. Hillocks can degrade device electrical performance and reliability, by perforating dielectrics or creating shorts between metal lines (Figure 1.a). Hillocks are particularly critical in RF high performances or power amplifier devices using thick (3-4 μm) copper lines, for which increased numbers of defects with increased deposited thickness have been highlighted. Besides, devices including MIM capacitors are sensitive to hillocks defect occurring during copper encapsulation by PECVD Si(C)N. The top of a hillock formed during copper line encapsulation is a potential area of concentration for electrostatic field, and that can result to MIM capacitors early failure by insulator breakdown (Figure 1.b).The Si(C)N capping includes several steps, mainly thermalization, surface cleaning, and deposition, with the wafer sitting on a heater at 370°C-400°C. Based on literature, compressive stress is introduced during capping in copper films and lines because of materials dilatation mismatch, between metals and dielectrics [3]. Hillocks are formed due to thermal plastic stress relaxation [2, 3]. Literature [2, 4] is indecisive about which diffusion mechanism favours hillocks growth during anneal: diffusion at copper grain boundaries, at interface with metal barrier, at line surface. It is probable that copper diffusion mechanism for hillocking is geometry dependent because of "sidewalls" contribution in compressive stress distribution [5].

Figure 1.a: A hillock on level n metal result in shortcut at level n+1.

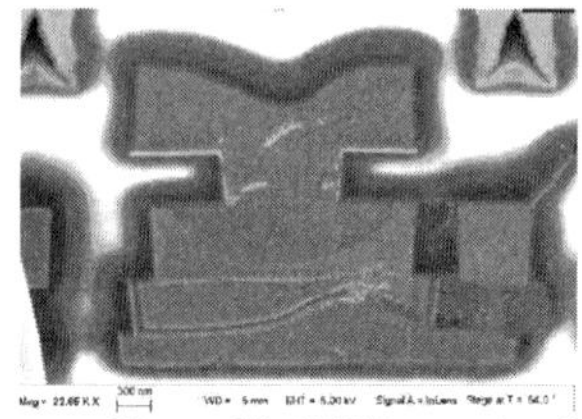

Figure 1.b: MIM capacitor breakdown due to a hillock.

First, the impact of copper lines geometry (width and thickness) and environment (copper local density) on post Si(C)N hillock formation is determined. Two seed layers (Cu and CuAl1%) have been studied. The aim is to propose a mechanism of hillock growth as function of line width.
In a second part, Si(C)N capping thermalization is simulated with a anneal on a hot plate with H_2 ambient, set at the same temperature and duration as Si(C)N deposition. This paper presents a characterization of hillocks as a function of capping thermal budget, and the activation energy (Ea) of hillocks formation on copper films is determined.

EXPERIMENT

The Cu lines samples studied in this paper are fabricated using electroplated Cu Damascene processing in a 300 mm semiconductor processing equipments. Copper lines are built in TEOS oxide dielectrics. Five series of two Cu-patterns (one after polishing, one after 40 nm Si(C)N capping), with different Cu-line thickness and width (**w**) are first analyzed. The first four series correspond respectively to a line thickness equal to 0.7μm, 2μm, 3μm and 4μm. The last one, has a CuAl(1 at-%) seed layer with a Cu line thickness equal to 3μm. The five series present a large range of line widths (0.4 μm to 30μm). Post polished 3μm thick-Cu patterns were analyzed using Electron Backscattered Diffraction (EBSD). Cu grain size, grain boundary density and the volume fraction of <001>,<111>, and <011> oriented grains are quantified as a function of line width. Post Si(C)N hillocks are counted and reviewed by scanning electron microscope defect review equipment.

Besides, 3μm thick Cu-pattern with varied Cu-line widths (0.4 μm to 40μm) and pattern densities (**d**) (20% to 80%) are analyzed after 60nm Si(C)N capping. Post Si(C)N hillocks are characterized using Atomic Force Microscope (AFM). A pattern recognition software permits to determine post Si(C)N hillocks volume, height and diameter by assuming capping layer thickness is uniform and hillock shape is conical. The number, volume, height and diameter of hillock are quantified as a function of Cu line width and pattern density.

Finally, a Design Of Experiment is carried out on copper thin films by using Statgraphics® software. Three factors are studied: thermalization temperature [180°C to 420°C] and duration [5s to 60s] and copper films thickness after CMP [0.5μm to 6μm]. The deposition stack on 300 mm Si wafers is the following: 100nm TEOS oxide, 25nm TaN/Ta diffusion barrier, 150nm seed Cu, and an electrochemically deposited (ECD) Cu layer of various thicknesses. After ECD Cu, all samples are annealed in furnace at 400°C for 30min, then 150nm of Cu is polished. Finally the anneal simulating Si(C)N thermalization is performed following the experiments table. Samples are analysed by AFM and SEM tools. Maximal hillock height and hillocks density are directly measured. Total hillock volume is measured from AFM scan.

DISCUSSION

Post Si(C)N hillocks formation in Cu lines

After capping Si(C)N at above 380°C, two shapes of hillocks are observed. First, large hillocks are single grains that extend from the free surface to the substrate (Figure 2). Such observations can be explained by grain boundary sliding [6]. A small grain is compressed by surrounding bigger grains during thermalization and extrude out of film by grain boundaries sliding to relax stress. In the second case, hillocks grow by copper atomic diffusion [2]. When subjected to thermal compressive stress, copper grains do not relax uniformly. There are local areas where the relaxation is larger. Once such localized differential relaxation occurs, a potential gradient is generated between the relatively relaxed grains. This leads to mass flow. The incoming atoms arrive to relaxed grain in a manner to remove the difference in stress, and the relaxed area grows out from the base as a hillock. Such hillocks are typically located at grain boundaries. Only 30% of hillocks are found at triple grain boundaries (Figure 2).

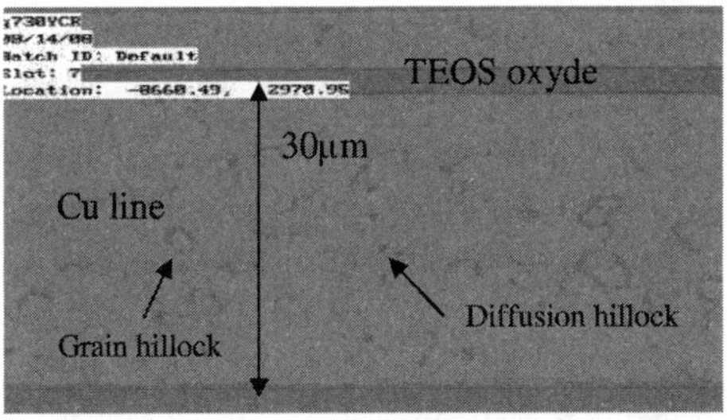

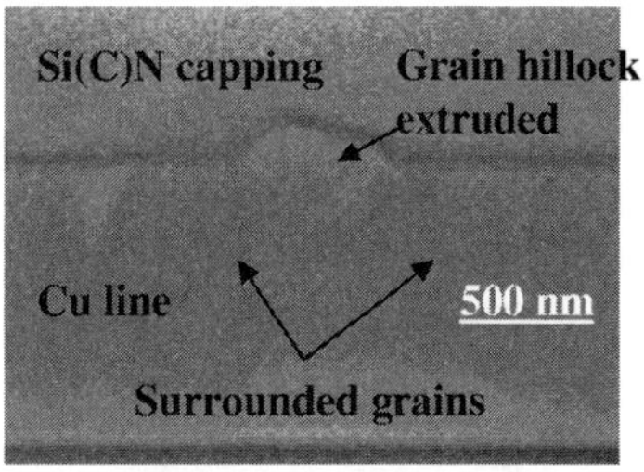

Figure 2: (a) Optical observation of two hillocks shapes (top view). (b) SEM observation of grain hillock section.

Post Si(C)N hillocks are observed and reviewed. SEM and optical pictures show that grain boundaries groove. But hillocks are clearly distinguished from grain boundaries ridge. Figure 3 shows hillocks density increases with Cu-pattern thickness until 2µm and saturates beyond.

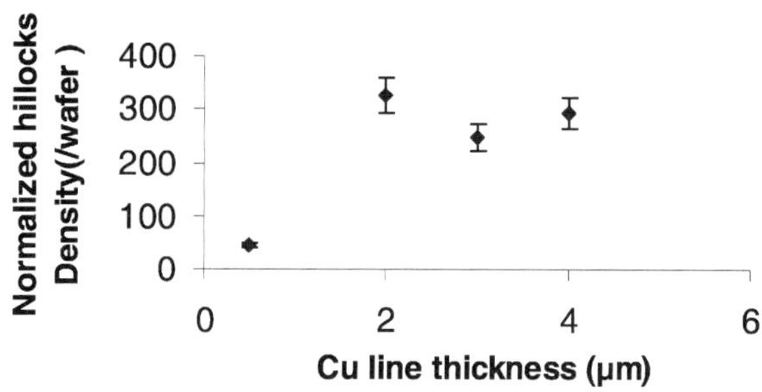

Figure 3: Hillocks density as a function of Cu-pattern thickness. Cu lines are capped with Si(C)N at 380°C.

Cu-pattern density and Cu-line width influences for 3µm thick Cu lines

Height, mean volume and diameter of a hillock increase with Cu line width until 10 µm and saturates beyond, as shown in Figure 4. It is suggested that thermal compressive stress during capping remains constant for copper lines wider than 10µm. Hillock height is density independent for lines wider than 10µm but increases linearly with density for the narrowest ones. Assume the increase of Cu lines density corresponds to the increase of sidewalls contribution in transverse thermal stress during capping. Interface between copper and TaN barrier impacts on hillocks growth in narrowest lines but do not for hillocks on widest ones. It is proposed that post Si(C)N hillocks mainly grow by interface diffusion on narrowest copper lines, whereas on widest ones a bulk mechanism is needed. Given post Si(C)N hillocks are formed only at grain boundaries on wide lines (see Figure 2 a), grain boundary diffusion is expected to be the main diffusion mechanism for hillocks growth on the widest lines.

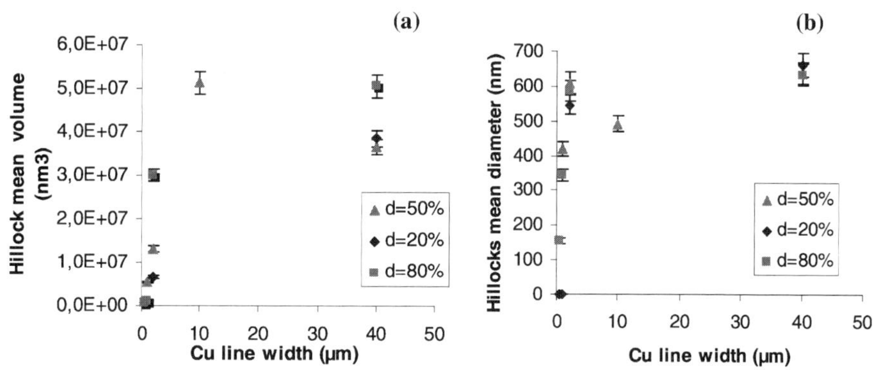

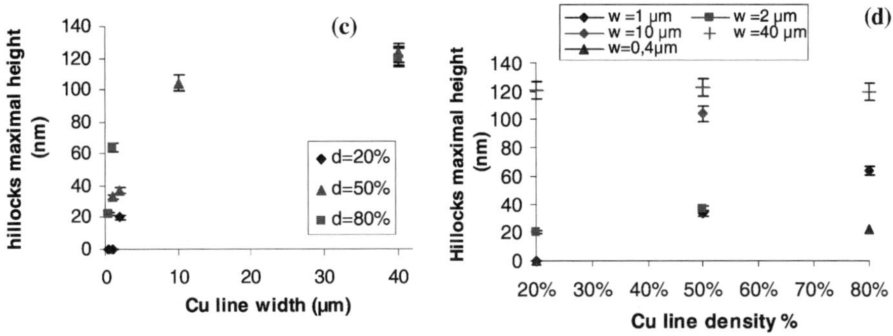

Figure 4: Characterization of a hillock as a function of Cu line width and density: (a) mean volume (b) mean diameter (c) maximal height (d) Cu pattern density effect.

Hillocks density decreases with Cu line until width equals 20 µm and saturates beyond as shown in
Figure 5 . Given post Si(C)N hillocks are always located at grain boundaries (double or triple), post CMP Cu grain surface boundaries density and post Si(C)N hillocks are expected to behave similarly as function of Cu line width. Figure 5 confirms that assumption: the more the grain surface boundary density, the more the hillocks density is.

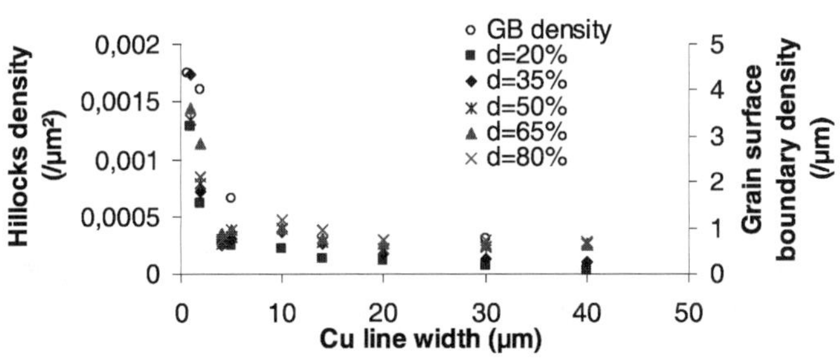

Figure 5: Hillocks density and copper grain surface boundary density as a function of Cu line width.

At a given Cu line width, hillocks density is independent of Cu pattern density for lines wider than 10 µm, while it linearly increases with Cu line pattern density for the narrowest ones (Figure 6).

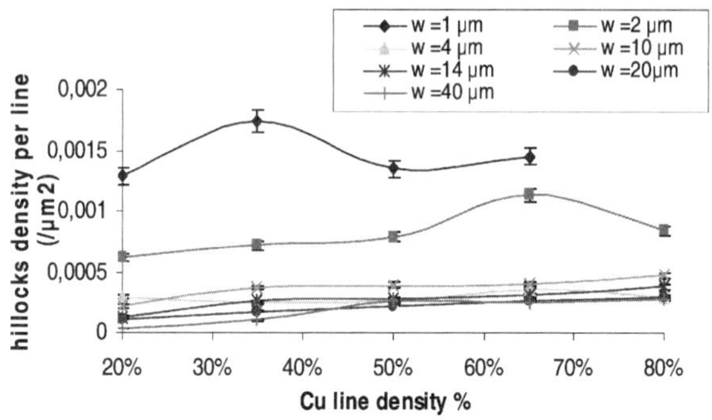

Figure 6: Hillocks density per line as Cu pattern width and density.

Cu grain size, volume fractions of <111>, <001>, <011> oriented grains, and twins are given as a function of Cu line width in Table 1. For narrow lines of 2µm, <111> orientation is minimum and <011> is the main orientation. For line widths above 5µm, <111> and twins increase at the expense of <001> and <011>. When the line width increases from 2µm to 30µm, the total volume fraction of twins, and grain size, both increase as well, suggesting that increasing line width favors stress relaxation by rotation of grains and grain growth. We propose that this stress relaxation is the main contributor to hillocks reduction.

After J.E. Sanchez [8], grain relaxation depends on orientation. When the three main volume fractions oriented grains are weak, stress gradient during Si(C)N capping is limited and flow mass transport is not efficient to form hillocks. Therefore, hillocks formation during Si(C)N deposition may be explained by the influence of Cu line density on hydrostatic stress behavior. The surrounding Cu lines exert a transversal stress on the line under study, contributing to the increase in hydrostatic stress. For narrow Cu lines, thermal hydrostatic stress increases with density [7]. For large Cu lines, contribution of sidewalls on stress is weaker [2], thus, hydrostatic stress is density independent. Consequently, post Si(C)N hillocks density also depends on flow stress in Cu line during capping.

	volume fraction				
Line width (µm)	**f<111>**	**f<001>**	**f<011>**	**ftwins**	**Grain size (µm)**
2	1.4%	12.8%	20.7%	65%	1.0
5	6.8%	10.3%	7%	76%	1.8
30	7.4%	3.1%	3.2%	80%	5.5

Table 1: Volume fraction and grain size as function of Cu line width. The twin fraction f_{twin} includes <221> (<001> Σ3 twin), <511> (<111> Σ3 twin), <210> (<011> Σ3 twin) fractions.

Estimation of activation energy of hillocks growth

There are three factors: Cu thickness, thermalization temperature and duration. Hillocks are characterized on thermalized copper films by AFM measurements. Hillocks are localized on AFM scan, and their volume, height, diameter are directly measured.

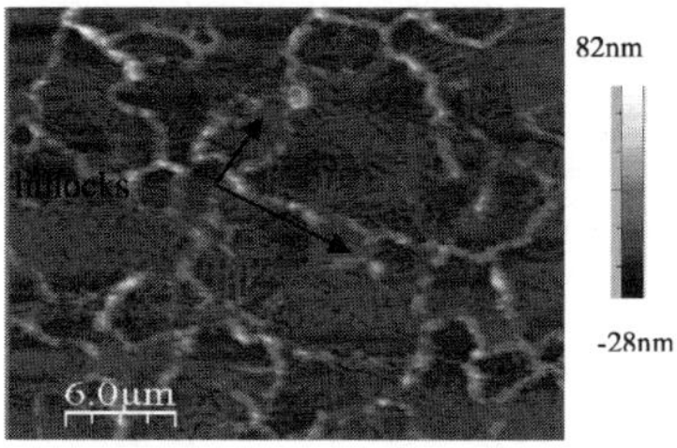

Figure 7: AFM scan after copper film thermalization.

The adjusted R-Squared statistic indicates that the model as fitted explains 91.7% of the variability in hillocks total volume. Hillocks total volume increases linearly with thermalization temperature and duration (Figure 8.a). Figure 8.b shows that hillocks total volume could be fit by an Arrhenius model. The corresponding activation energy for hillocks growth is 0.7eV +/-0.1eV. In previous paragraph, it has been demonstrated that hillocks grow on large Cu line by grain boundary diffusion mechanism. By considering blanket copper film as a very large line, it is possible to estimate that 0.7eV is the value of activation energy of copper grain boundary diffusion under thermal process.

Temperature (°C)	Thickness (μm)	Duration (s)	Hillocks density (/μm²)	Hillocks total volume (nm³)	Hillocks diameter (nm)	Hillocks maximal height (nm)
228	1.6	16	0.0627	5.4E+07	404	7.07
371	1.6	16	0.29	2.0E+09	417.5	60.6
228	4.8	16	0.00343	6.4E+07	0	6.2
371	4.8	16	0.262	1.4E+09	387.7	58.7
228	1.6	48	0.303	9.5E+07	202.2	14.3
371	1.6	48	0.307	1.9E+09	415.8	43.2
228	4.8	48	0.05	7.4E+07	413.6	21
371	4.8	48	0.281	2.8E+09	430.7	83
180	3.25	32.5	0.00667	3.7E+07	0	5
420	3.25	32.5	0.198	2.9E+09	487.1	81.2
300	0.5	32.5	0.256	1.9E+08	250.3	18
300	6	32.5	0.266	3.5E+08	301.2	21.8
300	3.25	5	0.151	1.2E+08	295.2	23
300	3.25	60	0.26	1.2E+09	351.5	65.4
300	3.25	32.5	0.321	7.6E+08	308.4	34.5
300	3.25	32.5	0.346	8.3E+08	312.8	36.2
300	3.25	32.5	0.316	7.2E+08	305.5	40.6

Table 2: Table of experiments.

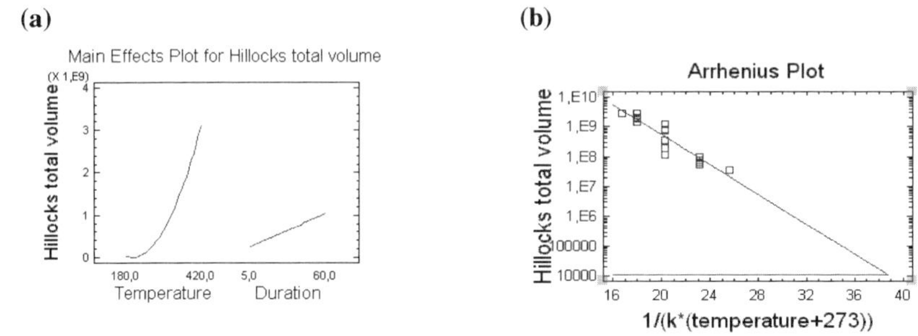

Figure 8: Model of total hillock volume : (a) Main effects plot (b) Arrhenius model.

CONCLUSIONS

After Cu line Si(C)N capping, two hillocks formation modes are observed, by grain boundary sliding or by diffusion. Diffusion can occur either at grain boundary or at interface. The activation energy of grain boundary diffusion has been estimated at 0.7eV, under thermal stress. It has been shown that when line width increases, grain boundary density decreases while grain size and twins increase and it is suggested that it contributes to stress reduction in Cu lines, which minimize hillocks. Results suggest that preferential diffusion mechanism for hillocks growth is interface for narrowest lines and grain boundary for wider ones. Cu line density plays also a role on hillocks formation by exerting a transversal mechanical stress on dielectric and then on sidewall of the studied Cu line. This transversal component increases the overall hydrostatic stress, which promotes hillocks formation.

REFERENCES

1. K. Sangchul and al.: "Copper hillock induced copper diffusion and corrosion behavior in a dual damascene process", Electrochemical and Solid-State Letters, Vol. 10, N° 6, June 2007.
2. P. Chaudhari: "Hillock growth in thin films", J.A.P., Vol. 45, N°10, Oct. 1974.
3. W. D. Nix and al.: "Creep-controlled diffusional hillock formation in blanket aluminium thin films as a mechanism of stress relaxation" J. Mater.Res., Vol. 15, N° 8, Aug. 2000.
4. M.S. Jackson and al: "Stress relaxation and hillock growth in thin films" Acta metal. Vol. 30, pp 1993-2000, 1982.
5. J.-Y Cho and al.: "Texture investigation of copper interconnects with a different line width"J.E.M., Vol.34, N°1,2005.
6. M. Ronay and al: "Hillock formation lead films by grain boundary sliding", Phi. Mag., Vol.42, N°2, 1980.
7. M. Gregoire and al: "Stress-void nucleation induced by Local delamination at Cu/Ta interfaces", Conferences Proceedings AMC 23, pp 679-685,2007.
8. J.E. Sanchez : "Effects of grain orientation on hillock formation and grain growth in aluminum films on silicon substrates" Script. Met., Vol. 27, pp 285-29,1992.

Conference Proceedings AMC XXIV © 2009 Materials Research Society

The Perspective Study of Cu Fill for Memory Devices

Hyunbae Lee, Tsukasa Matsuda, Hyun Park, Hyekyung Jung, Jungsun Choi, Byunglyul Park, Jongho Yun, Jongmyeong Lee, Gilheyun Choi, Hyeondeok Lee, and Jootae Moon
Process Development Team, SAMSUNG ELECTRONICS, SAN #16 Banwol-Dong, Hwasung-City, Gyeonggi-Do, 445-701, Korea, Republic of

ABSTRACT

The successful development of sub 40nm Cu interconnect technology for memory devices is described. The damascene pattern was fabricated using self-aligned DPT (double patterning technology) process. The sub 40nm Cu interconnect lines were completely filled by using optimized iPVD barrier/seed and electroplating process. The resistivity of Cu lines and the TDDB lifetime were investigated to confirm the reliability of Cu process integration.

INTRODUCTION

The low resistance Cu interconnects with low inter-metallic capacitance is now essential to compensate RC delay of high density memory device. However, as the metal line width shrinks down to below 40nm, the fill performance of Cu and patterning of narrow damascene structure become critical issues for the device shrinkage. For a narrow pattern filling, a novel PVD process or ALD/CVD process for barrier/seed metals have been widely investigated to obtain very thin and conformal step coverage [1]. For scaling down patterning, DPT process has been developed as a viable option to overcome limits of conventional lithography technique [2].

In this paper, we will discuss the critical issues associated with Cu fill process for sub 40nm lines. Optimizing iPVD barrier/seed and electroplating process was investigated. The resistivity of narrow Cu lines and TDDB reliability were also characterized for Cu interconnects in memory devices.

EXPERIMENTAL

The damascene pattern was fabricated using self-aligned DPT scheme [3]. For the damascene pattern filling, Ta(N) barrier metals and Cu seed layer were deposited by iPVD process. Cu electroplating was carried out in a commercial electroplating tool. After electroplating, the excess Cu in the field area was removed by a CMP process. The Si(C)N layer was deposited as a capping layer. Patterns were analyzed to investigate fill performance using SEM. TDDB test was performed using a snake-comb test structures.

RESULTS AND DISCUSSION

To fill the very narrow damascene pattern, minimization of asymmetry and overhang of iPVD barrier/seed Cu is essential, because such an asymmetry and overhang may generate void defects during electroplating process. We have optimized M+ (metal ion) based re-sputtering process to obtain robust sidewall coverage and minimize overhang. An appropriate fraction of high biased M+ and Ar+ was introduced at the bottom of trench with high RF bias, which enables simultaneous processes for deposition as well as resputtering of metals from the bottom of trench by M+ itself. Fig. 1 shows step coverage of barrier/seed layer on narrow trench. M+ based resputtering process reduced overhang and expanded the opening area compared to conventional deposition/etch/flash process. This lead to the sufficient filling capability of electroplated Cu.

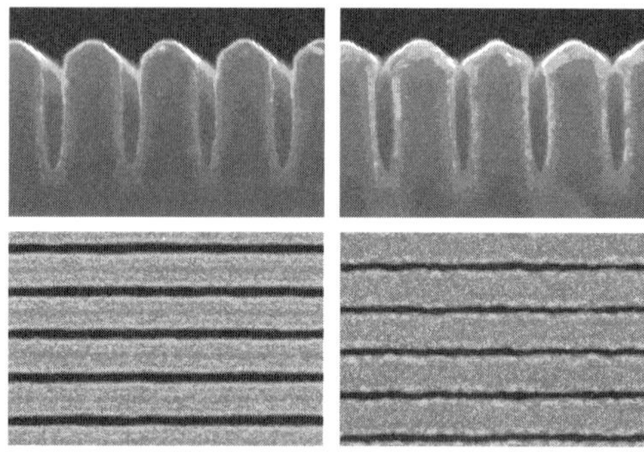

(a) (b)

Fig. 1 Step coverage and top opening size of barrier/seed deposited by (a) M+ resputter process and (b) conventional deposition/etch/flash process.

The Cu electroplating was optimized with respect to the current of initial electroplating step. As shown in Fig 2(a), under a low current condition, majority of defects were side voids, which might be attributed to the terminal effect during electroplating process or dissolution of thin seed layer due to high resistivity. However, under a high current condition most defects were center voids as shown in Fig. 2 (b). Center voids may be originated from overgrowth of overhang, resulting in a pinch off before completion of bottom-up fill. Fig. 3 shows the void frequency in metal lines depending on the electroplating current. As the current increases, the side voids decrease and center voids increase. Depending pattern density, we have optimized appropriate entry voltage and current density to minimize void defect.

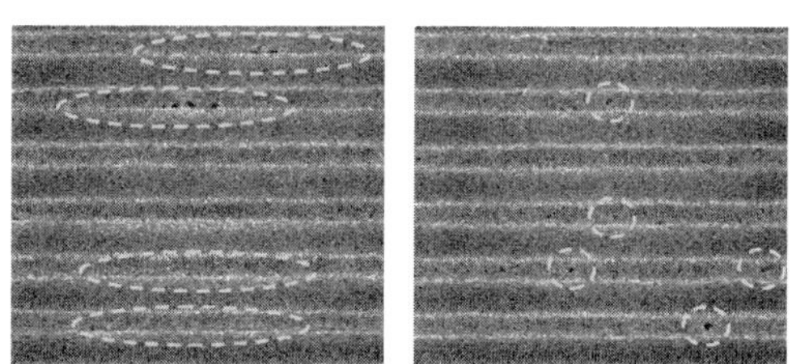

(a) (b)

Fig. 2 Defect void images (top) with respect to plating conditions; (a) Low initial current condition and (b) High initial current condition.

Besides optimizing of iPVD and electroplating process, the key integration issues related to void defects involve fabricating damascene pattern through self-aligned DPT scheme. If the shape and height of either sides of the trench are different, the barrier/seed layer shows asymmetric profile due to shadowing effect and resputtering angle, as shown in Fig. 4 (a). The asymmetry of sidewall and overhang can generate void defect during electroplating process. Thus, barrier/seed and electroplating process should have wider process capability window to overcome the pattern shape difference originated from self-aligned DPT process at sub 40nm trench. Fig. 4 (b) and (c) shows SEM images of trenches filled with optimized barrier/seed process and electroplating process on self-aligned DPT pattern after CMP. No voids or defect can be seen in Cu lines, in spite of variation of self-aligned DPT pattern.

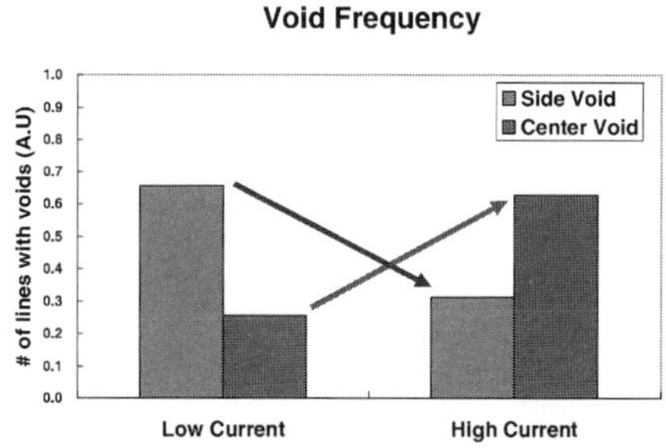

Fig. 3 Void frequency analysis depending on the electroplating process condition.

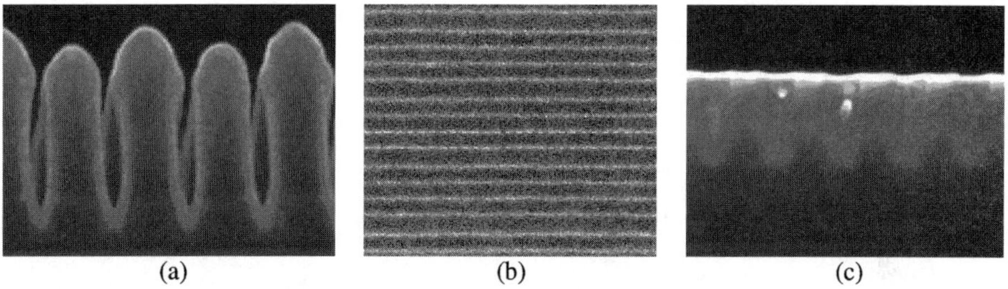

Fig. 4 (a) Step coverage of barrier/seed deposited on the self-aligned DPT pattern. (b) Top view and (c) cross-sectional images of metal lines after CMP.

Fig. 5 shows Cu resistivity measured on narrow damascene structures. The conventional equation for the scattering effects was used to fit experimental data [4]. The parameters of equation could be obtained to fit experimental data: p(Specularity parameter)=0.6, R(Reflective parameter)=0.22. The value of p and R were in similar range with previous results [5]. At ~40nm width of Cu lines, the resistivity was estimated to be ~ 4.5 μΩcm.

We have also compared the maximum electric field that satisfies the 10 year criteria of TDDB lifetime for sub 40nm space. At ~40nm spacing, the maximum electric field for 10 year criteria was estimated to be ~1.2 MV/cm with SiO_2 dielectric and optimized barrier/seed condition. We could confirm that the TDDB reliability for memory device can be satisfied by using optimized process conditions.

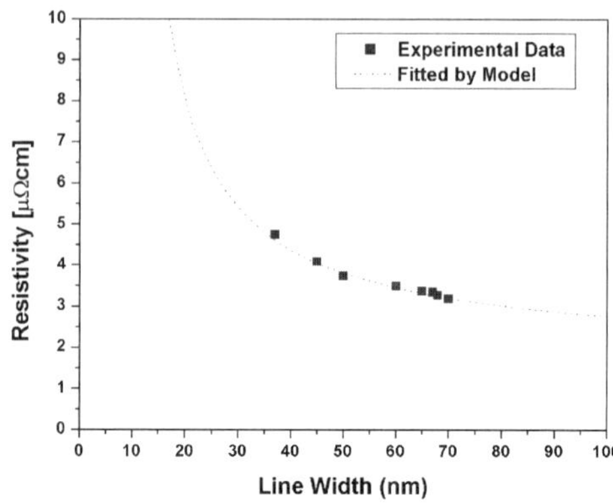

Fig. 5 Resistivity of Cu lines for sub 40nm interconnects. Squares indicate experimental data. Dot line indicates fitted results using conventional equation.

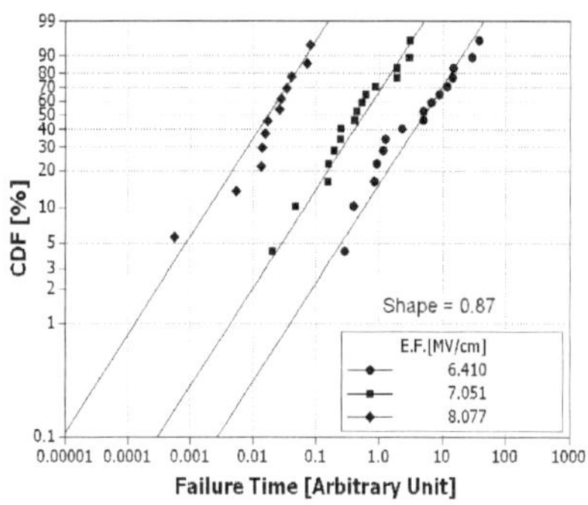

(a)

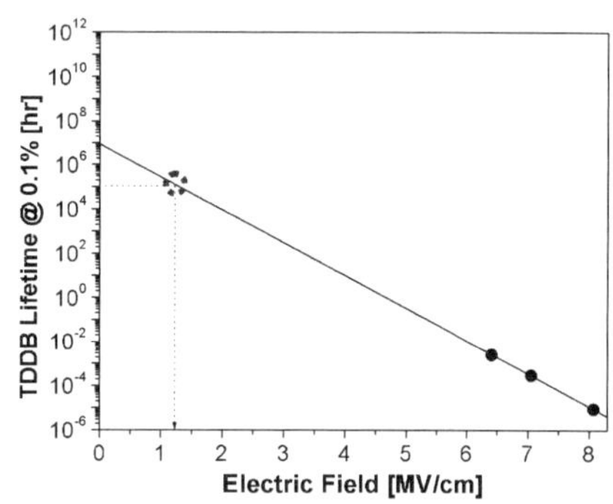

(b)

Fig. 6 (a) TDDB breakdown time distributions and (b) TDDB lifetime prediction based on E-model.

CONCLUSIONS

A progressive approach to fill Cu at sub 40nm trenches was performed for memory devices. The self-aligned DPT process was used for a single damascene pattern. A narrow trench pattern of sub 40nm width was successfully filled by optimizing iPVD barrier/seed and electroplating process. It was found that the resistivity of ~40nm Cu lines was ~ 4.5 μΩcm and the maximum electric field for 10 year lifetime was estimated to be ~1.2 MV/cm, which satisfy the requirements of memory devices.

REFERENCES

[1] P. H. Haumesser, et al., Advanced Metallization Conference, 2006, pp. 25~26, 2006.
[2] W.Y. Jung et al, Proceedings of SPIE 2007, Vol. 6520, pp. 65201C, 2007.
[3] B. J. Hwang, et al., Advanced Semiconductor Manufacturing Conference, 2007, pp. 356~358, 2007.
[4] Kapur, et al., Electron Devices, IEEE Transactions on., 49(4), pp. 590~597, 2002
[5] W.F.A Besling, et al, Microelectronic Engineering, 76, pp.167~174, 2004.

Conference Proceedings AMC XXIV © 2009 Materials Research Society

Spatial Correlation of Metal Line Variations

Jianhong Zhu, James Buller, David Wu, Scott Johnson, and Ravinder Rachala
Advanced Micro Devices, Inc., 7171 Southwest Parkway, Austin, TX, 78735

ABSTRACT

Statistical analysis on data collected from duplicated metal test structures within die reveals a roughly linear spatial correlation of metal line variations. The correlations are attributed to the planar feature in the CMP and etch height profiles. Considerable improvement is achieved in delay match analysis with the consideration of metal variation spatial correlation.

INTRODUCTION

Microprocessor fab process variations have become progressively more important [1] as design dimensions decrease. With continued scaling of IC technology, the number of relevant sources of variation and variation magnitudes have increased, causing design margins to shrink [1,2]. Variation-aware and variation-tolerant design has thus prevailed. Thorough characterization of the variations of design-related parameters is imperative for robust designs. Metal resistance and capacitance variations are usually closely monitored, but primarily for lot-to-lot, wafer-to-wafer, and die-to-die comparison. Understanding metal location variation or variation spatial correlation across a chip is important to the design of circuits that have strict delay match requirements.

In this paper, statistical analysis was performed on metal data collected from test structures duplicated within-die. A roughly linear spatial correlation of metal variation was revealed. Further analysis of metal geometries extracted from electrical data helped in understanding the correlation effects. Substantial performance improvements in delay matching circuits were seen once metal variation correlation was considered.

EXPERIMENT

Duplicated test structures were built on a 45 nm node test vehicle with a Cu low-k multiple layer BEOL stack. Stacked metal line-comb-serpentine test structures were duplicated with different separations within a die. Metal resistance and capacitances of each layer were then measured. Smaller metal Kelvin resistance test structures were duplicated within a 2.5 mm-length macro that was much smaller than the die. Electrical data was collected on a large number of lots for statistical analysis. Different die on the same wafer are treated as independent entities.

RESULTS AND DISCUSSIONS

For characterization of variation spatial correlation of, for example, metal resistance R, relative match variation (RMV) of R at a distance determined by locations B and A is defined as 3σ of (R(B)-R(A))/R(A), where R(B) and R(A) are the metal resistances at locations B and A, respectively. Normalized RMV of R is the result of RMV of R divided by relative 3σ (*i.e.*, 3σ/mean) of R. With this definition, a small normalized RMV of R indicates a high R variation correlation whereas a theoretical maximum number of 1.4 indicates no correlation. Figures 1(a), 1(b) and 1(c) show the within-die distance dependencies of the normalized RMVs of metal resistance, total metal capacitance and metal inter-level capacitance (Cdown), respectively. A roughly linear spatial correlation of variation is revealed for all three parameters. Clearly, two metal lines that are close to each other are almost fully correlated in resistance and capacitance, resulting in very small normalized RMVs. When the distance increases, they gradually lose such correlation and their normalized RMVs increase. Even at a distance close to 40 mm, their normalized RMVs are well below 1.4, indicating that metal variations are well correlated within the die. Normalized RMVs of M1 in Fig. 1 are smaller than those of other layers. This is due to larger M1 resistance and capacitance variations.

Figure 1(d) shows the normalized resistance RMV from the metal Kelvin resistance test structures that have distances less than 2.5 mm. The data roughly matches that in Fig. 1(a) in their overlapping distance range. When two metal lines are very close to each other (*i.e.*, at a distance less than 0.2 mm), the resistance match variation is below 10% of the resistance variation. This shows that treating close metal lines as independent is overly pessimistic.

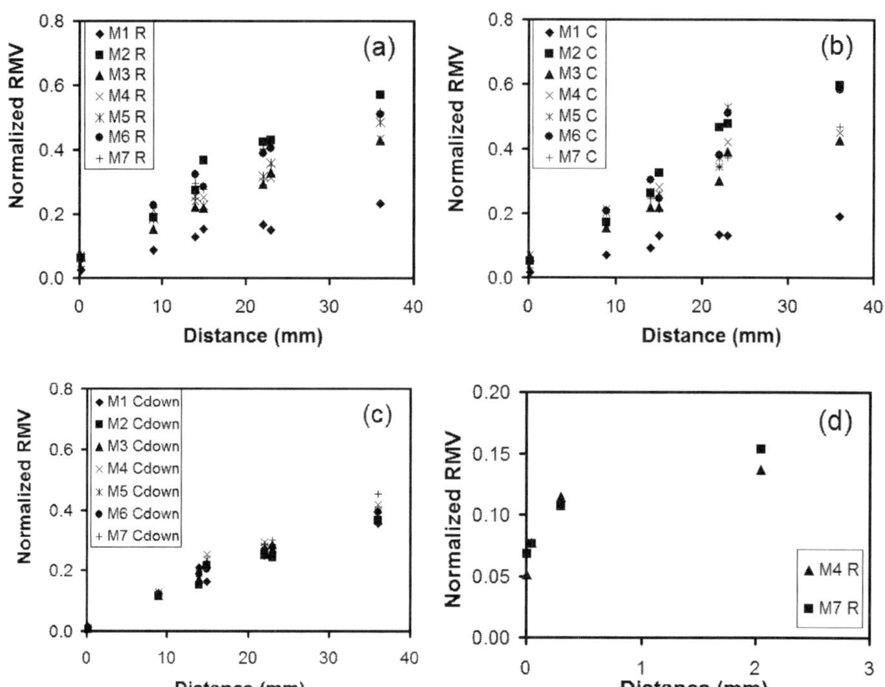

Figure 1. Variation spatial correlation of (a) metal resistance, (b) total metal capacitance, (c) inter-level metal capacitance, and (d) within-macro metal resistance.

To help understand the observed linear spatial correlation of metal resistance and capacitance variations, metal geometries of each individual metal line were extracted from measured electrical data using a recently developed method [3]. Similar statistical analysis was performed on these geometries. Figures 2(a) and 2(b) show the normalized RMVs of metal module height (metal height plus lower via height) and via height, respectively. Module height is determined by the CMP process while via height is determined by the metal etch process. Both module height and via height in Fig. 2 show a similar variation in spatial correlation as metal resistance and capacitance, indicating that CMP and etch have an intrinsic spatial correlation feature. Via height in Fig. 2 has smaller normalized RMVs than module height. This is due to a large via height variation. Match variations of module height and via height are comparable, with that of module height being a little larger. This shows that both CMP and etch processes contribute to the observed spatial correlation of metal variations.

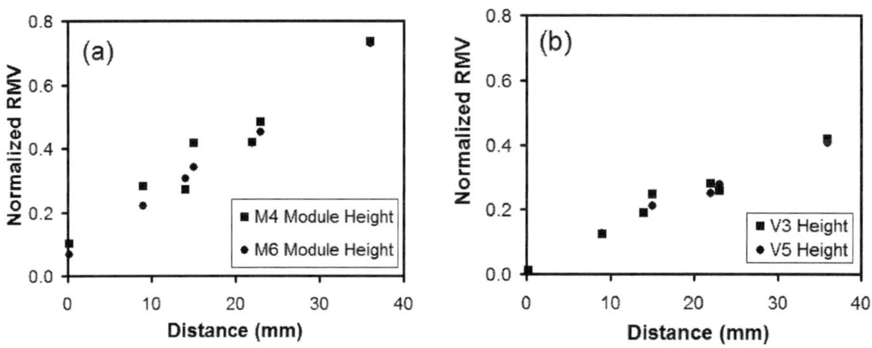

Figure 2. Variation spatial correlation of (a) metal module height and (b) via height.

Metal module height and via height spatial gradients at different distances but in the same direction were calculated and compared. Figures 3(a) and 3(b) show the comparison at distances of 11 and 22 mm. Both module height and via height show comparable and correlated gradients, indicating a near planar feature in CMP and etch height profiles. Such a local planar feature in process results in the observed roughly linear spatial correlation of metal variations.

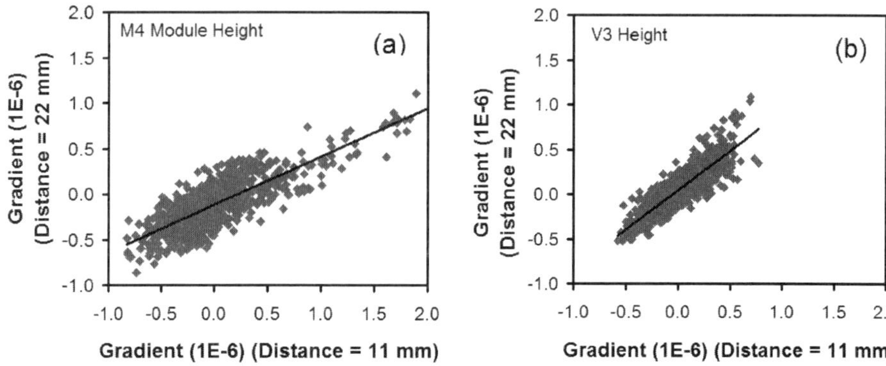

Figure 3. Comparison of spatial gradients of (a) M4 module height and (b) V3 height at distances of 11 and 22 mm. Lines are drawn for visual guidance of correlation.

To estimate the impact of the spatial correlation of metal variations on design, delay match variation analysis was exercised on two protocol circuits, including a 45 nm clock tree and a 45 nm source synchronous bus. These two structures were chosen because the circuits require good delay matching. Clock trees need good delay matching to deliver low clock skew to avoid race conditions and deliver good cycle times. Another circuit technique that needs good delay matching is a source synchronous bus, which delivers synchronous data with a forwarded clock far away from the original clock source. The source synchronous bus consists of a data bus that travels with a corresponding synchronous clock. The source synchronous bus frequencies and latency inherently depend on good delay matching

The simplified clock tree (Figure 4) consists of two parallel horizontal fingers separated by 5 mm. Each finger has ten M4 metal segments sequentially arrayed with repeaters inserted in between. Each metal segment has a metal width of 570 nm and a length of 0.415 mm. The clock signal is assumed to arrive simultaneously at the starting points S1 and S2 of the two fingers. Elmore delays at the ends E1 and E2 of the two fingers are statistically calculated with fixed device (repeater) parameters but varying metal parameters. The 20 metal segments are considered to be either spatially correlated in their resistance variations or independent of each other. For the calculation with correlation, resistance variation correlations in horizontal and vertical directions are considered independent and each is generated by a random factor combined with the data in Fig. 1(a). Metal capacitance varies with metal resistance based on the simulated C-R curve of the metals. The delay RMV at the two ends is calculated to be 11.5% when the metals are independent of each other. This delay RMV is reduced to 3.2% when the metals are spatially correlated in their variations.

The source synchronous bus circuit was mimicked by reducing the vertical separation to 0.3 mm from the 5 mm value in Figure 4. In this case, the delay RMV obtained is only 1.8% with the inclusion of metal correlation. The delay match variation analysis results are improved significantly by considering metal parameter variation correlation. Without these correlation improvements, the frequency of bus operation would be reduced. Alternatively, flops could be inserted to fix the frequency problem at the expense of data latency. Due to the correlation improvements, the bus will operate at better frequencies and lower latency.

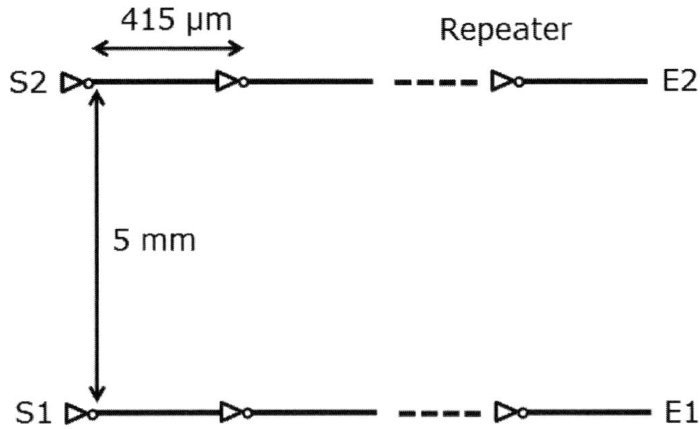

Figure 4. Schematic of a protocol circuit of clock tree.

CONCLUSIONS

A roughly linear spatial correlation of metal variations is revealed. These correlations are attributed to the planar feature in the CMP and etch height profiles. A significant improvement in circuit delay matching performance is achieved with the inclusion of metal variation spatial correlation.

ACKNOWLEDGMENTS

The authors thank AMD Fab 36 (Dresden, Germany) for strong support of data collection.

REFERENCES

1. *http://www.itrs.net/Links/2007ITRS/Home2007.htm* ITRS 2007 Edition.
2. D. Boning and S. Nassif, "Models of Process Variations in Device and Interconnect," *Design of High Performance Microprocessor Circuits*, A. Chandrakasam (ed.), 2000.
3. J. Zhu, J. Buller, D. Wu and P. Huebler, *in Proc. of* Advanced Metallization Conference 2007, pg. 577 (2007).

Metallization

Conference Proceedings AMC XXIV © 2009 Materials Research Society

Ionized-PVD Stacked Barrier Structure of TaN/TaRu for 32nm BEOL Integration

Takeshi Nogami[1,2], Chih-Chao Yang[1,2], Stephen M. Rossnagel[2], Christopher J. Penny[1], Donald F. Canaperi[1], James J. Kelly[1], Philip L. Flaitz[3], Patrick W. DeHaven[3], Thomas M. Shaw[2], Stephan A. Cohen[2], Chao-Kun Hu[2], Tuan A. Vo[1], Inka Zienert[4], Atsunobu Isobayashi[5], Hua Chung[6], Rongjun Wang[6], James Ren[6], Soon-Cheon Seo[1], Satyavolu S. Papa rao[1], Andrew H. Simon[3], Oscar van der Straten[1], Iqbal Ali[7], Joe Piccirillo[7], Sunny K. Chiang[7], Jean Wynne[1], Terry A. Spooner[1], and Daniel C. Edelstein[2]

[1]IBM in Albany Nano Technology Research Center, Albany, NY, 12203

[2]IBM T. J. Watson Research Center, Yorktown Height, NY, 10598

[3]IBM Microelectronics Center, Hopewell Junction, NY, 12533

[4]Advanced Micro Devices Inc., AMD Saxony LLC & Co. KG, Germany

[5]Toshiba America Electronic Components, Inc., Albany, NY, 12203

[6]Applied Materials Inc., Santa Clara, CA, 95054

[7]Applied Materials Inc. in Albany Nano Technology Research Center, Albany, NY, 12203

ABSTRACT

An ionized-PVD stacked barrier structure of TaN/TaRu was developed for 32nm BEOL integration. The TaN layer serves as a most robust diffusion barrier layer, while the TaRu barrier layer improves wettability of the ionized-PVD Cu seed layer. Observed benefits included enhanced Cu gap filling performance, resolving a line voiding problem, improvement in electromigration performance, and higher via chain yield than TaN/Ta controls at 32nm BEOL dimension. This stacked barrier structure, which uses two PVD chambers incorporating pure Ta and TaRu targets, was proposed based on our material studies showing that neither TaRu nor TaRuN performed adequately alone as Cu and O_2/H_2O diffusion barriers, but performed well when in conjunction with TaN or TaN/Ta backing layers. The newly proposed TaN/TaRu stacked barrier is a possible solution to extend ionized PVD to 32 nm node BEOL.

INTRODUCTION

At 32 nm node BEOL dimensions, the extendibility of ionized PVD (iPVD) for dual damascene liner/seed for Cu/ultra low-k (ULK) integration is reaching its limits. As will be described in detail below, the limits are coming not only from the shrunken dimensions, but also from degraded pattern integrity associated with porous ultra low-k dielectrics at these fine dimensions. Ruthenium has been investigated as a final liner layer under the Cu seed layer [1] or as an alternate (presumably non-etchable) material for direct Cu plating without a Cu seed layer [2]. While CVD and ALD Ru have been expected to serve as Cu gapfill enhancement layers because of their highly conformal deposition, iPVD TaRuN/TaRu has been reported to extend the iPVD technology [3].

In contrast to [3], our conclusions from the work done here are that neither TaRu nor TaRuN can function adequately as a Cu or O_2 diffusion barrier. The polycrystalline structure observed in TEM analysis was attributed to the poor diffusion barrier properties where presumably either or both of grain boundaries and Ru grains are possible diffusion paths for Cu and O_2, and grain boundaries are not stuffed by the addition of N to the process. Another finding shown in this paper is that the electroplated Cu gap filling performance increased with Ru %.

Based on these fundamental material data, we proposed and examined the TaN/$Ta_{10at\%}Ru_{90at\%}$ stacked liner structure. The TaN layer serves as a more robust diffusion barrier layer, while the TaRu barrier layer improves wettability of the PVD Cu seed layer. This stacked liner approach requires an additional barrier PVD chamber compared to the conventional TaN/Ta case, and Ru is a very costly material. However the cost of ownership can be reasonable, given the very thin TaRu layers being considered, and the fact that Ru can be > 90% recoverable from the PVD tooling and CMP slurries. Improvements in via-chain yield and electromigration performance, and resolved line corner voiding with the stacked TaN/TaRu are discussed to demonstrate the extendibility of ionized PVD to the 32nm node in ultralow-k dielectric.

EXPERIMENT

(1) Diffusion Barrier Properties and Microstructure of TaRu and TaRuN

Ta, TaN, TaRu and TaRuN films with thickness of 3nm were deposited by PVD with Ar or Ar/N_2 plasmas in a 300 mm commercial i-PVD system. PVD alloy targets of $Ta_{30at\%}Ru_{70at\%}$ and $Ta_{10at\%}Ru_{90at\%}$ (hereafter called TaRu70% and TaRu90%, respectively) were used for formation of TaRu and TaRuN. Diffusion barrier tests included triangular voltage sweep (TVS, [4]) for Cu diffusion, and a Cu oxidation test [5] for O_2/H_2O diffusion.

Figures 1 and 2 show the sample structure and raw data for the TVS tests. The peaks in the C-V plots represent changes in capacitance caused by Cu^+ ion diffusion into the SiO_2 layer during the voltage sweeps, through the barrier metals under test. As is shown in Fig. 2, neither of Ta and TaN showed any peaks indicating robust Cu diffusion barrier performances in this test. On the other hand, TaRu and TaRuN show peaks and indicate that neither of these films fully block Cu^+ diffusion, even at mild temperatures. Relative comparisons may be made from the heights of the peaks: (1) TaRu with higher Ru % has worse diffusion barrier performance; and (2) incorporation of nitrogen (TaRuN) somehow worsened the diffusion barrier performance.

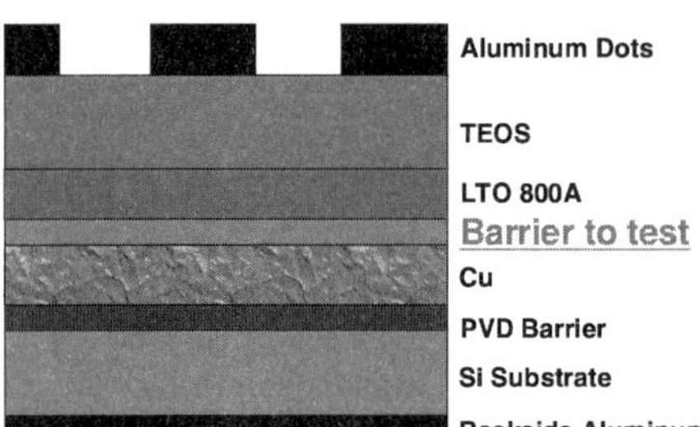

Fig. 1 Triangular Voltage Sweep (TVS) Test Structure

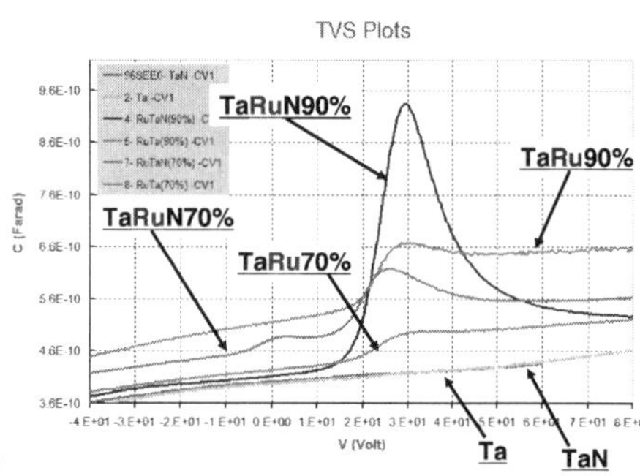

Fig. 2. TVS Plots

Figure 3 shows the Cu oxidation test sample structure and raw data for TaRuN/TaRu with 70% and 90% Ru vs. the TaN/Ta control. The blanket barrier metal films deposited on Cu were exposed to the air at 200C for eight days. Changes in surface appearance over time were observed in an optical microscope. As is shown in the pictures, a color change was observed with the TaN/TaRu samples, while TaN/Ta stack did not show any change over this stress time period. This and previous experience with this test indicate that these TaRuN/TaRu films have inadequate O_2/H_2O diffusion barrier performance for our BEOL integration requirements.

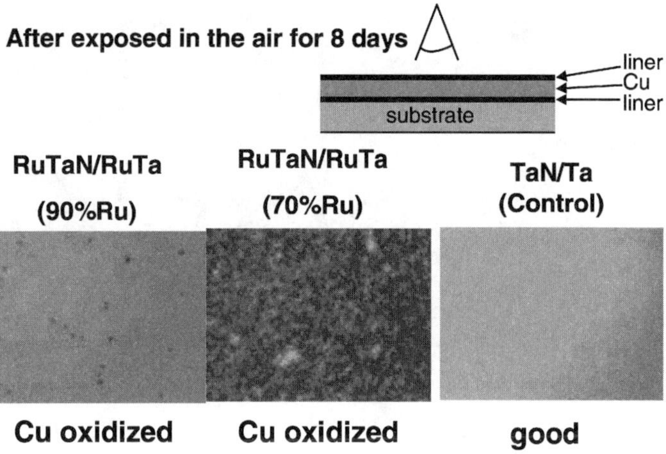

Figure 3. Anti Oxidation Test of TaRuN/TaRu

Figure 4 shows XRD spectra of a TaRu70%N/TaRu70% stacked film and a TaN/Ta reference sample. The TaN/Ta spectrum has broad peaks suggesting small-grained structure, while the sharp peaks in the TaRu spectrum indicate more highly crystalline morphology. Other TaRu films such as TaRu90% and TaRu monolayers showed similar spectra to the TaRu70%N/TaRu70% depicted here, which happened to represent the broadest peaks for any TaRu films studied.

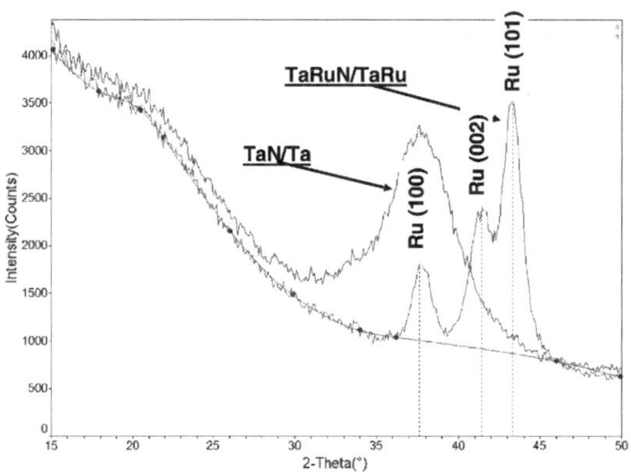

Figure 4 Thin Film XRD Spectra of Ta30at%Ru70at% and its Nitride Films in Stack

Figure 5 shows a TEM image and the electron diffraction pattern for the TaN/ TaRu70% stack. The polycrystalline structure is substantiated by these TEM results. Similar TEM images and EDX patterns showed similar polycrystalline morphology for TaRu90%, and its nitride.

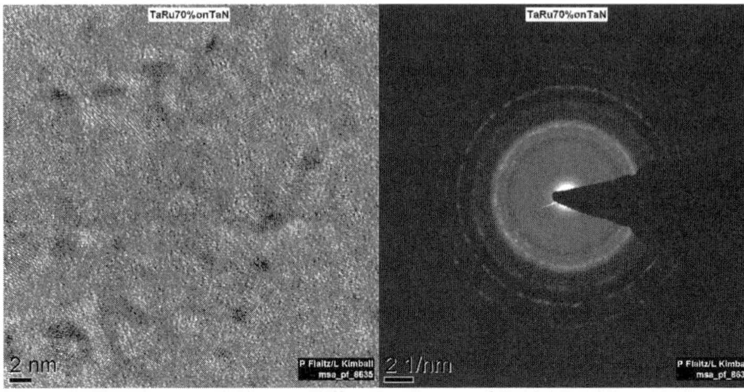

Figure 5 TEM Image and Electron Diffraction Pattern of Ta30%Ru70% on TaN

Figure 6 shows a cross-sectional TEM image for the TaN/TaRu70% stack. TaRu70% was polycrystalline and was not influenced by the underlying TaN film morphology.

Figure 6 Cross-sectional TEM Image of TaN/TaRu80% Stack

(2) Improvement in Cu Plating Gap-Fill Performance

Cu gap-fill performance by electroplating was compared among liner films of Ta, TaRu70%, and TaRu90% underneath a Cu seed layer in high aspect-ratio trenches (A/R>5:1). Figure 7 shows cross-sectional SEM images. Improvement by the use of TaRu is observed, in a similar trend as reported by K. Mori et al. in Ref. [3]. As is shown in Fig. 7, Cu gap-filling performance is improved with increased percentage of Ru in the composite film. This result seems quite reasonable from the view point of the direct plating taking place on Ru, and the enhanced wettability of Cu on Ru. However, it is not clear if the intermetallic Ta-Ru phase or the metallic Ru phase in the film helps the plating performance at this moment.

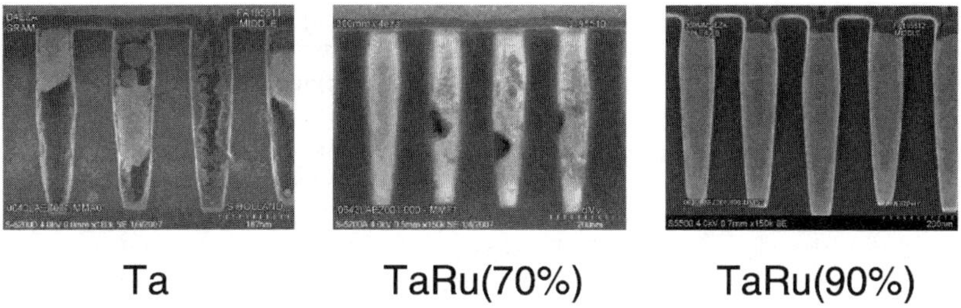

Figure 7 Cu Plating Fill as a function of Ru Content in TaRu

(3) TaN/TaRu90% Stacked Liner Structure

Because of substandard barrier performance with TaRu and TaRuN, and the more significant improvement in Cu gap-filling performance with higher Ru % in TaRu, the TaN/TaRu90% stacked liner structure was proposed as shown in Figure 8. A standard recipe was used to deposit the underlying TaN film, as in our TaN/Ta baseline. The Ta layer was simply replaced by TaRu90% prior to deposition of the Cu seed layer. The resistivity of the TaRu90% was 52 $\mu\ \Omega\cdot$cm.

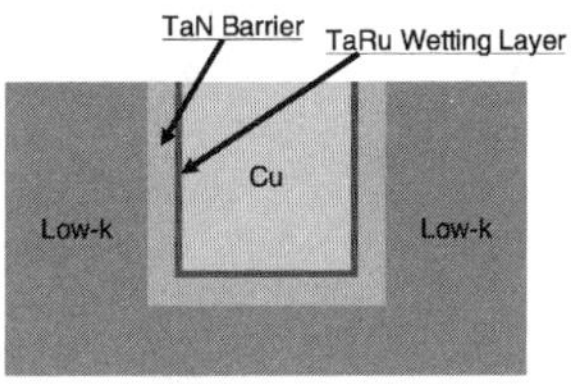

Figure 8 PVD Stacked TaN/TaRu

Figure 9 shows a cross-sectional TEM image and the EDX line profile through the stacked liner structure within a trench etched in k = 2.4 ultra low-k porous SiCOH at 32 nm node dimensions. The contrast between the TaN layer with no grains structure evident here, and the polycrystalline TaRu is clearly seen. The good conformality and continuous TaRu step coverage are confirmed along the trench sidewall, with thickness ranging from ~2 nm to ~4 nm.

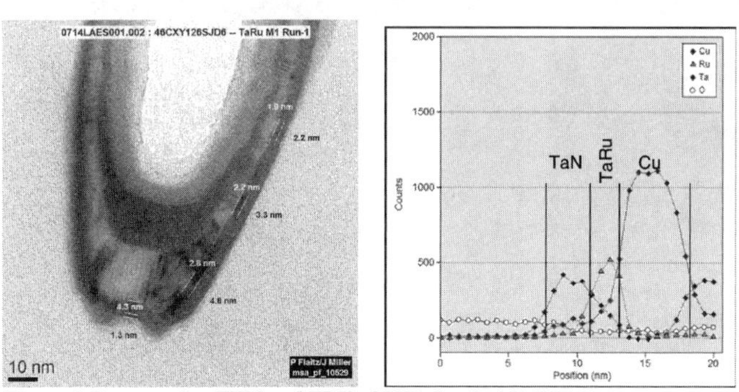

Figure 9 Cross-section TEM and EDX of PVD Stacked TaN/TaRu

The TaN/TaRu/Cu-seed PVD deposition process was optimized in thickness to avoid pinch-off at the entrance of the 32 nm node dual damascene trenches in the ULK SiCOH. The CMP process parameters were re-centered appropriately for complete removal of the TaRu layer. All other integration processes remained unchanged.

Figure 10 shows 3.5M via-chain yields for TaRu vs. POR liner splits. In this lot, the POR baseline was suspected of having had a defective Cu fill due to breaks in the Cu seed–layer. Remarkably higher chain yields were obtained for the TaN/TaRu splits. Fig. 11 shows electromigration performance of the TaN/TaRu vs. TaN/Ta liner splits. More than three times longer failure time were observed consistently from multiple runs. We believe this is due to increase of the process window for defect-free Cu seeding and Cu gap-filling.

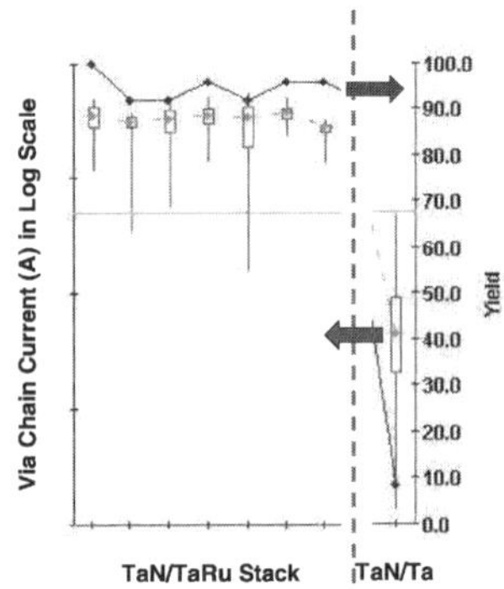

Fig. 10 3.5M Via Chain Yield Comparison between TaN/Ta and TaN/TaRu

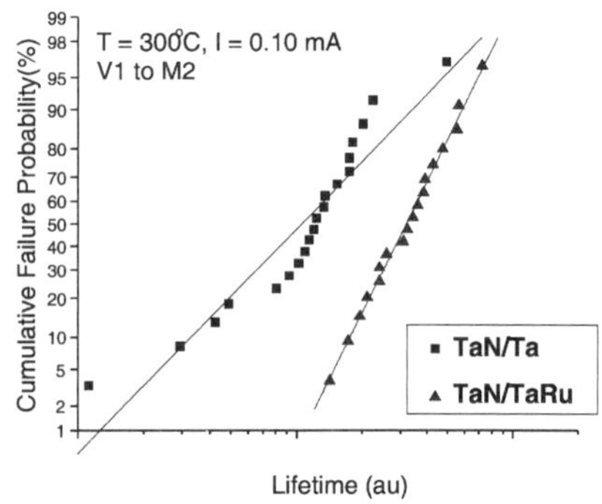

Fig. 11 EM Performance

The SEM image in Fig. 12a shows corner voiding caused by plasma damage in this particular sample, due to a non-optimized patterning process. This, in turn, can lead to Cu seed-layer discontinuities and subsequent Cu plating defects. The TaN/TaRu works to avoid formation of voids at this specific location as shown in Fig. 12b. The improved EM performance with TaN/TaRu in Fig. 11 is attributed to the improved Cu wettability of TaRu, leading to a continuous Cu seedlayer.

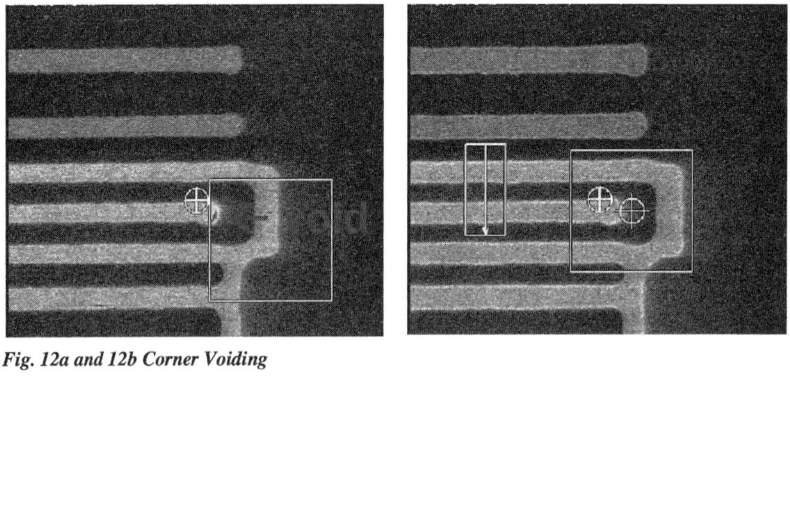

Fig. 12a and 12b Corner Voiding

With regards to the Cu microstructure and crystalline texture, no significant differences were observed by EBSD (electron backscatter diffraction) as shown in Table 1, and hence Cu microstructure is not attributed to the improved EM performance.

liner	treatment	med. grain size	{111} [%]
TaN/Ta40	with anneal	585nm	27.9
TaN/TaRu40	with anneal	506nm	26.8

Table 1 EBSD Analysis – Median Grain Size and (111) Orientation

CONCLUSIONS

The newly proposed iPVD stacked barrier of TaN/TaRu is a strong candidate to extend the use of iPVD for Cu metallization in the 32 nm node BEOL integration and beyond.

ACKNOWLEDGMENTS

This work was performed by the Research Alliance Teams at various IBM Research and Development Facilities.

REFERENCES

1. C-C. Yang, T. Spooner, S. Ponoth, K. Chanda, A. Simon, C. Lavoie, M. Lane, C.-K. Hu, E. Liniger, L.Gignac, T. Shaw, S. Cohen, F. McFeely, D. Edelstein, "Physical, Electrical, and Reliability Characterization of Ru for Cu Interconnects", IEEE IITC, 187, (2006)
2. M. W. Lane, C. E. Murray, F. R. McFeely, P. M. Vereecken, R. Rosenberg, "Liner materials for direct electrodeposition of Cu", Appl. Phys. Lett. 83, 2330 (2003)
3. K. Mori, K. Ohmori, N. Torazawa, S. Hirao, S. Kaneyama, H. Korogi, K. Maekawa, S. Fukui, K. Tomita, M. Inoue, H. Chibahara, Y. Imai, N. Suzumura, K. Asai and M. Kojima, "Effects of Ru-Ta Alloy Barrier on Cu Filling and Reliability for Cu Interconnects", IEEE IITC, 99, (2008).
4. S. A. Cohen, J. Liu, L. Gignac, T. Ivers, D. Armbrust, K.P. Rodbell, S.M. Gates, "Characterization of thin dielectric films as copper diffusion barriers using triangular voltage sweep", Mat. Res. Soc. Symp. Proc. 564, 551 (1999).
5. T.M. Shaw, D. Jimerson, D. Haders, C.E. Murray, A. Grill, D.C. Edelstein, "Moisture and oxygen uptake in Low K/ Copper Interconnect Structures", Adv. Metall. Conf., 77 (2003).

Conference Proceedings AMC XXIV © 2009 Materials Research Society

Material and Process for Advanced Cu/ULK Integration

Chih-Chao Yang[1], Daniel C. Edelstein[2], Terry A. Spooner[1], Ping-Chuan Wang[3], Kaushik Chanda[3], Ronald Filippi[3], Birendra N. Agarwala[3], Paul S. McLaughlin[3], Baozhen Li[4], Fen Chen[4], John Arnold[1], Stefan Schmitz[1], Hmayak A. Bagramian[1], Yunpeng Yin[1], Kaushik Kumar[3], Youngjin Choi[3], Samuel Choi[3], Andrew H. Simon[3], Wei-Tsu Tseng[3], Chao-Kun Hu[2], Eric G. Liniger[2], Lynne M. Gignac[2], Thomas M. Shaw[2], Stephan A. Cohen[2], and Robert Wisnieff[2]

[1]IBM, Albany Nanotechnology Research Center, 255 Fuller Road, Albany, NY, 12203

[2]IBM, T.J. Watson Research Center, Yorktown Heights, NY, 10598

[3]IBM, Semiconductor Research and Development Center, Hopewell Junction, NY, 12533

[4]IBM, Microelectronics Technology Reliability, Essex Junction, VT, 05452

ABSTRACT

Copper metallization at 32nm and beyond poses multiple technical challenges to the semiconductor industry for its continued evolution. It becomes necessary to consider alternative liner materials, deposition processes, and integration schemes together to successfully meet the technology extendibility, circuit performance and reliability requirements. This paper describes a new enhanced via integration scheme which includes a novel patterning technique coupled with a conformally deposited CVD Ru adhesion layer between the TaN diffusion barrier and the Cu interconnect. The new scheme shows potential as a replacement to the conventional "via punch-through" integration scheme with physical vapor-deposited (PVD) Ta adhesion layer.

MATERIAL EXTENDIBILITY

Thin PVD TaN/Ta bilayer liners with relatively thick PVD Cu seeds and electroplated Cu fill have been implemented in many advanced BEOL metallization schemes [1, 2]. In these stacks, one role played by the Ta layer is as an adhesion promoter between Cu and TaN, because Cu does not bond well to the TaN diffusion barrier [1, 3]. With continual shrinking of interconnect structures, scaling of the Cu seed thickness is required to avoid feature top pinch-off during subsequent Cu electroplating process. However, the constraint of forming a continuous Cu film on a Ta adhesion layer sets a fundamental limit on the minimum Cu seed thickness of ~30-40nm as measured on the field [4]. Figure 1 shows a cross-sectional SEM image taken from a comb-serpentine structure with PVD TaN/Ta liner and the minimum required thickness of Cu seed. The associated top view SEM images in Figure 2 are taken from the same structure following Cu plating and chemical mechanical polishing (CMP), showing highlighted feature-center voids. These voids are due to the severe Cu seed overhang in Figure 1.

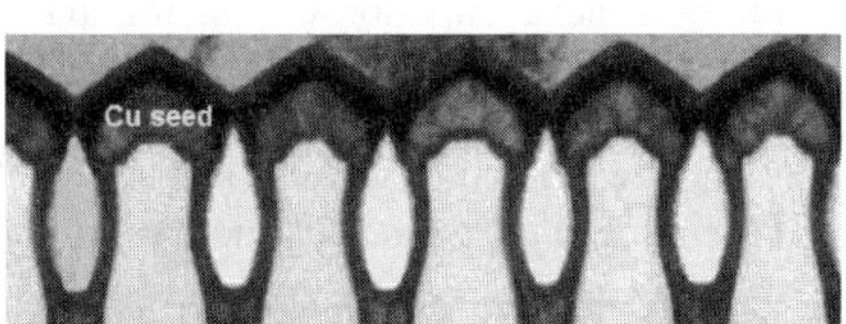

Fig. 1. SEM section from a line structure post PVD TaN/Ta liner and PVD Cu seed deposition.

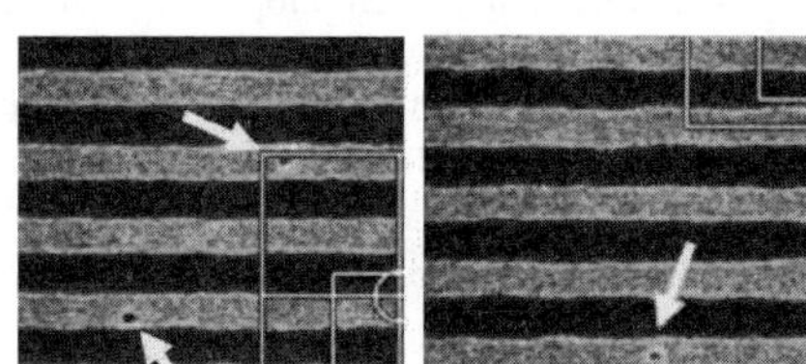

Fig. 2. SEM top-views from the structure in Fig.1 post Cu plating and CMP.

Any new candidate for an adhesion layer between the TaN diffusion barrier and the Cu interconnect must meet or exceed a number of properties of the conventional Ta adhesion layer. These include adhesion to TaN and Cu, thermodynamical stability with Cu, wettability of Cu to the adhesion layer, oxidation resistance, low resistivity, and thereby contributing to electromigration redundancy. Further, enhanced wettability and oxidation resistance of the deposited adhesion layer would help to prevent agglomeration of ultra-thin Cu seed layers, and the oxidation resistance may aid in Cu platability on ultra-thin (and hence discontinuous) Cu seed layers [5].

PROCESS

The presence of a TaN diffusion barrier and a Ta adhesion layer reduces the Cu volume fraction in the interconnects, and impacts wire resistances [5]. An additional resistance penalty comes from the non-conformal liner deposition process, which requires deposition of thicker liner at line bottoms to prevent discontinuity on sidewalls, especially inside vias. For this reason, it would be desirable to have a more conformal liner deposition process.

Conformal liners with good step coverage are therefore desirable for extending the existing technology. Recent advances in ionized physical vapor deposition (iPVD), chemical vapor deposition (CVD) and atomic layer deposition (ALD) techniques have enabled the deposition of nearly conformal liner layers in high aspect ratio features, but not all of these can do this for appropriates sets of materials or appropriate material quality. One positive example is an optimized PVD-TaN/ CVD-Ru that was recently developed [5]. SEM images of dual-damascene isolated via structures comparing TaN/PVD Ta and TaN/CVD Ru liner stacks [5] are shown in Figure 3. It can be observed that the CVD Ru liner appears to exhibit improved conformality and better capability for thin depositions than the PVD Ta liner.

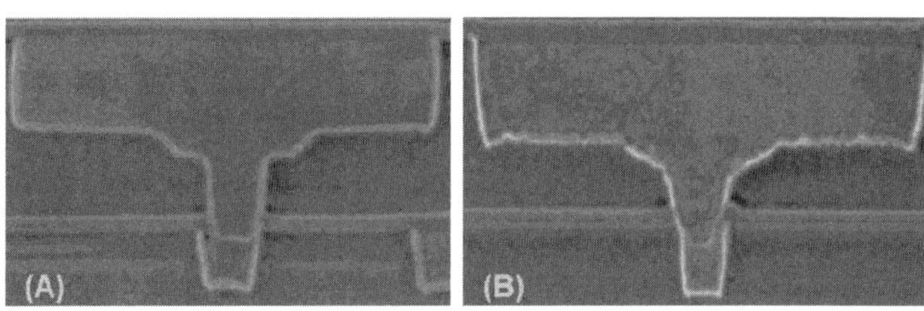

Fig. 3. SEM sections of isolated vias with (A) TaN/CVD Ru liner and (B) TaN/PVD Ta liner stacks.

INTEGRATION

In order to enhance interconnect reliability while employing copper and low-k dielectric materials in an interconnect structure, we pioneered a so-called **"via punch-through"** technique which has since been adopted by several companies [6-8]. The punch-thorough provides a via-gouging feature within the interconnect structure to achieve superior Cu wiring reliability through enhanced via-line contact geometry [7-9]. However, this process has drawbacks for extendibility to ultra-low k (ULK) dielectrics, due to dielectric damage following punch-through in features other than via bottoms. Detailed processing steps for the conventional approach are illustrated in Fig's. 4A-4E. Figure 4A shows a dual damascene interconnect structure, which includes an upper line and a lower contact via feature. Figure 4B shows the interconnect structure after depositing a TaN diffusion barrier over the exposed surface. Ar^+ sputtering, Figure 4C, is then performed to clean the via bottom surface and to form a gouged feature into the Cu interconnect which lies below. During the Ar^+ sputtering process, the TaN diffusion barrier is removed from the bottom of the line feature, and the lower dielectric layer is damaged. A Ta adhesion layer is then deposited as shown in Figure 4D. Figure 4E depicts the final interconnect structure after filling the features with electroplated Cu, and the subsequent chemical mechanical polishing (CMP) planarization step. Note that the final interconnect structure in Figure 4E has roughness and poor TaN diffusion barrier coverage at the line bottom surfaces. Both of these characteristics reduce the quality of the diffusion barrier and can degrade the overall Cu interconnect reliability. Associated SEM images taken from an actual line structure, Figure 5(A) and from an isolated via structure, Figure 5(B), reveal both line bottom roughness and via profile damage, respectively. As compared to dense low k dielectrics, the ULK damage from sputtering is much greater; this makes the conventional integration scheme of Fig's. 4A-4E quite challenging. Figure 6 shows the severe dielectric damage in a ULK/Cu interconnect as a result of Ar^+ sputtering.

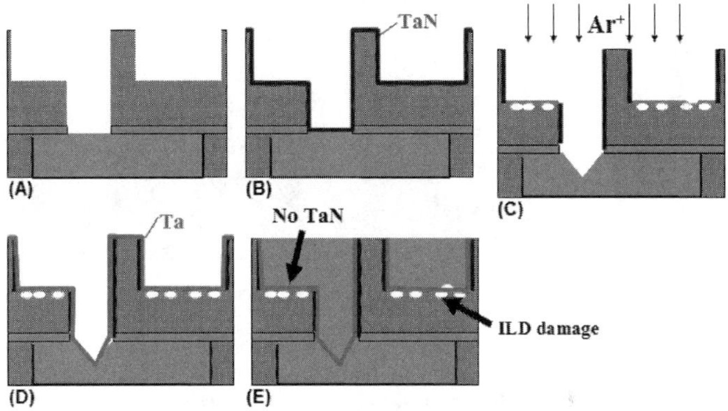

Fig. 4. Conventional "via punch-through" integration scheme.

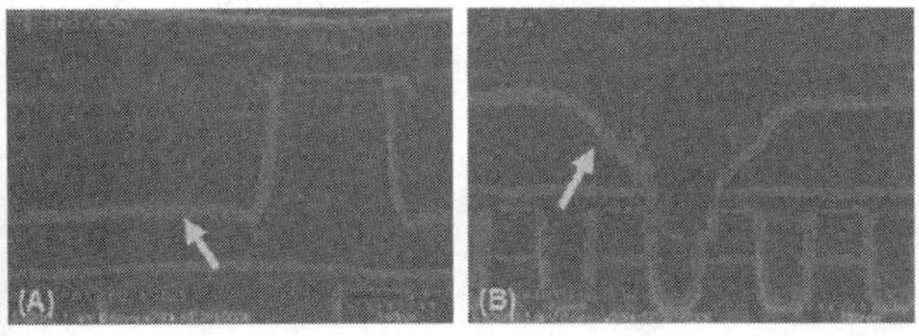

Fig. 5. SEM sections of (A) a line and (B) an isolated via in Cu/dense low-k, fabricated using the conventional "via punch-through" integration scheme.

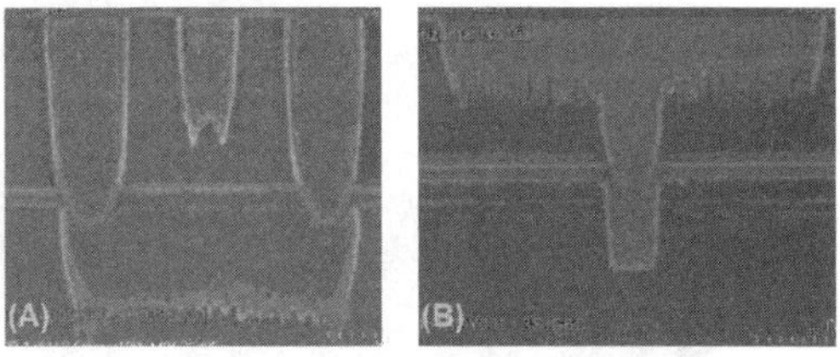

Fig. 6. SEM sections of (A) a via chain link and (B) an isolated via in Cu/porous low-k, fabricated using the conventional "via punch-through" integration scheme.

In view of the concerns with the current integration approach, a new enhanced via integration process has been developed which decouples via gouging from trench patterning, thereby eliminating the dielectric damage. This enhanced via integration approach begins with a TaN barrier deposition after via contact patterning, Fig's. 7(A), 7(B). The structure shown is then subjected to an Ar^+ sputtering process which removes the deposited TaN from bottom of the via and create a gouging feature, Figure 7(C). During the Ar^+ sputtering process, TaN located at the field area is removed, but the underlying dielectric is protected from sputtering damage by a hard mask layer. It's optional to form a thin liner layer on the wafer in order to prevent the exposed underlying Cu from subsequent process contamination. Next, the line level is patterned by conventional lithographic and etching processes, Figure 7(D), followed by liner/seed deposition, Cu plating and CMP, Fig's. 7(E), 7(F), to complete the interconnect structure.

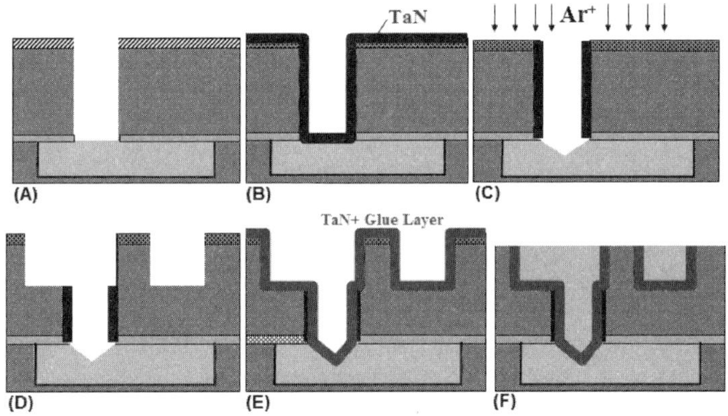

Fig. 7. Enhanced via integration scheme.

The novel enhanced via integration scheme described above overcomes dielectric/feature profile sputter damage because the line patterning process is carried out after forming the via gouging feature. SEM images of a via chain link, Figure 8(A), and of an isolated via, Figure 8(B) fabricated using this process sequence demonstrates the high integrity of the resulting cross section using this new approach. Figure 9 shows that this process can be carried to the extreme limit of via gouging, to the bottom of underlying interconnect, still with no observed dielectric damage. This further demonstrates technology extendibility of this enhanced via integration approach.

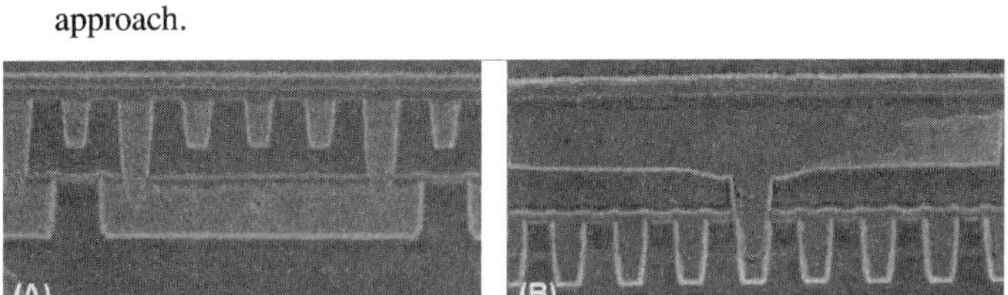

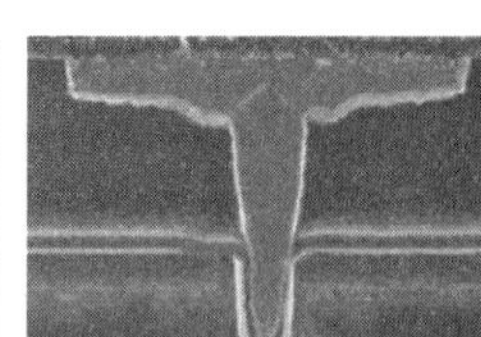

Fig. 8. SEM sections of (A) a via chain link and (B) an isolated via in Cu/dense low-k, fabricated using the enhanced via integration process.

Fig. 9. SEM section of an isolated via in Cu/porous low-k, fabricated using the enhanced via integration scheme.

EXPERIMENT

In this study, patterned wafers were prepared for electrical and reliability evaluations. CVD Ru liner was deposited in a 300mm deposition system integrated with degas, Ar^+ sputter, iPVD Ta(N), and iPVD Cu seed deposition chambers. Conventional Cu electroplating followed by CMP were then performed on the wafers. The feasibility of the enhanced via integration scheme including the conformal deposited CVD Ru adhesion layer for the Cu-low k system was evaluated by inline parametric measurements, electromigration (EM) and stress-migration (SM) tests.

RESULTS

Electrical Measurements

The line and via resistances from a representative lot are shown in Figure 10. In addition to a control split, this lot contained two experimental splits with the enhanced via integration process and conformal Ru adhesion layer. Split P2 used a more aggressive sputtering process than P1, which resulted in deeper via gouging. As compared to the control process, both experimental splits resulted in lower via resistances with tighter distributions. The deeper via gouging from the P2 process showed the lowest via resistances with the tightest distributions. Lower line resistance is also observed from both the experimental processes. Reasonable parametric yields were obtained from a comb-serpentine defect monitor structure as shown in Figure 11.

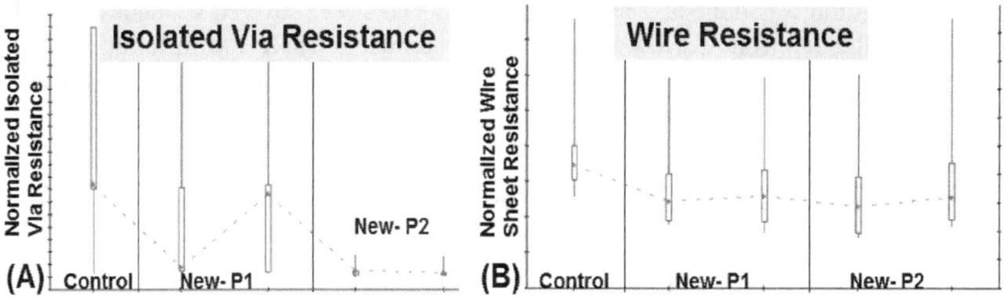

Fig. 10. Measurements of (A) via resistance and (B) line resistance. P1 and P2 are the new integration scheme with shallower (P1) and deeper (P2) via gouging, respectively.

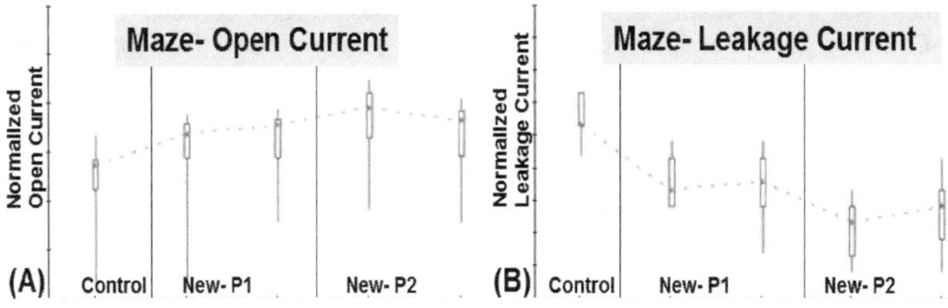

Fig. 11. Measurements of (A) open and (B) short at a comb-serp structure.

Electromigration Test

Electromigration testing was performed under a constant current density of 25 mA/μm^2 at 295°C. The test structure was a dual damascene multi-level via-line structure. The electron flow was forced through a single contact via to an 80nm wide line located either at the top for the upstream stress mode, or below for the downstream stress mode. Failure was defined as a 20% change in resistance, and the failure distribution was assumed to be lognormal. Figure 12 shows representative lognormal cumulative time to failure distributions for the two types of samples.

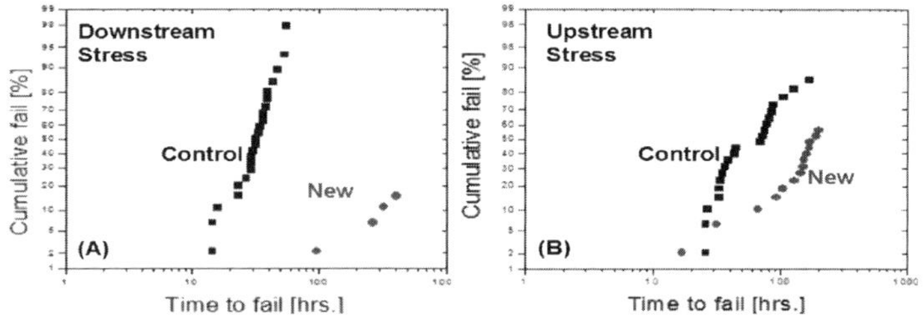

Fig. 12. Distribution of (A) a downstream and (B) an upstream EM fails.

Stress-Migration Test

Stress migration was done at 225°C for 300 hrs. The stress covered 4 major types of aggressive structures. Results from Cu/low-k hardware are shown in Table 1 with failure defined as a 20% change in resistance. The results shows similar failure rates for the control and the new process.

Table 1 Stress migration test results.

Stress Migration Test Results (# of fails) 67 structures/chip; 24 chips/wafer; 1 wafer/split					
Process \ Structure	**Via Chain**	**Plate Above**	**Plate Below**	**IsoVia-A**	**IsoVia-B**
Control	0	1	1	1	2
Experimental	0	1	1	0	3

CONCLUSIONS

To summarize, a new enhanced via integration scheme with a novel patterning approach coupled with a conformally deposited CVD Ru adhesion layer between TaN diffusion barrier and Cu interconnect was investigated for technology extendibility. This approach preserves the advantages of via/line contact interpenetration (gouging) while eliminating concomitant dielectric sputter damage, as especially beneficial for Cu/ULK integrations. Reduction of interconnect and via electrical resistances through this new approach have been demonstrated, along with supporting preliminary EM and SM performance. This new approach has potential to replace the current integration scheme with PVD Ta adhesion layer.

ACKNOWLEDGMENTS

This work was performed by the Research Alliance Teams at various IBM Research and Development Facilities.

The authors would like to thank TEL Technology Center, America, LLC in Albany, New York, for the support of CVD Ru metallization technology development.

REFERENCES

[1] D. Edelstein, *et al.*, IEEE Int. Interconnect Technology Conf., 9 (2001)
[2] D. Edelstein, *et al.*, IEEE Int. Electron Devices Meeting, 376 (1997)
[3] C.-C. Yang, *et al.*, MRS Adv. Metallization Conf., 213 (2004)
[4] D. K. Kwak, *et al.*, Electrochem. Soc. 502, 601 (2006)
[5] C.-C. Yang, *et al.*, IEEE Int. Interconnect Technology Conf., 9.5 (2006)
[6] D. Edelstein, *et al.*, IEEE Int. Reliability Physics Symp., 316 (2004)
[7] C.-C. Yang, *et al.*, IEEE Int. Interconnect Technology Conf., 7.7 (2005)
[8] C.-C. Yang, *et al.*, U.S. patent 7,122,462 (2006); R. Geffken, *et al.*, U.S. patent 5,985,762 (1998)
[9] A. Fischer, *et al.*, IEEE Int. Reliability Physics Symp., (2006)

Conference Proceedings AMC XXIV

Conformal CVD Co deposition for enhancement of Cu gapfill application

Jiang Lu, Hyoung-Chan Ha, Joseph AuBuChon, Paul Ma, Jennifer M Tseng, Sang-Ho Yu, and Murali Narasimhan
Applied Materials, Inc., Santa Clara, CA, 95054

ABSTRACT

Chemical vapor deposition (CVD) cobalt (Co) films were investigated as the Cu barrier seed enhancement layers for the application of 22nm node and beyond Cu interconnect technology. It was observed that the deposition of a 1.5nm-thick CVD Co layer between a physical vapor deposited (PVD) Ta barrier layer and a PVD Cu seed layer could effectively improve the electroplating (ECP) Cu gap-fill performance in the small geometry trench/via structures. The CVD Co enhancement layer could also significantly improve the electromigration (EM) resistance of the Cu interconnects.

INTRODUCTION

Cu-based interconnect has been used in the leading edge integrated circuits because of its low resistivity and improved electromigration resistance. In the current Cu metallization scheme, a thin continuous Cu seed layer is directly sputter-deposited on the barrier stack, i.e., PVD Ta/TaN, followed by ECP Cu [1]. However, with the aggressive shrinking of via and trench feature sizes beyond the 45nm technology node, it is increasingly more challenging to form a continuous, symmetric, and conformal sputtered Cu seed layer. This will likely lead to the formation of voids in ECP Cu interconnects. Transition and noble metals deposited by CVD and atomic layer deposition (ALD) methods can form a continuous and conformal seed enhancement layer in the narrow via/trench structures. The use of this enhancement layer can improve the adhesion [2] and surface nucleation of subsequent Cu metallization. Figure 1 illustrates the potential benefits of using an enhancement layer approach. For example, CVD/ALD Ruthenium (Ru) had already been widely studied for this application due to its many superior material properties [3-5]. It was also reported that CVD/ALD Co, a transition metal, could be used as enhancement layers to promote the adhesion and nucleation between Cu and its barriers [6, 7]. In this work, we propose to deposit a thin CVD Co enhancement layer between a PVD Ta/TaN barrier and a sputtered Cu seed layer to improve the gap-fill performance of ECP Cu metallization.

EXPERIMENT

The properties of various CVD/ALD Co films had been previously reported by many researchers [8-10]. In this work, the CVD Co film was deposited at temperatures below 200°C, which is essential to back-end metallization. Direct and remote NH_3 or H_2 plasma post-treatment was also performed to modify the films properties and minimize the carbon content.

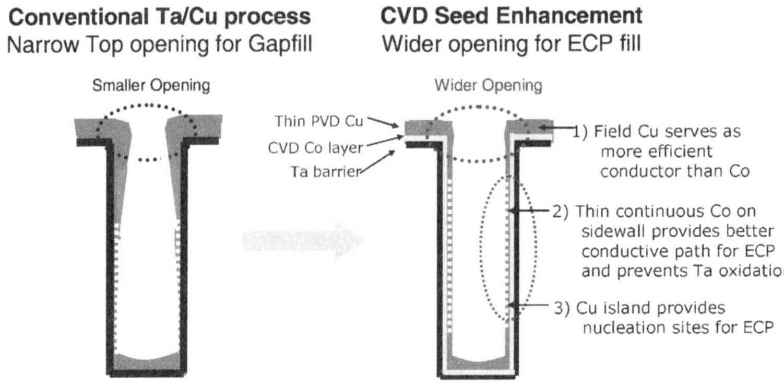

Figure 1: Schematic of conventional PVD barrier/seed gap-fill and the CVD seed enhancement approach

The effects of CVD Co enhancement layer on ECP Cu gapfill performance were examined using 50nm via and 35nm trench structured wafers. The step coverage of the CVD Co film was studied using 35nm trench structures wafers. CVD Co enhancement layers with different process conditions were in-situ deposited between a PVD Ta barrier and a sputtered Cu seed layer in an Applied Material Endura® Cu barrier seed system. Process parameters, including CVD Co thickness, CVD Co process temperature, direct and remote NH_3 or H_2 plasma post-treatment conditions, and sputtered Cu seed thickness, were varied to optimize the ECP Cu gapfill results.

CVD Co thickness and uniformity were determined by x-ray fluorescence (XRF), film density was determined by x-ray reflectometry (XRR), film crystal structure was investigated by grazing angle x-ray diffraction (XRD), and ECP Cu gapfill result was showed by cross-section scanning electron microscopy (SEM).

RESULTS AND DISCUSSION

CVD Co films with low bulk resistivity on the order of 30μΩ-cm were achieved. The bulk CVD Co film density was around 8.7g/cm^3 determined by modeling the XRR measurement data. Continuous thin (~1.5nm) CVD Co layer can be deposited on PVD Ta substrate without agglomeration as evidenced by the top-view SEM micrograph shown in Figure 2a. The step coverage of CVD Co deposited on PVD Ta is shown in Figure 2b and is close to 100% as the TEM illustrates. The excellent step coverage was obtained because the CVD Co process in this work was operated in a surface-reaction controlled regime.

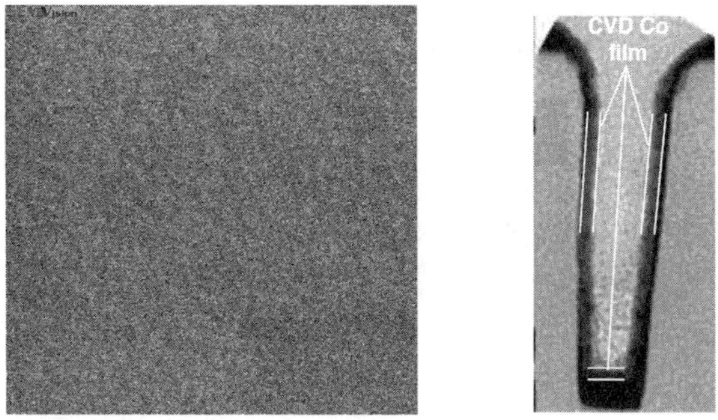

Figure 2: (a) Top-view SEM (field of view 1μm×1μm) of ~1.5nm CVD Co on 10nm PVD Ta and (b) TEM of CVD Co step coverage in 35 nm trench deposited on 10nm PVD Ta barrier.

The crystal structure of the 20nm CVD Co film was face-centered cubic (FCC) with preferred (111) phase orientation. An XRD spectrum is shown in Figure 3. Since the FCC Cu (111) phase is desired for interconnect applications due to low resistivity and better EM resistance[11], the formation of FCC Co (111) phase structure in CVD will be potentially beneficial for the Cu integration.

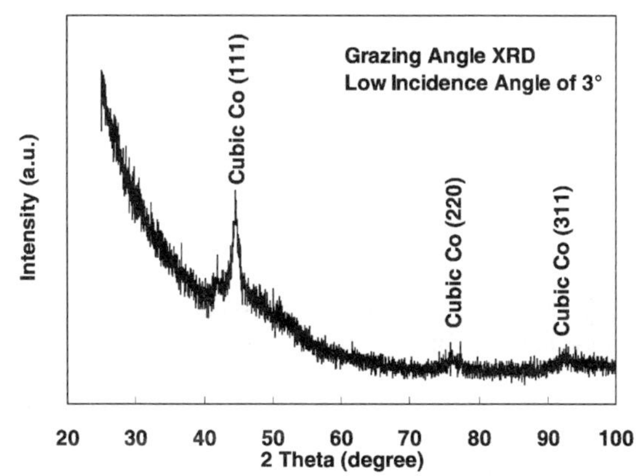

Figure 3: grazing angle XRD showed face centered cubic (FCC) Co (111) preferred crystal phase texture for a 20nm CVD Co film.

Figure 4 demonstrates the deposition of an ultra-thin, i.e., ~1.5nm, CVD Co layer between 10nm PVD Ta barrier and 30nm PVD Cu seed could dramatically improve the ECP Cu gap-fill performance in 50nm via structures. However, voids appeared if thinner, i.e., less than 1.0 nm, CVD Co layers were used, likely due to a discontinuous Co layer on the side-walls. With CVD Co enhancement layer, good gap-fill performance was also achieved using even thinner, i.e., <20nm, PVD Cu seed in 35nm trench structures. The presence of a continuous and conformal CVD Co enhancement layer may prevent the oxidation of the Ta barrier and provide a more conductive path on side-walls, which will help to enhance the bottom-up ECP Cu gapfill.

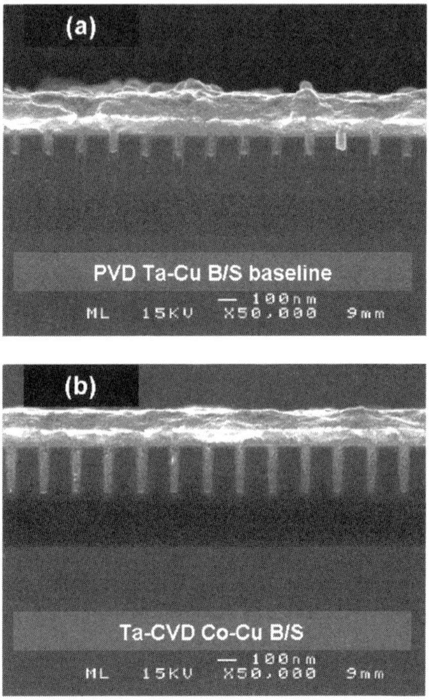

Figure 4: ECP Cu gap-fill in 50nm via, (a) 10nm PVD Ta-30nm PVD Cu barrier seed (B/S) baseline, and (b) 10nm PVD Ta-1.5nm CVD Co-30nm PVD Cu B/S.

In Figure 5, the electromigration performance using CVD Co is shown. As can be seen, the EM performance of the PVD-based TaN/Ta/Cu interconnects can be significantly improved with a CVD Co enhancement layer. The best EM performance was accomplished when the CVD Co layer was treated with optimized NH_3/H_2 plasma.

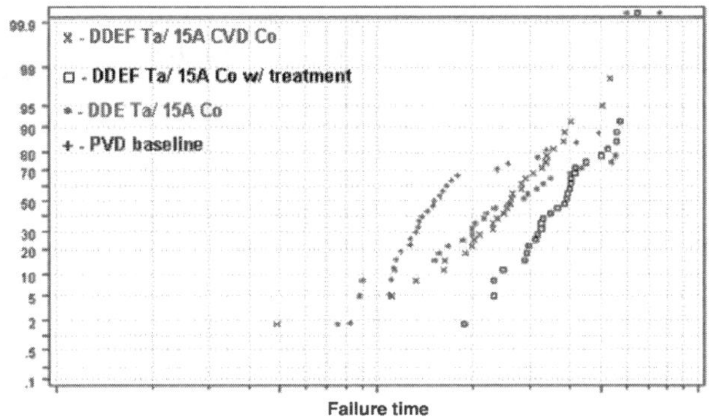

Figure 5: EM results with and without CVD Co

CONCLUSIONS

In summary, continuous and highly conformal CVD Co films were deposited and explored for the first time as Cu barrier seed enhancement layers. It was observed that an ultra-thin CVD Co enhancement layer could effectively improve the ECP Cu gapfill performance in small geometry trench/via structures. The presence of a CVD Co enhancement layer also significantly improve the electromigration resistance of the PVD-based TaN/Ta/Cu interconnects. CVD Co is a promising Cu barrier seed enhancement material for 22nm node and beyond Cu interconnects technology.

ACKNOWLEDGMENTS

The authors would like to thank Dr. Seshadri Ganguli for invaluable technical discussion, Ms. Ginger Nowlin for the top-view SEM analysis, Mr. Phillip Wang for the cross-section TEM analysis, Mr. Ming-Guang Liang and Ms. Haoyi Chen for the cross-section SEM analysis.

REFERENCES

1. P. Ho, U. Kelkar, A. Sundarrajan, J. Tseng, et al, *Advanced Metallization Conference*, p.421-427 (2005).

2. M. Lane, C. Murray, F.McFeely, P. Vereecken, and R. Rosenberg, *Appl. Phys. Lett.*, **83** (12), 2330 (2003).

3. S. Johnston, D. Feller, V. Dubin, and P. Moon, *Advanced Metallization Conference*, p.335-340 (2003)

4. O. Chyan, T. Arunagiri, and T. Ponnuswamy, *J. Electrochemical Soc.*, **150** (5), C347 (2003).

5. C.-C. Yang, T. Spooner, S. Ponoth, et al, *IEEE International Interconnect Technology Conference*, p.187-189 (2006)

6. Z. Li, R. G. Gordon, D. B. Farmer, Y. Lin, and J. Vlassakb, *Electrochemical and Solid-State Lett*, **8** (7), G182 (2005)

7. Y. K. Ko, B. S. Seo, D. S. Park, H. J. Yang, W. H. Lee, P. J. Reucroft, and J. G. Lee, *Electrochemical and Solid-State Lett.*, **6** (10), C141 (2003).
8. S.B. Kang, H.S. Kim, K.J. Moon, W.H. Sohn, et al, *IEEE Electron Devices Meeting, IEDM Technical Digest*, p. 20.6.1 - 20.6.4 (2003)
9. K. Kim, K. Lee, S. Han, T. Park, J. Kim, S. Yeom, and H. Jeon, *Japanese J. of Appl. Phys.*, **46** (8), p. L173 (2007)
10. H.-B.-R. Lee and H. Kim *Electrochemical and Solid-State Lett.,* **9** (11), G323 (2006)
11. S. Vaidya and A. K. Sinha, *Thin Solid Films*, **75**, 253 (1981).

Conference Proceedings AMC XXIV © 2009 Materials Research Society

New barrier metal selection principle for 32nm metallization and beyond

Koichi Motoyama[1], Akira Nakajima[1], Makoto Tohara[1], Tetsuya Kurokawa[2], Yasuaki Tsuchiya[1], Kunihiro Fujii[1], and Makoto Sekine[1]

[1]Process Technology Division, NEC Electronics Corporation, 1120, Shimokuzawa, Sagamihara, Kanagawa, 229-1198, Japan

[2]Test and Engineering Division, NEC Electronics Corporation, 1120, Shimokuzawa, Sagamihara, Kanagawa, 229-1198, Japan

ABSTRACT

This paper describes selection principles of new Cu barrier metals for 32nm node and beyond. Cu barrier properties for Zirconium (Zr), Hafnium (Hf), and Titanium (Ti) are evaluated and compared to explain three important characteristics of barrier metal, which are 1) heavier atomic weight, 2) low solubility in Cu, and 3) no compound formation with Cu at low temperature for performance improvement such as good step coverage and low wire resistance.

INTRODUCTION

Ta-based metals (Ta, TaN) have been widely used as diffusion barriers for Cu interconnects because of their high thermal stability and excellent barrier property against Cu diffusion. However, as devices shrink, Ta-based barrier metals suffer from reliability issues such as EM and SM [1]. To obtain highly reliable interconnect performance, Ti-based barrier metals are proposed [2, 3].

However, there are two drawbacks of Ti-based barrier metal deposited by ionized PVD. One is poor coverage compared with Ta-based barrier metals. Ti has a similar atomic weight to Argon (Ar) which is used as the sputtering gas for PVD, thus Ti is easy to be re-sputtered by Ar^+ during deposition and can lead to serious overhang (Fig.1). The other drawback is Ti has a high solid solubility in Cu which results in higher Cu resistance [4, 5].

To develop new barrier metal which has good coverage and high reliability, we have evaluated Zr and Hf which belong to the same group in periodic table as Ti. Both metals have heavier atomic weight and have less solid solubility in Cu.

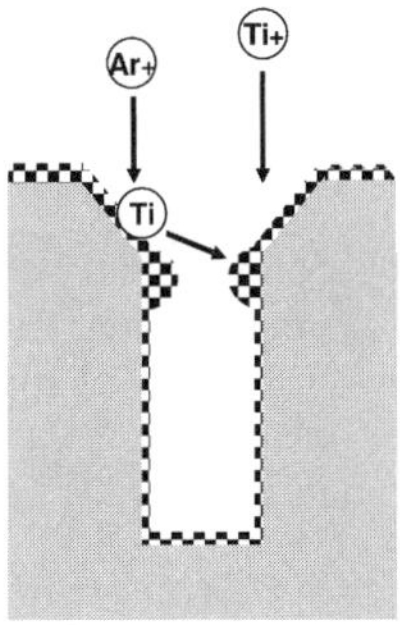

Fig. 1: Schematic figure of Ti coverage deposited by iPVD.

EXPERIMENT

Barrier metal deposition was performed with Applied Material's ionized PVD Extensa TTN chamber with 38kW target DC power, 750W wafer AC bias and low pressure.

To confirm the basic characteristics of Zr and Hf as the Cu barrier metal, resistivity and barrier property were evaluated. Cu40nm / barrier10nm / SiO_2 structures were prepared for sheet resistance (Rs) measurement and the barrier property evaluation by backside SIMS method. Annealing condition was 350°C-3hr. These samples were also examined by XRD after annealing.

Two-level low-k/Cu Dual Damascene (DD) structures were fabricated with our 55nm manufacturing process for e-Test. Wire to wire leakage current, wire resistance and via resistance of each barrier were measured.

DISCUSSION

Figure 2 shows the resistivity measurement result for each metal of 30nm thickness. All metals show lower resistivity than that of Ta (≈200uΩcm). Figure 3 is backside SIMS analysis result for Zr after 350°C-3hr annealing. No Cu diffusion in SiO_2 was evident, although Cu diffusion into Zr was observed. Ti and Hf have also been confirmed to have good barrier property by the same method.

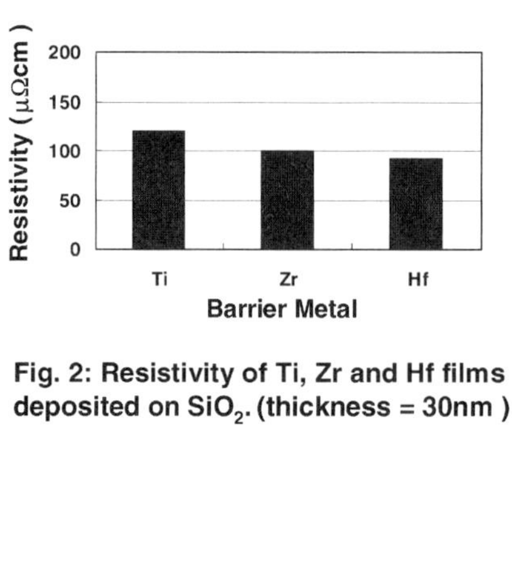

Fig. 2: Resistivity of Ti, Zr and Hf films deposited on SiO_2. (thickness = 30nm)

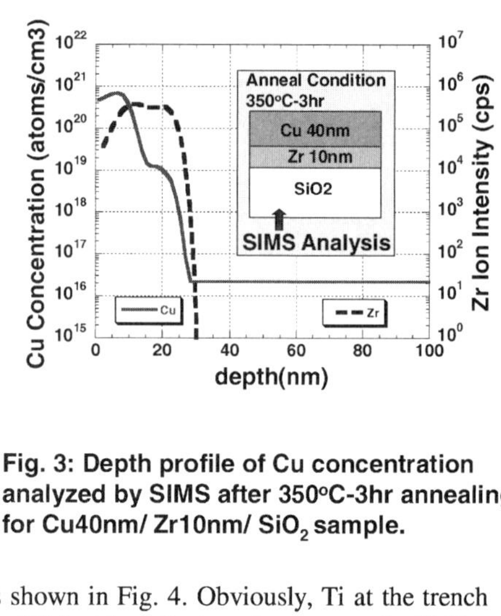

Fig. 3: Depth profile of Cu concentration analyzed by SIMS after 350°C-3hr annealing for Cu40nm/ Zr10nm/ SiO_2 sample.

Film coverage in trenches for each metal is shown in Fig. 4. Obviously, Ti at the trench corner is re-sputtered by Ar^+ and results in overhang. Dielectric at the trench corner under Ti film is also damaged. Dielectric damage can lead to reliability degradation such as TDDB. On the other hand, good coverage profiles were demonstrated by Zr and Hf due to their limited overhang and dielectric damage. Figure 5 shows the top view of via and via opening diameter after barrier deposition. Larger via opening was obtained as atomic weight increases (Ti→Zr→Hf). This result indicates that heavier atom makes wider process window against pinch off during Cu ECP.

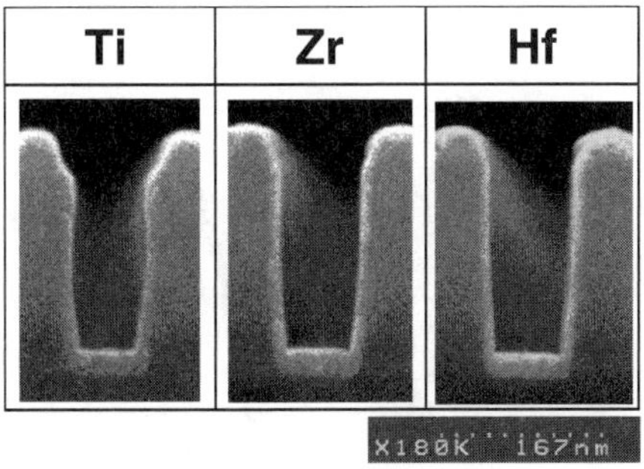

Fig. 4: Cross-sectional SEM image of film coverage in trench for each barrier metal.

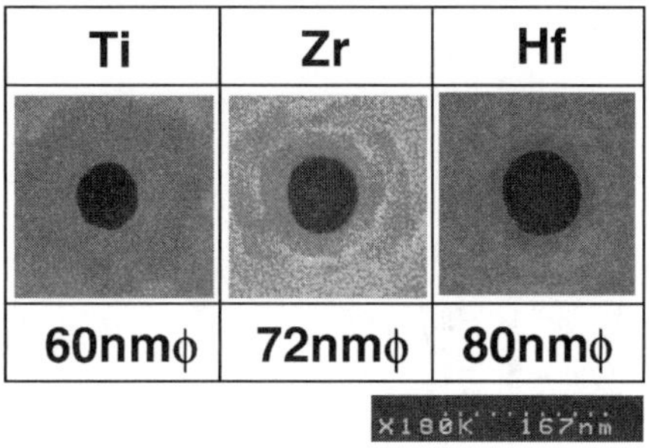

Fig. 5: Top image of via by SEM and via opening diameter after barrier metals deposition.

The cross-sectional view and bird's eye view of fabricated Cu wiring for each barrier after CMP are shown in Fig. 6. No defects such as void and scratch are observed and wiring height of each barrier is almost same.

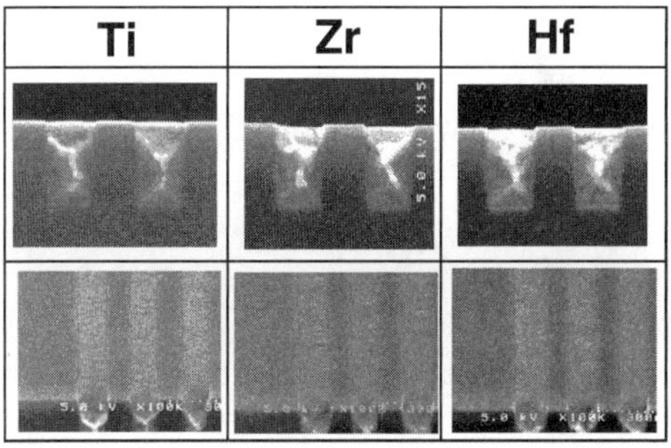

Fig. 6: Cross-sectional and bird's eye image of fabricated Cu wiring for each barrier after CMP.

Wire to wire leakage current for each barrier metal is shown in Fig. 7. Leakage current is low enough for all barrier metals. Based on these results, one can conclude that all barrier metals have adequate barrier property.

M1 wire resistance of 80nm line width for each metal type is shown in Fig. 8. Both Zr and Hf showed higher wire resistance than that of Ti. The distribution of 100k via-chain resistance for each barrier is compared in Fig. 9. All barrier metals show good via-chain yields. Via resistance of both Zr and Hf are higher than that of Ti.

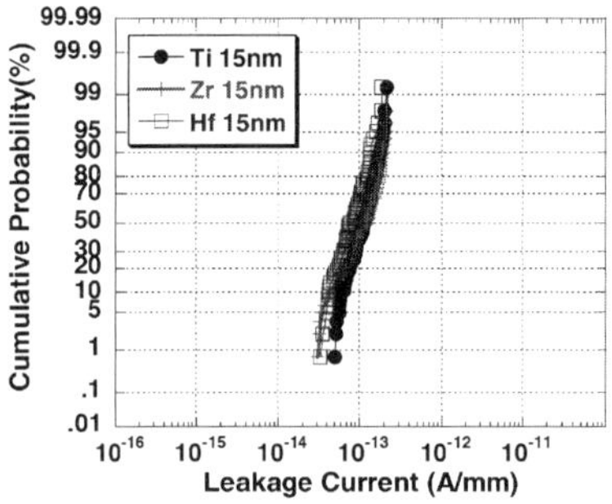

Fig. 7: Cumulative leakage current distribution of barrier metals. (L/S = 90nm/90nm)

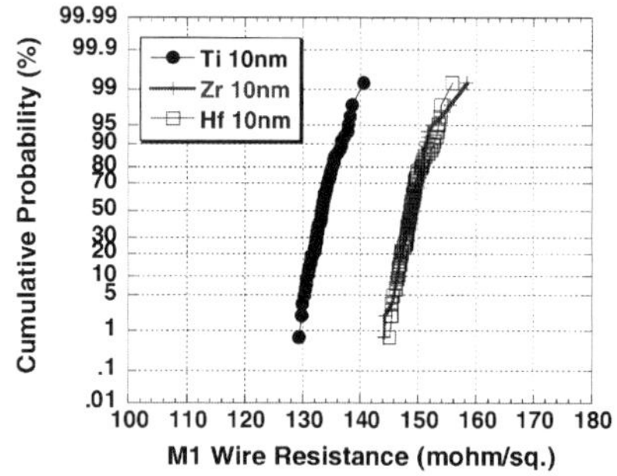

Fig. 8: Cumulative M1 wire resistance distribution of barrier metals. (Wire width = 80nm)

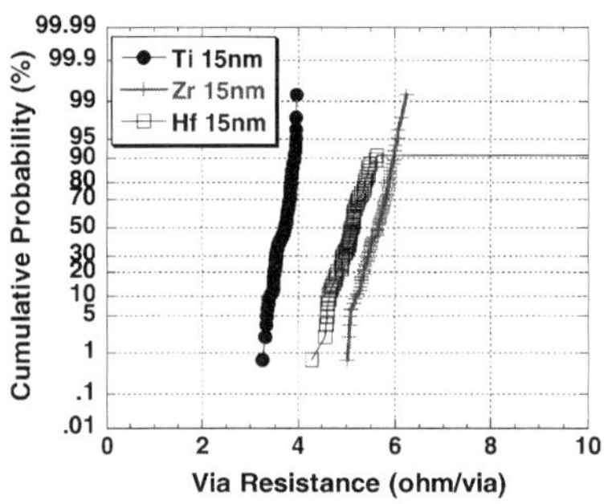

Fig. 9: Cumulative via resistance distribution of barrier metals. (via diameter = 75nm, via number = 100k)

Figure 10 shows the Rs shift of Cu on each barrier after 350°C-3hr annealing compared to Cu Rs shift on Ta. The slight increase of Rs for Ti is due to Ti doping in Cu film, because Ti can diffuse into Cu easily. On the contrary, both Zr and Hf show an excessive increase in Rs after annealing regardless their extremely low solubility in Cu. Thus, both Zr and Hf should have quite different reason of Rs increase from Ti case. This Rs result corresponds to the wire resistance result well.

To clarify the cause of this Rs increase, these samples were examined by XRD. Figure 11 is the XRD result of Cu / Hf structure with and without annealing. The Hf peak is changed to Cu_xHf_y peak after annealing indicated formation of Cu_xHf_y compound through annealing. The resistance of Cu_xHf_y is considered to be much higher than that of Cu. This is the root cause of the

Rs increase. Such a high resistance compound can be formed in via and trench by thermal budget during integration.

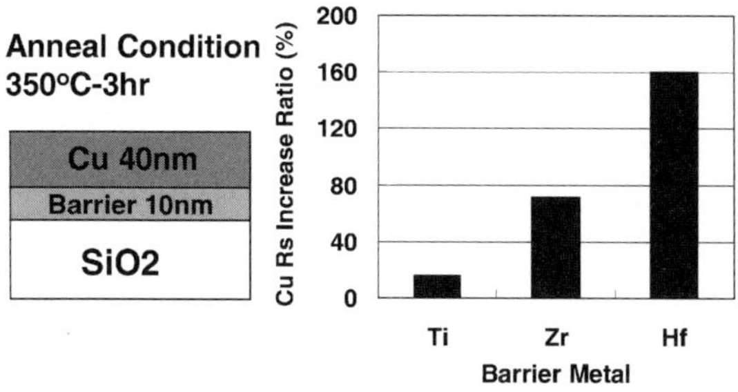

Fig. 10: Sheet resistance (Rs) increase ratio of Cu on each barrier after 350°C-3hr annealing compared to Cu Rs on Ta.

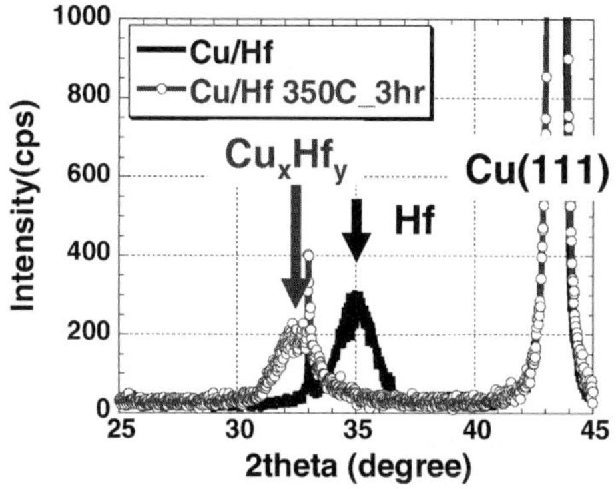

Fig. 11: XRD patterns for Cu / Hf structure with and without annealing.

Although it is well known that metal nitride has worse wettability to Cu than pure metal, HfN barrier was evaluated to suppress the reaction between Cu and Hf. Figure 12 is M1 wire resistance comparison between Hf and HfN. HfN shows much lower resistance than that of Hf. A similar trend is confirmed for the case between Zr and ZrN. Suppressing compound formation between barrier metal and Cu is important to maintain low wire resistance.

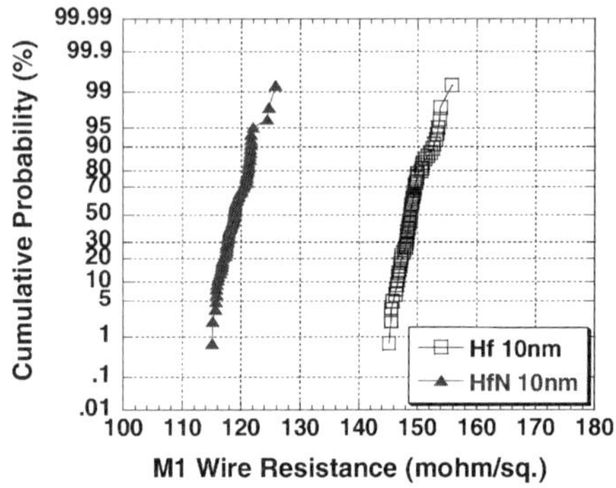

Fig. 12: Cumulative M1 wire resistance distribution of Hf and HfN used as Cu barrier metal with anneal. (Wire width = 80nm)

CONCLUSIONS

We have successfully fabricated 2-level low-k/Cu DD interconnects by applying Zr and Hf as the Cu barrier.

We have found there are three important characteristics of barrier metal required for 32nm metallization.

1) heavier atomic weight than Ar
2) low solubility in Cu
3) no compound formation with Cu at low temperature

ACKNOWLEDGMENTS

The authors would like to thank N. Sasaki and D. L. Diehl (AMJ), J. Lu and D. Saigal (AMAT) for their great support. The authors would like to express their gratitude to CMP group members of ACCRETECH for their help with Cu CMP.

REFERENCES

[1] Baozhen Li et. al., Microelectronics. Reliability 44, p365 (2004)
[2] A. Sakata et. al., Proc. of IITC2006, p101 (2006)
[3] B-S. Suh et. al., Proc. of IITC2005, p138 (2005)
[4] A. Sakata et. al., Proc. of IITC2008, p165 (2008)
[5] W. Wu et. al., Proc. of IITC2008, p202 (2008)

Conference Proceedings AMC XXIV © 2009 Materials Research Society

Electrochemical Cu Deposition in sub-100-nm Interconnects - Results for a New Model

Romy Liske[1], Axel Preusse[2], Susanne Wehner[2], Peter Kuecher[3], and Johannn W. Bartha[4]
[1]Fraunhofer Center for Nanoelectronic Technology CNT, Koenigsbruecker Strasse 180, 01099 Dresden, Germany
[2]AMD Saxony LLC & Co. KG, Wilschdorfer Landstraße 101, 01109 Dresden, Germany
[3]Fraunhofer Center for Nanoelectronic Technology (CNT), Koenigsbruecker Strasse 180, 01099 Dresden, Germany
[4]Semiconductor and Microsystems Laboratory, Dresden University of Technology, Noethnitzer Straße 64, 01187 Dresden, Germany

ABSTRACT

In high integrated circuits copper metallization is an established technology. In order to achieve the performance required for metallization schemes for the 65 nm technology node and beyond, void-free filling and a defined crystal structure of copper are crucial. Electrochemical copper deposition and the usage of plating bath additives is the process of choice. Although it is generally accepted that the plating bath additives are responsible for the bottom-up copper filling (so-called super filling), the mechanism of super filling is not completely understood. Existing models are mostly based on the mass transport of the additives including convection and diffusion which are -up to now- thought to play the main role in the super filling process.

In the present paper, a different view on the phenomenon of super filling is taken. Cross sectional micrographs reveal the individual effects of accelerator and suppressor on the resulting fill profiles. The change from an interrupted copper film profile to a continuous film deposition to super filling is shown. Simulation and calculation of electrical field lines suggest that a shielding of field enhanced sites is required to obtain a continuous copper film across a trench feature. The shielding is due to the adsorption behavior of suppressor molecules, which is verified by charge transfer measurements. Charge transfer measurements also show that the surface coverage of suppressor molecules can be adjusted by accelerator molecules.

INTRODUCTION

Copper metallization by electrochemical deposition of copper is an established process for the production to achieve ultra-pure, crystalline interconnects for high electromigration and mechanical stability. The typical electrochemical deposition process is characterized by two plating bath additives, suppressor and accelerator. Super filling with a higher rate of deposition at the bottom of the contact or trench occurs which is crucial to achieve void free copper interconnects (Figure 1).

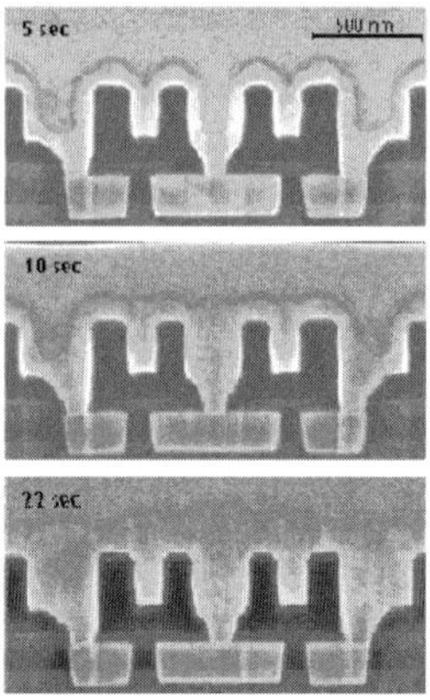

Figure 1. Partial fill experiments demonstrating the Cu super fill behavior in presence of suppressor and accelerator

The copper filling process is chemically driven by the active component of a suppressor which consists of polyethylene glycol, a long chained organic molecule with a mol mass of up to 10.000 g/mol. Due to its functional endgroups it is soluble in water and acids, while the ether groups possess free electron pairs which preferentially interact with Cu(I) ions [1,2]. Since Cu(I) ions are stabilized by chloride, adding small amounts of chloride to the Cu electrolyte is necessary for the suppressor to function as a surfactant [3,4].

The active component of the accelerator (bis(3-sulfopropyl)-disulfide or mercaptopropylsulfonic acid) is known to adsorb on Au and Ag surfaces by forming thiolate bonds. Controversy exists about the surface activity of accelerator on copper especially in acidic solutions [5]. Spectroscopic measurements did not identify a specific coordination of accelerator to a copper surface [6].

Describing the phenomena of super filling has been the focus of substantial research during the last years. The mass transport of plating bath additives is believed to play the main role in that process. Three models have been developed to explain the observed bottom-up fill behavior for electroplated copper.

(1) Concept of convection dependent surface blocking [7,8]. The concept considers a concentration gradient of a blocking additive (suppressor) across a feature. Due to convection, more suppressor is delivered to the feature top, while depletion of suppressor occurs in the feature interior. Thus, feature interiors are less protected against copper deposition than feature openings.

(2) Later on, the influence of additive diffusion and a second additive (accelerator) was incorporated into the super filling concept [9]. The so-called competitive adsorption model states that accelerator molecules diffuse faster into the feature interior than the bigger and less mobile suppressor molecules. Therefore, copper deposition is enhanced inside features by adsorbed accelerator and blocked at feature openings, where suppressor molecules are placed.

(3) The concept of curvature enhanced accelerator coverage [10] refers to the role of the accelerator. It states, that the growth velocity is proportional to the local accelerator surface coverage and the accelerator remains segregated at the metal/electrolyte interface during the copper deposition. Based on those assumptions the accelerator coverage increases at shrinking surfaces, i.e. feature bottom, which leads to an enhanced Cu deposition on those sections. At

expanding surface sections like shoulders, the accelerator coverage declines which results in a locally decreased Cu deposition.

All of those concepts are able to model super filling shapes using time-dependent mass transport calculations for accelerator and suppressor molecules or shape induced surface changes, but there exists discrepancies when applying those concepts to real copper filling of sub-100-nm structures.

In the present study the individual influence of accelerator and suppressor was investigated on the formation of three different copper fill shapes. Field line calculations and charge transfer measurements have been carried out, supporting the theory of a field shielding by adsorbed suppressor molecules. The surface coverage of suppressor is quantitatively determined and the individual impact of suppressor and accelerator as well as the synergetic effect between these two additives are discussed.

EXPERIMENT

The plating electrolyte used for all experiments was composed of 17 g/l $CuSO_4$, 170g/l H_2SO_4 and 50 ppm chloride. Two commercial additives called accelerator and suppressor were added to the electrolyte according to the referred concentrations. Partial fill experiments and charge transfer measurements were performed with a PAR galvanostat/potentiostat. All experiments were carried out at room temperature.

Partial fill experiments were performed potentiostatically in a Kocour Hull cell containing 250 ml of electrolyte. The solution was stirred magnetically at 50 rpm. A copper anode was used as counter electrode. The samples used for partial fill experiments possessed an area of 10 x 7.5 cm^2 and were cut from a 300 mm structured test wafer. The test wafers were processed at the AMD manufacturing site, Dresden. All samples shown in this paper were taken from the same batch and experienced the same pretreatment. The patterned wafers were physical vapor deposited with a conventional Ta based barrier layer and a copper seed layer. For the Hull cell partial fill experiments a current of 600 mA (dc) was applied.

SEM micrograph cross sections from via/trench features were taken to verify the plating fill behavior. Cross sections were prepared by focused ion beam technique. In order to conserve the fragile partially filled features, the specimens were covered with a chromium-platinum protection film deposited by e-beam evaporation prior to the FIB preparation. The shown micrographs correspond to plating current densities of 5 and 10 mA/cm^2 estimated by using a conventional Hull scale.

Charge transfer measurements were carried out using a conventional three-electrode cell (PAR, Model 616). The working electrode consisted of a platinum rotating disc electrode with a diameter of 5 mm. The reference electrode was a saturated Ag/AgCl electrode with a potential of 0.197 V versus the standard hydrogen reference electrode. A copper rod constituted the counter electrode. All potentials are reported with respect to the Ag/AgCl electrode. The charge transfer was estimated by measuring the current density during a potential sweep between -0.5 V to 1.5 V at a rate of 100 mV/sec. Rotation rates of 100 rpm were adopted.

DISCUSSION

Influence of individual additives on fill shapes

Figure 2 shows cross-sectional scanning electron micrographs of via/trench features. The features were half filled with copper using different combinations of accelerator and suppressor.

In absence of both additives a directional copper growth is observed (Figure 2a). Copper is deposited on horizontal surfaces, i.e. on the feature bottom and on the feature top, while the feature sidewalls stay uncoated. Adding accelerator alone has no influence on the filling behavior (Figure 2b). There is no accelerated copper deposition inside the features due to the presence of accelerator and again, the sidewalls are free of copper deposits. With the addition of suppressor but in absence of accelerator the fill behavior changes significantly (Figure 2c). Conformal copper deposition is obtained with equal deposition rates at bottom, sidewall and the trench top. The conformal fill behavior indicates that suppressor adsorbs uniformly across the feature. Only in presence of both additives suppressor and accelerator, super filling is achieved (Figure 2d).

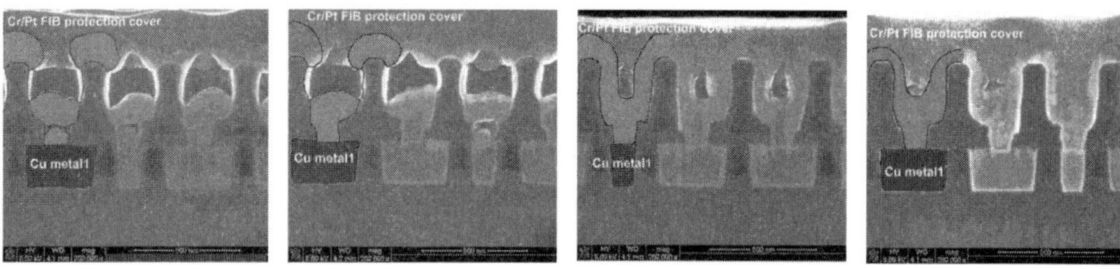

a) $CuSO_4 + H_2SO_4 + Cl^-$ b) $CuSO_4 + H_2SO_4 + Cl^- +$ Acc c) $CuSO_4 + H_2SO_4 + Cl^- +$ Supp d) $CuSO_4 + H_2SO_4 + Cl^- +$ Acc + Supp

Figure 2. Cu partial fill studies from indicated electrolyte compositions. Cu plating was performed in a Hull cell at 10 mA/cm² for 10 sec. (For more clarity copper deposits are traced in the left part of the pictures.)

Influence of additive adsorption on the electrical field

According to Faraday's law, the copper deposition rate is proportional to the local current density. So, super filling requires a higher current density at the feature bottom than at the feature top. Usually, that is not the case. The plots in Figure 3 show the ohmic potential drop in the electrolyte and the field streamlines for trenches with an aspect ratio of 2.

In the regular case, when a constant potential across a trench is considered, the electrical field lines sparely reach the trench bottom, the current density is not conformal and field enhanced sites occur at the trench opening. That case reflects the situation of an additive free electrolyte. In order to achieve a higher current density inside the trench, the field lines need to be distracted from the trench top towards the trench bottom. That distraction is supposed to be realized by adsorbing additives, e.g. suppressor molecules. Through adsorption, field enhanced sites can be electrically shielded, thereby increasing the current density inside the trench.

However, to simulate a conformal fill shape or super filling behavior requires a more complex model, e.g. the incorporation of a charge transfer resistance, which was not done at this stage.

1.0 0.9 0.5 0.2

Figure 3. Ohmic potential drop (colored surface) and electric field streamlines (white lines) inside a trench calculated for various degrees of potential fluctuations across trench bottom and trench top. Numbers indicate the potential ratios between trench top and bottom.

Surface coverage of adsorbed suppressor and the impact of accelerator

The conformal fill profile in presence of suppressor requires a sufficient bulk concentration of suppressor. Lower suppressor concentrations yield in a disrupted copper film formation on feature sidewalls, indicating an incomplete surface coverage. After a complete coverage is reached any further increase in suppressor concentration has no influence on the conformal fill behavior (Figure 4).

a) 0.1 ml/l Supp b) 0.5 ml/l Supp c) 1.0 ml/l Supp

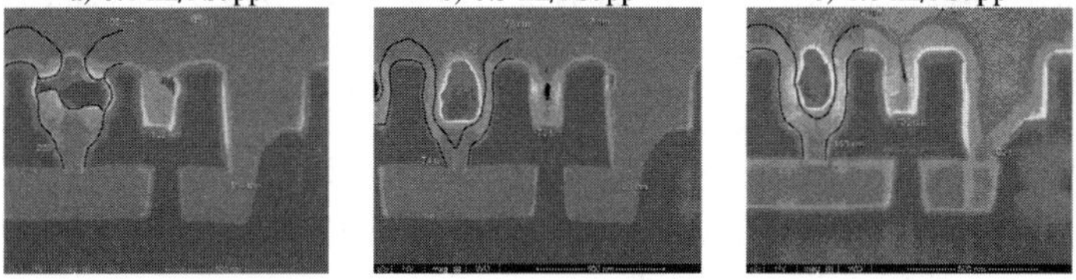

Figure 4. Impact of suppressor concentration on conformal fill behavior. Cu plating was performed in a Hull cell at approx. 5 mA/cm² for 15 sec. (For more clarity copper deposits are traced in the left part of the pictures.)

The adsorption behavior of suppressor in absence and presence of accelerator was examined by measuring the charge transfer versus copper deposition potential (Figure 5). The charge transfer is proportional to the amount of copper ions that pass the electrolyte/electrode interface and react to incorporated copper atoms. For reference, an accelerator and suppressor free base electrolyte was measured (solid line). In absence of suppressor, the addition of accelerator to the electrolyte does not show a remarkable change in charge transfer (Figure 5a). So, there is indeed no evidence that accelerator molecules influence the electrolyte/electrode interface, which is in good agreement with the missing proof of accelerator adsorption on copper surfaces.

a) b) c)

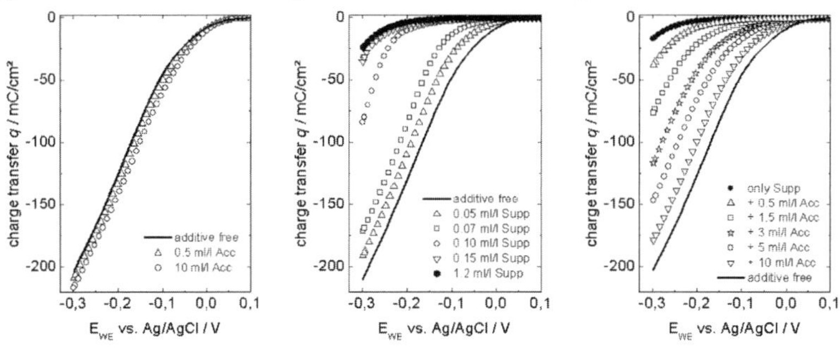

Figure 5. Impact of accelerator and suppressor on measured charge transfer in the copper deposition range: a) Comparison of accelerator free and accelerator containing electrolyte; b) Addition of suppressor to accelerator free electrolyte; c) Addition of accelerator to suppressor containing electrolyte

In contrast to the accelerator, suppressor alters the charge transfer significantly (Figure 5b). Adding suppressor leads to a decreased charge transfer which is explained by the formation of an inhibiting adsorbed film. Thereby, a saturation concentration exists where a minimal charge transfer is obtained. In presence of this charge transfer suppressing film the addition of accelerator shows an effect (Figure 5c) and increases the charge transfer.

Based on Frumkin's adsorption theorem [11] the electrode area that is covered by adsorbed additives can be calculated. The measured charge transfer through the electrolyte/electrode interface is modeled by two parallel capacitors. One capacitor C_0 is filled with additive free electrolyte which is considered to be not surface active, and the other capacitor C_1 is filled with a saturated suppressor electrolyte. At an applied potential E, the two parallel capacities correspond to a suppressor surface coverage of 0 and 1 at a maximum charge transfer q_0 and at minimum charge transfer q_1, respectively. Any measured charge transfer q between those extreme values can be converted into a surface coverage θ by the following equation:

$$q = (C_0 + C_1)E = q_0(1-\theta) + q_1\theta \qquad 1$$

As a result, the ratio of the electrode area that is covered by suppressor molecules to the total electrode area is obtained (Figure 6 left). There exists a dependence on potential for low suppressor concentrations, but the saturated suppressor concentration that leads to a complete surface coverage is independent of potential. Adding accelerator to an electrolyte that is already saturated with suppressor, results in a decreased surface coverage of adsorbed suppressor (Figure 6 right). Thus, the accelerator helps the suppressor to desorb from the surface which opens up surface sites for copper deposition.

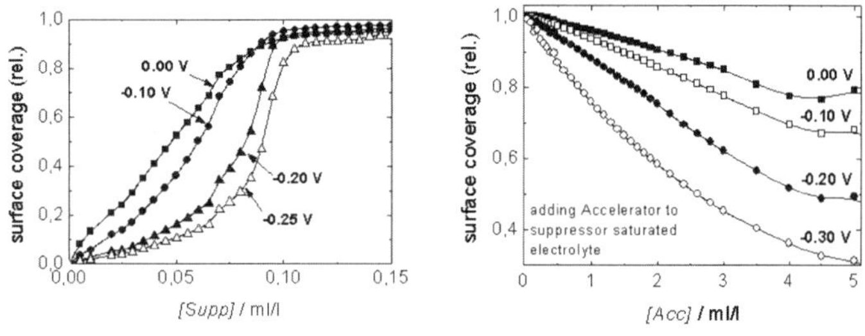

Figure 6. Formation of a closed adsorbing suppressor film (left) and the influence of accelerator on the decrease of the suppressor surface coverage (right)

CONCLUSIONS

Partial fill studies on copper interconnects showed that super filling is based on the synergetic effect of the plating bath additives suppressor and accelerator. Without additives a discontinuous copper deposition on the trench top and bottom occurs, while sidewalls are not coated. That plating profile is the result of an uneven current density distribution which exhibits field enhanced sites at the trench top. Adding only accelerator does neither influence the plating profile nor the measured charge transfer. That supports the thesis that accelerator does not change the electrolyte/electrode interface, i.e. there is no evidence for accelerator adsorption.

Upon the addition of suppressor to an accelerator free electrolyte, a conformal copper deposition is observed, which reflects the surfactant properties of suppressor. Both, partial fill experiments and charge transfer measurements indicate that a saturation concentration of suppressor exists, where a continuous surface film of adsorbed suppressor molecules is formed. From electric field calculations it can be concluded that suppressor molecules shield field enhanced sites.

The transition from a conformal filling to the required super filling is accomplished by accelerator but only in presence of suppressor. Accelerator decreases the suppressor surface coverage noticeably in dependence of accelerator concentration and electrode potential. From partial fill experiments it becomes obvious that suppressor desorption is enhanced at the trench bottom. Probably due to the electrical field a stronger suppressor adsorption takes place at the trench top than at the trench bottom and the accelerator impact becomes more feasible at the bottom.

ACKNOWLEDGMENTS

The authors would like to thank G. Grimm for SEM preparations and S. Bott for assisting in electrical field calculations. This work was financially supported by the Federal Ministry of Education and Research of Germany (BMBF).

REFERENCES

1. K. Doblhofer, S. Wasle, D. M. Soares, K. G. Weil, G. Weinberg, G. Ertl, *Z. Phys. Chem.*, 217, pp. 462-479 (2003)
2. M. Yokoi, S. Konishi, T. Hayashi, *Denki-kagaku-oyobi-kogyo-butsuri-kagaku*, vol. 52, pp. 218-223 (1984)
3. Z. V. Feng, X. Li, A.. A. Gewirth, *J. Phys. Chem. B*, 107, pp. 9415-9423 (2003)
4. D. Stoychev, C. Tsvetanov, *J. Appl. Electrochem.*, vol. 26/7, pp. 741-749 (1996)
5. A. Kudelski, *Langmuir*, 19, 3805-3813 (2003)
6. Z. D. Schultz, Z. V. Feng, M. E. Biggin, A. A. Gewirth, *J. Electrochem. Soc.*, 153, C97-C107 (2006)
7. P. C. Andricacos, C. Uzoh, J. O. Dukovic, J. Horkans, H. Deligianni, *IBM J. Res. Develop.*, vol. 42/4, pp. 567-574 (1998)
8. M. Georgiadou, D. Veyret, R. L. Sani, R. C. Alkire, *J. Electrochem. Soc.*, vol. 148/1, pp. C54-C58 (2001)
9. R. Alkolkar, U. Landau, *J. Electrochem. Soc.*, vol. 151/11, pp. C702-C711 (2004)
10. T. P. Moffat, D. Wheeler, D. Josell, *Electrochem. Soc. Interface*, vol. 4, pp. 46-52 (2004)
11. B.B. Damaskin, V.A. Safonov, O.A. Baturina, N.V. Safonov, *J. Electroanal. Chem.*, 550-551, pp. 3-11 (2003)

Conference Proceedings AMC XXIV © 2009 Materials Research Society

Copper Seed Feature Level Modeling

Umesh Kelkar, Daniel Lubben, Tza-jing Gung, Arvind Sundarrajan, Qian Luo, Yong Cao, Xianmin Tang, Maurice Ewert, and Murali Narasimhan
Applied Materials, Inc, Santa Clara, CA, 95054

ABSTRACT

This paper describes the feature level modeling of copper seed layer deposition. The feature level model is used in tandem with the reactor level models and the experimental data from Applied Material's EnCoRe II RF chamber. The model predictions were used to evaluate the extendibility and to provide guidelines to overcome the limitations of the current process and hardware. Both the model predictions and experimental data showed that the new process/hardware significantly improves the step coverage and gapfill performance for 30nm trenches.

INTRODUCTION

The physical vapor deposition (PVD) has been a workhorse of the semiconductor industry for several years to deposit high purity films used in forming multilayer interconnects. A thin stack of barrier (TaN, Ta) and seed (Cu) metal films is deposited before filling the feature with electroplating. It is extremely challenging to obtain a thin, conformal film inside the feature as the feature sizes shrink. Multiple deposition/removal steps and/or complex hardware are employed to extend the PVD to below 45nm technology node. Computational modeling and simulation is a powerful tool to enable faster product development at low cost.

Several authors performed reactor level modeling to enhance an understanding of the physics involved in the PVD plasma reactors [1-3]. These studies showed a correlation between process parameters such as power, pressure, bias, magnetic field on the fundamental plasma characteristics such as ion density, electron temperature, ion energy and angular distribution, etc. Feature level modeling predicts film profile evolution as a function of these fundamental plasma parameters [4-5]. A truly integrated reactor-feature level model is necessary to predict the influence of process/hardware parameters on the film profile evolution. Low pressure (<1mT), a three-dimensional magnetron geometry, and large gradients in the magnetic field present challenges for obtaining reliable quantitative results from the reactor level modeling of our copper seed chambers. This paper describes feature level modeling of copper seed deposition by partially taking inputs from the reactor level models and experimentally measured data. An iterative scheme is used to match the film profile predictions to the transmission electron microscope images. The validated the model is then used to assess the extendibility of the existing process/hardware as well as to develop new process/hardware for the below 32nm node.

EXPERIMENT/MODELING METHODOLOGY

Figure 1 shows a schematic of the EnCoRe II RF chamber used to deposit copper seed. High density plasma is generated near the target by using a rotating magnetron and by applying DC power to the target (typically 10kW-38kW). An inductively coupled coil (0.5kW-3kW) generates a secondary plasma to ionize the neutral atoms sputtered from the target and/or to create a pure argon plasma near the wafer. The wafer is placed on an electrostatic chuck, which can be independently biased by a RF power supply to control ion energy. The figure also shows a schematic of additional hardware used to control the neutral/ion angular distribution arriving at the wafer.

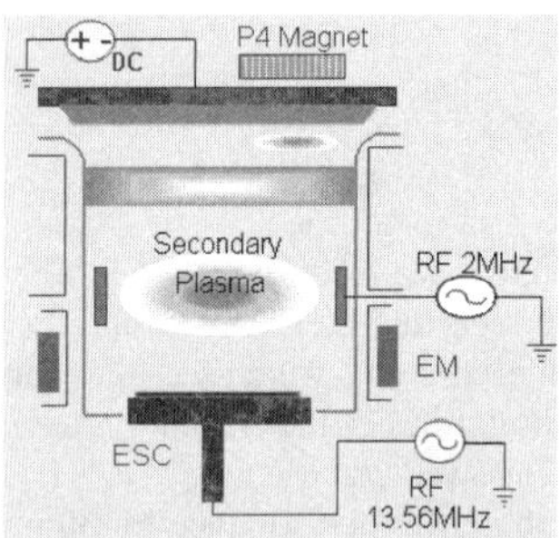

Figure 1: EnCoRe II RF Cu schematic

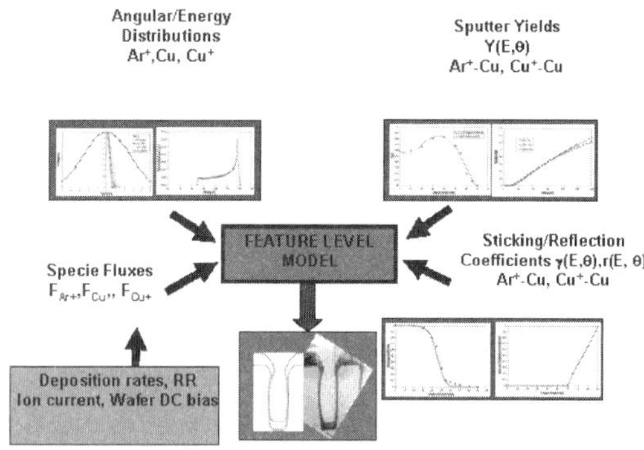

Figure 2: Feature profile modeling inputs

The particles arriving at the wafer consist of argon and copper neutrals as well as their ions. The film profile evolves due to a combination of angle and energy dependant surface processes such as sticking, reflection, re-sputtering processes. Figure 2 shows typical inputs necessary for simulating copper growth using the feature scale model. The energy dependent sputter yields for Ar^+/Cu and Cu^+/Cu are obtained from Matsunami et al.[7] The angular dependence of sputtered yield for Ar+/Cu and Cu+/Cu were obtained by curve fitting the data by Kress et al [8]. The angular and energy dependent sticking coefficients are taken from the molecular dynamics simulation by Kress et al. The Cu neutral angular distributions are obtained from line of sight models developed in-house. The Ar^+ and Cu^+ ion energy distribution are obtained by sheath models based on Miller and Riley [9]. The neutral/ion fluxes are approximated by using experimental data for deposition/etch rate, resputter ratio, wafer DC bias voltage and ion current measurement.

DISCUSSION

Figure 3 shows model-predicted and experimentally measured step coverage on a 65nm trench for two different wafer bias powers. The model accurately predicts bottom coverage as well as the amount of trench opening as a function of bias power. Note that the bottom coverage increases with increasing bias power due to narrowing of the metal ion angular distribution. The model was used to evaluate the extendibility of the current hardware/process.

Figure 4 shows the step coverage comparison with decreasing feature width keeping feature height the same. It is evident that the step coverage degrades substantially as the feature size shrinks. The top opening reduces substantially for the features less than 60nm in width and causes voids during subsequent electroplating.

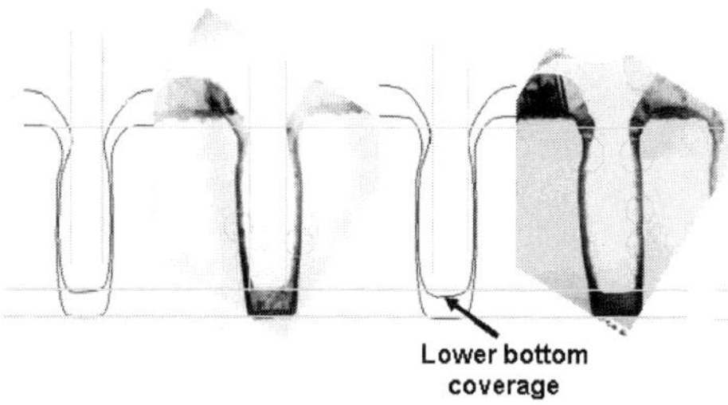

Figure 3: Model validate on 65nm trench

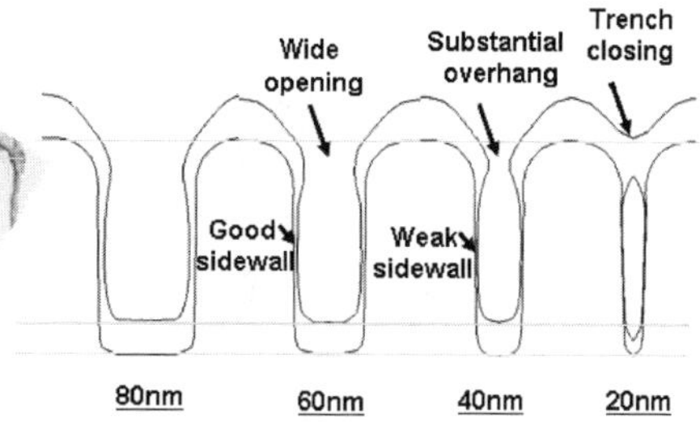

Figure 4: Step coverage degrades with decreasing feature size

Figure 5 shows that the sidewall coverage can be significantly enhanced by a pure Ar^+ etch process following the deposition for 65nm feature. The Ar^+ etch redistributes the material inside the feature and increases the sidewall coverage. It is found that there is an optimum ion energy that improves the step coverage without building the overhang. Figure 6 shows trench opening after etching as a function of amount of material etched for various values of Ar^+ energy. Note that, for ion energy of 75eV, the trench opening reduces as the material is etched. This overhang buildup is primarily due to re-deposition of the re-sputtered material from one side of the facet to the other side. For intermediate ion energy (150eV), the trench opening initially decreases and then increases once the etch amount exceed more than 60% of the initial deposition thickness. For energy above 250eV, the trench opening increases for etch amounts greater than 40% of the initial deposition thickness.

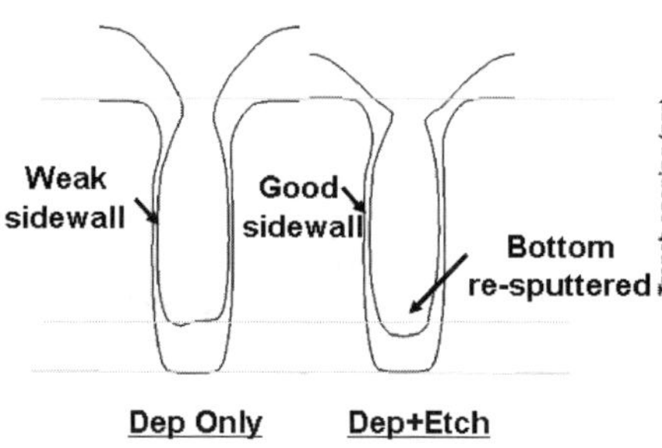

Figure 5: Re-sputtering improves step coverage

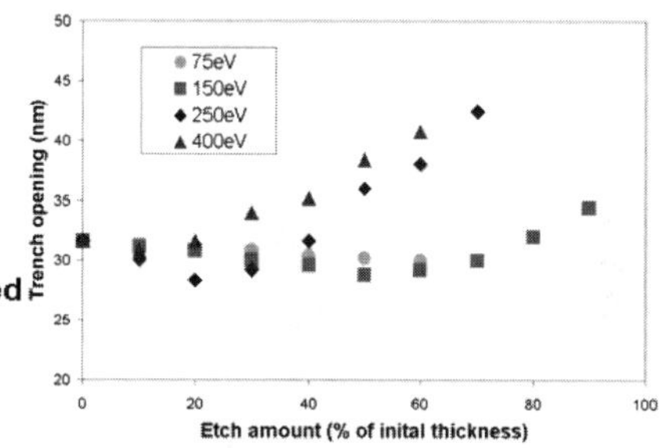

Figure 6: Effect of ion energy on trench opening

For the feature sizes below 45nm, the top opening significantly reduces due to re-deposition of the etched material on the opposite trench wall. Thus it is found that the step coverage can not be improved for reasonable etch amounts. Also, it is not feasible to use excessive etch amounts in dual damascene structures. Figure 7 shows that the excessive etching to keep the top open will damage the bevel between the trench bottom and via. The removal of seed and sometimes barrier film from the bevel region can cause serious stress and/or electro-migration failures.

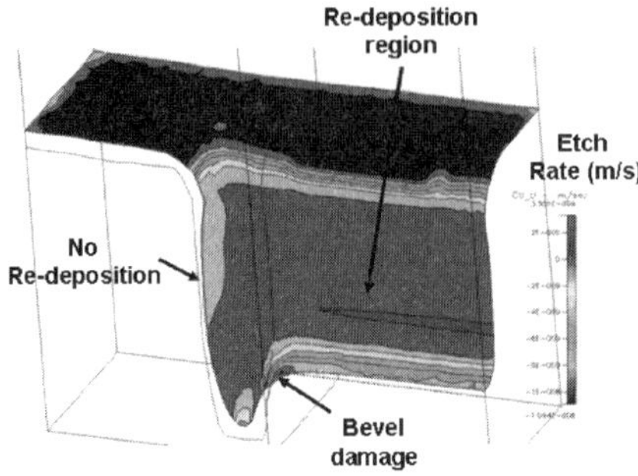

Figure 7: Bevel damage due to excessive etching

A new hardware was developed to extend the PVD copper seed process to sub 45nm feature sizes. The new hardware narrows the neutral angular distribution, increases the metal ion fraction, and reduces the deposition rate allowing higher bottom coverage and reduced overhang. Additionally, the new hardware enables novel process regime to significantly improve the amount of material that can be deposited inside very small features. Figure 8 shows that the new hardware substantially reduces the overhang and increases the bottom coverage on a 30nm (6:1 aspect ratio) trench.

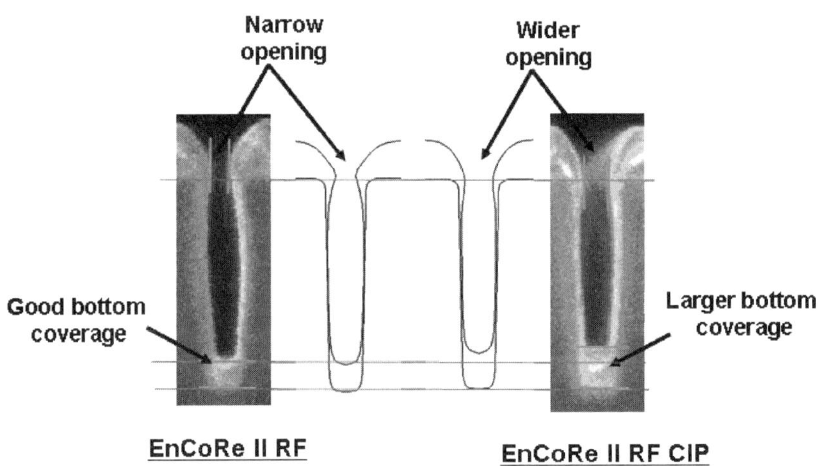

Figure 8: EnCoRe II RF CIP improves step coverage

The continuous thick sidewall results in very good gapfill performance. Figure 9 shows that the EnCoRe II RF CIP shows significantly improved post CMP defects compared to our baseline process.

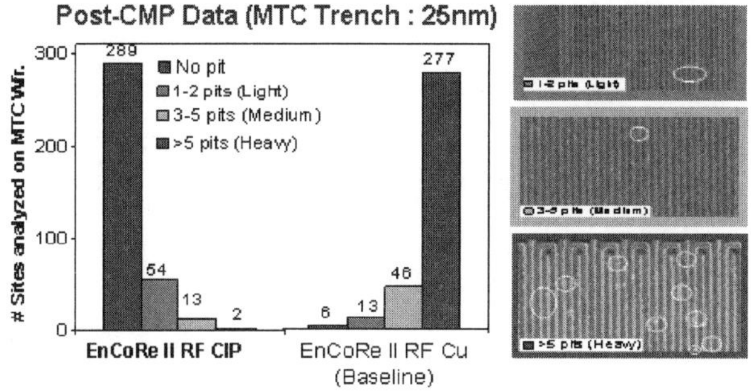

Figure 9: EnCoRe II RF CIP shows better defect and gapfill performance

CONCLUSIONS

The feature level modeling used in tandem with experimental data can be used to enhance understanding of the copper seed layer deposition. The feature level modeling was extensively used to provide guidelines for new process/hardware development and substantially reduced the developmental time and cost. EnCoRe II RF CIP showed superior gapfill and post CMP defect performance for 30nm trenches.

ACKNOWLEDGMENTS

The authors would like to acknowledge Jennifer Tseng and Mehul Naik from MTC group and Phillip Wang from TEM lab at AMAT for their valuable support.

REFERENCES

1. Stout et al., J. Vac Sci Tech B 20(6), Nov/Dec 2002, pg. 2421-2432
2. Arunachalam et al., J of Applied Physics, Vol. 90, No. 1, pg 64-73, July 2001
3. Kelkar et al., HT2005-72338, ASME/Interpak conference, July 17, 2005
4. Kersch and Hansen, J. Vac. Sci. Tech., A 20(4), July/Aug 2002, 1284-1294
5. Stout et al., J. Vac. Sci. Tech., A 21(3), May/Jun 2003, pg 596-606.
6. ESI-CFD, CFD-TOPO
7. Matsunami et al., Atomic Data and Nuclear Data Tables, Vol. 31, No. 1, pg 1-80, July 1984
8. Kress et al., J. Vac Sci, Tech A 17(5), Sep/Oct 1999, 2819-2825
9. Miller and Riley, J. Appl Phys, 82(8), Oct 1997, 3689-3709

Conference Proceedings AMC XXIV © 2009 Materials Research Society

Copper Contact Metallization using Ru-based Barrier Liners for 45 nm and 32 nm nodes : Reliability and Device Performance

Soon-Cheon Seo[1], Chih-Chao Yang[1], Chao-Kun Hu[2], Andreas Kerber[3], David Horak[1], Susan Fan[1], Bala Haran[1], Christian Lavoie[2], Veeraraghavan Basker[1], Donald Canaperi[1], Satyavolu S Papa Rao[2], Karen Petrillo[1], Steven Holmes[1], Matthew Smalley[1], Tuan Vo[1], James Kelly[1], Jason Cummings[1], Lahir Adam[2], James Demarest[1], Philip Flaitz[4], Zhibin Ren[4], Andreas Knorr[3], Sanjay Mehta[1], Charles Koburger[1], Dae-Gyu Park[2], Jean Wynne[1], Terry Spooner[1], Daniel Edelstein[2], Vamsi Paruchuri[1], Bruce Doris[1], and Dale McHerron[1]

[1]IBM at Albany Nanotech Research Center, Albany, NY, 12203

[2]IBM T.J. Watson Research Center, Yorktown Heights, NY, 10598

[3]Advanced Micro Devices, Yorktown Heights, NY, 10598

[4]IBM Semiconductor Research and Development Center, Hopewell Junction, NY, 12533

ABSTRACT

This work presents Cu contact metallization results using CVD Ru-containing liners for 45 and 32 nm nodes. CVD Ru-containing liner stacks utilized a non-halogen deposition process. CVD Ru-containing liners exhibited very good gap fill performance compared to PVD-based Ta liners. We selected CVD Ru-containing liners that passed barrier breakdown test and Cu diffusion test with 400°C 2 hours anneal. The selected CVD-Ru containing liners were then tested on fully integrated 45 nm and 32 nm node devices. Contact resistance of Cu using CVD Ru liners showed about one-third and one-fourth of W contact resistance for 45 nm and 32 nm nodes, respectively. No yield degradation was observed in 7.5-million contact chain current for 45 nm node and 100-thousand contact chain current for 32 nm node when BEOL thermal anneal was applied. The gate dielectric voltage breakdown test of Cu contact was comparable to that of W contact. Electromigration of Cu contact was also comparable to that of W contact. There was about 20% improvement in drive current with Cu contact metallization for 32 nm node device.

INTRODUCTION

Contact resistance becomes a key element in the device scaling as the total external resistance of CMOS devices becomes a limiting factor. As the contact size shrinks for the 45 nm or 32 nm nodes and beyond, tungsten contact resistance exhibits broad distributions and its magnitude increases significantly due to poor gap fill. It causes variations and delays in device performance. In order to avoid these issues, an alternative contact metallization scheme is required. This scheme involves lower resistivity material and void-free gap fill. Copper metallization for back-end-of-the-line (BEOL) applications using tantalum-based barriers[1] is well-established. Copper contact metallization was investigated using Ta- and Ti-based liners at 0.2 μm node and 90-150 nm contact holes with the aspect ratios of 2-3.[2-4] Metallization for high aspect ratio (AR>5) contact vias is significantly more challenging than that for BEOL dual damascene structures due to poor Cu seed coverage on the sidewalls. A Ru liner has been shown to be one of candidate liners for direct plating and a very good wetting layer for the thin copper seedlayer.[5,6] It provides a very good gap fill, electrical, and reliability performance on the BEOL structure.[5] In this paper, we present results of gap fill, parametric, and reliability tests on 45 nm and 32 nm device wafers using CVD Ru-containing liners.

EXPERIMENTAL RESULTS AND DISCUSSION

Blanket barrier breakdown test

Barrier breakdown tests were performed at the National Synchrotron Light Source (NSLS) beamline at Brookhaven National Laboratory. Wafers were ramped at 3°C/sec in purified He while we measured *in-situ* sheet resistance, light scattering, and evolution of x-ray diffraction (XRD) intensity of Cu peak during annealing. The blanket sample stacks were silicon / nickel silicide / barrier liner / copper seed. Fig. 1 shows the barrier breakdown test results. Barrier liners consisted of multi-stacks of TaN, Ti, Ti/TaN, Ru/TaN, TiN, or Ti/TiN followed by PVD Ta or CVD Ru. Thick TaN (2xTaN) increases the breakdown temperature for Cu diffusion. The barrier breakdown temperatures of some of the liners containing CVD Ru are comparable to that of the TaN/Ta stack.

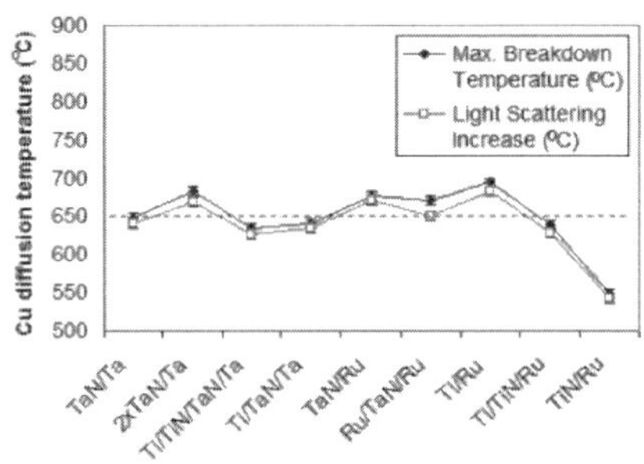

Fig. 1) Barrier breakdown test results for PVD Ta- and CVD Ru- containing liners

Cu diffusion test of Cu contact

Although blanket barrier property is reasonable, it does not guarantee Cu diffusion barrier performance in high aspect ratio contact holes. In general, the PVD process technique provides poor step coverage in high aspect ratio contact holes. In order to compensate sidewall and bottom liner coverage, one needs to deposit thick liner materials. However, it causes overhang at the top corner and results in poor gap fill. Cu diffusion test for 45 nm node contact holes on a blanket NiSi layer was studied with thermal anneals for four hours at 400°C in an N_2/H_2 ambient. Fig. 2 shows the TEM images of Cu contacts using a) liner A and b) liner B. Liner A did not show any indication of Cu diffusion to NiSi. However, severe Cu diffusion was observed with liner B indicating that liner B was not a good barrier.

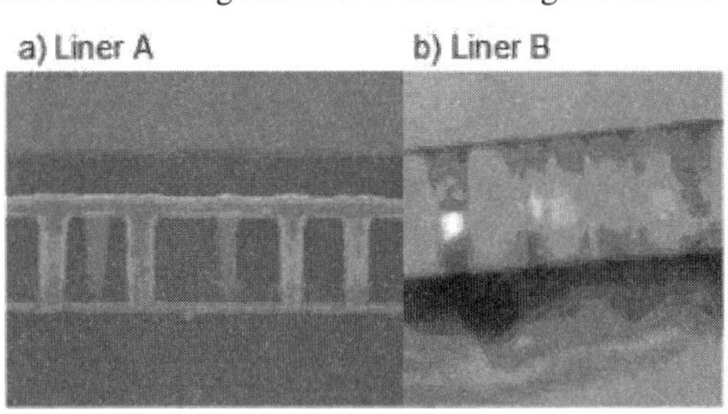

Fig. 2) SEM and TEM images of 45 nm node Cu contact holes using a) Liner A and b) Liner B on the blanket nickel silicide film after thermal stress (400°C for 4 hours).

Gap fill

Gap fill performance of PVD Ta- and CVD Ru-containing liners were compared in Fig. 3 using high aspect ratio structure where bottom CD was about 35 nm and via height was about 300 nm. These SEM cross sections were obtained after thermal stress at 400°C for 2 hours. CVD Ru-containing liner demonstrated a superior gap fill performance over PVD Ta-based liner in the high AR contact holes. No visible Cu diffusion to nickel silicide was observed.

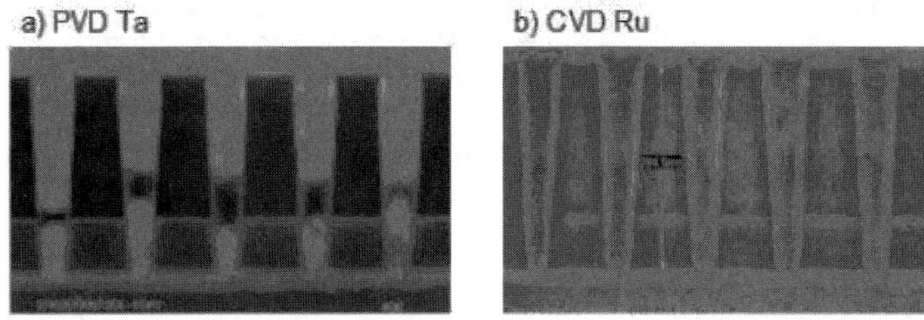

Fig. 3) Post-thermal anneal (400°C for 2 hours) gap fill performance for high aspect ratio contact holes (bottom CD ~35 nm and via height ~300 nm). CVD Ru-containing liners demonstrate the advantage of good gap fill in high AR contact holes as compared to PVD Ta-based liners.

Contact resistance

Based on Cu diffusion barrier test and gap fill tests, we have narrowed down CVD Ru-containing liner options for Cu contact metallization for processing a fully integrated FEOL device. The choice of CVD Ru-containing liner stacks was a halogen-free liner deposition process providing good barrier performance against Cu diffusion. Integrated Cu contact metallization using CVD Ru-containing liner was built into 45 nm and 32 nm devices to investigate the contact resistance difference between Cu and W. High aspect ratio contact resistance showed larger resistance reduction with Cu contact than that of low AR. In Fig. 4, the Cu contact resistance of a 45 nm node contact hole landing on active area (AR> 5-6) was about one-third of W contact resistance and that landing on gate (AR ~2-3) was about a half of W contact resistance. As the contact hole shrunk for 32 nm node, the contact resistance of Cu contact was reduced down further and it was only one-fourth of that of W contact (Fig. 5). Note that the resistance distribution of Cu contact was much tighter than that of W contact.

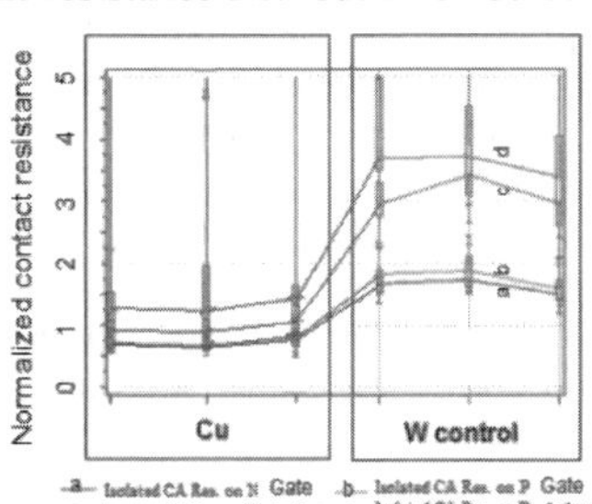

Fig. 4) 45 nm node contact resistance comparison between Cu and W contacts landing on gates (a & b) and active areas (c & d).

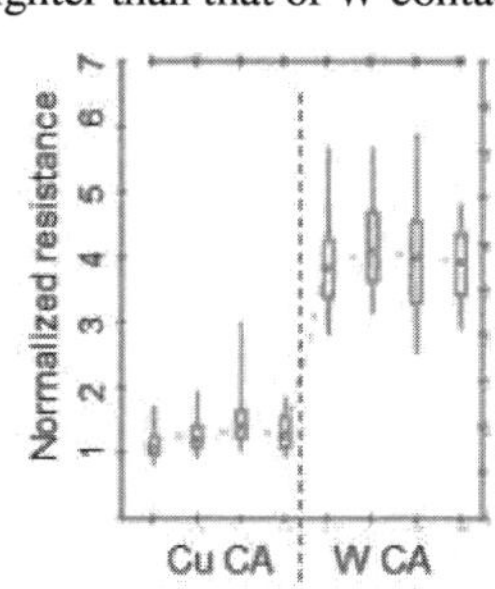

Fig. 5) 32 nm node contact resistance of Cu and W landing on active area.

BEOL thermal stress evaluation

The via chain resistance degradation of Cu contact was checked after building five BEOL metal levels and including a polyimide cure (375°C for 30 min). In Fig. 6, via chain currents (inversely proportional to resistances) for chains with 7.5 million contact links were measured at M1 and LT levels. This LT level was M5 with final passivation and included a polyimide anneal. We evaluated two different CVD Ru-containing liners. A1 liner showed some degradation in via chain current while A2 liner showed no degradation in via chain current following this multi-level build. A2 liner option enhanced gap fill performance.

A2 liner was used for 32 nm node contact metallization. Fig. 7 shows the pre- and post-thermal stress a) contact resistance and b) contact chain current of Cu contact at 32 nm node. The thermal stress was equivalent to building ten BEOL metal levels including a polyimide cure anneal (375°C for 30 min). There was no significant change in contact resistance and via chain current due to this thermal stress.

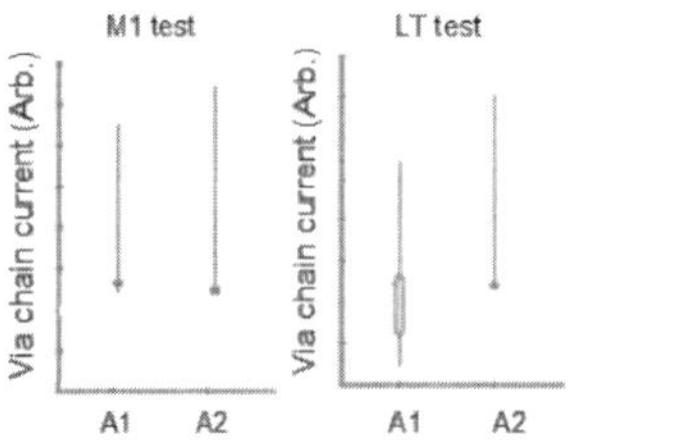

Fig. 6) 45 nm node 7.5-million chain yield tested at M1 and LT levels. The A2 liner option shows no visible signs of degradation after the multi-level build and a final polyimide anneal.

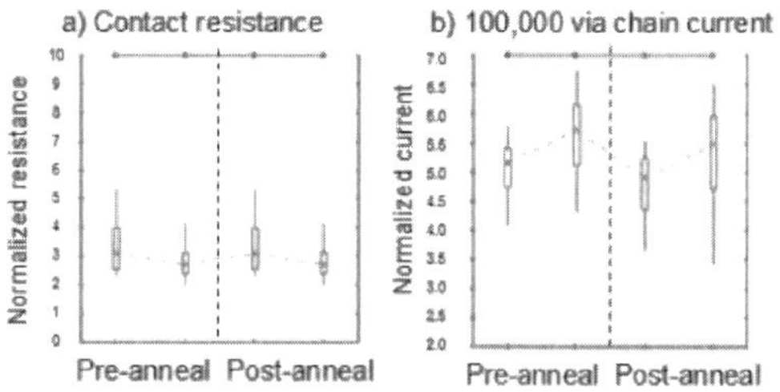

Fig. 7) Pre- and post-thermal stress a) contact resistance b) contact chain current of Cu contact at 32 nm node. The thermal stress is equivalent to 10-metal-level build.

Reliability tests

Two different types of reliability tests were executed, i.e. gate dielectric voltage breakdown (VBD) tests and electromigration tests. Fig. 8 represents gate dielectric voltage breakdown (VBD) between the contact and gates of 45 nm node thin and thick oxide gate devices, UG (upper curve) and DG (lower curve), respectively. The test was done at 125°C with the voltage ramp rate at 1 V/sec. The distances between the contact and gate were 82 nm and 45 nm for the DG and UG devices. Two different types of Cu contact metallization processes (Cu CA1 and Cu CA2) were compared to W contact (W CA). There were a few extrinsic failures on the negative voltage side (pFET), pointed by arrows, with the Cu CA2. Overall, the intrinsic VBD distributions of Cu and W contacts were comparable. Fig. 9 illustrates electromigration (EM) performance of Cu compared to W contacts on 45 nm node structure with electron flow from contact to polysilicon gate. EM lifetime to failure for the Cu contact was similar to that of the W contact although there were some early failures for the Cu contact. EM failure occurred most probably in nickel silicide. The failure analysis will be performed to understand the failure mechanism.

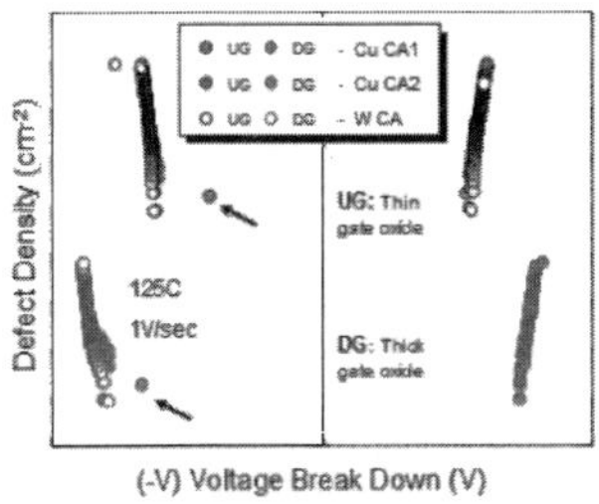

Fig. 8) Gate dielectric voltage breakdown tests show comparable results between Cu and W contacts at the 45 nm node.

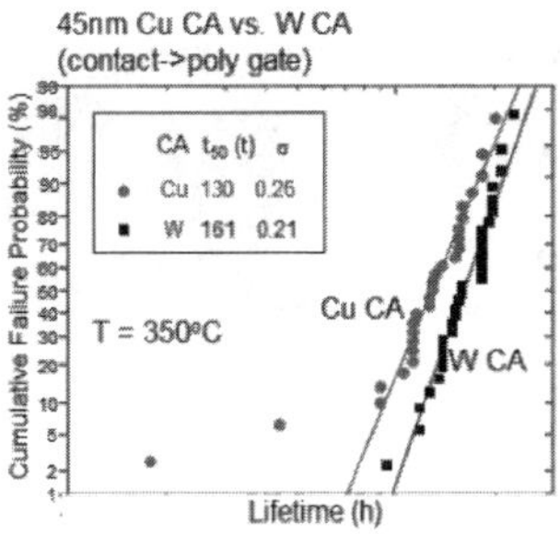

Fig. 9) Electromigration data for 45 nm node Cu vs. W contacts

Electromigration (EM) test results were obtained from 32 nm node Cu and W contacts with via depletion mode (electron flow from contact to M1) (Fig. 10a). They were comparable. The via depletion mode EM test result from 32 nm node Cu contact was also compared to that from 45 nm node Cu V1 (Fig. 10b). Although 32 nm node contact size is smaller than 45 nm node BEOL via size, they both exhibited equivalent EM performance.

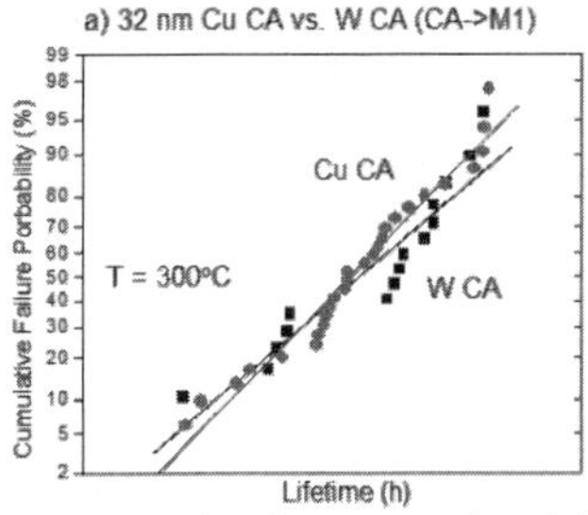

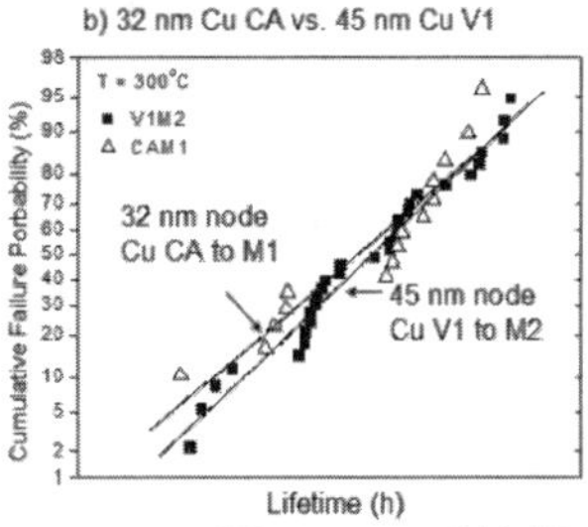

Fig. 10) Electromigration test results: a) 32 nm node Cu contact vs. W contact and b) 32 nm Cu contact vs. 45 nm Cu V1.

Device performance

Fig. 11 illustrates STEM images of a fully-integrated 32 nm node contact landing on a FET device. The aspect ratio of contacts landing on the silicided polysilicon gate was ~3 and that of contacts landing on a silicided active area was greater than 5. These contacts exhibited good gap fill. Device performance of Cu contact lot was compared to that of W contact lot at 32nm node in Fig. 12. There was about 20% improvement in drive current of Cu contact lot.

In another separate work, we also demonstrated functional 22 nm node 0.1 μm^2 SRAM cell and FETs at 90 nm gate pitch with Cu contact.[7]

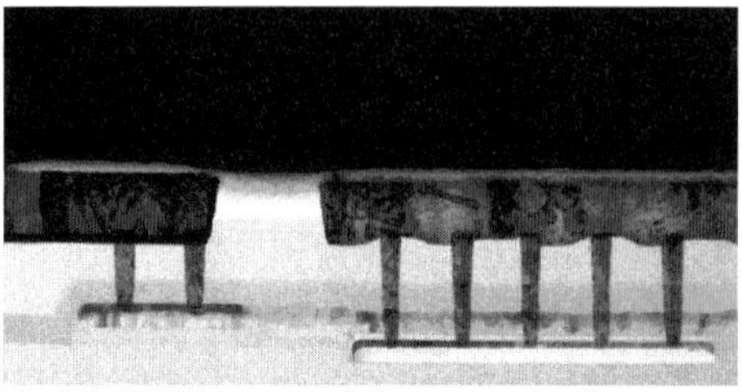

Fig.11) STEM images of Cu contacts landing on 32 nm node FET device

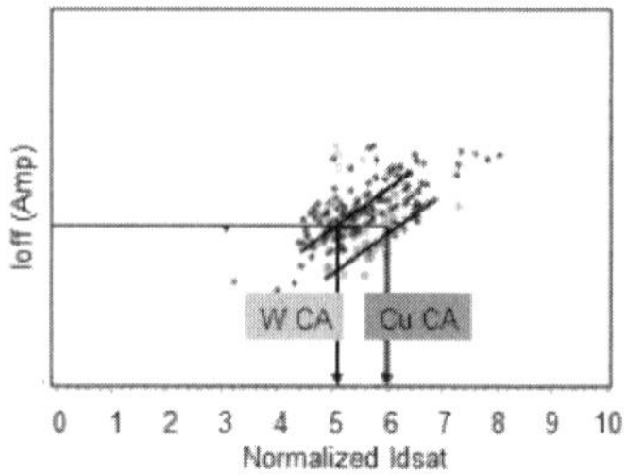

Fig. 12) 32 nm node device showed improved drive current with Cu contacts as compared to W contacts.

CONCLUSIONS

CVD Ru-containing liners exhibited superior gap fill performance compared to PVD-based Ta liners in contact holes with the bottom CD of 35 nm and AR of greater than 8. Cu contact liner choices were narrowed down using the results of the barrier breakdown test and the Cu diffusion test using Cu contact holes landing on blanket nickel silicide after 400°C for 2 hours anneal. Contact resistance of Cu using CVD Ru liner showed greater reduction than that of W for smaller contact holes with high AR. Cu contact resistance is about one-third and one-fourth of W contact resistance for 45 nm and 32 nm nodes, respectively. BEOL thermal stress was tested on 45 nm and 32 nm node contact chains and showed no degradation in contact chain currents. Gate dielectric voltage breakdown test and electromigration of Cu contact were comparable to those of W contact. We demonstrated a successful Cu contact metallization for 32 nm node device showing about 20 % improvement in drive current over that of W contact metallization. In another separate work, we also demonstrated functional 22 nm node 0.1 μm^2 SRAM cell and FETs at 90 nm gate pitch with Cu contact.

ACKNOWLEDGMENTS

This work was performed by the Research Alliance Teams at various IBM Research and Development Facilities. The authors would like to thank TEL Technology Center, America, LLC in Albany, New York, for their support of CVD Ru metallization technology for copper contact application.

REFERENCES

[1] D. Edelstein *et al.*, IITC, 9 (2001)
[2] R. Islam *et al.*, VLSI, 22 (2000)
[3] S. Demuynck *et al.*, AMC, 57 (2007)
[4] G. Van den bosch *et al.*, Proc. IEDM (2006)
[5] C. C. Yang *et al.*, IITC, 187 (2006)
[6] M. Lane *et al.*, Appl. Phys. Letters, 83, n12, 2330 (2003)
[7] B. Haran *et al.*, accepted to IEDM (2008)

Conference Proceedings AMC XXIV © 2009 Materials Research Society

Extendibility of Tungsten Technology for 32nm CA Contact Schemes with Innovative Process Improvements

Keith Kwong Hon Wong[1], Filippos Papadatos[1], Zhengwen Li[2], Hao Zhang[3], Valli Arunachalam[4], Jin-Woo Choi[5], Jay Strane[2], Jin Wallner[2], Haoren Zhuang[6], Doug Lee[4], Troy Graves-Abe[2], Andreas Knorr[4], Dan M. Mocuta[2], Shurong Liang[4], Stephan Grunow[2], and Xiaomeng Chen[2]

[1]IBM Corporation, 2070, Route 52, Hopewell Junction, NY, 12533

[2]IBM Corporation, 2070 Route 52, Hopewell Junction, NY, 12533

[3]Charter Semiconductor Manufacturing, 2070 Route 52, Hopewell Junction, NY, 12533

[4]Advanced Micro Devices, Inc., 2070 Route 52, Hopewell Junction, NY, 12533

[5]Samsung Semiconductor, Inc., 2070 Route 52, Hopewell Junction, NY, 12533

[6]Infineon Technologies, 2070 Route 52, Hopewell Junction, NY, 12533

ABSTRACT

This paper reports a systematic reduction of the contact resistance through a) identification of the key resistance contributors and b) resistance reduction of these components through process and material improvements without compromising the device yield. More specifically, a 75% reduction of contact resistance with tight distribution and comparable yield to the control samples has been achieved as demonstrated herein.

INTRODUCTION

Reduction of the RC delay at the contact level has been a key focus in 45 nm node and beyond. A 2 to 3% improvement of the Ion/Ioff ratio by replacing W with Cu has been reported (Ref 1), presumably attributed to the lower resistivity of Cu. However no semiconductor products with Cu Middle-of-the-line (MOL) contacts have been reported since the reliability of the barrier/seed/copper fill process is something to be proven and requires additional work. It is therefore important that the existing W process can be extended to future nodes.

Other novel approaches such as WN have recently been considered an alternative to Ti/TiN liners since the absence of Ti makes it less susceptible to F-attack. Thinner WN is believed to provide more "real estate" for the W fill which may potentially lower the contact resistance. Its introduction however is not trivial mainly because it requires efficient pre-cleaning processes to remove excess oxygen from the silicide interface. Failure to do so leads to high contact resistances and large data variation. Figure 1 shows normalized contact resistance data for Ti/TiN and WN splits. It is apparent that WN is not yet an optimized and mature process to replace traditional W. Extending the W process needs therefore to rely on optimization of all the factors that contribute to the CA (contact to silicide) resistance.

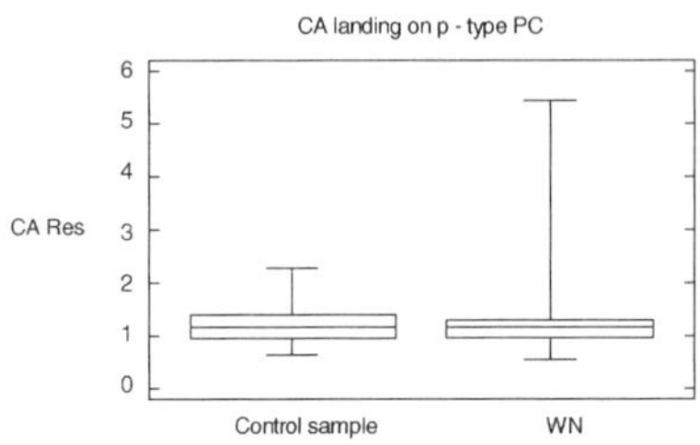

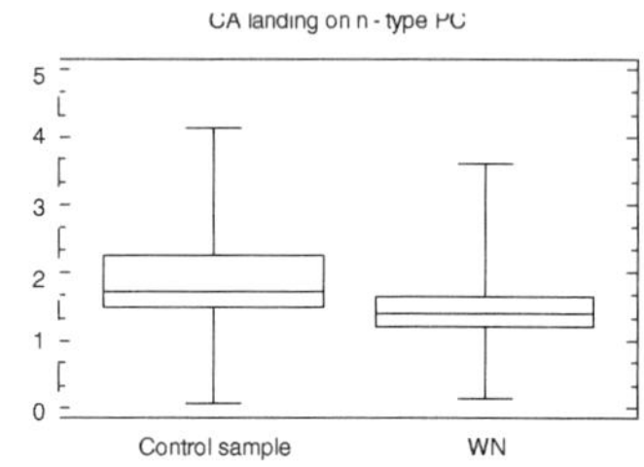

Figure 1 shows the normalized resistance of isolated CAs on p and n –type PC lines. The data demonstrates that there is no advantage of WN over the control sample.

EXPERIMENT

In this study, single level flop down wafers with 32nm feature size contacts were used for gapfill check, while fully integrated device wafers were employed for the electrical contact resistance measurements. The Ti metal was deposited by Ionized Metal Plasma (IMP) and TiN films were prepared by Chemical Vapor Deposition (CVD) and Atomic Layer Deposition (ALD) followed by plasma treatment. A thin W nucleation and a subsequent bulk CVD W fill completed the contact fill. Gapfill performance was evaluated by cross section Scanning Electron Microscopy (XSEM) imaging.

DISCUSSION

Profile tuning

Key contributors in MOL CA resistance are the liner pre-clean and the in-coming shape of the contact. In order to obtain low-resistance contacts, Ar sputter is commonly used to clean the metal silicide surface before the liner metal deposition. However, Ar sputter can make the top critical dimension (CD) smaller and consequently worsen the Ti overhang, which led to later W filling voids. A benign CD profile with wider top CD will facilitate Ti deposition over the trench, and due to the openness on the top, deposition of Ti on the bottom of the trench becomes easier, leaving the possibility of reducing the amount of Ti, but achieving the same amount of Ti on top of the silicide. The Ti pinch-off problem on top of the trench will be greatly reduced by using less amount of Ti. This improvement in overhang Ti metals improved later on TiN and W depositions. Figure 2 shows the bulk W fill improvement with additional CA trench shape adjustment of the CA shapes.

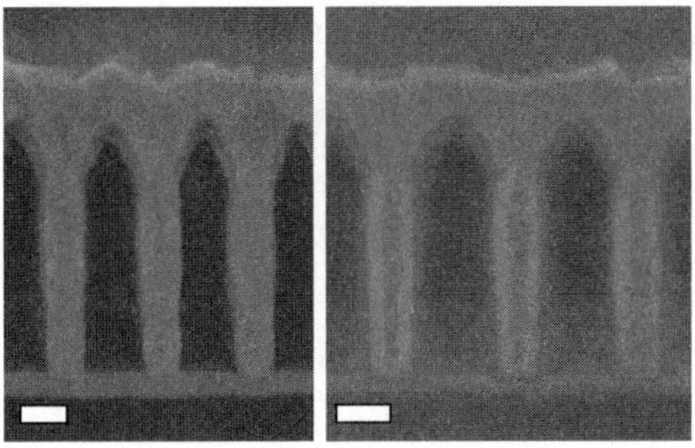

a)

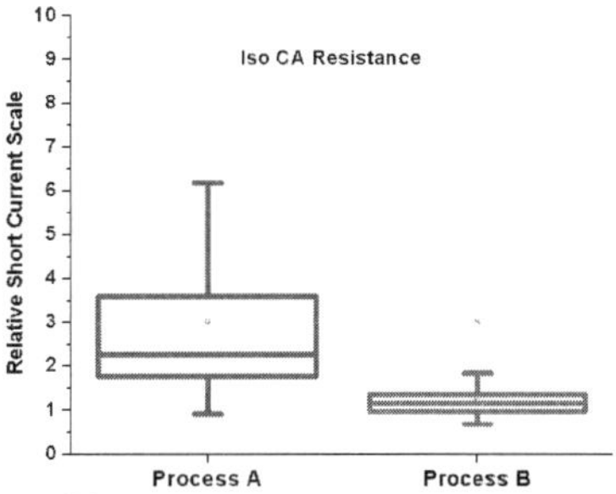

b)

Figure 2 (**a**) Process modification to improve metal fill: (left) without (right) with via adjustment. Both SEM cross-section have same Ti/TiN/W processes. (**b**) Electrical data showing advantage of profile tuning and other liner improvements (process B)

Ti Overhang reduction

Another way to reduce Ti overhang is to reduce the Ti field thickness. This may be achieved by applying a substrate bias so that the Ti field thickness is reduced while sufficient amount of Ti still remains at the bottom of the contact. Figure 3 shows the improvement of contact fill with reduced overhang by this approach. However, the wafer pedestal bias power process needs to be carefully examined because very high bias will make the overhang worse due to re-sputtering and re-deposition at the top corners of the contacts, while very low bias power will not improve bottom Ti coverage. Other process parameters such as chamber pressure and gas flows might be fine tuned to reduce the energy of the ions so that less re-sputtering occurs. One concern of this approach is that the wafer pedestal bias applied to the wafer will induce significant plasma damage to the substrate since the Ti is deposited right on the silicide. The voltage breakdown data (VBD) in Figure 4 suggests that there is no charging issue caused by applying this substrate bias to the wafer. The bias used herein to improved gapfill is typically lower than the commonly used Ar physical sputter pre-clean bias prior to contact metal liner.

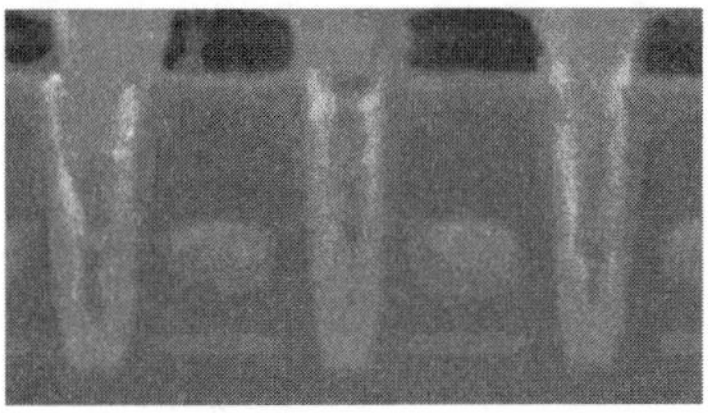

(A)

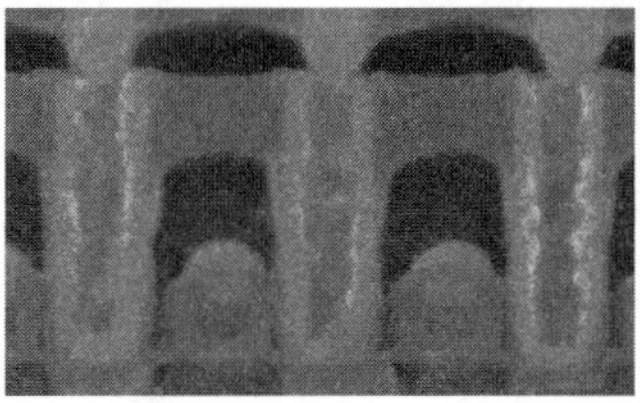

(B)

Figure 3 32nm wafer W gap fill comparison with (a) Thick Ti without wafer pedestal bias vs. (b) thin Ti with wafer pedestal bias. The Ti overhang was reduced greatly so that there is minimum pinch-off of the contact opening, thus gap fill was much improved

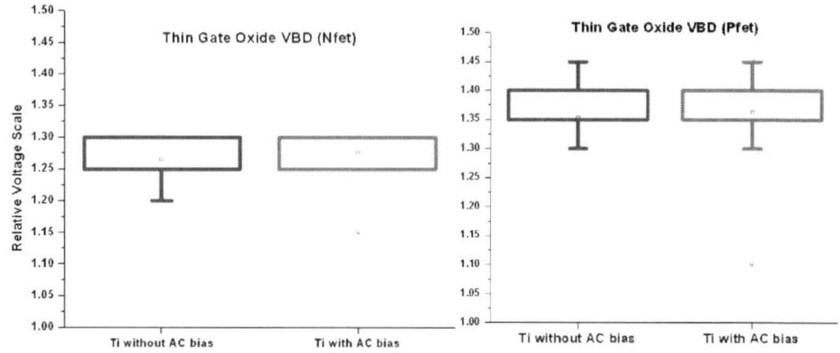

Figure 4 32nm M2 device data for thin gate oxide device VBD at both NFET and PFET for IMP Ti wafer substrate bias split: there is no degradation observed with applying substrate bias to the wafer during Ti deposition

TiN liner optimization

Effects of TiN on W nucleation have been reported (Ref. 2, 3). Rigorous process optimization and aggressive scaling of the CVD TiN and W nucleation layer (as much as 50% from last technology node) is required to reduce contact resistance and tighten the data distribution. The TiN process optimization was performed via design of experiment (DOE). Optimized responses were the Ti/N ratio and sheet resistivity of the TiN thin films. The optimal conditions within the process window were identified and later used for electrical performance evaluation. Figure 5 shows 3D surface plots for each response. Encouraging electrical data have been demonstrated for 32nm hardware with thin CVD TiN and ALD-like W nucleation processes. Thermal annealing at 400C has been employed to stress the metal liner and test the liner integrity. The results indicate that no contact resistance degradation is observed for the thinner liners which in turn suggests that continuous liner improvement may extend its viability in future nodes as shown in Figure 6.

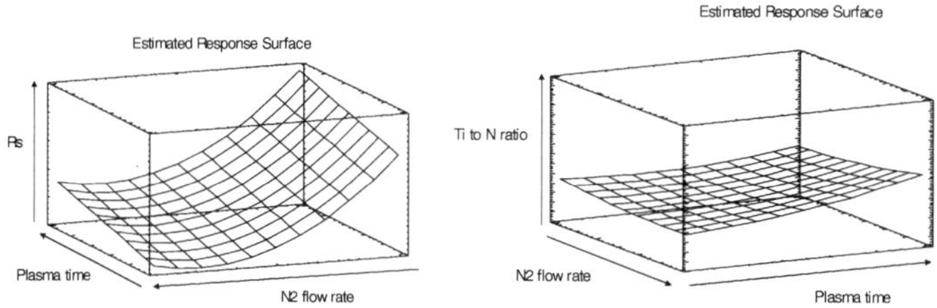

Figure 5 TiN Sheet resistivity and Ti/N ratio as a function of N_2 flow rate and plasma treatment exposure time

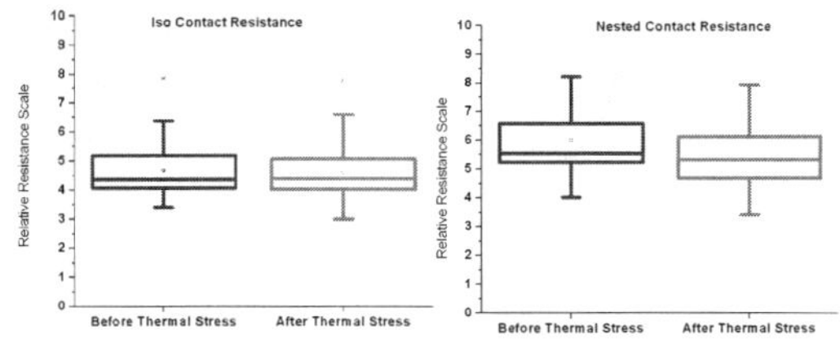

Figure 6 Contact Resistance comparison before and after thermal annealing stressing to test the thin metal liner integrity and reliability on 32nm hardware. No contact resistance degradation was observed after thermal stress, which demonstrated good liner integrity

Lately, ALD TiN has become a particularly attractive option, since ALD processes have inherently better step coverage and more efficient plasma treatment than CVD processes. In this work the impact of ALD and CVD TiN films on the resistance of 32nm node contacts has been studied. Figure 7 shows that the median iso-contact resistance can be reduced by as much as 29% with the use of ALD TiN as a barrier. Additionally, ALD TiN has a tighter resistance distribution than CVD TiN.

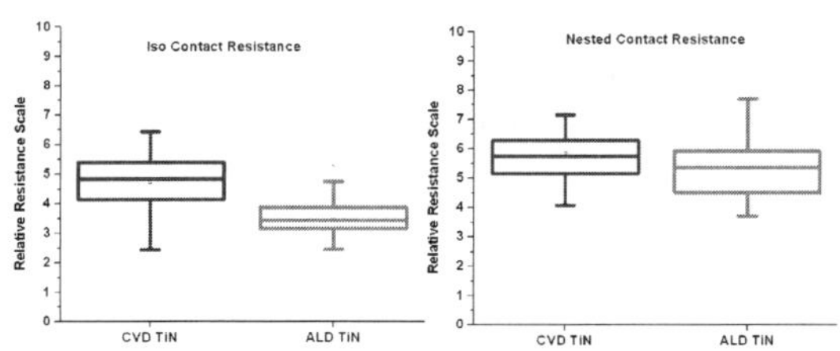

Figure 7. Contact resistance comparison for CVD TiN and ALD TiN on 32nm hardware with same target TiN thickness and all other processes are same. ALD TiN showed lower mean contact resistance value and tighter distribution than CVD TiN.

A top down image (Fig. 10a) and FIB cross-section (Fig. 10b) of an electrically measured structure indicates good fill with minimal seam void formation, thereby corroborating the electrical data.

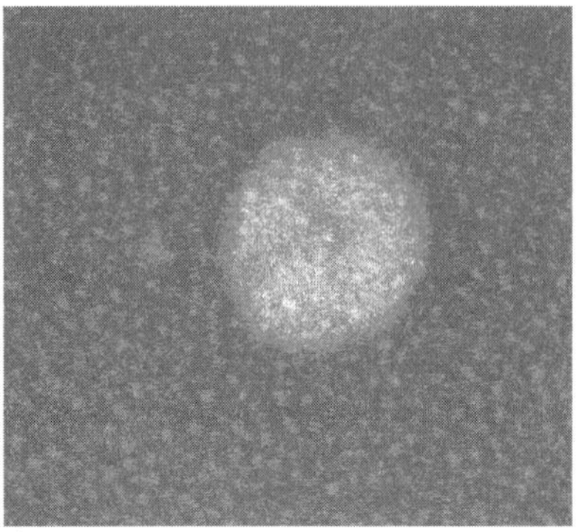

a)

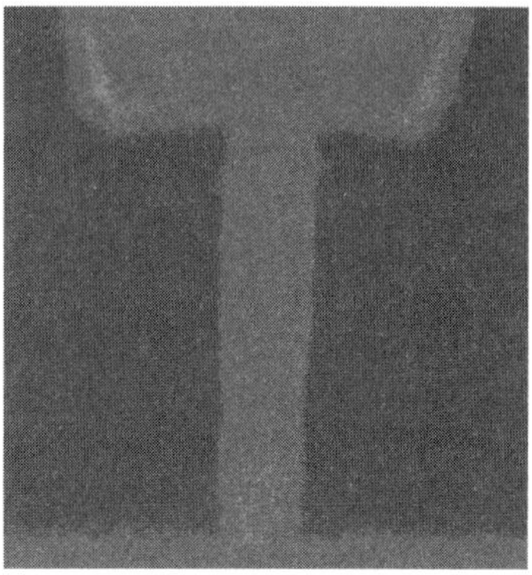

b)

Fig. 10 Left: top down image of an electrically measured isolated contact with ALD TiN liner, taken after de-layering, shows good fill with minimal seam void formation. Right: Cross-section of an isolated contact showing good fill with ALD TiN liner

<u>W fill</u>

The W process itself may also be fine tuned to lower the overall material resistivity. The fill consists primarily of two major components; the W nucleation layer and the W fill which also known as "bulk" deposition step. The nucleation component is the most resistive and important step because it acts as a template for the W fill. More specifically, it controls the grain size of the "bulk" W and hence the contact resistance and fill properties. Keeping the nucleation layer thin, reduces its contribution to the CA resistance allowing more room for the less resistive "bulk" W. Changes in the reactant gas sequence, combination of different reactant gases and/or flows may further improve the W properties. For example the replacement of silane with diborane in the nucleation step followed by a change in the gas flows, enhances the grain growth of the "bulk" W film(Ref 4). Reducing the nucleation thickness further improves the fill and the contact resistance as shown in Figure 8

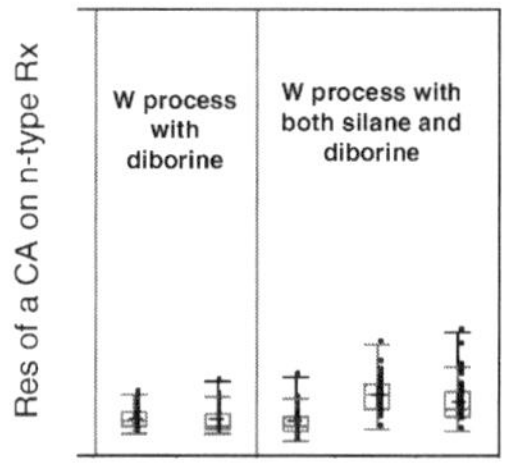

Figure 8 Resistance of isolated CAs on n-type active regions for two different W processes. The change in process conditions during the nucleation step may reduce the overall contact resistance

CONCLUSIONS

Careful fine tuning and process optimization of the CA profile, liner quality, Ti overhang and W deposition may lead to more than 75% reduction in contact resistance with tight distribution as demonstrated in Figure 9. These results suggest that Ti/TiN/W stacks can be extended for 32 nm node.

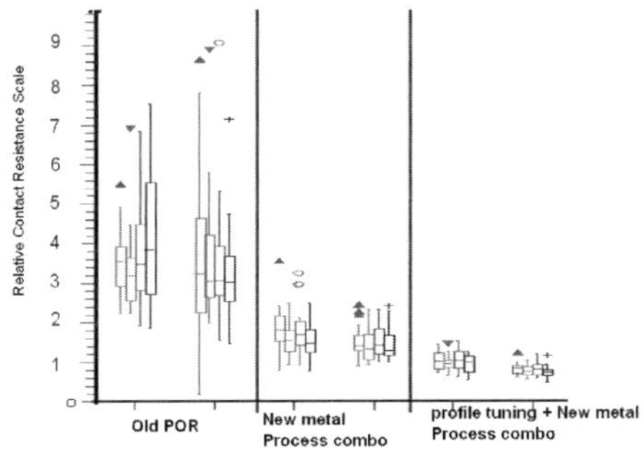

Figure 9 Contact Resistance mean value and distribution improvement with optimized CA metallization (Ti, TiN, W) process (Process A); and via adjustments enable further improvement and scaling.

ACKNOWLEDGMENTS

This work was performed at the IBM Microelectronics Div. Semiconductor Research and Development Center Hopewell Junction, NY 12533.

REFERENCES

1. Lower Resistance Scaled Metal Contacts to Silicide for Advanced CMOS - A.Topol et al.VLSI 2006.
2. Nucleation and Film Growth of Chemically Vapor Deposited Tungsten on Various Types of TiN Glue Layer , D.S. Kim et al., J. Research Institute for Catalysis , p69-74.
3 A new ALD-TiN/$CoSi_2$ contact plug process for reliable and low defect density bit-line integration in sub-quarter micron giga-bit, S.G .Park et al, Interconnect Tech. Conf, 2002 Proceeding, IEEE 2002, p282-284.
4 Effects of phase of underlying W film on chemical vapor deposited –W film growth and applications to contact-plug and bit line processes for memory devices, S.H. Kim et al., J. Vac. Sci. Technol. B. 2007, p 1574-1580.

Conference Proceedings AMC XXIV © 2009 Materials Research Society

Tungsten Nitride Barrier/Tungsten Via to Copper Interconnect for Memory Device Applications

Juwen Gao, Josh Collins, Wei Lei, Michal Danek, SungHyun Hong, Kaushik Chattopadhyay, Tom Mountsier, and Roey Shaviv
Novellus Systems Inc., San Jose, CA, 95134

ABSTRACT

In this paper, we report a novel, high productivity and low cost integration scheme for copper interconnects in memory devices using a tungsten nitride barrier and a tungsten via. Key integration challenges overcome in this work include zero sputter Cu_2O reduction, damage-free conformal WN barrier deposition, void-free W fill, and W CMP with minimal dishing, plug coring, or barrier erosion. Electrical results show high via chain yield with a tight distribution in via chain resistance. Electromigration results show comparable mean time to fail to that of the Cu-only control. Effects of WN film thickness, preheat time and WN deposition temperature were investigated.

INTRODUCTION

With increasing demand on speed and device density, copper interconnect has been introduced into Flash and DRAM devices. Tungsten still remains as the material of choice for the via as shown in figure 1. The tungsten via to the Cu interconnect brings new integration requirements and challenges to the conventional liner/barrier scheme. For example, in the conventional Ti/TiN liner/barrier approach, a bi-layer of PVD Ti/MOCVD (metal organic chemical vapor deposition) TiN is required [1]: PVD Ti is deposited at the contact bottom for the copper barrier followed by a MOCVD TiN barrier layer to protect Ti from fluorine attack during the subsequent CVD-W fill. This bi-layer scheme has several limitations: i) poor PVD-Ti step coverage, ii) PVD-Ti overhang, iii) MOCVD TiN film needs to be thick and dense enough to be an effective barrier, making this scheme more costly and less extendable. In addition, it was reported that SiH_4, which is typically used in CVD-W nucleation step, can easily diffuse through MOCVD-TiN [2, 3]. The penetration of SiH_4 to the underlying copper may lead to formation of high resistivity $CuSi_x$.

Pulse nucleation layer deposition (PNL) of WN/W vias can overcome these limitations. PNL is capable of depositing highly conformal WN/W films [4,5]. PNL-WN has good adhesion on dielectric and is an excellent barrier for copper and SiH_4 diffusion. Therefore, a single thin layer of PNL-WN can replace Ti/TiN as both the diffusion barrier and the liner.

In this paper, we report a W via integration scheme for copper interconnects, using PNL-WN as the barrier and PNL-W for via fill. Process characteristics are summarized and electrical performance of this process flow is also evaluated.

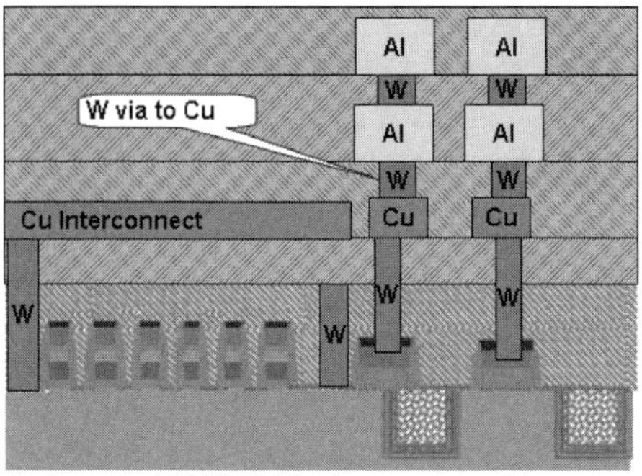

Figure 1. A typical NAND Flash structure with copper interconnect

EXPERIMENT

The PNL-WN/W via deposition was performed in a Novellus 300mm DirectFill™ system. The system consists of a preclean module, a PNL-WN deposition module and a PNL-W fill module. The multi-station sequential deposition (MSSD) architecture of PNL-WN module enables simultaneous processing of four wafers in each module, resulting in a high throughput. The MSSD architecture also allows individual control of gas flows, deposition temperatures, and step times at each station to optimize performance for each process step.

Ar/H_2 plasma was used to clean Cu_2O. The process pressure was 10mTorr to 100mTorr. The plasma power was 1000W to 3000W. The wafer bias power was 0 to 300W.

The PNL WN barrier deposition process is a thermal ALD process at temperatures between 200 to 400°C and a pressure of 1-10 Torr. One deposition cycle consists of sequential exposure of B_2H_6, WF_6, and NH_3. Each precursor exposure was followed by an Ar purge (Figure 2). The cycle is repeated until desired film thickness is obtained.

Tungsten fill is accomplished by a PNL-W nucleation (sequential introduction of SiH_4 and WF_6 reactive gases at 300 °C) followed by a conventional WF_6-H_2 CVD plug fill at 395 °C, both at 40 Torr. Details of PNL-W fill process can be found in the literature [5].

Tungsten chemical-mechanical-polishing (CMP) was carried out on a Novellus Xceda™ tungsten CMP system.

Cu_2O thickness was measured using reflectivity. WN film sheet resistance and thickness were measured by four-point probe and x-ray fluorescence (XRF), respectively. WN barrier property to SiH_4 diffusion was evaluated by SIMS (secondary ion mass spectroscopy) depth profile analysis. Both top-view and cross-section scanning electron microscope (SEM) were used for post-CMP defect inspection. Novellus internal test structure wafers (with M1 and M2 Cu interconnects connecting through 90 nm via) were used in the integration test.

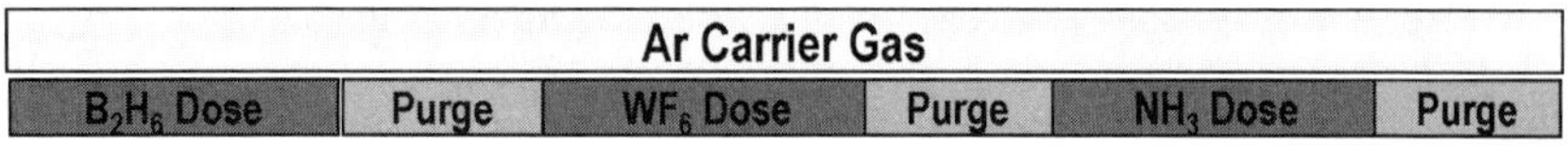

Figure 2. Schematic of PNL-WN deposition sequence

DISCUSSION

Via bottom Cu_2O clean

Cu at the via bottom can be easily oxidized in the atmosphere after via etch. Cu_2O will increase the electrical resistance of the via due to its highly insulating nature and therefore needs to be removed before barrier deposition. In this work, an Ar-H_2 plasma was used to reduce Cu_2O to copper. The plasma was carefully-controlled to avoid Cu sputter, which could degrade the reliability of the devices.

Cu_2O removal was characterized using a 120Å Cu_2O film grown by thermal oxidation of a 1000Å PVD-Cu film. This film was subjected to Ar-H_2 plasma to reduce Cu_2O at different clean times. Optical reflectivity at wavelength of 550nm was measured and converted to Cu_2O thickness on copper [7]. As shown in figure 3, an Ar-H_2 plasma can effectively remove Cu_2O. XRF measurement of the Cu substrate before and after preclean showed zero Cu sputter loss.

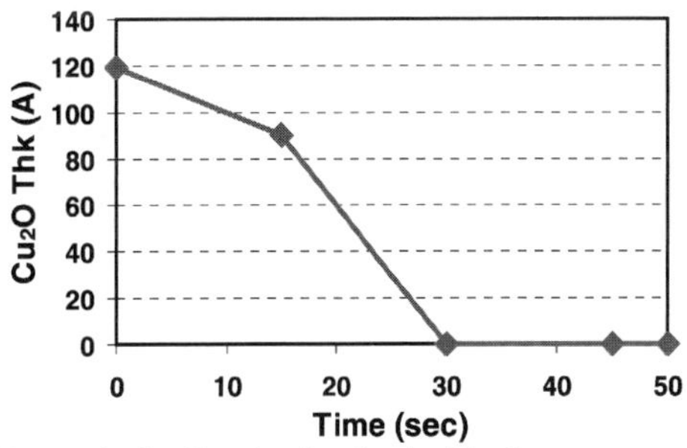

Figure 3. Cu_2O reduction by Ar-H_2 plasma

WN barrier/W process

The WN deposited by the PNL process has a growth rate of 2.1 Å/cycle. The ALD process nature produces a film of WN with excellent step coverage. XRD analysis shows that the WN barrier layer is polycrystalline β-W_2N, which provides a superior diffusion barrier property [5].

It should be pointed out that Cu itself is non-reactive with WF_6 and HF (positive Gibbs energy of reaction under typical deposition conditions) and thus requires no special protection. However, SiH_4 from the subsequent tungsten nucleation process may diffuse through the barrier into the copper layer and form highly resistive copper silicides. To evaluate the barrier performance of PNL-WN film for SiH_4 diffusion, 1000Å of PVD-Cu films with different thicknesses of PNL-WN were exposed to a PNL-W nucleation process which uses SiH_4-WF_6 at 300 °C. SIMS analysis (Figure 4) shows that without WN film, there is a significant amount of silicon diffused into the copper film. As thin as 20Å of WN film effectively blocks diffusion of SiH_4 into Cu.

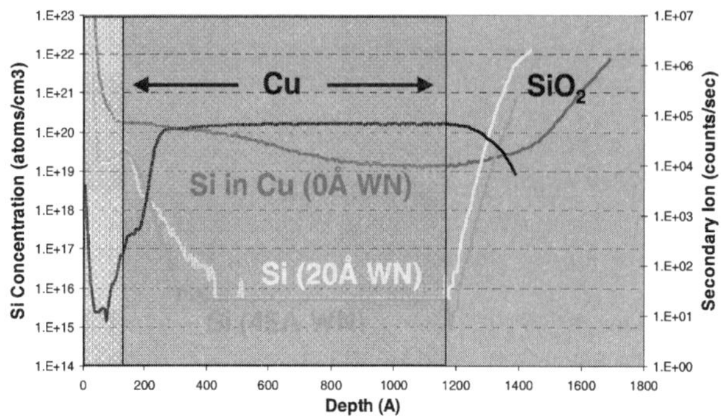

Figure 4. SIMS analysis showing Si in Cu with and without WN barrier.

After WN barrier deposition, the vias were filled with PNL-W. The PNL-W provides excellent via fill capability [6].

A TEM image of a single contact via showed a very clean copper-tungsten interface without any sign of fluorine attack or SiH_4 diffusion to the copper layer (Figure 5). Due to the conformality of WN there is no void at the Cu-WN or Cu-SiN interfaces.

Barrier adhesion to the dielectric and copper layer meets requirements as verified with scribe-tape test on blanket wafers with full WN/W (2000 Å) stacks.

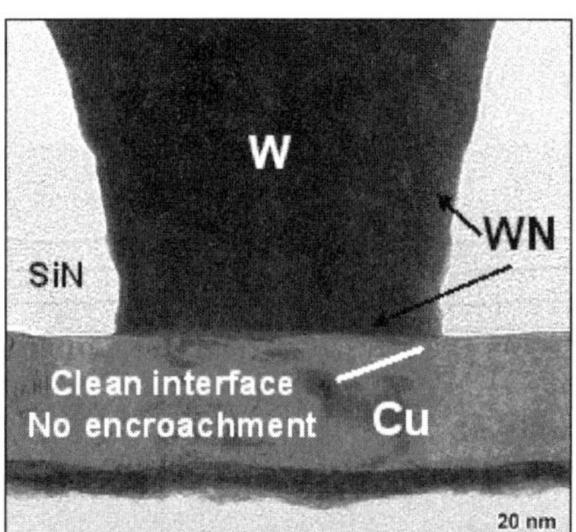

Figure 5. Cu/WN interface with PNL-WN/W fill

Tungsten CMP

Tungsten CMP is a key step in the integration. Top-view SEM inspection indicates that PNL-WN/W fill integrates well with CMP. No via coring or barrier erosion was observed. A cross-section SEM image after CMP (Figure 6) shows excellent CMP performance. Note that an oxide buffing step was applied after tungsten polishing to allow plug extrusion such that there was no post-CMP cleaning residue remaining at the top of the plug after CMP.

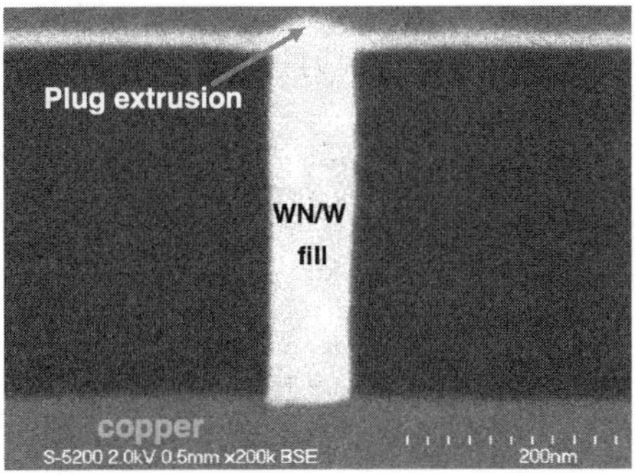

Figure 6. Cross-section SEM image of WN/W via to Cu after CMP and M1 deposition

Electrical performance

Electrical testing was carried out using Novellus internal test structures with 90nm tungsten vias between Cu metal-1 and metal-2 layers. Both Kelvin and chain contact resistances were measured. The effect of WN film thickness, preheat time and WN deposition temperature were explored.

As can be seen from figure 7a, there was no significant effect on Kelvin contact resistance related to preheat time and deposition temperature. As expected, a thinner (45 Å) WN layer results in lower via resistance. Figure 7b demonstrates the via chain parametric yield on 2 million vias for the baseline condition where 65 Å of PNL-WN was used as the barrier. Nearly 100% chain yield was achieved with very tight distribution.

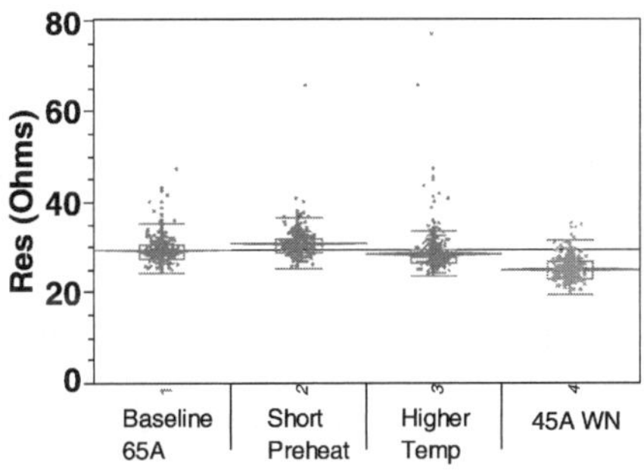

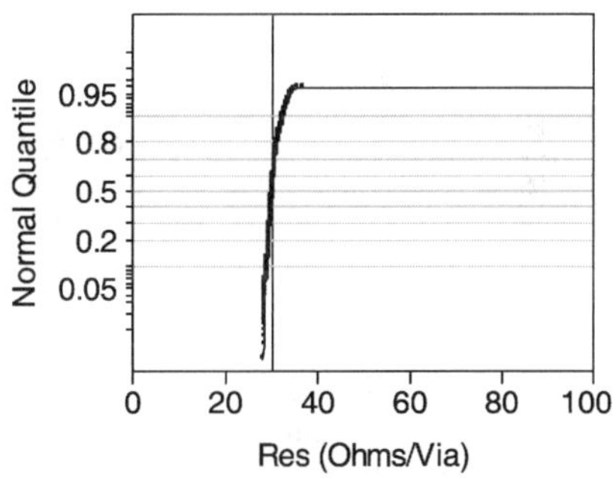

(a) (b)

Figure 7. Parametric performance for 90nm vias : (a) Kelvin via resistance ; (b) Rc yield for 2 million via chain (baseline WN 65 Å)

Electro-migration performance

Downstream and upstream electromigration (EM) results are presented in figure 8. These were compared to a dual damascene all-copper control. Package level EM testing was conducted at 325° C using a current density of 2.5 MA/cm^2. A wider distribution ($\sigma \sim 0.8$) in downstream EM time to fail is observed for the baseline process relative to control ($\sigma \sim 0.3$). Mean time to fail (t_{50}) for the baseline process is comparable to that of the control. Shorter preheat has no effect on downstream EM as both t_{50} and σ are similar to that of the baseline process. J_{max} with these WN conditions is about 3 MA/cm^2. Downstream EM of the higher temperature WN process has a t_{50} that is about 2 times longer than that of the control with a similar distribution ($\sigma \sim 0.3$). Consequently, with these conditions J_{max} is > 10 MA/cm^2 and is significantly larger than results for the control where $J_{max} \sim 7$ MA/cm^2. Short preheat appears to be detrimental to upstream EM where t_{50} is about half that of the control and J_{max} is about 7 MA/cm^2. The upstream EM results with all other processes are statistically equivalent to those of the copper control. For upstream EM, J_{max} > 10 MA/cm^2 for all these process conditions.

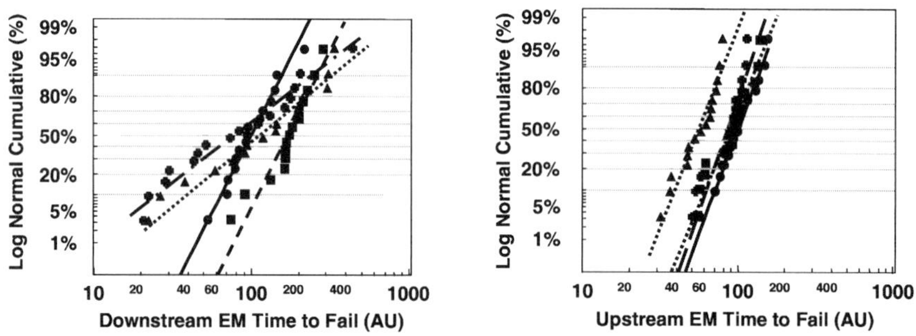

a. *b.*

Figure 8. Downstream (*a.*) and upstream (*b.*) EM time to fail for: Cu control (●), baseline WN/W (✚), short preheat (▲), and high temperature WN (■).

CONCLUSIONS

A PNL-WN/W via has been applied to Cu interconnect. The integration scheme includes plasma preclean, WN barrier deposition and W via fill. It was demonstrated that Ar/H_2 plasma clean can effectively remove Cu_2O. The highly conformal ultra thin PNL-WN film has excellent step coverage, superior barrier property to SiH_4 diffusion and good adhesion to both Cu and dielectrics. CMP integration was very successful with low defects counts. Parametric electrical tests showed that there was no significant impact of process temperature and preheat time. A thinner WN layer results in lower via resistance. Nearly 100% yield was achieved for 2 million via chains with 90nm via structure. Finally, electromigration performance was comparable to copper control. In summary, the low cost, high productivity PNL-W/WN via scheme in this work is a viable solution for the via to copper interconnect in memory applications.

ACKNOWLEDGMENTS

The authors wish to acknowledge many contributions from Wenhong Yan, Jing Guo and Kum Nam Kang.

REFERENCES

[1] N. Kumar, et al. *Semiconductor International*, May 1, 2008
[2] J. Bonitz, et al., *Microelectronic Engineering*, 82, (2005), p618-622
[3] C. Marcadal et al., J. Electrochem. Soc. 149 (1) (2002), C52–C58.
[4] S.H. Kim, et al. AMC 2005
[5] K. Ashtiani, et al. Semicon Korea 2006
[6] S.H Lee, et al. AMC 2001
[7] US Patent US 6,525,829

Conference Proceedings AMC XXIV © 2009 Materials Research Society

RuTaN Films by PEALD using N_2 / H_2 Plasma for Cu Diffusion Barrier Layer

Daekyun Jeong, Kunitoshi Namba, Hiroaki Inoue and Hiroshi Shinriki

ASM Japan, Process Development Department
23-1, 6-chome, Nagayama, Tama-shi, Tokyo, Japan
Tel.:+81-42-337-6314 Fax:+81-42-389-7555 *E-mail:daekyun.jeong@asm.com

Abstract

A novel formation technique of ultra-thin RuTaN films and its copper diffusion barrier property are reported in this paper. The RuTaN films are deposited by plasma enhanced atomic layer deposition (PEALD) by using N_2 / H_2 mixed gas plasma. Ru/Ta atomic composition ratio can be controlled by Ru/Ta precursor pulse ratio. Nitrogen composition in the RuTaN films is remarkably stable independent of N_2 flow rate in N_2 / H_2 mixed gas. However, single hydrogen plasma process can provide RuTa films having low resistivity and lower nitrogen composition below 1 at%, which is confirmed as good nucleation layer for CVD-Cu film. Including of Ru atoms into TaN film can improve adhesion property to Cu film while keeping Cu diffusion barrier property. 2 nm-thick RuTaN film deposited with Ru/Ta precursor pulse ratio of 2 shows sufficient Cu diffusion barrier property after annealing at 400°C. This RuTaN film by PEALD using N_2 / H_2 plasma is one of the most promising Cu barrier metal for advanced Cu interconnect.

Introduction

Beyond 32 nm node, reducing thicknesses of Cu barrier metal for higher Cu volume is one of the most effective approaches to improve RC delay [1]. Thus, conformal and extremely thin metal barrier formation technique is required while keeping good adhesion to Cu and sufficient Cu barrier property. A TaN barrier deposited by atomic layer deposition (ALD) has been reported as a Cu barrier metal, however, glue layer like PVD-Ta flush was required between ALD-TaN and bottom Cu interface to maintain sufficient adhesion property of ALD-TaN to Cu [2]. Fig.1 (a) shows stack films of PVD-Ta/ALD-TaN/PVD-Ta flush with Cu dual-damascene structure. However, this approach cannot be applied to high-aspect via filling because of the poor PVD-Ta step coverage. As shown in Fig.1 (b), RuTaN film was reported to be effective for improving adhesion to Cu layer [3]. Higher Ru/Ta precursor pulse ratio than 20 was required for obtaining Ru/Ta atomic composition ratio higher than 1. It is because including rate of Ru atoms into RuTaN film was extremely low.

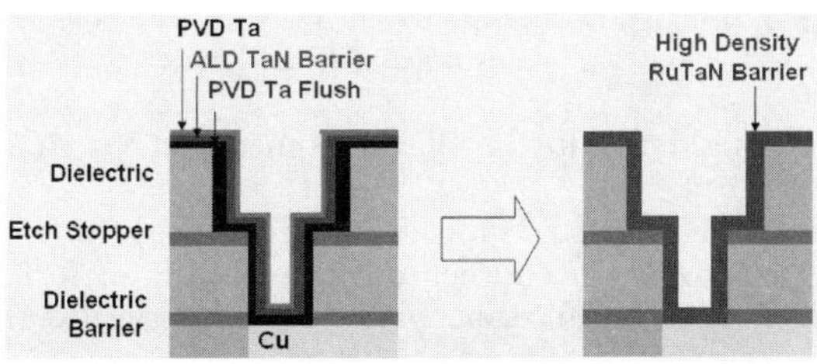

Fig. 1 Advanced Cu barrier films (A) PVD-Ta/TaN/PVD-Ta and (B) PEALD-RuTaN single layer structure

In this paper, we shows a novel PEALD-RuTaN process providing flexible control of Ru/Ta atomic composition up to higher than 3 by increasing Ru/Ta precursor pulse ratio from 1 to 5. This PEALD process can provide 2 nm-thick RuTaN film having a favorable Cu barrier property and low resistivity which is attractive for advanced Cu interconnect beyond 32 nm node.

Experimental

In this experiment, RuTaN films were grown by using plasma enhanced atomic layer deposition tool. Each Ru and Ta precursors were supplied to the same reactor by thermal evaporation of liquid metal–organic precursor. The deposition temperatures of Ru and Ta were fixed at same temperature below 300°C. In order to control the atomic compositions of Ru and Ta, after one Ta precursor pulse, a number of Ru pulse cycles ranged from 1 to 5 were inserted as shown in Fig.2. Decomposition of Ta and Ru precursor is achieved by H_2 plasma step [4] and N_2 / H_2 mixed gas plasma step. Combination of Ru precursor, Ta precursors and plasma processes can control atomic composition of Ru, Ta, and nitrogen in the film.

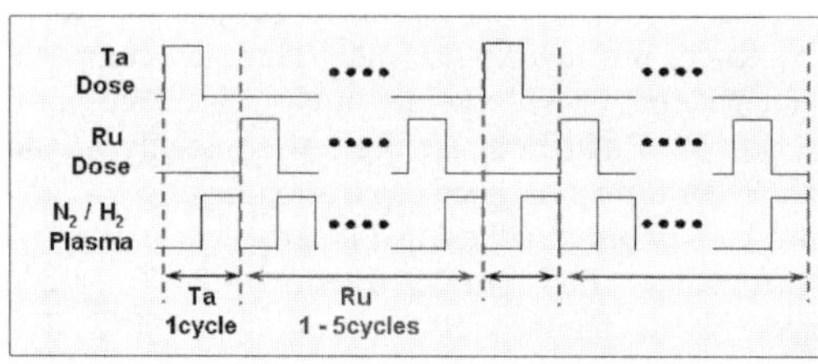

Fig. 2 RuTaN films deposition sequence

After TaNC, Ru/TaNC and RuTaN films were fabricated by PEALD, all of these films were analyzed by X-ray reflectometer (XRR) to measure the thicknesses. Sheet resistance was measured with 4-point probe. Adhesion property of these films to Cu and SiO_2 was evaluated by using tape test. Atomic compositions of their films were analyzed using 9.0 MeV He^{2+} Rutherford backscattering spectroscopy (RBS). Continuity of extremely thin CVD-Cu film deposited on Ru, RuTa and RuTaN films were evaluated by sheet resistance. CVD-Cu films were thermally deposited with Cu(hfac)tmvs precursor [4]. On Ru, RuTa and RuTaN films, incubation time of CVD-Cu formation was not observed. Cu barrier property of RuTaN films was evaluated by two methods. After annealing of samples having CVD-Cu/2 nm-thick RuTaN/Si structure, these samples were analyzed by X-ray diffraction (XRD) using Cu K-α radiation at 1.5405 Å with small contact angle mode. RuTaN films deposited with Ru/Ta precursor pulse ratios of 1:1 and 2:1 were evaluated. Backside secondary ion mass spectroscopy (BS-SIMS) analysis used as another method. After annealing sample having CVD-Cu/2 nm-thick RuTaN/SiO_2/Si structure, atomic depth profile was evaluated from Si substrate by SIMS analysis.

Results and Discussions

1. Atomic composition of RuTaN film as a function of N_2 / H_2 flow ratio

Fig. 3 shows atomic composition of RuTaN films deposited with Ru/Ta precursor pulse ratio of 2:1. For improving Cu diffusion barrier properties of RuTaN films, dependence of nitrogen composition of RuTaN films on N_2 flow rate is investigated. H_2 gas flow rate was fixed in every deposition. Without nitrogen gas, deposited film includes low Ta atomic composition below 5% and nitrogen composition below 1%, which is remarkable Ru rich film. As N_2 flow was included and increases from 1 to 3 (arbitrary unit), Ta/Ru and N/Ru atomic ratios become stable independent on nitrogen flow. Ta atomic composition is very similar to nitrogen atomic composition. This results means that nitrogen atom mainly make a chemical bond with Ta atoms. As N_2 flow increase from 1 to 3, Ta atomic composition slightly increases.

Therefore, RuTaN film having stable Ta and nitrogen composition can be formed by using N_2/H_2 plasma and RuTa film having Ru rich and low nitrogen composition below 1 at% can be formed by using H_2 plasma respectively. N_2 flow is fixed at 1 in the following results.

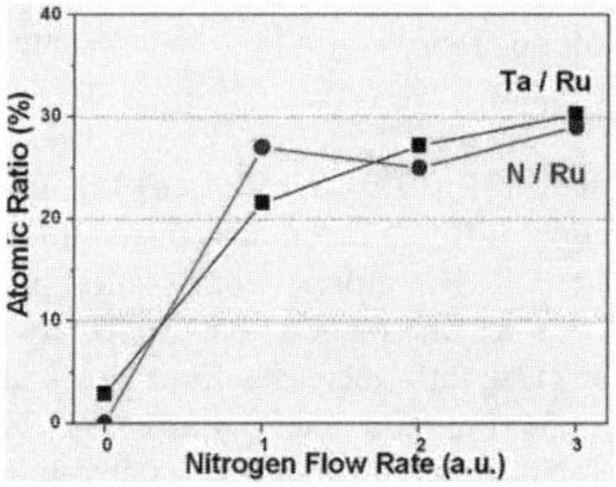

Fig. 3 Atomic ratio of Ta/Ru, N/Ru dependence on N_2 flow rate (a.u.)

2. RuTaN film properties as a function of Ru/Ta precursor pulse ratio

Growth rate, resistivity of RuTaN films and RuTa films are shown in Fig.4 and Fig.5. When Ru/Ta pulse ratio increased from 1:1 to 3:1, the typical growth rates of these RuTaN films also increased from 0.15 to 0.23 nm/cycle. On the other hands, the growth rate of RuTa films increased form 0.08 to 0.10 nm/cycle at the same deposition condition. Fig. 5 shows resistivity of PEALD-RuTaN films and RuTa films dependent on Ru/Ta pulse ratio. H_2 plasma process can provide RuTa film having lower resistivity than that of RuTaN films deposited by using N_2 / H_2 mixed gas plasma process. As Ru/Ta pulse ratios increased from 1 to 2, resistivity of RuTa films decreased from 170 to 60 μΩ-cm as shown in Fig.5.

As Ru/Ta precursor pulse ratio increased up to 3, resistivity of RuTa film became saturated at 50 μΩ-cm. As Ru/Ta precursor pulse ratio is higher than 2, resistivity of RuTaN films becomes lower than 160 μΩ-cm, which seems to be acceptable as a Cu barrier layer.

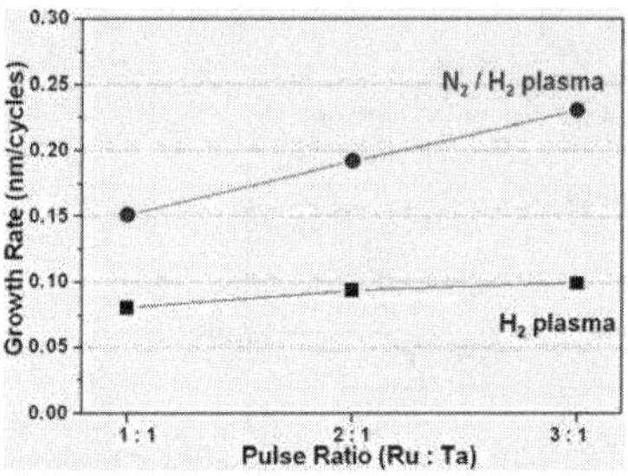

Fig. 4 RuTaN Growth rate by Ru/Ta pulse ratio

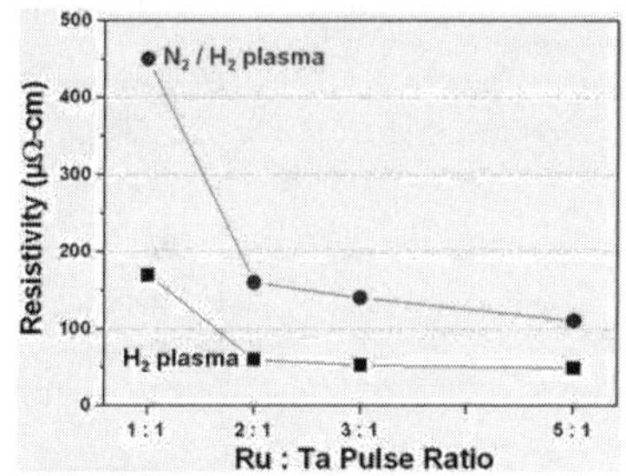

Fig. 5 Resistivity by Ru pulse ratio

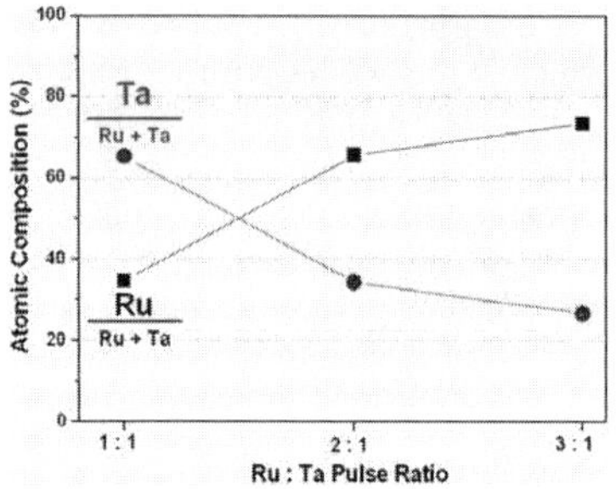

Fig. 6 Relation of atomic ratio and Ru/Ta pulse ratio

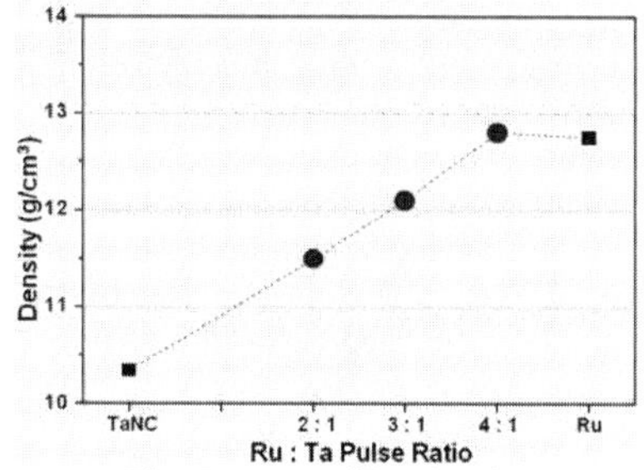

Fig. 7 RuTaN film density by changing of Ru Pulse Ratio

Fig.6 and Fig.7 show Ru/(Ru+Ta) and Ta/(Ru+Ta) atomic composition ratio and film density of RuTaN as a function of Ru/Ta precursor pulse ratio respectively. As Ru/Ta pulse ratios are increased from 1:1 to 3:1, Ru atomic composition rapidly increased and Ta atomic composition decreased. These results suggest that Ru/Ta atomic ratio is controlled by Ru/Ta pulse ratio. Ru/Ta precursor pulse ratio increases from 1 to 2 and 3, Ru/(Ru+Ta) atomic ratio of RuTaN films increases from 34.6 to 65.6 and 73.4 at%. Fig. 7 shows the film density measured by XRR. Film density of TaNC single layer was estimated to be 10.3 g/cm^2, and density of Ru film was estimated to be as same as Ru bulk. And as the Ru/Ta pulse ratio increased, densities of RuTaN films linearly increased. RuTaN film deposited with Ru/Ta precursor pulse ratio of 2 has higher density about 11.5g/cm^2. But Ru/Ta precursor pulse ratio increased to 4, density of the RuTaN film became 12.8 g/cm^2 which is very close to Ru bulk.

3. Adhesion property of RuTaN films to Cu and SiO_2

After CVD-Cu films (approximately 500 nm) were deposited on TaNC, RuTaN and Ru films, adhesion property were evaluated by tape test after and before annealing. Table.1 shows summary of tape test results. RuTaN films show good adhesion property both to Cu and SiO_2, but film peeling happened in CVD-Cu/Ru/SiO_2 films and CVD-Cu/TaNC/SiO_2 films.

To check the peeled off interface, we measured element existing on the upper and bottom peeled off surfaces by energy dispersion x-ray spectroscope (EDX). In the peeled interface of Cu/TaNC/SiO_2 films, high concentration of copper was detected from upper peeled interface, but high concentration of Ta was detected from the bottom peeled interface. These results suggest that a peeling happened between CVD-Cu and TaNC film interface. In the case of Cu/Ru/SiO_2 structure, EDX analysis also shows that peeling happened at Ru/SiO_2 interface.

<table>
<tr><th>Pulse Ratio</th><th>As deposition</th><th>After Annealing</th></tr>
<tr><td rowspan="2">TaNC</td><td>Fail</td><td>Fail</td></tr>
<tr><td colspan="2">Peeled off at Cu/TaN interface</td></tr>
<tr><td>Ru : Ta = 1 : 1</td><td>Pass</td><td>Pass</td></tr>
<tr><td>Ru : Ta = 2 : 1</td><td>Pass</td><td>Pass</td></tr>
<tr><td>Ru : Ta = 3 : 1</td><td>Pass</td><td>Pass</td></tr>
<tr><td rowspan="2">Ru</td><td>Fail</td><td>Fail</td></tr>
<tr><td colspan="2">Peeled off at Ru/SiO_2 interface</td></tr>
</table>

Table. 1 Summary of tape test with Cu/barrier films/ SiO_2 structure

However, peeling was not observed in Cu/RuTaN films/SiO_2 structures after and before annealing even if Ru/Ta precursor pulse ratio was ranged from 1 to 3. It is noted that RuTaN films have good adhesion both to Cu and SiO_2. Ru including in TaN film can improve adhesion property dramatically.

4. Effect of bottom materials to sheet resistance of extremely thin CVD-Cu films

Fig.8 shows plane view SEM photographs of (A) CVD-Cu(20 nm)/RuTaN/SiO_2 and (B) CVD-Cu/Ru/TaNC/SiO_2 structures which have the same Cu thicknesses. In these results, (A) Cu on RuTaN grain size is larger than (B) Cu on the Ru film. These results suggest that Cu nucleation density on the RuTaN film seems to be lower than on the Ru film at the same film thickness. So even though they have the same Cu thicknesses, grain size and density results are different.

Fig.9 shows the sheet resistance of extremely thin CVD-Cu film deposited on RuTaN films, RuTa films and Ru film. Thickness of CVD-Cu films was below 5 nm, which were deposited with same conditions. Considering that there is no incubation time on the RuTaN films, RuTa films and Ru film, sheet resistance is related with Cu film continuity. Lower sheet resistance of Cu film is obtained on Ru film and RuTa film than on the RuTaN films. But sheet resistance on the RuTaN films increases as Ru/Ta pulse ratio decreases from 4 to 2. Especially, as Ru/Ta pulse ratio is 2, sheet resistance value is much higher.

Theses results suggested that continuous CVD-Cu film can be formed on the Ru and RuTa film due to higher nucleation density, leading smaller Cu grain size as observed in Fig.8. Therefore, Ru film or RuTa film seems to be attractive as a glue layer before CVD-Cu seed formation, leading to thin and continuous Cu seed required advanced Cu interconnect.

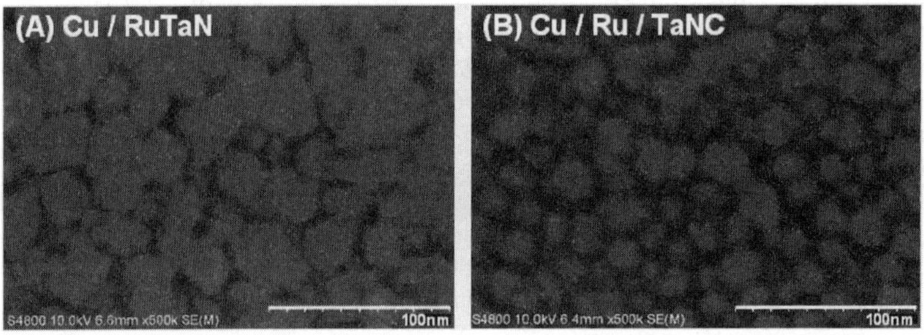

Fig. 8 FE-SEM results of Cu grain sizes on (A) RuTaN and (B)Ru/TaNC

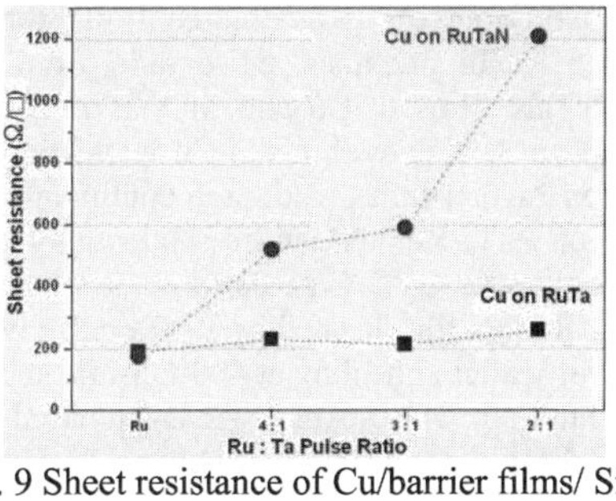

Fig. 9 Sheet resistance of Cu/barrier films/ SiO_2

5. Cu diffusion barrier properties of RuTaN films

Fig. 10 shows XRD spectrums of Cu/RuTaN/SiO_2/Si structures with Ru/Ta pulse ratio of 2 and 1 after rapid thermal annealing at 400°C for 1Hr. As shown in this result, main Cu and Cu_2O peaks are detected in this XRD spectrum because Cu surface was slightly oxidized during ramp annealing process. And small Ru and TaN peaks are also detected. But $CuSi_x$ peaks are not detected in both XRD spectrums of (A) RuTaN (2 : 1) and (B) RuTaN (1 : 1) films. This XRD results shows that 2 nm-thick RuTaN films deposited with Ru/Ta precursor pulse ratio of 1 and 2 is sufficient as Cu barrier film.

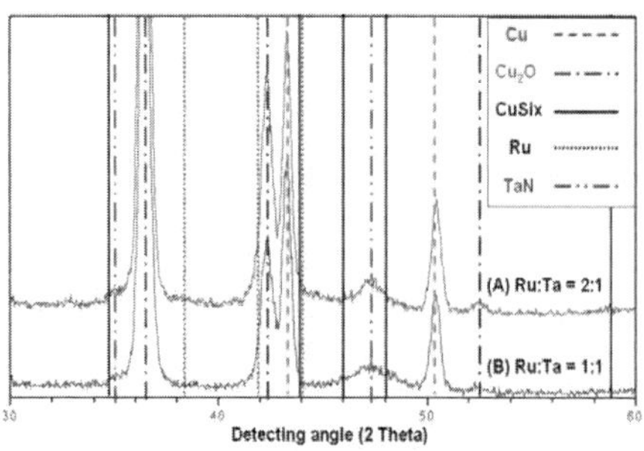

Fig. 10 XRD result of annealed Cu/RuTaN/Si structure

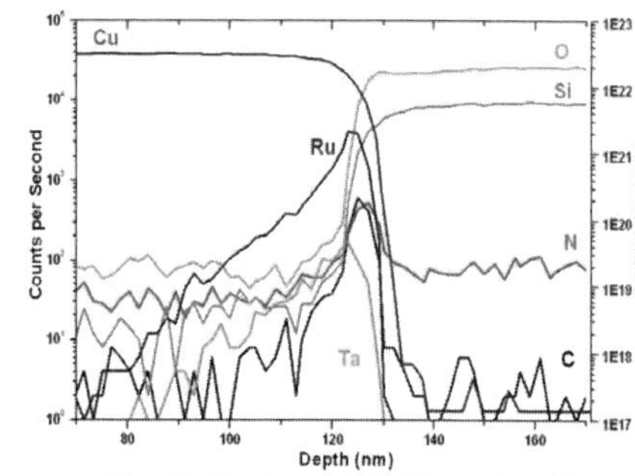

Fig. 11 Backside SIMS results of annealed Cu on RuTaN film

In addition to this Cu barrier evaluation, Cu barrier property of the RuTaN deposited with Ru/Ta precursor pulse ratio of 2 is investigated with backside SIMS for measuring the Cu diffusion to bottom Si. After annealing the sample which has Cu/RuTaN (2:1)/SiO_2/Si structure with the same condition, depth profile are analyzed by SIMS from Si substrate to upper Cu film as shown in Fig.11. Cu diffusion seems to be completely suppressed by the RuTaN film. It is because Cu concentration in SiO_2 film below the RuTaN film reduces down to below detection limit which is less than 10^{18}. This means that Cu diffusion can be prevented from 2 nm-thick RuTaN barrier metal.

Conclusion

A novel PEALD-RuTaN process with suitable metal precursors and N_2/H_2 plasma can provide remarkably 2 nm thick-Cu barrier film, which is attractive for advanced Cu interconnect. Advantage of PEALD-RuTaN process is flexible controllability of Ru, Ta and nitrogen atomic composition by adjusting Ru/Ta precursor pulse ratio from 1 to 3 and plasma conditions. Adhesion property of RuTaN films to Cu and SiO_2 is remarkably improved independent on Ru/Ta pulse ratio from 1 to 3. However, CVD-Cu nucleation density on Ru and RuTa films seems to be higher than on RuTaN films, leading to continuous Cu film deposition. Stacked film structures of Ru/RuTaN or RuTa/RuTaN can be sequentially deposited in single ALD reactor, which are effective for continuous CVD-Cu seed formation.

2 nm-thick PEALD-RuTaN film formed by Ru/Ta pulse ratio of 2 and N_2/H_2 plasma shows sufficient Cu barrier property after annealing at 400°C while keeping good adhesion property and low resistivity below 160 μΩ-cm. This novel PEALD-Ru based barrier process is one of the most promising Cu barrier metal formation techniques beyond 32 nm node.

Acknowledgments

The authors are thankful to Prof. Yukihiro Shimogaki in the University of Tokyo for discussion about CVD-Cu film property. And we are also thankful to Dr. Nobuyoshi Kobayashi of ASM for his continuous support.

Reference

[1] The International Technology Roadmap for Semiconductors 2007
[2] N. Ohashi, et al. The AVS 5th Topical Conference on Atomic Layer Deposition (2005)
[3] Seong Jun Jeong, Se Hun Kwon, Sang Won Kang, The AVS 6th International Conference on Atomic Layer Deposition (2006)
[4] D. Jeong, H. Inoue, H. Shinriki, International Interconnect Technology Conference, IEEE, 156 (2008)

Conference Proceedings AMC XXIV © 2009 Materials Research Society

Resistance of Ru barrier layer against oxidation by moisture in low-k film

Nobuaki Tarumi, Morio Shiohara, Mistuhide Abe, Hiroki Imamura, Seiichi Kondo, and Shinichi Ogawa
Semiconductor Leading Edge Technologies, Inc., 16-1, Onogawa, Tsukuba, Ibaraki, 305-8569, Japan
E-mail: tarumi.nobuaki@selete.co.jp

ABSTRACT

A ruthenium (Ru) barrier layer showed higher resistance against oxidation of Cu interconnect by moisture in a low-k film than a TaN/Ta barrier. Barrier properties of the Ru film did not degrade by moisture.

INTRODUCTION

It has been reported that a barrier property of a polycrystalline Ru (p-Ru) film against a Cu film was poor because of its column structure, while an amorphous Ru (a-Ru) formed by Ar+N2 atmosphere sputtering worked as a Cu diffusion barrier on an underneath dielectric layer (Fig. 1), and then an a-Ru / p-Ru stacked barrier structure was recently proposed [1]. It was reported that a Ta barrier has lower oxidation resistance to moisture uptake in a low-k film, and it resulted in lower yield because of Cu or Ta oxidation [2]. On the other hand, resistance against moisture in the low-k film of the Ru barrier film has never been reported.

In this work, it was demonstrated that resistance of the Ru film against oxidation by moisture uptake in a low-k film was higher than that of a TaN / Ta barrier film.

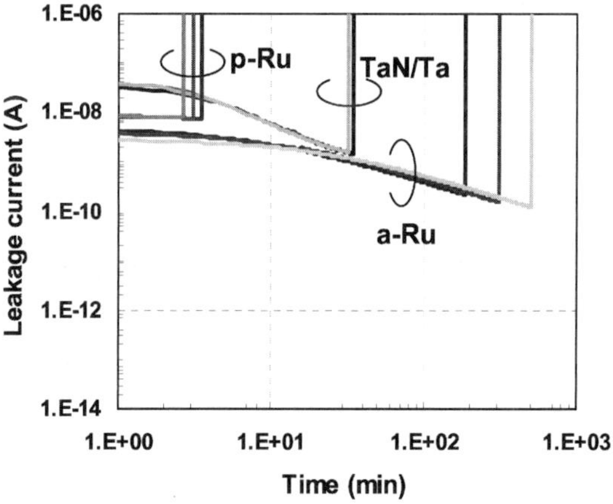

Fig. 1 An elapsed time change of leakage currents at 200 degrees C, 50V applied voltage for barrier metal. (From proc. 2008 IITC p.104 Fig.6)

EXPERIMENTAL

A p-Ru film and an a-Ru film was deposited by pure Ar sputtering and Ar+N2 sputtering, respectively, onto low-k (k=2.7) films with or without moisture uptake. Crystallinities of the Ru and Cu films were evaluated by X-ray diffraction (XRD). To characterize resistance of the Ru barrier film against the moisture in the low-k film, Si-sub. / 100 nm thick low-k / 10 nm thick p-Ru / Cu / Al-pad (passivated) diode structures (Fig.2) were examined at 200 degrees C, 12 V for up to 5 hours. Amount of moisture into the low-k film and damages onto the low-k films during the barrier metal deposition were evaluated by a micro beam FTIR method [3]. Cu / low-k interconnects of 140 nm pitch with 10nm p-Ru or 5nm TaN / 5nm Ta barriers were fabricated, and electrical properties (line resistance and capacitance between lines) were characterized. Moisture in the low-k film was adsorbed by ash and wet cleaning processes without anneal before the barrier metal deposition.

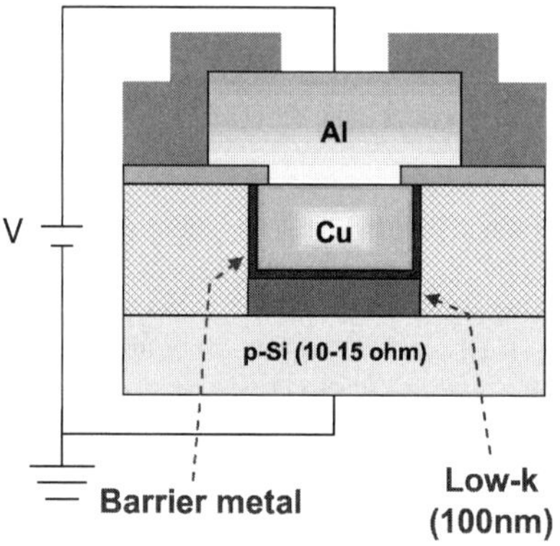

Fig. 2 A cross section structure of the examined diode in which barrier properties were characterized.

RESULTS AND DISCUSSION

As shown in Fig. 3, any XRD peak from Ru oxide and Cu oxide was not observed from the 10 nm p-Ru / 50nm Cu stacked film after 350 degrees C anneal in N2 atmosphere for 30 minutes, which indicated that the oxidation as being expected might not occur. The micro beam FTIR characterization of space areas did not show any spectra with enough intensity of moisture transmittance in these samples. In spite of the treatments for moisture uptake, difference in amount of moisture was not seen in these XRD and micro bam FTIR.

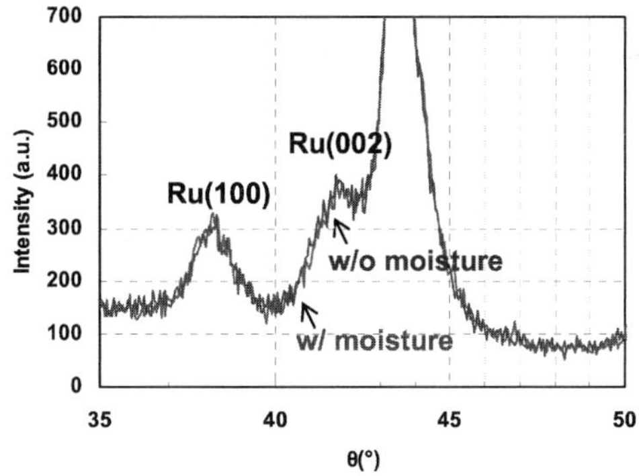

Fig. 3 XRD spectra of p-Ru with and without moisture after anneal at 350 degrees C, 30 mins.

Initial low-k capacitances with or without moisture were almost same, while initial currents-voltage properties of the a-Ru / p-Ru stacked barrier film and TaN/Ta stacked barrier were worse than that of the p-Ru film as shown in Fig.4. It seems that there might generate damages onto the low-k film during deposition of a-Ru or TaN in our experiments.

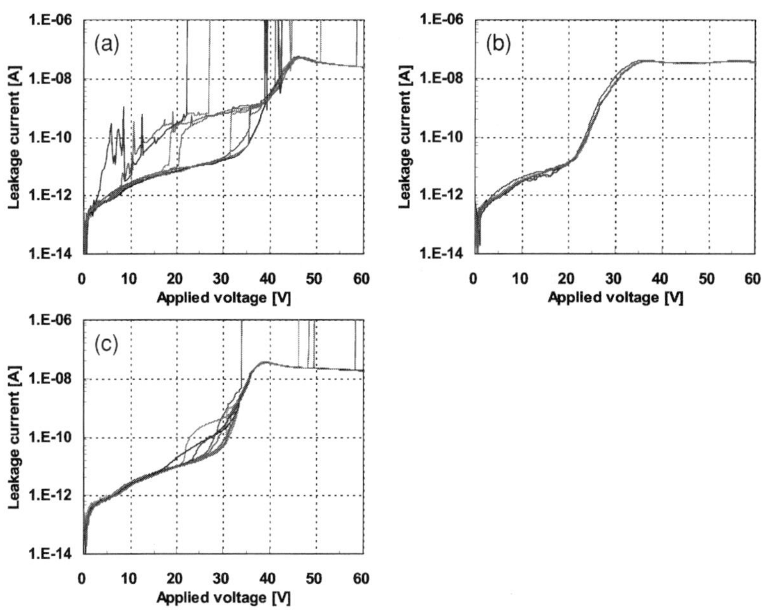

Fig. 4 I-V characteristics of the 100 nm low-k film.(a) a-Ru/p-Ru, (b) p-Ru, (c) TaN/Ta barrier.

Damages during the deposition of barrier metals were characterized by the micro beam FTIR [3]. Fig.5 shows dependency of Si-CH3/Si-O ratios on the deposited barrier metals calculated from the micro beam FTIR spectra, which showed that the damage into the low-k film during deposition of the a-Ru barrier and TaN / Ta stacked barrier might be worse than that during depositing the p-Ru barrier, and it seems that the difference might not depend on the difference in oxidation of Ru but the deposition damage. Then we focused on the p-Ru barrier behavior because of the lower damage into the low-k film to clarify the influence of the moisture on the barrier property, if any.

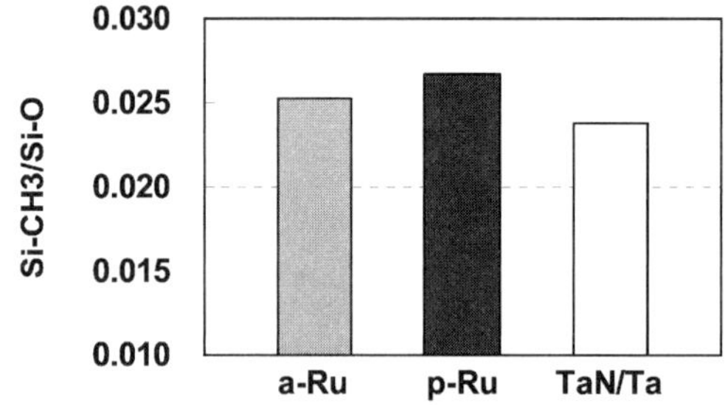

Fig. 5 Si-CH3/Si-O intensity ratio by the micro beam FTIR measurement of interconnect structure.

Fig.6 shows an elapsed time change of leakage currents at 200 degrees C, 12 V applied voltage for the 10 nm p-Ru film. As for time until breakdown, the p-Ru with moisture was same as without moisture. It indicates that barrier properties of the p-Ru film did not degrade by moisture in the low-k film.

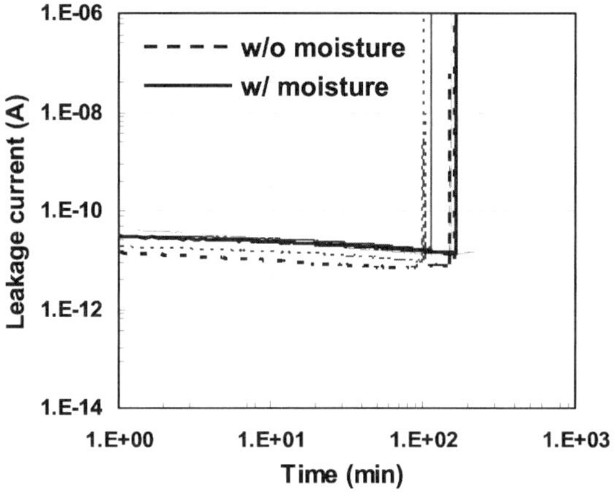

Fig. 6 An elapsed time change of leakage currents at 200 degrees C, 12V applied voltage for p-Ru with and without moisture.

Fig.7 showed a dependency of line resistance of 140 nm pitch interconnects (70nm line and space) on barrier materials with the moisture. A change in the p-Ru barrier interconnect was smaller than that of the TaN/Ta barrier. Fig.8 showed change in capacitance between lines with the p-Ru and TaN/Ta barrier interconnects, and no difference was seen between the Ru and TaN/Ta barrier. From these results, we suppose that moisture might not diffuse through the Ru barrier and Cu interconnect might not be oxidized. Behavior of moisture in the interconnects with barriers needs further investigation. In this experiment, the moisture of a small amount which was not seen by XRD and the micro beam FTIR might oxidize the Cu metal through the TaN/Ta barrier layers or the barrier layers themselves, however the moisture was not enough to affect the capacitance, while further investigation will be required.

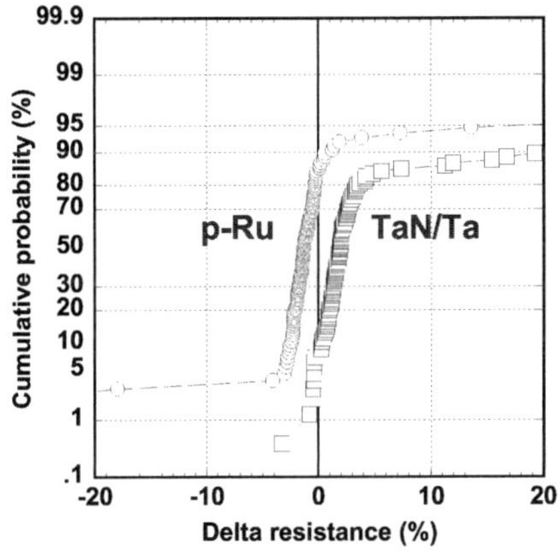

Fig. 7 Line resistance change by moisture in 140 nm pitch interconnect with p-Ru and TaN/Ta barrier.

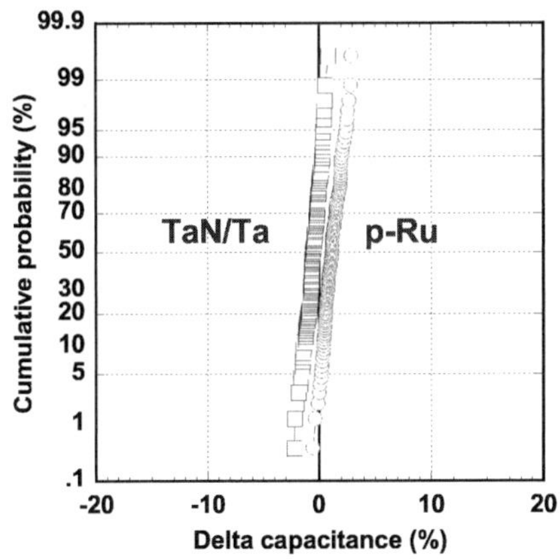

Fig. 8 Capacitance between lines change by moisture in 140 nm pitch interconnect with p-Ru and TaN/Ta barrier.

CONCLUSIONS

The Ru barrier layer showed higher resistance against oxidation of Cu interconnect by the moisture in the low-k film than the TaN/Ta barrier, and barrier properties of the Ru film did not degrade by moisture.

ACKNOWLEDGEMENTS

The authors would like to thank H. Seki and H. Hashimoto (TRC) for the micro beam FTIR measurements, colleagues in a Selete backend process program for discussions, and S. Saito for his encouragement throughout this work.

REFERENCES

1. S. Ogawa et al., Proc. of IITC (2008), p.102.
2. M. Okazaki et al., Proc. of IITC (2004), p.211
3. S. Ogawa et al., Proc. of IITC (2008), p.76

Conference Proceedings AMC XXIV © 2009 Materials Research Society

Formation of a low resistivity W nucleation layer using H_2 (hydrogen radical) reduction gas for W plug

Takashi Nishimori[1], Mitsuo Tanaka[1], Hideki Yuasa[1], and Mitsuhiro Tachibana[1]
[1]Tokyo Electron AT Limited,
650 Mitsuzawa, Hosaka-cho, Nirasaki City, Yamanashi 407-0192, Japan
E-mail:takashi.nishimori@tel.com

ABSTRACT

We have successfully formed a tungsten (W) nucleation layer with a low resistivity using hydrogen radical.

The conventional W nucleation layer, which is formed using SiH_4 or B_2H_6 as the reduction gas, has a higher resistivity than the main deposition layer, which is formed using H_2 as the reduction gas. It is difficult, however, to use H_2 as the reduction gas for nucleation layer at 350°C because Gibbs free energy of H_2 is too small to form a heteroepitaxial layer.Therefore, we experimented with hydrogen radical as the reduction gas for formation of W nucleation layer. The nucleation layer showed a low concentration of impurities, a low lattice strain, and a lower resistivity than the conventional W nucleation layer.

INTRODUCTION

Tungsten (W) film is an important metallic material for the applications of semiconductor. W film is usually used as contact metals or as interconnects. We focused on W film used as contact metals. One of the popular methods to form such W film is CVD (chemical vapor deposition). A typical CVD W film consists of two layers: the lower layer is called 'nucleation layer', which is deposited by reaction of WF_6 with SiH_4 or B_2H_6 on the substrate surface (ex. TiN). The upper layer is called 'main deposition layer £, which is deposited by reaction of WF_6 and H_2 on the nucleation layer.

Although the main deposition layer can hardly be formed without the nucleation layer, the resistivity of the nucleation layer is higher than the main deposition layer (see figure 1), which results in low electric properties at the contact holes. We assume that SiH_4 or B_2H_6 leaves a high impurity content in the W film, leading to the high resistivity. However, the reduction gas for a nucleation layer can not be changed to H_2 because Gibbs free energy of H_2 is too small (see table 1) to form W film as a heteroepitaxial film. In contrast, hydrogen radical's Gibbs free energy is higher than SiH4's one or B2H6's one, therefore we assumed that hydrogen radical's Gibbs free energy is enough to form W nucleation layer.

We report the resistivity of a nucleation layer formed using hydrogen radical as the reduction gas in this paper, as well as the impurities and crystallization of the film.

EXPERIMENT

Figure 2 shows a schematic diagram of Cat-SFD-W apparatus used in this experiment. "SFD" is one of the nucleation methods for the filling a deep hole with an alternating flow of a source gas (ex. WF_6) and a reduction gas (ex.SiH_4) [1]. SFD is an abbreviation of "sequential flow deposition", and "Cat" of "catalytic". Cat is one of the CVD methods and it makes hydrogen radical using a filament and H_2 gas. "Cat-SFD" is a combination of "Cat" and "SFD". This apparatus is a cold wall CVD chamber and it supplies hydrogen radical, which is highly reactive chemical.

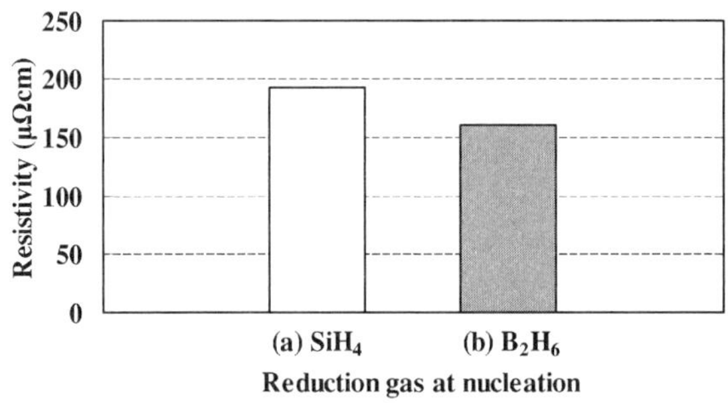

Figure 1. W nucleation layer's resistivity as a function of reduction gas at nucleation. Every layer is 10nm in thickness. In case of 50nm in thickness, nucleation layer's resistivity (about 150μΩcm, SiH_4 or B_2H_6-reduction) is higher than main deposition layer's one (about 30μΩcm).

Chemical Reaction	ΔG° (Kcal/mol)
$WF_6[g] + 3H_2[g] = W[s] + 6HF[g]$	-27
$WF_6[g] + B_2H_6[g] = W[s] + 2BF_3[g] + 3H_2[g]$	-195
$WF_6[g] + 1.5SiH_4[g] = W[s] + 1.5SiF_4[g] + 3H_2[g]$	-211
$WF_6[g] + 6H[g] = W[s] + 6HF[g]$	-295

Table I. The change in Gibbs free energy at 350°C (atmospheric pressure)

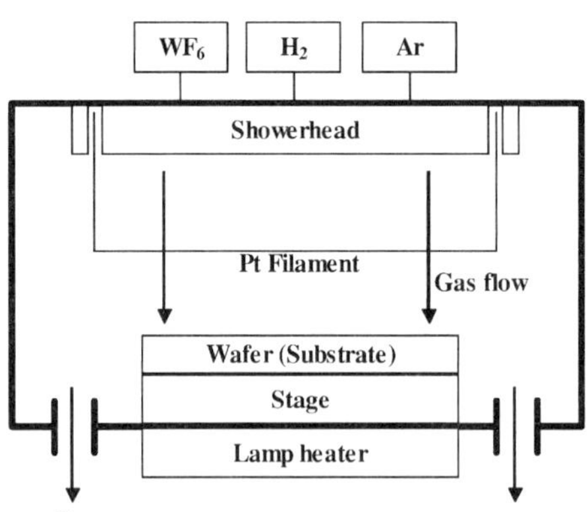

Figure 2. Schematic diagram of Cat-SFD-W apparatus.

H_2 (a reduction gas) changed to hydrogen radical with a 1000°C platinum (Pt) filament and WF_6 (a precursor gas) were alternately introduced into the vacuum chamber through a showerhead. Argon (Ar) was introduced into the chamber for purging purpose between H_2 and WF_6. For comparative experiments, SiH_4 or B_2H_6 was also used as the reduction gas in the apparatus without the filament. The gases were controlled by mass flow controllers (MFC). The chamber pressure was kept at a certain level (mainly at 1000Pa), using a throttle valve and a capacitance manometer. The filament was located between the showerhead and a substrate. The temperature of the filament is controlled to 1000°C while H_2 is being introduced into the chamber using a pyrometer and a voltage transformer. We used 20nm-thick CVD TiN films as an underlayer metal for W deposition are deposited onto thermal oxidized 300mm Si wafers. Each wafer was placed on the stage and heated to a high temperature (mainly 320°C) by a lamp system.

The resistivity of W films was calculated from sheet resistance measurements using a four-point probe and the film thickness measurements using x-ray fluorescence (XRF). The distribution of atoms such as silicon (Si), boron (B), and fluorine (F), and hydrogen (H) in the film was analyzed by secondary ion mass spectroscopy (SIMS). The crystallization of films was analyzed by an x-ray diffraction (XRD).

RESULTS AND DISCUSSION

We confirmed that nucleation film can be formed using hydrogen radical (H_2(Cat))as the reduction gas. The activation energy involved in the formation of the film can be found in figure 3 (Arrhenius plot). The graph shows the activation energies involved in the formation of nucleation film using other reduction gases, SiH_4, B_2H_6, H_2 as well as 'H_2 (Cat)'. In this report, hydrogen radical is referred to as 'H_2 (Cat)' to differentiate from 'H_2'. While 'H_2' was used for the main deposition layer, 'H_2 (Cat)' was used for the nucleation layer. Figure 3 indicates that the activation energy of 'H_2 (Cat)' is smaller than that of 'H_2'. Figure 3 also indicates that the activation energy of 'H_2 (Cat)' and that of 'SiH_4' and that of 'B_2H_6' are in the same range.

We also confirmed step coverage of the H_2 (Cat)-nucleation film in a deep hole (see figure 4). Figure 4(b) shows that the nucleation layer is formed to the bottom of the deep hole, indicating that hydrogen radical has reached to the bottom. Because the main deposition layer can hardly grow to the bottom of the deep hole without nucleation layer, H_2 (Cat)-nucleation film allows the deep hole to be filled with W (figure 4(a), (c)).

Figure 5 shows resistivity of various W nucleation layers. Resistivity of the H_2 (Cat)-nucleation layer is lower than the others. We assume that the low resistivity is caused by two principal factors.

The first factor contributing the low resistivity of H_2 (Cat)-nucleation layer is low impurity concentration in the film. We confirmed that the film contained 4 kinds of impurity: (Si(silicon), B(boron), F(fluorine), and H(hydrogen)). Figure 6 shows SIMS results on three nucleation layers (SiH_4, B_2H_6, and H_2(Cat)). Concentration of each impurity varies by the reduction gases. SiH_4-nuclation layer contains a high amount of Si as an impurity. B_2H_6-nucleation layer contains a high amount of B as an impurity. On the other hand, Si and B concentrations of H_2(Cat)-nucleation layer are below the lower detection limits. Furthermore, F concentration of H_2 (Cat)-nucleation layer is lower than the others because Gibbs free energy of H_2 (Cat) is large (see table I) and F is easily removed as F-H compounds.

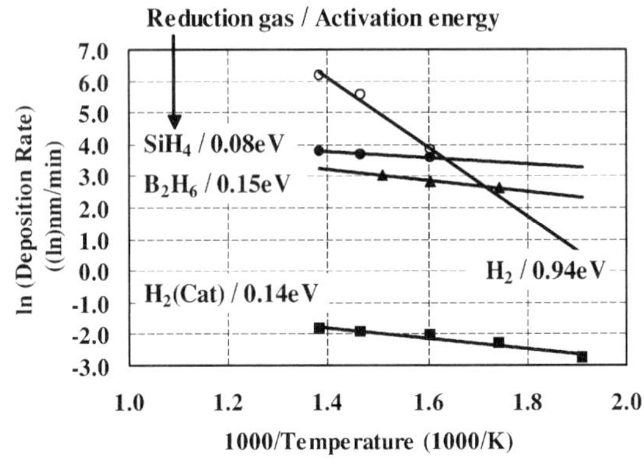

Figure 3. An Arrhenius plot of deposition rate of W films using various reduction gases. SiH_4, B_2H_6 and H_2(Cat) are reduction gas at nucleation. On the other hand, H_2 is reduction gas at main deposition.

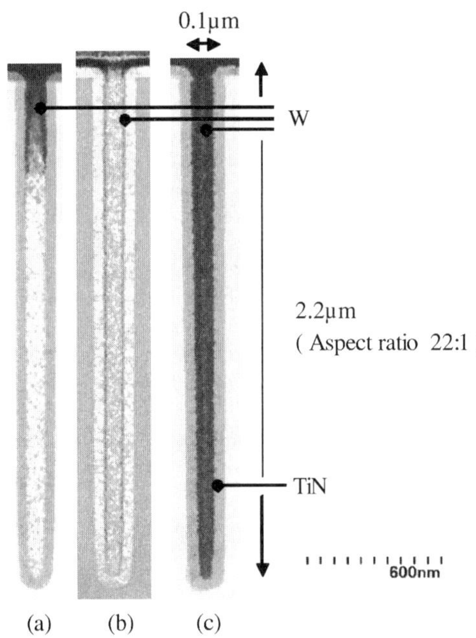

Figure 4. Step coverage of W film in deep hole.
(a): Main deposition layer only
(b): Nucleation layer only (H_2(Cat) reduction)
(c): Nucleation(H_2(Cat))+Main deposition layers

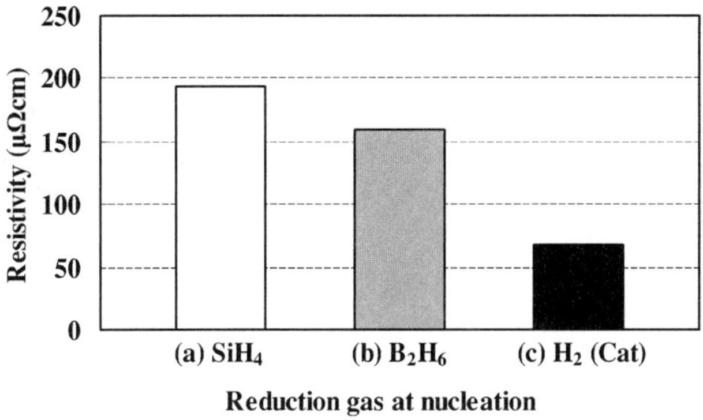

Figure 5. W film resistivity as a function of reduction gas at nucleation. Every layer is 10nm in thickness.

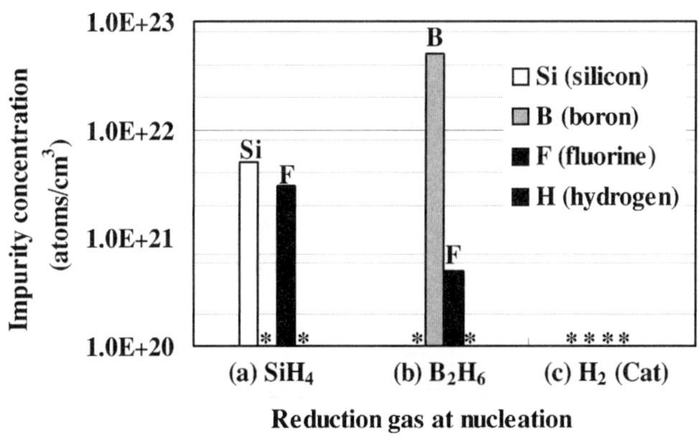

Figure 6. Impurity concentration in W film as a function of reduction gas at nucleation that were analyzed by SIMS.

* Lower than lower detection limit (Si:1.0E18 , B:1.0E18 , F:5.0E18 , H:1.0E20 atoms/cm^3).

The second factor contributing the low resistivity of H_2 (Cat)-nucleation layer is lattice strain. Figure 7 shows XRD spectra of the three nucleation layers (SiH_4, B_2H_6, and H_2(Cat)). H_2 (Cat)-nucleation layer shows higher peak intensity than SiH_4 and B_2H_6, therefore H_2 (Cat)-nucleation layer's crystallization is stronger than the others. XRD spectra also show the lattice strain of the nucleation layers. Figure 8 shows that lattice strain of H_2 (Cat)-nucleation layer is smaller than the others.

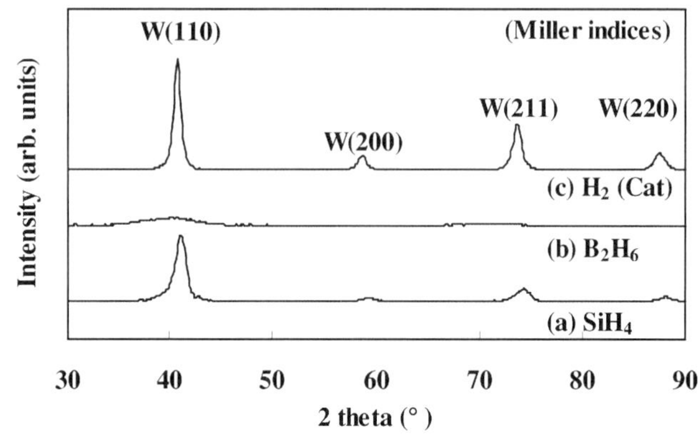

Figure 7. X-ray diffraction spectra of various W nucleation layers at different reduction gas. Every layer is 50nm in thickness.

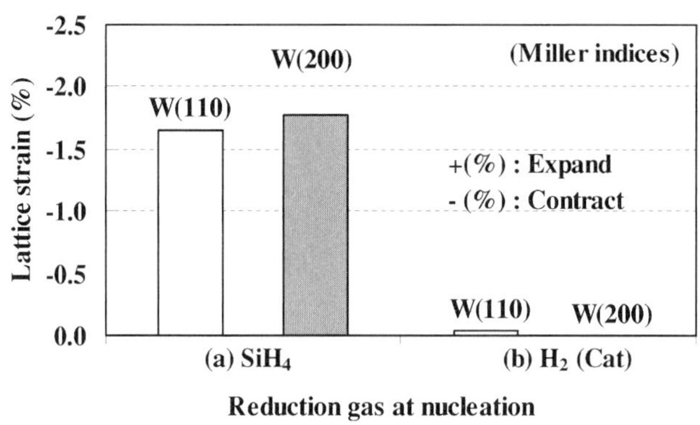

Figure 8. W film's lattice strain as a function of reduction gas at nucleation. Lattice strain is the ratio of this study's lattice constant to literature's one[2].

CONCLUSIONS

We have successfully formed W nucleation layer using hydrogen radical. This nucleation layer can be formed to the bottom of the 0.1μm hole with 22:1 aspect ratio. Resistivity of this nucleation layer (about 70μΩcm) is lower than that of conventional layer (resistivity of SiH_4 or B_2H_6 reduction's nucleation layer is about 160-190μΩcm. Every layer is 10nm in thickness.).

This layer also shows that impurity concentrations (Si , B , F) are below the lower detection limits (analyzed by SIMS) and lattice strain is low. These factors contribute to a low resistivity.

ACKNOWLEDGMENTS

The authors wish to thank the following people for their contributions to this work: T. Ueda, T. Mase, A. Matsuyama, and M. Moroi.

REFERENCES

1. K. Okubo, H. Ishizuka, K. Suzuki, K. Sato, and M. Tachibana, *Advanced Metallization Conference 2001:Asian Session*,76(2001).

2. Swanson, Tatge.Natl. Bur. Stand. (U.S.), Circ. 539, vI p28 (1953).

Conference Proceedings AMC XXIV © 2009 Materials Research Society

Influence of Hydrogen Plasma Treatment on Ni and NiPt Silicide Process

T. Tsutsumi , K. Kihara, T. Yamaguchi, T. Okudaira, K. Kashihara ,
K. Maekawa, S.Kudo, K. Asai, M. Kojima and M.Yoneda
Process Development Dept., Production Technology Development Div.,
Production and Technology Unit, Renesas Technology Corp.
4-1, Mizuhara, Itami-shi, Hyogo, 664-0005, Japan
Phone: +81-72-787-2516 E-mail: tsutsumi.toshiaki@renesas.com

ABSTRACT

Influence of hydrogen plasma treatment on electrical characteristics of Ni and NiPt salicide on CMOS were investigated. Using Ni silicide, the sheet resistance and the junction leakage current were increased only on PMOS devices after hydrogen plasma treatment. In the silicide layer, the pyramidal and the pipe shaped-Ni silicides were observed, and they were identified to $NiSi_2$. We found that the phase transition to $NiSi_2$ from NiSi increased the sheet resistance due to its high resistivity and pipe shaped $NiSi_2$ increased the junction leakage current due to the crossing junction. In order to suppress these defects, NiPt alloy silicide was introduced, and it is demonstrated that the degradation of electrical characteristics was successfully suppressed. It is revealed that the NiPt alloy silicide is effective not only to suppress the rapid diffusion of Ni atoms, but also to suppress the hydrogen induced defects.

INTRODUCTION

Ni silicide is the most suitable contact material for CMOS devices with sub-50 nm gate length because of its low resistivity, less silicon consumption and low reaction temperature compared with those of Co silicide. During conventional process of silicidation, source/drain regions of silicon are exposed and damaged by reactive ion etching (RIE) plasma in the formation of gate sidewall spacer, and the silicidation is strongly affected by this RIE plasma damage. In the case of Co silicide, it has been reported that the sheet resistance is increased by the RIE plasma damage [1]. In the case of Ni silicide, it has been reported that the sheet resistance of Ni silicide on P^+ Si surface etched by CF_4-CHF_3-Ar gas mixture is increased [2]. It also has been reported that the pyramidal $NiSi_2$ growth at the interface between Ni silicide and Si substrate increases the sheet resistance of Ni silicide. In addition, it is strongly suggested that the hydrogen ions (H^+) injected into silicon during RIE cause the pyramidal $NiSi_2$ and degrade the electrical properties. However, the quantitative analysis of influence of hydrogen plasma on the pyramidal growth and the electrical properties of Ni silicide has not been clarified yet. To stabilize the NiSi phase, it has been reported that NiPt alloy silicide is thermally stable and improve electrical characteristics[3,4,5]. However, it has not been investigated that NiPt alloy silicide is effective to suppress the electrical degradation induced by the hydrogen plasma damage.

In this paper, we report the influence of hydrogen treatment on the sheet resistance of Ni silicide and the junction leakage current of CMOS devices in detail. Especially, we focus on the role of hydrogen in silicon on the diffusion of Ni atoms during silicidation. Furthermore, the effect of NiPt alloy silicide on the reduction of hydrogen induced defect is also reported.

EXPERIMENT

Three groups of samples for sheet resistance evaluation were prepared by the following procedure shown in Fig.1. Firstly, undoped silicate glass(USG) layers were formed on heavily B or As doped Si(100) wafers. Secondly, contact windows for sheet resistance measurement were opened on the USG films in the size of (15×45) mm^2 using dry and wet etching methods. Sample A was etched by capacitively coupled plasma using hydrogen free gas mixture(CF_4-Ar system) and hydrogen contained gas mixture(CF_4- CHF_3-Ar system). Sample B was etched by diluted HF solution to avoid dry etching damage. Sample C was etched by diluted HF solution and then exposed to hydrogen plasma generated by inductively coupled plasma system. The exposure time of hydrogen plasma is varied from 5sec to 30sec, and the bias power was kept constant at 100W. Thirdly, 7nm thick Ni or NiPt films and TiN capping layers were sequentially deposited by DC magnetron sputtering method. Finally, Ni silicide films were formed by conventional two-step process[6]. The sheet resistances were measured by four-terminal probe method. The CMOS TEG devices for junction leakage current evaluation were prepared by the conventional procedure. These TEG devices were also exposed by hydrogen plasma prior to Ni or NiPt film deposition. The gate edge-leakage current, that is the largest component for MOSFET leakage current [6], was measured. The failure points in the TEG device, in which the large leakage current was detected, were observed by cross-sectional TEM.

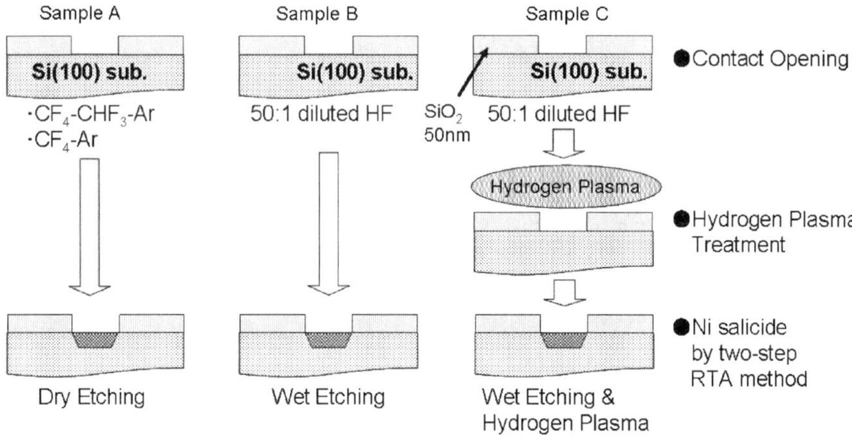

Figure 1. Process sequence of sheet resistance evaluation samples

RESULTS and DISCUSSION

Firstly, influence of etching gas mixture on sheet resistance was investigated. Sheet resistance of Ni silicide film formed on P^+ substrate etched by CF_4-CHF_3-Ar gas mixture was 32ohm/sq, which is higher than that etched by CF_4-Ar gas mixture of 18ohm/sq. Since this value is almost equal to the value of 17ohm/sq. of that etched by diluted HF solution, these results indicate that the etching process using CF_4-CHF_3-Ar gas mixture causes some damage on Si surface prior to silicidation. In order to clarify the cause of the difference in these sheet resistances, the outgas from the surface was

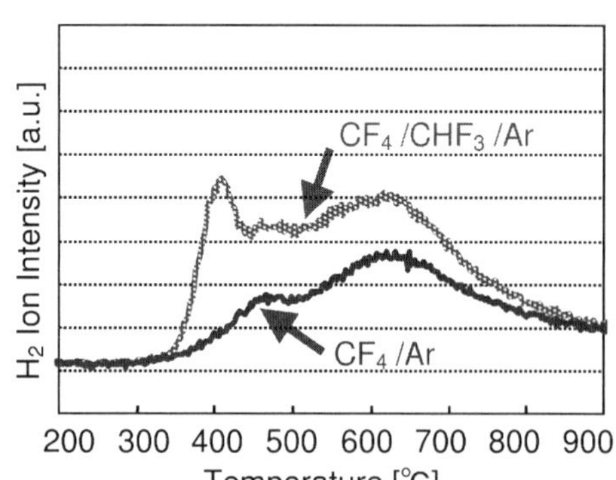

Figure 2. TDS spectra of H_2 from Si substrate exposed to CF_4-CHF_3-Ar and CF_4-Ar plasma

evaluated by TDS (Thermal Desorption Spectrometry). As shown in figure 2, H_2 desorption intensity from the substrate etched by CF_4- CHF_3-Ar gas mixture was higher than that from the substrate etched with CF_4-Ar gas mixture. Comparing the component of these gas mixtures, it is considered that hydrogen ions are injected in the Si surface by the CF_4- CHF_3-Ar plasma, and these injected hydrogen ions cause the increase in Ni silicide sheet resistance. In order to clarify this mechanism, the dependence of hydrogen plasma treatment time on junction leakage current and sheet resistance was investigated.

Figure 3 shows Ni silicide sheet resistance on P^+ and N^+ substrate as a function of hydrogen plasma treatment time. Although the sheet resistance of Ni silicide on N^+ substrate was not dependent on the hydrogen plasma treatment time, that on P^+ substrate increased drastically with the treatment time. It is strongly suggested that hydrogen plasma is main cause of the sheet resistance increase on the substrate exposed by CF_4- CHF_3-Ar gas mixture. Figure 4 shows the junction leakage currents of CMOS devices as a function of the hydrogen plasma treatment time. As the exposing time increased, the junction leakage currents increased only on PMOS devices.

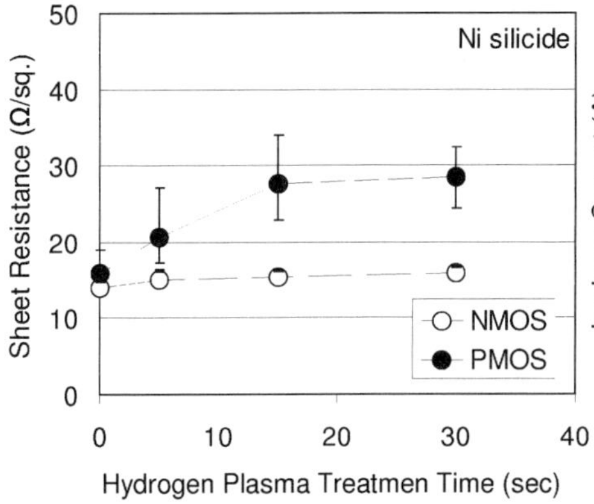

Figure 3. Dependence of Ni silicide sheet resistance on hydrogen plasma treatment time.

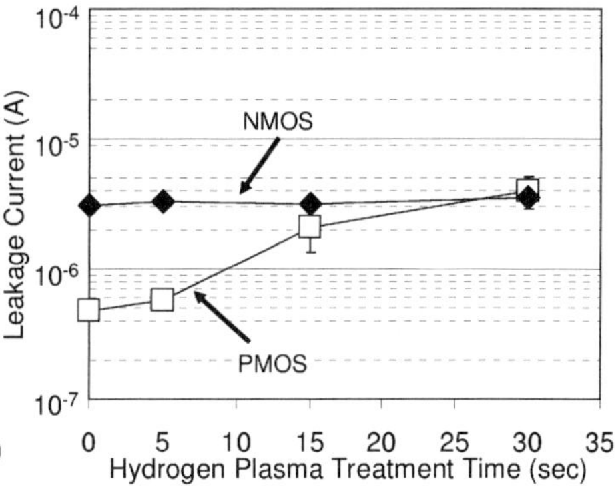

Figure 4. Dependence of junction leakage current of CMOS on hydrogen plasma treatment time.

It has been reported that implanted B reduces the $NiSi_2$ formation temperature [7]. Then the cause of the sheet resistance increase only on PMOS is considered to be the pyramidal $NiSi_2$ growth. In order to clarify the cause of junction leakage degradation on PMOS, failure point was observed by cross-sectional TEM. As shown in figure 5, pyramidal $NiSi_2$ growth and pipe shaped Ni silicide were observed. This pipe shaped Ni silicide crosses p/n junction and increases the leakage current. Using high resolutional TEM, it is revealed that the pipe shaped Ni silicide is identified to be single crystal $NiSi_2$ epitaxially grown on Si{111} plane along <$\bar{1}$10> direction. It is concluded that pipe shaped Ni silicide is formed by Ni atom diffusion along the <110> direction and epitaxial growth of $NiSi_2$ on Si{111} planes. It has been reported that hydrogen ions were injected deep into the Si substrate by the hydrogen plasma treatment and generated vacancies at Si surface[2]. These vacancies enhanced Ni diffusion along <110> direction, because this direction is the channeling direction of Si crystal[8]. Finally the $NiSi_2$ grew epitaxially on Si{111} planes and the pipe shaped $NiSi_2$ was formed.

Figure 5. Cross-sectional TEM picture of high leakage point on PMOS.

In order to reduce the leakage current caused by the pipe shaped $NiSi_2$, phase transition from NiSi to $NiSi_2$ should be suppressed. It has been reported that additional Pt (5atm %) increases the formation temperature of $NiSi_2$ and stabilizes NiSi phase [3,4]. Sheet resistances of Ni and NiPt alloy silicide film on PMOS as a function of hydrogen plasma treatment time were shown in figure 6. Sheet resistance of NiPt alloy silicide was not increased by hydrogen plasma treatment. Figure 7 shows the junction leakage currents at gate edge of CMOS devices with NiPt alloy silicide. Increase in leakage current was successfully suppressed.

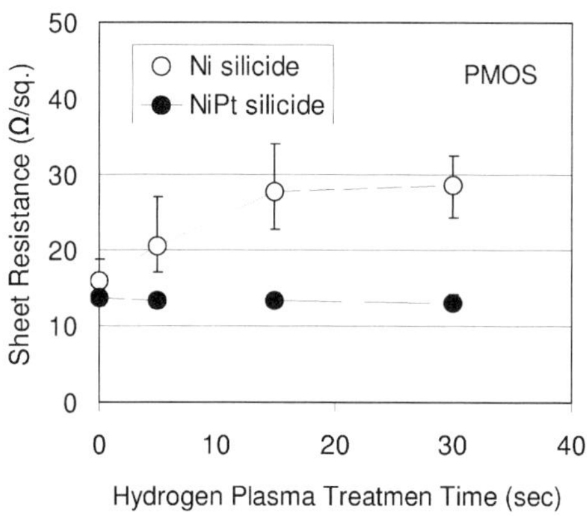

Figure 6. Dependence of sheet resistance of Ni and NiPt silicide on hydrogen plasma treatment time.

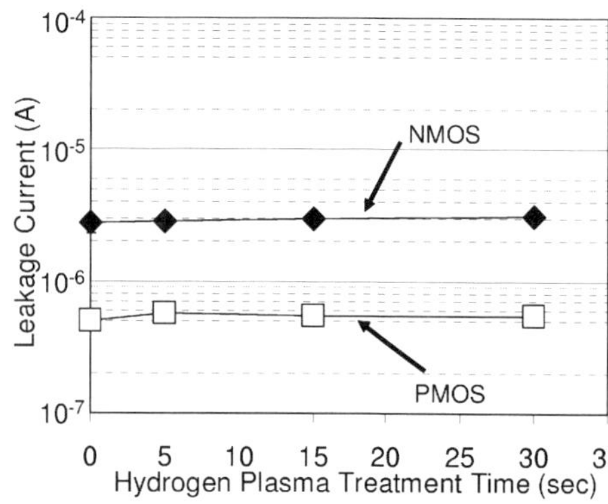

Figure 7. Dependence of junction leakage current on hydrogen plasma treatment time on CMOS devices with NiPt silicide.

These results indicate that the Pt atoms suppress the anomalous $NiSi_2$ growth on the substrate after hydrogen plasma treatment. We propose a following model as shown in figure 8. Ni atoms diffuse into Si substrate by exchange with neighboring vacancies induced by hydrogen plasma. When Ni silicide is formed from NiPt alloy, some Pt atoms segregate at the NiSi/Si interface and are trapped by the vacancies induced by implanted hydrogen. Consequently, the Ni diffusion via the vacancies is suppressed and abnormal $NiSi_2$ growth is not occurred.

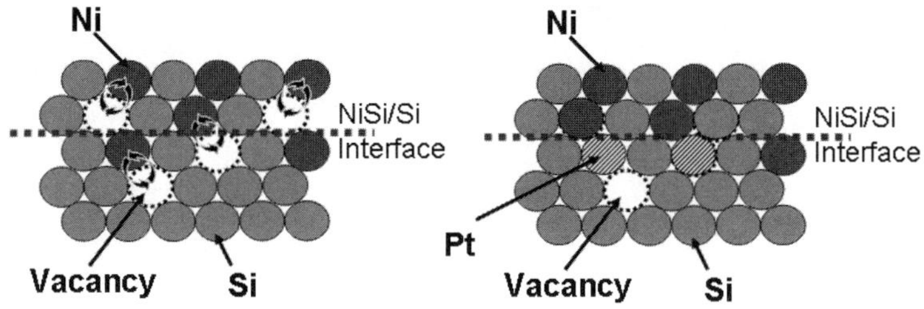

Figure 8. Suppression model of Ni diffusion by Pt segregation.

CONCLUSIONS

We revealed the influence of hydrogen plasma damage to Ni silicide electric characteristics. Hydrogen plasma enhances anomalous $NiSi_2$ growth on Si{111} planes along <110> direction which causes the increase in leakage current on PMOS device. It is also revealed that the degradation of leakage current by hydrogen plasma on PMOS is successfully suppressed by adopting NiPt alloy silicide. NiPt alloy silicide is a leading candidate with resistance to plasma damage.

REFERENCES

1 T.Kimura et al.;J.VacSci.Technol. A25 P1068 (2007)
2 K.Kashihara et al.; Ext.Abst.8th IWJT, P154 (2008)
3 D.Mangelinck et al; Appl. Phy. Lett., 75 P1736 (1999)
4 C. Detavernier et al; ECS Transactions, 3 P131 (2006)
5 Y.Chiang et al.;AMC 2005, P193
6 T. Yamaguchi et al., IEDM Tech.Digest, P855 (2006)
7 L.J.Chen et al;J.Appl.Phys, 62 P2789 (1987)
8 S.Kudo et al.; 46th IRPS, P580 (2008)

Conference Proceedings AMC XXIV © 2009 Materials Research Society

Extending Al Interconnect Technology Beyond the Classical Performance Limits

William J Murphy, David C Thomas, Tom C Lee, Jonathan Chapple-Sokol, Jeffrey P Gambino, Timothy D Sullivan, Daniel A Delibac, Daniel S Vanslette, Zhong-Xiang He, Stephen A Mongeon, and Stephen E Luce
IBM Microelectronics, Essex Junction, VT, 05452

ABSTRACT

In this paper, we report on a novel Al interconnect metallurgy that addresses the current density requirements of advanced ULSI IC technologies. We report on the effects of bottom liners on electrical upstream electromigration performance of three metallization stacks: (I) Ti/TiN/AlCu/Ti/TiN (conventional Al wiring), (II) W/Ti/TiN/AlCu/Ti/TiN and (III) W/TiN/AlCu/Ti/TiN. Stack II shows a 6x lifetime improvement in the electrically open fail mode and a 2-3x current density improvement (see Fig. 9) over conventional Al wiring for the upstream electromigration mechanism. Stack III showed a ~20x improvement in open fails, and a 4-5x improvement in current density.

INTRODUCTION

Despite the introduction of copper (Cu) interconnect technology, aluminum (Al) interconnects are still used in a wide portfolio of advanced products.. When using a typical fail criteria (for example, a 20% resistance shift as reported in this paper), Cu has a longer electromigration lifetime than Al, because self-diffusion in Cu is slower than self-diffusion in Al. While Cu interconnect provides improved current density performance to Al, the downstream via fail mode is the limiter for Cu application due to the absence of a redundant layer to the via landing on the Cu wire (Figure 1). As shown, only enough void volume needs to accumulate to separate the Cu wire from the via. The R-t plot for Cu dual damascene (DD) interconnect (Figure 2) shows the pop up fail resulting when the connection fails through voiding. A pop up fail is defined as an abrupt increase in resistance caused by a high resistance or even open circuit.

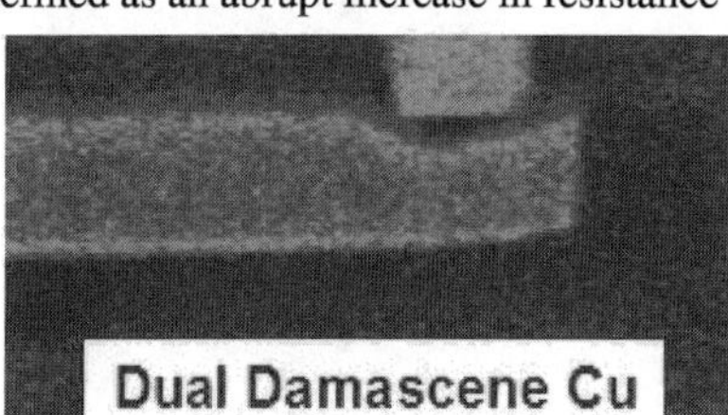

Figure 1 Cu EM fail by via downstream

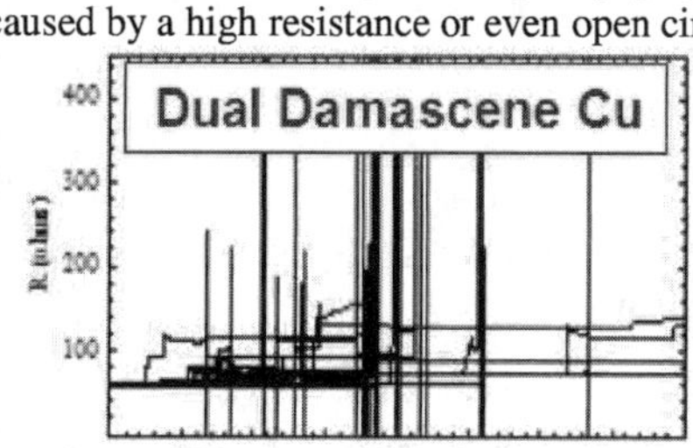

Figure 2 R-t plot of Cu via downstream

Al interconnect wiring levels have been limited by their current carrying capability, with shorter electromigration lifetimes compared to copper. The limiting fail mode in Al interconnect is the upstream via when the vacancies accumulate over the W stud (Figure 3). The reliability of traditional Al interconnects is greatly improved by adding metal layers, such as titanium (Ti), titanium nitride (TiN) and/or tungsten (W) above or below the Al [1-3]. While the void in Figure 3 and Figure 4 is well defined in that it the entire volume of Al is swept away over the via, the void shape could be of any configuration [4]. Ti/TiN underlayers are frequently used with Al metallization, and provide improved electromigration lifetimes. However, the conductivity of Ti/TiN redundant layers is degraded by Ti reactions with silicon dioxide (SiO_2) resulting in a pop fail. Therefore, we propose an improved Al metallization with a W/Ti/TiN metal stack under the Al. Figure 4 shows the bottom redundancy intact of stack II.

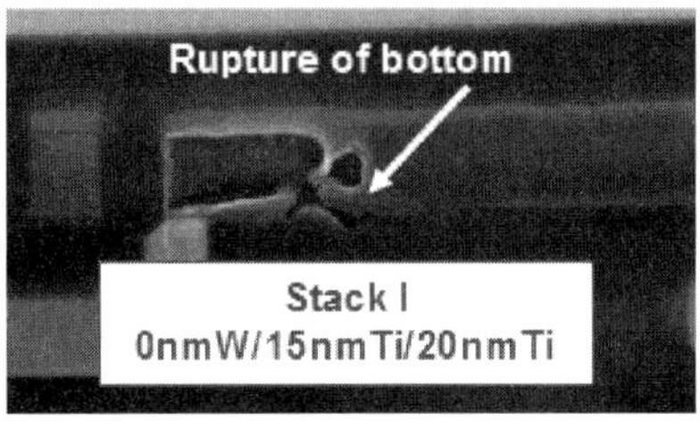

Figure 3 Al EM fail, via upstream of stack I.

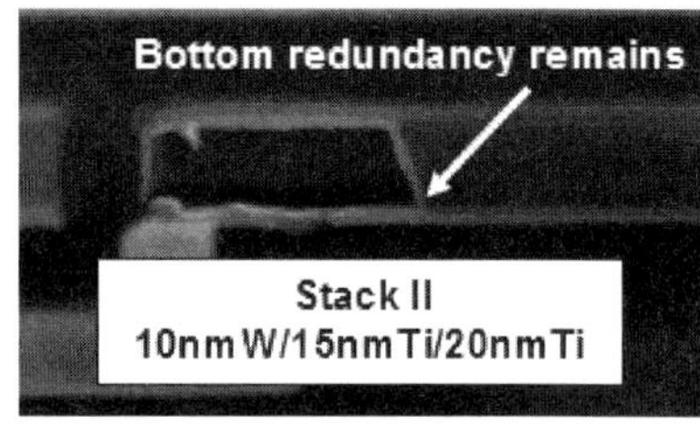

Figure 4 Al EM fail, via upstream of stack II.

EXPERIMENT

Using a combined Cu and Al 180nm technology, electromigration test structures were fabricated on 200mm wafers. Wafer processing was started at poly-silicon (PC) through CA (W) contact, then M1 was processed with Cu damascene. A V1 (W) via in SiO_2 connects to M1. M2 metallization experiments with the three metallization (I-III) stacks were run. Table 1 provides the process descriptions for each stack. The aluminum composition is Al-0.5%wt Cu. All films were deposited by PVD in conventional DC Magnetron PVD equipment using planar sources. The aluminum deposition was temperature controlled using an electrostatic chuck. All other metal films have no temperature control. The wafers were initially degassed using lamp heating in a vacuum for water desorption. Stack II, III used a 5 nm oxide equivalent preclean prior to W deposition. Figure 5 shows the schematic diagram of stacks I, II, and III

Stack	Process Description
I	150c degas, 15nm Ti, 20nm TiN, 370nm Al @200c, 5nm Ti/70nm TiN
II	**250c degas, 5nm preclean, 10nm W, + air break +** **250c degas**, 15nm Ti, 20nm TiN, 370nm Al @250c, 5nm Ti/70nm TiN
III	**250c degas, 5nm preclean, 25nm W, + air break +** **250c degas**, 20nm TiN, **370nm Al @250c**, 5nm Ti/70nm TiN

Table 1, Stack I-III description

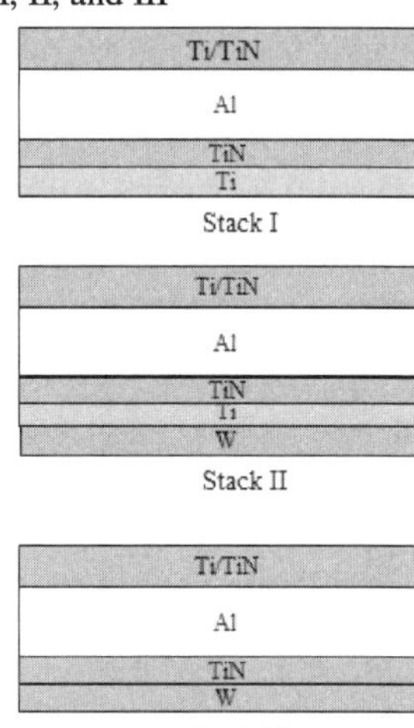

Figure 5, Stack I-III schematic

The top Ti/TiN is deposited in the same chamber with no preconditioning of the Ti target between wafers. After M2 pattern definition, wafers continued through V2/M3 and final passivation to replicate typical manufacturing conditions.

Electromigration was measured at M2 on minimum dimension, single via structures with no metal extensions over the vias. The nominal M2 line is 0.28 μm wide and 200 μm long. The V1 diameter is 0.24 μm. Electromigration evaluations were carried out on all three stacks at 250°C stress temperature and stress current densities between 10 to 15 mA/μm^2. Current flow is M1-V1-M2. Using 20% resistance increase (dR/R) as the failure criterion, the lifetime distributions of all three stacks are shown in Fig. 9. The results show the improved t50 times for stack II and III over stack I.

DISCUSSION

Electromigration Performance

Referring Figures 6-9, Figure 6 shows the R-t plot for stack I. Clearly, the bottom redundancy fails as shown in the failure analysis in Figure 4. The bottom redundancy failure is driven by the thermal runaway condition when the void occurs at the upstream via. A void in any other location in the wire is shunted by the top and bottom redundant layers. This provides a

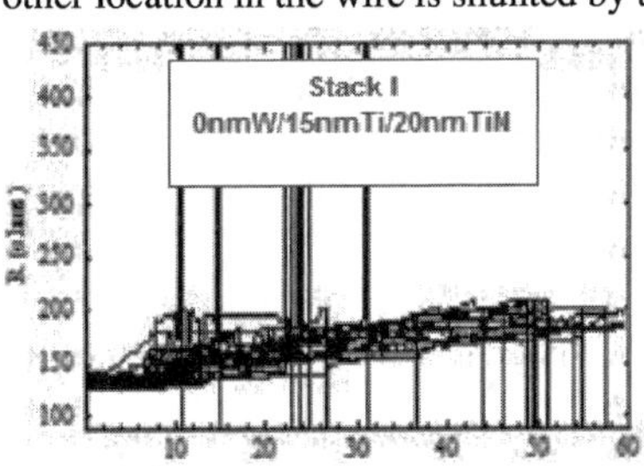

Figure 6 R-t plot, Stack I via upstream

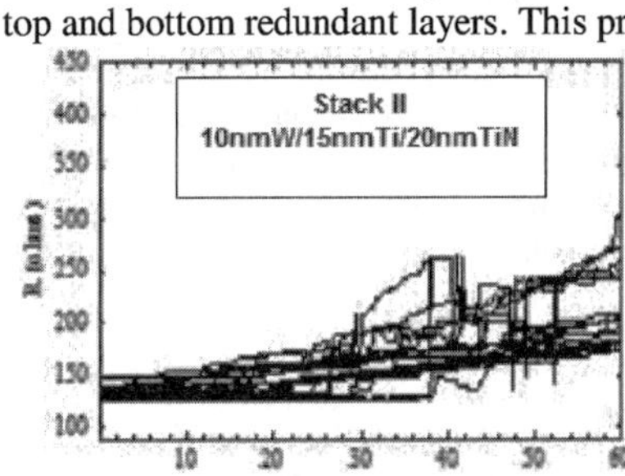

Figure 7 R-t plot, Stack II via upstream

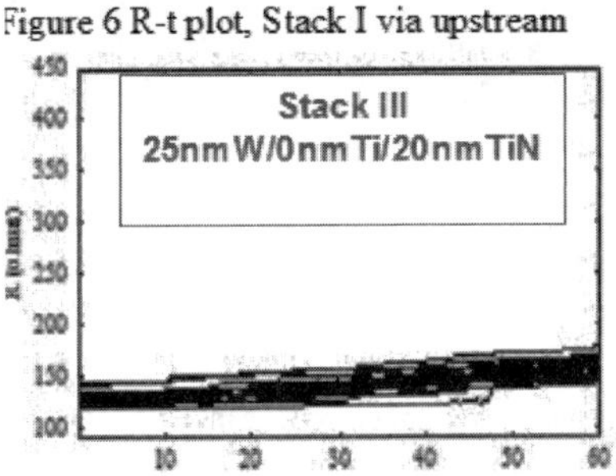

Figure 8 R-t plot, Stack III via upstream

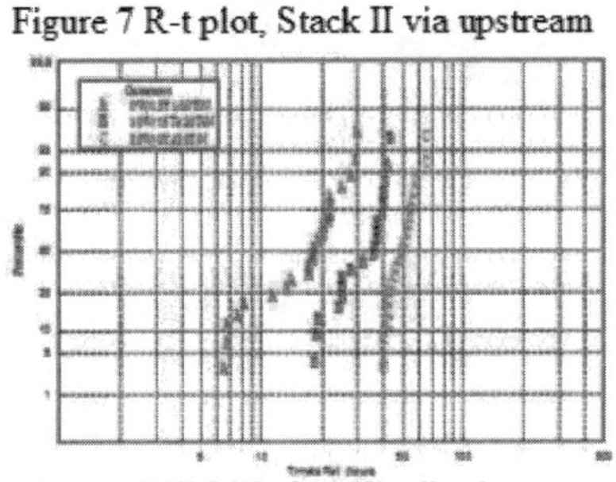

Figure 9 EM lifetime distributions of stack I, II, III (A,B,C in the plot)

parallel resistance path for the current and the power through each layer is dissipated through the ILD (inter-level dielectric). When the void occurs at the upstream via, as soon as the void fully exposes the via, the current is in a series connection to the Al wire. As the temperature of the bottom redundancy increases through joule heating, the Ti reaction with the SiO_2 proceeds. While the manufacturing processes are sufficient to drive some SiO_2 interaction with the Ti, the joule heating provides significant local temp increase to react more Ti/SiO_2. This only further increases the bottom redundancy series resistance. Ultimately, the bottom redundancy fails through thermal overload. An analysis of the bottom redundancy reaction kinetics is as follows. For the general case, at 600K,

$$\text{Metal} + SiO_2 \rightarrow M_xSi_y + M_xO_y \quad (1)$$

Using (1), with stack I and (II or III) shows in Table 2 that W reacting with SiO_2 is not

Metal + SiO_2	$M_xSi_y + M_xO_y$	ΔG
2 Ti + 1 SiO_2	TiSi + TiO2	-162 kJ/mole
3 W + 2 SiO_2	WSi_2 + 2 WO_2	557 kJ/mole

Table 2, Reaction kinetics for bottom redundancy

energetically favorable, while Ti reacting with SiO_2 proceeds. Clearly, this will result in the bottom redundancy of stack I having increasing resistance over stress conditions.

A one dimensional analysis of the temperature, Figure 10, of the bottom redundancy where the void occurs at the upstream via shows that Ti as a first layer contact to SiO_2 is not sufficient to provide low resistance shunt to the void. The high resistivity, the low melting temperature and as shown in table 2, reactivity with SiO_2, contribute to the EM issues associated with Al metallurgy.

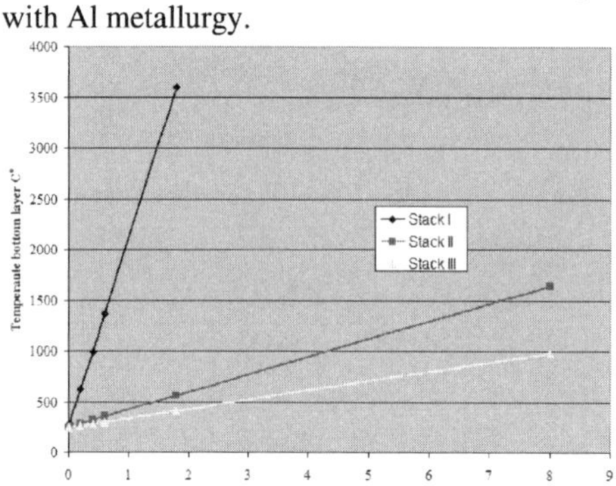

Figure 10, Bottom redundancy temp vs. void length

	Ti	W	SiO_2
Resistivity [μΩ-cm]	75	13	
Thermal Conductivity [W/μm-C°]			9.6e-7
T melt [C°]	1668	3422	

Table 3, constants for 1-dimensional analysis

Table 3 shows the constants used in the analysis. The direction of heat flow is from the bottom conductor through the SiO_2 to the substrate. Figure 10 is plotted against void length from and including the via. Plotting against void length provides for analysis independent of void shape. The model considers that the aluminum is voiding away from the bottom redundancy. A use current of 1.29mA is used. Stack I shows thermal runaway as void length increases, whereas W as a first layer to SiO_2 shows excellent current shunting do to its low resistance, therefore less joule heating.

Further analysis of the R-t plot in Figure 6-8 show a shallower slope as W is used, stack III being the best. Figure 9, a log-normal plot of the t50 times for 20% resistance shifts, shows that not only is the fail mode associated with the upstream via effectively solutioned, but the late mode fail is extended out in time through the low resistance shunt provided by the bottom redundancy. Figure 11 shows the dR/R vs void length where the void occurs away from the via.

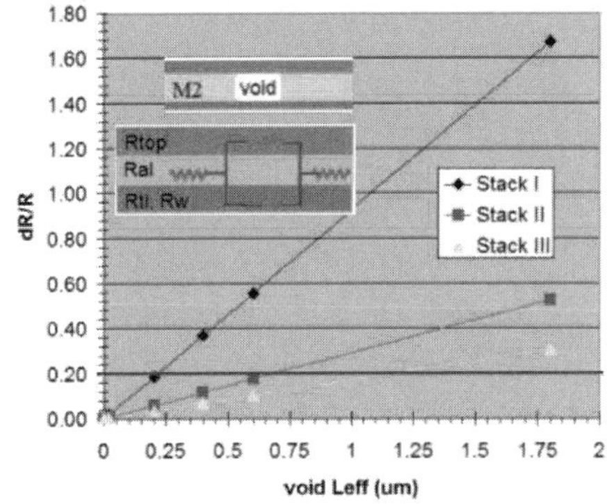

Figure 11, dR/R vs void length for Stack I-III

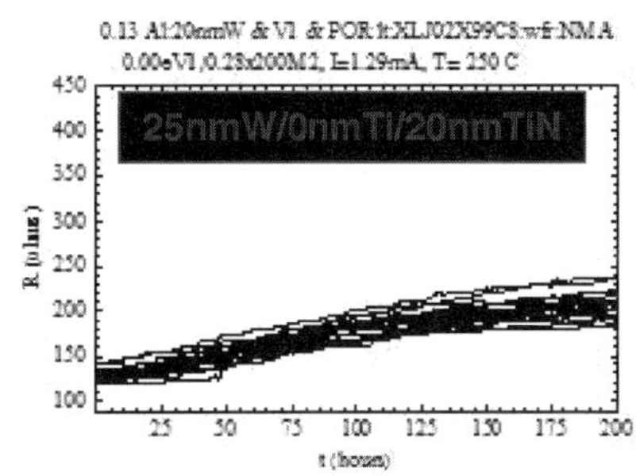

Figure 12, R-t plot of stack III to 200 hours

As a result, the W bottom redundancy allows 5x more void length in stack III compared to stack I. Allowing the EM stress conditions to proceed for more hours reveals a saturation limit in vacancy accumulation as shown in Figure 12. After 200 hours of stressing the R-t plot slope approaches zero, indicating a limit to vacancy accumulation. The resistance upper bound correlates to a void volume of ~4%. The coefficient of thermal expansion (c.t.e.) difference between Al and SiO_2 is 5x. At processing conditions in typical Al interconnect, this accounts for 2.8% of the total volume. The balance would be in process variables such as line roughness and RIE processing.

CONCLUSIONS

A novel Al interconnect metallurgy has been developed to address the current density performance requirements of advanced ULSI IC technologies. It incorporates a thin W film in the bottom redundancy to provide robust EM resistance. It corrects the issues associated with Ti based redundant systems by providing a low resistance shunt, a high melting temperatue, and low reactivity with SiO_2. Using a W redundant system eliminates chip failure associated with pop up failure.

ACKNOWLEDGMENTS

The authors would like to thank Carl Farris for support in wafer fabrication, Chuck Cox for SEM and Chad M Burke, Kevin Lindstam and Randy Austin for EM testing support.

REFERENCES

1. J. Tao et al., IEEE Trans. Elec. DEv., 1996, p. 1819.
2. M. Sekiguchi et al., Japan J. Appl. Phys., 1995, p. 978.
3. K. Kamoshida , J. Vac. Sci. Tech., B19, 2001, p.2190
4. J.C. Doan et al., IEEE Intl. Physics Reliability Symposium, 1999, p. 206.

Conference Proceedings AMC XXIV © 2009 Materials Research Society

Cu(Mn) seed layers in single damascene trenches with dimensions down to 30 nm

Henny Volders[1], Richard Olivier[1], Laureen Carbonell[1], Roger Palmans[1], Patrick Verdonck[1], Nancy Heylen[1], Kristof Kellens[1], Silvia Armini[1], Hugo Bender[1], Larry Zhao[2], and Zsolt Tőkei[1]

[1]RDO, IMEC VZW, Kapeldreef 75, Heverlee, 3001, Belgium

[2]RDO, Intel Corporation, Kapeldreef 75, Heverlee, 3001, Belgium

ABSTRACT

Cu-8 at% Mn seed layers with thickness values ranging between 10 and 60 nm were deposited by Physical Vapor Deposition (PVD) in single damascene trenches for dimensions down to 90 nm (180 nm pitch) in SiCOH (k=2.5) films, for dimensions down to 30 nm in SiO_2 films and on planar capacitors on 30 nm SiO_2 with feature sizes of 60 micron width. Good filling was obtained in 30 nm trenches in SiO_2 and a 1.4 nm self-formed barrier was evidenced by Transmission Electron Microscopy (TEM). The Median Time To Failure (MTTF) at 100°C of a 2.2 nm self-formed barrier on planar capacitors in SiO_2 was comparable to the MTTF of a standard 6 nm Ta(N)Ta. The resistivity of a Cu line in SiCOH, when using the self-forming barrier process, is not higher than the traditional Ta(N)Ta and Cu combination. However, any residual Mn in the Cu line increases the resistivity of the Cu line and should therefore be avoided.

INTRODUCTION

Alternative routes, such as self-forming barriers, are being investigated to replace the standard PVD Ta(N)Ta barriers for Cu interconnects. The process of self-forming barriers, when deposited by PVD, entails the deposition of a Cu alloy seed layer directly onto the dielectric, followed by plating of the Cu bulk. The barrier formation is induced by a thermal treatment during which the alloy element is migrating through the Cu to the oxygen containing dielectric to form a thin barrier layer at the Cu-dielectric interface. Self-forming barriers have been reported with thicknesses ranging between 2 – 7 nm [1]. Omitting the conventional Ta-based barrier layer opens the process window for copper seed. By depositing PVD seed layers directly onto the dielectric helps to tackle the challenge of filling increasingly narrower features, because the seed layer does not have to cope with the overhang caused by the barrier process. On the other hand, in the case of Cu(Mn) self-forming barriers, Mn is resistive, so any residual Mn should be prevented in order to preserve the benefits of pure Cu [2]. The formation of the barrier is highly dependent on the oxygen availability in the vicinity of the top dielectric surface. The surface sensitivity of the barrier formation [3] is also directly related to the adhesion of the barrier which has a significant effect on the performance during Chemical Mechanical Polishing (CMP).

EXPERIMENT

In the present work, we have studied the integration of Mn-based self-forming barriers by depositing Cu-8 at% Mn seed layers with minimal overhang for thicknesses ranging from 60 nm to 10 nm (Figure 1) in single damascene trenches with dimensions down to 30 nm.

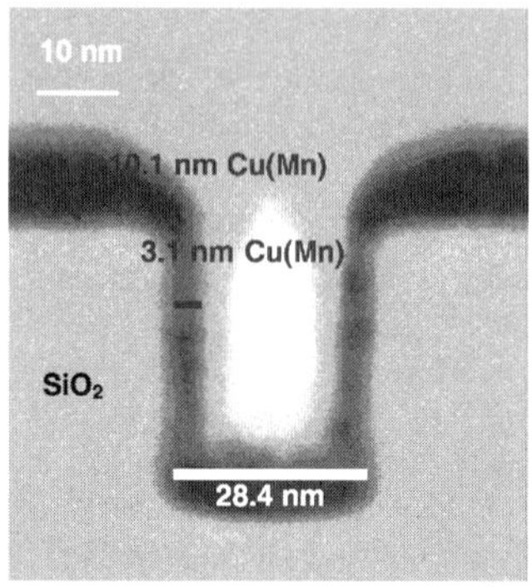

Figure 1. 10 nm PVD Cu-8 at% Mn deposition on field in 30 nm wide trenches in oxide AR 2:1

SiCOH (k=2.5) was used as Inter Metal Dielectric (IMD) for dimensions down to 90 nm (180 nm pitch) to study the resistivity of Cu(Mn) based metal lines. The extendibility to narrower features with dimensions down to 30 nm was studied on metal lines embedded into SiO_2 [4]. In order to study the intrinsic barrier properties of the self forming barrier, 60 nm Cu-8 at% Mn seed layers were also deposited on planar MOS capacitors with 30 nm SiO_2 thickness and with a feature size of 60 micron width. Cu fill of the trenches was enabled through a 2- step plating approach to overcome the terminal effect encountered when plating on thin resistive Cu(Mn) layers of 10 nm thickness. Conformal Cu(Mn) seed layer enhancement was done first by using a low Cu content alkaline plating chemistry [5] and was followed by conventional "bottom-up" fill of the trenches using a 3-component medium acid chemistry. A Ta capping layer was applied for the narrower features with dimensions down to 30 nm to promote barrier formation at the interface and to prevent the segregation/oxidation of Mn atoms at the surface of the Cu layer during the thermal treatment. Formation of the barrier layer was finally obtained by applying an anneal at 430°C for one hour in Ar. The Ta capping layer and excess of Cu and barrier on the field areas were removed by CMP. Film thickness was studied in oxide and SiCOH (k=2.5) trenches using TEM along the [110] cross-sectional direction. The elemental composition of the dielectric/seed interface was analyzed by using either Energy Dispersive X-ray Spectroscopy (EDS) and/or Electron Energy Loss Spectroscopy (EELS) analysis. Time Dependent Dielectric Breakdown (TDDB) measurements were performed at 100°C on planar MOS oxide capacitors to demonstrate the intrinsic barrier properties of the self-formed barriers. The Median Time To Failure (MTTF) was extracted from the experimental current vs. time curves and for describing the field acceleration an E-model was used.
Splits combining and comparing Ta(N)Ta and Cu(Mn) were also deposited in order to evaluate the barrier properties of the self-formed layers and their impact on the resistivity of Cu. In order to take process variability into account (e. g. dishing and erosion during CMP), the specific resistivity and cross-sectional area of Cu in damascene features were obtained separately using a calculation based on

Matthiessen's Rule [6]. As the specific resistivity and cross-sectional area depend essentially on the line resistance and its variation with temperature, any factor influencing these relevant parameters (CD variation, line height, impurity incorporation, grain size, etc…) will reflect in changes of the Cu specific resistivity in the lines. The specific resistivity of Cu (ρ_{Cu}) for different line widths was extracted from line resistance vs. temperature measurements and plotted as a function of the cross sectional area of the Cu.
Although the microstructure and the impurity levels of the copper might be different in trenches as compared to blanket film layers (different deposition mechanisms are involved), the analysis of unpatterned layers could establish some basic insight in the relationships between the Cu resistivity and performance of the self-forming barrier and seed in patterned features. Therefore also Back Scattering Electron (BSE) imaging was performed when combining and comparing Ta(N)Ta and Cu(Mn).

DISCUSSION

TEM analyses show that good Cu fill was achieved in trenches as small as 30 nm wide (Figure 2A). Moreover, a 1.4 nm thin barrier layer was formed at the bottom of the trench, on the sidewalls and on the field areas as evidenced by the detection of Mn by EDS at the interface between Cu and oxide (Figure 2B).

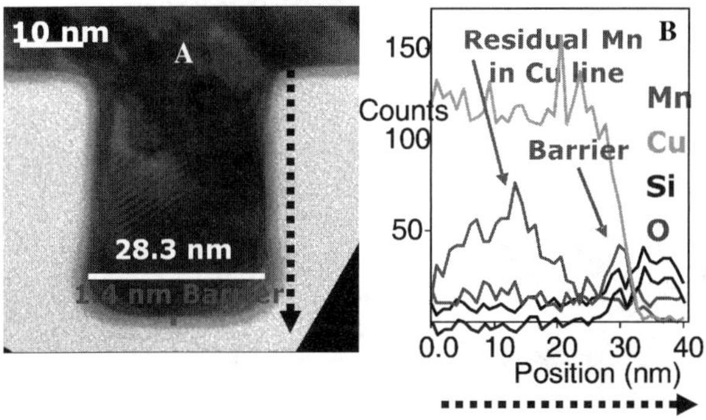

Figure 2A. 10 nm PVD Cu-8 at% Mn deposition on field in 30 nm wide trenches in oxide AR 2:1 with a 2-step plating and 5 nm Ta cap after a 1 hour anneal in Ar at 430°C
Figure 2B. EDS analysis of a 30 nm wide trench in oxide with a 1.4 nm barrier layer formed at the bottom

One drawback of using Ta capping layers during the self-forming barrier process is the presence of unreacted Mn atoms inside the Cu trench after completion of the anneal, thus leading to a resistance increase in the Cu line compared to pristine Cu. The resistivity of Cu was measured for interconnects with dimensions down to 90 nm in SiCOH (k=2.5) (Figure 3).

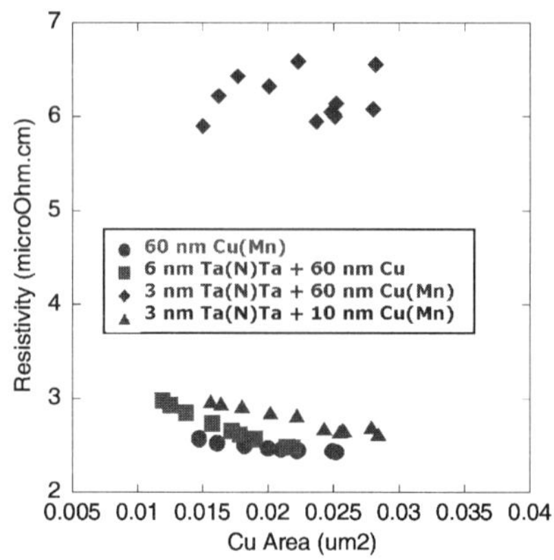

Figure 3. Resistivity of Cu for interconnects with dimensions down to 90 nm in SiCOH (k=2.5)

When using a 60 nm Cu-8 at% Mn seed layer directly deposited onto the dielectric (Figure 4) and combining it with an optimized anneal, the resistivity of the Cu line is comparable to the resistivity of the standard 6 nm Ta(N)Ta and 60 nm pure Cu.

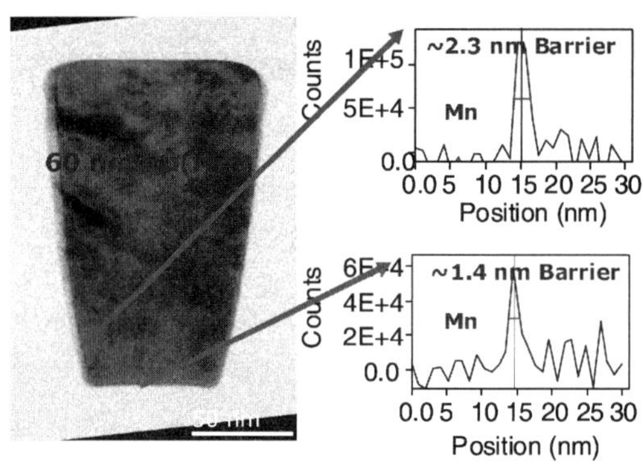

Figure 4. TEM and EELS analysis of a trench in SiCOH (k=2.5) filled with 60 nm Cu(Mn) and standard acidic plating after an anneal of 1 hour in Ar at 430°C and CMP

The observed slight changes in resistivity could be related to changes in the Cu grain size, changes in electron scattering at the interfaces and impurities. BSE imaging shows a significant increase of the amount of small Cu grains when using 3 nm Ta(N)Ta and 60 nm Cu-8 at% Mn (Figure 5). Indeed, a significantly larger grain size is observed for Cu(Mn) layers deposited onto the low-k dielectric as compared to Cu(Mn) deposited onto Ta based barriers.

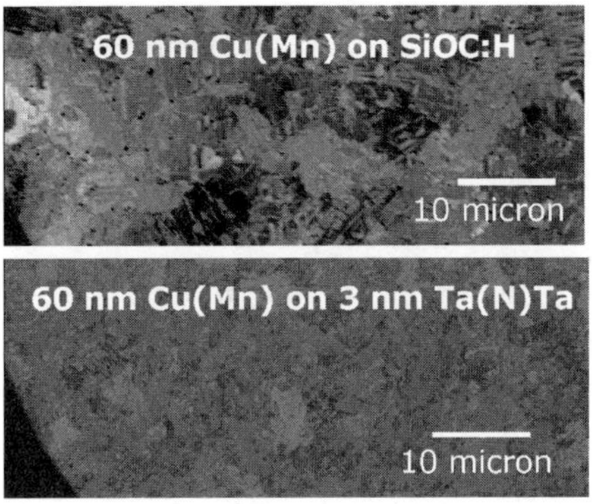

Figure 5. Back Scattering Electron imaging of the Cu grains for Cu (Mn) on SiCOH (top) and on 3 nm Ta(N)Ta (below)

Though, the significant increase in resistivity for 3 nm Ta(N)Ta and 60 nm Cu-8 at% Mn can not only be explained by the increased amount of small Cu grains. Large amounts of unreacted Mn atoms remaining in the trenches contributed mainly to the resistivity increase of Cu. In the case of 3 nm Ta(N)Ta and 10 nm Cu–8 at% Mn only a slight increase in resistivity is observed. In the case of such thin Cu(Mn) layers, a much lower concentration of unreacted Mn atoms is present in the trench (below the detection limit of EELS, 1 at. %), thus the observed resistivity increase is more limited.

TDDB measurements were performed at 100°C on planar MOS oxide capacitors (Figure 6) to demonstrate the intrinsic barrier properties of the self-formed barrier. They were compared to TaN(Ta) reference barriers as well as to structures that do not contain any barrier. As expected, the lifetime of the structures without a barrier layer is significantly decreased as compared to standard Ta-based barriers. The self-forming barrier, with a thickness of around 2.2 nm evidenced by TEM (Inset Figure 6) exhibits a MTTF similar to that of the reference of 6 nm Ta(N)Ta, indicating the 2.2 nm thick self-formed layer is a barrier.

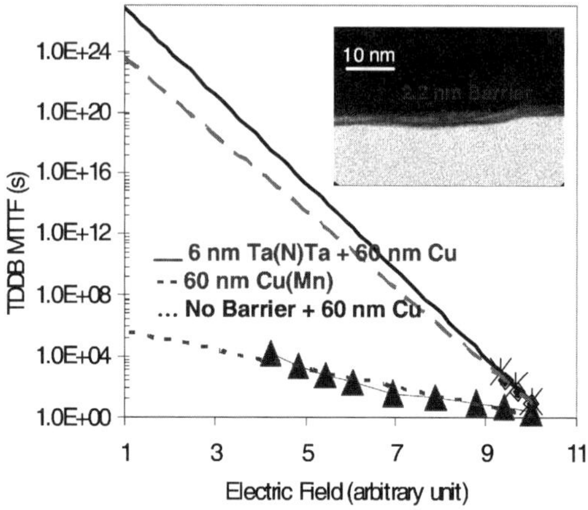

Figure 6. TDDB performance of planar capacitors at 100°C for 60 nm PVD Cu-8 at% Mn , 60 nm pristine Cu and the standard 6 nm Ta(N)Ta with 60 nm Cu Inset. TEM of the interface between Cu and oxide for a planar capacitor

CONCLUSIONS

Filling of 30 nm features is enabled by combining direct PVD Cu alloy seed deposition and a 2-step plating process. A 1.4 nm self-formed barrier has been detected with TEM in these narrow oxide trenches. Good TDDB MTTF performance of the self-forming layer indicates that a barrier is formed and low Cu resistivity can be achieved if no residual Mn is present in the Cu line.

REFERENCES

1. J. Koike et al, J. Appl. Phys. 102, 043527 (2007)
2. J. Koike and M. Wada, Applied Physics Letters 87, 041911 (2005)
3. S.-M. Chung, J. Koike and Zs. Tökei, Proceedings of the MRS Conference 2008 (2008)
4. J. Van Olmen et al, Proceedings of the IEEE International Interconnect Technology Conference 2007 (2007)
5. Y. Lantasov, R. Palmans, and K. Maex, Proceedings of the Advanced Metallization Conference 2000 (2000) 145
6. Hua Li et al., Proceedings of the AMC 2000 Conference (2000).

Conference Proceedings AMC XXIV © 2009 Materials Research Society

Two Component Additive Plating for Through-Silicon-Via Filling

Ole Lühn[1,2], Alex Radisic[1], Philippe M Vereecken[1], Chris Van Hoof[1,3], Jean-Pierre Celis[2], Bart Swinnen[1], and Wouter Ruythooren[1]

[1]IMEC, Kapeldreef 75, Heverlee, B-3001, Belgium

[2]MTM, Katholieke Universiteit Leuven, Kasteelpark Arenberg 44, Heverlee, B-3001, Belgium

[3]ESAT, Katholieke Universiteit Leuven, Kasteelpark Arenberg 44, Heverlee, B-3001, Belgium

ABSTRACT

We report on a method that allows the use of simple, two component electrolytes for filling of Cu through-silicon-vias. A thin Ta layer was deposited on top of the seed layer. By inhibiting the growth of Cu at the top surface of the wafer, the Ta acts as a replacement to the leveler otherwise needed in the electroplating process. For vias with a diameter of 5 μm and a depth of 25 μm, complete filling was obtained in less than 15 minutes.

INTRODUCTION

Electrical interconnects through the Si substrate (through-silicon-vias, TSV) are an essential building block for 3D integrated circuits [1]. To keep resistance low, the preferred metallization of the vias is Cu. Since the volume of the TSV metal is fairly large, electroplating is the preferred deposition technique due to high deposition rates. Vias have to be filled completely and defect free to avoid reliability issues. At the same time, the thickness of the material deposited on the field or top surface of the wafer has to be limited for compatibility with further processing (e.g. CMP). The duration of the plating process is an important determining factor of the overall 3D process cost and thus needs to be minimized.

To achieve bottom-up plating, the deposition rate at the top surface has to be reduced compared to the deposition rate inside the structure. Typically this is not possible when using a two additive component system. Fill experiments conducted with such electrolytes tend to result in the inclusion of large voids in the via structure (Figure 1).

Still, a number of successful implementations for bottom up TSV plating have been reported, but these are mostly based on electrolytes containing 3 different additive components [e.g. 2, 3]. These components are usually referred to as 'accelerator', 'suppressor' and 'leveler'. The chemical complexity of the electrolytes makes them harder to monitor and maintain. Oxygen purging has been indicated as a way to avoid the use of leveler [4], but application of this technique may require modifications to existing hardware.

As an alternative approach, we have earlier reported on the application of a thin Ta layer to block electroplating at the wafer surface [5]. We reported that this could be successfully combined with a 3 component chemistry for void-free via plating. Based on these promising results, we have further extended our work and can now report that even a simple two component chemistry without leveler can be successfully applied.

Figure 1. *Typical result for TSV Cu plating using a two component chemistry when no surface blocking is applied. A large void is present in the via and the Cu thickness on the top surface of the wafer is significant. All vias shown have a diameter of 5 µm and a depth of 25 µm.*

EXPERIMENTAL TECHNIQUES

Our work has been focused on a 'via first' approach, i.e. the vias are created before wafer thinning. The via geometry was 5 µm diameter by 25 µm depth for use in the 3D-Stacked IC approach (3D-SIC) currently developed at IMEC.

The vias were etched in Si wafers by deep reactive ion etching. After etching, the via sidewalls were passivated by a 100 nm thick oxide liner deposited by chemical vapor deposition.

Barrier and seed layers consisting of 80 nm thick tantalum and 300 nm thick copper respectively were deposited by sputtering on top of each other (Applied Endura CuBS (Cu Barrier/Seed) PVD (Physical Vapor Deposition) with the SIP EnCoRe™ process. After deposition of the barrier/seed layer stack, a thin Ta-layer (~5 nm) was deposited directly on the copper seed with the same process. The thickness values indicated are the ones obtained on the wafer surface only. Since their deposition is mostly a line-of-sight process, the thickness of the three layers dimishes very rapidly along the sidewalls of the vias.

The wafers were diced into coupons (2 cm × 2 cm) and were mounted on a rotating-disk-electrode (RDE) and placed in a three-electrode-cell of 100 ml cell-volume which was filled with an electrolyte (Figure 2). A platinum mesh was used as counter electrode and separated from the main cell compartment by a glass frit. The flat area of the rotating disk electrode was 1.5 cm^2. The true area of the patterned wafer samples inclusive the 5 µm × 25 µm vias, was about 2 times larger, approximately 3 cm^2. The applied geometric (1.5 cm^2) current density was $|i|$ = 2.5 mA/cm^2. The rotation speed of the RDE was 400 rpm.

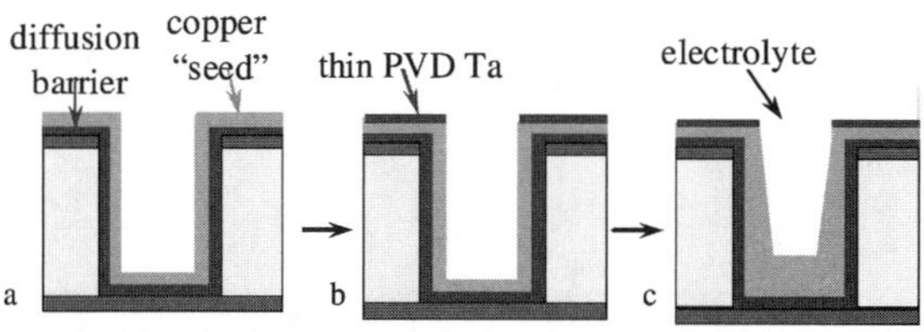

Figure 2. *Before electroplating, a PVD Ta layer is applied on top of the seed layer. The Ta layer blocks the deposition of Cu in the electroplating process.*

The electrolyte used for filling experiments contained 0.8 M $CuSO_4{\cdot}5H_2O$, 0.7 M H_2SO_4, 200 ppm polyethylene glycol (PEG) with a molecular weight of 8000 (Sigma-Aldrich) as suppressor, and 60 ppm Cl^- (NaCl, Sigma-Aldrich), and 7.5 ppm bis(3-sulfopropyl) disulfide (SPS, Raschig).

The overpotential needed for Cu electrodeposition on Ta is much larger than this on Cu [6]. The copper electrodeposition potential (or current) can be chosen in such way that copper deposits in the vias only.

RESULTS AND DISCUSSION

Figure 3 shows SEM images obtained by FIB after partial filling of the vias by depositing copper for 450 s (Figure 3; left), and a completely filled via after 850 s (Figure 3; right). The vias show no voids. Deposition was stopped before the vias were overgrown with copper, which would lead to mushroom-like structures.

Figure 3. *Cu filled TSV obtained by plating from a two component electrolyte. Left: Partial fill, right: full fill. The electrolyte consisted of 0.8 M CuSO4, 0.7 M H2SO4, 7.5 ppm SPS, 200 ppm PEG and 60 ppm chloride.*

At the top corner of the vias, the electroplated Cu slightly overgrows the Ta blocking layer (Figure 4). Note that the effect of the blocking layer on the adhesion of the Cu at that location and further down the via should be studied to assess the implications on long term reliability of the structures.

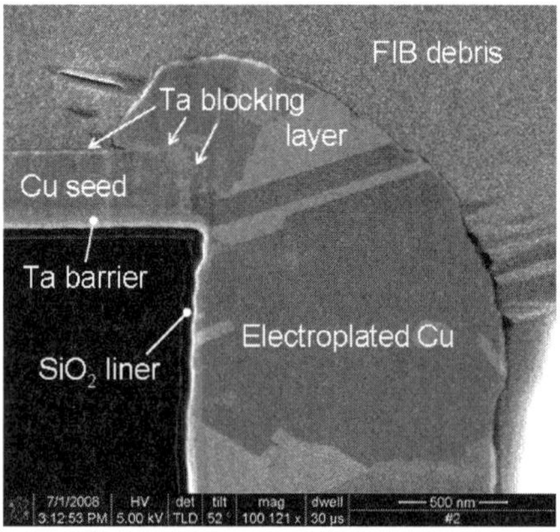

Figure 4. *Cu deposition near the top of the via opening. The Cu seed layer on the field of the wafer is separated from the electroplated Cu by a thin line indicative of the Ta blocking layer. The Ta layer is only apparent on the seed layer itself and does not seem to extend into the via opening itself.*

Figure 5 shows a top-view of a via that is filled completely with copper by electrodeposition. The remaining topography can be removed by CMP that is anyway required to remove the copper seed and barrier layers in the field area. Some copper islands are present due to defects in the Ta blocking layer.

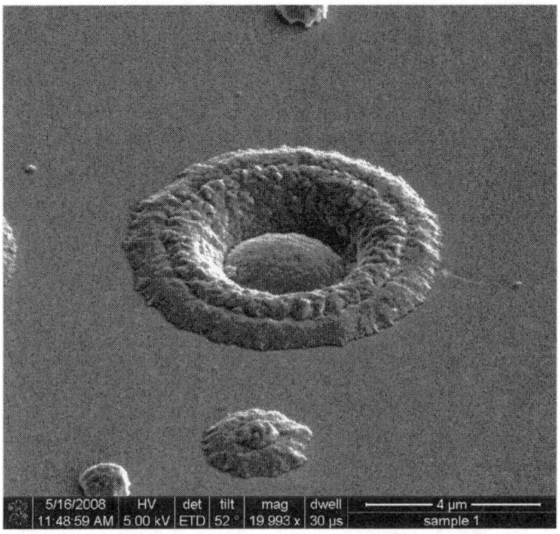

Figure 5. *Wafer surface after via filling. Next to the main feature indicating the presence of a via below, a number of parasitic deposits have been formed. These spurious outgrowths are probably indicative of defects in the Ta blocking layer.*

CONCLUSIONS

Through-silicon-vias were filled void-free with copper by electrodeposition using a thin Ta blocking layer on the copper seed-layer. An electrolyte contained only 2 additive components. The filling time for 5 µm diameter vias with a depth of 25 µm (aspect ratio 5) was about 15 min.

REFERENCES

[1] E. Beyne, *Proc. IEEE International Interconnect Technology Conference*, Burlingame, 4-7 Jun., 2006, 1-5.

[2] B. Kim, C. Sharbono, T. Ritzdorf, and D. Schmauch, *Proc. IEEE Electronic Components and Technology Conference*, San Diego, 30 May – 2 Jun., 2006, 838-843.

[3] O. Lühn, J.-P. Celis, C. Van Hoof, K. Baert, and W. Ruythooren, *ECS Transactions,* **6** (8), 2007, 123-133.

[4] D.P. Barkey, J. Callahan, A. Keigler, Z. Liu, A. Ruff, J.. Trezza, and B. Wu, *57th Electronic Components and Technology Conference (ECTC)*, Reno, May 29 - June 1, 2007.

[5] O. Lühn, A. Radisic, P.M. Vereecken, B. Swinnen, C. Van Hoof, W. Ruythooren, J.-P. Celis, *58th Electronic Components and Technology Conference (ECTC)*, Lake Buena Vista, May 27 - 30, 2008.

[6] Radisic, A., Oskam, G., and Searson, P. C., *Journal of The Electrochemical Society*, **151** (2004), C369-C374.

Conference Proceedings AMC XXIV © 2009 Materials Research Society

A new approach to reduce the Titanium deposition overhang: Optimized deposition thermal budget

Chong Jiang, Tommy Gu, Anthony C-T Chan, A.F. Bello, Charles C Wang, Phillip Wang, Ashish Bodke, Xiaoxi Guo, Keyvan Kashefizadeh, Miller Allen, James Ren, Jim van Gogh, and Mei Chang

Metal Deposition Products Division, Applied Materials Inc, 974 E.Arques Ave., M/S 81339, Sunnyvale, CA, 94085

Abstract

A new approach to reduce the Titanium physical vapor deposition overhang was evaluated. A remarkable result was obtained by this new method. Controlling the deposition thermal budget is the key point in this new procedure. A few models were introduced to explain the experiment results. Minimum free energy and thermal diffusion theory can explain the phenomenon. AFM and XRD analysis were also performed to understand the mechanism of this new approach.

Introduction

Titanium (Ti) has been widely used in the semiconductor industry as part of aluminum interconnect schemes and is now seen as a material of interest in copper (Cu) physical vapor deposition (PVD) interconnect applications as well. Previously used as an aluminum (Al) film wetting layer [1], Ti is now being considered to replace the industry standard Ta (N), for copper barrier layer [2] applications. Cost and electrical performance are the two major drivers for this change. With the via geometries shrinking, overhang created during the Ti deposition process can present a critical challenge for subsequent gap fill processes. New processing schemes to minimize the creation of overhang during Ti deposition have been evaluated. High temperature deposition and substrate bias control have produced remarkable results, providing extendibility of PVD processing. In this paper, we discuss the mechanism of high temperature processing and its impact on the overhang evolution.

Experiment set up and data

For these experiments, wafer temperature was modulated with the use of a preheat step in a Degas chamber prior to Ti deposition processing. Wafer transfer time was controlled so that the Ti film deposition temperature was dominated by the Degas chamber pedestal temperature set point. It was confirmed that the Ti deposition temperature was about 50ºC lower than the Degas chamber temperature set point when the Degas chamber pedestal temperature was set at 450ºC.

Figure 1 is a TEM picture of the Ti film deposited at high temperature. Another wafer deposited at room temperature was used as a control (Figure 2). A clear distinction between the overhang developed can be observed. When depositing the Ti at high temp, the Ti film was very smooth at the via corner; while with the room temp Ti deposition, a distinct overhang was observed. A 0.14um via was used as the test vehicle. Similar results were obtained on 45nm structures.

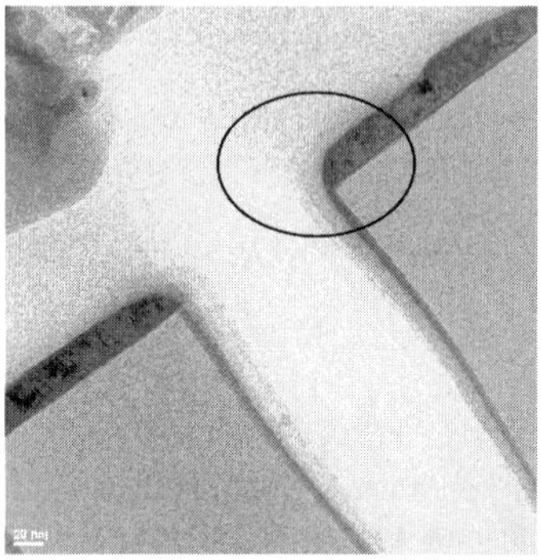
Figure 1: Ti deposited at high temp

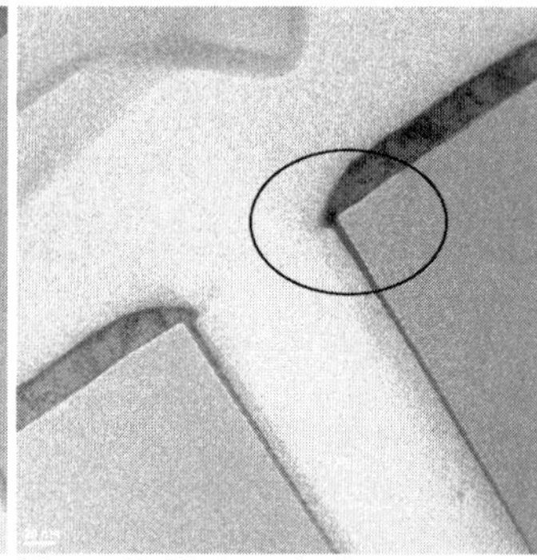
Figure 2: Ti deposited at room temp

Data analysis

Fundamental studies were completed to understand the impact of Ti deposition temperature and deposition bias voltage on film properties.

Ti film orientation was evaluated using XRD. Figure 3 is the XRD analysis result of the Ti film deposited at room temp and high temp. Ti deposited at room temp shows (111) and (101) orientation while deposited at high temp it has strong (002) orientation rather than the (111) and (101). The (002) peak is getting smaller when the bias is increased during the deposition. It is common knowledge that overhang will degrade when the bias is increased. Table 1 is the summary of the correlation between the Ti film orientation and overhang. We conclude that Ti film orientation might be an important factor impacting overhang.

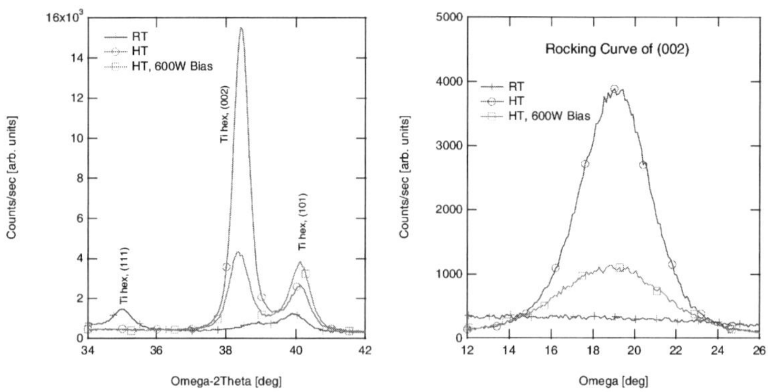

Figure 3: XRD analysis of the Ti film orientation

The Ti film orientation-temperature effect was also observed by our another experiment: three wafers were deposited with Ti film, two of them at room temp and one at high temp, then followed by high temp PVD Al film deposition on the top. Those three wafers Al (111) orientation XRD analysis results are shown on Figure 4. A significant difference was observed between the room and high temp Ti deposition. We will discuss this phenomenon details in a separate paper.

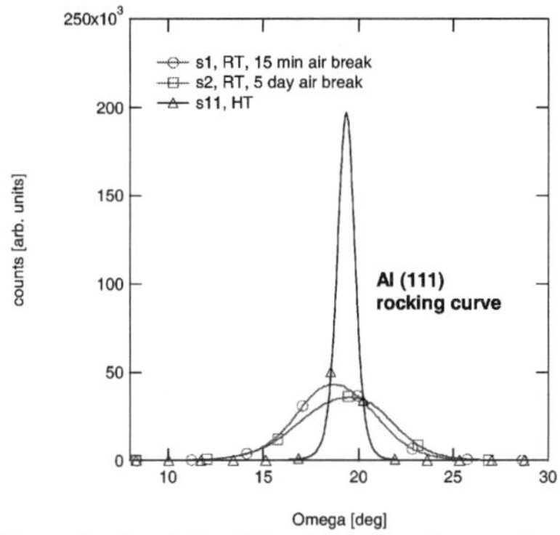

Figure 4: XRD analysis of the Ti orientation impacting the Al film orientation

Table 1: Orientation and Overhang Summary.

Room Temp Normal Bias	High Temp High Bias	High Temp Normal Bias
no (002)	weak (002)	strong (002)
large overhang	some overhang (worse than the normal bias case)**	limited overhang

** This conclusion is based on common knowledge. Although data is not available on the same 0.14um via structure, similar phenomenon is observed on smaller structures.

It has been observed that (002) Ti assists in Al (111) growth and provides better EM resistance in the Al application [3,4,5,6]; for the Cu application, further work is required to understand if (002) or (111) Ti is better for the Cu growth and EM prevention.

Surface roughness was the next effect studied. Figure 5 is the AFM analysis pictures of high temp and room temp cases. Ti film deposited at room temp is smoother than the case of high temperature deposition. Table 2 is the AFM data summary. There is no correlation between the overhang and the Ti film surface roughness. However, ignoring the high bias data it may introduce another mechanism, conclusions can be drawn that the smoother the surface, the larger the overhang.

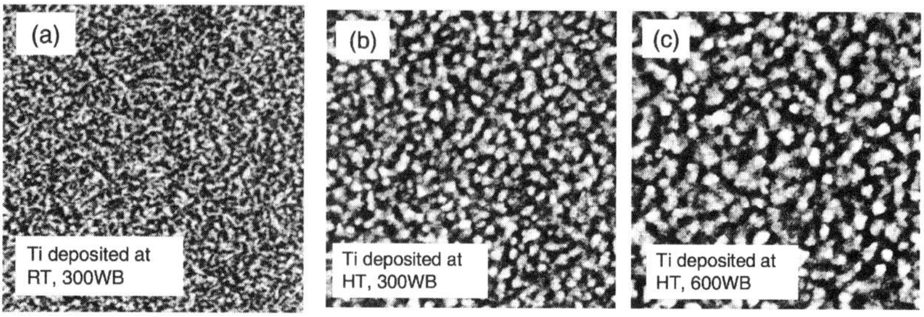

Figure 5: AFM result at RT, HT and HT with high bias

Table 2: Ti film surface roughness vs. overhang

Ti condition	Ti Rms	Ra	Rmax	overhang
RT, 300WB	0.48	0.38	3.79	large
HT, 300WB	0.63	0.51	4.61	small
HT, 600WB	0.7	0.56	5.58	larger than 300WB case **

Discussion

Interface stress could be a root cause of the overhang. Interface stress between the Ti film and the dielectric substrate, is present because of the lattice mismatch and thermal expansion coefficient differences between the two materials. This stress strongly depends on the deposition conditions and deposition temperature can make a significant difference. At the via corner this difference could be exaggerated. As the Ti film grows, arriving atoms always search for conditions of the lowest stress state. The stress tension difference between the corner and the nearby area is small in the room temp case compared to the high temp case. This could produce a lower growth rate at the via corner in the high temp over the room temp, eventually creating a smoother corner for the high temp case.

This phenomenon can be predicted from the minimum free energy point of view (Figure 6). High temperature deposition can be thought of as a simultaneous deposition/anneal process. The high overhang case can be thought of as a sub-stable state, and smooth corner/less overhang case should have lower free energy level and a more stable state. When deposition happens at high temperature, the material atoms have high kinetic energy so they have higher probability to overcome the potential wall and go to lower free energy state, eventually, resulting in a small overhang. The process condition can be described as the following equation:

$$E_k - E_b > U$$

Where E_k is atom kinetic energy which is proportional to the temperature; E_b is the rest potential energy that is against the atoms moving around; U is the potential barrier level.

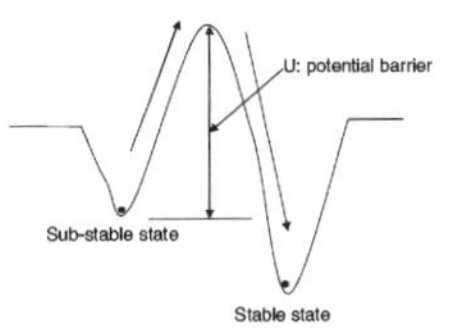

Figure 6: Free energy at overhang case and rounding corner case.

Thermal diffusion could be another explanation of thermal budget controlling the overhang. The diffusion process follows Fick's law:

$J=-D\partial\varphi/\partial x$

And $D=D_0e^{(-Ea/kT)}$

Where J is the diffusion flux; φ is the concentration; x is the position; D is the diffusion coefficient; D_0 is the diffusion coefficient at infinite temperature; Ea is the activation energy; k is the gas constant and the T is the temperature.

Rounded surface (high temp case) has lower surface tension and lower free energy because rounded surface has larger curvature radius. At room temperature, the diffusion coefficient is small so that the material has less probability to be restructured. When deposited at high temperature, the material has higher probability to diffuse due to a large diffusion coefficient. This may result in a system with a lower free energy state producing a rounded surface (less overhang).

The film orientation data also shows a correlation between the overhang and film orientation. Ti (111) film is denser than the (002) Ti film. Denser films may have higher binding energy creating a material that is more difficult to be re-sputtered. Reduction in re-sputter rate would provide a higher net growth rate that may contribute to increased overhang on these (111) films.

Conclusion

We proposed a new method to reduce overhang development during PVD deposition of a Ti film. Controlling the deposition temperature is shown as an approach to minimize overhang. In the high overhang case, the system has higher free energy and it is a sub-stable state compared to the rounded corner, lower overhang case. In the high temp deposition, the atoms have higher kinetic energy so they have more probability to overcome the potential barrier and go to stable state. Our preliminary data is very promising. Further work is needed to verify the result on different structures, particularly on structures with rounded via corners.

Acknowledgment

The Authors thank Applied Materials Silicon System Group (SSG) Metal Deposition Products (MDP) Division application laboratory co-workers' supports. We could not make those experiments done without their help. Thank Applied Materials Defect and Thin Film Characterization Laboratory for the XRD and AFM analysis. Thank Applied Materials TEM laboratory for the TEM samples preparation.

Reference

1. G. Steinlesberger, G. Schindler, M. Engelhardt, W. Steinhogl, M. Traving, " Aluminum Nano Interconnects" p51, IITC 2004
2. A. Sakata et al "Reliability Improvement Adopting Ti-barrier Metal For Porous low-k ILD Structure", p101, IITC 2006
3. Semicond. Sci. Technol. 21 No 5 (May 2006) 633-637
4. Toyoda H et al 1994 IEEE Int. Reliab. Phys. Symp. (IRPS) pp 178-84.
5. Sasaki Tsutomu and Dohnomae Hioshi 1998 Japan. J. Appl. Phys. 37 6544-51
6. D-S Park and Y-H Kim, 2002, Journal of Electronic Materials. 31 1009-1015 .

Conference Proceedings AMC XXIV © 2009 Materials Research Society

High-quality Ru thin film grown on TaN by plasma enhanced atomic layer deposition(PEALD)

qi xie[1,2], Xin-Ping Qu[1], Yu-Long Jiang[1], Guo-ping Ru[1], Bing-zong Li[1], Jan Musschoot[2], Christophe Detavernier[2], and Roland Van Meirhaeghe[2]

[1]Department of Microelectronics, Fudan University, Handan Road 220, Shanghai, China, Shanghai, 200433, China, People's Republic of

[2]Department of Solid State Science, Ghent University, Ghent, B9000, Belgium

ABSTRACT

Ru thin films were sequentially deposited onto TaN(5nm) by plasma enhanced atomic layer deposition using $Ru(EtCp)_2$ and NH_3 as precursors. The effect of growth temperature on the electrical resistivity and morphology of the Ru films were studied. It was found that the Ru films can achieve a low resistivity of 14μΩcm and a low root-mean-square roughness at a growth temperature of 270°C. The thickness of the underlying TaN film was found to affect the Ru film growth. The oxidation of the very thin TaN film was correlated with the island growth of Ru. Ex and in-situ XRD was employed to verify the copper diffusion barrier properties of a Ru(3nm)/TaN(5nm) bi-layer structure.

INTRODUCTION

Sputter deposited Ta/TaN bi-layer barriers with a Cu seed layer have been successfully employed for building Cu interconnect structures. However, with continual scaling of integrated circuit feature size, the step coverage provided by the physical vapor deposition (PVD) technique will become poor and the effective resistivity of the interconnect line will also increase. Seedless copper interconnect technology becomes increasingly appealing, since it will simplify the process and reduce the fabrication cost. Ruthenium recently has got much research attention, due to its good adhesion with copper, high thermal stability, and the capability of direct electrodeposition of copper on ruthenium. Many techniques including atomic layer deposition (ALD) have been used to grow the Ru film. [1-23]. ALD is a promising technology for the conformal growth of high quality thin films by sequential self-limited surface reactions. In comparison with thermal ALD, plasma enhanced ALD (PEALD) offers potential advantages such as improved film purity, and the capability for deposition at reduced substrate temperature. Since the Ru film does not adhere well to the SiO_2-based dielectric material [18] and furthermore, it is not an effective diffusion barrier, [8] an effective barrier like TaN is required. There are many reports to ALD grow Ru on TiN [16], grow mixed Ru/TaN [19], but only few reports to grow Ru on TaN because of difficulty of grow Ru on TaN.[13, 21] In this paper, we report on the growth kinetics and properties of Ru films which were deposited by PEALD onto ALD-grown TaN. The thermal stability of Ru(3nm)/TaN(5nm) bi-layer structure was characterized by in-situ XRD techniques.

EXPERIMENT

A home-made temperature-controllable ALD chamber was pumped by a turbomolecular pump, resulting in a base pressure in the range of 10^{-7} Torr. NH_3 gas was used as plasma enhanced reactants. The substrates were heated using a resistive heating plate, providing temperatures up to around 400°C. The bubbler containing bis(ethylcyclopentadienyl)ruthenium [$Ru(C_2H_5C_5H_4)_2$; $Ru(EtCp)_2$] liquid (99.999%, Sigma-Aldrich) was heated at 80°C to develop adequate vapor pressure while the delivery lines were heated to 100°C to prevent condensation of the precursor. The bubbler containing PDMAT [Pentakis(dimethylamino)tantalum] solid (99.9%, Sigma-Aldrich) was heated to 80°C while the temperature of the delivery line was sustained as 100°C. Ar gas with 99.999% purity was used as a carrier gas for the precursors. P-type Si (100) substrates were loaded into the vacuum chamber through a loadlock after standard chemical cleaning without further HF dipping. It should be noted that this cleaning method is known to result in the growth of a thin SiO_2 layer on the Si surface. The TaN layer was first deposited onto the substrates by using PEALD. The details for this process have been reported earlier. [25] The pulse time of the PDMAT precursor for the ALD TaN growth is 2s and the exposure time to the NH_3 plasma is 30s. The evacuating time for both the PDMAT precursor and NH_3 plasma reactants was 10 seconds. C-free TaN film was deposited on the Si substrate at a growth temperature of 200°C and the atomic ratio of Ta/N is close to 1:1 according to XPS measurement. Then Ru film was deposited without breaking the vacuum. A typical ALD cycle for growing Ru consists of, first 5s of the precursors pulse time followed by 10s to evacuate the chamber, then 10s NH_3 plasma burning time to provide the second reactant, and 10s for evacuation. The growth temperature varied from 100°C to 300°C and the pulse time was also varied to optimize the growth condition. Then a 50nm thick Cu film was deposited on some selected samples in a magnetron sputtering system attached to the ALD chamber without breaking the vacuum. The thermal stability for the barriers against Cu diffusion was characterized by ex and in-situ X-ray diffraction (XRD). For in-situ XRD, the samples were heated from room temperatures to 950°C at a fixed ramp rate of 0.2°C/s in a home-built heating chamber mounted on a Bruker diffractometer. There is a ±3°C error bar of the temperature calibrations. The film thickness was monitored by X-ray reflectivity (XRR). The sheet resistance, phase information, surface morphology, chemical composition and microstructure of the samples were investigated by four-point-probe, ex-situ XRD, Atomic force microscopy (AFM), X-ray Photoelectron Spectroscopy (XPS) and cross-sectional transmission electron microscopy (XTEM), respectively.

DISCUSSION

The growth rate per cycle (GPC) of the ALD Ru films on the ALD TaN(5nm)/Si was measured when the growth temperature changed from 100°C to 300°C. A constant growth rate of 0.2Å/cycle was observed, as can be seen in Fig.1. This value is much lower than what was reported in Ref 15. It was also found that the electrical resistivity of the Ru films grown at different temperatures was different. The thickness of these films is comparable, around 7nm. For films deposited at 100, 200°C and 300°C, the films' resistivity is around 60 - 80μΩ·cm. The film deposited at 270°C has the lowest resistivity of 14μΩ·cm.

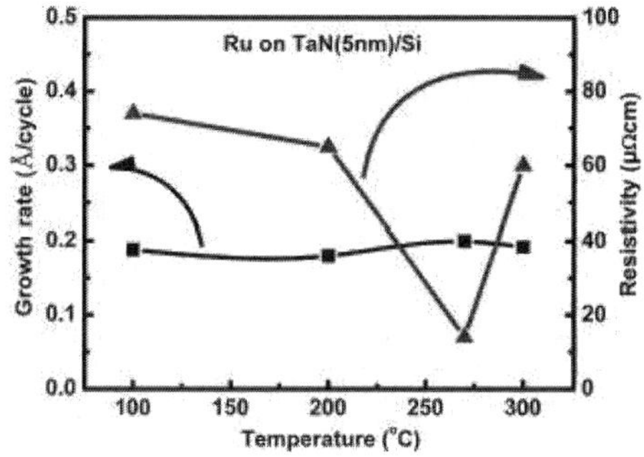

Figure 1 The GPC and resistivity of Ru films grown on TaN(5nm)/Si as a function of growth temperatures

As shown in Fig. 2, there is good linearity for the Ru films grown on the TaN(5nm)/Si substrate at 270°C. The fitted line does not go through the origin, indicating an incubation time before the film starts to grow. The saturation behavior was verified by increasing the pulse time of $Ru(EtCp)_2$ from 5s to 10s. The growth rate per cycle increased from 0.2 Å to 0.26Å for the 10s exposure, indicating a non-ideal saturation behavior.

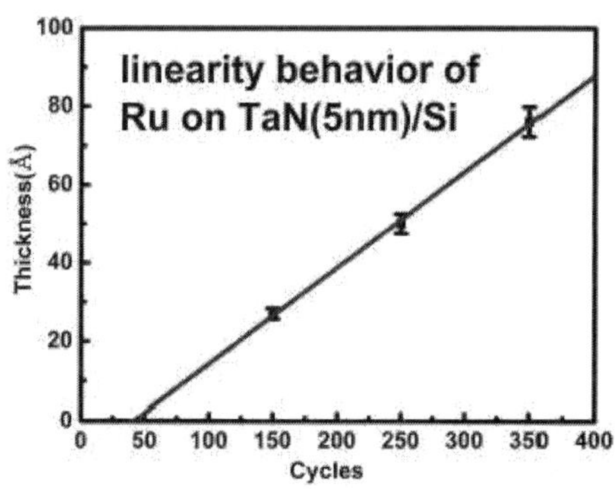

Figure 2 Linearity features of Ru films grown on TaN(5nm)/Si at 270°C.

As we know, electron mobility of the metallic films is affected by the presence of impurities and by grain boundary, surface and interface scattering. To investigate the reason for the different resistivity for the films grown at different temperatures, the crystallinity, surface roughness and film composition were measured using XRD, AFM and XPS measurement.

Figure 3 shows the XRD spectra of the as-deposited Ru films grown at different temperatures. The crystallinity of the Ru films deposited at 100, 200, and 300°C is very poor. Only the film deposited at 270°C has a diffraction peak of Ru (002). The better crystallinity may explain the low resistivity of the film deposited at 270°C. After deposition, the films were subjected to annealing at 600°C for 60 seconds. The crystallinity of the films deposited at 100, 200 and 270°C was improved by post-annealing, with the Ru(002) peaks full width at half medium

(FWHM) becoming narrower, as shown in Fig.3(b). The narrowest Ru(002) peak from the sample originally deposited at 270°C and then annealed at 600°C among all samples indicates that the grain size of this sample is the largest. The crystallinity of the Ru film deposited at 300°C remained very poor even after annealing at 600°C.

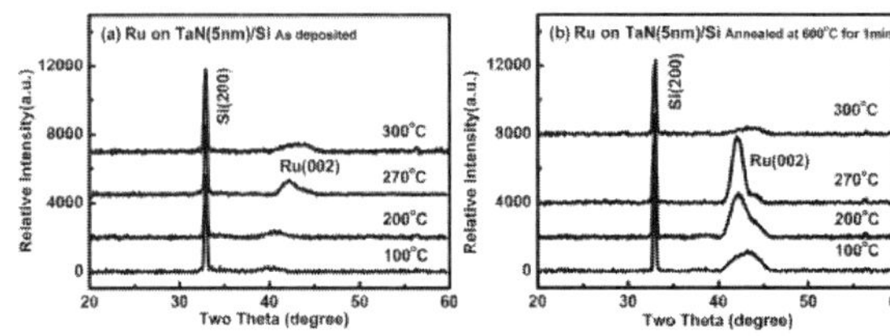

Figure 3 XRD spectra of the Ru films on TaN(5nm)/Si with different growth temperatures (a) as deposited, (b)annealed at 600°C for 1 minute.

The surface morphology of the Ru films deposited at different temperatures was compared by AFM, as shown in Fig.4. The film deposited at 100°C has a much rougher surface (root mean square (RMS) roughness of 1.8 nm) than the one deposited at 270°C (RMS of 0.6 nm). Fig. 4(c) indicates island growth of Ru at a growth temperature at 300°C.

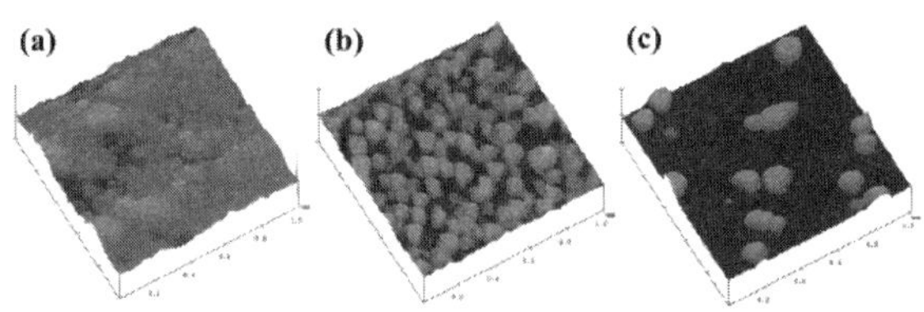

Figure 4 AFM images for the as deposited Ru(5nm)/TaN(5nm)/Si sample grown at (a) 100°C, (b) 270°C, and (c) 300°C.

XPS was employed to analyze the chemical composition of the Ru films grown at 270°C. Fig.5(a) gives the XPS spectra of the Ru3d core levels of the Ru films deposited on a TaN(5nm)/Si substrate. It shows that the Ru peak is composed of two peaks at 279.9eV for Ru 3d5/2 and 284.12eV for Ru 3d3/2, and the area ratio of Ru 3d5/2 and Ru 3d3/2 is close to 3:2. It is very difficult to determine C contamination in Ru films using XPS since the C1s core level is at 284.5eV, which is very close to the Ru 3d5/2. However, peak fitting has shown that the Ru 3d5/2 and Ru 3d3/2 peaks match the theoretical spin-orbit splitting of Ru 3d, suggesting that the amount of carbon is very small.[26-28] As also shown in Fig.5 (b), no N related binding energy peak was detected. The lack of nitrogen incorporation is due to the fact that ruthenium does not have any stable nitride phase. However, around 2% O was detected in the Ru films.

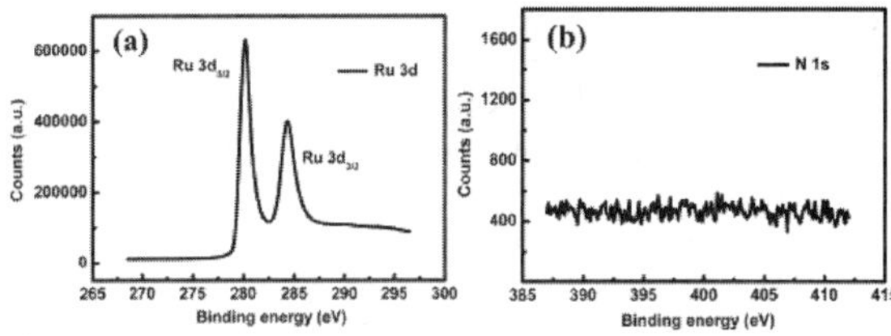

Figure 5 XPS spectra of (a)Ru3d and (b)N 1s for Ru film deposited on TaN(5nm)/Si substrate at 270°C.

The substrate effect on the ALD growth of the Ru film was also studied. As discussed in the above, high quality Ru films could be grown on a TaN(5nm)/Si substrate. However, it was found that when the thickness of the ALD TaN film was decreased to 2nm, only very thin and discontinuous Ru islands could be grown on the TaN/Si substrate. XTEM was used to characterize the microstructure of these films and check the reason. After ALD deposition (Growth temperature 270°C), the sample was transferred under vacuum into an adjacent PVD chamber, and a 50 nm Cu film was sputtered as a protective layer. Figure 6 gives the XTEM images for the as-deposited Cu(50nm)/TaN(2nm)/Si and Cu(50nm)/Ru(3nm)/TaN(5nm)/Si samples. Seen from both Fig.6(a) and (b), there is an amorphous interface layer with thickness of around 3.7 nm between the TaN layer and the Si substrate. This interface layer is SiO_2 layer caused by the chemical cleaning as we mentioned before. From Fig.6(b), we can also see that the PEALD deposited Ru film has a columnar grain structure with a grain size of 4~10 nm and the interface between Ru and TaN is rather sharp. The ultrathin Ru film is continuous.

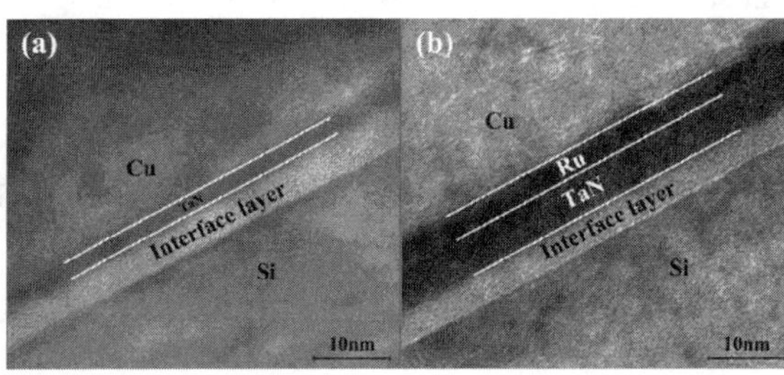

Figure 6 XTEM images for (a) Cu(50nm)/TaN(2nm)/Si and (b) Cu(50nm)/Ru(3nm)/TaN(5nm)/Si samples.

The brighter contrast for the 2nm thick TaN film (Fig.6(a)) as compared with the 5nm thick TaN film (Fig. 6(b)) suggests a larger amount of O incorporation in the thinner film. In principle, one does not expect oxidation of the TaN, since no oxygen was intentionally introduced into the chamber (as opposed to thermal ALD with the $RuCp_2$ precursor [14]). Moreover, the ALD growth of TaN and Ru were performed in the same chamber, without breaking the vacuum. However, after chemical cleaning of the Si substrate, there surface is covered with a thin silicon oxide film. The oxygen from this native silicon oxide layer may react with the ultrathin TaN layer, and cause it to become partially oxidized [29, 30]. The native oxide present on the Si

substrate after cleaning may provide enough O to oxidize the 2nm thick TaN film. For increasing TaN thickness, oxidation of the TaN layer becomes less significant. The real composition of the oxidized TaN film is difficult to analysis because of its ultrathin thickness. It is possible that the film is composed of Ta, O, N and Si. Apparently, the nucleation and growth of Ru on the oxidized TaN films were suppressed, which is might be due to the limited number of reaction sites available for the Ru precursors and the low chemisorption probability of the precursor on this kind of surface. These results demonstrate that for proper PEALD growth of Ru film, the TaN film on oxidized substrates (such as low k dielectrics, SiO_2) can be too thin.
The diffusion barrier properties of the Ru/TaN barrier stack were studied by using in-situ and ex-situ XRD [31]. The growth temperature is still fixed at 270°C for Ru deposition. Fig.7shows the in-situ XRD graph for the sample with same structure annealed from room temperature to 950°C at a fixed ramp rate of 0.2°C/s. It can be seen that the formation temperature of Cu_3Si, which indicates catastrophic failure of the barrier, occurs around 660°C. It is worth mentioning that during in-situ annealing, the Ru(3nm)/TaN(5nm) diffusion barrier withstands more than 20 minutes at annealing temperatures higher than 400°C.

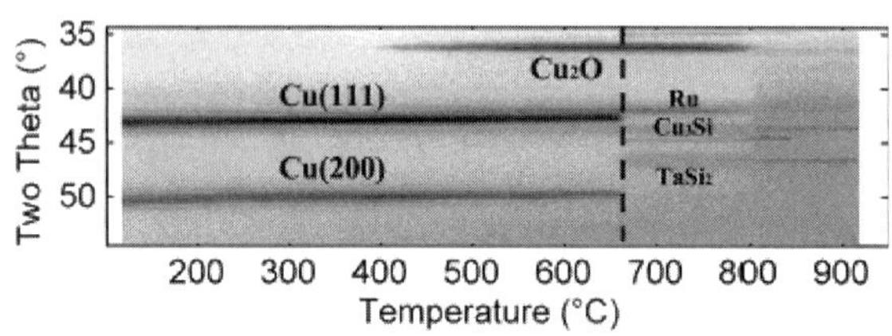

Figure 7 In-situ XRD graph with ramp annealing at 0.2°C/s for the Cu(50nm)/Ru(3nm)/TaN(5nm)/Si sample

CONCLUSIONS

In summary, PEALD grown Ru films were successfully deposited on TaN(5nm)/Si substrates. The effect of growth temperature and the TaN thickness on the growth of Ru films were studied. The resistivity of the Ru was correlated with crystallinity and surface roughness. The thickness of the underlying TaN film was found to affect the Ru film growth. The occurrence of Ru islands grown on a 2nm TaN film was related to the oxidation behavior of very thin TaN films. Good diffusion barrier properties were observed for the Ru(3nm)/TaN(5nm) bi-layer structure.

ACKNOWLEDGMENTS

This work was supported by National Basic Research Program of China (No. 2006CB302703), Bilateral Scientific and Technological Cooperation Project Flanders-China (B/06086/01), National natural science foundation in China (NSFC-60476010, NSFC-90607018), International cooperation program of Shanghai (075007034), SRC (2006-KJ-1467) and IRTG (Materials and Concepts for Advanced Interconnects). The authors would like to acknowledge the support from Micro/Nanoelectronics Science and Technology Innovation Platform in Fudan University and IWT-Vlaanderen for financial support of the SBO-METACEL project.

REFERENCES

1. 1. C. C. Yang, T. Spooner, S. Ponoth, K. Chanda, A. Simon, C. Lavoie, M. Lane, C.-K. Hu, E. Liniger, L.Gignac, T. Shaw, S. Cohen, F. McFeely, D. Edelstein, IITC pp187, 2006.
2. M. Abe, M. Ueki, M. Tada, T. Onodera, N. Furutake, K. Shimura, S. Saito and Y. Hayashi, IITC pp4, 2007.
3. O. Chyan, T. N. Arunagiri, and T. Ponnuswamy, *J. Electrochem. Soc.* **150**, C347 (2003)
4. D. Josell, D. Wheeler, C. Witt, and T. P. Moffat, *Electrochem. Solid-State Lett.* **6**, C143 (2003)
5. R. Chan, T.N. Arunagiri, Y. Zhang, O. Chyan, R. M. Wallace, M. J. Kim, and T. Q. Hurd, *Electrochem. Solid-State Lett.* **7**, G154 (2004)
6. S. K. Cho, S. K. kim, H. Han, J. J. Kim, and S. S. Oh, *J. Vac. Sci. Technol. B* **22**, 2649 (2004)
7. Q. Wang, J. G. Ekerdt, d. Gay, y. M. Sun, and J. M. White, *Appl. Phys. Lett.* **84**, 1380 (2004)
8. T. N. Arunagiri, Y. Zhang, O. Chyan, M. El-Bouanani, M. J. Kim, K. H. Chen, C. T. Wu, and L. C. Chen, *Appl. Phys. Lett.* **86**, 083104 (2005)
9. J. Liu, J. Lei, N. Magtoto, S. Rudenja, M. Garza, and J. A. Kelber, *J. Electrochem. Soc.* **152**, G115 (2005)
10. X. P. Qu, J. J. Tan, M. Zhou, T. Chen, Q. Xie, G. P. Ru and B. Z. Li, *Appl. Phys. Lett.* **88**, 1 (2006)
11. H. Li, D. B. Farmer, R. G. Gordon, Y. Lin, and J. Vlassak, *J. Electrochem Society,* **154**, D642-D647 (2007).
12. J. J. Tan, X. P. Qu, Q. Xie, Y. Zhou, G. P. Ru, *Thin Solid Films*, **231**, 504 (2006)
13. M. Zhou, T. Chen, J. J. Tan, G. P. Ru, Y. L. Jiang, R. Liu, X. P. Qu, *Chin. Phys. Lett.*, **24**, 1400 (2007)
14. T. Aaltonen, P. Alen, M. Ritala, and M. Leskela, *Chem. Vap. Depo.*, 9, 45 (2003)
15. O. K. Kwon, S. H. Kwon, h. S. Park, and S. W. Kang, *Electrochem. Solid-State Lett.* **7**, C46 (2004)
16. K. Kwon, S. H. Kwon, H. S. Park, and S. W. Kang, *J. Electrochem. Soc.,* ***151***, C753 (2004)
17. O. K. Kwon, J. H. Kim, H. S. Park, and S. W. Kang, *J. Electrochem. Soc.*, **151**, G109 (2004)
18. X. Zhao, N. P. Magtoto, and J. A. Kelber, Mater. Res. Soc. Symp. Proc. **812**, 129 (2004)
19. H. Kim, T. Koseki, T. Ohba, Y. Kojima, H. Sato and Y. Shimogaki, *J. Electrochem. Soc.*, **152**, G594 (2005)
20. S. H. Kwon, O. K. Kwon, J. S. Min and S. W. Kang, *J. Electrochem. Soc.*, **153**, G578 (2006)
21. H. Kim, Surface & Coatings Technology **200**, 3104 (2006)
22. J. Heo, S. Y. Lee, D. Eom, C. S. Hwang, and H. J. Kim, *Electrochem. Solid-State Lett.* **11**, G5 (2008)
23. S. K. Kim, S. Y. Lee, S. W. Lee, G. W. Hwang, C. S. Hwang, J. W. Lee and J. Jeong, *J. Electrochem. Soc.*, **154**, D95 (2007)
24. S. H. Kwon, O. K. Kwon, J. H. Kim, H. R. Oh, K. H. Kim, and S. W. Kang, *J. Electrochem. Soc.*, **155**, H296, (2008)
25. Q. Xie, J. Musschoot, C. Detavernier, D. Deduytsche, R. L. Van Meirhaeghe, Y. L. Jiang, Guo-Ping Ru, Bing-Zong Li and Xin-Ping Qu, *Microelectron. Eng.* (2008), doi:10.1016/j.mee.2008.05.02
26. L. C. Reldman, J. W. Mayer, Fundamentals of Surface and Thin Film Analysis, North-Holland, New York, 1986

27. H. Kim, I. R. D. Moraes, G. Tremiliosi-Filho, R. Haasch, A. Wieckowski, *Surf. Sci.* **474**, L203 (2001)
28. C. Willocq, A. Delcorte, S. Hertnans, P. Bertrand, M. Devillers, *J. Phys. Chem.* **B 109(19)**, 9482 (2005)
29. X. Zhao, N. P. Magtoto, and J. A. Kelber, *Thin solid films,* **188**, 478 (2004)
30. Y. Y. Wu, A. Kohnm and M. eizenberg, *J. Appl. Phys.*, **95**, 6174 (2004)
31. H. Kim, C. Detavenier, O. van der Straten, S.M. Rossnagel, A. J. Kellock, D.G. Park, *J. Appl. Phys.* **98**, 014308 (2005)

Conference Proceedings AMC XXIV © 2009 Materials Research Society

Reducing Effective Tungsten Line Resistivity with Advanced Pulsed Nucleation Layer

Anand Chandrashekar, Raashina Humayun, Feng Chen, Lana Chan, Kaihan Ashtiani, and Michal Danek
Direct Metals Business Unit, Novellus Systems, Inc., 4000 North First Street, San Jose, CA, 95134

ABSTRACT

This paper describes a new tungsten deposition (W-CVD) process – called Advanced Pulsed Nucleation Layer (PNL^{TM}) – on a Novellus ALTUS platform that yields low effective resistivity and excellent fill in narrow linewidths and interconnects. Advanced PNL provides sufficiently conformal nucleation layer at thicknesses down to approximately 10 Å. The nucleation layer is followed by low resistivity tungsten (LRW^{TM}) growth initiation and bulk W-CVD fill. A thinner nucleation film allows more of the feature to be filled with bulk tungsten thereby reducing line and contact resistance, while low resistivity tungsten fill has the potential to extend W plug fill (to 2x nm nodes) by reducing size effects associated with grain boundary scattering in the bulk W. Consequently, the advanced PNL process results in ~40% lower median line resistivity while the ALTUS multi-station architecture enables process optimization at distinct deposition conditions.

INTRODUCTION

Chemical Vapor Deposition of tungsten (W-CVD) continues to be the process of choice for contact metallization in logic and memory integrated circuits, mostly due to a combination of excellent fill performance and relatively low contact resistance[1]. Despite the gradual transition to copper metallization, W-CVD also remains the predominant conductor for bit-line and local interconnect applications. Pulsed Nucleation Layer (PNL) technology [2] has extended the conventional W-CVD fill by providing a thin, conformal nucleation layer. However, recent studies have shown that as feature sizes shrink, increase in resistivity of tungsten damascene interconnects starts at relatively larger feature sizes (300 nm), than for Al or Cu.[3] This phenomenon has been attributed to (a) the larger mean free path of tungsten causing more charge carrier collisions in smaller features and (b) formation of seams in the trench resulting in increased electron scattering. This is a critical issue in need of a solution, since the technological trend towards smaller feature sizes could otherwise limit tungsten applications in integrated circuits.

Conventional W plugfill techniques use a two-step process approach – a Pulsed Nucleation Layer (PNL) at lower temperatures followed by CVD plugfill at higher temperatures. The PNL film is vital as it promotes conformal tungsten fill inside the trench. Bulk resistance is primarily determined by grain size of the CVD W film,[4] and is typically higher than the resistivity of single crystal bulk tungsten. Due to its small grain nature, the nucleation film is not expected to provide an ideal template for large W grain growth. Therefore, a process that deposits large grain tungsten inside high aspect ratio (HAR) features is key to maximizing electrical performance.

EXPERIMENT

The Novellus C3 ALTUS tungsten deposition system consists of four heated pedestals, or stations that are isolated from each other using an inert gas curtain. Different steps of the W deposition process are performed sequentially at each station. This multi-station architecture can sequentially deposit films using alternate chemistries and temperature on a wafer; while maintaining a high throughput.

DISCUSSION

In this paper, we describe the advanced Pulsed Nucleation Layer (PNL) process yielding excellent tungsten step coverage and low resistivity at narrow linewidths. The new Advanced PNL process replaces the conventional silane based PNL nucleation with (i) a new boron based nucleation step at low temperature followed by (ii) low resistivity tungsten (LRW) treatment[5] at higher temperature that promotes growth of large W grains. While the conventional plugfill

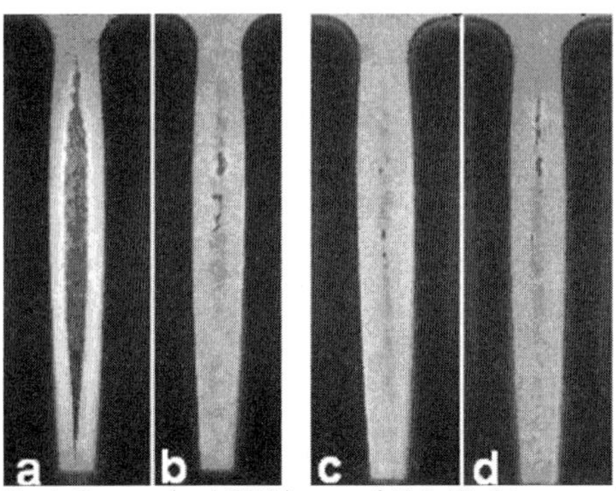

Fig. 1 Cross sectional SEM images of 10:1 AR, 0.1 μm opening contact with CVD TiN barrier. Plugfill failed for (a) 10 Å conventional PNL film while good step coverage is observed for (b) 23 Å conventional PNL film, (c) 12 Å advanced PNL film and (d) 23 Å advanced PNL film.

method requires a nucleation layer thickness of approx. 25 Å (Fig. 1 b), it has been found that approx 10 Å of the advanced nucleation film is sufficient for good plugfill (Fig.1c). The larger void in Fig. 1(a) is due to insufficient coverage of conventional nucleation layer near the feature bottom, causing delay in the H_2-WF6 CVD reaction that follows. The ability to achieve good plugfill with a thinner nucleation film is especially important when filling small critical dimensions like high aspect ratio features (characteristic of the 2x nm technology nodes) as it allows more of the feature to be filled with bulk tungsten thereby reducing line and contact resistance. For the advanced PNL nucleation film, it is believed that excellent plugfill is possible with minimal nucleation film thickness due to its amorphous nature as XRD studies indicate (Fig. 2). Also, due to the absence of grain boundaries such an amorphous film can be an excellent barrier to fluorine diffusion[6] and attack on underlying layers. For consistent device performance in a production environment, twice the minimum thicknesses are typically recommended – 25 Å for the advanced nucleation film and 50 Å for conventional nucleation

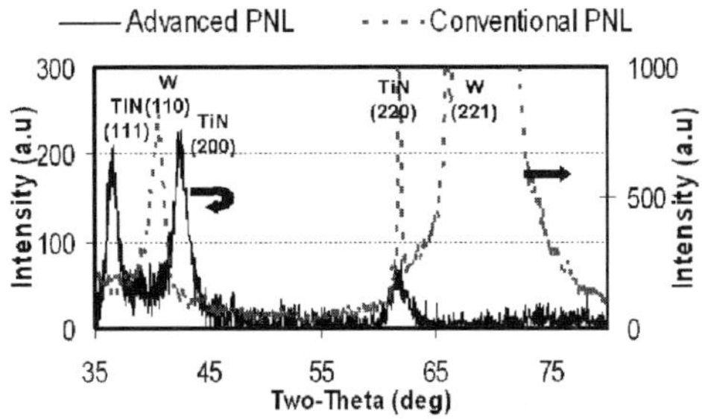

Fig. 2 Glancing angle XRD pattern of nucleation films for conventional PNL and advanced PNL processes, deposited on PVD TiN barrier. Conventional PNL nucleation film shows peaks corresponding to crystalline W, while advanced nucleation film appears to be amorphous.

film. Aside from occupying a lesser volume of the HAR feature, it has been found the resistivity of the advanced PNL nucleation film (55 μΩcm for 25 Å on PVD TiN) is lower than that for the conventional nucleation film (76 μΩcm for 55 Å on PVD TiN).

Another significant difference in the new process is the low resistivity tungsten (LRW) treatment performed between deposition of the nucleation layer and the bulk CVD. The low resistivity treatment deposits a LRW template with alternating boron-based and tungsten layers, at a higher process temperature than the nucleation sequence. This results in larger W grain size (and hence lower resistivity) in the subsequent CVD bulk film. In the case of thin films, it is well known that interfacial and grain boundary scattering can increase the total resistivity significantly above the bulk resistivity - an effect commonly known as the size effect [4, 7]. The total resistivity can then be calculated by adding to the bulk resisitivity (ρ_{bulk}), the contributions due to interfacial scattering ($\Delta\rho_i$), surface roughness scattering ($\Delta\rho_r$), and grain boundary scattering ($\Delta\rho_{gb}$). Bulk resistivity results from the scattering of conduction electrons with phonons. Interfacial scattering is associated with the barrier and is typically not a parameter that can be controlled in W deposition. Surface roughness scattering does not increase the resistivity significantly for W films below 100nm. This leaves grain boundary scattering as the key process controller for W resistivity improvement. The effect of grain boundary scattering on resistivity for polycrystalline films with average grain size D is estimated [8] as:

$$\frac{\Delta\rho_{gb}}{\rho_{bulk}} \approx \frac{3}{2}\frac{\lambda}{D}\frac{R}{(1-R)} \tag{1}$$

where λ is the materials-dependent electron mean free path (λ = 41nm for W)[9] and R is the reflection coefficient denoting the fraction of electrons reflected from the grain boundary. These reflected electrons contribute to higher resistivity. Equation 1 shows that the larger the grain size, the lower the resistivity. Thus the larger grain W film deposited using the advanced PNL process has lower resistivity than the conventionally deposited W-CVD films, although it is still higher than the resistivity of single crystal tungsten (5.3 μΩ-cm)

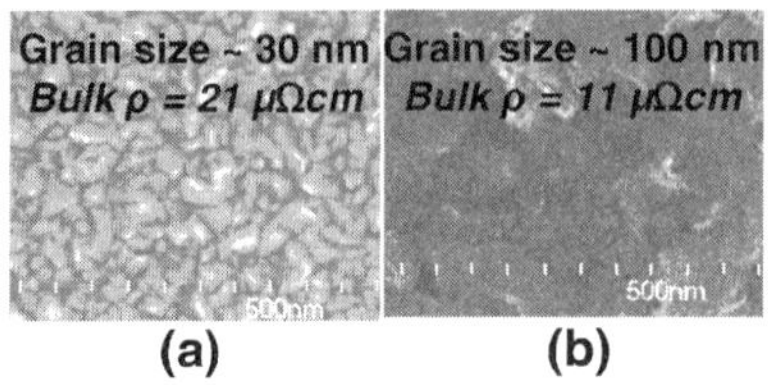

Fig.3. 500 Å Blanket film SEM images for (a) conventional PNL + W-CVD, and (b) advanced PNL + LRW treatment + W-CVD

Figs. 3(a) and 3(b) compare W grain sizes for (a) conventional PNL process and (b) advanced PNL process with LRW treatment. Similar CVD-W processes were used in both cases. It is clear that for 500 Å W film deposited on PVD TiN, W grain size on the blanket wafer is three times larger with the advanced PNL process.

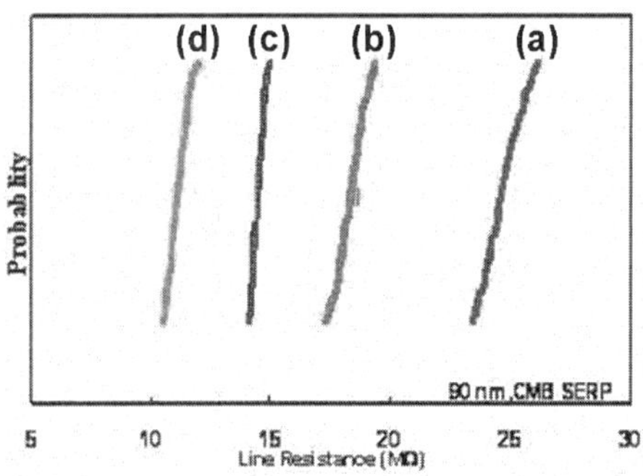

Fig. 4 Measured line resistance on 90 comb-serpentine lines with CVD Ti/TiN liner/ barrier. The four processes shown here are (a) conventional PNL + W-CVD plugfill (b) conventional PNL + LRW + W-CVD plugfill (c) advanced PNL + W-CVD plugfill (d) advanced PNL + LRW + W-CVD plugfill.

In order to validate the significance of the new process on tungsten grain size differences and on electrical performance, line resistance measurements were performed on 90nm comb-serpentine lines. 75 Å of Ti and 120 of Å CVD-TiN were used as liner and barrier respectively. In this study, four processes were used to evaluate the effect of thin amorphous low resistivity nucleation film and LRW treatment: (a) conventional PNL + W-CVD plugfill (b) conventional PNL + LRW + W-CVD plugfill (c) advanced PNL + W-CVD plugfill (d) advanced PNL + LRW + W-CVD plugfill. Similar CVD tungsten fill process conditions were maintained for all four splits. As-measured line resistance is plotted in Fig. 4 for the four different processes. Line resistivity was calculated for the die with median line resistance, using actual cross sectional area of the line from post-CMP SEM images. From Fig. 4, (d) advanced PNL + LRW + W-CVD plugfill, (c) advanced PNL + W-CVD plugfill and (b) conventional PNL + LRW + W-CVD plugfill show a 42%, 32% and 22% drop in median line resistivity respectively, compared to (a) conventional PNL + W-CVD plugfill. From these electrical test wafers, in-trench SEM images (Fig. 5) were also obtained after decorating the grain boundaries using KOH etch. This inspection technique involved a Focused Ion Beam (FIB) to cut the trench lines parallel to the sidewall (lengthwise). The grain boundaries were decorated using 100 % KOH treatment at room temperature, followed by 45 sec. water rinse and dry. Cross sectional SEM imaging (across the trenchline) was also attempted but was found not to be an ideal method. The cross-sectional area is very small; so grain growth from the sidewalls cannot be clearly observed.

It is clear that compared to a conventional PNL nucleation process, the advanced PNL process results in reduced line resistivity due to (i) larger in-trench W grain size due to LRW treatment, resulting in less electron scattering at the grain boundaries (ii) lower resistivity of the nucleation film and (iii) higher percentage of CVD W fill due to thinner advanced PNL nucleation.

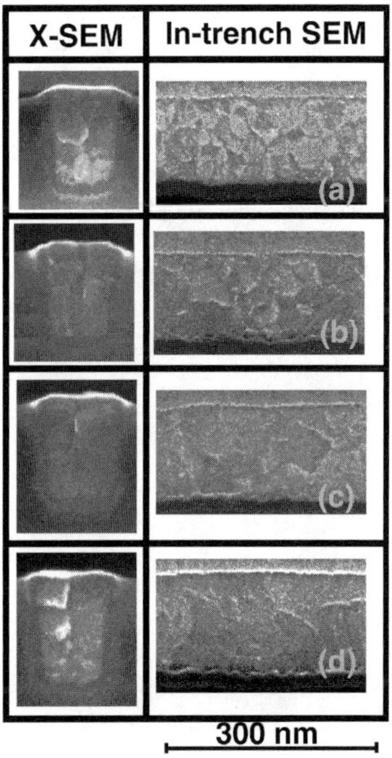

Fig. 5 Tungsten texture in 90nm trench. (a) conventional PNL + W-CVD plugfill (b) conventional PNL + LRW + W-CVD plugfill (c) advanced PNL + W-CVD plugfill (d) advanced PNL + LRW + W-CVD plugfill.

CONCLUSIONS

Advanced PNL provides an ultra-thin nucleation layer, down to ~10 Å and promotes growth of large, low resistivity W-CVD grains within features. These benefits translate into low resistance single damascene lines which enable improved performance in memory and logic circuits.

ACKNOWLEDGMENTS

The authors thank the Customer Integration Center at Novellus Systems for providing test structures and for data analysis.

REFERENCES

1. R. A. Levy and M. L. Green, J. Electrochem. Soc. Vol. 134, 2, 37C (1987)
2. S.H. Lee, L. Gonzalez, J. Collins, K. Ashtiani, K. Levy, Advanced Metallization Conference 2001 (AMC 2001), Editors McKerrow, Shacham-Diamand, Zaima, and Ohba, 649 (2001)
3. W. Steinhogl, G. Steinlesberger, M. Perrin, G. Scheinbacher, G. Schindler, M. Traving, M. Engelhardt, Microelec. Eng. 82, 266 (2005)
4. S. M. Rossnagel and T. S. Kuan, J. Vac. Sci. Technol. B, 22, 240 (2004)
5. S Smith et al., "Low Resistivity Tungsten for Contact Metallization," in Microelectronic Engg., Vol. 82 (3-4), 261-265, Dec 2005
6. J. Schmitz, Chemical Vapor Deposition of Tungsten and Tungsten Silicides for VLSI/ULSI Applications, William Andrew Publishing (1991)
7. I.P. Ivanov, I. Sen, and P. Keswick, J. Vac. Sci. Technol. B, 24, 533 (2006)
8. A.F. Mayadas and M. Shatzkes, Phys. Rev. B, 1, 1382 (1970).
9. E. Fawcett and D. Griffits, J. Phys. Chem. Solids, 23, 1631 (1962).

Conference Proceedings AMC XXIV © 2009 Materials Research Society

Specific Grain Boundary Resistivity Measurements in Thin Film Copper Bicrystals

Nagraj S Kulkarni[1,2], Boyd M Evans[2], Tae Hwan Kim[2], An-Ping Li[2], Edward A Kenik[2], and Don M Nicholson[2]
[1]Center for Materials Processing, University of Tennessee, Knoxville, TN, 37996-0816
[2]Oak Ridge National Laboratory, Oak Ridge, TN, 37831

ABSTRACT

A technique to estimate the specific grain boundary resistivity contribution in Cu lines using nanoscale four-probe STM measurements across thin film bamboo grain boundaries is presented. FIB milling of evaporated copper films was utilized to create Cu lines of various widths (400-4 μm). BSE and OIM microscopy was used to characterize the microstructure and texture of annealed Cu films containing these lines. Preliminary measurements using a fixed probe spacing with the four-probe STM technique indicated the sensitivity of the technique to detect variations in resistances along the length of the line though conversion to resistivities resulted in abnormally high values likely caused by material redeposition and/or Ga ion impregnation during FIB milling.

INTRODUCTION

The resistivity of copper interconnects increases appreciably with reducing feature sizes due to a variety of scattering mechanisms. However, the relative importance of the various scattering mechanisms remains an issue of debate. Most reports including ITRS have pointed to the importance of both grain boundary (gb) and interface scattering [1,2], while others [3] have suggested that interface scattering is the dominant contributor. The source of these differing analyses can be traced to difficulties in determining the experimental resistivity in narrow lines with varying cross-sectional area, and the subsequent fitting of this data using the Fuchs-Sondheimer and Mayadas-Shatzkes formalisms. Since a variety of fitting parameters involving the different scattering mechanisms can fit the experimental data equally well, it is difficult to state with confidence the relative role of the various scattering contributions.

In the approach that is the broader focus of our efforts, we seek to obtain estimates of the specific grain boundary resistivity contribution in copper lines by measuring the individual grain boundary resistivities in patterned lines containing bamboo grain boundaries (Fig.1). In this preliminary work, however, we have investigated lines whose microstructure still resemble that of an annealed blanket Cu film since post-patterning annealing to yield bamboo grain boundaries was not yet carried out. Cu lines or trenches about 25 μm in length and ranging in width from 400-4000 nm were patterned from annealed blanket films (250 nm thickness) using a dual-beam focused ion beam (FIB) milling technique.

Orientation Imaging Microscopy (OIM) and Backscatter Electron Microscopy (BSE) were used to characterize the texture of the films containing the lines. A four-probe scanning tunneling microscope (STM) (Fig. 2) [4] developed at the Center for Nanophase Materials Science (CNMS) at ORNL was used to measure resistances at various locations on the line in order to determine if the technique was sensitive enough to detect variations due to the presence of grain boundaries.

Future efforts will focus on post-pattern annealing of these lines in order to obtain grain boundaries that span both the width and thickness of the line in a manner similar to Al interconnect lines in order to obtain a direct correspondence between an individual grain boundary resistivity measurement and its type.

BICRYSTAL GB RESISTIVITY

Consider a thin film consisting of bamboo grains in which the grain boundaries (gb's) extend through the thickness of the film with the grain boundary plane normal to the surface of the film. This situation is typically seen in annealed Al films that are less than 500 nm thick. Assuming a similar situation exists in Cu, a film thickness of <500 nm in Cu is expected to result in a bamboo grain microstructure when annealed at a suitable temperature and time. Typically the average grain size observed is about twice the film thickness. A polycrystalline grain structure will be seen when the annealed film is observed from the top surface (plan view). However, if lines or strips are generated from the film (e.g., by etching lines), and these lines are subsequently annealed, the grain boundaries may be expected to also span the width of the film as well giving a true bamboo microstructure that is equivalent to a series of single crystals separated from each other by bicrystal grain boundaries [5].

In Fig. 1 we consider the situation where the grain boundaries when viewed from the top surface of the film extend across the width of the film as well. Assume the gb's are normal to the current flow and the length of the line. For the present example, assume the grain size for the three crystals are equal to each other and to the average grain size *d*. First measure the resistivity in the single crystals 1, 2, 3, etc. by measuring resistances R_{AB}, R_{CD}, etc. by placing the probes at AB and CD in the interior of the crystals 1 and 2 respectively. L_{AB} and L_{CD} should be normal to the grain boundary and should be known. Assume the width w and thickness t are constant throughout the length of the sample. Then the single crystal resistivities for crystals 1 and 2 along AB and CD are:

$$R_{AB} = \rho_{AB} \cdot \frac{L_{AB}}{Ao_{AB}} = \rho_{AB} \cdot \frac{L_{AB}}{w_{AB} . t_{AB}} = \rho_{AB} \cdot \frac{L_{AB}}{w.t} \quad \text{and} \quad R_{CD} = \rho_{CD} \cdot \frac{L_{CD}}{Ao_{CD}} = \rho_{CD} \cdot \frac{L_{CD}}{w_{CD} . t_{CD}} = \rho_{CD} \cdot \frac{L_{CD}}{w.t}$$

where Ao is the cross-sectional area. Now after locating the grain boundary between crystals 1 and 2 (gb_{12}), consider the situation if the resistance is measured between B in crystal 1 and C in crystal 2. This resistance R_{BC} includes the resistance due to gb_{12}. In order to compute the resistivity of gb_{12}, resistance R_{BC} can be expressed as:

$R_{BC} = R_{hyp} + R_{gb}$, where R_{hyp} is the resistance of a hypothetical ideal system between BC with no grain boundary resistance contribution, and R_{gb} is that attributed to the grain boundary, which is in effect the difference between the resistance of the actual system between BC and that of the hypothetical system. Hence:

$$R_{hyp} = \rho_1 \cdot \frac{L_1}{Ao} + \rho_2 \cdot \frac{L_2}{Ao} \quad \text{and} \quad R_{BC} = \rho_{BC} \cdot \frac{L_{BC}}{Ao} = \rho_{BC} \cdot \frac{L_3}{Ao} = \rho_{BC} \cdot \frac{2L_1}{Ao}$$

Assuming $L_1 = L_2$ so that $L_3 = L_{BC} = 2L_1$, the resistivity of the grain boundary ρ_{gb12} can be expressed in terms of the specific grain boundary resistivity γ_{gb12} which is the grain boundary resistivity/surface area per unit volume for that particular grain boundary type in the line.

$\rho_{gb12} = \gamma_{gb12}.S/V$, where S is the surface area of the gb ($S = w.t$) assuming a sharp grain boundary, and V the volume between BC ($V = w.L3.t$). Hence R_{BC} can be written as:

$$R_{BC} = \rho_{BC}.\frac{2L_1}{Ao} = \rho_1.\frac{L_1}{Ao} + \rho_2.\frac{L_2}{Ao} + \gamma_{gb12}.S/V.\frac{L_1 + L_2}{Ao}$$

Since $L_1 = L_2$, it follows from the above equation that

$$\rho_{BC}.\frac{2L_1}{Ao} = \rho_1.\frac{L_1}{Ao} + \rho_2.\frac{L_1}{Ao} + \gamma_{gb12}.S/V.\frac{2L_1}{Ao}$$

The expression for the specific gb resistivity given by Nakamichi [6] is thus obtained:

$$\gamma_{gb12}.S/V = \rho_{BC} - \frac{\rho_1 + \rho_2}{2} \quad \text{or} \quad \gamma_{gb12}.\frac{w.t}{2L_1.w.t} = \rho_{BC} - \frac{\rho_1 + \rho_2}{2}$$

$$\gamma_{gb12} = \left(\rho_{BC} - \frac{\rho_1 + \rho_2}{2}\right).2L_1$$

If the distances L_1 and L_2 are not equal, the specific grain boundary resistivity can be show to be:

$$\gamma_{gb12}.S/V = \frac{L_1.(\rho_{BC} - \rho_1) + L_2(\rho_{BC} - \rho_2)}{L_1 + L_2} \quad \text{or} \quad \gamma_{gb12} = L_1.(\rho_{BC} - \rho_1) + L_2(\rho_{BC} - \rho_2)$$

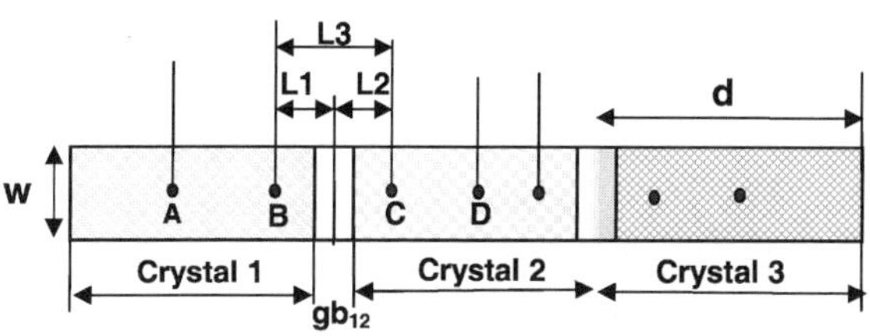

Fig. 1. Top view of a bamboo microstructure in a copper interconnect line that is used to measure bicrystal grain boundary resistivities using a 4-probe STM technique.

Fig. 2. A photo of four-probe STM

EXPERIMENTAL

The substrate used for these experiments was a four inch (100) Si wafer with 5 μm of thermal oxide on the surface. A Ti film 10 nm thick was first deposited on a wafer followed by a 250 nm Cu film both using electron beam evaporation. Film thicknesses were determined by a quartz crystal microbalance. The sample was subsequently annealed at 400°C for 4 hours on a PECVD chuck. A dual-beam focused ion beam (FIB) system (FEI Company™ Nova 600 Nanolab DualBeam™ -SEM/FIB) was used to create nanoscale conductors in the copper films. Conducting regions were isolated from the surrounding material by sputtering trenches in the sample with a focused stream of Ga ions from the FIB system. The final lines were written at a magnification of 4,800X with settings of 30 kV and 50 picoamps and a beam dwell time of 2 μs. A series of lines contained conductors 25 μm long and 0.4 - 4 μm in width was created (Fig. 3a, b). The depth of the trenches were approximately 400 nm, however the ultimate milling rate was dependent on the orientation of the particular grain. It is estimated that the beam diameter was 19.0 nm by the system software. BSE and OIM scans of the film are shown in Figs. 4a-c indicating both (111) (dominant) and (100) orientations.

An alternate set of FIB-milled trenches were also created as shown in Fig. 3c in order to avoid completely isolated conduction paths as in Fig. 3a (beam diameter 23 nm, 30 kV, 0.1 nA) This was believed to prevent charging effects during SEM imaging and measurements using the four-probe STM. The width of these trenches was also increased to about 200 nm.

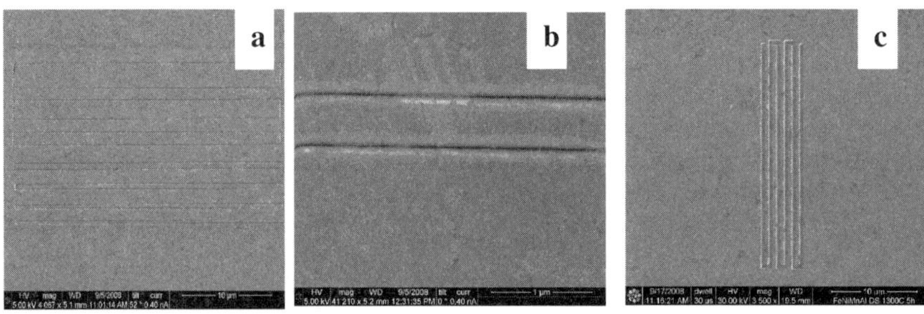

Fig. 3. (a) FIB-milled trenches (lines 0.4 – 4 μm wide, 25 μm long) in an annealed Cu film; (b) 400 nm wide line from Fig. (a); (c) serpentine trenches (lines 500 nm wide, 25 μm long).

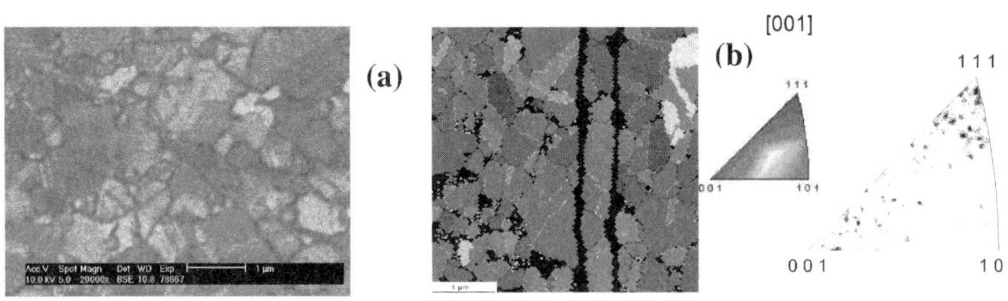

Fig. 4. (a) BSE image (b) Inverse pole figure (IPF) grain map (c) IPF map of annealed Cu film.

4-PROBE STM

To carry out four-probe resistivity measurements, the probes were monitored and located on a surface with nanometer scale features using a SEM and STM as shown in Fig. 5a. The STM motion controller was used to locate the probes just above the surface without unwanted crash between the probes and the surface. In order to make contact, first the tunneling current (0.1-0.3 nA) was achieved by adjusting the heights of the STM probes with the STM feedback loops on. Then, the gain of current preamplifier was changed from high to low in order to measure a dramatically increased current (1-10 uA) when the probe made physical contact with the surface. The lateral spacing between the probes was controlled with nanometer precision by the piezo-driven motors. The minimal spacing was only limited by the radius of the probes (approx. 100 nm). When all four probes were located on the surface, current was sourced through the two outer probes using a Keithley 2400 source meter and the voltage drop was measured through the two inner probes using a Keithley 6430 source meter as shown in Fig. 5a.

RESULTS

With a spacing of 200-700 nm between the inner voltage probes, a total of 12 resistance measurements were carried out at room temperature along the length of one of the 500 nm serpentine lines. Fig. 5b shows the voltage drop between the inner probes with respect to sourcing current. A linear fit was used to obtain a resistance value for a given spacing between voltage probes. Assuming a width of 500 nm and a height of 250 nm for the Cu line, this resistance was converted to resistivity (using $\rho = RA/L$). By fitting the data, a mean value of 4.63 ± 0.35 μohm-cm was obtained (Fig. 5c). These resistivity values varied from location to location (see Fig. 5d). This mean value compared quite favorably to the values of 4.73 and 4.53 μohm-cm obtained for a much higher probe spacing of 20 μm along the length of the same line and another line respectively. To compare with the above measurements, the van der Pauw method was applied to measure the resistivity in an unpatterned area of the same sample. A film thickness of 264 nm was obtained with measurements at 300 K and 350 K using the TCR method [1] and hence a film resistivity of 1.53 μohm-cm was obtained at room temperature.

DISCUSSION

The four-probe measurements made using a small probe spacing appeared to detect variations in resistances and hence resistivities along the length of the 500 nm wide line and the mean was comparable to that obtained using a much larger probe spacing; however, we suspect that the high resistivity values are the result of FIB-induced redeposition during milling (causing errors in the height) and/or Ga ion penetration because the resistivity in an unpatterned area of the same film was comparable to the bulk resistivity. Additionally, because the current probe tip diameters (about 10 nm) were much smaller than the line width (500 nm), it is not clear if the usual equation for converting resistances to resistivities is accurate. Hence, our results at the present time must be regarded as inconclusive with regard to gb resistivity effects. Measurements at two different temperatures to obtain the resistivity of lines using the TCR method [1] will help us remove the uncertainty in the thickness or cross-sectional area dimensions. Future work will focus on improvements in FIB milling to control redeposition and Ga ion penetration along side-walls, chemical mechanical planarization (CMP) to remove excess material on the lines, and development of suitable standards using our 4-probe STM technique. Additionally, we plan low temperature plasma etching experiments of Cu for narrow lines in order to avoid the issues encountered in FIB patterning. Optimization of processing conditions to obtain bamboo gb's are needed to establish a relation between a specific gb type and specific resistivity measurement. Gb resistivity measurements on bulk bicrystals in Cu using strain annealing [6], and resistivity measurements in thin Cu films of various grain sizes in order to obtain an average specific grain boundary resistivity [7,8] are also planned.

CONCLUSIONS

A dual-beam FIB system was used to create Cu lines of various widths in an evaporated annealed Cu film. An approach to measure the specific grain boundary resistivity in thin film bicrystals was presented. Preliminary, four-probe resistivity measurements indicated the potential of the technique to detect resistivity variations along the length of a 500 nm wide line, though redeposition and/or Ga ion contamination may have resulted in abnormally high values.

ACKNOWLEDGMENTS

This research was sponsored by the Laboratory Directed Research and Development Program of Oak Ridge National Laboratory, managed by UT-Battelle, LLC for the U. S. Department of Energy under Contract No. DE-AC05-00OR22725. A portion of this research was conducted at the Center for Nanophase Materials Sciences and the SHaRE User Facility, which are sponsored at Oak Ridge National Laboratory by the Division of Scientific User Facilities, Office of Basic Energy Sciences, U. S. Department of Energy. We are also grateful to C. Michael Garner and S. Sadasivan at Intel Corporation for their interest and encouragement.

REFERENCES

1. W. Steinhögl, G. Schindler, G. Steinlesberger, M. Traving, and M. Engelhardt, "Comprehensive study of the resistivity of Cu wires with lateral dimensions of 100 nm and smaller," J. Appl. Phys. 97, 023706, 2005.
2. International Technology Roadmap for Semiconductors (ITRS), 2007 edition.
W. Zhang, S. H. Brongersma, Z. Li, D. Li, O. Richard, and K. Maex, "Analysis of the size effect in electroplated fine copper wires and a realistic assessment to model copper resistivity," J. Appl. Phys. 101, 063703, 2007.
3. T.-H. Kim, Z. Wang, J. F. Wendelken, H.Weitering, W. Li, and A.-P. Li, "A cryogenic Quadraprobe scanning tunneling microscope system with fabrication capability for nanotransport research," Rev. Sci. Instr. 78, 123701, 2007.
4. Y.-C. Joo and C. V. Thompson, "Analytical model for the grain structures of near-bamboo interconnects," J. Appl. Phys. 76, 7339, 1994.
5. I. Nakamichi, "Electrical Resisitivity and Grain Boundaries in Metals," Mat. Sci. For. 207-209, 47, 1996.
6. R. A. Brown, "A dislocation model of grain boundary electrical resistivity," J. Phys.F: Metal Phys. 7, No.8, 1477, 1977.
7. P. V. Andrews, M. B. West, and C. R. Robeson, "The effect of grain boundaries on the electrical resistivity of polycrystalline copper and aluminium," Phil. Mag. 19:161, 887, 1969.

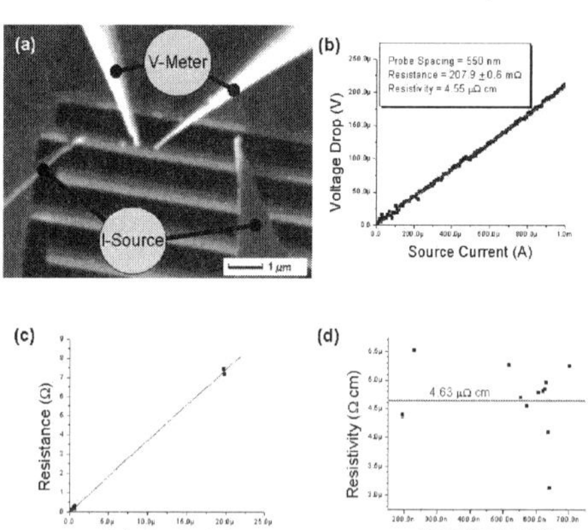

Fig. 5. (a) An SEM image after making contacts with four probes on one of Cu lines. Schematic drawings indicate measurement configuration. (b) Four-probe measurement of (a). (c) Probe spacing dependence of resistance along Cu lines. Red line indicates linear fitting. (d) Distribution of calculated resistivities with respect to probe spacing. Red line indicates average value of resistivity from linear fitting of (c).

Conference Proceedings AMC XXIV © 2009 Materials Research Society

In-situ CVD CoSix Deposition for Direct Contact Applications

See-Eng Phan[1], Seshadri Ganguli[1], Hyoung-Chan Ha[1], Sang Ho Yu[1], Michael Jackson[1], Sanghyeob Lee[1], Abraham Anapolosky[2], and Phillip Wang[3]
[1]MDP, Applied Materials, 3320 Scott Blvd, Santa Clara, CA, 95054
[2]DTCL, Applied Materials, 3535 Garrett Drive, Santa Clara, CA, 95054
[3]MDP, Applied Materials, 3340 Scott Blvd, Santa Clara, CA, 95054

ABSTRACT

A CVD $CoSi_x$ process has been developed to replace $TiCl_4$ Ti/TiN for direct contact applications, with the potential to reduce the number of total steps in the manufacturing process. This process deposits CoSi from an organo-metallic precursor and SiH_4, which is then plasma-treated to reduce C. After RTP, this CoSi transforms to $CoSi_2$. AES and XRD are used to show this transformation, with TEM providing further confirmation. Step coverage of ~70-80% is demonstrated on structures with aspect ratio ~5:1 and 100nm opening. Potential to form $CoSi_2$ directly via a CVD process is also shown.

INTRODUCTION

Current memory devices use $TiCl_4$ Ti/TiN for direct contact, but as linewidth shrinks, $TiSi_2$ sheet resistance dramatically increases because $TiSi_2$ depletes dopants from the junction. In addition, the $TiCl_4$ process leads to junction leakage due to rough silicide interfaces, agglomeration and lateral encroachment. CVD $CoSi_x$/Co is an attractive material because of its low resistivity, thermal stability and excellent step coverage.[1,2] Also, excessive silicon consumption during thermal process is a critical issue for shallow junction formation. Si consumption ratio for Co alone is quite high at ~3.64 atoms Si per atom Co. In-situ CVD $CoSi_x$ can reduce silicon consumption and CVD Co serves as a fluorine and oxygen barrier. CVD $CoSi_x$ may also have a more simplified integration scheme with one less RTP step and possibly no wet etch or $TiCl_4$ Ti/TiN barrier layer (Figure 1). This paper describes some results with in-situ CVD CoSi. CVD $CoSi_x$/Co can also be used for S/D cobalt salicide with PVD Co and 3-D transistors for 22nm and beyond

Current: Co/RTP1/Wet etch/ RTP2/$TiCl_4$ TTN/W **New: $CoSi_2$/RTP/WN/W**

Figure 1, showing a contrast between current Co integration scheme and optimized CVD CoSi integration scheme, removing one RTP step, the wet etch and an ex-situ liner.

EXPERIMENT:

This work was done an Applied Materials ALD TaN chamber, modified with two extra gas lines and a RF subsystem which involves a generator, RF match and RF-hot showerhead. The RF-hot showerhead and process kit were borrowed from the Applied Materials TxZ MOCVD TiN. The chamber was mounted on an Applied Materials Endura2 mainframe, permitting in-situ cleaning and capping.

The precursor was used as received. It is stored in a "hot can" in the chamber's gasbox, where it is heated to a temperature determined to be an optimal compromise between vapor pressure and decomposition. The films as deposited have a high concentration of C (~50-60%), so a cyclical process involving multiple hydrogen treatments was used to reduce the carbon concentration to below 5%, as measured using AES. One major advantage to this precursor over other precursors that give a more pure Co film is that, under these conditions, the Co reacts with SiH_4 to give CoSi. This can be seen in Fig 2a, where a Si:Co ratio of ~1:1 (measured with AES) is demonstrated. Fig 2b shows the same wafer after RTP, where the Si:Co ratio varies from 2.2-1.5:1. Fig 3 shows XRD from $CoSi_x$ films, where Figure 3a (before RTP) does not show a $CoSi_2$ peak, whereas Fig 3b (after RTP) shows the appearance of a $CoSi_2$ peak.

The process recipe for deposition of these films involves a cyclical process where co-flow of the Cobalt precursor and SiH_4 together is alternated with a H_2 plasma treatment to reduce C. The Si:Co ratio as deposited under these conditions is ~1:1. Table I summarizes a variety of process conditions and resistivities, both before and after RTP.

			Before RTP				After RTP		
Description	Wafer	CoSix Time	Calc Thk [Å]	Center Rs [Ω/sq]	Calc r [μΩ·cm]	Density [g/cm3]	Calc Thk [Å]	Center Rs [Ω/sq]	Calc r [μΩ·cm]
	Ox	10s×18	122	782	957	6.71	Same	499	611
CoSix LD 1	Ox	5s×18	54	1812	973	6.42	Same	1032	554
CoSix LD 2	Ox	5s×12	36	4121	1495	5.99	Same	19402	7040
CoSix LD 3	Ox	5s×8	22	NM	NM	5.99	Same	NM	NM
CoSix	BS	10s×18	122	245	299		220	23	51
CoSix LD 1	BS	5s×18	60	180	108	5.99	108	50	54
CoSix LD 2	BS	5s×12	40	209	83	6.32	72	114	82
CoSix LD 3	BS	5s×8	25	258	65	6.23	45	279	127

Table I: $CoSi_x$ properties for a variety of conditions. Wafer types are oxide ("ox") and bare silicon ("BS"), $CoSi_x$ time has the deposition time (5 sec or 10 sec) and the number of dep/treat cycles (from 8 to 18). Baseline : 10s deposition and 18 cycles; LD1 : 5s deposition and 18 cycles, LD 2: 5s deposition and 12 cycles, LD3 : 5s deposition and 8 cycles. The calculated thickness comes from a measurement of wafer thickness using the Co signal from XRF multiplied by the CoSi:Co thickness ratio (2.5). Using this CoSi thickness and the center-point Rs, the resistivity can be calculated as reported. The density reported is measured by XRR. Similar numbers are reported after RTP. As can be seen from the data, the films on oxide do not get thicker after RTP, while those on bare Si do get thicker as $CoSi_2$ is formed after RTP

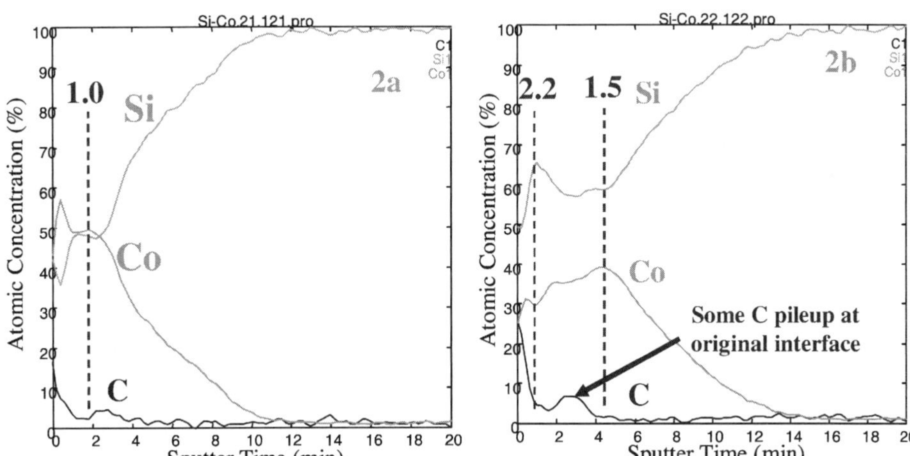

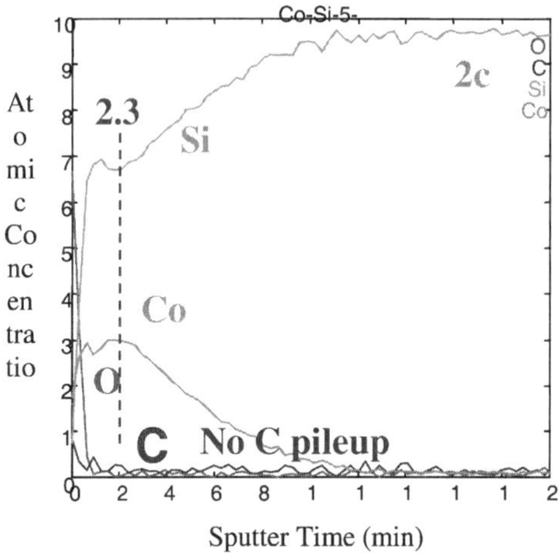

Figure 2: AES depth profile spectra showing transformation of as-deposited CVD Co (Fig 1a, left) to $CoSi_2$ (Fig 2b, right) after RTP. There is some variation in the Si:Co ratio as well as some indication of C pile-up at the original interface between Si and $CoSi_x$ \Fig 2c shows no C pile-up at the interface after recipe modifications.

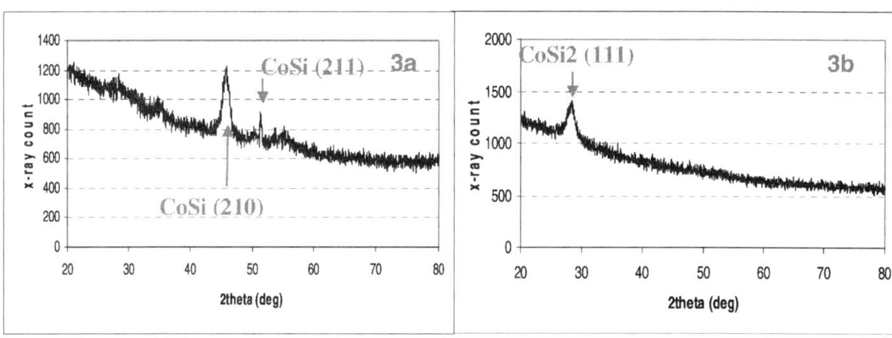

Figure 3 Comparison of XRD on $CoSi_x$ films before (3a) and after (3b) RTP, confirming that the as-deposited CoSi transforms to $CoSi_2$ after RTP.

Further evidence for the transformation of CoSi into $CoSi_2$ can be seen in Figure 4. Figure 4a shows the CoSi before RTP, where the film is quite homogenous. Fig 4b shows the film after RTP, where the $CoSi_2$ film appears as two layers, although both layers are crystalline $CoSi_2$. The top layer is slightly more porous than the bottom layer. The interface between these two layers is likely the original Co:Si interface. Using the measured thicknesses to calculate the Si consumption from the substrate, the number works out to be between 1.7 and 1.9 atoms Si per atom Co, much less than the ~3.6 for Co alone.

The step coverage is illustrated in Fig 5 (a composite of two TEM images), where, depending on the details of measurement, the step coverage is from 71-80% on a ~5:1 structure with a 100nm opening. The images were composited as neither TEM image had the combination of field-of-view and resolution desired.

The majority of this work has concentrated on CoSi, however Fig.6 shows that with a combination of process and hardware modifications, we can deposit $CoSi_2$. In this case, it was only deposited at the edge of the wafer, but improvements are ongoing to deposit $CoSi_2$ across the entire wafer. These improvements have also reduced C below the AES detection limit.

DISCUSSION

The main requirements for success in a CVD $CoSi_x$ process are the ability to form good ohmic contact as measured on an electrical test or device wafer, with acceptable step coverage for future device generations. We have focused on XRD (showing that we have polycrystalline $CoSi_2$) and TEM (showing a transformation after RTP). For a $CoSi_x$ process, the ability to reduce amount of Si consumed is critical. As shown in Fig 4, the CoSi process reduces Si consumption notably, but still consumes Si from the substrate. Fig 6 shows evidence that we have the potential to further reduce Si consumption towards our ultimate goal of depositing CVD $CoSi_2$ film. In order to minimize C pile-up at the interface (which is shown to some degree in Fig. 2b), our recipe has been carefully optimized with a variety of minor changes in nucleation steps and reactant introduction order. The results of these modifications are shown in 2c, where C is below the detection limit for AES, even after RTP.

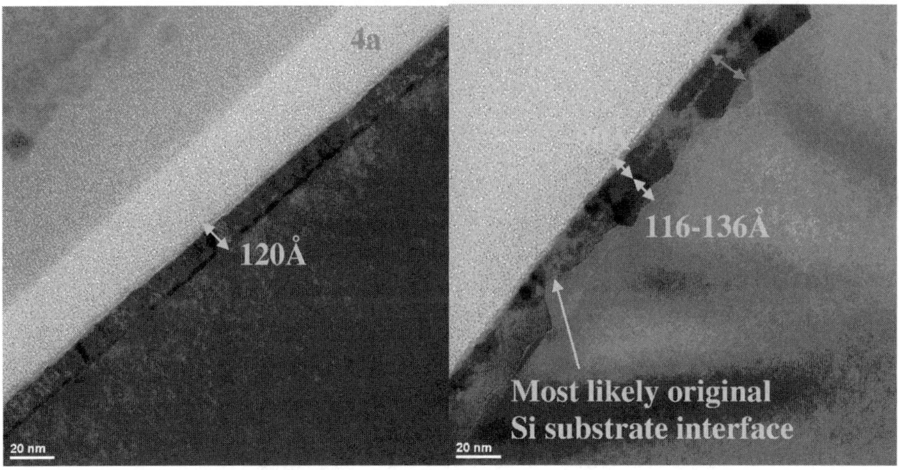

Figure 4: (4a) As deposited TEM showing homogenous CoSi film. ($b) After RTP, some of the silicon substrate reacts with CoSi to form CoSi2.

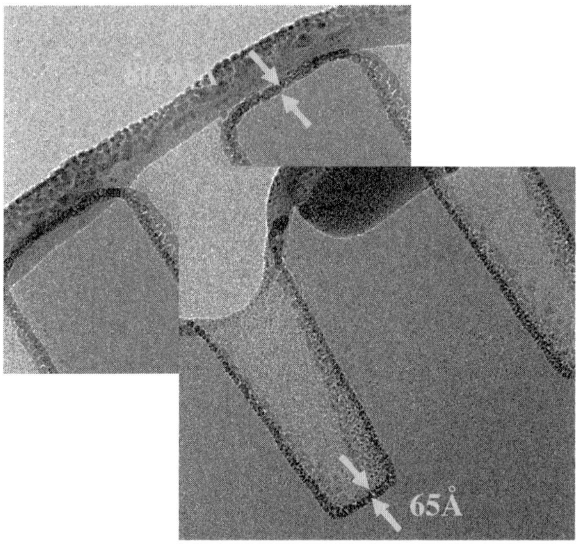

Figure 5: Step coverage of CVD CoSix is about 71-80% for structures with a 5:1 aspect ratio and 100nm opening

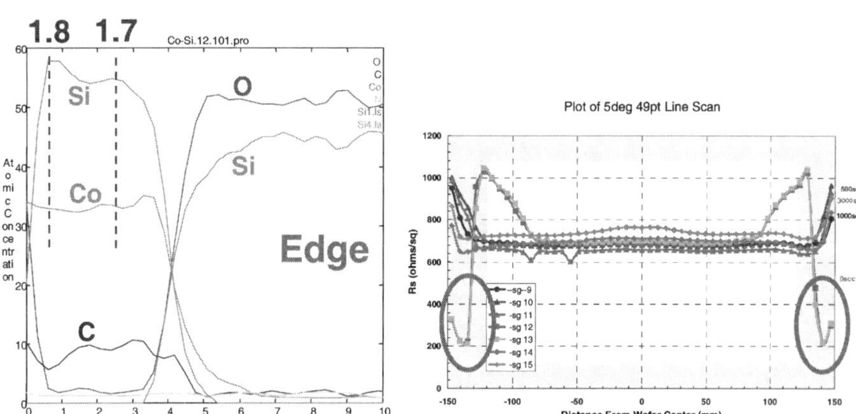

Figure 6 AES depth profile showing Si:Co ratio of ~1.8:1 at the edge of the wafer.

The resistivity of CVD $CoSi_x$, as measured in this paper, is still higher than that of PVD Co/RTP $CoSi_2$. As seen in table 1, the resistivity of CVD CoSi2 after RTP plateaus ~50 μΩ·cm, still higher than PVD $CoSi_2$ resistivity ~14-20 μΩ·cm. The density of the as-deposited CVD

CoSi (Table I) is similar to that of PVD CoSi. [3] For thinner films (<50A), the CVD $CoSi_x$ resistivity is quite high. Since no capping layer (e.g. TiN) was used in this work, we believe oxygen was incorporated into the thinner $CoSi_2$ layers at a substantial concentration, resulting in higher resistivity.

CONCLUSIONS

A process for the deposition of a CoSi film has been demonstrated, along with a variety of characterization data (AES, XRD and TEM) showing that this film has a very low level of C and can be transformed into $CoSi_2$ using RTP. Evidence was shown that there is a route to deposit $CoSi_2$ directly via a CVD reaction.

REFERENCES

[1] S. Wolf, Silicon *Processing for the VLSI Era Vol. 4* (Lattice Press, Sunset Beach, CA,2002).

[2] G.J.M Dormans, G.J.B.M. Meekes and E.G.J. Staring, J Cryst Gr 114, 364 (1991)

[3] L.J. Chen, *Silicide Technology for Integrated Circuits,*(IEE London 2004*)*

Conference Proceedings AMC XXIV © 2009 Materials Research Society

Grain Growth of Electroplated Copper Film by Alternative Annealing Methods

Kazuyoshi Ueno[1], Kouichiro Shimotani[1], Yuji Shimada[1], Shigeru Yomogida[2], Takato Takeshita[1], Akiko Hashimoto[2], and Takashi Yata[2,]

[1] Shibaura Institute of Technology, 3-7-5 Toyosu, Koto, Tokyo 135-8548, Japan
[2] KISCO Limited, 4-11-2 Nihonbashi Honcho, Chuo, Tokyo 103-8410, Japan
E-mail: ueno@shibaura-it.ac.jp

ABSTRACT

Alternative annealing methods using current stress or supercritical fluid (SCF) are investigated for electroplated copper (Cu) films. The current stress method promotes grain growth at room temperature, mostly through the Joule heating effect. The grain size resulting from the SCF annealing method was more uniform; therefore, it is a potentially better method for improvement of the microstructure in electroplated Cu films.

INTRODUCTION

Cu interconnects are used for ultra large scale integrated circuits (ULSI) due to low resistivity and high endurance against electromigration (EM). However, increase of resistivity induced by electron scattering is becoming a critical issue for width narrower than 100 nm. In order to suppress the increase of resistivity, it is necessary to reduce the electron scattering at Cu interfaces and grain boundaries.[1] An effective method to achieve this is to reduce grain boundaries by enhancing grain growth. Microstructure is also an important factor affecting reliability.[2,3] Electroplating is conventionally used for Cu deposition in the Cu damascene process. Cu films deposited by electroplating initially consist of very fine Cu grains that are unstable, which leads to room-temperature grain growth referred to as “self-annealing”.[4,5] Conventionally, thermal annealing is adopted for grain growth and stabilizing the Cu after electroplating; however, it is difficult to control Cu microstructure in the conventional thermal annealing method with the limited process parameters.

In this study, two alternative annealing methods were studied for control of the Cu microstructure. One is current induced grain growth, and the other is annealing in a supercritical fluid (SCF). The grain growth characteristics of electroplated Cu films subjected to these alternative annealing methods were investigated.

EXPERIMENT

Cu films were deposited on 2×2 cm^2 Cu/Ta/TaN/SiO_2/Si substrates by electroplating using a Cu sulfate bath with polyethylene glycol (PEG) 4000 and sulfopropyl-sulfonate (SPS) as additives. The thickness of the Cu seed layer was 50 nm. The thickness of the electroplated Cu film was varied between 250 and 500 nm. The current density used for electroplating was 10 mA/cm^2.

After electroplating, the new annealing methods were adopted as follows. For the current induced grain growth experiments, current stress was applied to the Cu plated films in a N_2

atmosphere using connecting terminals at the opposite ends of the film. The DC current density was varied between 0.02 and 0.2 MA/cm^2 at the room temperature. The sample temperature was raised during the current stress by the Joule heating effect; therefore, the surface temperature of the sample was measured using a thermocouple. In order to compare grain growth between the current stress and the thermal annealing by Joule heating, samples were also annealed at the same temperature without current stress.

The SCF annealing method was performed in a high pressure cell with variation of the gas species, pressure, and temperature conditions. CO_2 and N_2 were investigated as gases for use in this method. Typical gas pressure and temperature were 15 MPa and 300 °C for the gases to be in the SCF phase. Samples annealed in a N_2 gas atmosphere were also fabricated for comparison.

The Cu microstructure was observed using focused ion beam (FIB) microscopy. The sheet resistance of the films was measured by the four-point probe method. The change in sheet resistance over time during storage at room temperature (self-annealing) was also measured. X-ray diffraction (XRD) measurements were performed for film texture analysis.

RESULTS & DISCUSSION

Grain growth by current stress

The current stress effect on the sheet resistance was investigated as shown in Fig. 1, which shows the change in sheet resistance for a 500 nm thick electroplated films stressed at 0.1 MA/cm^2 for 1, 2, and 5 h. After the current stress, the sheet resistance was reduced by approximately 20%. On the other hand, the change in resistance for the sample without current stress was as small as 5%. The reduction in resistance was almost the same for different stress times, which indicates that the resistance reduction was saturated after the 1 h stress treatment.

Figure 2 shows a comparison of the change in sheet resistance for 250-nm-thick electroplated films stressed at different current densities of 0.02, 0.1, and 0.2 MA/cm^2 for 2 h. During the stress at 0.1 and 0.2 MA/cm^2, the sample temperature became 50 and 160 °C, respectively. As the current density was increased, the reduction in resistance after the current stress was also increased. The reduction in resistance was approximately 20% after the 0.2 MA/cm^2 current stress treatment. Comparing the reduction in sheet resistance between the samples subjected to 0.2 MA/cm^2 current stress and 160 °C annealing, the reduction was almost the same at around 20%, as shown in Fig. 2. After the reduction in resistance, further changes in resistance were not observed during room temperature storage. The reduction in resistance after the 0.1 MA/cm^2 stress shown in Fig. 2 is smaller than that observed in Fig. 1. It is considered that the 250-nm-thick film of Fig. 2 was more difficult to grow than the 500-nm-thick film of Fig. 1, due to the thickness dependence of the grain growth.[4]

Figure 3 shows the microstructure of the 250-nm-thick Cu films after current stress at different current densities. After the 0.02 MA/cm^2 current stress, the grain size remained very small, although some grain growth was observed as indicated by the dotted line areas in Fig. 3(a). When the current density was increased to 0.1 MA/cm^2, such grain growth sites increased, as shown in Fig. 3(b). After the 0.2 MA/cm^2 current stress, larger grains were observed (Fig. 3(c)). The observed grain size difference is in good accordance with the sheet resistance reduction given in Fig. 2.

Figure 4 shows the grain size distribution and microstructure of the 250-nm-thick Cu film after annealing at 160 °C, which was the same temperature obtained during the 0.2 MA/cm^2

current stress by Joule heating. There appears to be no distinct difference in microstructure when Figs. 4(b) and 3(c) are compared. There is a little difference in the grain size distribution (Fig. 4(a)), which indicates that the 0.2 MA/cm^2 current stress leads to larger grains than 160 °C annealing. Since there is no distinct difference in the microstructure and the sheet resistance reduction, current induced grain growth is considered to be driven mainly by Joule heating during the stress. Both films have high {111} textures after annealing, as indicated by XRD data summarized as follows. The XRD peak area intensity of the 0.2 MA/cm^2 stressed film is approximately 20% more than that of the 160 °C annealed film. The full width at the half maxima (FWHM) is 0.11° for the 0.2 MA/cm^2 stress and 0.12° for the 160 °C annealing, respectively, indicating that the {111} texture improved a little bit by the current stress.

The results suggest that current stress can promote grain growth at room temperature, mainly due to the thermal annealing effect from Joule heating. The grain size and the texture were slightly improved compared to those obtained by conventional thermal annealing at the same temperature attained by Joule heating during the current stress. This may be an indication of enhanced Cu diffusion by the EM phenomenon. However, it is expected that EM potentially enhances Cu diffusion and leads to enhanced motion of the grain boundary. Further studies under different current stress conditions are necessary to clarify the effect of EM.

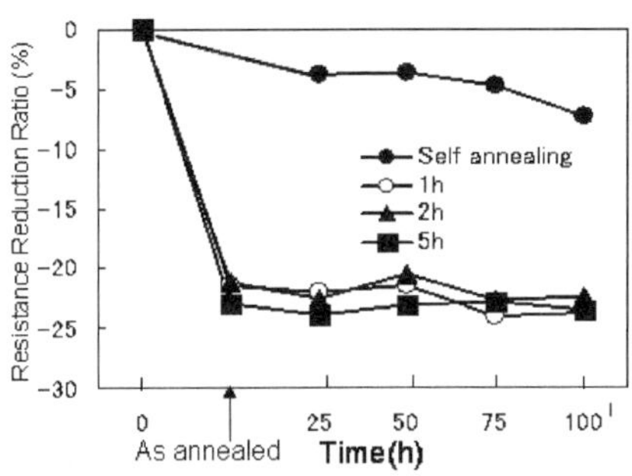

Fig. 1. Change in sheet resistance after current stress at 0.1 MA/cm^2 for 1, 2, and 5 h, and during storage at room temperature.

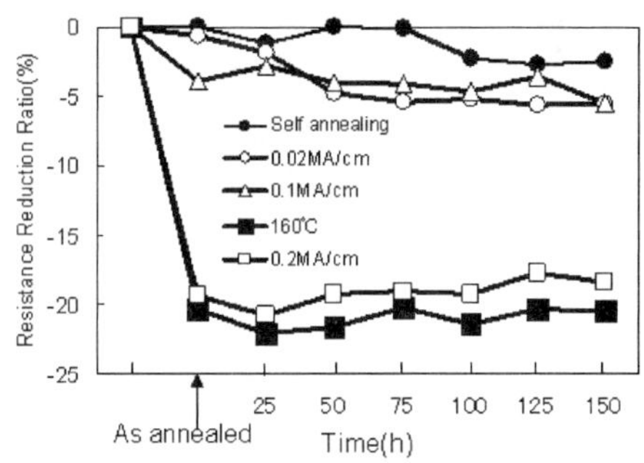

Fig. 2. Change in sheet resistance after current stress at different current densities, after 160 °C annealing, and during storage at room temperature.

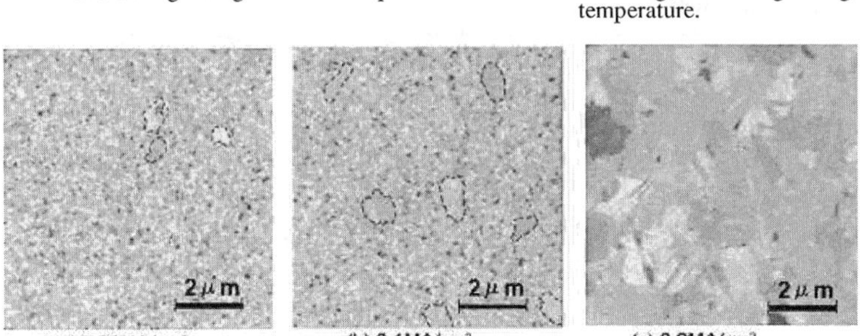

(a) 0.02MA/cm^2 (b) 0.1MA/cm^2 (c) 0.2MA/cm^2

Fig. 3. Microstructures obtained by FIB microscopy after different current stress at (a) 0.02, (b) 0.1, and (c) 0.2 MA/cm^2.

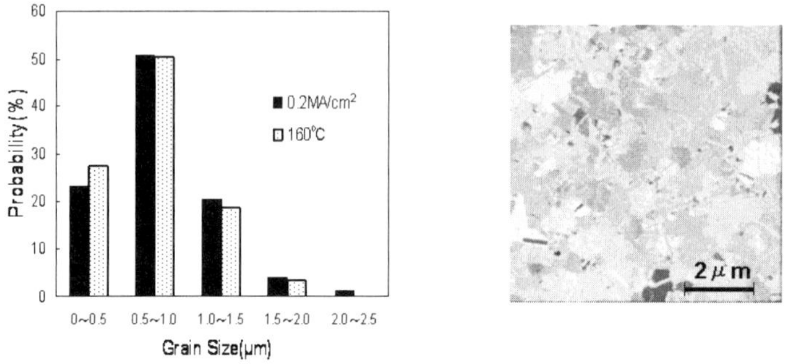

(a) Grain size distribution. (b) Microstructure after 160 °C annealing.

Fig. 4. Grain size distribution and microstructure after 160 °C annealing.

Grain growth by SCF annealing

Figures 5 and 6 show the changes in sheet resistance and the microstructure of the 500 nm thick Cu films annealed in a supercritical (SC)-CO_2 (H_2) atmosphere (300 °C, 15 MPa) and in an atmospheric pressure N_2 at 300 °C, respectively. There is no distinct difference in the resistance changes after the annealing and during the subsequent room temperature storage. On the other hand, it is noted that after the SCF annealing the grains have a more uniform shape than that after N_2 annealing, as shown in Fig. 7. In addition, grooving at the grain boundaries is observed more clearly for the SCF annealed samples.

In order to reveal the factors that affect the microstructure during SCF annealing, the dependency of the microstructure on the base gases used, and the effect of hydrogen addition were investigated. Figure 7 compares the microstructures for N_2 and CO_2 atmospheres at 300 °C and 15 MPa. The difference in microstructure is not clear, as both films contain small and large grains, and they appear to have bimodal grain size distributions. This indicates that abnormal grain growth occurred under both conditions. The grain boundaries formed without H_2 are difficult to distinguish. On the other hand, the grain boundaries formed in the presence of H_2 are more clearly observed in the microstructure. It is also noted that grooving at the grain boundaries is clearer for SCF annealing in the presence of H_2. The grain size distribution became relatively narrower for SCF annealing with H_2 as shown in Fig. 8.

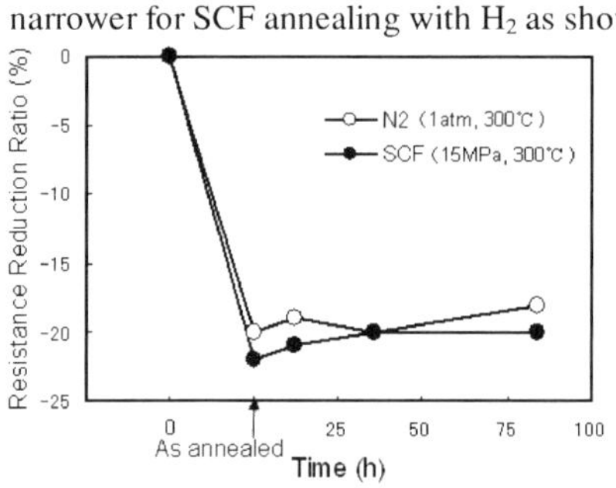

Fig. 5. Change in sheet resistance for SC-CO_2 (H_2) annealing and atmospheric N_2 annealing at 300 °C.

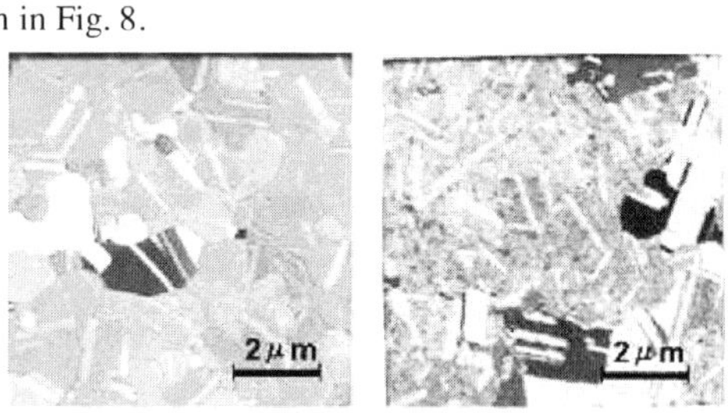

(a) SCF-CO_2(H_2) annealing (b) Atmospheric N_2 annealing

Fig. 6. Microstructure for (a) SC-CO_2 (H_2) annealing, and (b) atmospheric N_2 annealing at 300 °C.

It has been considered that abnormal grain growth in electroplated Cu films is due to anisotropic grain boundary energies combined with impurities from the electroplating bath. [6,7] Impurities have been considered to potentially pin the grain boundary movement. [4] SC-CO_2 is well known as an excellent cleaning solution, due to its high solvating power for contaminant species. [8] Therefore, SC-CO_2 should remove impurities in the electroplated films during grain growth. H_2 may also enhance the removal of the impurities due to its chemical reactivity. Further investigation, including secondary ion mass spectrometry (SIMS) analysis, of the impurities in the annealed films is necessary to confirm this speculation.

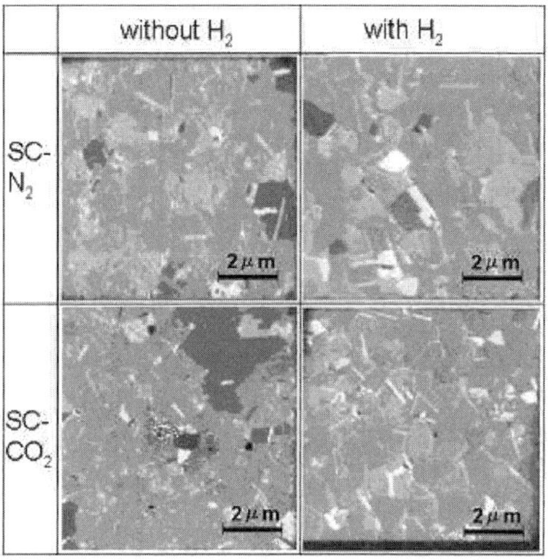

Fig. 7. Dependency of the microstructure on the base gas species and the effect of H_2 addition.

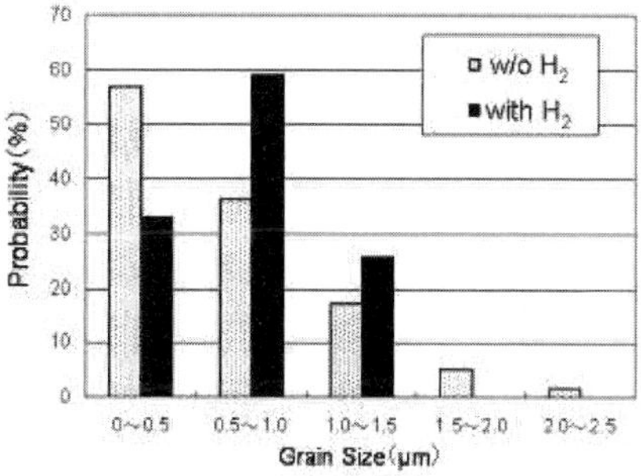

Fig. 8. Comparison of the grain size distribution between the SC-CO_2 annealing with and without H_2.

CONCLUSIONS

Alternative annealing methods, current stress or SCF, were investigated for control of the microstructure in electroplated Cu films. Current stress annealing can promote grain growth, mainly by thermal effects from Joule heating. SCF annealing promotes different grain growth characteristics, such as improved uniformity of grain size. Therefore, SCF annealing was determined to be a potentially better method for control of the microstructure in electroplated Cu films.

REFERENCES

1. *ITRS Roadmap, Interconnect*, Semiconductor Industries Association, **2007 ed.**, available from <www.itrs.net> (2007).
2. J. M. E. Harper and K. P. Rodbell, *J. Vac. Sci. Technol. B*, **15**, 763-779 (1997).
3. C. K. Hu, L. Gignac, and R. Rosenberg, *Microelectronics Reliability*, **46**, 213-231 (2006).
4. J. M. E. Harper, C. Cabral. Jr., P. G. Andricacos, L. Gignac, I. C. Noyan, K. P. Rodbell, and C. K. Hu, *J. Appl. Phys.*, **86**, 2516-2525 (1999).
5. K. Ueno, T. Ritzdorf, and S. Grace, *J. Appl. Phys.*, **86**, 4930-4935 (1999).
6. V. Wehnacht and W. Br ckner, *Thin Solid Films*, **418**, 136-144 (2002).
7. J. M. Paik, Y. J. Park, M. S. Yoon, J. H. Lee, and Y. C. Joo, *Scripta Materials*, **48**, 683-688 (2003).
8. G. L. Weibel and C. K. Ober, *Microelectronic Engineering*, **65**, 145-152 (2003).

Conference Proceedings AMC XXIV © 2009 Materials Research Society

Feasibility study of a multi-wafer reactor system by supercritical fluid deposition of Cu for mass production

Takeshi Momose, Masakazu Sugiyama[1], and Yukihiro Shimogaki
Department of Materials Engineering, The University of Tokyo
[1]Institute of Engineering Innovation, The University of Tokyo
7-3-1 Hongo, Bunkyo-ku, Tokyo 113-8656, Japan

ABSTRACT

Feasibility of Cu supercritical fluid deposition (SCFD) in a multi-wafer reactor system (MWRS) for mass production was evaluated based on the productivity and uniformity of the deposited film thickness on a 12-inch stack of wafers. The dependence of Cu deposition kinetics as a function of Cu precursor, H_2, and by-product concentration was experimentally evaluated for both of productivity and uniformity. Results revealed that the Cu precursor profile affected the uniformity of the deposited film thickness, because both the change in the H_2 concentration profile in the reactor during deposition and the effect of by-product concentration on the kinetics of Cu deposition were negligible. Diffusion coefficient of the precursor in $scCO_2$ was also investigated for evaluating transport property. Then, productivity and uniformity of the MWRS was evaluated based on the balance between consumption by Cu deposition and transportation of the precursor for determining the optimal structure and size of a MWRS reactor capable of high throughput required for mass production. The optimal reactor shape and the film thickness uniformity were estimated by simulation using a finite element method. In conclusion, the most suitable choice for mass production of Cu deposition is a flow channel type MWRS, where $scCO_2$ dissolving the precursor flows horizontal to the stacked wafers.

INTRODUCTION

ULSI interconnects are assembled through a damascene process, which first involves Cu seed-layer formation by sputtering and then gap-filling by electro plating into high aspect ratio features [1]. The poor step coverage of sputtering is unacceptable for features with high aspect ratio of over 5. Supercritical fluid deposition (SCFD), which is reduction of metal organics with hydrogen in supercritical carbon dioxide ($scCO_2$), reportedly is a promising technology for superior gap filling and conformal deposition [2, 3]. For example, we previously demonstrated formation of a 10nm-thick, smooth, continuous film and complete gap-filling of Cu into ultra narrow vias (50nm Φ and 1μm depth) [4]. However, the productivity of SCFD has not yet been evaluated. Our current study focuses on the design of a deposition system capable of high throughput and uniform film formation on a 12-inch stack of wafers for mass production. Single wafer reactor systems (SWRS) are well suited for a wide variety of low-volume manufacturing as well as for uniform film formation if its throughput is high enough, for example, 20 wafers/hr. Considering the time to load and unload a single wafer into/from the reactor in SWRS, the deposition time for such throughput of a 1μm-thick overburdened Cu film is 1 min at most, which means a required growth rate (G.R.) of over 1μm/min for SWRS. The G.R. in Cu-SCFD, however, is typically 10-20 nm/min [5, 6]. Although stacking multiple wafers in a single reactor (MWRS) should improve the productivity, the system should be carefully designed based on the uniformity of the deposited film thickness profile on a 12-inch of wafers. This uniformity can be estimated from the kinetics of the Cu deposition and reactant concentration distribution on the wafer surface. This distribution is controlled by the balance between the transport property of the reactant and consumption on the wafer surface by Cu deposition. In addition, the $scCO_2$ flow direction in the reactor of an MWRS (horizontal/perpendicular to the stacked wafers) affects the transport phenomena of the precursor as shown in Fig. 1. In a diffusive transport type reactor, Cu precursor between the stacked wafers is transported by diffusion, whereas in a flow channel type reactor, Cu precursor is transported by the $scCO_2$ flow.

Figure 1. Cross-sectional schematics of two types of multi-wafer reactor systems (MWRS).

In this study, first, the kinetics of Cu deposition and diffusion coefficient of the precursor in $scCO_2$ were experimentally evaluated to obtain deposition parameters. Then, the optimal structure and size of a MWRS reactor capable of high throughput for mass production were determined based on simulation using a finite element method and the obtained deposition parameters.

EXPERIMENTS

Deposition and analysis

Surface reaction kinetics of Cu-SCFD were experimentally investigated using a cold-wall flow type reactor. Deposition was carried out by introducing precursor dissolved in $scCO_2$ onto a heated 2×2cm Si substrate placed in the reactor as shown in Fig. 2 [7]. $Cu(tmhd)_2$ was the precursor [8], high purity CO_2 was the supercritical solvent, and H_2 was the reducing agent. The substrate had a 60nm-thick sputtered Cu film that acted as a seed-layer for the Cu-SCFD process, which helps extract the real growth kinetics by removing the initial incubation period to form Cu nuclei. In the deposition, first, a mixture of $scCO_2$ and H_2 in the syringe pump was continuously introduced into the reactor, and then $Cu(tmhd)_2$ dissolved in acetone was injected into the flowing $scCO_2/H_2$ mixture by a plunger pump. During deposition, the substrate temperature was kept at either 200 or 220 °C, the side wall of the reactor was kept at less than 160 °C to prevent decomposition of the precursor in $scCO_2$, the pressure in the reactor was kept at 15 MPa, and the deposition time was 30 min.

The diffusion coefficient of the precursor was determined based on the measured film thickness profile as follows using the macro cavity method [9]. In the macro cavity, the gap between the two substrates was only 0.1 mm, and consequently the precursor was transported only by diffusion because the Grashoff number was 10^6, indicating no convection. The deposited film thickness profile in the macro cavity was measured using FE-SEM (JEOL, JSM6340F), and then used to extract the experimental diffusion coefficient of the precursor.

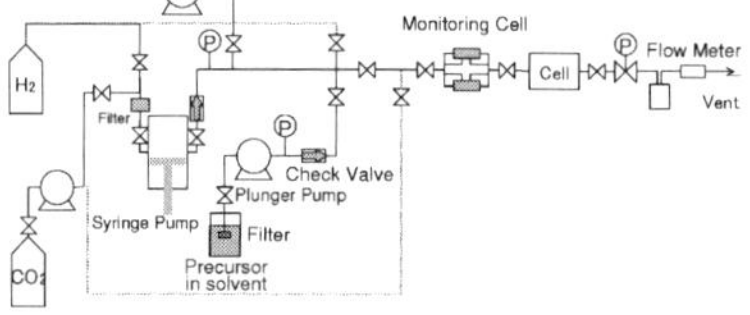

Figure 2. Schematic of the apparatus of Cu-SCFD.

Simulation

The deposited thickness profile in the macro cavity was also estimated by numerical simulation under steady state condition using finite element simulation software (COMSOL Multiphysics, FEMLAB). Then, based on these simulation results, the optimal MWRS size for mass production was determined. Here, only the Cu precursor was considered as a deposition species because Cu precursor concentration was the dominant factor and because the effect of H_2 concentration and by-product concentration was negligible (discussed in Results and Discussion section). In the simulation of the macro cavity, Cu precursor concentration at the inlet of the macro cavity was fixed as the Cu precursor concentration in the reactor, and the outgoing flux on the substrate surface was defined as Cu precursor consumption by deposition. The estimated Cu precursor concentration profile in the cavity determined by this simulation was then converted to the growth rate (G.R.) profile in the cavity using Cu deposition kinetics. Experimental film thickness profile was thus analyzed to extract the diffusion coefficient of the Cu precursor.

RESULTS AND DISCUSSION

Surface reaction kinetics of Cu-SCFD

Kinetics of Cu-SCFD were experimentally studied to obtain the deposition parameters required for estimating (using simulation) the optimal MWRS structure and size for mass production. The G.R. dependence on precursor concentration was investigated for two different temperatures, 200 °C and 220 °C, where precursor concentration was varied from 1×10^{-4} to 2×10^{-3} mol/L, and H_2 concentration was 0.4 mol/L in all deposition runs, which was over 200 times higher than the precursor concentration. The measured G.R. dependence on precursor concentration was found to exhibit nonlinear surface reaction kinetics (Fig. 3). The G.R. dependence on H_2 concentration was also investigated, where H_2 concentration was varied from 0.04 to 0.4 mol/L and precursor concentration was 4×10^{-4} mol/L in all runs. The measured G.R. dependence on H_2 concentration was also found to exhibit nonlinear surface reaction kinetics (Fig.4). The G.R. dependence on by-product concentration was examined. Here, we defined by-product (H(tmhd)) concentration as the added H(tmhd) concentration and H(tmhd) concentration generated during deposition. When the measured apparent G.R. was 5 nm/min (Fig.5), considering the by-product concentration yielded during deposition was about 1×10^{-4} mol/L, suppressed G.R. by this by-product concentration was estimated as about 0.1 nm/min, indicating that the by-product concentration has a negative, yet negligible impact on G.R.

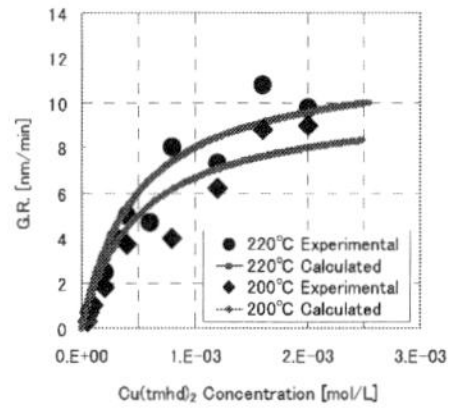

Figure 3. Dependence of G.R. of Cu-SCFD on $Cu(tmhd)_2$ concentration. H_2 concentration was 0.4mol/L. Surface reaction constant ks was 10 and 12 nm/min at 200 °C and 220 °C, respectively. Absorption/ desorption coefficient K was 2 m^3/mol at both temperatures.

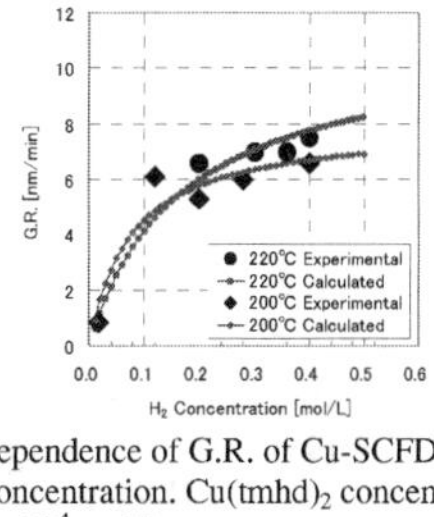

Figure 4. Dependence of G.R. of Cu-SCFD on H_2 concentration. $Cu(tmhd)_2$ concentration was 4×10^{-4} mol/L.

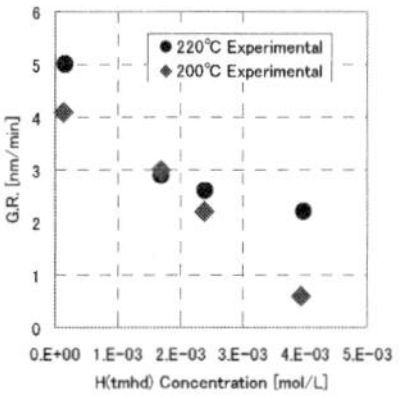

Figure 5. Dependence of G.R. of Cu-SCFD on by-product (H(tmhd)) concentration defined as added H(tmhd) concentration and H(tmhd) concentration generated during deposition. $Cu(tmhd)_2$ concentration was 4×10^{-3} mol/L, and H_2 concentration was 0.4 mol/L.

These non-linear kinetics imply Cu-SCFD is in a reaction-limited regime. When two different reactants exhibit non-linear reaction kinetics simultaneously, a reaction model should be carefully considered. When two different species compete for adsorption at the same site, a competitive reaction model should be introduced. Contrarily, when two different species adsorb on independent sites, reaction kinetics should be introduced independently for each species. Here, to simplify the reaction model, we introduced this latter model. However, excess H_2 compared to the precursor could be assumed constant and uniform in the reactor, and, consequently, has no effect either on the G.R. profile on the stacked wafer or on the complex structure. Thus, Cu precursor distribution is the only factor to control the uniformity of the G.R. The measured G.R. dependence on Cu precursor concentration was, therefore, approximated using Langmuir-Hinshellwood-type (LH) kinetics, which is expressed as

$$G.R. = \frac{ksKC}{1+KC} \qquad (1)$$

where k_s is surface reaction rate constant, K is adsorption equilibrium constant, and C is precursor concentration. In our approximation, k_s = 10 and 12 nm/min at a substrate temperature of 200 °C and 220 °C, respectively, and K = 2 m^3/mol at both temperatures.

In summary, although both the Cu precursor and H_2 concentrations exhibit LH type kinetics and affect the G.R. in Cu-SCFD, the Cu precursor distribution is the only factor to control the uniformity of the G.R., i.e., control the film thickness on the stacked wafers in an MWRS.

Diffusion coefficient of the Cu precursor in $scCO_2$

Diffusion coefficient of the precursor in the 200 °C $scCO_2$ was estimated by combination of experimentally obtained thickness profile in the macro cavity as follows, and calculation by a finite element method. First, deposition was carried out for 60 min at 1.7×10^{-4} mol/L $Cu(tmhd)_2$ and 0.17 mol/L H_2. Figure 6 shows the deposited Cu film thickness profile and calculated Cu film thickness profile as a function of diffusion coefficient in the macro cavity. In the calculation, only the Cu precursor concentration was considered, and the obtained kinetics in the previous section were used. Based on comparison between the deposited profile and calculated profiles in Fig.6, the diffusion coefficient of the Cu precursor was 3×10^{-8} m^2/s.

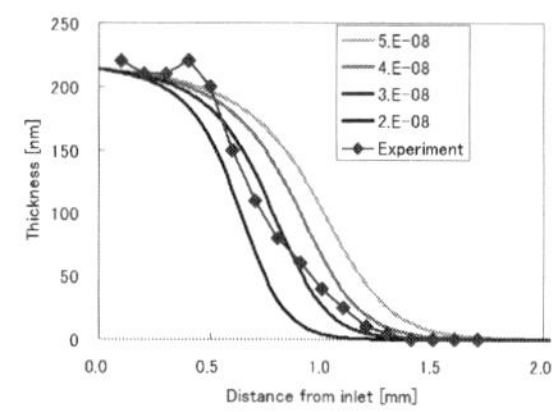

Figure 6. Deposited and calculated Cu thickness profile in a macro cavity. Calculation was done using a finite element method with different values of diffusion coefficient. $Cu(tmhd)_2$ concentration was 1.7×10^{-4}mol/L. H_2 concentration was 0.17mol/L.

Design of a MWRS in Cu-SCFD

The optimal structure and size of the MWRS for mass production were examined using simulations with a finite element method. In the simulation, only the Cu precursor concentration was considered, and obtained kinetics were used. The reactor temperature, including the flowing $scCO_2$ and substrate, was assumed to be 200 °C, i.e., a hot-wall reactor. Two different types of reactors were simulated, namely, diffusive transport reactor and flow channel type reactor. In a diffusive transport type reactor, the precursor is transported by diffusion to the growing surface of the stacked wafers. Figure 7 shows the G.R. profile on stacked wafers simulated by changing the interval between wafers for G.R. of 3.7 nm/min. When the interval reached 10 cm, G.R. at the center of the 12–inch stack of wafers was 99.1% of that at the edge, thus indicating that an interval of 10 cm achieves acceptable uniformity. However, an assumed G.R. of 3.7 nm/min was too slow to achieve a throughput of 40 wafers/hr, which is the throughput for electro plating. Considering that a Cu film thickness of 1μm is required as overburdened, the deposition time will be 4.5hrs with this slow G.R. Thus, 180 wafers will need to be stacked to achieve productivity of 40 wafers/hr. The reactor height therefore will need to be 18 m.

In a flow channel type reactor, the precursor is transported by the $scCO_2$ flow on the growing surface of the stacked wafers. In this case, the G.R. profile was calculated from the mass balance of the introduced precursor and consumed precursor by deposition. When the interval between wafers was 6 mm for G.R. of 10 nm/min and for $scCO_2$ flow rate of 1 mm/min, the G.R. profile was uniform. Considering the G.R. used here was 10 nm/min, 67 wafers have to be stacked to achieve productivity comparable to that of electro plating, and therefore the reactor height will need to be 40 cm. In conclusion, the optimal MWRS for mass production by Cu-SCFD is a flow channel type.

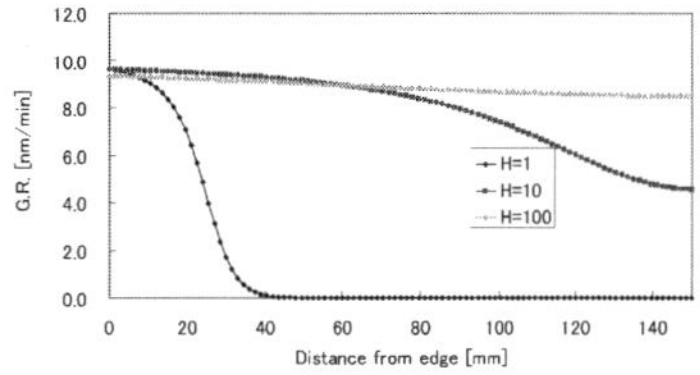

Figure 7. Calculated G.R. profile on stacked wafers determined using finite element method with different intervals (H) between stacked wafers. 150 mm in X-axis indicates the center of the 12-inch stack wafer.

CONCLUSIONS

The feasibility of a Cu-SCFD MWRS reactor for mass production was evaluated based on the productivity and uniformity of the deposited film thickness on a 12-inch stack of wafers. To evaluate the productivity, the kinetics of Cu deposition was studied, revealing that the Cu precursor was the only factor to control film thickness profile on the stacked wafers. Diffusion coefficient of the precursor in $scCO_2$ was also studied using the macro cavity method. To evaluate the uniformity, the optimal structure and size of a MWRS reactor capable of high throughput equal to that of electroplating for mass production and the film thickness uniformity were examined using simulation with a finite element method using the obtained kinetics. By considering the balance between the consumption by deposition and transportation of the precursor, a flow channel type MWRS, where $scCO_2$ dissolving the precursor flows horizontal to the stacked wafers, is suitable for mass production of Cu deposition by Cu-SCFD.

ACKNOWLEDGMENTS

The authors gratefully thank M. Yoshimaru, O. Yamazaki, and N. Shimizu at the Semiconductor Technology Academic Research Center and thank E. Kondoh at Yamanashi University for intensive discussions.

REFERENCES

1. International Technology Roadmap for Semiconductors (ITRS) 2007 edition.
2. J. M. Blackburn, D. P. Long, A. Cabanas, and J. J. Watkins; *SCIENCE* **294** (2001) 141-145.
3. E. Kondoh, and H. Kato; *Micro. Eng.* **64** (2002) 495-499.
4. T. Momose, M. Sugiyama, and Y. Shimogaki; *Appl.Phys. Express*, 1 (2008) 097002.
5. Y. Zong, and J. J. Watkins; *Chem. Mater.* **17** (2005) 560-565.
6. E. Kondoh, M. Hirose, and J. Fukuda; *Proc. Advanced Metallization Conference 2006*, San Diego, C.A., Oct. 17-19, 2006, p.91-95 (2007).
7. T. Momose, M. Sugiyama, and Y. Shimogaki; Jpn. J. Appl. Phys., 47 (2008) 885-890.
8. T. Momose, M. Sugiyama, and Y. Shimogaki; *Jpn. J. Appl. Phys.*, **44** (2005) L1199-1202.
9. L. S. Hong, Y. Shimogaki, and H. Komiyama; *Thin Solid Films*, **365** (2000) 176-188.

Conference Proceedings AMC XXIV © 2009 Materials Research Society

Nanometer-scale Four-Point Probe Resistance Measurements of Cu Wires Using Carbon Nanotube Tips

Y. Kitaoka,[1*] S. Yoshimoto,[1**] T. Hirahara,[1] S. Hasegawa,[1] T. Ohba[2]
1 Dept. of Physics, University of Tokyo, 7-3-1, Hongo, Bunkyo-ku, Tokyo 113-0033, Japan
2 School of Engineering, University of Tokyo, Tokyo 113-0033, Japan
**Present address: ISSP, University of Tokyo, Kashiwa 277-8581, Japan
*Tel/Fax: +81-3-5841-4209 kitaoka@surface.phys.s.u-tokyo.ac.jp

ABSTRACT

In the present study, the four-terminal conductivity measurements of damascene copper (Cu) wires with various widths have been done at room temperature using Pt-coated carbon nano-tube (CNT) tips in a four-tip scanning tunneling microscope. The physical stability and high aspect ratio of the CNT tips made it possible to reduce the probe spacing down to 70-nm, which is the shortest among the measurements of interconnects wires for semiconductor devices performed so far. The measured resistivity of Cu increased as the line width decreased and direct evidence of grain boundary scattering was observed.

INTRODUCTION

Nano-electronics such as semiconductor devices requires low and stable electrical resistance of interconnects to maintain device performance. Copper (Cu) is now widely used for interconnects, but as the wire dimension reaches down to the sub-micrometer scale, on the order of the mean free path of conduction electrons, an increase in the resistivity is observed. This is speculated as due to the increased surface and grain boundary scatterings. As the wire width scales down, electrons will undergo more reflections at the surfaces, so the collisions with the surfaces will become a significant fraction of the total number of collisions [1]. In addition, grain boundaries in polycrystalline wires/films may act like partially reflecting planes located perpendicular to the electric field [2], so they also contribute to the increase of resistivity. To investigate these effects directly, the technique of nanometer-scale conductivity measurements is necessary. Therefore, in this work, we have successfully demonstrated nanometer-scale four-probe conductivity measurements on Cu wires at room temperature using four Pt-coated carbon nanotube (CNT) tips [3] in a four-tip scanning tunneling microscope[4].

EXPERIMENT

Copper interconnects with line width between 70-nm and 1-μm were prepared using a Cu/Low-k damascene process. Figure 1 shows the cross-sectional SEM images of the damascene structure. Tantalum (Ta) was used as a barrier metal (BM) for Cu interconnects in this experiment. Cu damascene line was formed using conventional Cu process such as seed Cu, electrochemical plating (ECP) Cu for trench filling, and chemical-mechanical polishing. Figure 2 shows the grain size and its distribution along the Cu damascene line at 70-nm of line width, measured by the Electron Backscatter Diffraction (EBSD) method. In the left figure, each color represents different crystalline direction. The estimated average grain size is about 100-nm.

The four-tip scanning tunneling microscope enables four-point probe conductivity measurements with various probe arrangements and probe spacing as shown in Fig. 3. By using CNT tips, the probe spacing can be reduced down to the order of several ten nm routinely [3]. The minimum probe spacing in the present study was 70-nm. The notable feature of the CNT tips is that direct electrical contacts causes no electrically or mechanically damage on the sample as well as on the tips even after repeated contacts. On the other hand, even using the Pt-coated CNT tips, the contact resistance between the tip and sample could not be reduced to less than 50-kΩ because of its very small contact area. This means that it is impossible to measure conductive materials whose resistance is less than 50-kΩ by the two-point probe method. Only with the four-point probe method, resistances much smaller than the contact resistance (as small as 0.1-Ω in the present case) can be measured. Therefore, the combination of the CNT tips and the four-tip STM is very powerful for studies in nanoscience and nanotechnology.

Four-terminal *I-V* measurements were performed by sweeping the bias voltage between tips 1 and 4 and recording the current flow *I* and the voltage drop *V* between tips 2 and 3 (Fig. 3). The probe spacing between the voltage probes was reduced while measuring the *I-V* characteristics. The two current probes (tips 1 and 4) and one of the voltage probes (tip 2) were fixed at the same place during the measurements, and only tip 3 was moved between tip 2 and tip 4.

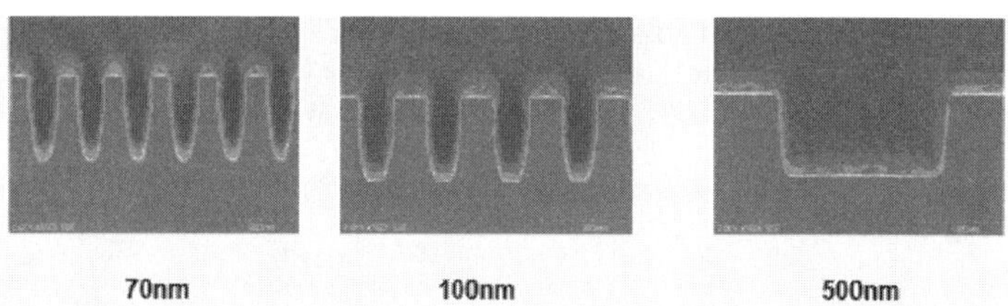

Figure 1. Cross-sectional SEM image of damascene trench.

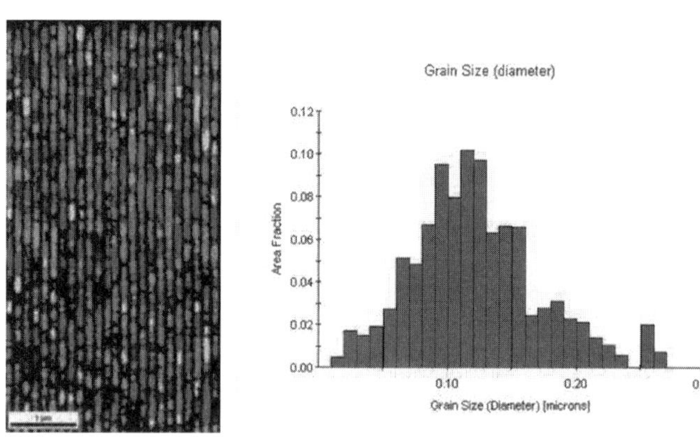

Figure 2. Grain mapping and distribution of grain size of the Cu wire with 70-nm width The different colors in the left figure show the difference of the crystalline direction.

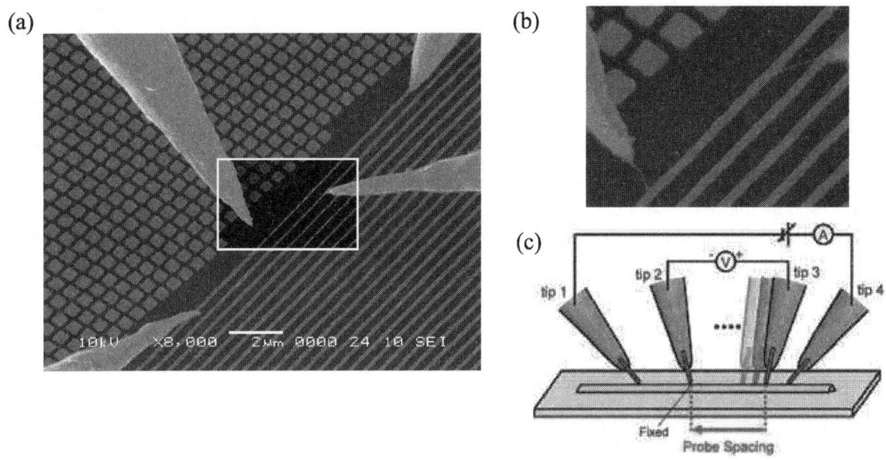

Figure 3. (a) An SEM image of the Cu wire being contacted with four probes. (b) The enlargement of the area of rectangle in the (a). (c) The illustration of the four-terminal *I-V* measurement.

RESULTS and DISCUSSION

The measured values of four-terminal resistance as a function of the probe spacing between the contact points of the voltage probes on the Cu wires are shown in Figs. 4 and 5. In the case of 70-nm wide wire as shown in Fig. 4 (a) and (b), the minimum probe spacing was 70-nm, which is the shortest among the measurements of interconnects wires for semiconductor performed so far. The probe-spacing dependence of resistance basically showed a linear one-dimensional feature down to 70-nm spacing, meaning a diffusive transport. For the wide Cu wires, tungsten (W) tips were used for the measurements, so the minimum probe spacing was limited down to 500-nm due to the whole width of the W tips. The solid line shows the fit to all of the data. By multiplying the gradient of the fitted straight lines and the cross section of Cu wires, the three-dimensional resistivity was calculated as 4.6-μΩ·cm, 3.7-μΩ·cm, and 3.4-Ω·cm, for the 70-nm, 500-nm, and 1-μm wide Cu wires, respectively. The measured resistivity of Cu increased as the line width decreased as shown in Fig. 5 (b), which can be ascribed to the increased surface scattering. The details concerning the line width dependence will be reported elsewhere. No change in the measured resistance was found in each measurement before and after repeated contacts of CNT tips. It means no significant damage on the sample by the probe contacts.

In this experiment, the probe spacing was reduced to the scale which is comparable with the grain size. Therefore, it can be expected that there will be some change in the resistance when the probe spacing gets so short that electrons do not undergo grain boundary scattering. This has indeed been observed. Figure 4 (b) shows the enlarged view of the data shown in Fig. 4(a) for probe spacing smaller than 600-nm. There is a slight change in the resistance when the probe spacing is shorter than 200-nm. An extrapolated plot using the resistance at the spacing shorter than 200-nm shows lower resistance, as represented by a dotted line. There is about 15% difference compared to the solid line, and we speculate that this directly corresponds to the grain boundary scattering.

CONCLUSIONS

An electrical measurement using Pt-coated CNT tips in a four-tip STM has been demonstrated for the Cu damascene wires. Since the apex of CNT tips is small and sharp, it is able to measure nano-scale surface areas. The minimum probe spacing in the four-point probe resistance measurement was reduced to 70-nm, which is similar or less than the Cu grain size. The resistance along the Cu line increased linearly with the measured length, and the resistivity increased as the line width decreased. From the estimation of resistance using small region less than 200-nm in length, the resistance is approximately 15% lower. It is believed that higher resistance is caused by electron scattering at the grain boundaries.

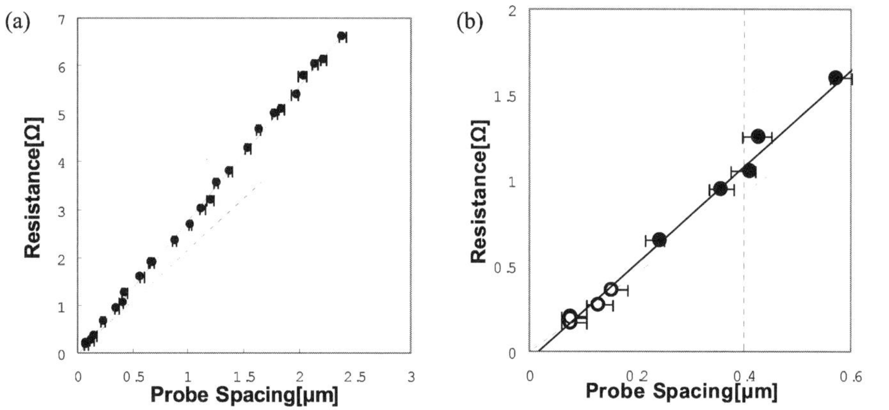

Figure 4. (a) The measured resistances of the 70-nm wide Cu wire are shown as a function of probe spacing. (b) The enlargement of the area under 600-nm of probe spacing.

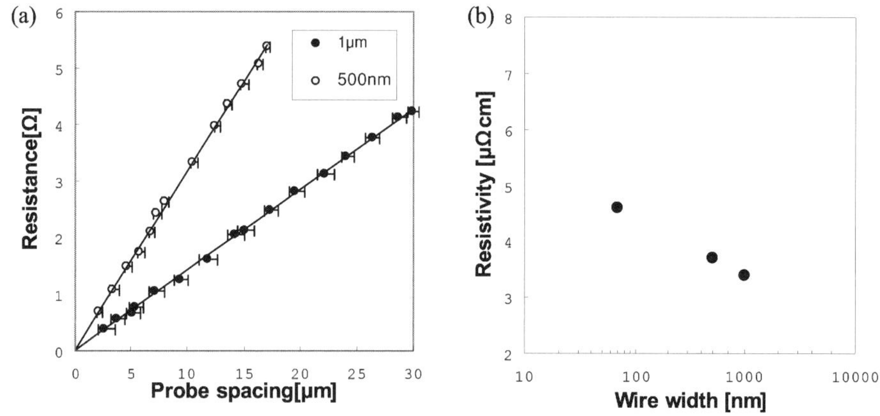

Figure 5. (a) The measured resistances of Cu wires are shown as a function of probe spacing for the 500-nm and 1-μm wide wires. (b) Plot of the resistivity vs Cu line width

ACKNOWLEDGEMENTS

We would like to thank N. Shimizu and H. Kitada, FUJITSU LABORATORY LTD. for their sample preparation and grain analysis.

REFERENCES

[1] J. R. Sambles and T. W. Preist, J. Phys. F: Met. Phys. **12**, 1982, (1971).

[2] A. F. Mayadas, M. Shatzkes, and J. F. Janak, Appl. Phys. Lett. **14**, 345, (1969).

[3] S. Yoshimoto, Y. Murata, R. Hobara, I. Matsuda, M. Kishida, H. Konishi, T. Ikuno, D.Maeda, T. Yasuda, S. Honda, H. Okado, K. Oura, M. Katayama, S. Hasegawa, Jpn. J. Appl. Phys. **44**, L1563 (2005); Nano Letters **7**, 956 (2007).

[4] S. Hasegawa, I. Shiraki, F. Tanabe, R. Hobara, Curr. Appl. Phys. **2**, 465, (2002); I. Shiraki, F. Tanabe, R. Hobara, T. Nagao, S. Hasegawa, Surf. Sci. **493**, 633, (2001),

Conference Proceedings AMC XXIV © 2009 Materials Research Society

In-die Cu Thickness Monitoring of Memory Chip

Heung-Soo Lee[1], Chulhong Kim[1], Hyungwon Yoo[1], Tae-Kyung Kim[1], Chang Hwan Lee[1], Sanghoon Son[1], Kwansoon Park[2], Priya Mukundhan[2], Cheolkyu Kim[2]

[1]Hynix Semiconductor Inc., San 136-1, Ami-ri, Bubaleub, Icheon-si, Kyunggi-do, 467-701, Korea

[2]Rudolph Technologies Inc., One Rudolph Road, Flanders, NJ 07836, USA
Cheolkyu.kim@rudolphtech.com

ABSTRACT

Picosecond Ultrasonic Metrology (PULSE™) has been successfully adopted in copper process monitoring by logic device manufacturers. In this article, we present the capability of this technique to measure in the in-die cell area of a memory chip. Using picosecond ultrasonic technology, we measured in-die Cu thickness of Electroplated (EP) Cu and post-CMP Cu wafers. Cu thickness measurements were also made on solid pads located in the scribe lines for comparison. EP Cu data showed that Cu was thicker on the in-die cell area demonstrating the well known overburden effect. For CMP wafers, Cu thickness at cell area was significantly different from the thickness at pad box. PULSE results confirm the necessity and benefits for monitoring the Cu thickness in the in-die cell area versus relying on pad measurements for process monitoring.

INTRODUCTION

Rapid device scaling has governed the growth of the semiconductor industry resulting in integrated circuits with faster speed and lower power consumption. In order to minimize resistance-capacitance (RC) delays, logic device manufacturers have transitioned from using aluminum interconnects to copper interconnects and from oxide to low dielectric constant materials. Over the last few years, high speed logic chipmakers have successfully adopted Cu process for metallization. The main advantage of changing from aluminum to copper is the reduction of RC delay, resistance to electromigration and savings in manufacturing cost. Aluminum still has been the metal of choice for most memory chipmakers, because memory devices have fewer levels of interconnect and can operate at slower speed than logic devices.

As technology node shrinks down to 45nm, memory chipmakers are also making the transition to Cu process. Unlike logic chipmakers, the main driving force for the move to Cu process seems to be derived mostly in the savings of manufacturing cost and higher yield. Some memory chipmakers have already adopted Cu process in their volume production, and others are developing the process in their pilot lines.

Film thickness control in semiconductor process is mostly done on product wafers by measuring the thickness on test pad in scribe line to reduce the cost for monitor wafers. As technology node shrinks, some critical processes require in-die monitoring due to the lack of correlation between the thickness in the cell area and the thickness in the test pad on scribe lines. In-die cell area of memory chip is repeated patterns of memory units. At metallization level, the whole area of cell is repeats of metal lines and spacing. This relatively simple structure of memory cell area makes in-die monitoring less difficult.

Copper damascene process represents a typical example that requires in-die measurement. Pattern-dependent electroplating effects have been extensively studied and are well understood[1]. During Cu electroplating process, excessive copper build up occurs on sub-micron line arrays resulting in "super filled" structures. This overburden of Cu on line arrays need to be controlled to minimize subsequent problems with CMP. During Cu CMP, thickness of Cu in the line array of cell area becomes significantly different from Cu thickness on pad box, because the removal rate of Cu is very different from that of Inter Level Dielectric (ILD). Therefore, direct monitoring of Cu thickness on in-die cell area becomes crucial in successful thickness control of Cu process.

Picosecond Ultrasonic Metrology (PULSE™) has been successfully adopted in copper process monitoring by logic device manufacturers[2,3]. In this paper, we present the evaluation result of picosecond ultrasonic technology to measure in-die Cu thickness of memory chip. We will also demonstrate the significance of in-die monitoring by comparing the thicknesses between in-die cell area and the test site on scribe lines.

METHOD

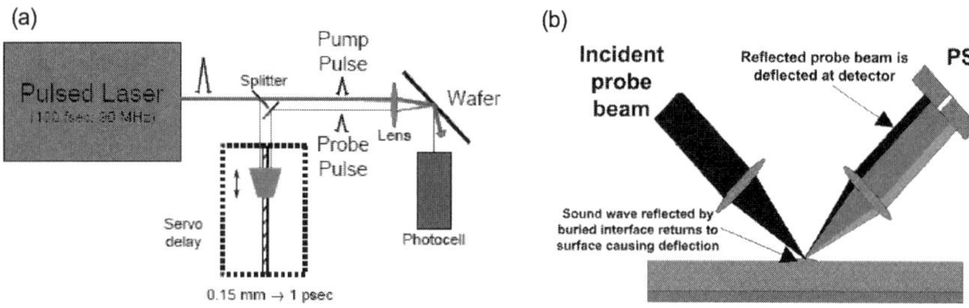

Figure 1. (a) Schematic representation of picosecond ultrasonic technology. Time delay of probe beam arrival on the wafer surface with respect to pump beam allows measurement of surface reflectivity change as a function of time delay. (b) Schematic of PSD measurement.

Picosecond ultrasonic technology[4,5] is a unique way of measuring opaque film thickness using ultrashort pulsed laser. The optics schematic of this technology is shown in Figure 1(a). A 0.1ps laser pulse (pump) is focused to about $7\times5\mu m^2$ spot onto a wafer surface to create a sharp acoustic wave. The acoustic wave then travels away from the surface through the film at the speed of sound. At the interface with another material, a portion of the acoustic wave gets reflected and comes back to the surface while the rest is transmitted. When the reflected acoustic wave reaches the wafer surface, it is detected by another focused laser pulse (probe) which was diverted from the pump pulse by the beam splitter. There are 2 different methods of detecting the reflected acoustic wave at the surface. The first method is to detect the change of optical reflectivity that is caused by the strain of acoustic wave. The second method is to detect the deflection of reflected probe beam that is caused by the deformation of surface due to the acoustic wave. The second method requires a position sensitive detector (PSD), which is described in Figure 1(b). The PSD method shows better sensitivity and signal to noise for measuring thick copper and line array structures. All the data reported in this paper were measured using the PSD mode. The servo-delay controls the time difference between pump and probe, which allows us to measure the round trip travel time of the acoustic wave within the film. Combining this with speed of sound of the material, one can calculate the thickness of the film using the following equation;

Thickness = (Speed of sound in the film) × (Round trip transit time) / 2

EXPERIMENT

EP Cu Measurement

As-plated wafers of target thickness 5000Å, 7000Å, and 9000Å were measured using picosecond ultrasonic metrology. Measurements were made on pad boxes in the scribe line. The pad box was 120μm × 100μm in size and was covered by blanket Cu film on flat ILD. Measurements were also made on 5mm × 5mm cell area, where Cu was deposited on line array whose lines are ~0.05μm wide and 50% dense. The two measurement sites were ~2.5mm apart.

Figure 2(a) shows the PSD signals from the center die of all the three electroplated wafers. Measurements shown are from the pad box in the scribe line. The first large peak of each signal is due to the large deflection of reflected probe beam that was caused by the arrival of reflected sound wave on the surface. This corresponds to the round trip travel time of the sound wave in Cu film. As film thickness increases, the transit time for the sound wave in the film is expected to increase. One can clearly see the shift in the peak position to the right (from ~200ps to ~400ps) as thickness increases. Copper thickness is calculated by using the position of the echo and assuming 45Å/ps for the speed of sound in copper. Figure 2(b) summarizes the result of measuring Cu thickness on pad box located in the scribe line on all the 3 wafers on various die. Average thickness of the 3 wafers are ~4900Å, 6490Å, 8050Å and within-wafer thickness range is ~250Å, 355Å and 305Å , respectively. Typical dynamic repeatability of these measurements is about 0.8% or less for 1 sigma standard deviation.

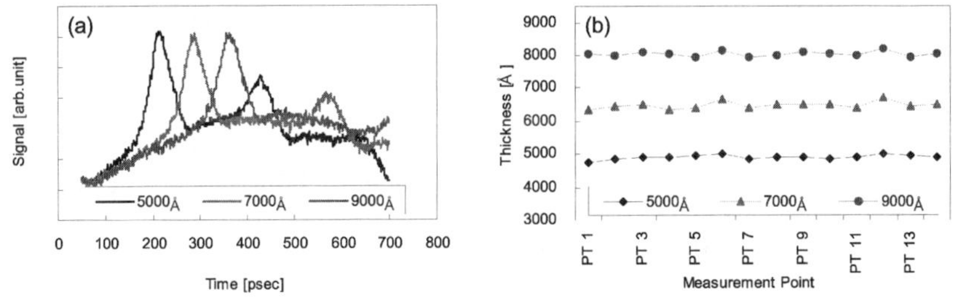

Figure 2. (a) Picosecond ultrasonic signals measured on pad box in scribe line. The position of the first main peak corresponds to the round trip travel time of sound wave in Cu film. (b) EP Cu thickness on pad box in scribe line measured by picosecond ultrasonics

Figure 3(a) shows PSD signals measured on in-die cell area at the center die of the same wafers. The shape of the signals measured on cell area looks different from that of the pad box.

This is because the sound wave generated on the surface of the pad box gets reflected at the interface between the pad and the ILD, whereas in the case of the cell area, part of the sound wave gets reflected at the top of the trench and part of it at the bottom of the trench. Figure 3(b) shows Cu overburden thickness and trench depth on 13 different points in cell area of the 3 wafers. The average copper overburden thickness and trench depth of the three wafers are 5560Å , 7090Å , 8720Å and 1650Å 1660Å 1690Å , respectively. Dynamic repeatability of these measurements is also about 0.8% or less.

It should be noted that there is a noticeable difference (~700Å) between the Cu thickness measured in the pad box in the scribe line and the copper thickness measured in the in-die cell area. This result is consistent with known overburden or "Superfill" effect. EP Cu is deposited more on the in-die cell area (i.e. line array) than on pad box. Accuracy of the PULSE measurements was verified by correlating pad box measurements with Transmission Electron Microscopy (TEM) data, and the linear correlation factor between the two was about 0.99. This excellent correlation between the two techniques demonstrates the accuracy of the picosecond method for such measurements.

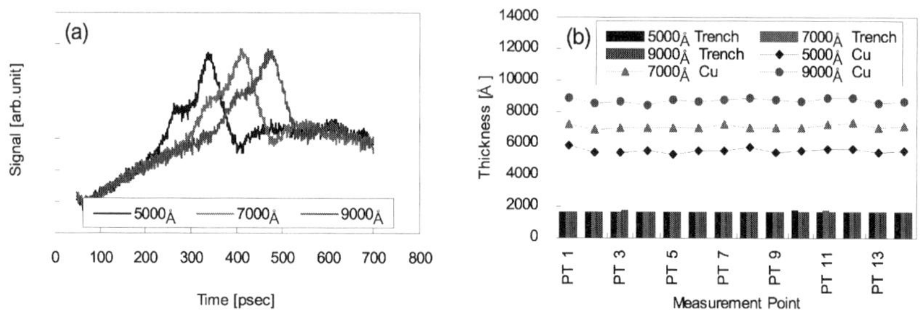

Figure 3. (a) PSD signals measured on in-die cell area. One can see the main peak is moving to the right as can be expected from the increase of round trip travel time in thicker Cu film. (b) EP Cu thickness of in-die cell area measured by picosecond ultrasonic. Bar chart in the graph shows the trench depth. Line graphs represent the Cu overburden thickness.

Post CMP Measurement

Post CMP wafers were measured for 3 different polish (CMP) times; $t - \Delta t$, t, $t + \Delta t$, where t represents nominal polishing, t-Δt represents under polishing and t+Δt represents over polishing conditions. From the CMP polish times, one would expect that the Cu thickness on pad box would decrease as CMP time increases. However, our measurements revealed that this was not the case. Figure 4(a) shows thickness of 13 pad box for each 3 wafers. The thickness variations across the three wafers were not consistent with the trend that was expected for the different polish times.

Figure 4(b) shows 13 point measurement data of in-die cell area. In sharp contrast to the data of pad box, the within wafer thickness variation is consistent and also the overall thickness trend is consistent with (CMP) polish times. This clearly demonstrates that the Cu CMP process needs to be monitored by measuring in-die cell area, not by measuring bond pad in scribe line. Standard deviation (1 sigma) of 9 dynamic (load/unload) repeatability for this measurement was less than 0.3%. PULSE measurement results showed excellent correlation with TEM (R2> 0.9) confirming the accuracy of the measurements.

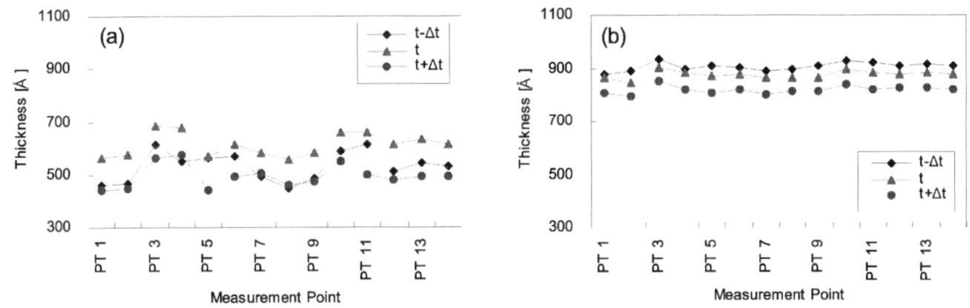

Figure 4. (a) Cu thickness of the 3 wafers measured on the pad box after CMP. (b) Cu thickness of in-die cell area after CMP. Note that pad box thickness data do not show a consistent trend with CMP polish time, while thickness data of in die cell area shows a consistent and expected trend with CMP time.

CONCLUSION

We have successfully demonstrated that picosecond ultrasonic technology is well suited for Cu process monitoring on the in-die cell area of memory chip for both electroplate and CMP process. For EP copper measurements in the in-die cell area, overburden and trench thickness are reported simultaneously. Comparison with pad box measurement showed about 700 Å overburden effect. For post-CMP wafers, measurements on the cell area was consistent with CMP times while thickness on pad box were not. The pad box showed a much lower thickness than the in-die cell area for the underpolished wafer. These results confirm the necessity and benefits of monitoring on the actual device area instead of relying on pad measurements. The repeatability, accuracy and throughput of the measurements are well within the requirements for a metrology tool that is production worthy.

REFERENCES

1. D. Boning, B. Lee, T. Tubawa, and T. Park., "Models for Pattern Dependencies: Capturing Effects in Oxide, STI, and Copper CMP", *Semicon/West Technical Symposium: CMP Technology for ULSI Manufacturing*, July 17, 2001
2. M. Hsieh, J. Yeh, M. Tasi, K. Wang, J. Tan, S. Leary, "In-line monitoring of advanced copper chemical mechanical planarization processes with picosecond ultrasonic metrology," Proc. SPIE Vol 6152, 61522C, Metrology, Inspection and Process Control for Microlithography, March 2006
3. P. Mukundhan, G.Tas, J. Dai, S.P. Leary, J. Clerico, " Characterizing CMP Processes using Picosecond Ultrasonic Metrology," ISTC 2007
4. C. Thomsen, H.T. Grahn, H. J. Maris, and J. Tauc, "Surface generation and detection of phonon by picosecond light pulse", Physical Review B, Vol 34, No. 6, pp.4129-4138, 1986
5. C. J. Morath, G.J. Collins, R.G. Wolf, "Ultrasonic multi-layer metal film metrology", Solid State Technology, June 1997

Conference Proceedings AMC XXIV © 2009 Materials Research Society

Influence of moisture on the CVD formation of a MnO_x barrier layer

K. Neishi[1], K. Matsumoto[2], H. Sato[2], H. Itoh[2], S. Hosaka[2] and J. Koike[1]
[1] Dept. of Material Science, Tohoku University
6-6-11 Aoba, Aramaki, Aoba-ku, Sendai 980-8579, Japan
[2]Technology Development Center, Tokyo Electron Ltd., Nirasaki 407-0192, Japan
Phone +81-22-795-7339, Fax +81-22-795-7338, E-mail: neishi@material.tohoku.ac.jp

ABSTRACT

We investigated the effects of moisture on the deposition behavior of the MnO_x diffusion barrier layer on TEOS-oxide substrates. The moisture effects were investigated in two ways: (1) with and without the introduction of H_2O gas in the reaction chamber and (2) with and without pre-annealing of the TEOS-oxide substrates without introducing H_2O gas. A thick crystalline MnO_x layer was formed under the presence of H_2O gas. A thin amorphous MnO_x layer was formed without H_2O gas. Without H_2O gas in the reaction chamber, the thickness of the MnO_x layer was slightly thinner on the substrate with pre-annealing than on the substrate without annealing.

INTRODUCTION

As Cu interconnect is shrunk in its size in advanced large-scale integrated circuits, barrier-layer thickness should be reduced to minimize the increase of effective resistivity [1]. In this aspect, conventional PVD barrier layers of Ta/TaN have been facing substantial difficulties to reduce its thickness without sacrificing the diffusion barrier property. Recently, we reported the advantages of a self-forming barrier process by sputter deposition of a Cu-Mn alloy seed layer on TEOS-SiO_2 substrates and subsequent annealing in an oxygen containing atmosphere [2-6]. This process can be used to form a thin and uniform barrier layer of MnO_x with a good thermal stability and diffusion barrier property. Additionally, a low resistivity of 2.0 μΩcm of the Cu film can be obtained in the optimized annealing condition. However, as the technology node is advanced from 45 nm to 32 nm and beyond, a physical vapor deposition process for the barrier layer formation will encounter increasing difficulties because of its poor coverage. Since chemical vapor deposition (CVD) is well known for its good step coverage, we investigated the possibility of utilizing CVD to form a MnO_x barrier layer and successfully demonstrated its excellent thickness uniformity and barrier property [7,8].

Meanwhile, dielectric materials tend to absorb moisture which may influence the deposition behavior of CVD-MnO_x. Since the control of thickness and quality of the barrier layer is an essential part of the barrier formation process, in the present work, we investigated the influence of moisture on MnO_x layer formation with thermal-MOCVD.

EXPERIMENT

Substrates were p-type Si wafers having plasma-TEOS oxide of 100 nm in thickness. The $(EtCp)_2Mn$ precursor was vaporized and introduced into a hot-wall reaction chamber with H_2 carrier gas. In this study, two types of experiments were performed. In the first, influence of additional moisture in the CVD reaction process was investigated by introducing H_2O gas at a flow rate of 0.2 sccm. In the second, influence of absorbed moisture in the substrate was investigated by depositing a MnO_x layer on the TEOS-oxide substrate with or without pre-annealing the substrates at 500 °C for 2 h in vacuum of 1×10^{-5} Pa.

Gas components with the TEOS-oxide substrate were analyzed with thermal desorption spectroscopy (TDS). Thickness of the MnO_x layer was investigated by X-ray fluorescence (XRF) analysis. The cross sectional microstructure was observed with transmission electron microscopy (TEM). Chemical bonding states of the CVD-Mn layer was analyzed with X-ray photoelectron spectroscopy (XPS).

RESULTS and DISCUSSION

Figure 1 shows a cross sectional TEM image and a diffraction pattern of the as-deposited CVD-Mn/TEOS with CVD condition at 100 °C for 15 s with the H_2O gas flow of 0.2 sccm. A continuous layer having a thickness of 45 nm is observed on the TEOS oxide layer. The diffraction pattern and the image contrast of a CVD-Mn layer indicate that the CVD-Mn layer has a crystalline structure. It is noted that the existence of moisture in the reactor causes a rapid growth of MnO_x layer.

Figure 2 shows a depth profile of XPS spectra of Mn 2p and O 1s of the same CVD sample as shown in Fig. 1. Depth profile was obtained by etching from the CVD-Mn layer surface to the TEOS oxide layer. The peak position of Mn 2p indicates that the CVD-Mn layer is in oxide states. These results suggest that the Mn oxide is deposited on a TEOS substrate by reaction between the Mn precursor and the H_2O gas. It is noted that the CVD-MnO_x layer formed with H_2O gas may show a poor diffusion barrier property, because the crystalline layer has a fast diffusion path of a grain boundary.

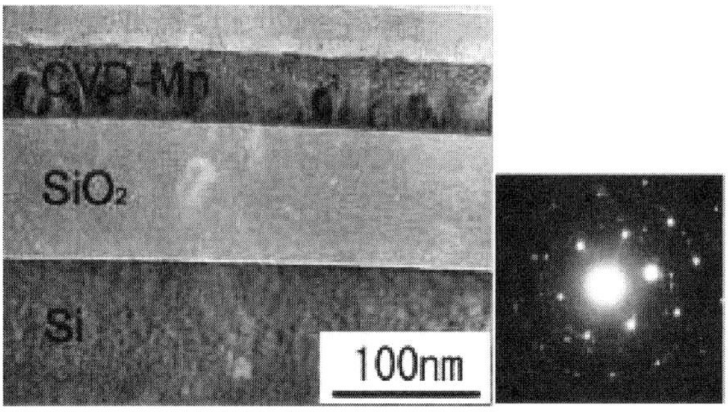

Fig.1 A Cross sectional TEM image and a diffraction pattern of as-deposited CVD-Mn/TEOS deposited at 100 °C for 15 sec. with H_2O gas.

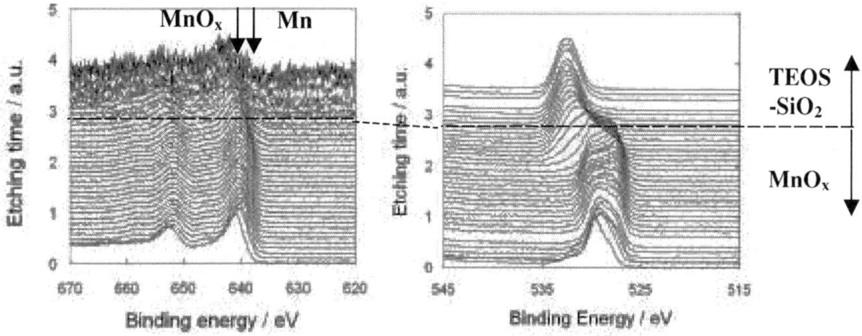

Fig.2 XPS spectra of Mn 2p and O 1s for CVD-Mn/TEOS deposited at 100°C for 15sec. with H_2O gas.

Figure 3 shows the change of the Mn intensity from the MnO_x layer with substrate temperature. The MnO_x was deposited for 1 min. with the H_2O gas flow of 0.2 sccm. The amount of Mn contained in the MnO_x was measured with XRF. The increase of the Mn intensity with temperature corresponds to the increase of the MnO_x thickness.

Figure 4 shows the TDS result of the TEOS-oxide substrates. A Sample size was 10 mm square. A heating rate was 1 K/ sec. This result indicates that H_2O, CH_4 and NH_3 are expelled from the TEOS-oxide substrate with the peak desorption temperature of around 150 °C. The TDS peak temperature of H_2O gas corresponds the physical absorption temperature on plasma-TEOS /Si substrate [9]. This result implies that these gas species are physically absorbed in the TEOS substrate during air exposure after CVD deposition. Additionally, it is considered that the pre-annealing condition used in the present work (500 °C for 2 hours in vacuum) is sufficient for moisture removal from the TEOS-oxide substrate.

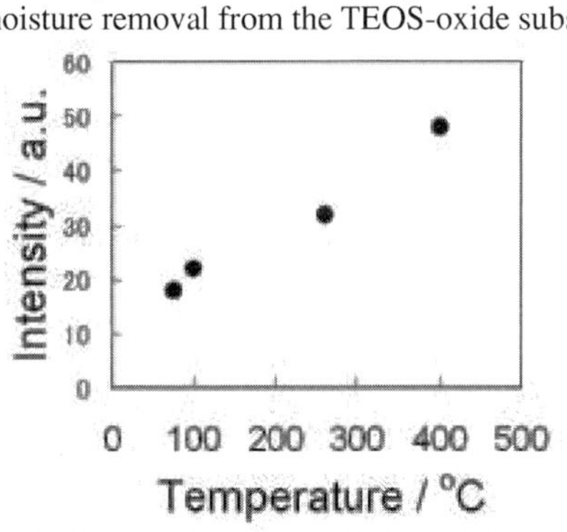

Fig.3 The change of the Mn intensity from the MnOx layer with substrate temperature.

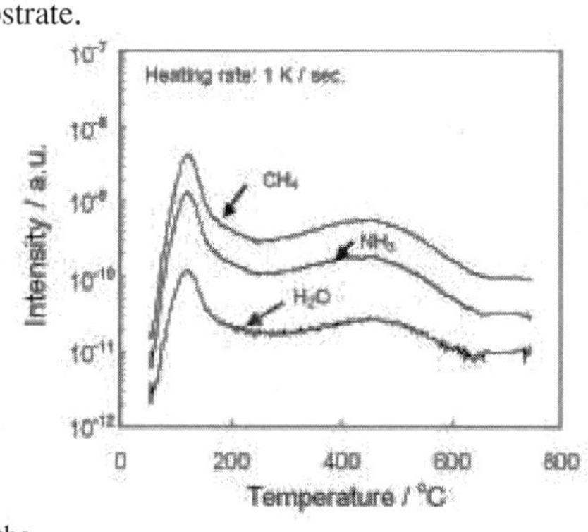

Fig.4 The TDS result of the TEOS-oxide substrates.

Figure 5 shows cross-sectional TEM images of the MnO_x/TEOS sample with and without pre-annealing. The CVD MnO_x layer was deposited at 100 °C for 30 min. with H_2 carrier gas flow rate of 25 sccm. A thin continuous MnO_x layer having an amorphous structure is observed on the TEOS oxide layer in both samples. The thickness of MnO_x on the pre-annealed TEOS substrate is slightly thinner than that on the non-annealed TEOS substrate. It is considered that the amount of chemisorbed and/or physisorbed H_2O molecules on the surface influences the MnO_x layer thickness. This result implies that the management of moisture in substrates is important to ensure a good thickness reproducibility of the MnO_x barrier layer.

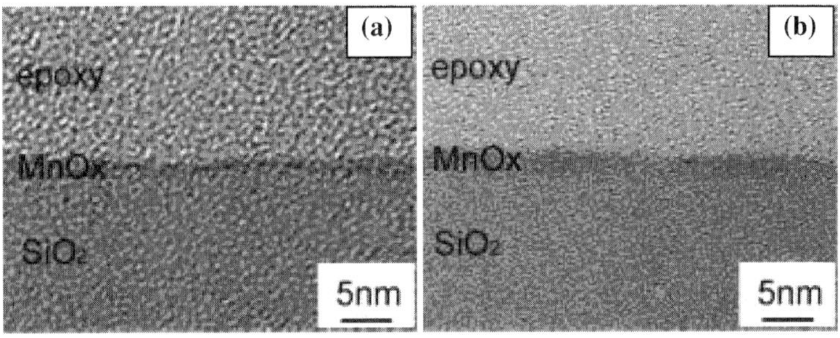

Fig.5 Cross sectional TEM images of CVD-MnO_x/TEOS deposited at 100 °C (a) with and (b) without pre-annealing.

Figure 6 shows cross-sectional TEM images of the MnO_x/TEOS sample with and without pre-annealing. The CVD MnO_x layer was deposited at 400 °C for 30 min. with H_2 carrier gas of 25 sccm. A thin continuous MnO_x layer having an amorphous structure is observed on the TEOS oxide layer in both samples. A Mn particle is observed on the MnO_x/TEOS layers in the pre-annealed sample. This result implies that the decomposition of $(EtCp)_2Mn$ precursor occurs around 400 °C, which is similar to the decomposition behavior of $(MeCp)Mn(CO)_3$ precursor [10]. It is considered that the interaction between the thermally decomposed Mn precursor and the desorbed moisture from the substrate leads to stable oxide formation without metallic Mn deposition.

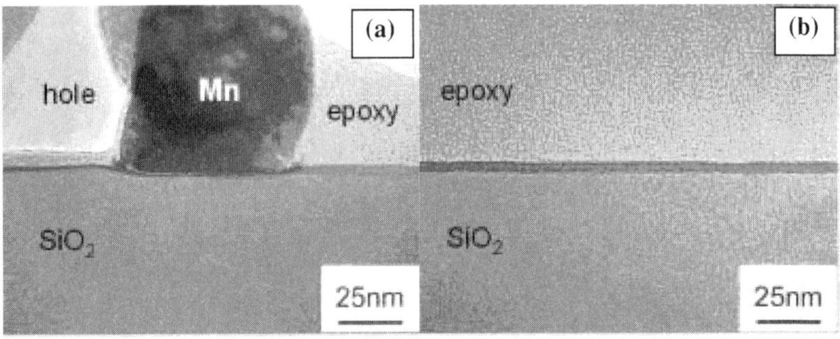

Fig.6 Cross sectional TEM images of CVD-MnO_x/TEOS deposited at 400 °C (a) with and (b) without pre-annealing.

Figure 7 shows the thickness variation of the MnO_x layer with substrate temperature. Open and solid circles correspond to the films deposited on the substrates with and without pre-annealing, respectively. The thickness gradually increases with increasing the substrate temperature. It is also found that the thickness is thinner on the substrate with pre-annealing than without pre-annealing, especially at high temperatures. It should be noted that the thickness of the MnO_x layer on the pre-annealed TEOS substrate is in the range of 1.5 nm to 3.7 nm, which is within the range of the thickness values required in the ITRS road map.

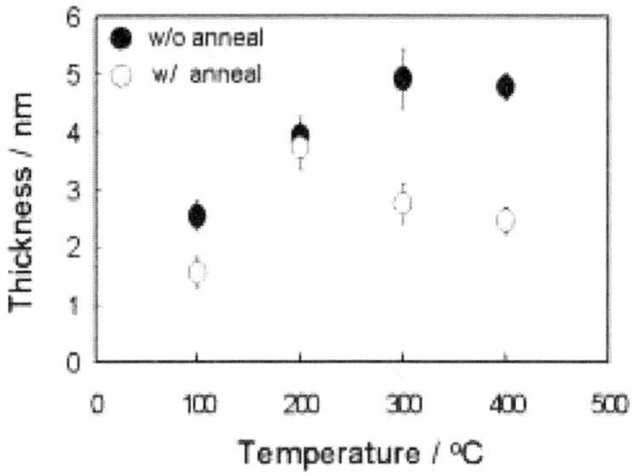

Fig.7 The thickness variation of MnO_x layer with substrate temperature.

CONCLUSIONS

Influence of moisture on the CVD formation of MnO_x was investigated. With the introduction of H_2O gas in the reaction chamber, a crystalline MnO_x layer was formed and the thickness was more than 45 nm. Without H_2O gas, an amorphous MnO_x layer was formed with a thickness of less than 4 nm. The MnO_x layer was slightly thicker when the substrate contained moisture. The obtained results implied that the management of moisture in the reaction chamber and in the substrate is needed to obtain a thin MnO_x barrier layer with a good reproduciblilty.

REFERENCES

1. ITRS 2006 update.
2. J. Koike and M. Wada, Appl. Phys. Lett. **87**, 041911(2005).
3. M. Haneda, J. Iijima, and J. Koike, Appl. Phys. Lett. **90**, 252107 (2007).
4. T. Usui, H. Nasu, S. Takahashi, N. Shimizu, T. Nishikawa, M. Yoshimaru, H. Shibata, M. Wada and J. Koike, IEEE Trans. Electron Devices **53**, 2492(2006).
5. J. Koike M. Haneda, J. Iijima, Y. Otsuka, H. Sako and K. Neishi, J. Appl. Phys. **102**, 043527 (2007).
6. J. Koike, J. Iijima and K. Neishi, Proc. AMC, 3-9(2007).
7. K.Neishi, S. Aki and J. Koike, Proc. MRS **1079**, N03-11(2008).
8. K.Neishi, S. Aki, K. Matsumoto, H. Sato, H Itoh, S. Hosaka and J. Koike, Appl. Phys. Lett. **93**, 032106(2008).
9. N. Hirashita, S. Tokitoh and H. Uchida, Jpn. J. Appl. Phys., **32**, 1787-1793(1993).
10. S. Wen-bin, K. Durose, A. W. Brinkman, and J. Woods, J. Cryst. Growth, **113**, 1 (1991).

Conference Proceedings AMC XXIV © 2009 Materials Research Society

Copper diffusion barrier properties of a 10 nm W-Mo alloy films on porous SiOC:H

Kuo-Chung Hsu[a], Dung-Ching Perng[a], Jia-Bin Yeh[a,*], Ruo-Ping Chang[a], Jia-Feng Fang[a], and Climbing Huang[b]
[a]Institute of Microelectronics, Department of Electrical Engineering, National Cheng Kung University
No. 1 University Road, Tainan 70101, Taiwan, Rep. of China
[b]Advanced Thin Film Dept., United Microelectronics Corp.
No. 18, Nanke 2nd Road, Sinshih Township, Tainan County, Taiwan, Rep. of China

* Corresponding author Tel: +886-6-275-7575 ext.62400-1260, E-mail: q1895137@mail.ncku.edu.tw

ABSTRACT

The thermal properties of 10 nm W-Mo alloy films as copper diffusion barrier on porous SiOC:H were studied. The GIXRD results show that this layer has great barrier capabilities against copper penetration. Cross-sectional TEM image and EDS line profiles taken from TEM sample indicate Mo and W still accumulate at the interface. Sharp decline of Cu and Si concentrations at interface and lack of Cu penetrations to porous SiOC:H, indicating that W-Mo barrier was effective for sample annealed at 600°C for 30min. Cu agglomeration occurred after 5 min 700°C annealing, suggesting that Cu had poor adhesion to underneath barrier. The 10nm W-Mo layer could be a potential barrier candidate for advanced copper interconnects.

INTRODUCTION

The integration of copper interconnects with porous ultra low-k (ULK) dielectric has been implemented for advanced integrated circuits (IC). The refractory transition metals such as W [1], WN [1, 2], WCN[3,4], Mo/WN [5] were widely investigated as copper barrier for several benefits. First, quite low bulk resistivity and high thermal stability are suitable for barrier's requirements. Additionally, desired properties of non-solute with Cu and enhancement of EM (electromigration) reliability were reported to meet advanced copper interconnect [4]. For the viewpoint of manufacture in IC industry, the usage of W metal was a mature technology and Mo was inexpensive material. This consideration of low cost is desirable for manufacturing in IC industry. In this work, diffusion barrier and thermal properties of a co-sputtered 10nm W-Mo refractory transition metals layer on porous low-k film is reported.

EXPERIMENTS

The 200 nm porous ULK, SiOC:H (k=2.55), film was prepared by Applied Materials Inc.'s plasma enhanced chemical vapor deposition (PECVD) system. A 10 nm W-Mo alloy films were deposited on the porous ULK film by co-sputtering with pressure of 1×10^{-5} torr then followed by in-situ 70 nm Cu deposition without breaking vacuum. The substrate was rotated with 15rpm without heat treatment during barrier's co-deposition. The stack of 300nm W-Mo / Si was used to determine composition ratio by EDS (Energy Dispersive Spectrometer). The result shows the composition ratio between W-Mo film is about 85:15 at. %. In order to evaluate thermal stability of the Cu/W-Mo/ULK stacked films, rapid thermal annealing (RTA) was used for annealing with chamber pressure below $3x10^{-6}$ torr. Four-point probe was employed to measure the sheet resistance of the stacked films. The surface morphology of the samples was observed by scanning electron microscopy (SEM). Phase identification of the stacked film was characterized by glazing incident x-ray diffraction (GIXRD) with 2θ from 20° to 80°. The XPS depth profile analysis was used to observe Cu diffusion behavior through 10nm W-Mo barrier. The XPS survey scan was employed to detect interfacial chemical state between Cu and W-Mo liner after Cu removal. TEM and attached EDS were used to observed microstructures and composition depth profiles across the interface of the Cu/W-Mo/ULK stack films.

RESULTS AND DISCUSSION

GIXRD Identification:

In figure 1, only Cu peaks was observed in XRD patterns for the as-deposited stack films. As the annealing temperature increased, the Cu grains growth resulted in sharper peaks. No copper silicide peaks were found when samples annealed up to 700°C for 5 min, indicating robust barrier capability of W-Mo layer against Cu diffusion to porous SiOC:H. However, MoO_2 formed during the heat treatment at 500 °C. Formation of tungsten oxide also started at 700°C. These results were confirmed by the higher sheet resistance measured. Nonetheless, it demonstrated that W-Mo alloy barrier could provide great thermal and barrier properties against Cu penetration.

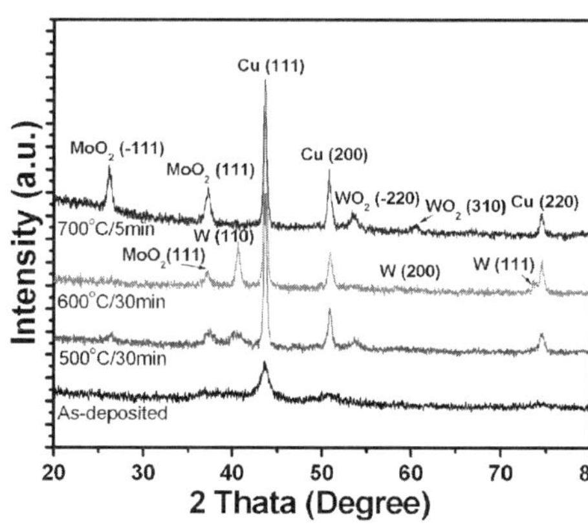

Figure 1. The GIXRD patterns of Cu/W-Mo/ULK stacked films after annealing at various temperatures.

Sheet resistance measurements:

Figure 2 shows the sheet resistance of the Cu/W-Mo/ULK stack films decreased after annealing at 400°C for 30 min in vacuum, indicating that Cu grains growth at elevated temperature. Grain growth leads to less grain boundaries and results in decrease of sheet resistance due to lower electron scattering between grain boundaries. Sheet resistance of the stack films increased while annealing above 700°C due to the formation of WO_2 as well as Cu agglomeration. This was further confirmed by XRD (fig. 1) and SEM results.

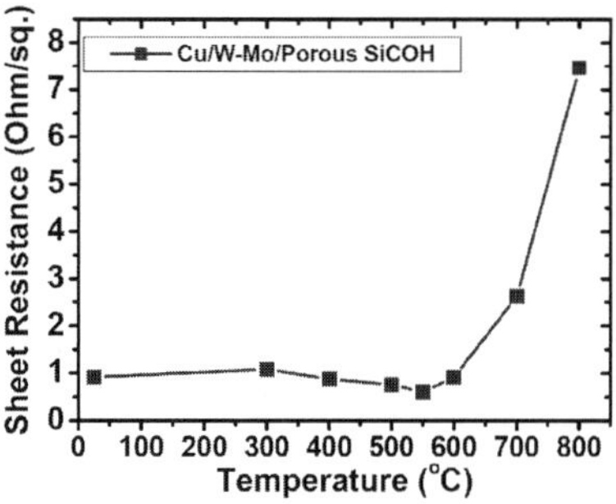

Figure 2. The measured sheet resistances of Cu/W-Mo/ULK stacked films as function of temperature.

The SEM top views (not shown) of the 400°C and 600°C annealed samples had smooth surfaces, indicating absence of Cu diffusion through W-Mo barrier. In the case of 700°C annealing, Cu agglomeration was observed and caused higher sheet resistance. This phenomenon and XRD results (Fig. 1) demonstrated that adhesion is critical rather than copper penetration if W-Mo is adopted as diffusion barrier.

Cross-sectional TEM observations and EDS line scan profiles

The TEM micrograph and EDS line profiles taken on cross-sectional TEM image of the Cu/W-Mo/ULK stack after annealed at 600°C for 30min are shown in figure 3(a) and (b), respectively. The TEM image shows a clear interface between ULK and barrier, indicating no Cu penetration to porous SiOC:H film. Furthermore, the W-Mo layer transformed to bi-layer after annealed at 600°C. The 600°C XRD result indicates that MoO_2 top layer was self formed. The standard free energy of Mo oxide formation is more negative than that of W and Cu. [6] This implies Mo prefer to react with oxygen than W and Cu illustrating Mo oxides formation instead of W oxide and Cu oxide. However, the standard free energy of Si oxide formation is more negative than W oxide and Mo oxide. The self-formed MoO_2 layer would exist between interface of Cu and W-Mo alloy. According to XPS depth profile, the oxygen source is possibly owing to residual 10 at.% oxygen content in as-deposited 70nm Cu film. In figure 3(b), Cu signal drops sharply near the diffusion barrier, indicating its excellent barrier properties against Cu diffusion. The MoO_2 top layer of the transformed bi-layer structure enhanced the barrier's functions.

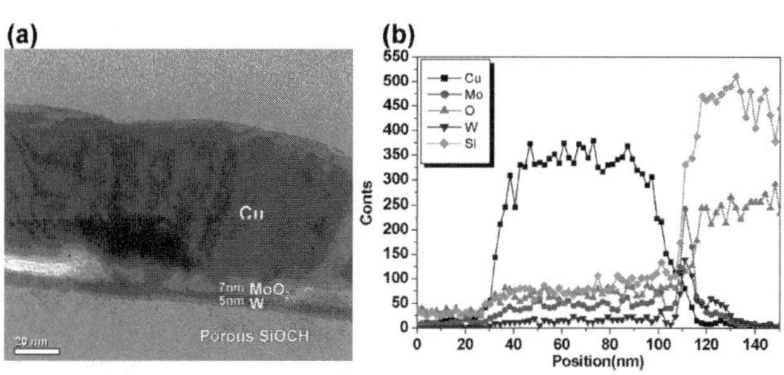

Figure 3. (a) Cross-sectional TEM image of Cu/W-Mo/ULK after annealing at 600°C for 30min (b) The EDS line profiles taken on cross sectional TEM of Cu/W-Mo/ULK after annealing at 600°C for 30min in vacuum.

XPS depth profiles

The XPS depth profiles of the as-deposited and post annealed samples are shown in Fig. 4. The 10 nm W-Mo barrier is clearly observed for the as-deposited film (fig.4a) and Cu atoms did not penetrate barrier to porous SiCO:H for 30 min 600°C annealed sample (fig.4b). They indicate the barrier's excellent thermal stability on porous material. This result is also confirmed by TEM line scan profiles and GIXRD results. For sample annealed at 700°C for 5 min (fig.4c), the distribution of W and Mo atoms broadened and diffused into Cu layer. In addition, some Cu atoms penetrate through 10 nm W-Mo barrier and diffuse into porous SiCOH implies the failure of the barrier. In the case of 5 min. 800 °C annealing, significant amount of Cu diffusion to porous SiCO:H was observed in figure 4(d).

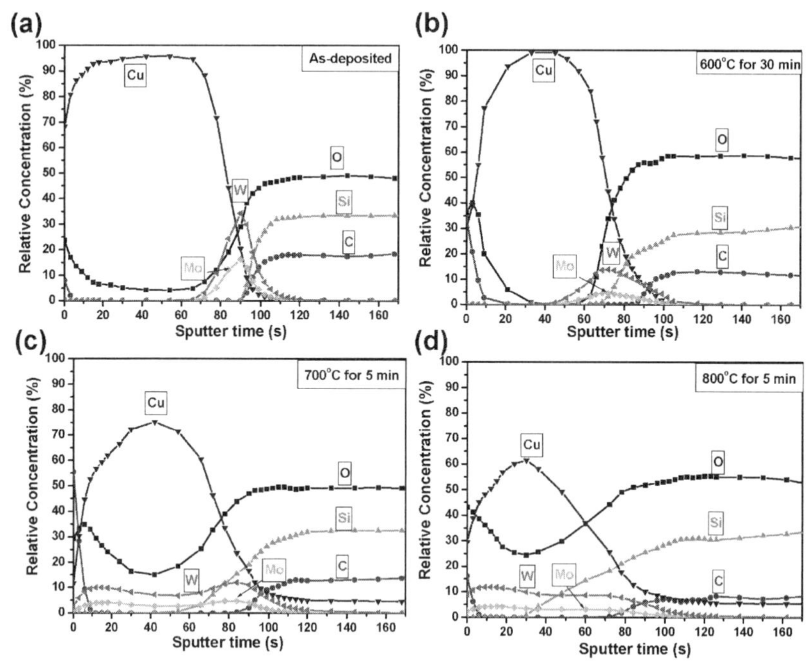

Figure 4. The XPS depth profiles of Cu/W-Mo/Porous SiCOH for various annealing conditions (a) as-deposited (b) annealed at 600°C for 30 min (c) annealed at 700°C for 5 min (d) annealed at 800°C for 5 min.

XPS photo-line scan analysis:

The chemical bonding states of interfacial MoO_2 layer was evaluated by XPS survey scan analysis. The reversed Cu electroplating with $CuSO_4$ bath was used to remove the Cu layer. Figure 5(a) shows the presence of 231.88eV for Mo $3d_{3/2}$ and 228.68eV for Mo $3d_{5/2}$ core-level spectra, which indicate the presence of stoichiometric MoO_2.[7] In addition, the figure 5(b) was observed the peak of O1s at 530.58eV. Both Mo 3d and O 1s signals were detected at the interface after annealing at 600°C for 30min indicating the formation of MoO_2. The XPS photo-line scan results confirm the MoO_2 layer shown in Fig. 3(a) and also consistent with GIXRD patterns. The self-forming MoO_2 layer enhanced W-Mo liner's capability against Cu penetration.

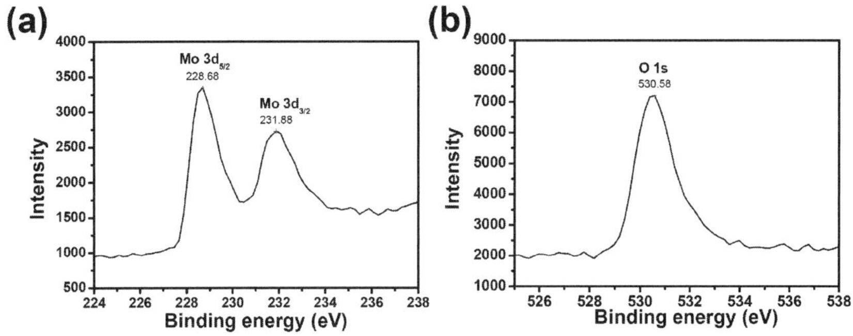

Figure 5. The XPS survey spectrums taken from interface of Cu-striped and W-Mo after annealed at 600°C for 30 min (a) Mo 3d spectra (b) O 1s spectra

CONCLUSION

In summary, we propose a 10nm refractory W-Mo diffusion barrier for Cu/ULK applications. The GIXRD results revealed that no copper silicide formations at least to 700°C for 5 min. TEM image and EDS line profiles show clear interface, indicating that there is no Cu penetration to porous SiOC:H film. The XPS depth profiles clearly demonstrate that the 10 nm W-Mo barrier is thermally stable at 600°C for 30 min. Small amount of Cu diffuse to porous SiOC:H film occurred at 700°C. The Mo atoms migrated and self-formed MoO_2 layer at high temperature which provides additional barrier functions against Cu diffusion for the cost of higher resistance. Poor adhesion of Cu to W-Mo alloy barrier was observed and needs improvement. A 10 nm W-Mo alloy layer has excellent thermal stability and diffusion barrier properties. It is a potential barrier candidate for advanced copper metallization.

ACKNOWLEDGMENTS

The authors gratefully acknowledge Taiwan's National Science Council for the financial support of this work under contract no. NSC 96-2221-E-006-291. This work is also partially supported by National Cheng Kung University under Landmark Project No. C004.

REFERENCES

[1]. K. M. Chang et al., *J. Appl. Phys.* ,vol 83 No.3, pp. 1469-75, (1997)
[2]. M. Hecker et al., *Microelectron. Eng.*, vol 64 pp 269–277 (2002) .
[3]. S. H. Kim et al., *Appl. Phys. Lett.*, Vol. 82, No. 25, (2003)
[4]. W. M. Li et al., *Proceeding of IITC* pp.191 – 193 (2002)
[5]. P. Majumder et al., *Appl. Phys. Lett.* Vol 91, 162108 (2007)
[6]. W. F. Gale et al. , *Smithells Metals Reference Book*, (Elsevier Butterworth-Heinemann, 2004) Ch. 11 p.36
[7]. J.G. Choi et al., *Appl. Surf. Sci.* Vol 93, pp.143-149 (1996)

Materials and Characterization

Conference Proceedings AMC XXIV © 2009 Materials Research Society

Chemical Vapor Deposition (CVD) of Manganese Self-Aligned Diffusion Barriers for Cu Interconnections in Microelectronics

Roy G. Gordon, Hoon Kim, Yeung Au, Hongtao Wang, Harish Bhandari, Yiqun Liu, Don Keun Lee, and Youbo Lin
Chemistry and Chemical Biology, Harvard University, 12 Oxford Street, Cambridge, MA, 02138

Keywords: manganese, chemical vapor deposition, diffusion barrier, copper, adhesion

ABSTRACT

Barriers to prevent diffusion of copper, oxygen and water vapor were formed by CVD using a manganese precursor vapor that reacts with silica surfaces. The manganese metal penetrates only a few nanometers into the silica to make conformal amorphous manganese silicate layers. This $MnSi_xO_y$ was found to be an excellent barrier to the diffusion of Cu, O_2 and H_2O vapor. The adhesion strength of Cu to the $MnSi_xO_y$ was found to be sufficiently strong to satisfy the semiconductor industry requirements for interconnections in future microelectronic devices. CVD Mn dissolves into copper surfaces and then diffuses to increase adhesion to SiCNO capping layers.

INTRODUCTION

Copper (Cu) is now becoming the standard material for connecting transistors in microelectronic devices. However, transistors cannot function properly if Cu diffuses into them. Cu also reduces the resistance of the insulating material between the Cu wires. For both these reasons, Cu must be confined inside diffusion barriers that surround the Cu wires. Also, barriers to oxygen and water must separate these materials in the environment from the copper to prevent its oxidation. In current technology, the diffusion barriers are formed by sputtered tantalum nitride (TaN). Although TaN is an effective diffusion barrier, Cu does not adhere strongly to TaN. The adhesion of Cu to tantalum (Ta) metal is much stronger than to TaN, so Ta is sputtered on top of the TaN to enhance the durability of interconnect structures. Next in the process flow a thin Cu "seed" layer is sputtered, and then the trenches and holes are filled with Cu by electroplating.

As the dimensions of microelectronic circuits are being reduced, the non-conformal nature of sputtered barrier (TaN), adhesion (Ta) and seed (Cu) layers has caused problems. These layers are thicker near the top openings, so that the remaining openings are narrower at the top than lower inside the features. Then the electroplating process for Cu can form voids inside the features, which later can lead to failure of the interconnections. Another difficulty arises from roughness on the sidewalls of features, which may prevent deposition of the layers on recessed regions that are shielded from the line of sight to the sputtering target. If these areas are missing TaN, then Cu may diffuse out through these holes in the diffusion barrier. If Ta is missing from an area, then reliability may be reduced because of the reduced adhesion of the Cu to the TaN. If these areas are missing the Cu seed layer, then Cu electroplating will not take place on the oxidized surface of the exposed Ta, producing voids in the Cu. Thus there is a need for methods to produce more conformal diffusion barriers, adhesion layers and seed layers. CVD and atomic layer deposition (ALD) are methods that can produce conformal diffusion barriers, adhesion layers and seed layers.[1]

Another problem with current diffusion barriers and adhesion layers is that TaN and Ta have a much higher resistivity than Cu. Thus by displacing Cu metal, they increase the resistance of interconnects. The time delay due to interconnects increases with their resistance, and thus the volume of barrier material should be minimized in order to form the fastest circuits. Ideally, a barrier should not occupy any of the volume reserved for Cu. Such a "zero-thickness barrier" can be realized by sputtering a manganese-Cu alloy, from which manganese diffuses into the surface of silica insulation at 450 °C.[2] The resulting self-aligned barrier layer of amorphous manganese silicate ($MnSi_xO_y$), just a few nm thick, remains part of the surface of the insulator, whose dimensions are not significantly changed by in-diffusion of the manganese. The resulting interconnects have lower resistance than ones made with conventional TaN barriers, and have proven to be even more durable as well.[3]

The self-aligned $MnSi_xO_y$ barriers start with sputtering of a Mn-Cu alloy, so they are still subject to the conformality problems of sputtered barriers. Voids in electroplated Cu may form under overhangs near the tops of features or inside rough areas on the sidewalls In this paper we propose a way to make conformal self-aligned $MnSi_xO_y$ barriers using CVD of manganese metal. We find that vapors of a manganese compound react with silica surfaces to form a thin amorphous layer of manganese silicate mainly within the surface of the silica. The $MnSi_xO_y$ layer is shown to be a good barrier against diffusion of Cu, O_2 and H_2O. CVD Mn also greatly increases the adhesion of copper to capping layers of SiCNO. The layer is completely conformal inside the surfaces of holes and trenches.

EXPERIMENTAL

The compound that serves as a precursor for the manganese is called bis(*N,N'*-diisopropylpentylamidinato)manganese(II), whose chemical formula is shown in Figure 1. The compound was synthesized by methods similar to those described previously.[4] It is a pale yellow crystalline solid that melts at about 60 °C to a clear liquid. For the CVD experiments, the manganese precursor was evaporated from the liquid in a bubbler at a temperature of 90 °C into a 60 sccm flow of highly purified nitrogen (concentrations of water and oxygen less than 10^{-9} of the N_2). The manganese precursor reacts immediately with oxygen or water to form a black material when it is exposed to air. The vapor pressure of the precursor is estimated to be around 0.1 mbar at 90 °C.

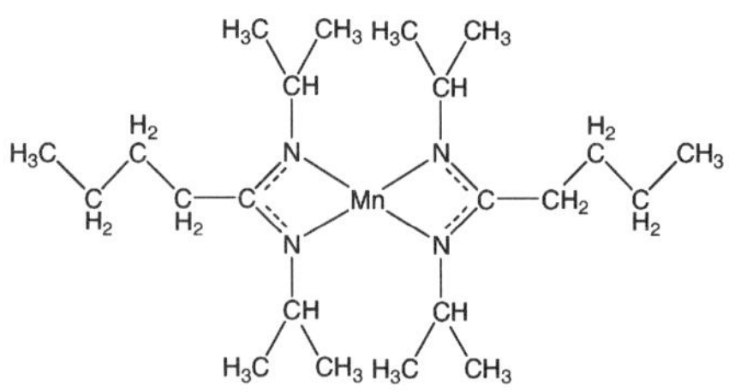

Figure 1. Chemical formula of bis(*N,N'*-diisopropylpentylamidinato)manganese(II).

The substrates of silica were either thermally oxidized silicon or silica deposited by ALD. The CVD was carried out in a hot-wall tube reactor (diameter 36 mm) within a tube furnace at temperatures between 200 and 400 °C and a total pressure of about 5 Torr. The amount of manganese deposited was measured by Rutherford backscattering spectroscopy (RBS). The $MnSi_xO_y$ formation was evaluated by cross-sectional high-resolution transmission electron microscopy (HRTEM). The effectiveness of the $MnSi_xO_y$ as a barrier to diffusion of Cu was tested in four ways: optical appearance, sheet resistance, Cu silicide formation and capacitance-voltage (CV) analysis of capacitors. For Cu diffusion tests, layers of SiO_2 8 nm thick were grown on HF-etched silicon wafers by ALD at 215 °C, followed by CVD Mn at 350 °C for 10 min, which formed 2.3 nm of Mn metal. Control samples of SiO_2 omitted the CVD Mn treatment. Then Cu layers about 200 nm thick were deposited on top of the CVD $MnSi_xO_y$ or SiO_2 layers. Anneals in a pure nitrogen atmosphere were carried out at temperatures of 400, 450 and 500 °C for one hour. For CV analysis, CVD Mn layer is deposited on 300 nm thermal SiO_2. Cu pad (500 μm diameter circle) is formed by thermal evaporator using shadow mask. CV was measured with a HP 4275A meter in a shielded probe station at room temperature. The Cu pad samples were used for Bias-Temperature Stress (BTS) test that was conducted at 250°C under 1 MV/cm bias condition in Ar ambient. After this stress condition, CV was measured at room temperature.

RESULTS AND DISCUSSION

Thin Mn layers (2.3 nm) deposited on SiO_2 did not have any conductivity by 4-point probe (> 10^6 ohm/square). It may form $MnSi_xO_y$, which is an insulator, and any remaining metal Mn may be oxidized when the sample is taken out into the air for measurement. Thus, it is not clear from this measurement whether the CVD process deposited manganese metal or some insulating manganese compound. To confirm Mn metal deposition, CVD was carried out on substrates of Cu films 50 nm thick that had been evaporated onto SiO_2/Si substrates. The resulting structure was examined by cross-sectional HRTEM. Figure 2 shows that the CVD metal Mn diffused through the Cu layer and reacted with the SiO_2 to form an amorphous $MnSi_xO_y$ layer about 2~5 nm thick between the Cu and the SiO_2. The $MnSi_xO_y$ layer is thicker near grain boundaries in the Cu, along which Mn diffusion is faster. This result is evidence of Mn metal deposition.

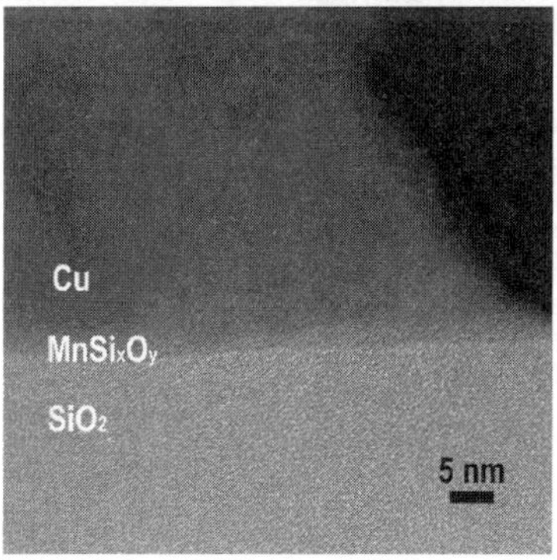

Figure 2. Cross-sectional HRTEM of CVD Mn / PVD Cu / SiO_2

The effectiveness of $MnSi_xO_y$ as a copper barrier was evaluated using a sample structure PVD Cu(200 nm)/CVD Mn (2.3 nm)/ALD SiO_2 (8 nm)/Si. A $MnSi_xO_y$ layer was formed between the Cu and ALD SiO_2 layers. The shiny Cu color and sheet resistances of these samples were unchanged by anneals in nitrogen at 400 or 450 °C. After a 500 °C anneal, the control sample, which does not have any $MnSi_xO_y$ layer, turned black and its sheet resistance increased by a factor of 200 because of massive diffusion of the Cu through the thin ALD SiO_2 into the silicon. The CVD Mn sample, by contrast, retained its shiny Cu color and increased its resistance only slightly.

To analyze for Cu diffusion, the remaining Cu layers were dissolved in nitric acid, and then the manganese silicate and silica were removed by dilute HF. The etched surfaces were then analyzed by an energy-dispersive X-ray spectrometer (EDX) and scanning electron micrographs (SEM). Figure 3 shows the SEM results after a 500 °C for 1 hr anneal. The few Cu-containing spots appear to be Cu silicide crystallites oriented by the crystal directions of the silicon. The control sample shows that the majority of its surface is covered by Cu silicide. The control sample showed a large Cu signal in EDX analysis that was stronger than the silicon signal, confirming that the thin ALD SiO_2 allowed diffusion of Cu. The CVD Mn-treated samples did not show Cu by large-area EDX. A few small areas of the SEM image did show some Cu by EDX, indicating some localized breakdown of the $MnSi_xO_y$ barrier at 500 °C. These spots might arise from dust or other defects in the films, which were not processed in a clean-room environment.

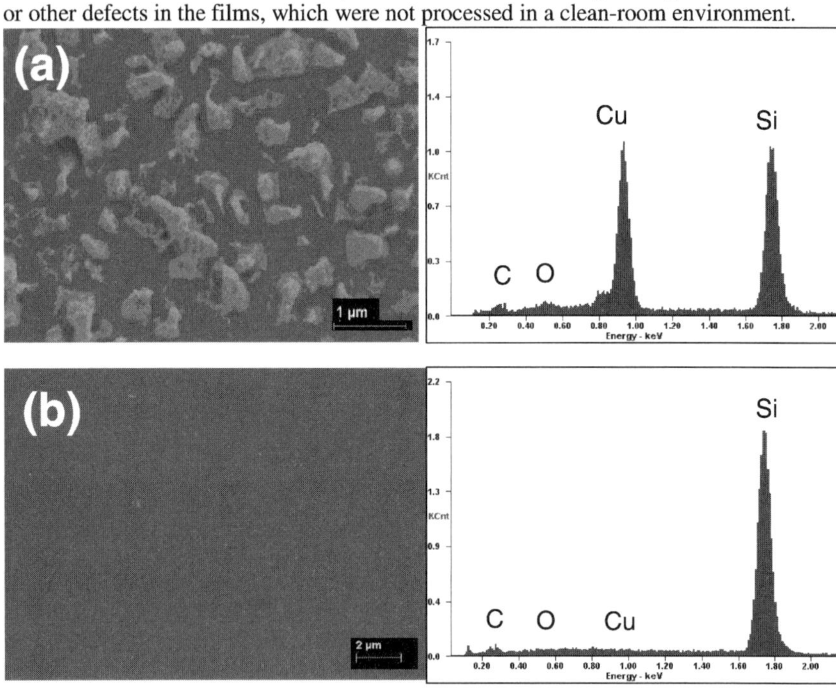

Figure 3. SEM of etched Si surface after annealing at 500 °C (a) with SiO_2 alone, and (b) with CVD $MnSi_xO_y$ on SiO_2.

An electrical test of barrier properties was carried out by patterning the Cu into capacitor electrodes. CV curves for samples annealed at 450 °C for 1 hr are shown in Figure 4. The large shift (-4.9 V) to negative voltages in the control sample is caused by positive Cu ions diffusing into the silica insulator.[6] In contrast, the silica protected by $MnSi_xO_y$ shows only a very small shift (- 0.1 V). This electrical test is more sensitive to diffusion of small amounts of Cu, than the other tests. These CV curves also demonstrate that the capacitance of the SiO_2 is not changed significantly by the CVD Mn treatment. Thus, we can expect this process is acceptable in low-k integration.

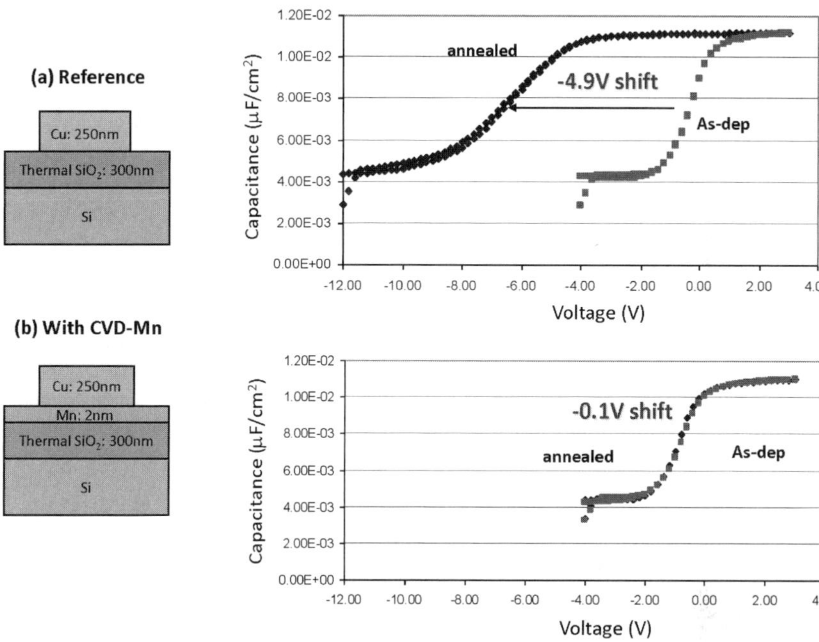

Figure 4. CV curves of samples before and after annealing at 450 °C (a) with SiO_2, and (b) with CVD $MnSi_xO_y$ on SiO_2.

Anneals of similar capacitors were also conducted under an applied voltage of 1 MV/cm at 250 °C. Bias temperature stress (BTS) test is more sensitive method for Cu diffusion into SiO_2, because Cu ion diffusion in the SiO_2 is accelerated by electric field.[6] BTS condition is closer to the environment of chip operation. Thus, BTS testing gives more relevant results for evaluating Cu diffusion barriers.[6] The control sample lost the capacitance behavior just after 2 min in the BTS condition, implying that a large amount of Cu diffused into the Si, so that the Si would not work as a semiconductor. However, the CVD Mn treated sample had no significant change in its CV curve (Figure 5). The results of this BTS test confirm the good Cu barrier properties of the $MnSi_xO_y$ layers.

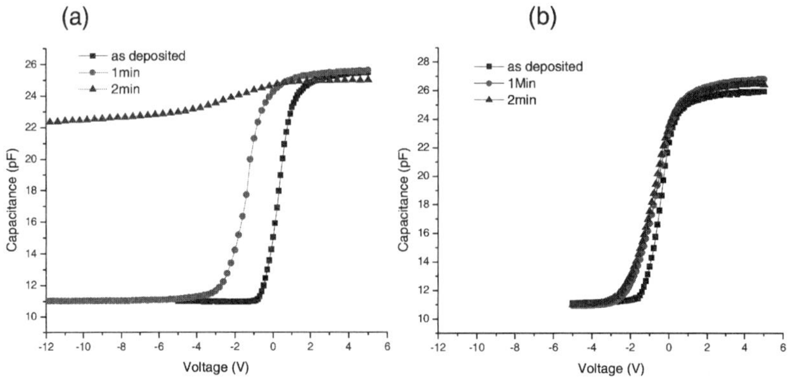

Figure 5. CV curves of samples before and after annealing at 250 °C under a 1 MV/cm field (a) with SiO_2, and (b) with CVD $MnSi_xO_y$ on SiO_2.

$MnSi_xO_y$ layers were also found to be good barriers to oxygen and water, which can corrode copper layers. To test how well the metal silicate layers protected copper, commercial low-k porous insulator layers from Applied Materials were coated with manganese as described above, followed by PVD copper. The top surface of the copper was protected with 40 nm of ALD silica.[7] The sample was cut into pieces to expose the edges of the low-k insulator so that oxygen or water vapor could diffuse into the low-k layer by the pathways shown in Fig. 6.

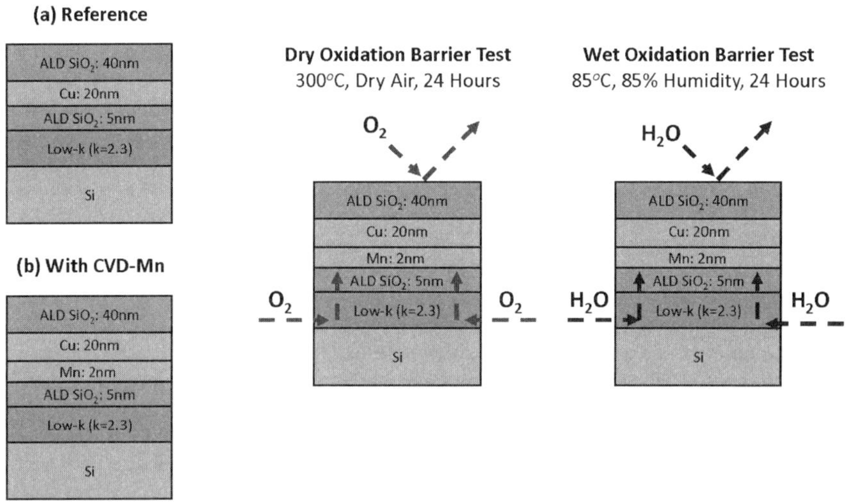

Figure 6. Sample structures and pathways for tests of $MnSi_xO_y$ as an oxidation barrier.

After exposure to dry air at 300 °C for 24 hours, the CVD-Mn-treated sample maintained its shiny copper color. A control sample without the CVD manganese treatment was corroded near its edges to dark copper oxide by the same exposure. This test shows that the manganese silicate layer is a good barrier to oxygen. Similar tests in a humid atmosphere (85% humidity at 85 °C for 24 hours) showed that the manganese silicate layer is a good barrier to water vapor. These exposures to water and oxygen also significantly increased the sheet resistance of the unprotected control copper layers, whereas the resistance of the Mn-protected copper layers remained nearly constant (Fig. 7.).

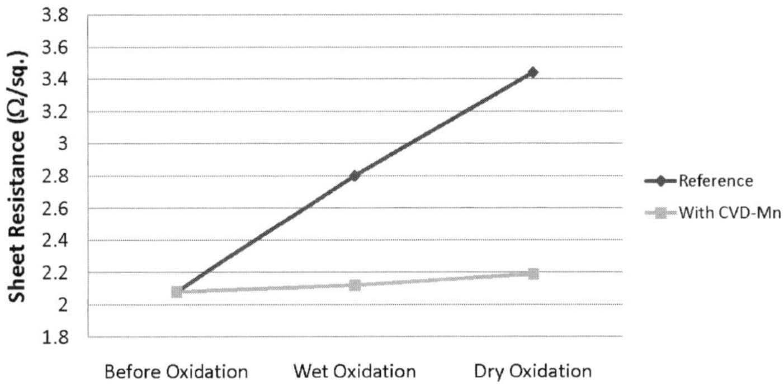

Figure 7. Sheet resistance of copper before and after oxidation.

The formation of the $MnSi_xO_y$ layer remarkably increased the adhesion of the Cu/SiO_2 interface, which failed the tape adhesion test prior to the CVD of Mn, but easily passed after the CVD of Mn. Adhesion strength was measured by 4-point bend tests.[5] The samples were prepared by CVD of Mn onto thermal SiO_2 on silicon wafers. Then CVD at 160 °C was used to form Cu oxynitride (CuON), which was then reduced to Cu metal by a hydrogen plasma at room temperature.[8] The adhesion energy was found to be 10.1 $\pm$ 1 J m^{-2}, which is higher than the value of 5 J m^{-2} that is considered to be sufficient for making durable interconnections.[9]

The CVD Mn process can also be used to strengthen the adhesion between copper and the commonly-used capping layers of SiCNO. To test this adhesion effect, 50 nm of copper was evaporated onto SiCNO layers (BLOK™, Applied Materials). The Cu showed very poor adhesion, with adhesion energy less than 3 J m^{-2}. Next, similar Cu/SiCNO layers were treated by CVD Mn at 350 °C for 10 minutes. This process increased the sheet resistance from 0.5 ohms/square to 1 ohm per square because of the manganese impurity in the copper. Then the structure was annealed for 1 hour at 400 °C in a nitrogen atmosphere. The sheet resistance then returned to 0.5 ohms per square because the manganese diffused to the surfaces or the interface. The out-diffusion of the manganese from the Cu film was confirmed by SIMS analysis. After the manganese treatment, the adhesion energy was remarkably increased to greater than 12 J m^{-2}, because manganese diffused to the interface, and made an interface layer. XPS confirmed the presence of manganese at both surfaces created by delamination of the 4-point bend test sample. The exact compounds formed at the interface are not known.

This manganese metal diffusion process could be used to strengthen the bond between copper and the usual SiCNO capping layers. To do this, CVD Mn would be applied just after chemical-mechanical polishing, thereby dissolving Mn into the upper portions of the Cu below its polished surface. After SiCNO deposition, this dissolved Mn will diffuse up to the surface of the SiCNO and strengthen its bonding to the copper, probably increasing the electromigration lifetime.

Complete conformality of the CVD CuON depositions in holes with aspect ratios up to 40:1 was confirmed by cross-sectional SEM studies.[8] Because the $MnSi_xO_y$ layers are too thin to image by SEM, the distribution of CVD $MnSi_xO_y$ was tested by Energy Dispersive Analysis by X-rays (EDAX) for Mn. The substrates were holes in silicon with an aspect ratio of 52:1 and a native oxide surface with which the Mn precursor reacted. The results plotted in Fig. 8 show how the Mn content decreased with distance into the hole. The amount of Mn at the bottom of the hole was 18% of the amount at the top entrance to the hole.

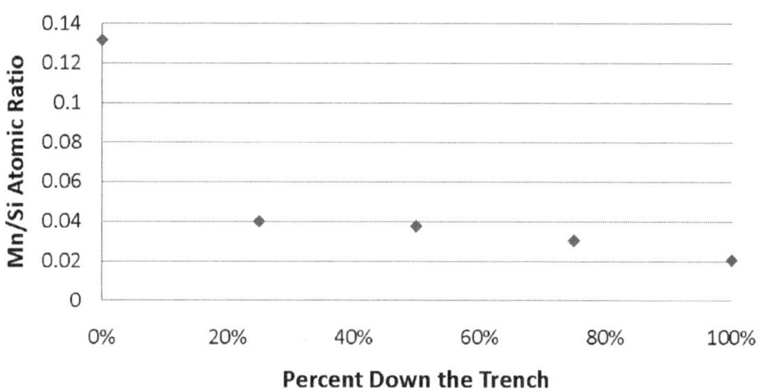

Figure 8. Distribution of CVD Mn down the length of a trench with aspect ratio 52:1.

This result shows that the Mn precursor has a fairly slow reaction with silica surfaces at 350 °C. Thus interconnect trenches and vias with typical aspect ratios less than 5:1 can be coated fairly uniformly with CVD $MnSi_xO_y$ at 350 °C. At lower temperatures, the surface reactions are expected to be slower, and the conformality should be even higher. Even when roughness is present on sidewalls, the CVD process should form a complete $MnSi_xO_y$ diffusion barrier without pinholes or gaps.

CONCLUSIONS

In summary, we presented a simple CVD method to make conformal self-aligned diffusion barriers of manganese silicate. This $MnSi_xO_y$ is an effective barrier to diffusion of copper, oxygen and water. Because the $MnSi_xO_y$ is formed within the insulator, it does not decrease the volume available for Cu, as happens with conventional Ta/TaN barriers. Also, the CVD Mn dissolves into copper surfaces, such as at the tops of vias, so no resistive barrier exists between the tops of vias and the copper in the next higher level. The CVD process can also be used to strengthen the interfacial bonding between copper and a capping barrier layer of SiCNO. Thus more conductive and more durable interconnections are possible using the self-aligned $MnSi_xO_y$ barriers. The CVD Mn-CuON process should allow formation of conformal seed layers for void-free electroplating of Cu in even the narrowest interconnects.

ACKNOWLEDGMENTS

This work was performed in part at the Center for Nanoscale Systems (CNS), a member of the National Nanotechnology Infrastructure Network (NNIN), which is supported by the National Science Foundation under NSF award no. ECS-0335765. CNS is part of the Faculty of Arts and Sciences at Harvard University. D. K. Lee was supported by a Korea Research Foundation Grant from the Korean Government (MOEHRD, KRF-2005-214-C00206). Substrates and SIMS analysis were supplied by IBM. This work was supported in part by Intel Corporation and the National Science Foundation.

REFERENCES

1. The International Technology Roadmap for Semiconductors, http://public.itrs.net (2007).
2. J. Koike and M. Wada, Appl. Phys. Lett. **87** (4), 041911/1 (2005); M. Haneda, J. Iijima, and J. Koike, Appl. Phys. Lett. **90** (25), 252107/1 (2007); J. Koike, M. Haneda, J. Iijima, Y. Otsuka, H. Sako, and K. Neishi, J. Appl. Phys. **102** (4), 043527/1 (2007).
3. J. Iijima, M. Haneda, and J. Koike, IEEE Int. Interconnect Technol. Conf., Proc., 155 (2006); J. Koike, M. Wada, T. Usui, H. Nasu, S. Takahashi, N. Shimizu, M. Yoshimaru, and H. Shibata, AIP Conf. Proc. **817** (Stress-Induced Phenomena in Metallization), 43 (2006); T. Usui, H. Nasu, S. Takahashi, N. Shimizu, T. Nishikawa, M. Yoshimaru, H. Shibata, M. Wada, and J. Koike, IEEE Trans. Electron Devices **53** (10), 2492 (2006); T. Usui, K. Tsumura, H. Nasu, Y. Hayashi, G. Minamihaba, H. Toyoda, H. Sawada, S. Ito, H. Miyajima, K. Watanabe, M. Shimada, A. Kojima, Y. Uozumi, and H. Shibata, IEEE Int. Interconnect Technol. Conf., Proc., 216 (2006); H. Haneda, T. Tabira, H. Sakai, H. Kudo, M. Sunayama, N. Ohtsuka, A. Tsukune, and N. Shimizu, Conf. Proc. AMC XXIII, 59 (2008).
4. B. S. Lim, A. Rahtu, J.-S, Park, and R. G. Gordon, Inorg. Chem. **42**, 7951 (2003).
5. Y. Lin, J.J. Vlassak, T.Y. Tsui, and A.J. McKerrow, Materials Research Society Symposium Proceedings **795**, 93 (2004).
6. A.L.S. Loke, C. Ryu, C. P. Yue, J. S. H. Cho, and S. S. Wong, IEEE Elec. Device Lett. **17**, 549 (1996).
7. D. Hausmann, J. Becker, S. Wang, and R. G. Gordon, Science **298**, 402 (2002).
8. H. Kim, H. B. Bhandari, S. Xu, and R. G. Gordon, J. Electrochem. Soc. **155**, H496 (2008).
9. L. Peters, Semiconductor International (July 1, 2001).

Conference Proceedings AMC XXIV © 2009 Materials Research Society

Effects of Current Density, Environment, and Temperature on Adhesion and Debonding Kinetics of Cu / Barrier Interfaces

Ryan Birringer[1], Roey Shaviv[2], and Reinhold Dauskardt[1]
[1]Department of Materials Science and Engineering, Stanford University, Durand Building, 496 Lomita Mall, Stanford, CA, 94305
[2]Novellus Systems, Inc., 4000 North First Street, San Jose, CA, 95134

ABSTRACT

Thin film adhesion techniques are commonly used to characterize the critical interface fracture energy of barrier / copper film stacks. In this study, modified versions of these thin film techniques are used to study subcritical (or time dependent) interface debonding as a function of environment, temperature, and current density through the Cu film. We believe these are the first reported measurements of subcritical interface debonding in Cu / barrier films.

INTRODUCTION

Cu interconnects at ≤ 32 nm technology nodes pose unique fundamental and technological challenges. The need for ultra-thin liners and selective capping layers that provide superior diffusion and dielectric barrier properties will require the integration of new materials into future interconnect structures. It is widely accepted that electromigration in modern Cu interconnects is closely related to adhesion and bonding between the Cu line and the top capping layer [1,2]. While four-point bend thin film adhesion techniques are commonly used to characterize adhesion in Cu / barrier layer interfaces, the kinetics of sub-critical debonding in these films has been largely ignored.

In the present study, we describe modified versions of well-known thin film adhesion techniques used to quantify adhesion and time-dependent subcritical debonding between Cu and selected barrier layers in the presence of variable current density through the film, testing environment, and temperature. We use these techniques in an attempt to simulate the growth kinetics of EM voids.

EXPERIMENT

A 50 nm PVD Cu seed layer was deposited onto silicon substrates containing a Ta/TaN barrier film on a 1000 nm dense oxide. 500 nm of Cu was electroplated onto the seed layer before the Cu film was reduced down to 250 nm total thickness by chemical mechanical polishing. A reducing plasma treatment was used to reduce the surface of the copper film and 50 nm dielectric barriers were then deposited using CVD. The film stack was capped with 1000 nm dense SiO_2. The wafers were then bonded to blank Si substrates to make a symmetric "sandwich" structure. A high-temperature, low viscosity epoxy was used to bond the wafers, and a 200 nm Cr/Al standoff film was deposited prior to bonding in order to isolate the thin films of interest from the bonding layer. A cross section of the final structure can be seen in Figure 1. Fracture mechanics specimens were fabricated by dicing this "sandwich" structure into beams using a high precision dicing saw.

A modified version of the well-known four-point bend technique [3] was used to measure critical fracture energies of the Cu / barrier interfaces. Additionally, a modified version of the standard DCB technique [3] was used to probe the debonding kinetics of the copper / barrier interface in the presence of varying current density through the film, environment, and temperature (Figure 2). Load relaxation methods were employed using the elastic compliance of the silicon beams to characterize the debond length as a function of time. This allows the debond growth rate, da/dt (m/s), to be measured as a function of the applied debond driving force expressed in terms of the strain energy release rate G (J/m^2). The value of G measured at high growth rates corresponds to the interface fracture energy G_c. Debond growth rates were measured in a range of oxidizing moist air and inert N_2 environments with controlled amounts of O_2 and reducing H_2. In order to clarify the fracture path and rationalize the measured adhesion values, surface sensitive characterization techniques such as AR – XPS and AES were performed on the resulting fracture surfaces.

DISCUSSION

Time dependent subcritical debonding of the Cu / SiN interface was measured and exhibited three distinctive regimes as discussed below. The kinetics of Cu / SiN interface debonding were significantly influenced by the test environment. Specimens tested in oxidizing environments, such as lab air or an O_2/N_2 mixture, exhibited a pronounced plateau region where debond growth velocity was relatively insensitive to applied driving force (Figure 3). Debond growth in this region may be attributed to oxidation of the Cu film and the growth velocity was limited by either diffusional transport of the oxidizing species to the crack tip or the kinetics of Cu oxidation at the crack tip. The growth velocity of the plateau region decreased as the activity of the oxidizing species was decreased. The kinetics of wet oxidation in Cu thin-films have been shown to be significantly faster than for dry oxidation at temperatures below 600 °C [4], which may account for the difference in plateau growth velocities in humid and dry (O2/N2) environments. This plateau region was followed by an abrupt threshold region where debond growth decreased dramatically. It is thought that this threshold region occurs at an applied driving force below which the oxidizing species is sterically hindered from reaching the crack tip. Similar steric hinderance effects have been reported previously for subcritical debonding of metal films in oxidizing environments [5].

Specimens tested in a 5 mol% H_2 forming gas environment exhibited a pronounced reaction region where debond growth depends strongly on applied driving force, indicating that debonding was limited by a stress-enabled chemical reaction at the SiN / Cu interface. As the activity of hydrogen is decreased, curves shift towards lower velocity and higher driving force and appeared to exhibit a small transport limited region (Figure 4).

The effects of increased temperature in a 0.45 mol% H_2 forming gas environment are shown in Figure 5. Tests performed at 50 and 75°C show a slight shift upwards and a pronounced shift to the right. The acceleration of growth velocity with increasing temperature is expected based on previous studies of reaction-controlled subcritical crack growth [6]. The pronounced shift to the right with increased temperature may be due to increased plasticity near the debonding region or crack tip blunting effects in the Cu layer. Figure 6 shows the effect of current flow through the Cu film during testing. In both cases, current was flowing in the same direction as crack propagation. One can see that testing at 75 °C and a current density of 130 KA / cm^2 results in no measurable change in the subcritical growth curve; however, when the current density was increased to 350 KA / cm^2, the curve shifts to the right. This shift is similar to that seen with increased temperature, and may indicate local joule heating near the debond tip or the effects of increased Cu atom mobility at the interface in the presence of the EM current.

CONCLUSIONS

Subcritical interface debonding in SiN / Cu films and the effects of environment, temperature, and current flow through the Cu film were studied using a modified version of the well-known DCB thin film adhesion technique. Subcritical debond growth due in a humid oxidizing environment is much faster than in a dry oxidizing environment. In reducing forming gas environments, subcritical growth is limited by a chemical reaction between the environment and the strained bonds at the interface debond tip. Increasing the test temperature results in a slight increase in debond growth rates and a pronounced shift towards higher applied debond driving force. As current density through the Cu film is increased, the curves shift to the right in a manner similar to that seen with increasing temperature.

ACKNOWLEDGMENTS

In addition to our collaborators at Novellus, Inc., we would like to acknowledge the support of Dr. Paul Besser (AMD, Inc.) for participating in useful discussions.

REFERENCES

1. M. Lane, E. Linger, and J. Lloyd, J. Appl. Phys. 93, 1417 (2003).
2. K. Chattopadhyay, B. van Schravendijk, T. W. Mountsier, G. B. Alers, M. Hornbeck, H. J. Wu, R. Shaviv, G. Harm, D. Vitkavage, E. Apen, Y. Yu, and R. Havemann, *Proceedings of The International Reliability Physics Symposium* (IEEE, San Jose, CA, 2006) pp. 128–130.
3. R. H. Dauskardt, M. Lane, Q. Ma, and N. Krishna, Engng. Fract. Mech. 61, 141 (1998).
4. Z. H. Hu, R. Sharangpani, and S.-P. Tay, J. Vac. Sci. Technol. A. 18, 2527 (2000).
5. M. Lane, R. H. Dauskardt, N. Krishna, and I. Hashim, J. Mater. Res. 15, 203 (2000).

6. S. M. Wiederhorn, Int. J. Fract. Mech. 4, 171 (1968).

FIGURES

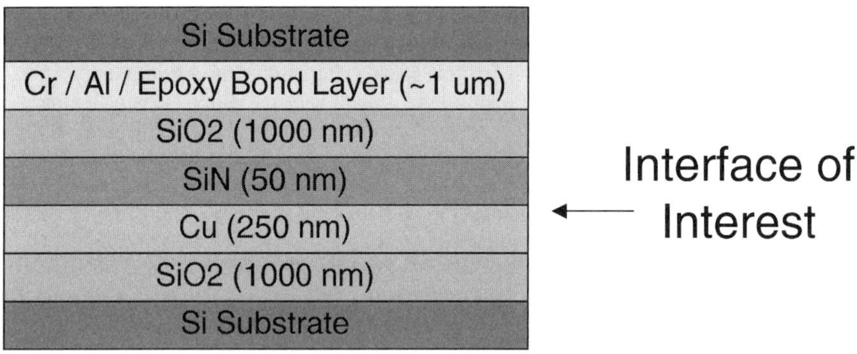

Figure 1: Schematic cross section of test specimens.

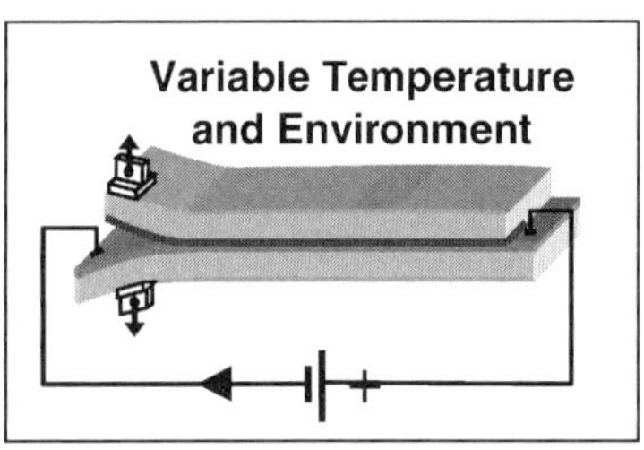

Figure 2: Schematic of subcritical testing technique.

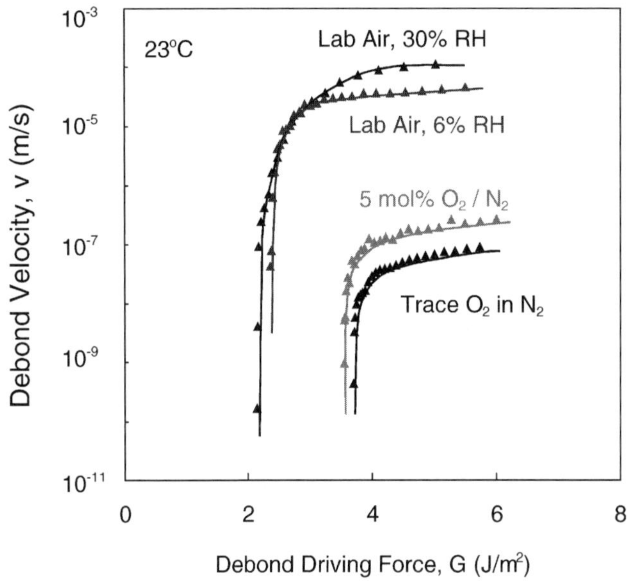

Figure 3: Subcritical debonding curves in oxidizing environments.

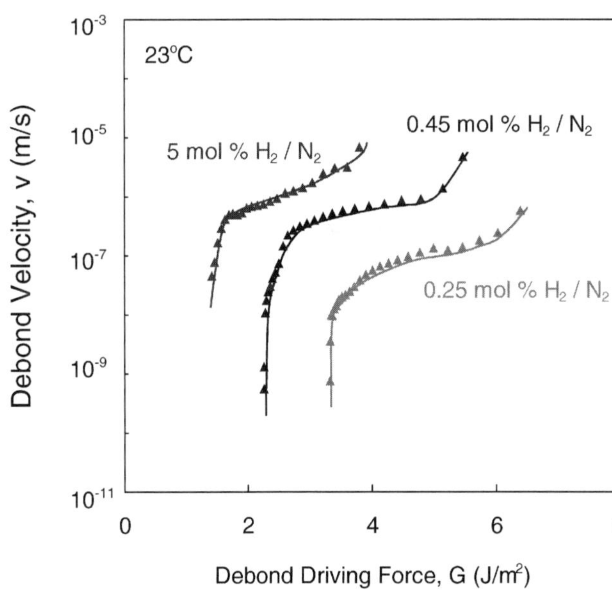

Figure 4: Subcritical debonding curves in forming gas environments.

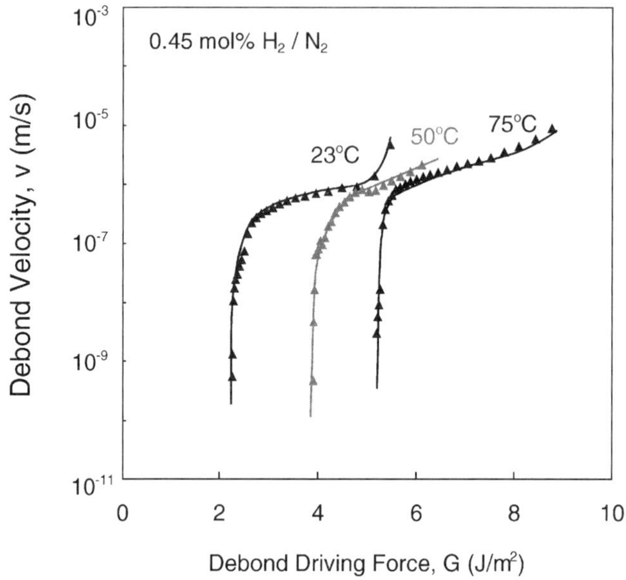

Figure 5: The effect of increased temperature on subcritical debonding behavior in a forming gas environment.

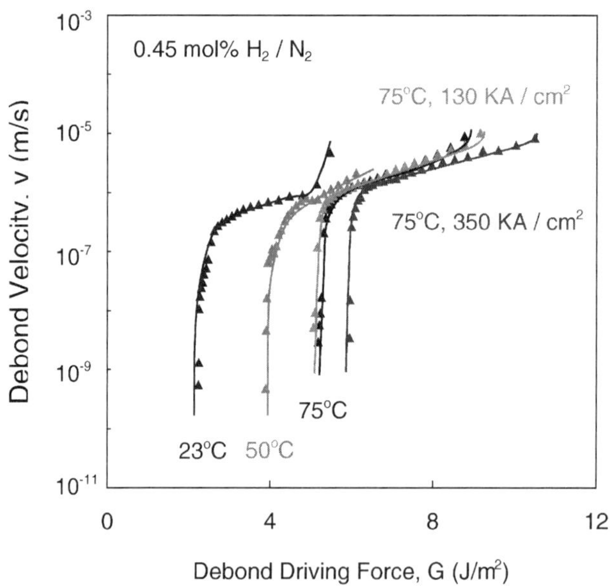

Figure 6: The effect of current density in the Cu film in a forming gas environment.

Conference Proceedings AMC XXIV © 2009 Materials Research Society

Tuning Depth Profiles of Low-k Dielectrics with UV Curing

Taek-Soo Kim[1], Naoto Tsuji[2], Johan van der Hilst[2], Kiyohiro Matsushita[2], Nobuyoshi Kobayashi[2], Dmytro Chumakov[3], Holm Geisler[3], Ehrenfried Zschech[3], and Reinhold H. Dauskardt[1]

[1]Stanford University, Stanford, CA, 94305

[2]ASM Japan K.K., Tokyo, 206-0025, Japan

[3]AMD Fab36 LLC & Co. KG, Dresden, D-01109, Germany

ABSTRACT

This study demonstrates that ultraviolet (UV) radiation curing can control depth profiles of organosilicate films. Fracture energies at both top and bottom interfaces of OSG were significantly increased by the use of an UV standing wave and an underlying optical spacer layer that helps to locate the maximum UV intensity at the low-k interfaces. SiN and SiCN underlying layers were tested for the optical spacer and dramatic differences for the effects of UV curing were observed. Depth profiles of the low-k layers were characterized by force modulation atomic force microscopy (FM-AFM). The implications of the tuning depth profiles of low-k dielectrics with UV curing for the next technology nodes are discussed.

INTRODUCTION

We recently reported depth dependence of UV curing of a series of experimental organosilicate low-k thin film glasses (OSG) which has implications for the variation of mechanical properties through the thickness of the films [1-4]. Specifically, an oscillating elastic modulus depth profile following UV cure was measured by FM-AFM as shown in Fig. 1(a). The oscillation is associated with UV light interference, which we modeled using a new standing wave equation that accurately predicts the resulting depth dependent stiffness variations considering changes of film absorption, shrinkage, and refractive index (Fig. 1(b)). Significant increase of the fracture energy at the top interface of the OSG was observed. However, the fracture energy at the bottom interface was insensitive to UV curing. We speculated that is due to either minimum UV light intensity by standing wave at the interface with highly reflecting Si, or strong UV light absorption by the underlying SiCN layer. In the current study, we present an innovative method to increase fracture energies dramatically at both "top" and "bottom" interfaces of OSG by controlling the film thickness and reflectivity of an underlying layer to locate the maximum UV light intensity of the standing wave at the top and bottom of OSG.

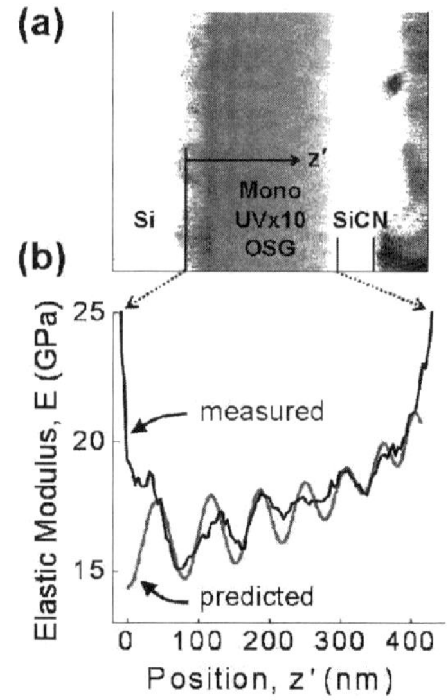

Figure 1. (a) The FM signal image of a cross-section of a specimen containing a UV cured OSG. (b) An elastic modulus depth profile measured by FM-AFM is compared with predicted one by the standing wave simulation.

EXPERIMENT

Experimental OSG films were deposited by plasma enhanced chemical vapor deposition (PE-CVD) on 20 nm thick SiN and SiCN underlying layers, respectively. After deposition, they were exposed to monochromatic UV radiation with a wavelength less than 200 nm for 5 minutes in a controlled environment. Following the UV cure, the films were capped with PE-CVD SiCN barrier layers. Four-point bend (FPB), double cantilever beam (DCB), and FM-AFM specimens were fabricated and tested by previously reported techniques [1-4]. The FM-AFM technique was used to investigate the variations in the elastic modulus on the cross-section of specimens containing UV cured 500 nm organosilicate films. Two thin film structures were bonded with epoxy, and the specimen consisting of the two thin film structures were diced and polished by chemical mechanical planarization (CMP) to produce a flat cross-section. During force modulation scans on the cross-section the AFM tip always remained in contact with the surface. The quantitative estimation of the elastic modulus is based on the modeling of the mechanics of the tip-specimen elastic interaction. The calculations rely on Hertzian contact theory to consider the deformation of both the surface and the tip. Present FM-AFM measurements require the knowledge of the reduced modulus at any two locations on the sample that possesses different elastic modulus values. These values are needed for the calibration of tip and cantilever parameters that play a determinant role in delivering the quantitative information. The corresponding calibration was performed by using a nanoindentation tool with in-situ scanning probe microscopy (SPM) imaging capability on sample regions which are still large enough to be measured with the tip of the nanoindenter.

DISCUSSION

Striking differences between the effects of SiN and SiCN underlying layers were observed. The SiN layer dramatically increased fracture energies at both top and bottom interfaces of OSG films (Fig. 2 and 3). The 20 nm thick SiN layer is transparent to the UV light and helps to locate the maximum UV curing intensity at the bottom of the OSG films, while the SiCN layer strongly absorbs the UV light and therefore decreases UV curing intensity. Consequently, the SiN layer results in stronger UV curing intensity at the top of the OSG film

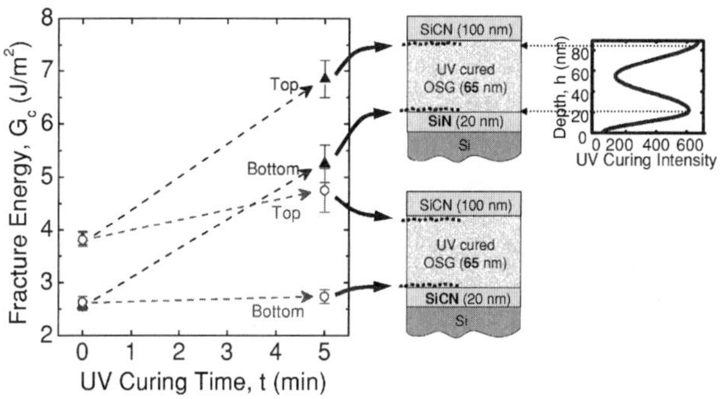

Figure 2. Fracture energies at the top and bottom interfaces of a 65 nm thick OSG before and after UV curing. The effects of SiN and SiCN underlying layers are investigated. Maximum UV curing intensity is located at the top and bottom interfaces of the OSG that has a SiN underlying layer.

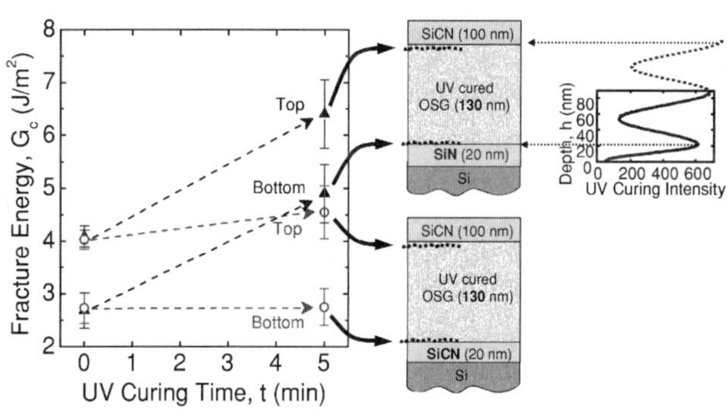

Figure 3. Fracture energies at the top and bottom interfaces of a 130 nm thick OSG before and after UV curing. The effects of SiN and SiCN underlying layers are investigated. Maximum UV curing intensity is located at the top and bottom interfaces of the OSG that has a SiN underlying layer.

than the SiCN layer. The thicknesses of 65 nm (Fig. 2) and 130 nm (Fig. 3) of OSG films were selected to locate the maximum UV curing intensity at top of the OSG films. The measured fracture energy and stiffness variation through the film thickness were consistent with the predicted by the standing wave model [4]:

$$U(z,t) = \int_0^t I\left(h(t), n(t), z\right) dt \tag{1}$$

where resulting UV cure intensity, U, in terms of the film depth z following cure duration t is simulated involving the change of film thickness h, and refractive index n.

In the presentation, details of a new standing wave model, UV curing depth profiles by FM-AFM, and multi-functional ultra-low-k (MF-ULK) layers with a single post-deposition UV curing process will be discussed. MF-ULK layers have high adhesion and O_2-ashing tolerance at the bottom, ultra-low dielectric constant through the film thickness, and high elastic modulus and fracture resistance at the top to endure CMP processing.

CONCLUSIONS

This study demonstrates that depth profiles of low-k dielectrics can be tuned to have dramatically improved fracture energy at both top and bottom interfaces by the use of the UV standing wave and the underlying optical spacer layer that helps to locate the maximum UV intensity at the low-k interfaces. SiN and SiCN underlying layers were tested for the optical spacer. The SiN layer is transparent to the UV light and resulted in the maximum UV intensity at the both interfaces of the low-k layer, and therefore significantly increased interface fracture energy. However, SiCN had no effect on the bottom interface fracture energy because it strongly absorbs the UV light. Significant difference in the depth profiles of the low-k layers were characterized by FM-AFM.

ACKNOWLEDGMENTS

The authors acknowledge assistance with sample preparation by Yvonne Ritz of AMD in Dresden. This work was supported by the Director, Office of Energy Research, Office of Basic Energy Sciences, Materials Sciences Division of the U.S. Department of Energy, under Contract No. DE-FG02-07ER46391.

REFERENCES

1. T. Kim, N. Tsuji, N. Kemeling, K. Matsushita, and R. H. Dauskardt, Proceedings of Advanced Metallization Conference 2006, 389-394 (2007).
2. T. Kim, N. Tsuji, N. Kemeling, K. Matsushita, N. Kobayashi, D. Chumakov, E. Zschech, and R. H. Dauskardt, Proceedings of Advanced Metallization Conference 2007, 455-458 (2008).
3. T. Kim, N. Tsuji, N. Kemeling, K. Matsushita, D. Chumakov, H. Geisler, E. Zschech, and R. H. Dauskardt, Journal of Applied Physics **103,** 064108 (2008).
4. T. Kim, N. Tsuji, K. Matsushita, N. Kobayashi, D. Chumakov, H. Geisler, E. Zschech, and R. H. Dauskardt, Journal of Applied Physics, In press (2008).

Study of Chip-Package Interaction and Crackstop Structures for Cu/Ultra low-k Interconnects

Xuefeng Zhang[1], Scott Smith[1], Rui Huang[2], and Paul S. Ho[1]
[1]Microelectronics Research Center, UT, Austin, TX, 78758
[2]Aerospace Engineering & Engineering Mechanics Dept., UT, Austin, TX, 78758

ABSTRACT

In this paper, the impact of CPI on the mechanical reliability of Cu/Ultra low-k interconnects in a flip chip package was investigated using a multi-level sub-modeling method. The crack growth behavior in a six-level Cu/low-k interconnect was simulated. The crack propagation path was calculated using the crack tip opening displacement method (CTOD) and based on a maximum hoop stress criterion. The crack was found to propagate from interconnect towards the Si substrate driven by shear stress and the calculated energy release rate (ERR) increases as crack grows. The result agrees well with our experimental observations. The effect of structure reinforcement such as crack stop on fracture resistance was also investigated. A decrease in the energy release rate was achieved by adding dummy Cu patterns into the low-k interconnect. The fracture toughness of crack stop test structures was measured using a Modified Edge Liftoff Test (MELT) method. The results showed a significant increase in the fracture resistance demonstrating that the crack stop structure can effectively suppress the crack propagation to improve the mechanical reliability of Cu/ultra low-k interconnects.

INTRODUCTION

The exponential growth in device density requires continuous scaling of the interconnect structure for high-performance microprocessor chips. This leads to the implementation of new material, process and design for interconnect and packaging structures. The current effort of the semiconductor industry is focused on implementing ultra-low k (ULK) dielectric material with k < 2.5 into Cu interconnects to reduce the RC delay[1]. The weak mechanical properties of the ULK dielectrics and the increase in the interconnect level raise serious reliability concern for Cu/ultra low k interconnects. Interfacial delamination caused by chip-packaging interaction (CPI) is commonly observed in low k interconnects after die assembly into a flip-chip package [2,3]. A previous study revealed that the CPI-induced crack driving forces for interfacial delamination can increase substantially when the dielectric material is switched from fully dense low-k to ultra low-k materials as shown in Fig. 1 [4,5]. The effect was found to depend more on the elastic modulus, E, of the low-k material than the coefficient of thermal expansion (CTE).

In this paper, we extended the study of CPI to a 6 level Cu/low-k structure and calculated the crack propagation path using a CTOD method based on a maximum hoop stress criterion. Local crack stops were added into the model and their effect on crack propagation was studied. Finally, the fracture behavior of Cu/low-k interconnect was measured using the MELT test and the effect of crack stop structures was investigated. The reliability impact on ULK interconnects and their design rules for 45nm technology and beyond will be discussed.

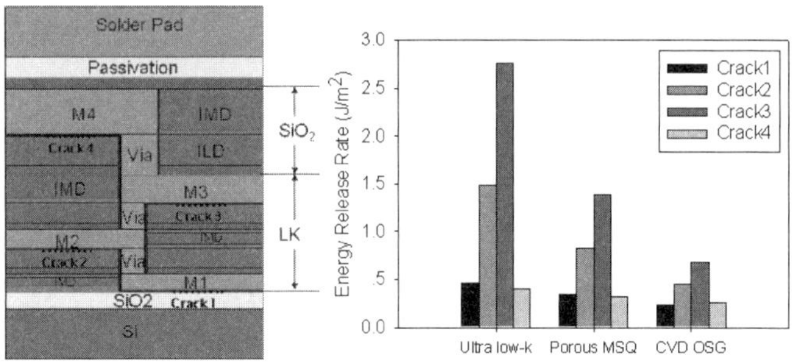

Fig. 1. ERR comparison between different low-k dielectrics under CPI [4]

SIMULATION

Sub-modeling technique

Multi-level sub-modeling technique was employed in this study in order to bridge the gap in feature size between the package and the chip [2-5]. We first developed a 2D FEA model based on a 3-level sub-modeling technique as shown in Fig.2. Level 1 is the package level for investigating the overall thermal deformation for the flip-chip package. Simulation results for this package level were verified with experimental results obtained from moiré interferometry. Level 2 is a sub-model with much finer meshes and focus on the critical solder bump region at the outermost chip corner. Level 3 contains detailed features for the interconnect structure with six metal levels. In this structure, the pitch and line dimensions in the first two metal levels (M1 and M2) are doubled in the third level (M3) while the fourth level (M4) is 1.5x the dimension of M3. This simulates approximately the hierarchical levels in real interconnect structures. Crack tip opening displacement method and the maximum hoop stress criterion were employed to study the crack propagation behavior in low-k interconnect [6,7,8]

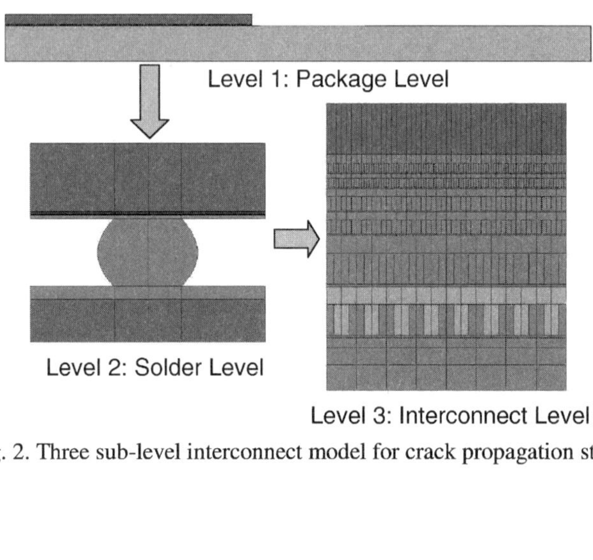

Fig. 2. Three sub-level interconnect model for crack propagation studies

EXPERIMENT

Modified Edge Liftoff Test (m-ELT) was used to determine the effect of fracture toughness due to the implementation of crack-stop structures into low k interconnects. To prepare the specimen, a thick layer of epoxy was deposited on top of Si wafer and then cured for one hour at 177°C (Fig. 3). After curing, the sample was diced into 1 cm x 1 cm coupons. These diced specimens were placed in a cooling chamber where debonding can be observed upon cooling with liquid nitrogen. The temperature at which debonding occurred was recorded and the corresponding residual stress can be extrapolated using a calibrated residual stress vs. temperature curve. The strain energy release rate can be calculated by Eq. 1 [9,10], where σ_e is the residual stress in the epoxy coating layer; h is is the epoxy thickness and E and v are Yong's modulus and Poisson's ratio.

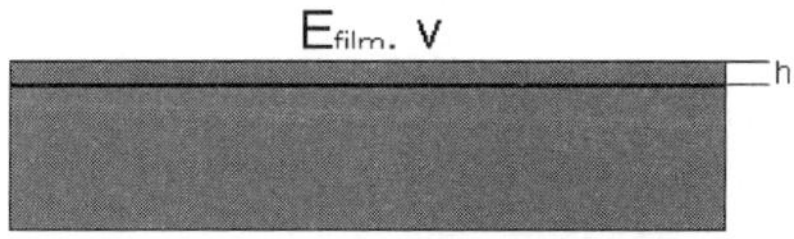

Fig. 3. Schematics of m-ELT sample

$$G_c = \frac{\sigma_e^2(T)\cdot h}{2\overline{E}_{film}}$$

$$\overline{E}_{film} = \frac{E_{film}}{1-\nu^2} \qquad \text{Eq. (1)}$$

RESULTS AND DISCUSSION

Crack Propagation and Crack Stop

An example of crack propagation in a real interconnect structure due to CPI is shown in Fig. 4. In general, the crack propagates along a path of maximum G/Γ, the ratio between the energy release rate and the fracture toughness [7]. Therefore, the crack propagation not only seeks a path with the largest energy release rate, but also favors a path with the lowest fracture toughness, either interfacial or cohesive. Depending on the local material combination and geometry, an interfacial crack may kink out of the interface, causing cohesive fracture of low-k materials. As shown in Fig.4, the crack propagated from the upper levels to the lower levels, eventually causing failure by die cracking [11]. Due to the complexity in the materials and structures, modeling of crack propagation in multilevel interconnects has not been well developed. In this paper, a 2D multi-level structure was used to study the crack propagation. The crack was assumed to initiate at an upper-level interface, which had been shown to have a higher energy release rate compared with that in a lower-level interface. The CTOD method was employed to calculate the mode mixity at each crack tip as crack growed and then the maximum hoop stress criteria was used to predict the crack propagation direction. The crack propagation path obtained by simulation was plotted in Fig. 5, propagating from the upper levels to the lower levels. The actual path may not be exactly as shown since it can be affected by process defects and material inhomogeneity in the low-k interconnects. Nevertheless, the overall crack behavior can still be deduced from the simulation. The result showed that as the crack propagates toward the lower levels and the total crack length increases, the energy release rate increases, indicating an unstable crack growth.

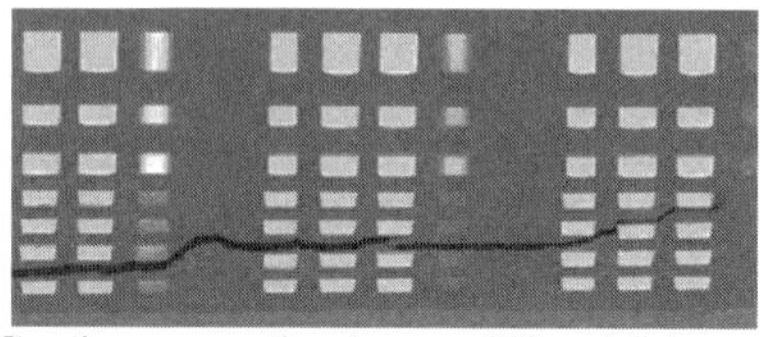

Fig. 4.Crack propagation in a multilevel interconnect

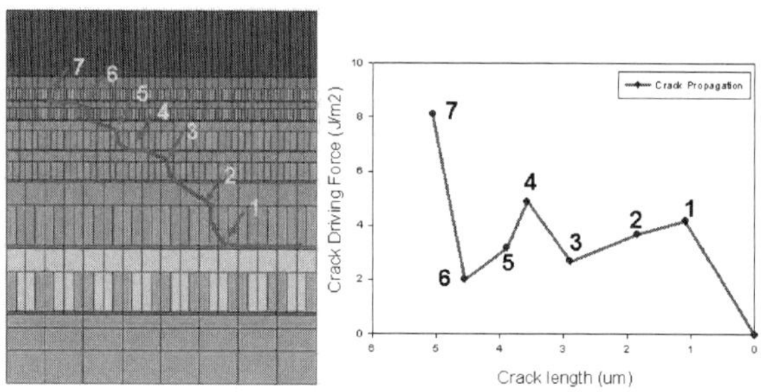

Fig. 5. ERR for crack propagation

A major challenge in packaging Si die is to prevent cracks propagating from the die edge to the active area of a chip during packaging assembly . One way to suppress crack propagation is to implement crack stop structures. The effect of crack stop structures was analyzed in this study. Dummy Cu barriers at via levels were added into Cu/low-k interconnect as local reinforcements. The simulation result showed that the crack driving force was suppressed by the crack stop structure as shown in Fig. 6. Our experiments confirmed the toughness increase due to the implementation of the crack stop structure, thus demonstrating its effectiveness in improving the mechanical reliability of low-k interconnect.

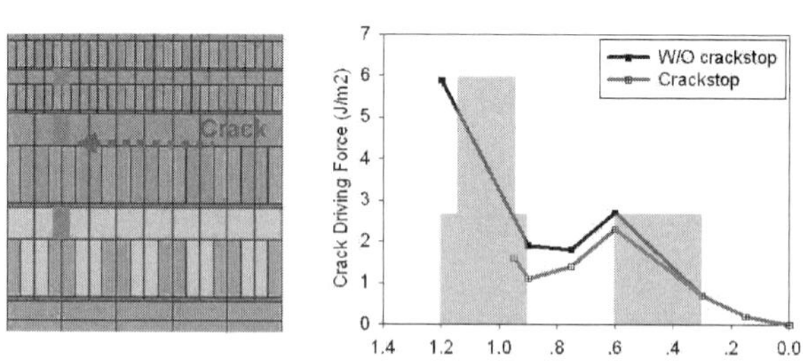

Fig. 6. Effect of crackstop structure on ERR

Fracture Toughness Measurement

Several crack stop designs were tested and the results were plotted in Fig. 7. Significant increase in the fracture resistance was observed for chips with crack stop structures to about 20 J/m2 as compared with chips without crack stop structures, usually about 1~4 J/m2 [12]. This demonstrated the effectiveness of the crack structure in preventing dicing cracks at the die edge from propagating into the active interconnect structure. Failure analysis of these samples yielded a crack propagation path from upper interconnect level to lower level as shown in Fig.8, which confirmed our simulation results.

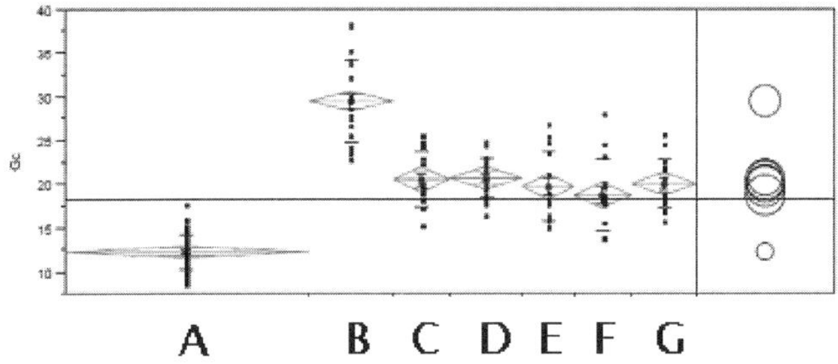

Fig. 7: Fracture toughness measurement of crackstop by MELT

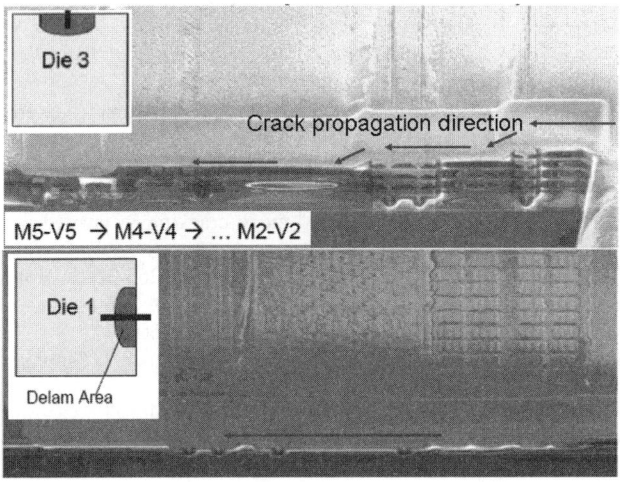

Fig. 8: Failure analysis of failed samples

CONCLUSIONS

The impact of CPI on the mechanical reliability of Cu/Ultra low-k interconnects in a flip chip package was investigated using a multi-level sub-modeling method. Modeling showed that crack propagated from the upper-level interconnect towards Si substrate under CPI, which agrees well with our experimental observations. A decrease in the energy release rate was achieved by adding dummy Cu structure at the via level into the low-k interconnect. The effect of crack-stop structure on fracture toughness was measured by m-ELT tests and the result confirmed that such structures can be very effective in suppressing crack propagation to improve the mechanical reliability of Cu/ultra low-k interconnect.

ACKNOWLEDGMENTS

The authors would like to thank Prof. H. F. Nied and Dr. Murat from Lehigh University for discussions of the crack propagation calculations. We'd also like to thank the Semiconductor Research Cooperation for support of this project.

REFERENCES

1. ITRS Roadmap 2007
2. L. Mercado, S.-M. Kuo, C. Goldberg, S.-M. Kuo, T.-Y. Lee, Proc 53rd Electronic Components and Technology Conference, 1784 (2003)
3. G. Wang, P. S. Ho and S. Groothuis, Microelectronics Reliability, Vol. 45, 2005.
4. X. Zhang, S. Im, R. Huang, P. S. Ho, *Electrical, Optical, and Thermal Interconnections for 3D Integrated Systems, chapter 2, 2008*
5. C. J. Uchibori; X. Zhang, P. S. Ho and T. Nakamura, International Interconnect Technology Conference, 196 (2006)
6. D. Broek, in *Elementary engineering fracture mechanics*, (Sijthoff & Noordhoff International Publishers, The Netherlands, 1978), p. 215
7. J.W. Hutchinson and Z. Suo, Mixed-mode cracking in layered materials, Advances in Applied Mechanics 29, 63-191 (1992).
8. Z. Suo, " Reliability of interconnect structures." pp. 265-324 in Volume 8: Interfacial and Nanoscale Failure (W. Gerberich, W. Yang, Editors), Comprehensive Structural Integrity (I. Milne, R.O. Ritchie, B. Karihaloo, Editors-in-Chief).
9. J. Im, E. Shaffer, T. Stokich. A. Strandjord, J. Hetzner, J. Curphy, etc., J. of Electronic Packaging, Volume 122, Issue 1, pp. 28-33
10. M. Chiang, W. Wu, J. He and E. J. Amis, Thin Solid Films, Volume 437, Issues 1-2, 2003, pp. 197-203
11. T. M. Shaw, E. Liniger, G. Bonilla, J. P. Doyle, B. Herbst, X. H. Liu, M. W. Lane, "Experimental determination of the toughness of crack stop structures," IEEE International Interconnect Technology Conference, 2007.
12. X. H. Liu, T. M. Shaw, M. W. Lane, E. G. Liniger, B. W. Herbst, D. L. Questad, "Chip-package interaction modeling of ultra low-k/copper back end of line," International Interconnect Technology Conference, 2007.

Conference Proceedings AMC XXIV © 2009 Materials Research Society

A FAST NEW APPROACH FOR EVALUATING THE CONNECTIVITY OF MICROPORES (<2 nm) AND SUPERMICROPORES (<0.5 nm) IN LOW-K THIN FILMS

Eiichi Kondoh[1*], Shosaku Aruga[1], Fuminori Ito[2], and Yoshihiro Hayashi[2]
[1] Department of Mechanical System Engineering, Interdisciplinary Graduate School of Medicine and Engineering, University of Yamanashi, 4-3-11 Takeda, Kofu 400-8511, Japan.
[2] LSI fundamental Research Laboratory, NEC Electronics Corporation, 1120 Shimokuzawa, Sagamihara, Kanagawa, 229-1198, Japan
* Corresponding author

ABSTRACT

This paper reports a novel method for evaluating the pore connectivity of porous low-k thin films. In this technique, in-situ ellipsometric measurements were carried out in a supercritical medium that dissolves a probe organic solute, and optical changes of low-k films upon the condensation of the probe solvent were detected. Different low-k films having a dielectric constant of 2.2–3.1 were examined using ethanol or acetone as a probe solute at a pressur of 10 MPa and temperatures of 40–100 °C. Large changes in ellipsometric parameters were observed, in any condition studied, when spin-on-dielectric films that have micropores were used. Measurement conditions for films with a smaller pore size were explored and the detection of supermicropores in plasma-enhanced chemical-vapor-deposited SiOCH was succeeded.

INTRODUCTION

Currently, the standard method for reducing the dielectric constant (k) of inter-metal dielectrics is to lower film density and/or introduce porosity. Both of these approaches may invoke water uptake, poor chemical resistance, and etching gas permeation. These issues are critical for "open" and "connected" pores, and a great effort has thus been made to embed isolated micro-/supermicropores [1] in the matrix [2,3]. For instance, molecular pore-stacked SiCOH (MPS) film, which is fabricated by plasma enhanced polymerization and contains supermicropores surrounded by siloxane rings, has been developed [2].

A limited number of methods are currently available for evaluating the pore structures of low-k thin films. Ellipsometric porosimetry (EP) [4] has high sensitivity and pore-size resolution but supposedly cannot easily detect supermicropores (<0.5 nm) [5]. Positronium annihilation lifetime spectroscopy (PALS) [6] is capable of detecting pores of such small sizes, but its instrumentation is too complicated to enable its use as a common inspection tool.

In-situ ellipsometry in supercritical media has been developed in one of the author's laboratory (EK) and an investigation has been conducted to detect adsorptive condensation in the pores of low-k thin films [7]. Supercritical fluids possess both good penetration capability and high density. We thus expect that using a supercritical fluid (SCF) as an adsorbent medium will allow the evaluation of supermicropores in porous low-k films. In this paper, we report results for PECVD, porous SOD, and MPS films.

PRINCIPLES

Ellipsometry is a technique used to determine the thicknesses and refractive indices of thin films. When the pores of a porous thin film are filled with a certain substance, the film's refractive index is changed from its original value. An ellipsometer itself determines the ellipsometric parameters, Ψ and Δ

$$\Psi = \arctan\left|\frac{R_p}{R_s}\right|$$

and

$$\exp(j\Delta)\tan\Psi = \frac{R_p}{R_s}$$

where R_p and R_s and are the complex reflectivity of p-polarized and s-polarized waves, respectively. Ψ represents the absolute ratio of p-polarized and s-polarized wave amplitudes, and Δ represents the phase shift between p-polarized and s-polarized waves. Therefore, Ψ and Δ change when adsorption occurs.

The refractive index of a porous thin film will not change after penetration of a supercritical fluid because the refractive index of the fluid is as low as that of the open pores. For this reason, a probe solute (adsorbate) is dissolved in the fluid so that the changes in Ψ and Δ upon solute condensation can be monitored (see Fig. 1). Note that the incorporation of the solute does not change the refractive index of the medium [7].

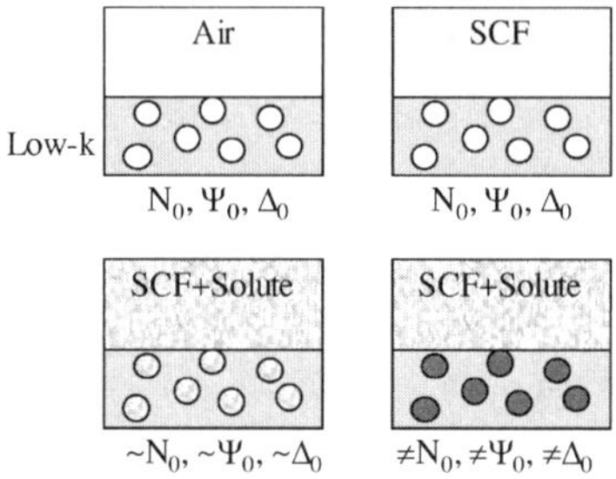

Fig. 1. Change in refractive index due to pore condensation

PROCEDURE

Figure 2 shows the experimental setup used in this study. A spectroscopic ellipsometer with a CCD spectrometer and a high-pressure optical cell was used. The incident angle was fixed at 70°. Pressurized CO_2 and a solute chemical were supplied to the cell; the pressure was fixed at 10

MPa using a back pressure regulator (BPR). Acetone, methanol, and ethanol were used as solutes. Table 1 lists the low-k films tested in this study. All films were formed on a Si wafer. In experiments, the wafers were cut into small pieces.

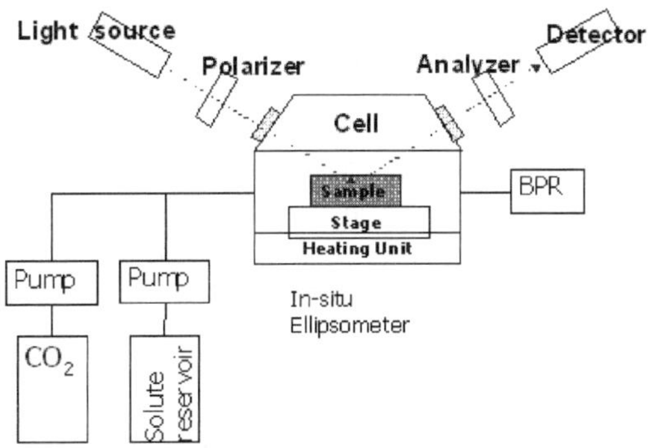

Fig. 2. Experimental setup

Table 1 List of films used in this work

Material	Thickness (nm)	Dielectric constant
PECVD SiOCH (MPS)	200	2.5
PECVD SiOCH#1	120	2.7
PECVD SiOCH#2	200	3.1
SOD-SiOCH #1	400	2.3
SOD-SiOCH #2	200	2.2

RESULTS AND DISCUSSION

Figure 3 shows the change in Δ during the supply of acetone as a solute at 100 °C. The specimen used was SOD-SiCOH #2 and the wavelength was 633 nm. We can see that Δ changed more as solute concentration increased, and that Δ traced the same value paths when solute concentration was reduced. Ψ behaved in the same manner besides its change of direction. (The experiment of Fig. 3 conducted for a rather long time, however, ten minutes is long enough to confirm whether a change occurs.) No remarkable differences were observed when thermally-grown oxide was used instead under the same conditions. These observations clearly indicate that the probe solute (acetone in this case) condensed in the film's open, connected pores.

A series of tests were carried out in the same manner on the other specimens listed in Table 1, with different solutes and temperatures, and the results are summarized in Table 2. Plus signs (+) indicate that changes in ellipsometric parameters were confirmed, whereas minus signs (–) indicates that no change was observed. It is clear that solute penetration and condensation occur more easily on SOD films. Plasma CVD SiCOH films were more penetration resistant. MPS films exhibited excellent penetration resistance.

Figure 4 shows the relationship between pore diameter and pore connectivity determined by using PALS analysis, for some of the materials used in the present work. The original graph was presented in reference 5. Material C is MPS, E and D correspond to PECVD-SiCOH, and SOD indicates SOD#1. The ease with which penetration and condensation were ellipsometrically detected (Table 2) accords well with the order of pore connectivity shown in Fig. 4. MPS has near zero pore connectivity, which fairly agrees with our observation.

It is seen in Table 2 that a lower temperature was needed to detect smaller pore size and connectivity. The SCF is a dense compressive medium and its density is very sensitive to pressure and temperature around the critical point. The critical point of CO_2 is 7.4 MPa and 31 °C, and even at 10 MPa, the CO_2 density changes greatly when the temperature is decreased from 100 °C to 50 °C. Therefore, it is presumed that a higher number density of the solute molecule is needed to fill small pores. Figure 5 show the relationship between the pore connectivity and the minimum fluid density required for pore detection. Obviously, the detection of largely connected pores is easier and does not require a high (molecular) density. The effect of the kind of solute is more or less reasonable, because ethanol has a smaller molecular size than acetone.

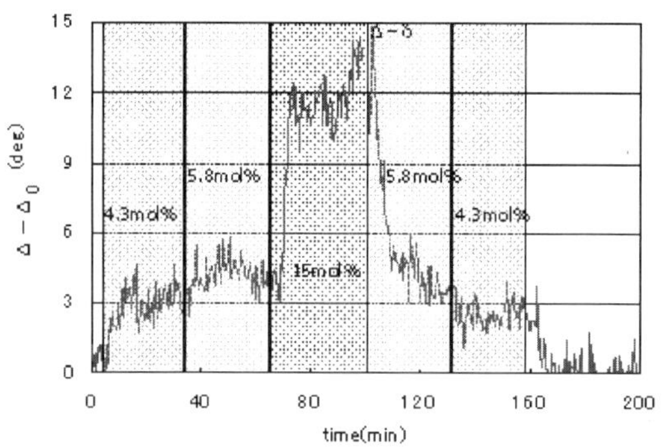

Fig. 3. Δ as a function of time during the supply of a probe chemical. Data at 633 nm.

Table 2 Summary of in-situ ellipsometric study in SCF. +/– indicates detected/undetected.

Probe	Temp. (°C)	MPS k=2.5	PECVD#1 k=2.7	PECVD#2 k=3.1	SOD#1 k=2.3	SOD#2 k=2.2	t-SiO2
Ethanol	100	–	–	–	+	+	–
Acetone	100	–	–	–	+	+	–
	50	–	+				
	40	–	+	+	+		–

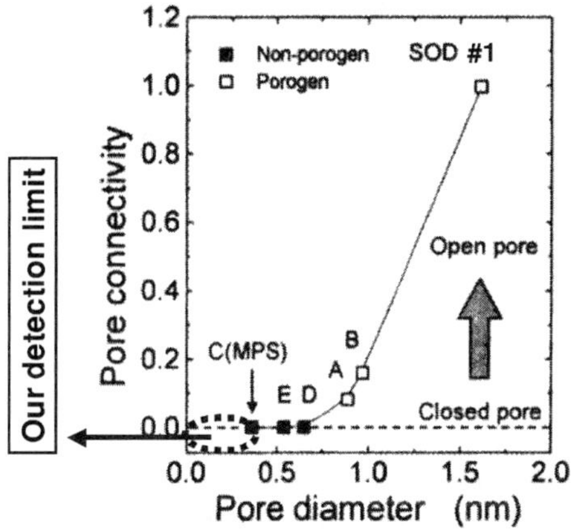

Fig. 4. Relationship between the pore diameter and pore connectivity of used materials. Adopted from ref. [5].

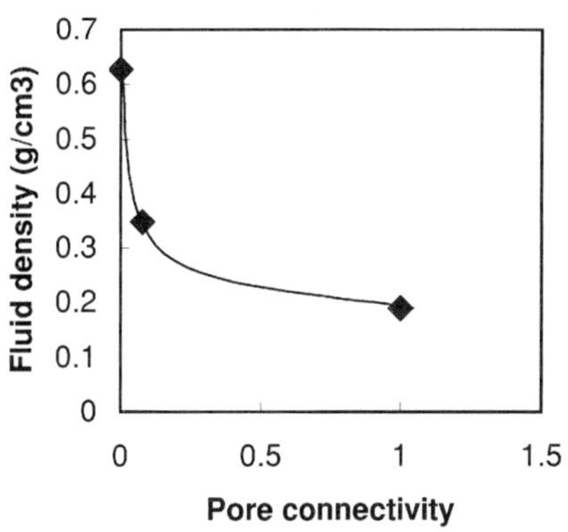

Fig. 5. Relationship between the pore connectivity and the minimum fluid density required for pore detection.

CONCLUSION

We studied the penetration and condensation of solute chemicals in different porous low-k films using in-situ ellipsometry within supercritical media. We confirmed changes in the ellipsometric parameters originating from the condensation of solutes in the pores and resulting in increases in the refractive index. Condensation was more readily observed in films with higher pore connectivities and larger pore diameters. MPS exhibited the best penetration resistance, suggesting its better chemical resistance. In conclusion, in-situ "supercritical ellipsometry" is a fast and sensitive technique for evaluating the connectivity of supermicropores in low-k thin films.

REFERENCES

1. IUPAC classification.
2. Y. Hayashi, F. Itoh, Y. Takeuchi, M. Tada, M. Tagami, H. Ohtake, K. Hijioka, S. Saito, T .Onodera, D. Hara and K.Tokudome, in Proc. IEEE IITC 2004.
3. T. Nakamura and A. Nakashima, in Proc. IITC 2004.
4. M.R. Baklanov, K. P. Mogilinikov, V. G. Polovinkin, and F. N. Dultsev, J. Vac. Sci. Technol. B18, 1385 (2000).
5. F. Ito, T. Takeuchi, H. Yamamoto, T. Ohdaira, R. Suzuki, and Y. Hayashi, in Ext. Abst. ADMETA/AMC 2007.
6. D. W. Gidley, E. Frieze, T. L. Dull, J. Sun, A. F. Yee, and C. V. Nguyen, Appl. Phys. Lett. 76, 1282 (2000).
7. E. Kondoh, E. Ukai, and S. Aruga, phy. stat. solidi c5, 1219 (2008).

Conference Proceedings AMC XXIV © 2009 Materials Research Society

Structural Analysis of Low-k porous SiOC Film by Electron Energy Loss Spectroscopy (EELS)

M. Shimada[1)], A. Kojima[2)], Y. Otsuka[3)], Y. Shimizu[3)], H. Seki[3)], N.Nakamura[2)], T. Yoda[1)]
1) Process & Manufacturing Engineering Center, Semiconductor Company, Toshiba Corporation
2) Center for Semiconductor R&D, Advanced BEOL Technology Dept., Semiconductor Company, Toshiba Corporation
3) TORAY RESEARCH CENTER, Inc..
8, Shinsugita-cho, Isogo-ku, Yokohama, Kanagawa 235-8522, Japan
Tel:+81-045-776-4580 Fax:+81-45-776-4100 e-mail:miyoko.shimada@toshiba.co.jp

ABSTRACT

In this paper, we investigate plasma induced damage of porous SiOC film in a patterned sample by electron energy loss spectroscopy (EELS) combined with scanning transmission electron microscopy (STEM). By using optimum conditions, it is possible to distinguish plasma process induced damage from TEM sample preparation and observation damage. In a trench patterned porous SiOC film, the Si-L spectra show the threshold peak shift to the higher energy side while approaching to the sidewall. It seems that the peak shift concerns with C bonding to Si.

INTRODUCTION

For a high-end logic device inter layer dielectric (ILD) film in Cu interconnects beyond the 45nm node, various porous low-k films (k=2.0-2.5) have been studied. However, porous low-k films have an inherent disadvantage in that the film property is degraded due to the plasma process induced damage. The "damage" here means deterioration of film properties affected by changes of film structures. The damages induced by plasma processes which decompose Si-CH_3 and C-H bonds in the SiOC film have been widely known to cause moisture absorption. Consequently the dielectric constant increases and it might affect performances and reliability of the interconnects. An analysis of damaged film was carried out by using X-ray photoelectron spectroscopy (XPS), Fourier transform infrared spectroscopy (FT-IR)[1] and so on, but they could not analyze nanostructure. Also, TEM-EELS for high spatial resolution analysis is used to study damaged porous low-k SiOC film with no distinction between observation damage and process induced damage. In addition, the process induced damage is estimated by the measurement of elemental ratio [2], and the damaged structure has not yet been explained. In this paper, the plasma process induced damage of porous SiOC film in a patterned sample is investigated by structural analysis using the TEM-EELS method as TEM sample preparation and observation damages are distinguished.

EXPERIMENTAL

Plasma induced damage of porous SiOC film was investigated in a patterned sample, as shown in Figure 1. The sample structure consists of porous polyarylene (PAr) (k=2.3) for the upper layer, and porous SiOC (k=2.3) for the lower layer [3]. A trench pattern was formed especially in both layers. The optimum sample preparation and observation conditions for TEM-EELS were used to detect the plasma process induced damage. Si-L and C-K spectra were measured at the different positions from the trench sidewall to study the process damage of porous low-k SiOC structure.

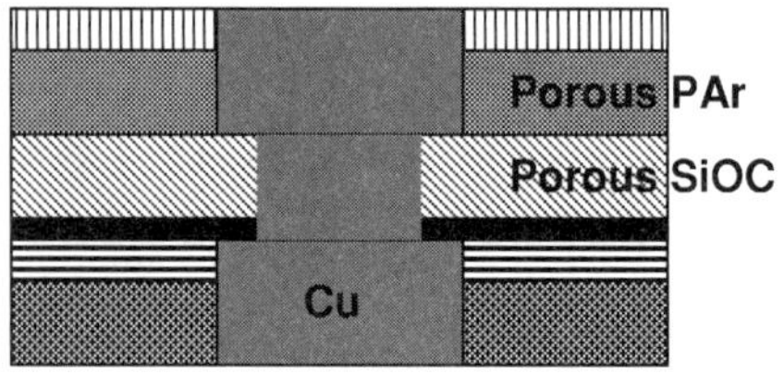

Figure 1 The schematic cross section of the patterned sample.

RESULTS AND DISCUSSION

Figure 2 shows the atomic concentration of C/Si and the C-K spectra of porous SiOC blanket films using the optimum TEM sample preparation and observation conditions compared with conventional conditions. The C/Si values are almost equal between the two conditions. However the peak around 285eV in the C-K spectrum did not appear clearly by using optimum conditions. This peak in C-K spectra suggests that the structure has a carbon double bond, however porous low-k SiOC materials in this study have no carbon double bond originally. Therefore we speculated that the peak had been generated by TEM sample preparation and observation. By using optimum conditions, it is possible to suppress the peak generation and distinguish plasma process induced damage from TEM sample preparation and observation in porous SiOC film.

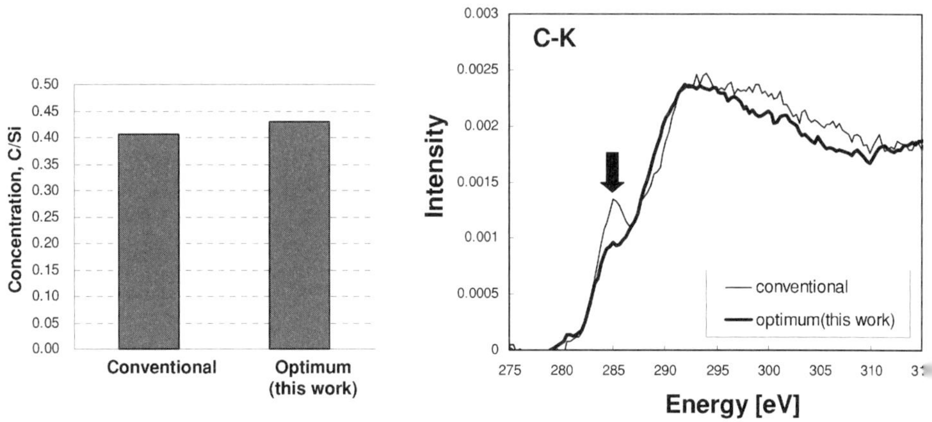

Figure 2 C/Si and C-K spectra of porous SiOC branket film using optimum conditions compared with conventional conditions.

Figures 3 show the atomic concentration of C/Si, O/Si, the Si-L and C-K spectra obtained by TEM-EELS analysis as a function of distance from the sidewall in patterned samples. It was found that the atomic concentration of C/Si decreased and O/Si increased gradually to the side-wall area. The peak around 285eV in C-K spectra appears clearly in the area close to the sidewall (5-50nm), though damages due to TEM sample preparation and observation are suppressed as shown in Fig.2. From C/Si and C-K spectra in Fig.3, it is obvious that the plasma process induced damage was introduced from the sidewall to at least 50nm distance in porous SiOC film. While the Si-L2,3 peak at about 107eV has not changed, the threshold position moved towards higher energy (side). It is proven that the porous SiOC structure was changed to SiO2 structure in the area close to the sidewall (5-20nm) by comparison with figure 4 that shows Si-L spectra from various ILD blanket films. In Fig.4, the peak at about 107eV seems to be broader and shift to the low energy side as O bonding with Si decreases. The increase of O/Si close to the sidewall region is related to increase of oxygen that bonds with Si in SiOC film, but does not bond with H in H2O.

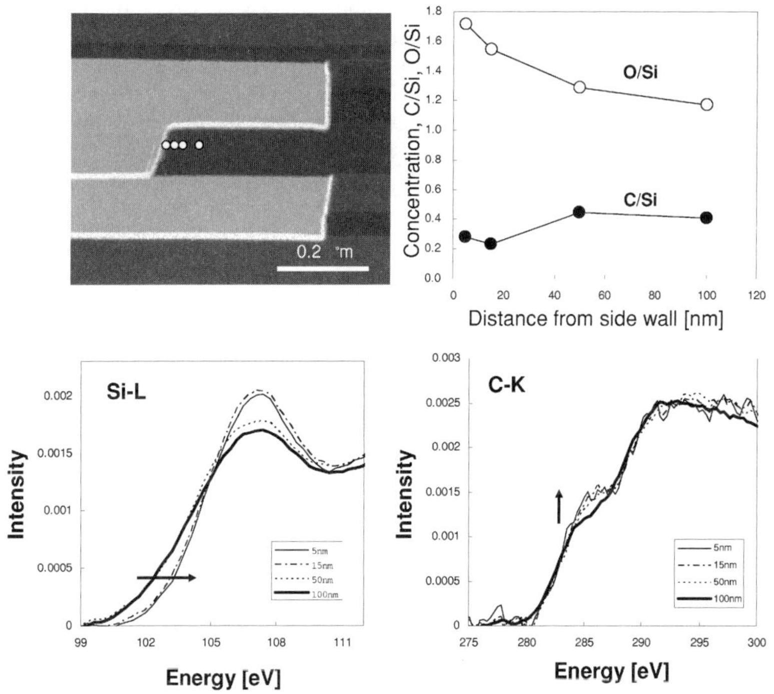

Figures 3 C/Si, O/Si concentrations, the Si-L and C-K spectra as a function of distance from sidewall.

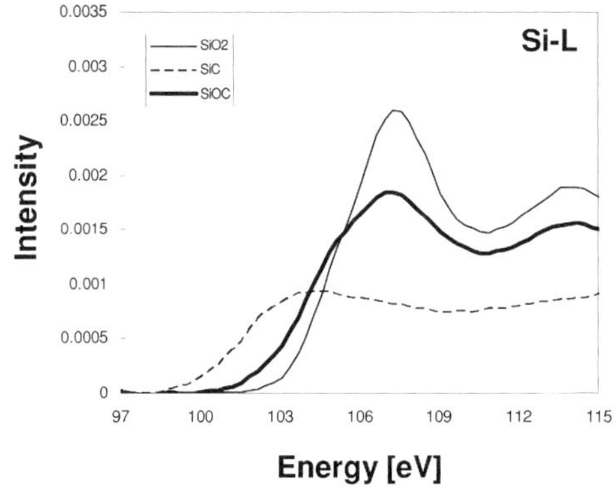

Figure 4 Si-L spectra of ILD films.

Figures 5 show atomic concentration of C/Si, O/Si and the Si-L spectra obtained by TEM-EELS analysis as a function of distance from the sidewall in a patterned sample with restoration treatment after dry etching. The atomic concentration of C/Si increases in the area close to the sidewall (5-20nm) compared with Fig.4 and is nearly the same at all regions. The peaks at 285eV in the C-K spectra are suppressed compared with Fig.3 due to the restoration treatment. However the only Si-L spectrum in the sidewall shifts to the higher energy region. Figure 6 shows Si-L spectra in the sidewall with and without restoration treatment. It is suggested that carbons by restoration treatment do not quite bond to Si as Si-CH3 in SiOC film.

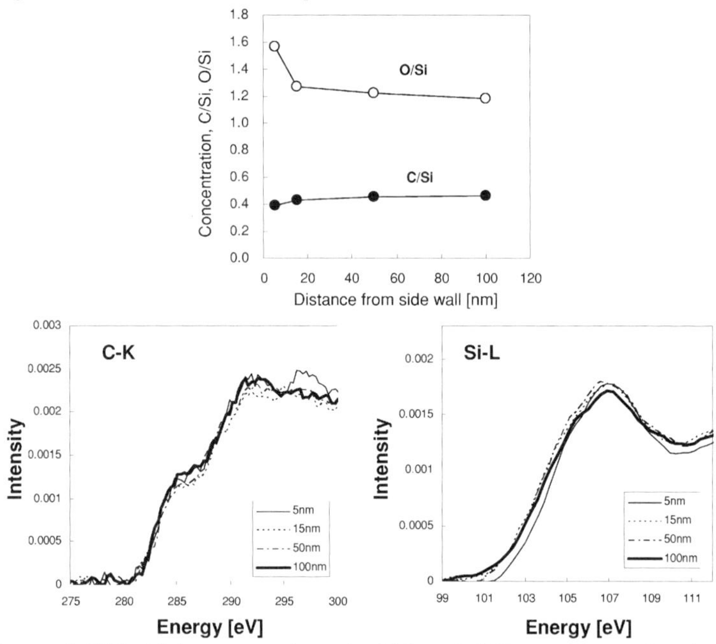

Figures 5 C/Si, O/Si concentrations, the Si-L and C-K spectra as a function of distance from sidewall in a patterned sample with restoration treatment.

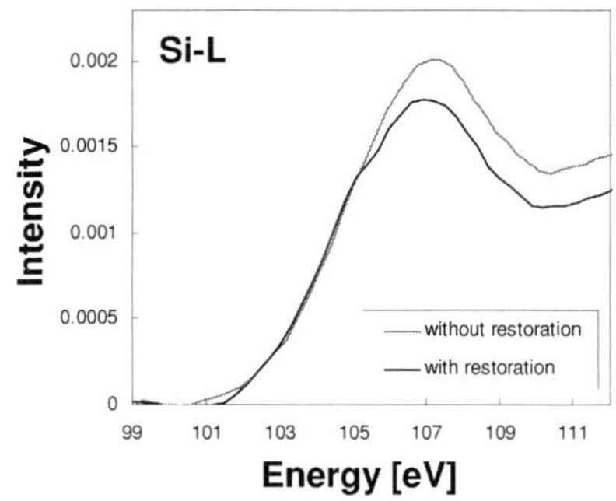

Figure 6 Si-L spectra in sidewall with and without restoration treatment.

SUMMARY

Porous low-k SiOC film damage was analyzed by TEM-EELS using optimum conditions. The spectra of Si-L were shifted by C bonding to Si. It is suggested this method is effective for analyzing the damage structure of low-k films in patterned sample and useful for improving the plasma process development.

References

[1] H.Seki et al., in Proc. ADMETA2004, p.34
[2] T.J.Dalton et al., in Proc. 2004IITC, p.154
[3] N.Matsunaga et al. in Proc. 2005IITC, p.6

Conference Proceedings AMC XXIV © 2009 Materials Research Society

Characterization of Defects Generated in SiCN Dielectrics for Copper Diffusion Barriers

Kiyoteru Kobayashi, Taketoshi Ide, and Yusuke Takahashi
Tokai University, 1117 Kitakaname, Hiratsuka, Kanagawa 259-1292, Japan
PHONE: +81-463-58-1211, E-mail: kkbys@keyaki.cc.u-tokai.ac.jp

ABSTRACT

We have characterized defects generated in silicon carbon nitride (SiCN) films deposited using PECVD by electron spin resonance (ESR) and electrical measurements, and have also studied current conduction through the films. Paramagnetic defects were generated via exposing the SiCN films to 4.9-eV UV illumination. On the other hands, the variation of the paramagnetic defect density after exposing the films to 3.4-eV UV illumination was small. The characteristics of the ESR signals suggest that the generated defects are neutral silicon dangling bonds. The defect generation can be explained by the "silicon charge conversion" model.

Due to 4.9-eV UV illumination, the build-up of negative charge in the SiCN film and the substantial increase in the leakage current through the film also occurred. The negative charge creation is attributed to electron injection from the silicon substrate. The generated paramagnetic defects are responsible for the increase in the leakage current.

INTRODUCTION

Silicon carbon nitride (SiCN) films deposited using plasma-enhanced chemical vapor deposition (PECVD) techniques have been receiving significant recent attention as copper diffusion barriers and etching stoppers for copper multilevel interconnects in ultra-large scale integration (ULSI) [1-3]. The dielectric constant of the SiCN films is about 5, which is larger than those of low-dielectric constant (low-k) interlayer dielectrics (~2.3). Therefore, the SiCN films used in copper interconnects need to be as thin as possible for minimizing the R-C time delay. In order to achieve excellent dielectric strength in such thin films, the origin and nature of defects in the films are crucial issues.

Electron spin resonance (ESR) is a powerful tool to investigate the microscopic nature of defects in dielectric films. In a previous study using ESR [4], the carbon-related defects, with a g value of 2.003 and ESR linewidth of 0.07 mT, were observed in thick SiCN ceramics derived from the pyrolysis of polymer precursors above 1000 °C. Unfortunately, there are few reports on defects in low-temperature PECVD SiCN films. In this work, we have characterized defects in PECVD SiCN films by ESR and electrical measurements, wherein ultraviolet (UV) light was used to activate the defects. Current conduction through the films is also studied.

In addition, ultra-violet (UV) illumination curing of the copper diffusion barrier dielectrics has emerged as a promising technique for suppressing stress-induced voiding failures [3]. Characterization of the defects generated in the SiCN films subjected to UV illumination is very important to establish the UV cure technology.

EXPERIMENTAL

P-type (100)-oriented silicon substrates were used in this study. Two types of samples were prepared: (1) SiCN-Si substrate structure and (2) SiCN-SiO_2-Si substrate structure. SiCN films with 200 nm in thickness were grown by a PECVD technique at 400 °C using $Si(CH_3)_4$ and NH_3 gases. The SiO_2 film with 10 nm in thickness was grown by thermal oxidation of the silicon substrates. UV-light irradiation was carried out at room temperature using two kinds of Hg arc lamps with monochromatic lights of 254 nm (4.9 eV in energy) and 365 nm (3.4 eV in energy) in wavelength.

ESR measurements were carried out at room temperature using a JEOL JES-FA200 X-band spectrometer. The external magnetic field was modulated by 0.3 mT at a frequency of 100 kHz. We took proper caution to ensure that the resonance was not saturated. Defect densities were derived by comparing the double numerical integral of the measured derivative power absorption spectrum with that obtained for a calibrated standard. The g value is defined as the zero-crossing magnetic field and the width of the spectrum, ΔH_{pp}, is defined as a peak-to-peak width.

A mercury probe was used to provide a temporary metal electrode onto both the SiCN-Si and SiCN-SiO_2-Si structures. Capacitance-voltage (C-V) measurements were done at a frequency of 100 Hz. Current-voltage (I-V) characteristics were measured using an Agilent 4157B semiconductor parameter analyzer.

RESULTS AND DISCUSSION

Defect Generation and Negative Charge Accumulation in SiCN films

Figure 1(a) shows ESR derivative power absorption spectra observed before and after exposing the SiCN-Si structure to 4.9-eV UV illumination. The intensity of the ESR signal was increased after the SiCN film was exposed to 4.9-eV UV illumination. The characteristics, g value and a peak-to-peak width (ΔH_{PP}), of the signals are 2.003 and 1.2 mT for the UV-illuminated films. Figure 2 shows the paramagnetic defect density derived from ESR signals as a

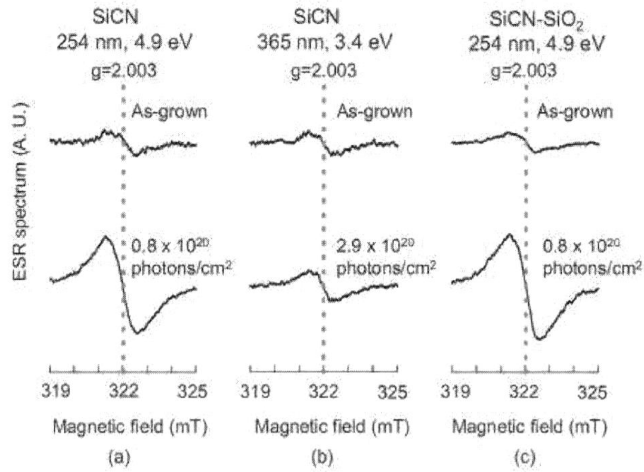

Fig. 1 ESR spectra of SiCN-Si structure after 4.9-eV UV illumination (a), SiCN-Si structure after 3.4-eV UV illumination (b), and SiCN-SiO_2-Si structure after 4.9-eV UV illumination (c).

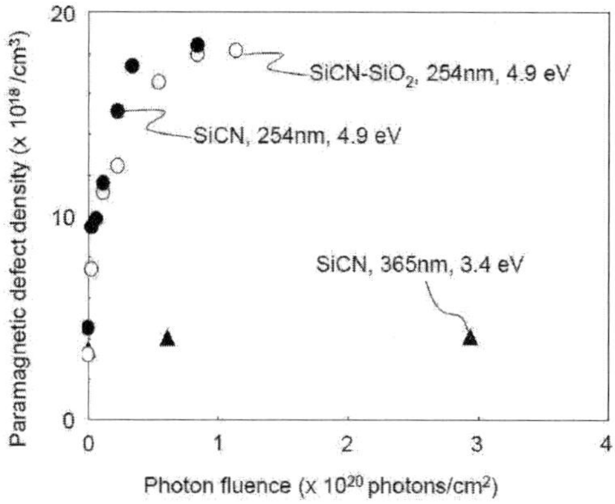

Fig. 2 Paramagnetic defect density as a function of UV radiation dose.

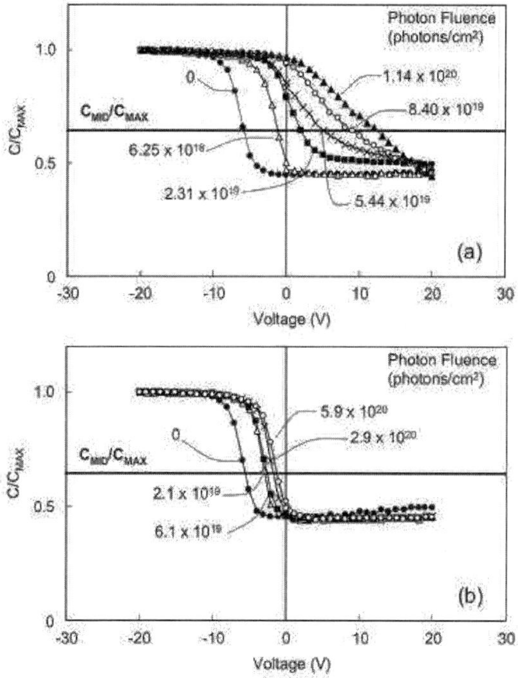

Fig. 3 C-V characteristics of SiCN-Si structure after 4.9-eV UV illumination (a), and SiCN-Si structure after 3.4-eV UV illumination (b).

function of UV radiation dose. The paramagnetic defect density of the SiCN film was substantially increased due to 4.9-eV UV illumination and then saturates at 1.8 x 10^{19} cm^{-3}. On the other hands, as can be seen in Fig. 1(b), the variation of the ESR signals after exposing the SiCN-Si structure to 3.4-eV UV illumination was small. The paramagnetic defect density was increased by only 7 x 10^{17} cm^{-3}, as shown in Fig. 2.

In the previous studies on N-rich PECVD silicon nitrides [5, 6], it has been reported that paramagnetic defects, characterized by a g value of 2.0028 and a ΔH_{PP} of 1.3 mT, were generated by UV illumination at room temperature, which are identified to be neutrally charged silicon dangling bonds ($\cdot Si \equiv N_3$). This silicon dangling bond center is often termed K^0 center. It has been also investigated that the defect generation occurs due to the UV lights with photon energies higher than 4.1 eV [5].

The characteristics of the defect center observed from the SiCN films are very similar to those of the K^0 center, because of the following: (1) a large number of paramagnetic defects appear due to 4.9-eV UV illumination at room temperature, (2) the g value of the observed ESR signals is 2.003, which is close to that of the K^0 center, (3) the ΔH_{PP} values of the observed signals are 1.2 mT, which are close to that of the K^0 center. These results suggest that the observed paramagnetic defect center in our specimens is the neutral silicon dangling bond.

Next, Figures 3(a) and 3(b) show C-V characteristics measured before and after exposing the SiCN-Si structures to UV illumination of 4.9 and 3.4 eV. The as-grown SiCN films without UV exposure were positively charged. As shown in Fig. 3(a), the C-V curve was shifted toward the positive voltage direction after 4.9-eV UV exposure. This shift indicates the appearance of net negative charge in the SiCN film. We also observe the stretch-out in the C-V curve, due to the generation of interface traps. In the case of 3.4-eV UV illumination, the curve shifts were small, as can be seen in Fig. 3(b).

Figures 4(a) and 4(b) show the leakage current in the SiCN-Si structure versus applied voltage. The leakage current density at a constant gate voltage (-84 V) is plotted in Fig. 5 as a function of UV radiation dose. The current through the SiCN film was substantially increased

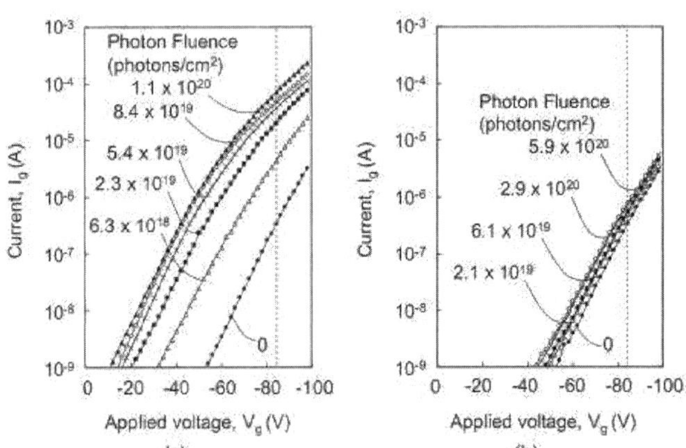

Fig. 4 Conduction current versus applied voltage. The current through the SiCN film was substantially increased due to 4.9-eV UV illumination (a), while the variation of the current was small in case of 3.4 eV (b).

due to 4.9-eV UV illumination, while the variation of the current was small in the case of 3.4-eV UV illumination. We consider that the paramagnetic defects generated by 4.9-eV UV illumination are responsible for the substantial increase in leakage current [7].

Defect Generation in SiCN-SiO_2 Stacked Films

Figure 6(a) illustrates a possible generation process of the negative charge in the SiCN-Si structure. High-energy UV illumination will produce electron-hole pairs in the silicon substrate. The electrons that are generated with enough energy to overcome the potential barrier at the SiCN-Si interface will be emitted into the conduction band of the SiCN film. The emitted electrons can be trapped by the defect centers, which will create a negative charge. The emission probability of electrons can be a function of the energy of the hot electrons generated in the silicon substrate, which is a function of the incident photon energy. Thus the emission probability of electrons may depend on the photon energy of UV light.

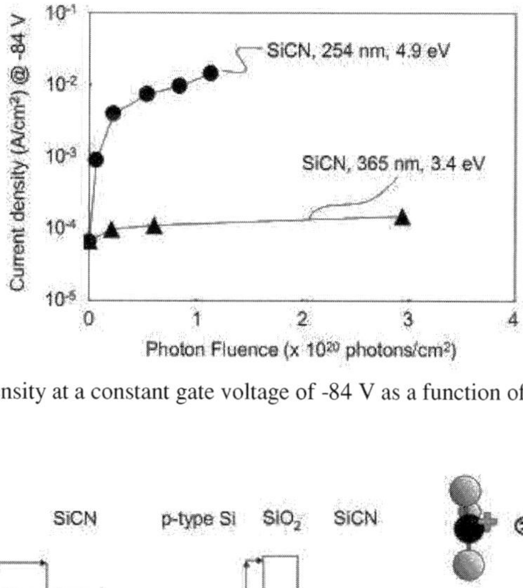

Fig. 5 Current density at a constant gate voltage of -84 V as a function of UV radiation dose.

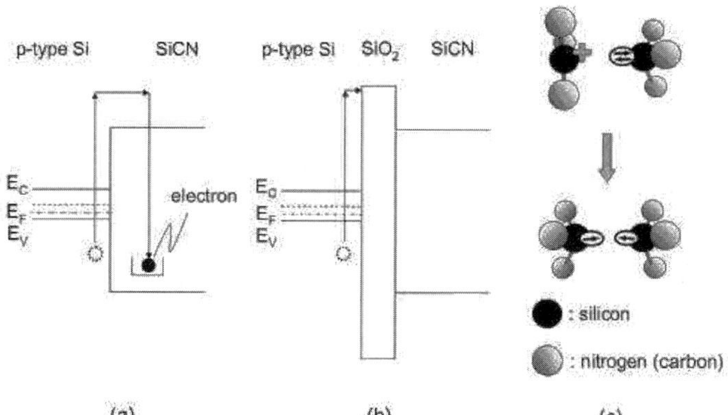

Fig. 6 Possible generation processes of the negative charge and the paramagnetic defects due to high-energy UV illumination.

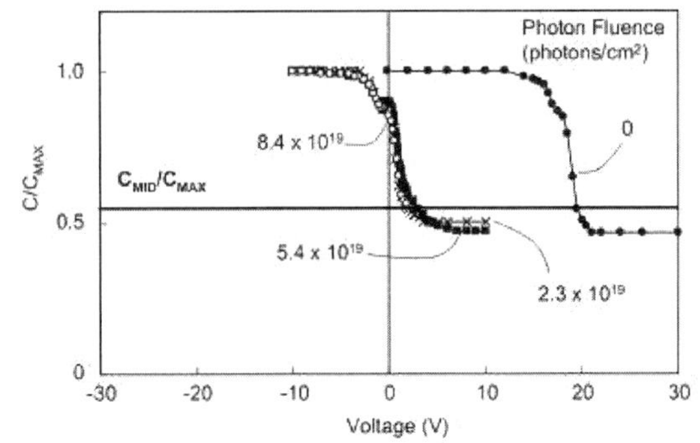

Fig. 7 C-V characteristics of $SiCN\text{-}SiO_2\text{-}Si$ structure before and after 4.9-eV UV illumination.

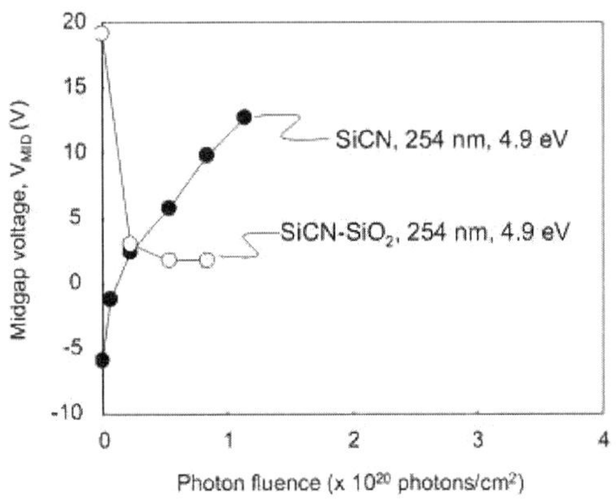

Fig. 8 Midgap voltage obtained from C-V characteristics as a function of UV radiation dose.

In order to examine the above assumption, we have investigated C-V characteristics of the $SiCN\text{-}SiO_2\text{-}Si$ structure subjected to UV illumination. The SiO_2 layer will become a higher energy barrier for the electron emission from the silicon substrate, as denoted in Fig. 6(b). In Fig.7, we illustrate C-V characteristics before and after exposing the $SiCN\text{-}SiO_2\text{-}Si$ structure to 4.9-eV UV illumination. We can see from the C-V curve before UV exposure that the $SiCN\text{-}SiO_2$ stacked film was negatively charged. This indicates that, during PECVD of the SiCN film onto the SiO_2 film, a lot of electrons were captured in the stacked film. The C-V curve was shifted toward the negative voltage direction due to the UV radiation dose of 2.3×10^{19} cm^{-2}. A large portion of the captured electrons seems to have been emitted from the stacked film. However the curve was not shifted by the subsequent UV exposure. In Fig. 8, we demonstrate the midgap voltage obtained from C-V characteristics shown in Fig. 3(a) and Fig. 7 as a function of UV radiation dose. The midgap voltage is defined as the gate voltage at which the middle of the silicon band gap at the insulator-silicon interface corresponds to the Fermi level of the silicon.

It is evident from the results indicated in Fig. 8 that negative charge accumulation did not occur in the SiCN-SiO_2-Si structure. Thus we claim that, in the SiCN-Si structure, the electrons emitted from the silicon substrate due to high-energy UV exposure were captured at neutral traps in the SiCN film.

Finally, we refer to ESR spectra, shown in Fig. 1(c), observed before and after exposing the SiCN-SiO_2-Si structure to 4.9-eV UV illumination. In Fig. 2, we also plot the paramagnetic defect density in the SiCN-SiO_2-Si structure versus UV radiation dose. As mentioned in the above discussion, the SiO_2 film plays a role of the energy barrier for the electron emission from the silicon substrate. However, a large number of paramagnetic defects were generated after 4.9-eV UV illumination even in this structure. This result means that the generation of the paramagnetic defects is due to changes of charge state of defects in the SiCN films. Figure 6(c) illustrates a possible generation process of the paramagnetic defects. This model is based on the change in spin and charge state of pre-existing positively and negatively charged silicon centers which have no unpaired electrons and are ESR inactive [6]. Due to high-energy UV illumination, both the charged centers will change into the neutral silicon dangling bonds which are ESR active. Thus we suggest that the generation of the paramagnetic defects is attributed to such a "silicon charge conversion" model.

CONCLUSION

Paramagnetic defects are generated via exposing the SiCN films to 4.9-eV UV illumination. The characteristics of the ESR signals suggest that the generated defects are neutral silicon dangling bonds. The defect generation can be explained by the "silicon charge conversion" model. The generated paramagnetic defects are responsible for the substantial increase in the leakage current due to 4.9-eV UV illumination. On the other hands, the accumulation of negative charge simultaneously occurs in the SiCN films. This phenomenon is due to electron trapping in the films. The hot electrons produced by high-energy photons are emitted from the silicon substrate, and are captured at traps in the SiCN film.

ACKNOWLEDGMENTS

We are indebted to Mr. M. Matsuura, Dr. K. Goto, and Dr. K. Asai for providing the SiCN films. We would like to express our sincere thanks to Mr. Y. Miyamoto, Mr. T. Haraki, Prof. Y. Kamigaki, and Prof. T. Izumi, for their helpful discussions. We would also like to thank Mr. T. Osawa, Mr. H. Yokoyama, Mr. K. Ukai, and Mr. Y. Mori for their cooperation. We are grateful to the Technical Service Coordination Office and the Advanced Analysis Center in Tokai University for the ESR measurements.

REFERENCES

1. C. C. Chiang *et al.,*: Jpn. J. Appl. Phys., **42** (2003) 5246.
2. H. Cui and P. A. Burke: J. Electrochem. Soc., **151** (2004) G795.
3. K. Goto *et al.,*: 2006 *International Interconnect Technology Conference , 2006, p.95.*
4. S. I. Andronenko, I. Stiharu, and S. K. Misra: J. Appl. Phys., **99** (2006) 113907.
5. D. T. Krick, P. M. Lenahan, and J. Kanicki: J. Appl. Phys., **64** (1988) 3558.
6. W. L. Warren *et al.,*: J. Electrochem. Soc., **143** (1996) 3685.
7. K. Kobayashi, H. Yokoyama, and M. Endoh, Applied Surface Science, **254** (2008) 6222.

Conference Proceedings AMC XXIV © 2009 Materials Research Society

Wideband frequency in-situ characterization of ultra thin ZrO_2 films for integrated MIM capacitors

Thomas Bertaud[1,2], Thu-Trang Vo[1], Cédric Bermond[1], Thierry Lacrevaz[1], Christophe Vallée[2], Yves Morand[3], Bernard Fléchet[1], Alexis Farcy[3], Mickaël Gros-Jean[3], and Serge Blonkowski[3]
[1]IMEP-LAHC Université de Savoie, Le Bourget du Lac, 73376, France, Metropolitan
[2]LTM UMR CNRS 5129, 17 avenue des Martyrs, Grenoble Cedex 9, 38054, France, Metropolitan
[3]STMicroelectronics, 850 rue Jean Monnet, Crolles Cedex, 38926, France, Metropolitan

ABSTRACT

High-K dielectrics are largely investigated as promising candidates to increase the capacitance density of Metal Insulator Metal (MIM) capacitors in advanced circuits: densities above 5 fF.μm^{-2} are expected in 2010 [1]. However, High-K properties are expected to vary with manufacturing process and also frequency as relaxation and resonance mechanisms occur [2]. It is necessary to analyze and evaluate High-K behavior from DC to microwave frequency. This paper deals with the characterization of zirconium oxide (ZrO_2) for the realization of MIM capacitors. The extraction of the ZrO_2 complex permittivity is performed from DC up to 5 GHz using dedicated test vehicles allowing an "in-situ" characterization. The oxide ZrO_2 is deposited by PEALD and four thicknesses are investigated: 10, 20, 32 and 45 nm.

INTRODUCTION

High permittivity insulators (High-K) are progressively introduced in high-speed integrated passives in order to optimize circuit performances. Metal Insulator Metal (MIM) are widely embedded for Radio Frequency (RF) blocks, Analog Mixed Signals (AMS) circuits, decoupling capacitors, or Dynamic Random Access Memory (DRAM), thus making MIM capacitors key components for the development of System on Chip (SoC). High-K dielectrics with a very stable permittivity versus frequency are needed to cover such a wide range of applications. As a consequence, low frequency and high frequency characterization techniques have to be mixed in order to extract insulator properties on a wide range of frequencies.

From SiO_2 (K ~ 3,9), the High-K dielectric in MIM has evolved using Si_3N_4 (K ~ 7) and Al_2O_3 (K ~ 10) to HfO_2 (K ~ 22) or Ta_2O_5 (K ~ 25) [3]. This paper focuses on the "in-situ" characterization of the zirconium oxide (ZrO_2, K ~ 35) integrated in MIM capacitors according to a damascene architecture, i.e. with ZrO_2 integrated in the same stack as the final MIM component one using industrial processes [4]. First, low frequency measurements have been carried out, such as I(V) and C(V) in order to determine the properties of this material: voltage breakdown, quadratic and linear voltage coefficients α and β. Then, frequency dependent measurements have been performed in two ranges to extract the permittivity and loss tangent of ZrO_2: i) in low frequency domain (from 10^{-3} Hz to 10^{+5} Hz) and ii) in microwave frequency domain (from 40 MHz to 5 GHz) thanks to dedicated MIM test devices.

EXPERIMENTAL PROCEDURE

The ZrO_2 films were grown on a TiN (60nm) /SiO_2 (200 nm) /Si substrate by Plasma Enhanced Atomic Layer Deposition (PEALD). Four ZrO_2 thicknesses are investigated: 10, 20,

32 and 45 nm. Top electrode of the MIM capacitors and dedicated test vehicles, allowing an "in-situ" microwave characterization, were realized in patterned aluminum deposited on W/TiN top of ZrO_2 or SiO_2 else (Figure 1).

Current versus voltage measurements were performed on a HP4155 Analyzer. C(V) and low frequency measurements (i.e. from 1 mHz to 100 kHz) were realized on a Novocontrol dielectric spectrometer coupled with a heated chuck for measures in temperature. Parasitic pads are de-embedded. High frequency measurements (i.e. from 40 MHz to 5 GHz) are carried out using an Anritsu 37387C Vector Network Analyzer and a de-embedding procedure as proposed in [5].

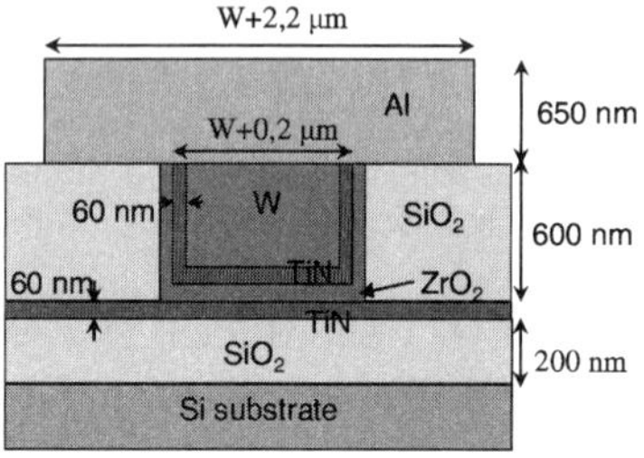

Figure 1: Sectional view of the stack

RESULTS AND DISCUSSION

Low frequency result

First, I(V) electrical characterization have been carried out in order to determine the material voltage breakdown, which is in the range of 3,5 $MV.cm^{-1}$ (Figure 2).

Next, capacitor voltage linearity, which is an important parameter for MIM capacitor, is revealed with measurements performed for two different temperatures (25°C and 125°C). Figure 3 shows the normalized capacitance of the MIM capacitor with applied voltage for three thicknesses, at the frequency of 10 kHz and 25°C. The normalized capacitance is fitted by:

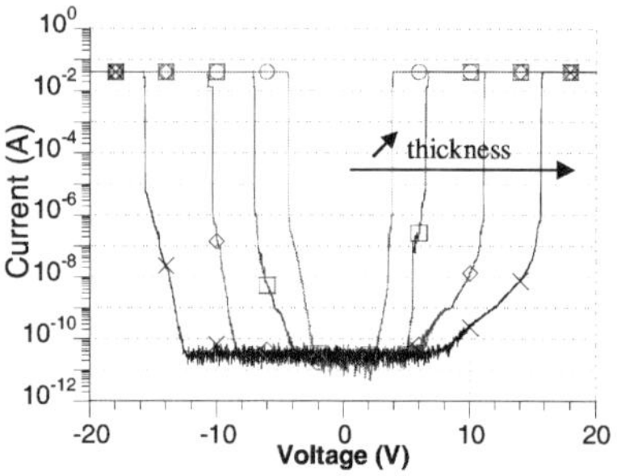

Figure 2: I(V) measurements at T=25°C

$$\frac{\Delta C}{C_0} = \frac{C(V) - C_0}{C_0} = \alpha V^2 + \beta V$$

where C_O is the zero-biased capacitance and α and β the quadratic and linear coefficients, respectively, in $ppm.V^{-2}$ and $ppm.V^{-1}$ (the voltage coefficients capacitance, VCC). Results are in line with specifications for thicknesses of 32 nm and above (Table 1). However, the normalized capacitance variations with temperature are shown in the inset of Figure 3 at the frequency of 10 kHz for the thickness of 45 nm. The temperature coefficient of capacitance (TCC or α_T, in $ppm.°C^{-1}$) can be determined with the fit:

$$\frac{\Delta C}{C_0} = \frac{C(T) - C_0}{C_0} = \alpha_T T.$$

Values / Thickness	Breakdown Voltage [V]	Surface Capacitance [$fF.\mu m^{-2}$]	VCC α [$ppm.V^{-2}$] 25°C	VCC α [$ppm.V^{-2}$] 125°C	VCC β [$ppm.V^{-1}$] 25°C	VCC β [$ppm.V^{-1}$] 125°C	TCC α_T [$ppm.°C^{-1}$]	α_E [$m^2.V^{-2}$]
10 nm	-4,3 & 3,8	30,2	-	-	-	-		-
20 nm	-7,0 & 6,5	16,6	1451	1143	2831	1457	403	$5,8.10^{-19}$
32 nm	-10,4 & 11,1	10	443	515	605	1118		$4,5.10^{-19}$
45 nm	-15,7 & 15,7	6,7	215	201	51	41		$4,3.10^{-19}$

Table 1: VCC and TCC for the ZrO_2

The last column shows the coefficient alpha function of the electrical field, α_E, calculated according to [6]: $\alpha_E = \alpha \cdot t^2$. This coefficient is constant and in accordance with results of [6] for 32 and 45 nm ZrO_2 thicknesses; and it rises for the 20 nm thickness. This effect has to be linked with the $\Delta C / C$ curves asymmetry (whereas the MIM is symmetrical) and could be a consequence of a material difference (upper leakages currents) or an interfacial TiO_x layer at the bottom electrode level, result of a TiN oxidation.

All these results are obtained at 10 kHz and values at 100 kHz show a generally observed evolution: a decrease of the TCC and VCC when frequency increases [3] and [7].

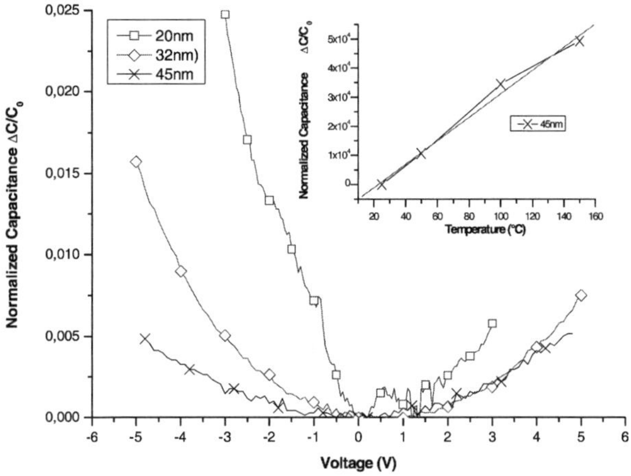

Figure 3: C(V) and C(T) measurements at 10 kHz.

RF measurement procedure

High frequency measurements were carried out on dedicated test devices: one-port MIM including contact pads, access line and the MIM capacitor with an S area, as shown in Figure 4. The equivalent circuit model is established and presented in Figure 5: the MIM capacitor (Device Under Test, DUT) is modeling with a capacitor basic model C and a shunt conductance G.

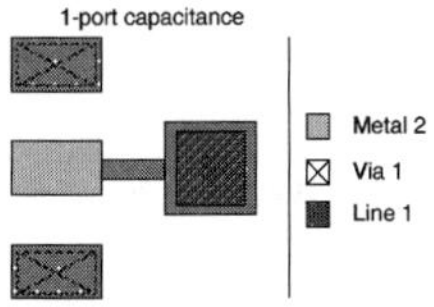

Figure 4: Top view of the device under test in microwave frequency domain.

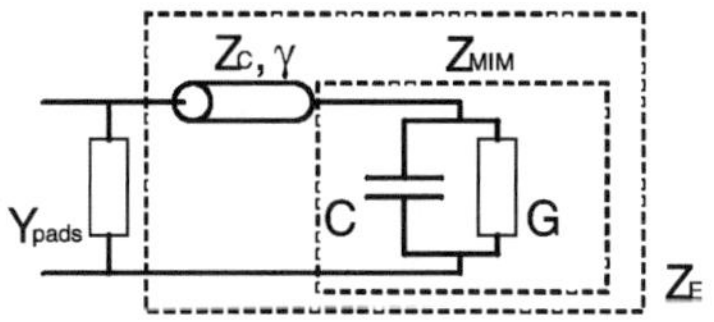

Figure 5: Equivalent circuit of the complete device under test.

First, an on-wafer OSTL (Open Short Thru Load) calibration is achieved and the scattering parameters of the complete device and of the de-embedding structures were measured using an Anritsu 37387C Vector Network Analyzer (VNA) from 40 MHz to 5 GHz.

The MIM capacitor impedance is given thanks to a de-embedding procedure proposed in [5]. After a de-embedding of the pads (Y_{pads}) and with the measurement of the line cascade ABCD parameters ($[ABCD]_{line}$ matrix), we find the characteristic impedance Z_C and the propagation exponent γ of the access line. The input impedance Z_E, which is the impedance of the MIM capacitor seen through the access line, is determined with scattering parameter S_{11}:

$$\textit{line characteristics}: \begin{cases} \gamma = ar\cosh\left(\dfrac{D_{line}}{l_{line}}\right) \\ Z_C = \sinh\left(\dfrac{\gamma l}{C_{line}}\right) \end{cases} \quad and \quad Z_E = \left[\frac{1-S_{11}}{50\cdot(1+S_{11})} - Y_{pads}\right]^{-1}$$

Finally, the MIM impedance (Z_{MIM}) and the values C and G of our model are determined by:

$$Z_{MIM} = \frac{Z_C \tanh(\gamma l_{access}) - Z_E}{Z_E \dfrac{\tanh(\gamma l_{access})}{Z_C} - 1} \quad and \quad Y_{MIM} = \frac{1}{Z_{MIM}} = G + j\omega C$$

Wideband frequency results

By mixing the high frequency extraction method, previously described, with low frequency measurements, we can get a characteristic of the capacitance C and the conductance G of the MIM device over a wide range of frequencies: from 1 mHz to 5 GHz. The permittivity and the loss tangent of the material are determined on all this frequency range by applying the relations:

$$\tan\delta = G/_{C\omega} \text{ and } \varepsilon_{ZrO_2} = {C \cdot t_{ZrO_2}}/_{\varepsilon_0 S}.$$

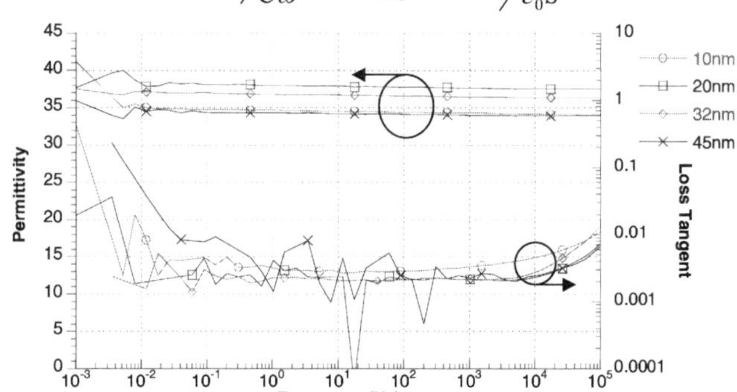

Figure 6: Permittivity and loss tangent from 1 mHz to 100 kHz for different thicknesses

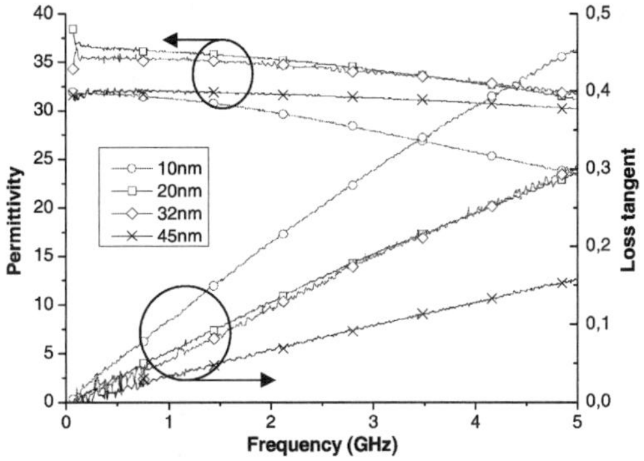

Figure 7: Extraction results (permittivity and loss tangent) from 40 MHz to 5 GHz

Frequency results on Figure 6 and Figure 7 show an excellent agreement between low and high frequencies and a stable capacitance with a rise towards the very low frequencies and a decline in high frequency for the smallest thickness. This is a consequence of a relaxation effect of the material in very low frequency [8]: measurements as a function of temperature verify this thermally activated phenomenon. In low frequency domain, loss tangent kept stable around 10^{-2} and in microwave domain, losses remain within 10^{-2} to 10^{-1} and are in agreement with low frequency results. These high values of dielectric losses could be a consequence of the thin TiN layer as the bottom electrode that induces leakages.

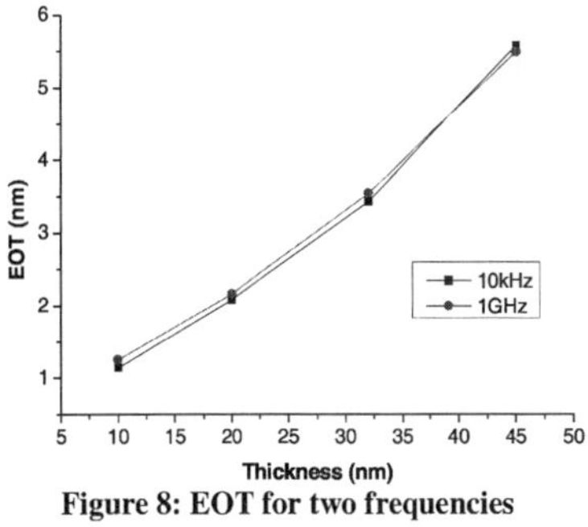

Figure 8: EOT for two frequencies

Then, the curve of the Equivalent Oxide Thickness (EOT) as a function of the ZrO_2 thickness is plotted for two frequencies (Figure 8), 10 kHz and 1 GHz, thanks to the relation:

$$\frac{\varepsilon_0 \varepsilon_{SiO_2} S}{C} = EOT = \frac{\varepsilon_{SiO_2}}{\varepsilon_{ZrO_2}} t_{ZrO_2}$$

However, the slope of this straight line is linked to the permittivity of the oxide ZrO_2 which is 31 at 10 kHz and 32 at 1 GHz. The dispersion in the permittivity results (Figure 6 and Figure 7) is linked to the uncertainty of the values of thicknesses and Figure 8 can escape this uncertainty.

CONCLUSIONS

To conclude, wideband frequency "in-situ" characterization of the oxide ZrO_2 has been performed using dedicated test devices directly integrated in the back-end of line of industrial process. ZrO_2 exhibits good performance for frequencies up to 5 GHz, with a capacitance density of 6 $fF.\mu m^{-2}$ for the thickness of 45 nm, making it an excellent candidate for MIM capacitors.

ACKNOWLEDGMENTS

The authors would like to acknowledge the CEA-LETI for carrying out the devices and the "Rhône Alpes" region, under the "Mini-π" project, for supporting these researches.

REFERENCES

[1] International Technology Roadmap for Semiconductors, 2006 Update.

[2] P. Robert, Electrical and Magnetic Properties of Materials, Artech House, 1988.

[3] K. C. Chiang, C. H. Lai, A. Chin, T. J. Wang, H. F. Chiu, J.-R. Chen, S. P. McAlister, C. C. Chi, "Very High-Density (23fF/μm2) RF MIM Capacitors Using high- TaTiO as the dielectric", IEEE Electron Device Letters, Vol. 26, N° 10, October 2005.

[4] M. Thomas, J. Piquet, A. Farcy, C. Bermond, J. Torres, T. Lacrevaz, B. Flechet, N. Casanova, C. Perrot, P. Caubet, B. Chenevier, "Characterization and optimization of a new Cu/SiN/TaN/Cu damascene architecture for metal-insulator-metal capacitors", in Microelectronic Engineering, Vol. 82, N° 3-4, p 521-528 (MAM 2005).

[5] Vandamme, E.P., Schreurs, D.M.M.P., Van Dinther, G.: "Improved Three-Step De-Embedding Method to Accurately Account for the Influence of Pad Parasitics in Silicon On-Wafer RF Test-Structures", IEEE Transactions on Electron Devices, Vol. 48, N° 4, april 2001, pp. 737-742.

[6] S. Blonkowski, "Nonlinear capacitance variations in amorphous oxide metal-insulator-metal structures", Applied Physics Letters 91 (2007).

[7] H. Hu, C. Zhu, Y. F. Lu, M. F. Li, B. J. Cho, W. K. Choi, "A High Performance MIM Capacitor Using HfO2 Dielectrics", IEEE Electron Device Letters, Vol. 23, N°9, September 2002.

[8] A. K. Jonscher, "Dielectric relaxation in solids", J. Phys. D:Appl. Phys. 32 (1999) R57-R70.

Conference Proceedings AMC XXIV © 2009 Materials Research Society

Cu Alloy and Cu Bilayer Adhesion Studies on PEALD Ta-C-N Using Al, Ag, Ru and Ta

Henry Wojcik[1], Steffen Strehle[1], Martin Knaut[1], Rainer Kaltofen[2], Ulrich Merkel[1], Mirela Pletea[2], He Jian[1], Irina Fiering[2], Christian Wenzel[1], Heidrun Hiemann[1], Johann W. Bartha[1], and Axel Preusse[3]

[1]Semiconductor & Microsystems Techn. Lab. (IHM), TU Dresden, Noethnitzer Str. 64, Dresden, 01187, Germany

[2]Leibniz Institute For Solid State & Materials Research (IFW), Helmholtzstr. 20, Dresden, 01069, Germany

[3]AMD Saxony LLC & Co.KG, Wilschdorfer Landstraße 101, Dresden, 01109, Germany

ABSTRACT

The scope of this paper is to investigate diverse combinations of PVD and PEALD to produce Cu alloy or Cu bilayer structures using Ag, Al, Ru and Ta as alloying or gluing elements, respectively, on top of PEALD TaCN diffusion barriers (approx. 5 nm). In particular, CuAg(<0.1 at%...2 at%) and CuAl(1 at%...5 at%) thin films were co-sputtered using two magnetrons at a time and verified by GDOES. Ru PEALD films were deposited using Ru(Etcp)2 and a N2/H2 plasma. Before and after annealing the samples at 600°C for 30 min under UHV conditions, in situ XPS and AFM were performed to get detailed information on Cu surface composition and topography. Sheet resistance, (4pb) and SEM investigations were used to obtain further Cu adhesion criterions. While Al addition lowers film agglomeration risk, CuAg(<0.1 %) films do not show an equivalent behavior on PEALD TaCN. In the bilayer concept, a 5-7 nm PVD Ta adhesion layer turned out to be most effective and seems to be mandatory even when a 10 nm Ru layer is inserted, for both PVD and PEALD Ru.

INTRODUCTION

For barrier and seed layer deposition, PVD probably continues to be the method of choice for 45 nm and 32 nm nodes. Nevertheless, there are great challenges in filling through silicon vias (TSV) having aspect ratios of 10:1 and more, as well as contact vias in the 22 nm technology node, even when using ionized PVD. As PEALD Ta-C-N films have proven to be highly conformal, conductive and effective in preventing Cu-diffusion down to a few nm in thickness [1], one approach is therefore to ensure at least a continuous barrier against Cu diffusion via PEALD, and to work on with more relaxed PVD requirements during the following seed layer stack deposition.

In any case, Cu adhesion is important with regard to electromigration (EM) and stress induced voiding (SIV) phenomena at the barrier interface. So far, this issue is a weak point of PEALD TaCN films and consequently one major integration draw back. To overcome this, intentions for some materials investigated here are as follows: Besides the standard material Ta, Ru is in discussion as a gluing element that could enable direct plating of Cu. A CuAg alloy offers a lower resistance as compared to CuAl in conjunction with a possible similar behavior.

EXPERIMENT

Sample Preparation

PEALD TaCN films (5 nm thickness) deposited on blanket SiO2 & SiCOH wafers represent the starting point for all investigations. TBTDET and Ar/H2 rf plasma were used as precursor and reducing gas, respectively, at 250°C and a working pressure of 200 Pa inside a parallel plate chamber during the PEALD process. Hereafter, diverse combinations of PEALD/PVD thin films of Ru, Ta and Cu form a series of glue and seed layers, respectively, as shown in table I. All sputtered PVD films were deposited at approx. 0.5 Pa in an Ar atmosphere. To some part, Cu seed was deposited at $3x10^{-5}$ Pa using evaporation. Ru PEALD films were deposited using Ru(Etcp)2 and N2/H2 rf plasma as precursor and reducing gas, respectively, at 270°C and a working pressure of 150 Pa. Additionally, references on PVD TaN films were prepared via reactive sputtering. In another series CuAl(1 at% + 2 at%) and CuAg(<0.1 at%, 1 at%, 2 at%) alloys were co-sputtered in a PVD chamber having two independent magnetrons and a baffle to control small amounts of material flow. Film compositions are verified through GDOES measurements.

Analysis

Before and after annealing of the samples at 600°C for 30min under UHV conditions, in situ XPS and AFM, SEM and sheet resistance measurements were performed to get detailed information on Cu surface coverage and topography.

Table I Design of experiment, development of sheet resistance and surface roughness (AFM) during annealing at 600°C for 30min. in UHV

TaN Barrier	Glue/Seed layer	Rs in Ω/□ @ RT	Rs in Ω/□ after 30min @600°C	AFM rms a.d.	600°C
Option 1	*Direct Cu*				
PVD 6nm	20nm Cu sputtered in-situ with barrier	2.1	1.9	0.7	1
PVD 6nm	20nm Cu evaporated in-situ w. b.	3.9	760	0.9	9.4
PVD Ta 6nm	20nm Cu sputtered in-situ w. b.	2.1	1.8	0.7	1.1
PEALD 4nm	20nm Cu sputtered in-situ w. b.	2.6 … 2.9	800…1000	0.5	37
Option 2	*Direct Cu alloy*				
PEALD 4nm	20nm CuAl1% sputtered	4	3.5	0.6	1.3
PEALD 4nm	20nm CuAl2% sputtered	4.6	3.6	0.6	0.5
PEALD 4nm	20nm CuAg1% sputtered	3.7	1170	0.5	21
PEALD 4nm	20nm CuAg0,1% sputtered	4.2	1900	0.6	23
Option 3	*Ta adhesion layer*				
PEALD 5nm	7nm Ta / 20nm Cu sputtered	2.3	1.7	0.2	2.2
Option 4	*Ru seed*				
PEALD 5nm	10nm PEALD Ru / 20nm Cu evap.	3.6	1.9	0.8	18
PEALD 5nm	10nm PVD Ru / 20nm Cu evap.	3.4	4	0.5	31
PVD 6nm	10nm PVD Ru / 20nm Cu evaporated in-situ	3.6	14	0.6	67
Option 5	*Ru seed + Ta adhesion layer*				
PEALD 5nm	10nm PVD Ru / 5nm Ta / 20nm Cu evap.	2.2	1.8	0.6	1.6
PEALD 5nm	10nm PEALD Ru / 5nm Ta / 20nm Cu evap.	2.5	1.8	0.8	2.3

RESULTS & DISCUSSION

Direct Cu

A 20 nm Cu seed layer sputter-deposited in-situ onto the PEALD barrier, leads to de-wetting at 600°C, indicating weak adhesion as compared to PVD TaN, which remains smooth and conductive (figure 1). We found that there is no difference in this behaviour when (porous) SiCOH or SiO_2 are used as underlying substrates. On PVD TaN, the slightly increased Cu/Ta ratio revealed by XPS after annealing (see table II) is probably due to crystallization effects and the movement of vacancies to triple junctions, which in turn make "pinholes" more evident. On the contrary, if Cu is in-situ evaporated on PVD TaN, agglomeration of Cu does occur. That originates from the small kinetic energy of evaporated Cu particles, hence little surface bombardment and therefore adhesion is weaker. However, this evaporated state is also interesting, as it is possibly more representative for sputtered particles hitting the sidewalls of vias or trenches, as their energy transfer is not comparable to planar surfaces.

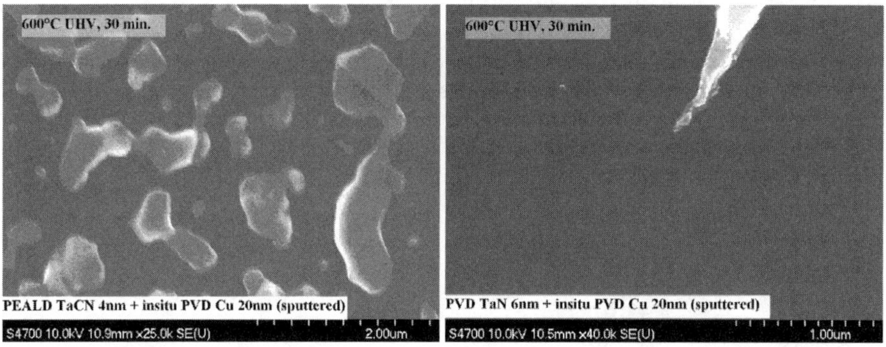

Figure 1 Left: SEM picture of annealed PEALD TaCN sample with a direct Cu seed layer. Right: SEM picture of annealed PVD TaN sample with a direct Cu seed layer.

Bilayer structures

Comparable to the conventional TaN/Ta stack, one option is to sputter a Ta adhesion layer between the PEALD barrier and Cu seed layer. A 7nm Ta layer greatly improves adhesion, from the standpoint of sheet resistance and AFM roughness (figure 2 + table I) and is comparable to PVD TaN/Cu. It is assumed that the carbon remains localized in the TaCN barrier and does not diffuse to the surface. The increased Cu/Ta ratio from XPS after annealing (see table II) is interpreted as it was in the section before. When only a 3nm Ta layer is inserted between TaCN and Cu seed, agglomeration behaviour is not yet consistent, i.e. it varies on the substrate.

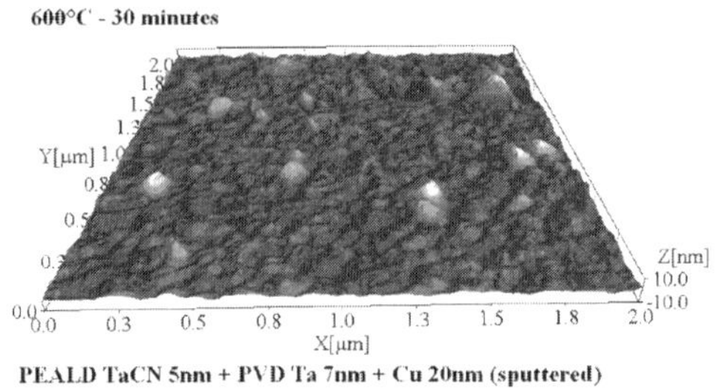

Figure 2 AFM picture with a Ta interlayer between PEALD TaCN and Cu. Area 2x2 μm^2, z-scale ±10nm, rms 2.2 nm. For rms comparison see table I.

Another option investigated here is a 10 nm Ru seed/glue layer on top of the PEALD TaCN barrier. In particular, PEALD Ru could be interesting because it offers a better conformity than PVD Ru. For both PVD and PEALD Ru glue layers Cu de-wetting is less pronounced in terms of an almost unchanged sheet resistance, but not completely avoided, given the XPS and AFM values (figure 3 + tables I,II). It is noteworthy that Cu has been evaporated here, since in-situ sputtering of Cu was not available for this material combination. The kinetic energy of Cu particles is therefore quite small, as seen before on PVD TaN. Nevertheless, in the next section we will find that on Ta, Cu agglomeration will not occur, even when it is evaporated. As mentioned above, the evaporated state is also interesting, since it is possibly more representative for sputtered particles hitting the sidewalls of vias or trenches.

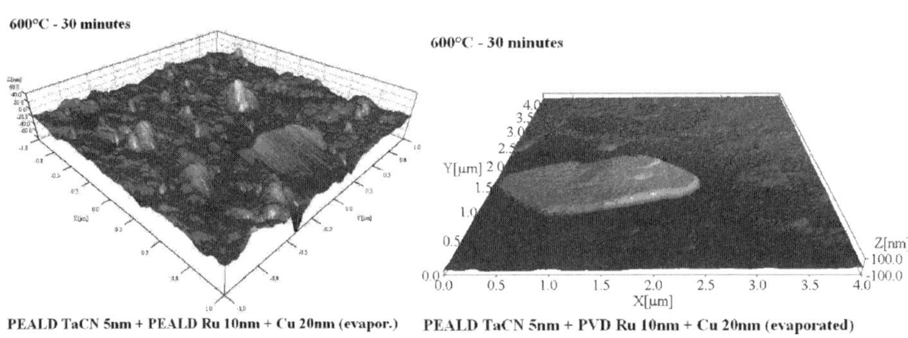

Figure 3 In situ AFM images of annealed samples with a Ru interlayer and evaporated Cu seed. Left: PEALD Ru, 1x1 μm^2, z-scale ±60 nm, rms 18 nm. Right: PVD Ru, 4x4 μm^2, z-scale ±100 nm, rms 31 nm. For rms comparison see table I.

So, in a final approach a Ta flash layer of 5nm is inserted between PEALD/PVD Ru and evaporated Cu, which leads to a significant improvement of Cu adhesion in both cases. After annealing only underlying Ta can be observed by XPS, but no Ru (figure 5, 6 + table II). Interestingly, Ta should not have a better adhesion to Cu than Ru does, according to its higher lattice mismatch. On the other hand, this result is in agreement with the common knowledge that Ta has better Cu adhesion properties than TaN, which was used as reference in the section above.

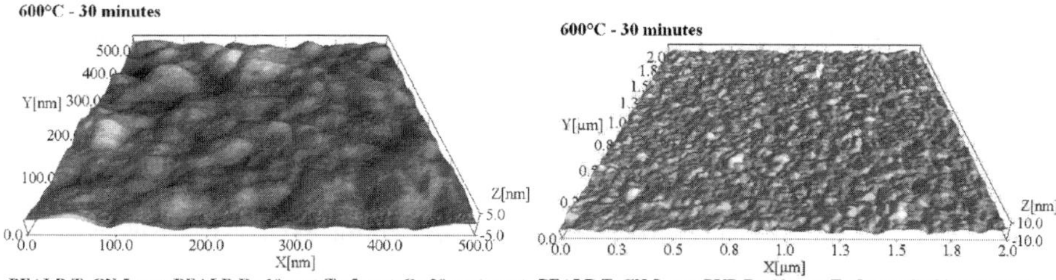

PEALD TaCN 5nm + PEALD Ru 10nm + Ta 5nm + Cu 20nm (evap.) PEALD TaCN 5nm + PVD Ru 10nm + Ta 5nm + Cu 20nm (evap.)

Figure 4 In situ AFM pictures of annealed samples with a Ru/Ta interlayer and evaporated Cu seed. Left: PEALD Ru, 0.5x0.5 µm², z-scale ±5 nm, rms 2.3 nm. Right: PVD Ru, 2x2 µm², z-scale ±10 nm, rms 1.6 nm. For rms comparison see table I.

Table II Results of the in-situ XPS surface analysis of all samples before and after annealing at 600°C for 30minutes in UHV. * All Al Peaks are superposed by stronger Cu signals and have not been quantified therefore. Exact values were obtained by GDOES.

TaN Barrier	Glue/Seed layer	XPS as deposited, @RT					XPS after 600°C 30min.				
		Cu	Al	Ag	Ta	Ru	Cu	Al	Ag	Ta	Ru
Option 1	*Direct Cu*										
PVD 6nm	20nm Cu sputtered in-situ	10			0		88			8	
PVD 6nm	20nm Cu evaporated in-situ	8.3			0.6		20			34	
PVD Ta 6nm	20nm Cu sputtered in-situ	10			0		80			9	
PEALD 5nm	20nm Cu sputtered in-situ	14			0		12			40	
Option 2	*Direct Cu alloy*										
PEALD 4nm	20nm CuAl1% sputtered	44	0*		0		86	0*		0	
PEALD 4nm	20nm CuAl2% sputtered	39	0*		0		82	0*		0	
PEALD 4nm	20nm CuAg1% sputtered	9		0.1	0.9		7.7		5.1	36	
PEALD 4nm	20nm CuAg0,1% sputtered	14		0	0.8		15		0.1	25	
Option 3	*Ta adhesion layer*										
PEALD 5nm	7nm Ta / 20nm Cu sputtered	14			0		57			30	
Option 4	*Ru seed*										
PEALD 5nm	10nm PEALD Ru / 20nm Cu evap.	16			0	0.4	65			0	17
PEALD 5nm	10nm PVD Ru / 20nm Cu evap.	19			0	0	61			0	17
PVD 6nm	10nm PVD Ru / 20nm Cu in-situ ev.	6.2			2	0	53			3	42
Option 5	*Ru seed + Ta adhesion layer*										
PEALD 5nm	10nm PVD Ru / 5nm Ta / 20nm Cu	17			0	0	51			33	0
PEALD 5nm	10nm PEALD Ru / 5nm Ta / 20nm Cu evap.	11			0	0	53			14	0

Cu alloys

The effect of alloying addition of 1 at% and 2 at% Al was investigated and led to a significant adhesion improvement, as proposed on ALD TaN already by Vanypre et al., who also presented promising reliability data [2, 3]. No Ta signal can be detected by XPS (table II). Best results are achieved with addition of 2 at% Al, but at the expense of a higher resistance. Incorporation of Al increases film resistivity, even though it decreases during annealing (table I). It will be interesting therefore to see whether a smaller amount of aluminium, i.e. < 1 at% is capable to be as beneficial as a CuAl(1 %) alloy with a smaller resistivity increase, which is under current work.

On the other hand, it is known from theory that Ag can be added to Cu up to 2 at% - without increasing its resistance - and was alloyed 0.5…2 at% to Cu in a next step. Cu agglomeration turned out even worse as Ag diffuses to the surface, confirmed by XPS (table II). This was also found on PVD TaN [3]. As the solubility of Ag in Cu is only 0.08 % according to the phase diagram, the final approach here was a CuAg(<0.1 at%) alloy, which did not change Cu de-wetting, according to sheet resistance, SEM, AFM and XPS (see figure 5 and tables I,II).

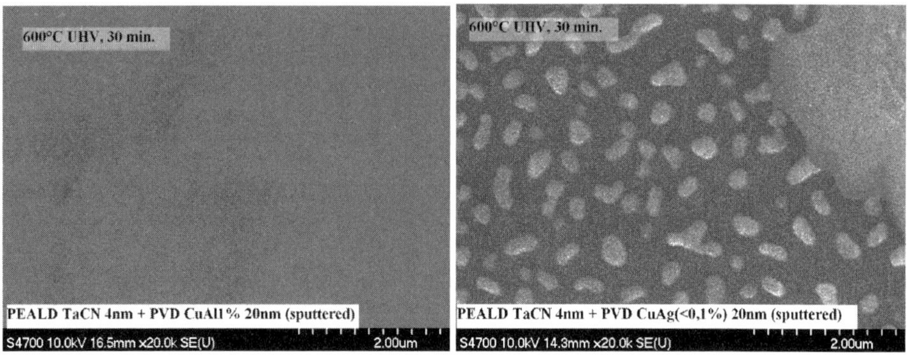

Figure 5 Left: SEM picture of annealed PEALD TaCN sample with a direct CuAl(1 %) seed layer. Right: SEM picture of annealed PEALD TaCN sample with a direct CuAg(<1 %) seed layer.

CONCLUSIONS

In equivalence to the conventional TaN/Ta stack it is possible to work with a PEALD TaCN/Ta bilayer in high aspect ratio structures, with the benefits of a highly conformal, reliable diffusion barrier and relaxed PVD requirements during Ta/Cu seed deposition.

When Ru shall be applied, which offers the possibility of direct plating in particular for PEALD Ru, a 5-7 nm PVD Ta adhesion layer seems to be mandatory. Hence, RuTa or RuTaN alloys should become interesting in future as they might offer a combined benefit on both Cu diffusion barrier properties and Cu adhesion.

CuAg(< 0.1 at%) films are not an equivalent option to CuAl(1 at%) as a direct Cu seed. Here, it will be of interest how far the adhesion improvement effect can be scaled down to minute fractions of Al, i.e. 0.5 at% or even less, minimizing the increase of resistivity.

REFERENCES

1. H. Kim, C. Detavenier, O. van der Straten, S. M. Rossnagel, A. J. Kellock, D.-G. Park: Journal of Applied Physics Vol. 98, 014308, 2005.
2. W. F.A. Besling , X. Federspiel, T. Vanypre, J. Torres: Microelectronic Engineering 82 (2005) 254–260.
3. Vanypre-T; Cordeau-M; Mourier-T; Besling-W-F-A; Dupuy-J-C; Torres-J: Microelectronic Engineering 83 (2006) 2373–2376.
4. Zhao, H. Kim and Y. Shimogaki: Japanese Journal of Applied Physics Vol. 44, No. 41, 2005, pp. L 1278–L

Conference Proceedings AMC XXIV © 2009 Materials Research Society

Impact of organic precursors on the deposition of advanced a-SiC:H low-k copper diffusion barrier layer

Cedric Charles-Alfred[1], Vincent Jousseaume[2], Agnès Granier[3], and Jean-Paul Barnes[4]
[1]STMicroelectronics, 850, rue Jean Monnet, Crolles, F-38926, France
[2]CEA-LETI-Minatec, 17 rue des martyrs, Grenoble, F-38054, France
[3]IMN/PCM, 2 rue de la houssiniere, Nantes, F-44322, France
[4]CEA-LETI-Minatec, Grenoble, F-38054, France

ABSTRACT

This study deals with the deposition of an amorphous hydrogenated silicon carbide (a-SiC:H) as a low-k barrier against copper diffusion. To perform this deposition, an organic precursor is added to trimethylsilane and helium in a Plasma Enhanced Chemical Vapor Deposition (PECVD) reactor. Three molecules are tested (toluene, ethylene and propylene) with the aim of investigating the impact of organic precursor chemistry on the deposited film properties. For the three precursors, soft plasma conditions and low substrate temperature favor carbon incorporation and the more carbon incorporated the lower the dielectric constant. However, the free volume generated by the phenyl rings of the toluene does not seem to impact the reduction of the dielectric constant, compared to the other precursors. This work also shows that toluene is the only of the three precursors that enables incorporation of double bonded carbon visible on infrared spectra. All the chemistries lead to the deposition of films that are stable up to 400°C, and the SIMS results show that increasing the carbon incorporation up to the ratio C/Si = 1.5 does not affect the barrier against copper diffusion properties of the material.

INTRODUCTION

Amorphous hydrogenated silicon carbon-nitride (a-SiCN:H) is the most used dielectric barrier against copper diffusion in microelectronics interconnections. The improvement of the devices performances leads to the reduction of the dielectrics permittivity and a-SiCN:H, with a dielectric constant around 5, is the material that limits this global reduction. Therefore, an amorphous hydrogenated silicon carbide (a-SiC:H) is developed as a low permittivity (low-k) dielectric barrier. This material, known to be a good barrier against copper diffusion [1-3], was deposited thanks to plasma enhanced chemical vapor deposition (PECVD) with trimethylsilane (3MS) diluted in helium as gas precursors. In order to reduce the dielectric constant, the carbon content of the a-SiC:H is increased by adding an organic gas precursor to the 3MS. Three different organic precursors were tested: toluene, propylene and ethylene. The differences in their chemical structures are not only expected to highlight the impact of their composition on the reduction of the dielectric constant, but also on the thermal stability of the deposited materials and on the barrier properties against copper diffusion.

EXPERIMENTAL DETAILS

The materials were deposited on silicon substrates using a capacitive rf-PECVD reactor (13.56 MHz). The gaseous precursors were injected in the chamber through the upper shower electrode. All the materials were analyzed by Fourier Transformed Infrared spectroscopy (FTIR) using a Biorad spectrometer. The composition of the deposited materials was determined by Rutherford Backscattering (RBS) for Silicon, Nuclear Reaction Analysis (NRA) for Carbon and Oxygen and Elastic Recoil Detection Analysis (ERDA) for Hydrogen. Additional surface analyses were carried out by X-ray Photoelectron Spectroscopy (XPS). The thickness was determined thanks to spectroscopic ellipsometry. The dielectric constant was evaluated using C(V) curves measured at 100 kHz with a mercury probe. In order to observe copper diffusion, stacks of SiO_2/a-SiC:H/Cu/TiN were deposited on silicon wafers and annealed at 400°C during one hour under nitrogen flow. Then, they were analyzed by Secondary Ion Mass Spectroscopy (SIMS) to see if copper has diffused through the barrier (40 nm) into SiO_2. The thermal stability of materials was determined thanks to thermogravimetry analysis. To perform this analysis, a part of the film deposited on the wafer was scraped and the obtained powder was heated on a balance from 20 to 500°C (10°C/min) under nitrogen flow.

DISCUSSION

Organic precursors incorporation

First, the impact of the organic precursors on the chemical structure of the deposited film was studied. The use of infrared spectroscopy allows to follow the evolution of the chemical bonds of the film with the variations of the deposition conditions. Actually, whatever the organic precursor used – and even if there is no organic precursor added -, the spectrum presents three peaks regions (cf. figure 1). The first one, between 500 and 1300 cm^{-1}, can be mainly attributed to silicon bonded to carbon groups (Si-C bonds), such as Si-CH_3 around 1245 cm^{-1}[4]. The second band around 2140 cm^{-1} can be attributed to silicon-hydrogen bonds (Si-H) [4]. Then the last one, located between 2700 and 3100 cm^{-1}, can be attributed to carbon-hydrogen bonds (C-H) such as CH_2 (2850 and 2925 cm^{-1}) and CH_3 (2900 and 2960 cm^{-1}) involving sp^3 hybridized carbon [5].

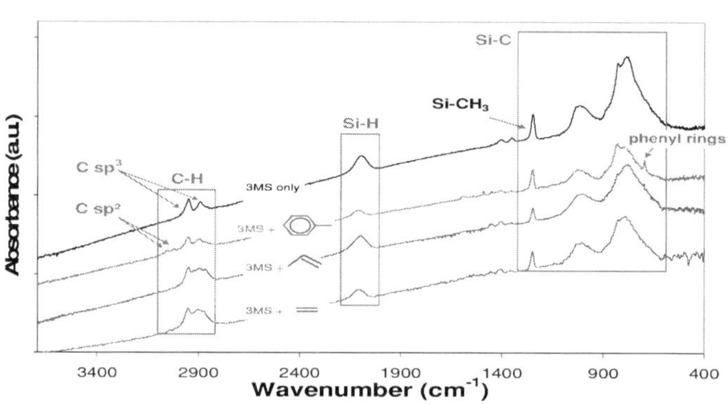

Figure 1: Comparison of the infrared spectra of the materials deposited with the different mixes of precursors (normalized absorbance)

The global incorporation of carbon from the organic precursor can be monitored by comparing the areas of the Si-C and C-H groups of peaks. In figure 1, the Si-C areas are lower for the mixed precursors than for the 3MS-only deposited material whereas, the areas of the C-H peaks are unchanged or slightly larger. This evolution of the peaks areas leads to think that the silicon content decreases while the carbon content increases when organic precursors are added. Therefore, the ratio of C-H to Si-C FTIR peak areas, denoted C-H/Si-C, was first used to follow the evolution of carbon incorporation as a function of the plasma parameters and substrate temperature. Then, to compare the a-SiC:H films deposited with different precursors, quantitative RBS measurements were performed in order to convert C-H/Si-C FTIR peak area ratios into C/Si ratio.

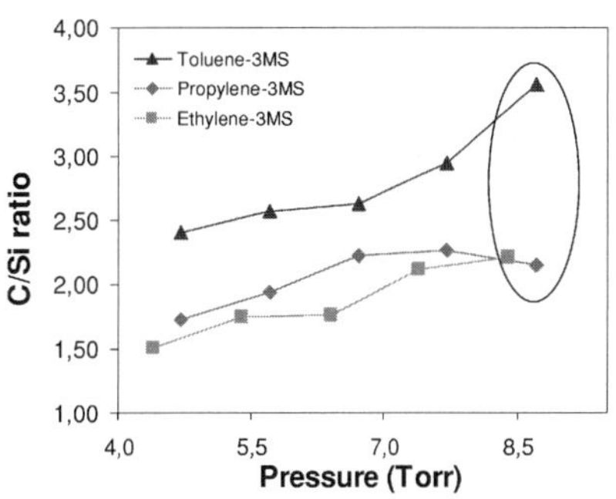

Figure 2 : Evolution of C/Si ratio with the pressure

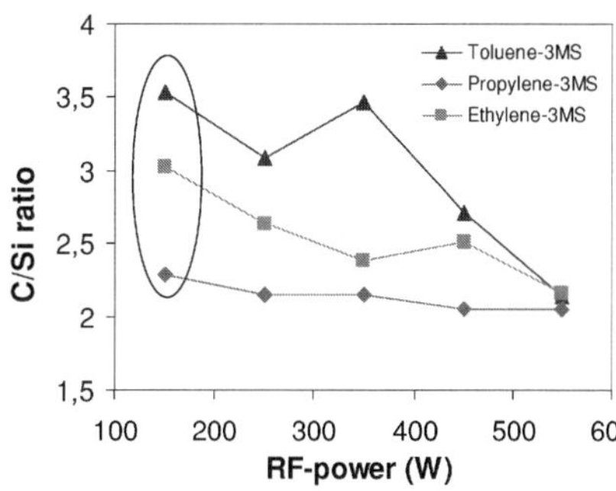

Figure 3: Evolution of C/Si ratio with the dissipated RF-power

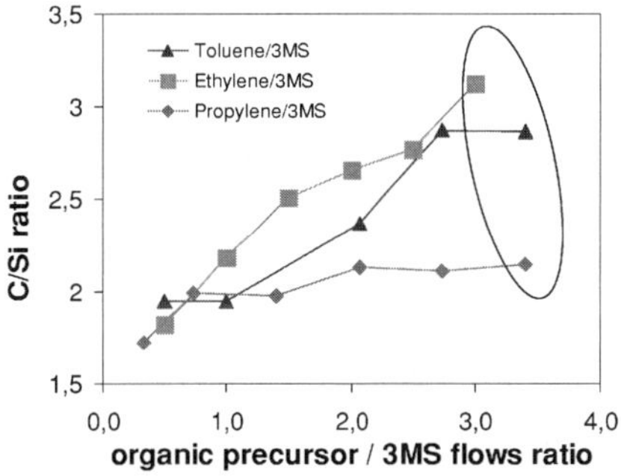

Figure 4: Evolution of C/Si ratio with the precursors flow ratios

plasma conditions pressure rf-power ethylene/3MS flows ratio	4,4Torr 650W 1	8,4Torr 550W 1	8,4Torr 550W 2.7	8,4Torr 350W 2.7
C/Si @ 250°C	**1.77**	**2.21**	**2.64**	**3.07**
C/Si @ 350°C	1.43	1.52	1.69	1.86

Table 1: Influence of the substrate temperature on the C/Si ratio

For each precursor, it can be noticed that the C/Si ratio increases with the pressure and the organic precursor flow rate (cf. figures 2 and 4, respectively) and decreases when the rf-power is increased (cf. figure 3), which means that soft plasma conditions seem to favor the carbon incorporation. Table 1 shows that, for fixed plasma conditions, the C/Si ratio increases when the substrate temperature is decreased, which means that low substrate temperature seems to favor carbon incorporation too.

Specific carbon incorporation can also be noticed on the infrared spectra (cf. figure 1). In the case of toluene, new peaks appear on the spectrum when the film is deposited under soft plasma conditions (low power, high pressure and high organic precursor flow). The first one is located around 700 cm^{-1} [6] and can be attributed to C-H bending vibrations out of the plan in phenyl cycles. It means that intact phenyl rings have been incorporated into the film. The others are located around 3025 and 3059 cm^{-1} and can be attributed to C-H stretching involving sp^2 hybridized carbon in aliphatic and cyclic chains respectively [5], which means that double-bonded carbons have been incorporated in the film.
Though all the tested organic precursors have double-bonded carbon in their molecular structure, sp^2 hybridized carbon can only be noticed in the infrared spectra for the material deposited with toluene precursor.

Decrease of the dielectric constant

To determine the impact of the organic precursor on the reduction of the permittivity, the dielectric constant of all the deposited films is measured and plotted as a function of the C/Si ratio (cf. figure 5). As expected, for each mix of precursor the dielectric constant of the a-SiC:H deposited decreases when the C/Si ratio increases. However, the fact that the three curves are almost superimposed leads to the conclusion that, the size and the steric hindrance of the phenyl rings in the toluene molecule do not seem to play an important role in the dielectric constant reduction when compared to the propylene and ethylene precursors (which are linear molecules).

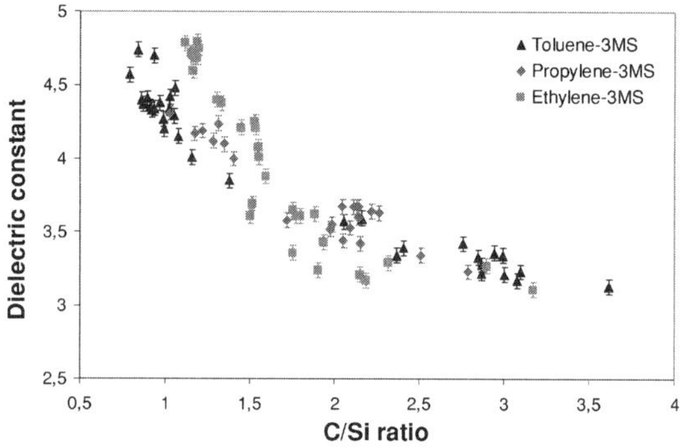

Figure 5: Evolution of the dielectric constant with the C/Si ratio

This figure shows that incorporating more carbon in an a-SiC:H helps to decrease the dielectric constant of this material (from 4.8 to 3.1). Actually, a film with a C/Si ratio around 6 and with a dielectric constant of 3 (not represented in figure 5) could only be deposited by using toluene precursor, which means that it is easier to increase the C/Si ratio above 3.5 with toluene than with the other organic precursors.

Thermal stability

The thermal behavior of the deposited films is determined using thermogravimetry analysis. Figure 6 shows for each organic precursor and for three different carbon contents in each case, the percentage of sample weight left after heating on vertical oven (10°C/min, from 20 to 400°C).

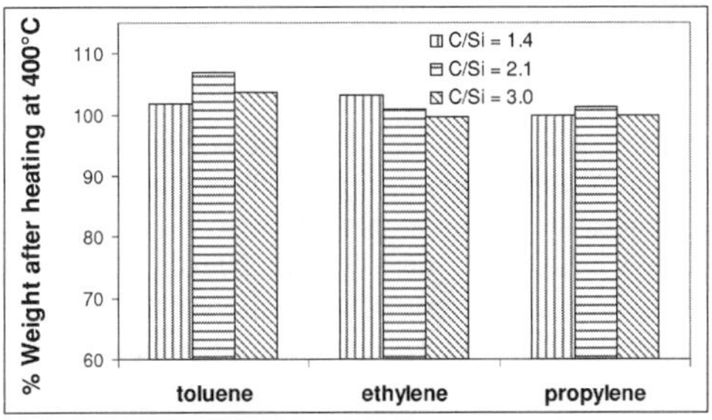

Figure 6 : percentage of the sample's weight left after heating from 0 to 400°C (Thermogravimetry)

This figure shows no significant weight loss after heating at 400°C, whatever the organic precursor used and the carbon content in the film. On the contrary, a little increase of weight can be observed. This is due to the fact that the analysis chamber is not free of oxygen so that a slight oxidation of the material can appear leading to a slight increase of weight.

Copper diffusion

In order to follow the copper diffusion through the new a-SiC:H deposited films, SIMS measurements were performed on SiO_2/a-SiC:H/Cu stacks after 1 hour of annealing at 400°C. As an example, figure 7 shows the SIMS results for a standard a-SiCN:H (k=5) considered as a good barrier against copper diffusion (on the left), and an a-SiC:H (k=4.1) deposited with 3MS and ethylene as precursors with a C/Si ratio around 1.5 (on the right).

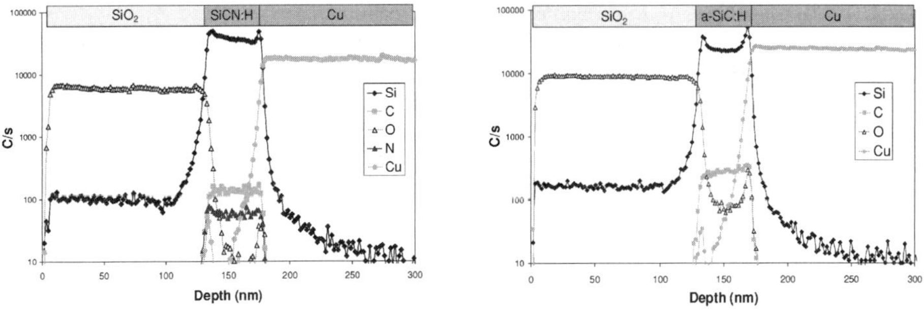

Figure 7: SIMS results for copper diffusion tests
Left: a-SiCN:H known to be a good barrier
Right: Ethylene-made a-SiC:H barrier film, C/Si=1. 5

The comparison of both graphs leads to the conclusion that the ethylene-made a-SiC:H is also a good barrier against copper diffusion, since, no copper can be detected in the SiO_2 layer after 1 hour of annealing at 400°C. The incorporation of carbon up to C/Si =1.5 does not seem to have modified the capability of this material to be a good barrier. Complementary experiments should be performed to have better insight of the C/Si limit above which the ability of a-SiC:H to be copper diffusion barrier is no longer effective.

CONCLUSION

In this work, new amorphous hydrogenated silicon carbides were deposited thanks to PECVD with three different organic precursors, toluene, ethylene and propylene, added to trimethylsilane. For each precursor, global carbon incorporation can be noticed in the infrared spectra, but toluene is the only molecule that leads to double bonded carbon in the film, as detected by infrared spectroscopy. It was shown that whatever the organic precursor, soft deposition conditions such as low substrate temperature, high pressure, low rf-power and high precursor flow rate favor the global carbon incorporation. Moreover, the more carbon incorporated, the lower the dielectric constant. For all the precursors tested, the permittivity can be reduced from 4.8 to 3.1. This study also showed that, every precursor lead to the deposition of materials that are resistant to thermal degradation up to 400°C. Moreover, it was shown that an increase of the carbon content in the material up to a C/Si ratio around 1.5 does not seem to affect its ability to prevent copper diffusion.

ACKNOWLEDGMENTS

The author would like to thank Claire Sourd (CEA-Leti) for the thermogravimetry analysis and François Pierre (CEA-Leti) for the RBS/NRA/ERDA measurements.

REFERENCES

1. M. J. Loboda, Microelec. Eng., **50** (2000).
2. L. Favennec, V. Jousseaume, A. Zenasni, D. Bouchu, G. Passemard, in *New low-k a-SiC:H dielectric barrier for advanced interconnects*, (IEEE, Proc. Burlingame, CA, 2006), pp.110-112.
3. C. Charles-Alfred, V. Jousseaume, Surf. Coat. Technol., **201** (2007).
4. N. B. Colthup, L. H. Daly, S. E. Wiberly, in *Introduction of infrared and Raman spectroscopy*, Academic Press Inc., Boston (1990).
5. G. Lucovsky, Adv. Mater. Opt. Electron., **6** (1996).
6. Vrtis R. N., O'Neill M. L., Vincent J. L. et al., *Plasma Enhanced Chemical Vapor Deposition of Porous Organosilicate Glass ILD Films wth k<2.4,* (Mat. Res. Soc. Symp. Proc. **766**, Warrendale, PA 2003), pp. 259-264.

Conference Proceedings AMC XXIV © 2009 Materials Research Society

Investigation of etch processes of dense and porous low-k dielectrics using OES and QMS as in situ diagnostic methods

Sven Zimmermann[1], Frieder Blaschta[1], Matthias Schaller[2], Hartmut Rülke[3], Stefan E. Schulz[1], and Thomas Gessner[4]

[1]Back-end of Line, Fraunhofer Research Institution for Electronic Nano Systems (ENAS), Reichenhainer Strasse 88, Chemnitz, D-09126, Germany

[2]AMD Saxony Limited Liability Company & Co. KG, Wilschdorfer Landstrasse 101, Dresden, D-01109, Germany

[3]AMD Fab 36 Limited Liability Company & Co. KG, Wilschdorfer Landstrasse 101, Dresden, D-01109, Germany

[4]Fraunhofer Research Institution for Electronic Nano Systems (ENAS), Reichenhainer Strasse 88, Chemnitz, D-09126, Germany

ABSTRACT

The focus of this paper is to find correlations between the sidewall profiles as well as etch rates of SiCOH based low dielectric constant (low-k) films in fluorocarbon based plasma etch processes and specific plasma parameters, which were measured using optical emission spectroscopy (OES) and quadrupole mass spectrometry (QMS) as *in situ* diagnostics. SEM cross-sections were prepared to investigate the trench profiles after the etch processes.

We found, that the change between CHF_3 and CF_4 base chemistry has an impact on the etch rate. In the SEM cross-section no distinctive impact on the sidewall profiles becomes visible. *In situ* measurements show a higher concentration of H ions and the presence of HF in the etch plasma using CHF_3 as base chemistry. With the introduction of argon, no distinctive impact on the sidewall profiles was observed in SEM cross-sections. But argon enhances significantly the concentration of CH and CN groups in the plasma. The additive C_4F_8 is a promising candidate to alter the sidewall profiles and the etch rate. SEM cross-sections show a minor reduced material removal at the trench sidewalls using C_4F_8 as additive. C_4F_8 has a smaller influence on the concentration of CH and CN groups than argon but it decreases significantly the concentration of C_2 bands (swan bands). This indicates, that the C_2 band concentration is a measure for the polymerization efficiency of the etch process. Finally, no differences in the chemical etch mechanism between dense and porous SiCOH films are observed as result of the *in situ* measurements. The porous films have pronounced higher etch rates than those of dense SiCOH and a weaker mechanical strength, which makes the etch process more difficult.

INTRODUCTION

In the last years the low-k material SiCOH has been replaced the SiO_2 as interlevel dielectric in CMOS logic ICs. Commonly used SiCOH materials consist of CH_3 groups, which are embedded in a Si_xO_y matrix. The hydrophobic CH_3 groups are necessary to avoid moisture adsorption, which increases the effective dielectric constant (k-value) [1]. The introduction of pores into SiCOH films results in an additional decrease of the k-value below 3.0 [1]. To produce the pores a porogen is introduced during the film deposition. During the following UV curing process the porogen is diffusing out of the films and pores are formed in the matrix. These new materials result in various integration challenges, especially for further scaling of the interconnect dimensions.

In this work the trench patterning of these materials in a dual damascene via first integration approach using reactive ion etching (RIE) was investigated. During these processes the SiCOH materials are damaged, which results in an increase of their k-values. The damage can originate from penetration of indiffusing active radicals [1], radiation and ion bombardment [2]. The combination of radicals and radiation as well as the ion bombardment were identified as strongest mechanisms [2]. The consequences of these mechanisms are moisture absorption as well as degradation of morphology [2]. With scaling down the interconnect dimensions to 45 nm or lower the damage of the sidewalls in the patterned trenches and vias becomes more and more critical [3]. Oxygen-free fluorocarbon based etch plasmas are able to reduce the damage of the low-k material during the RIE significantly [1]. This plasma chemistry provides a combination of etch reaction and plasma enhanced deposition of polymer films at exposed surfaces in the RIE reactor. These polymer films can be used to protect the sidewalls against damaging ions and radicals. Another critical issue is, to find a measure to quantify the sidewall damage. In some works models were generated to detect this quantity electrically [4].

The dependence of the combined etch and deposition mechanism of the fluorocarbon plasmas on process parameters, e.g.: etch chemistry, process pressure, ICP and bias power, gas flows, and the SiCOH material properties, e.g.: thickness, porosity, morphology, as well as the surface film stacks are far from being fully understood at present. However, the strong dependence of etch process parameters, e.g. selectivities or rates, on plasma parameters, e.g. ion fluxes, radical concentration, ion densities and so on, was already shown [5]. OES and QMS are common methods to measure plasma characteristics during the etch process [5-7]. However, the evaluation of the OES and QMS spectra is quite difficult. New computer based methods using complex search routines were evaluated to find optimum process conditions [7]. Another work demonstrates the control of the etch process in real time using OES and QMS results as actuating variables [8].

In this work the etch behavior of dense and porous SiCOH in CHF_3 and CF_4 plasmas was investigated. Argon and C_4F_8 were introduced into the CF_4 plasma to alter the etch process and its results. We analyzed the process *in situ* using OES and QMS. SEM cross-sections were prepared to estimate the sidewalls profiles of the trenches after etching and to evaluate the etch rate.

EXPERIMENTAL

The etch processes were performed in an OXFORD Plasmalab system 100 with ICP etch reactor. The ICP power was adjusted to 1000 W and the bias power to 60 W. All processes were performed at a pressure of 7 mTorr. Both different base chemistries CHF_3 and CF_4 with gas flow rates of 16 sccm and 12 sccm, respectively, were analyzed. The flow rates of the additives argon and C_4F_8 were 1 sccm and 3 sccm, respectively. The mass spectra were collected with a differentially pumped quadrupol spectrometer UTI 100 C in the range of 1 to 150 amu. The OES spectra were measured with an optical fiber spectrometer Ocean Optics USB 4000. The measurement range was defined from 190 to 890 nm.

45 nm node damascene structures with dense and porous SiCOH films were provided by AMD Fab 36. On these structures we performed the trench etch process. The stack configuration for the dense samples is shown in Figure 1 left and in Figure 5 left for the porous ones, respectively. For the etch process the provided 300 mm wafers were cleaved in coupons of 100 x 100 mm.

RESULTS

Dense SiCOH Films

Figure 1 shows SEM cross-sections of samples etched in CHF_3 plasma after different process durations. Figure 2 shows the corresponding OES and QMS spectra, which were collected during the same process. In the QMS spectra the removal of the different films of the stack can be identified using the CO and the O peaks at 28 amu and 32 amu. At the beginning the mask cap and the dielectric cap were etched. This results in pronounced CO and O peaks in the start mass spectrum. After the removal of the cap layers, the organic patterning layer (OPL) as well as the SiCOH material were reached and the CO and O peaks start to decrease. The etching of the organic OPL releases some complex organic components. It is supposed, that these compounds will be splitted in small fractions in the etch plasma. On the way between the plasma and the QMS these fractions form new compounds, visible in the QMS spectra at 72 amu, 74 amu and 92 amu. After 160 s process time the OPL was removed and the dielectric cap beneath the resist was etched. This results in a renewed increase of the CO and the O peaks.

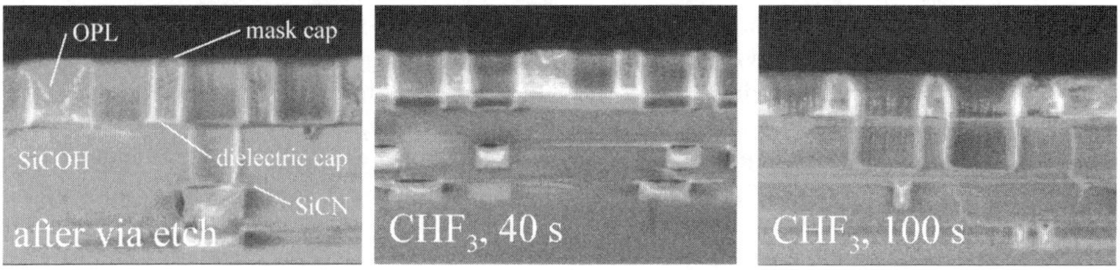

Figure 1. SEM cross-sections of samples prior and after exposure to CHF_3 based trench etch process

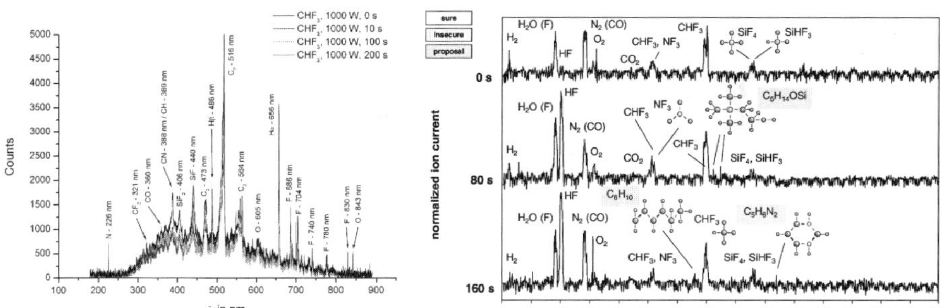

Figure 2. Results of *in situ* measurements during an etch process in CHF_3 base chemistry (left: OES results, right: QMS results)

The main reaction product is a $Si_xF_yH_z$ compound, which was identified as SiF and SiF_2 (lines at 406 nm and 440nm) in the OES and SiF_4 and $SiHF_3$ (peak at 86 amu) in the QMS spectra. The high H_α lines at 656 nm in the OES and the pronounced HF peaks at 20 amu in the QMS spectra are typical for an etch process in CHF_3 ambient. This indicates a higher concentration of H ions and HF compounds in the reactor in comparison to a process using CF_4 as base chemistry cf.

Figure 4. Another difference between CHF_3 and CF_4 is the higher etch rate of the CF_4 plasma, see Figure 1 and 3.

Figure 3 shows the results of three different etch processes for 40 s in a CF_4 plasma, without and with argon and C_4F_8 as additive. The respective OES and QMS spectra are shown in Figure 4. The main product of the reaction between CF_4 and SiCOH is also a $Si_xF_yH_z$ compound. The addition of argon results in an higher etch rate but has no influence on the sidewall geometry. In the OES spectra the addition of argon increases significantly the CH and CN lines but has no influence on the swan bands and the CF_2 line at 321 nm. This is in contrast to the observation of Rao *et.al.* [6], who reported that the argon bombardment enhances the production of CF and CF_2 radicals, which act as precursor to the formation of polymer films. In the QMS spectra the argon addition results in the disappearance of peaks from complex organic compounds. It is assumed, that these compounds are destroyed by argon bombardment.

Using C_4F_8 as additive, a slower etch reaction at the sidewalls of the trenches can be observed in the SEM cross-section, see Figure 3 right. Therewith C_4F_8 is a promising candidate to reduce the sidewall damage. The etch rate of the combined CF_4/C_4F_8 plasma is nearly the same like in the CF_4 process without additives. In the OES spectra the C_4F_8 addition results in a large decrease of the C_2 lines (swan bands at 473 nm, 516 nm and 564 nm). The CN and CH lines are also higher but the effect is not so pronounced in comparison to the argon addition. Just like argon C_4F_8 has no influence on the CF_2 line at 321 nm. This suggests, that the swan bands are an indicator for the polymerization efficiency of the etch process. The peaks at mass numbers higher than 100 amu in the QMS spectra in Figure 4 right are an indication for big macromolecules, which assumedly were formed between the plasma and the QMS. If the higher polymerization addiction of the combined CF_4/C_4F_8 plasma supports the formation of these molecules is not yet understood at the moment.

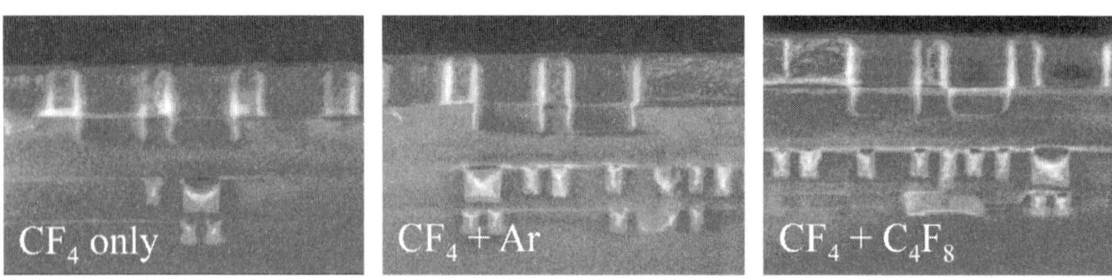

Figure 3. SEM cross-sections of an etch process for 40 s in CF_4 with argon and C_4F_8 as additives

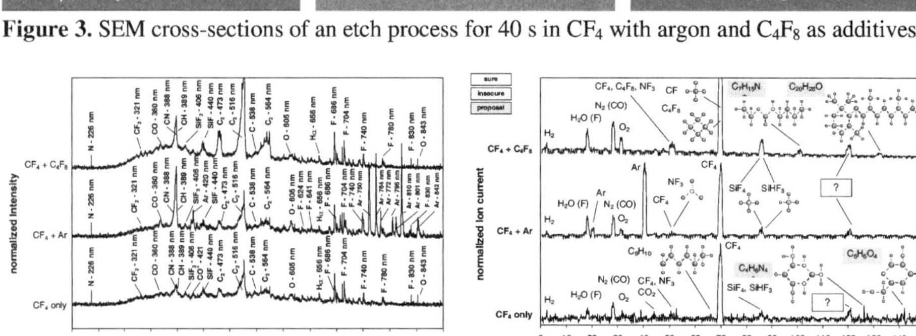

Figure 4. Results of the *in situ* measurement during etch processes of 40 s in CF_4 with argon and C_4F_8 as additives (left: OES results, right: QMS results)

Porous SiCOH Films

The major difference between dense and porous SiCOH material is the etch rate, which is nearly doubled for the porous films. So the difference between the etch rates of the SiCOH and the dielectric cap is more pronounced in comparison to the dense samples. The SEM cross-sections in Figure 5 show, that after 20 s etch time the etch reaction always takes place in the dielectric cap layer. After 50 s the SiCOH films were nearly etched completely down to the SiCN film. The microloading effect, which results in different etch rate between small and broad trenches, produces a significant difference in the etch depth in dependence from the trench geometry, see Figure 5 right.

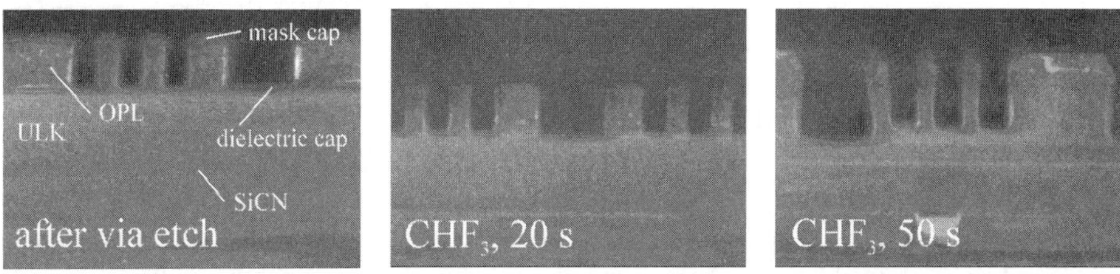

Figure 5. SEM cross-sections of samples prior and after exposure to CHF_3 based trench etch process

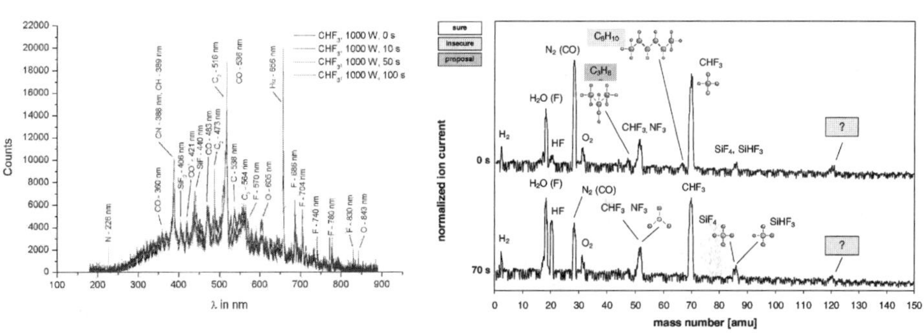

Figure 6. Results of *in situ* measurements during an etch process in CHF_3 base chemistry (left: OES results, right: QMS results)

The OES and QMS spectra derived from porous SiCOH etch in CHF_3 in Figure 6 show no differences to dense SiCOH (Figure 2). The process time is not long enough to remove the mask cap and to expose the OPL. For this reason no complex molecules are found in the QMS spectra at higher mass numbers, see Figure 6 right.

Figure 7 shows the results of *in situ* measurements during etch processes for 20 s with CF_4 as base chemistry. In this processes argon and C_4F_8 were introduced additionally. The only difference between porous and dense samples in the OES spectra (Figure 7 left) is the more pronounced CH and CN lines for etch of the porous films with argon and C_4F_8 addition. The interpretation of this observation needs more detailed investigations. The QMS spectra in Figure 7 right hand show no complex molecules, which consists of parts from the OPL, because the etch time was too short to expose this layer. The appearance of the measured fragments of $C_8H_{12}O$ and $C_{10}H_{16}$ at 81 amu and 93 amu did not completely understood yet.

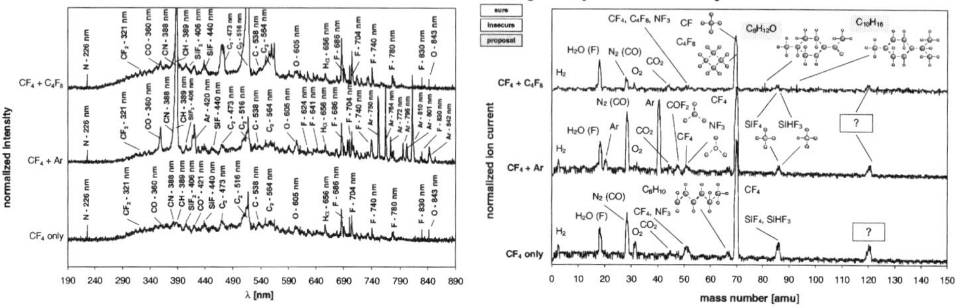

Figure 7. Results of the *in situ* measurement during etch processes of 20 s in CF_4 with argon and C_4F_8 as additives (left: OES results, right: QMS results)

CONCLUSIONS

In this work some correlations between results of etch processes and plasma conditions were described. CHF_3 and CF_4 base chemistries were tested and a higher etch rate for CF_4 was found. The CHF_3 plasma includes a higher concentration of H ions and the HF compound. For the addition of argon a higher etch rate and an increase of CH and CN groups were observed. The introduction of C_4F_8 hampers the etch reaction at the trench sidewalls and shows a significant increase of the swan bands in the OES spectra. So the swan bands were indentified as measure for the polymerization efficiency of the etch process. Finally no difference between the chemical etch mechanism of dense and porous SiCOH films was observed in the measured spectra. The microloading effect was identified more challenging during the etch of porous SiCOH.

The long term goal of our investigations is to find a mathematical correlation between the results of the *in situ* analysis and analyzed etch profiles and to estimate the sidewall damage using *in situ* measurements.

ACKNOWLEDGEMENTS

The project described in this publication is funded by the German Federal Ministry of Education and Research, registered under funding number 01M3173. The authors named in the publication are responsible for all published contents.

REFERENCES

[1] M.R. Baklanov *et.al.*, Proceedings of the 8th Intenational Confernce on Solid-State and Integrated Circuit Technology, 2006 (ICSICT'06), pp. 291
[2] S. Uchida *et.al.*, Journal of Applied Physics, vol. 103 (2008), pp. 073303-1
[3] R.M. Badam *et.al.*, Japanese Journal of Applied Physics, vol. 43, no. 11A (2004), pp. 7381
[4] F. Iacopi *et.al.*, Electrochemical and Solid-State Letters, vol. 7, no. 4 (2004), pp. G79
[5] F. Gaboriau *et.al.*, Journal of Physics D: Applied Physics, vol. 39 (2006), pp. 1830
[6] M.V.V.S. Rao *et.al.*, Plasma Sources Science and Technology, vol. 11 (2002), pp. 69
[7] J. Al-Kunzee *et.al.*, Plasma Sources Science and Technology, vol. 13 (2004), pp. 612
[8] V. Milosavljevi´c *et.al.*, Journal of Applied Physics, vol. 103 (2008), pp. 083302-1

All graphics in this publication have been generated by Fraunhofer ENAS.

Conference Proceedings AMC XXIV © 2009 Materials Research Society

Grain Growth in Sub-50 nm Cu Lines

R. L. Graham[1], G. B. Alers[1], N. Shamma[2], T. Mountsier[2], G. Kooi[2], A. Rairkar[2], I. Karim[2], L. Hakim[2], T. Ponnuswamy[2], W. B. Chou[3], S. Dhuey[4], S. Cabrini[4], S. Peddeti[5], and R. H. Geiss[6]
[1]Department of Physics, University of California, Santa Cruz, CA, 95064
[2]Novellus Systems, San Jose, CA, 95134
[3]Lam Research Corporation, Fremont, CA, 94538
[4]Molecular Foundry, Lawrence Berkeley National Laboratory, University of California, Berkeley, CA, 94720
[5]Department of Chemical and Biomolecular Engineering, Clarkson University, Potsdam, NY, 13676
[6]National Institute of Standards and Technology, Boulder, CO, 80305

ABSTRACT

Extending copper interconnect technology will be of critical importance for achieving continued increases in density for both memory and logic circuits. As feature size shrinks, smaller grain size can lead to increased resistivity and reduced reliability. Electromigration and stress migration are expected to worsen with small grain sizes and nonuniform texture. The goal of this work is to understand grain growth in sub-50 nm lines, where surface constraint plays an increasing role. It is commonly assumed that average grain size scales with critical dimensions (CDs) in small Cu wires. This examination demonstrates that grain size can be much larger than line width, even for widths less than 30 nm. Electron-beam lithography and a dual hard mask approach were used to fabricate copper damascene interconnects, with CDs between 25 nm and 55 nm. Conventional metallization with a 300 mm tool set was used to form the copper lines and vias. Electron backscatter diffraction (EBSD) was used to characterize the crystallographic structure of Cu grains in 35 nm to 55 nm lines. A simple model is presented for calculating resistivity due to surface scattering, which is valid for lines with CDs between 10 nm and 100 nm.

INTRODUCTION

Grain growth in wires is influenced by parameters of metallization [1], the geometry of the trench, and energy minimization. Thompson and Carel suggested that growth of grains with a <111> texture is generally favored because it reduces energy per unit area at the grain boundary interface [2].

Understanding grain growth is especially interesting to the semiconductor industry because grain boundaries contribute to increased resistance as feature sizes shrink. In bulk copper at ambient temperature, resistivity is dominated by electron-phonon scattering according to the Bloch-Grüneisen model. In features with dimensions less than the mean free path, grain boundary scattering and surface scattering are the two major contributors to resistivity. The commonly used model for surface scattering was developed by Fuchs [3], and the grain boundary scattering model was developed by Mayadas and Shatzkes *et al.* [4]. Mayadas and Shatzkes added a regularly spaced delta-function perturbation to Fuchs's four-wall scattering.

Mayadas and Shatzkes noted that the derivation of resistivity due to grain boundaries simplifies if it is assumed that the distance between grains is equal to the dimension of the wire cross-section. The implicit assumption is that during annealing the grain grows to the size of the trench.

The problem of increased resistivity generates the desire for further investigation into scattering mechanisms. Many investigations begin with a combination of the Fuchs model and the Mayadas and Shatzkes model, and simplify the parameters (interface reflectivity, distance between grain boundaries, and area of trench cross-section) by assuming grain size is equal to line width or height. With so many parameters, however, the solution that fits the data is not unique. This work attempts to separate the multiple scattering mechanisms that contribute to copper resistivity in sub-50 nm lines.

EXPERIMENT

Electron-beam lithography and a dual hard mask approach were used to fabricate copper damascene interconnects. Figure 1 shows both lines and vias patterned into Poly[methyl methacrylate] (PMMA) resist with a minimum CD of <17 nm. This pattern was transferred to a 35 nm SiN hard mask, and the resist was removed. The SiN pattern was then transferred into 100 nm of ashable hard mask (AHM). The use of a dual hard mask reduced line edge roughness after the final etch into 100 nm of TEOS-SiO_2, as shown with and without AHM in Figure 2. Conventional 300 mm etch and metallization processes were used to form single-damascene copper lines and vias, as shown in Figure 3. The dielectric stack was deposited using plasma-enhanced chemical vapor deposition (PECVD) with a Novellus Vector™. The barrier/seed layers were deposited using physical vapor deposition (PVD) on a Novellus Inova™. Copper plating was done by electrochemical deposition (ECD) on a Novellus Sabre™. Samples were annealed after ECD at 150 °C for 60 minutes. Chemical-mechanical polishing (CMP) was performed to remove the copper overburden and produce electrically testable sub-30 nm lines.

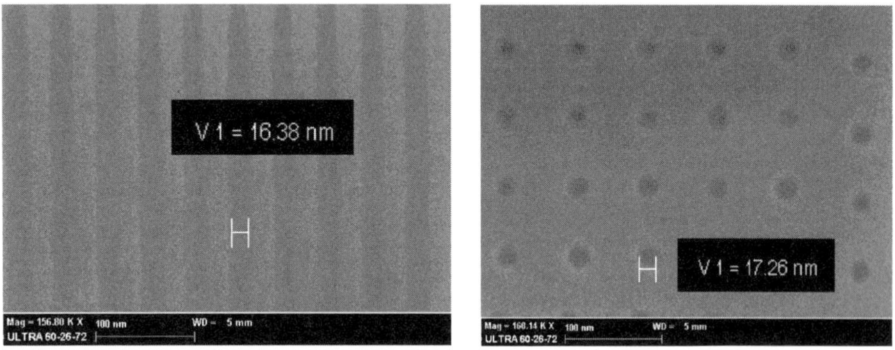

Figure 1: Optimized electron-beam lithography for trenches (left), with 16 nm width and 40 nm pitch, and vias (right) as small as 17 nm.

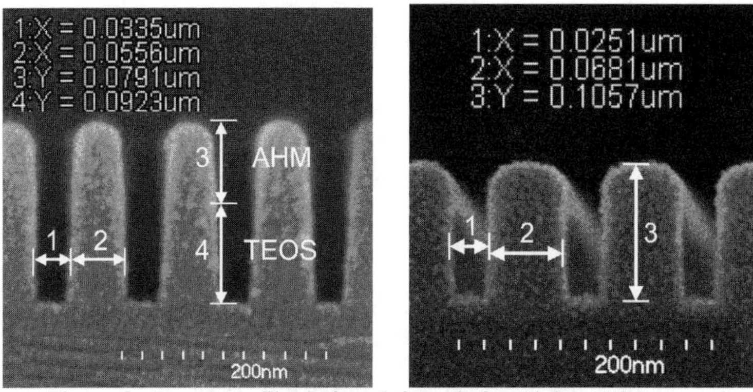

Figure 2: Trench profiles before (left) and after (right) AHM removal.

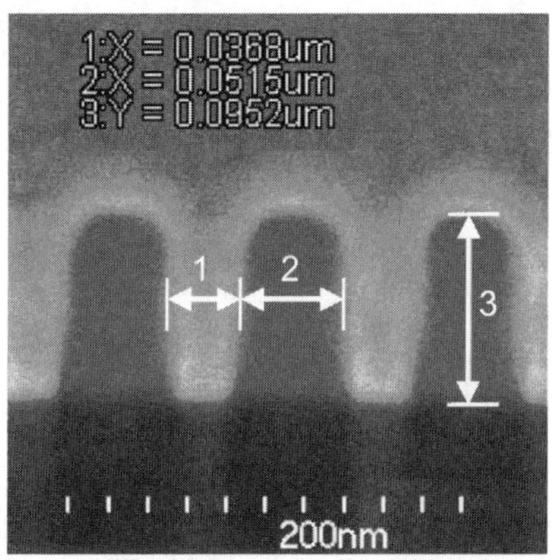

Figure 3: Copper fill of 36 nm trenches with ~ 3:1 aspect ratio.

RESULTS

Grain size and crystal orientation were characterized by electron backscatter diffraction (EBSD), a high-resolution diffraction technique performed in a scanning electron microscope. EBSD was applied to characterize the entire volume of the wire. For a 20 kV incident electron-beam, the penetration depth is between 100 nm and 200 nm, as drawn in Figure 4. The narrow Cu lines analyzed here have a depth of 100 nm, therefore EBSD samples the entire depth of the Cu lines. Each EBSD scan mapped 0.026 μm x 1.0 μm in 10 nm steps. Grains with dimensions ranging from 30 nm to 800 nm along the lines were observed. Based on all the collected data, it is reasonable to assume that the grains had a bamboo structure, with grain boundaries perpendicular to current flow.

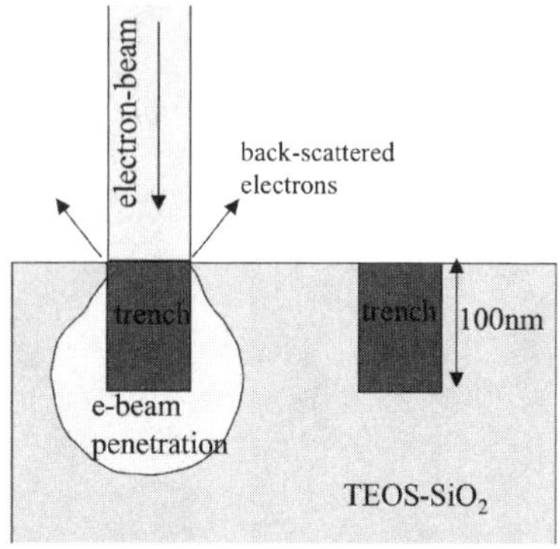

Figure 4: Electron beam penetration in EBSD analysis of Cu trenches.

The results show that some grains are much longer than the lines are wide, and grains in adjacent lines often have the same orientation, as shown in Figure 5. This suggests that the grain structures in the lines are remnants of the larger grain structure in the overburden after ECD and annealing. Figure 6 shows a (111) pole figure and a [001] normal direction (ND) inverse pole figure (IPF) from merged data in seven 50 nm Cu lines. The data in the (111) pole figure and the [001] IPF indicate that there is a preferred normal grain orientation near to <111>. Further, the data in the IPF show there are also some grains with normal orientations near to <001>. Note that the <111> orientation arises due to energy minimization at the surface.

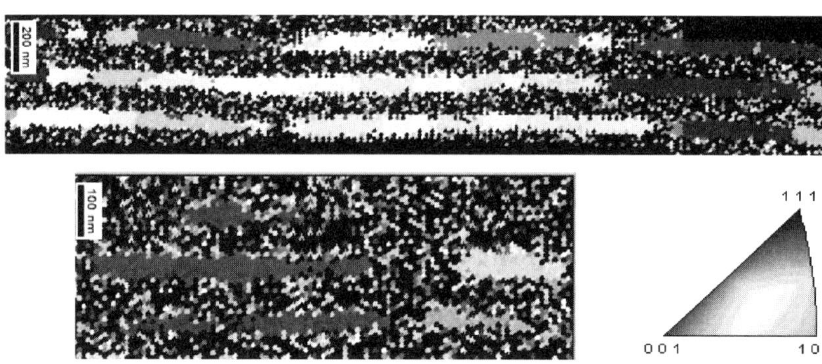

Figure 5: EBSD data of adjacent 55 nm Cu lines (top) and adjacent 35 nm (lower left) Cu lines. Line colors (colors online) correspond to the orientation normal to the sample surface, according to the triangle.

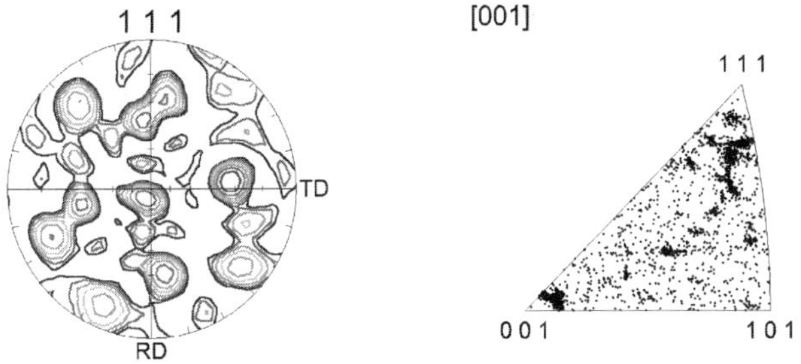

Figure 6: (111) pole figure (left) and [001] normal direction, ND, inverse pole figure (right) showing grain orientation in adjacent 55 nm Cu lines.

According to the model by Mayadas and Shatzkes, the contribution to resistivity from grain boundary scattering depends on the ratio of the mean free path (MFP) of electrons to the average grain size. If the grain size is larger than the MFP, then grain size is not important for scattering, and resistivity increase is from other mechanisms including electron-phonon scattering and surface scattering. Figure 7 shows examples of some long grains seen by the EBSD line scans performed on 35 nm to 50 nm lines. Both the 35 nm and 50 nm line scans show very large grains, with <111> normal orientation depicted in blue.

Figure 7: Images of large grains in 35 nm, 40 nm, 45 nm, and 50 nm Cu lines.

According to the Fuchs model, resistivity due to completely diffuse surface scattering is given by the following integral according to the geometry of Figure 8 [5, 6]:

$$\left(\frac{\rho_0}{\rho}\right) = 1 - \frac{6}{4\pi * hw}\int_0^w dx \int_0^h dy \int_{-arctan\frac{y}{x}}^{arctan\frac{h-y}{x}} d\phi \int_0^\pi d\theta * \left(sin\theta * cos^2\theta * exp\left(\frac{-x}{\lambda * sin\theta * cos\phi}\right)\right)$$
$$- \frac{6}{4\pi * hw}\int_0^w dx \int_0^h dy \int_{-arctan\frac{x}{h-y}}^{arctan\frac{w-x}{h-y}} d\phi \int_0^\pi d\theta * \left(sin\theta * cos^2\theta * exp\left(\frac{-(h-y)}{\lambda * sin\theta * cos\phi}\right)\right),$$

where *w* and *h* are the width and height, respectively, and λ is the MFP. This integral was evaluated numerically, for a range of heights and widths, using λ of 40 nm and assuming inelastic scattering. The simple model is as follows:

$$\left(\frac{\rho}{\rho_0}\right) = 1 + K * \left(\frac{\lambda}{w} + \frac{\lambda}{h}\right),$$

with fitting constant $K = 0.4584$, as determined from numerical integrations and averaged for line heights between 10 nm and 100 nm and aspect ratios between 0.25 and 4. Resistivity measurements are in progress and will be compared to the Fuchs and simple models.

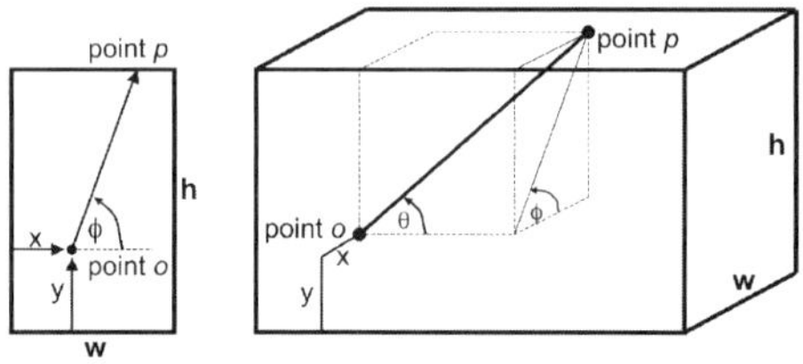

Figure 8: The geometry considered in the integral for wire resistivity due to surface scattering. The electron path starts from point *o* (defined locally by *x* and *y* in the cross-section), travels freely in the azimuthal angle θ and the radial angle ϕ, and scatters off the surface at point *p*.

DISCUSSION

It is reasonable to believe that for small wires the average grain cross-section is equal to the wire cross-section. Regarding length, however, the EBSD results reported in this paper show that the distance between adjacent grain boundary planes, along the length of the wire, is generally much larger than the wire width or height. The results discussed here suggest that the crystallographic behavior and grain size are determined by the copper overburden, after ECD and annealing but before CMP. The crystal structure matches in adjacent lines, which is a result of those grains coming from the same grain in the Cu overburden. With some grains as large as 20 times the CD, the resistance increase due to grain boundaries may not be as large as the Mayadas and Shatzkes model predicts. The large grains reported here suggest that grain boundary scattering is not a dominant scattering mechanism. Other factors that likely increase resistivity are the sidewall roughness, top surface roughness, and surface corrosion.

CONCLUSIONS

This work has extended our understanding of copper damascene grain growth to sub-50 nm lines, which is critical for the extendibility of current industry practices in producing ever smaller interconnects. Conventional metallization methods are shown to be extendible to sub-30 nm feature sizes. It was shown that grain size along the length of the Cu line could be much larger than the dimension of the line cross-section. A simple model was presented, which approximates the integration in the Fuchs model for line heights between 10 nm and 100 nm and aspect ratios between 0.25 and 4. Measurements of line resistance will be used to calculate resistivity, which will be compared to existing models for surface scattering. Some other scattering mechanisms include surface roughness, corrosion, and sidewall roughness. Further investigation, including statistical data of resistivity measurements, would help to better characterize resistivity in the 25 nm to 50 nm size regime.

ACKNOWLEDGEMENTS

The authors thank the following people for their helpful input regarding this project: Bob Havemann (Novellus), Don Schlosser (Novellus), Neil Mackie (Novellus), Vamsi Devarapalli (Clarkson U.), and Dr. S. V. Babu (Clarkson U.).

This work is a contribution of the US Department of Commerce and is not subject to copyright in the USA. Certain commercial equipment, instruments, or materials are identified in this document. Such identification does not imply recommendation or endorsement by the National Institute of Standards and Technology, nor does it imply that the products identified are necessarily the best available for the purpose.

FUNDING ACKNOWLEDGEMENT

Portions of this work were performed at the Molecular Foundry, Lawrence Berkeley National Laboratory, which is supported by the Office of Science, Office of Basic Energy Sciences, of the U.S. Department of Energy under Contract No. DE-AC02—05CH11231.

REFERENCES

1. J. M. E. Harper, C. Cabral, Jr., P. C. Andricacos, L. Gignac, I. C. Noyan, K. P. Rodbell, and C. K. Hu, *J. Appl. Phys.* **86**(5), 2516 (1999)
2. C. V. Thompson and R. Carel, *J. Mech. Phys. Sol.* **44**, 657 (1996)
3. K. Fuchs, *Proc. Cambridge Philos. Soc.* **34**, 100 (1938)
4. A. F. Mayadas, M. Shatzkes, and J. F. Janak, *Appl. Phys. Lett.* **14**, 345 (1969)
5. W. Zhang, S. H. Brongersma, Z. Li, D. Li, O. Richard, and K. Maex, *J. Appl. Phys.* **101**, 063703 (2007)
6. D. Josell, C. Burkhard, Y. Li, Y.-W. Cheng, R. R. Keller, C. A. Witt, D. R. Kelley, J. E. Bonevich, B. C. Baker, and T. P. Moffat, *J. Appl. Phys.* **96**(1), 759 (2004)

CMP and Cleaning

Conference Proceedings AMC XXIV © 2009 Materials Research Society

Recent Topics in CMP-Related IC Design for Manufacturing

Andrew B. Kahng[1], Kambiz Samadi[2], and Rasit Onur Topaloglu[3]
[1]CSE & ECE, University of California, San Diego, 9500 Gilman Drive, La Jolla, CA, 92093-0404
[2]ECE, University of California, San Diego, 9500 Gilman Drive, La Jolla, CA, 92093-0409
[3]Technology Development Group, Advanced Micro Devices, One AMD Place, MS 79, Sunnyvale, CA, 94085

ABSTRACT

CMP is a planarization technique for multilevel VLSI metallization processes. CMP fills are inserted as a design for manufacturability (DFM) methodology to improve pattern-dependent post-CMP topography, i.e., to improve interconnect planarity. Design-driven fill synthesis seeks to optimize CMP fill with respect to objectives beyond mere density uniformity. Design-driven fill synthesis minimizes the impact of CMP fill on function-driven performance and parametric yield metrics, while satisfying manufacturing-driven density criteria. This paper reviews CMP fill techniques as a DFM process, then describes a power-aware methodology wherein CMP fill synthesis is integrated within a holistic design- and manufacturing-driven flow. Experimental validation with production 65nm CMP models shows that this proposed methodology obtains power reductions on the design side, while satisfying density requirements imposed by manufacturing.

INTRODUCTION

In the past decade, *chemical-mechanical polishing* (CMP) has emerged as the predominant planarization technique for shallow trench isolation (STI) and back-end-of-the-line (BEOL) metallization. Failure to meet surface topography variation constraints leads to degradation of transistor characteristics in the case of STI, and electrical shorts or increased wire resistance in the case of BEOL interconnects, all of which are detrimental to circuit yield. As technology scaling advances into 45nm and beyond, variations due to lithography and CMP processes have to be accurately considered in the design flow. As an example, on-chip variation (OCV) from the interconnect thickness variation due to CMP becomes relatively larger and needs to be taken into consideration in the post-layout RC extraction and timing flow [10].

Design for manufacturability (DFM) targets design analysis, modeling, optimization and automation to enable manufacture of working chips, and improve chip performance and reliability, in the face of mounting challenges (variability, leakage, etc.) in semiconductor manufacturing. As the number of metal layers increases and line widths shrink, tolerance for topographical imperfections decreases. This is due to tight depth-of-focus variation requirements and high sensitivity of resistance to metal thickness. Despite improvements in CMP technology, layout pattern sensitivities are significant, causing certain regions to have higher topographies than others due to differences in underlying densities. One DFM technique that improves surface planarity in today's advanced technologies is the insertion of CMP *fill* shapes, which are non-functional features that do not directly contribute to the logic implementation and can either be grounded or left floating (Figure 3). Fill shapes impact electrical performance of the designs and hence should be optimized for circuit performance in addition to layer topography uniformity. Efficient characterization, modeling and optimization of CMP fills are necessary to help reduce design pessimism and improve suboptimal design performance.

The remainder of this paper is organized as follows. In the next section, we present a brief overview of CMP modeling, including for copper dual-damascene / low-k processes. We then review the state of the art in density control through fill insertion. Next, we present an overview of design-driven CMP fill synthesis. Specific examples of design-driven fill synthesis, namely, timing- and power-driven fill optimization methodologies are presented next, followed by a summary of CMP fill synthesis design flow issues and the paper's conclusions.

CMP MODELING

CMP is the technique of choice to satisfy local and global planarity constraints imposed by today's advanced photolithography methods, as well as interconnect performance margins. However, CMP heavily depends on the underlying pattern density and therefore, various degrees of metal dishing and dielectric erosion happen at different metal densities (Figure 1) [10]. To reduce the post-CMP thickness variation dummy fill shapes (CMP fill) are routinely inserted into the layout. Due to complex nature of CMP process (pad, interaction between slurry and metal/oxide, multi-level effects, etc.), a physical model is required to simulate CMP process on a full chip level. Such a model needs to address the CMP process at three scales of particle-, feature- and die-/wafer-level.

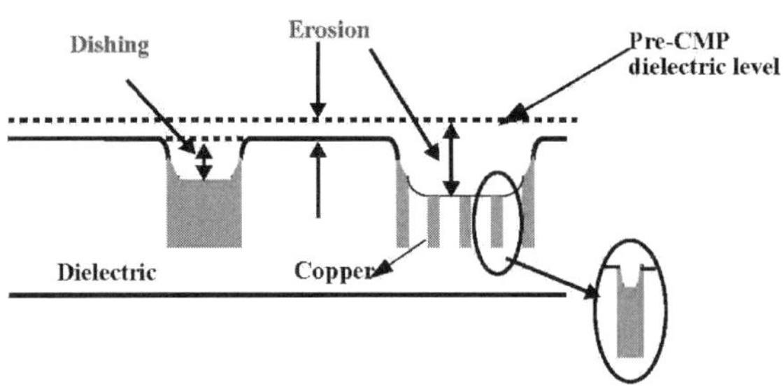

Figure 1. Metal dishing and dielectric erosion [2].

Particle-scale modeling addresses the roles and interactions of slurry particles, slurry chemicals, polishing pads and wafer materials. The two most important issues at the particle scale are the material removal rate (MRR) and the surface quality (surface roughness). The material removal rate determines the production rate of the process. Surface qualities determine the yield of the process. A widely known model is the experimental Preston's equation which was initially introduced for glass polishing [11]. Models at the feature and die scales are needed to address the topography evolution of IC chips as a function of pattern density, line width, and polish time. The MIT semi-empirical model is based on a framework first proposed by Stine *et al.* [5]; it characterizes and models the polishing behavior of inter-layer dielectric (ILD) as a function of structure pattern density. Experimental data indicates that the effective pattern density, which is defined as the ratio of the raised area to the total area for a window of a given size and shape, is the dominant factor for within-die nonuniformity, while the structure area, pitch, and perimeter/area play only minor roles. The effect of pattern density on within-die nonuniformity is due to pressure nonuniformity. In a high-density area, where the oxide area contacted by the pad is larger, the effective pressure is lower and hence, the material removal rate is smaller. Analytical equations to address this are developed by Stine *et al.* [6] based on Preston's equation of material removal rate. While the semi-empirical models of Stine *et al.* [6] and Ouma *et al.* [7] are the most successful at feature and die scales, others have tried to develop physical models to explain the topography evolution. Ouyang *et al.* [8] developed equations based on fluid mechanics to predict pressure at any given location on the die. Preston's equation is then used to evaluate the topography evolution [11]. Another fluid mechanics based feature-scale topography model was developed by Yao *et al.* [9].

With dual-damascene copper interconnects, the low mechanical strength of low-k films can cause defects when using conventional CMP. An alternative solution to this problem is to use *stress-free polishing* (SFP). Stress-free polishing is a low viscosity fluid contact alternative to CMP that enables the removal of copper overburden without exposing barrier or dielectric films to mechanical stress. SFP applies a voltage across the wafer with an electrolyte held as anode. The process removes copper through control of wafer rotation, recirculating electrolyte flow, and voltage and current without the use of slurry. Since the SFP process is inherently isotropic with no planarizing capability, a partial CMP step will be added at the end to achieve surface planarity [12]. Hence, the above-mentioned models can be applied to predict the resultant topography after the partial CMP process.

LAYOUT DENSITY CONTROL THROUGH FILL INSERTION

Pattern-density uniformity is achieved by insertion ("filling") or, to a lesser extent, partial deletion ("slotting") of features in the layout. The foundry's design rules specify upper and lower bounds on feature density for $w\times w$ windows of the layout. In practice, these bounds are verified and enforced only for overlapping windows of size $w\times w$ within a so-called *fixed dissection* of the layout. More precisely, windows of size $w\times w$ are stepped across an $n\times n$ layout region, with step size of w/r; this induces a dissection of the layout region into *tiles* of size $(w/r)\times(w/r)$, and density bounds are verified in all windows of $r\times r$ tiles having top-left corners at locations $(i\cdot(w/r), j\cdot(w/r))$, for $i, j = 0, 1, \ldots, r((n/w)-1)$, as shown in Figure 3a. Kahng *et al.* [16] developed an optimal solution approach for fill synthesis, using linear programming (LP). The LP approach finds the optimum total area of fill shapes to be inserted into each tile, such that the difference between maximum and minimum post-fill window densities in the layout is minimized. In a follow-up work, Tian *et al.* [17] developed a minimum-fill LP formulation, incorporating an elliptical weighting function (from the planarization model of Ouma [7]) which has better correspondence to the pad deformation. The works of [16] and [17] contrast (1) a 'manufacturer's objective', which is to minimize the post-fill density variation across the layout to improve manufacturing uniformity, versus (2) a 'designer's objective', which is to minimize the total amount of fill inserted into the layout so as to minimize added capacitance and coupling, timing, noise and power impacts on the design. Manufacturability goals also dictate the creation of "smooth" fill, e.g., through application of the LP approach with explicit upper bounds on the density difference between any pair of adjacent windows. Table 1 [19] compares relevant metrics of minimum-variation, minimum-fill, and maximum-smoothness solutions for a challenging image sensor chip. Figure 2 additionally shows layout density histograms for the original layout, a minimum-fill solution, and a minimum-variation solution.

Table 1. Comparison of different fill insertion formulations and post-fill layout metrics.

	Min. D	Max. D	Delta D	# fill	Avg. Smoothness
Original Solution	0.1652	0.4717	0.3065	-	0.0508
minVar	0.4153	0.5448	0.1295	784,968	0.0234
minFill	0.3234	0.4717	0.1483	416,773	0.0317
maxSmoothness	0.3945	0.5243	0.1298	711,429	0.0174

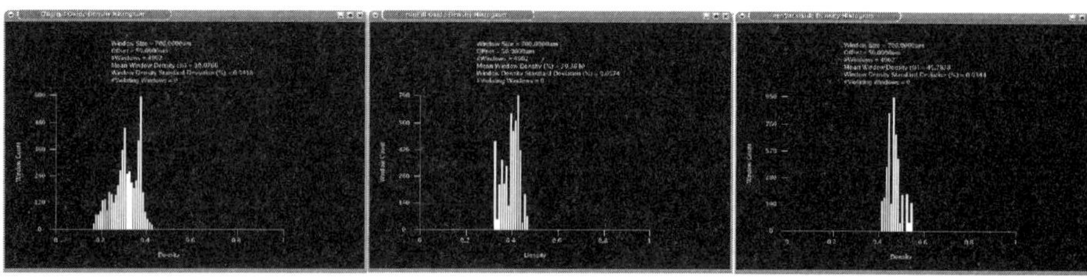

a) Original Density Histogram (ΔD = 31%) (b) Min-Fill Density Histogram (ΔD = 15%) (c) Min-Variation Density Histogram (ΔD = 13%)

Figure 2. Post-fill layout density histograms for original, minimum-fill, and minimum-variation solutions.

DESIGN-DRIVEN FILL SYNTHESIS

As the industry moves into the 45*nm* node and beyond, traditional fill synthesis methods have reached their limits of usefulness. One indication of this is the emergence of "recommended rules" in design rule manuals, e.g., "it is better to have a small difference between the density values of adjacent windows" (a smoothness objective), or "it is better to maximize the overlap of fill shapes on adjacent layers to enable dummy via insertion" (since dummy vias enhance low-k dielectric material stability). The impact of fill synthesis on parasitics and timing also continues to be a key concern for the designer. It is increasingly difficult for a design rule check (DRC) platform to obtain an *optimal, design-driven* fill synthesis solution that meets all basic CMP design rules and as many recommended rules as possible, while minimizing impact on timing, signal integrity and power. With this in mind, we now sketch the capabilities of more sophisticated, "design-driven" fill syntheses. Such techniques can potentially reduce engineering effort while enhancing manufacturability through increased process and design latitudes. Design-driven fill embodies such features as the following [2].

- *Global Optimization:* Typical foundry rules specify upper and lower bounds on density in windows of the layout region. These windows are "stepped" to improve density uniformity, e.g., a physical verification run deck might verify density in 200μm × 200μm windows, stepped in increments of 50μm (Figure 3(a)). In this scenario, windows have w = 200μm, each window is divided into $r = 4$ "steps", and the step distance of $w/r = 50\mu m$ defines a *tile* size. A large ASIC 20mm on a side would have 160000 tiles on each layer, and a 10-layer metal process implies 1.6 million tiles in the chip. Modern design-driven fill should be able to compute optimal amounts of fill to be added into each tile, simultaneously for all tiles on all layers of the chip. This is enabled by, e.g., highly scalable nonlinear optimization technology.

- *Model-Based Fill Synthesis:* Rule-based fill synthesis is based on concepts such as density or *keep-off distance* rules (i.e., no fill shape is inserted less than this distance away from any layout feature, as shown in Figure 3(b)), which may be applied to wiring segments belonging to nets with less than given threshold amounts of timing slack. Model-based fill synthesis, on the other hand, would use CMP models to, e.g., identify regions where maintaining planarity and hence wire cross-section is important (e.g., next to timing-critical segments and below critical segments). The model-based approach has implicit tight coupling to extraction and timing engines, particularly to model the impact of fill on post-polish wire thickness and RC values (Figure 4).
- *Timing-Driven Fill Synthesis:* One of the largest concerns in fill synthesis, apart from meeting the CMP design rules, is the impact of fill insertion on capacitances of signal nets. Excessive increase in wire capacitance can cause a net to violate setup timing constraint (or, on the other hand, added capacitance can increase hold timing slack). A large value for keep-off distance reduces the setup slack danger but reduces available area for fill insertion and therefore can make it impossible to meet minimum density constraints. With timing-driven fill, the impact of inserting fills on timing is continually assessed, and the minimum keep-off distance for each net to meet the setup time constraint can be computed to avoid a wastefully large "one size fits all" keep-off distance. In a more advanced, timing-driven fill flow, the impact of fill insertion on both wafer topography and timing would be analyzed and optimized concurrently. One additional advantage of timing-driven fill is that it can improve the hold timing slack of a net by deliberately and selectively introducing capacitance to that net.
- *Power-Driven Fill Synthesis:* Power consumption is also a very important consideration in today's integrated circuit (IC) designs. As CMP fill changes interconnect capacitance it significantly affects the dynamic power consumption of these nets. Power-driven fill synthesis consists of identifying the power-critical nets – i.e., those with high switching activity and large capacitance – and optimizing the fill shapes around them to minimize power consumption [3].

In the next section, we present our recently proposed timing- and power-driven CMP fill insertion methodologies as examples of design-driven fill synthesis. Experimental assessments in a production 65nm technology suggest that using our proposed methodologies can reduce performance guardbanding during the design process, and increase product performance.

TIMING AND POWER-DRIVEN FILL

Fill shapes must be placed to improve not only CMP uniformity, but circuit performance as well. Insertion of fills needs to be automated utilizing available fill insertion guidelines. For timing- and power-critical circuits, timing- and power-aware fill synthesis methodologies are needed. We target these problems and provide an automated fill synthesis framework by which circuit performance can also be improved. In our fill synthesis framework, which is based on a physical-analog paradigm originating in [13, 14], the possible locations for fill insertion have allocated energies. When a fill is to be inserted, the location with lowest energy is sought. Fills are inserted into a given region one by one, until a prescribed density constraint is satisfied while performance (timing or power) is improved.

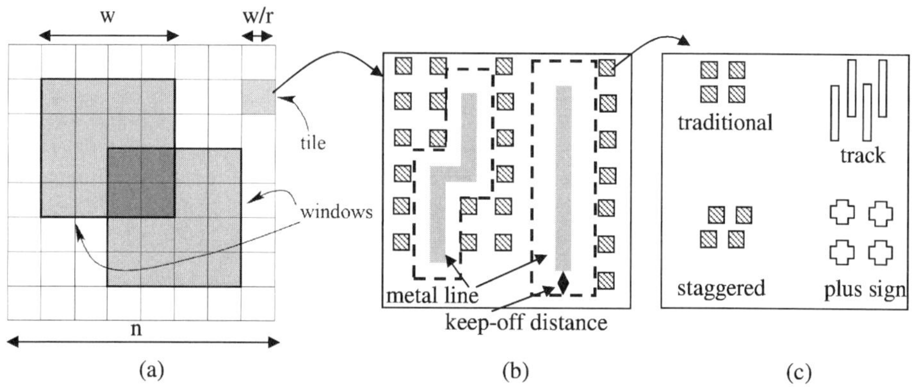

Figure 3. (a) Windows and tiles used in fill synthesis; (b) sample layout within a tile, showing keep-off distance and inserted floating fill shapes; and (3) different shapes and patterns of floating fill.

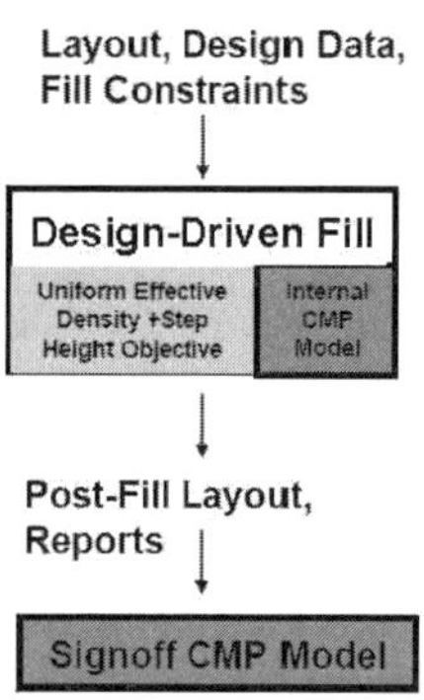

Figure 4. Design-driven fill with embedded CMP model.

Fill optimization methodology

We now describe our fill optimization methodology, covering the steps of region definition, grid definition, energy modeling within a grid, and fill insertion within a grid.

Adaptive Regions. Given a metal layer, we use a scanline algorithm to determine regions between facing interconnects with no other interconnect in between. A region consists of a maximal rectangular area such that two interconnects directly face each other in the horizontal (vertical) direction, when the main routing orientation in the given interconnect layer is vertical (horizontal). To obtain a uniform post-fill metal density while also considering performance impact of fill, we use a uniform target density for each region. Our region definition is adaptive in the sense that it follows the specific configuration of the interconnect design, in contrast to methods that simply dissect the layout into uniform square regions.

Converting a Region to a Grid. In converting a region into a grid, we can define possible grid rectangles for fill insertion. Each such region is converted into a grid after keep-off distances are stripped (i.e., made unavailable for filling). For each region, we form a grid of rectangles into which fills will be placed. A single region is shown in Figure 5, where square boxes are the grid rectangles into which metal fills can be placed. The auxiliary frame forms the outer edge of the grid. The grid rectangles and the frame are connected together with bonds, which have adjustable energy values between them. Bonds touching a grid rectangle are its incident bonds.

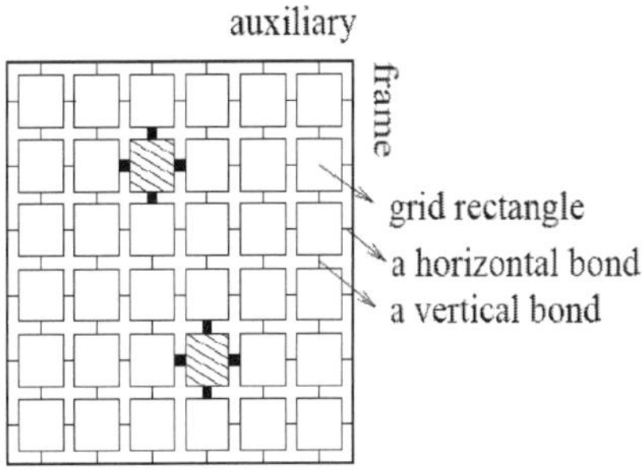

Figure 5. A grid inside a region consisting of grid rectangles to be filled.

Energy Modeling in a Grid. We have developed a parameterized model [3] to choose bond values according to a given fill insertion guideline. Fills may need to be inserted away from the critical interconnects. We design the bond energies so that they decrease as we move away from critical nets.

Insertion of Fills into a Grid. We map the fill insertion problem to an energy-minimization problem. Fills are placed in available locations to minimize an energy criterion. The rectangles are fixed in the grid, and hence the locations of bonds are fixed. We find the grid rectangle for fill insertion with minimum sum of energies for its incident bonds. We implement metal fill insertion using greedy optimization.

Timing-oriented optimization

Interconnect delay has been catching up with gate delay and even exceeding it in certain contexts. While STA (Static Timing Analysis) with gate delays alone would be sufficient previously, sub-90nm technologies require accurate incorporation of critical nets into the STA.

Hence, similar to critical gates being important in a critical path, critical nets are important during fill optimization and require special attention. In fact, chip performance can be improved by optimizing the fills around the critical nets. Table 2 shows our critical net-aware fill flow. The first step in timing critical net-aware fill is to identify the timing critical nets. We handle this step by using the STA timing reports and extracting critical net names. Fills should be placed away from critical nets. To account for this, we update bond energies so that the energies favor locations away from the critical nets.

Power-oriented optimization

Power consumption is a dominant consideration in integrated circuits. Dynamic power increases as larger capacitances are charged and discharged at every clock cycle. Power can be reduced by reducing the capacitances of power-critical nets. Similar to the critical-aware timing optimization, we determine the power-critical nets and optimize the fills around them. A net is power-critical if its dynamic power (proportional to the product of switched capacitance and switching activity factor) is high. To determine such nets, we rank the nets according to their power consumption. We select a number of the most power-critical nets and optimize the fills around these nets so that the capacitances seen by such nets are minimized.

Table 2. Timing critical net-aware fill flow.

1. Place, synthesize clock network, and route the design.
2. Extract SPEF parasitics from post-routing DEF file.
3. Run static timing analysis using SPEF file from Step 2.
4. Use Perl scripts to obtain top critical net names.
5. Input net names to metal fill optimization (MFO) for energy calculation.
6. Insert timing critical net-aware fills.

Experimental setup, protocol and results

Our experiments validate the efficiency and relevance of the proposed fill synthesis method by comparing layouts filled with the proposed method against those filled with a traditional fill method. We have implemented the proposed method using approximately 4000 lines of custom C++ code. Input is a GDSII (Graphic Data Stream) file and output is the fill-optimized GDSII and DEF (Design Exchange Format). In our flow, we use Cadence SOC Encounter v5.2 for placement, clock tree synthesis and routing, and Synopsys Star RCXT 2007.06 to extract post-fill parasitics in SPEF format. We compare layouts filled with the proposed scheme versus Blaze DFM, Inc. Intelligent Fill (IF) or Mentor Graphics Corp. Calibre-implemented fill scheme representing the merits of traditional fills; these tools are otherwise powerful industrial tools. We use Synopsys Primetime 2005.12 for static timing analysis. For CMP prediction, we use Cadence CCP 1.4 with models optimized for the TSMC 65nm process. Our analysis flow is given in Table 3. For the optimized case, fills are inserted using our code. The SPEF file on Line 4 is in Standard Parasitic Exchange Format, and the DEF file is in Design Exchange Format. For the power-oriented fill, instead of running static timing analysis, we run scripts to compute the interconnect switching dynamic power. We have used TSMC 65nm GPlus 8-layer metal technology in our experiments. We have used the ISCAS89 S38417 and

OpenCores ALU and AES benchmark circuits and a 2.2GHz server to conduct our experiments. The testcases are summarized in Table 4.

Table 3. Analysis flow.

1. Obtain filled (i.e., post-fill insertion) GDS
2. Use Perl scripts to obtain DEF file containing fills.
3. Extract SPEF parasitics from DEF containing fills.
4. Run static timing analysis using SPEF file from Step 3.
5. Use Mentor Calibre scripts to obtain density histograms.
6. Use Perl scripts to obtain timing slack histograms.

Figure 6 shows the worst 200 critical path setup slacks for the S38417 benchmark. Slack is a parameter which gives the difference between the actual signal arrival time at a flip-flop input or primary output, and the corresponding required arrival time. A distribution towards the right implies better circuit timing. We present the results using histograms to show that not only a single path is modified by optimization, but rather that the whole distribution shifts and its shape changes in most cases. As can be observed from the figure, our MFO approach results in improved timing as compared to the fill options available from the Blaze IF tool.

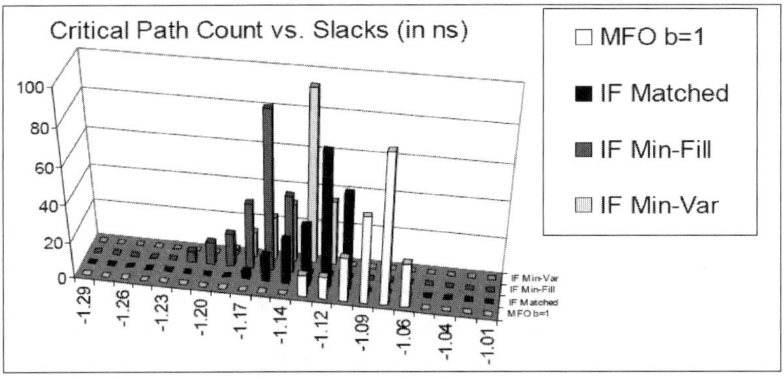

Figure 6. Timing slacks for S38417 ISCAS'89 benchmark with 43,076 interconnect segments for MFO, Blaze IF matched to the fill number of MFO, Blaze IF min-fill, and Blaze IF min-var cases for the case of 30% target density. Runtime is 172 seconds including printing out the GDS.

Table 4 summarizes our experimental results for power-oriented fill. We use 100 power-critical nets for each testcase. Power is given in the table for 1GHz operating frequency. For large circuits, a non-negligible The S38417 testcase contains many flip-flops with short critical paths, hence its optimization results (an observed 35.8% dynamic power savings) are highly optimistic. With longer critical paths, there will be more interconnects involved, not all of which can be optimized simultaneously, and fill optimization will result in smaller timing and power impact on average.

Table 4. Power-oriented fill results.

Circuit	Power with fills	Power with power-oriented fills	% Improvement
S38417	1746	1121	35.8
ALU	6337	6185	2.4
AES	11670	11350	2.7

CMP FILL SYNTHESIS: FLOW ISSUES AND FUTURES

As we look to the future, several aspects of the CMP fill synthesis flow are worth reviewing. We first consider the "platform" for delivery of future CMP fill syntheses that enhance manufacturability, as well as parametric yield, of the IC product. A natural scenario is for fill synthesis and timing to be integrated within the detailed routing tool. This is intuitively reasonable for such reasons as: (1) routers create geometries and close timing, and hence are the natural candidates to perform fill synthesis; (2) timing closure is more certain when fill is synthesized by the design team before handoff to manufacturing; and (3) grounded fill, which reduces timing uncertainty and improves IR drop, is a natural extension of power/ground routing. At the same time, there are significant barriers to emergence of such a fill synthesis solution: (1) complicated (wraparound, full-chip, width-distribution-dependent, etc.) density analyses that support high-quality CMP modeling are not easily performed by a router; (2) tools that are optimized for batch solution of full-chip detailed routing may not be able to deliver high-quality fill without a significant runtime hit; (3) with the possible exception of hold-time slack and coupling-induced delay uncertainty issues, grounded fill can be harmful to chip performance, and the more preferable floating-fill approach is less natural for a router; and (4) at the design-manufacturing handoff, better passing of design intent [15] (and, acceptance of the foundry's role in performing litho-driven reticle enhancement) could allow responsibility for CMP fill to remain with the foundry, and reduce any need for the router to assume this role.

Second, the industry will require new techniques to mitigate the impact of CMP fill on manufacturing closure. For example, depth of focus (DOF) is a major contributor to lithographic-process margin, and a major cause of focus variation is imperfect planarization of fabrication layers. Currently, optical proximity correction (OPC) methods are oblivious to the predictable nature of focus variation arising from wafer topography. As a result, designers suffer from manufacturing-yield loss, as well as loss of design quality through unnecessary guardbanding. Figure 7 shows how post-CMP-thickness variation can result in loss of critical dimension (CD) control [18]. To minimize the impact of pattern-dependent effects of the CMP process, the OPC methods should be aware of the post-CMP topography to assign appropriate defocus value for all the features with the same topography. Gupta *et al.* [18] propose a flow and methodology to drive OPC with a topography map of the layout that is generated by CMP simulation. The experimental results show that the proposed topography-aware OPC can yield up to 67% reduction in edge-placement errors and 12% reduction in timing uncertainty at the cost of little increase in mask cost.

Finally, we observe that any future CMP fill methodology will involve some subset of four potential elements: (1) CMP simulation, (2) topography-aware RC extraction, (3) timing and signal integrity awareness, and (4) multilayer fill synthesis. A "smart" fill synthesis, if it accurately optimizes planarity and accounts for multilayer topographic effects, minimizes the need for (2). And, while efforts toward (1) and (2) only address the analysis side of fill pattern design, a combination of (1), (3) and (4), with planarity assumptions validated by best possible CMP simulation, may provide a very strong replacement for today's physical-verification-based floating fill or router-based grounded fill [2].

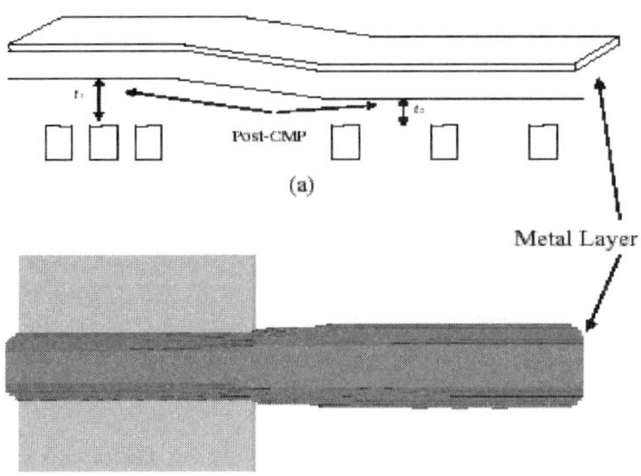

(b)

Figure 7. (a) Side view showing thickness variation over regions with dense and sparse layout. (b) Top view showing CD variation when a line is patterned over a regions with uneven wafer topography, i.e., under conditions of varying defocus [18].

CONCLUSIONS

We have reviewed the role of CMP fill in reducing manufacturing variation (specifically, post-polishing wafer topography). We have also pointed out the recent trend toward design-driven fill synthesis that integrates scalable global optimization, handling of a rich array of constraint types, CMP process modeling and simulation, and incremental parasitic extraction and timing analysis. In this paper, we introduced a fill synthesis framework that heuristically accepts and realizes complex, performance-driven CMP fill guidelines. Our heuristic approach uses an energy minimization framework to achieve metal fill insertion in a given region. We have used our tool to optimize fill in both timing criticality-aware and power-aware use contexts. With large 65nm industrial testcases, we recover up to 96.1% of timing slack that otherwise would have been lost due to critical net capacitance increase caused by added floating fills. We have also demonstrated up to 35.8% reduction in total interconnect switching power using power-aware fill insertion.

REFERENCES

1. A. B. Kahng, K. Samadi and P. Sharma, "Study of Floating Fill Impact on Interconnect Capacitance," in *Proc. ISQED*, 2006, pp. 691-696.
2. A. B. Kahng and K. Samadi, "CMP Fill Synthesis: A Survey of Recent Studies," in *IEEE Trans. on CAD*, 27(1), 2007, pp. 3-19.
3. R. O. Topaloglu, "Characterization, Modeling and Optimization of Fills and Stress in Semiconductor Integrated Circuits," *Ph.D. thesis, Dept. of Computer Science and Engineering, UC San Diego*, 2008.
4. D. White and A. Gower, "Impact of Multi-Level Chemical Mechanical Polishing on 90nm and Below," in *Proc. VMIC*, 2005.
5. B. Stine, D. Ouma, R. Divecha, D. Boning, J. Chung, D. L. Hetherington, I. Ali, G. Shinn, J. Clark, O. S. Nakagawa and S. -Y. Oh, "A Closed-Form Analytic Model For ILD Thickness Variation In CMP Processes," in *Proc. CM P-MIC*, 1997, pp. 266.273.
6. B. Stine, D. O. Ouma, R. R. Divecha, D. S. Boning, J. E. Chung, D. L. Hetherington, C. R. Harwood, O. S. Nakagawa and S.-Y. Oh, "Rapid Characterization and Modeling of Pattern-Dependent Variation in Chemical-Mechanical Polishing," in *IEEE Trans. on Semiconductor Manufacturing,* 11(2), 1998, pp. 129-140.
7. D. Ouma, "Modeling of Chemical-Mechanical Polishing for Dielectric Planarization," *Ph.D. Dissertation*, Dept. of Electrical Engineering and Computer Science, MIT, Cambridge, MA, 1998.
8. C. Ouyang, K. Ryu, L. Milor, W. Maly, G. Hill and Y.-K. Peng, "An Analytical Model of Multiple ILD Thickness Variation Induced by Interaction of Layout Pattern and CMP Process," in *IEEE Trans. on Semiconductor Manufacturing*, 13(3), 2000, pp. 286-292.
9. C.-H. Yao, D. L. Feke, K. M. Robinson and S. Meikle, "Modeling of Chemical Mechanical Polishing Processes Using a Discretized Geometry Approach," in *J. Electrochemical Society,* 147(4), 2000, pp. 1502-1512.
10. H. Liao, L. Song, N. Jakatdar and R. Radojcic, "Integration of CMP Modeling in RC Extraction and Timing Flow," in *Proc. CICC*, 2007, pp. 249-252.
11. J. Luo and D. Dornfeld, *Integrated Modeling of Chemical Mechanical Planarization for Sub-micron IC Fabrication: From Particle Scale to Feature, Die and Wafer Scales*, Springer, 2004.
12. D. Wang, W. Barth, W. Catabay, S. Lakshminaryanan, P. Burke and J. Pallinti, "Removing Copper Over low-k Films Using Stress-Free Polishing," in *Solid State Technology Magazine*, 2004.
13. A. B. Kahng and R.O. Topaloglu, "Performance-Aware CMP Fill Pattern Optimization," **Invited Paper**, in *Proc. VMIC,* 2007.
14. R. O. Topaloglu, "Energy-Minimization Model for Fill Synthesis," in *Proc. ISQED*, 2007, pp. 444-451.
15. Y. Chen, P. Gupta and A. B. Kahng, "Performance-Impact Limited Area Fill Synthesis," in *Proc. DAC*, 2003, pp. 22-27.
16. A. B. Kahng, G. Robins, A. Singh, H. Wang and A. Zelikovsky, "Filling and Slotting: Analysis and Algorithms," in *Proc. ISPD*, 1998, pp. 95–102.
17. R. Tian, D. F. Wong, R. Boone and A. Reich, "Dummy feature placement for oxide chemical–mechanical polishing manufacturability," *Technical Report*, Dept. Computer Science, Univ. Texas, Austin, TX, pp. 9-19, 1999.
18. P. Gupta, A. B. Kahng, C.-H. Park, K. Samadi and X. Xu, "Wafer Topography-Aware Optical Proximity Correction," in *IEEE Transactions on CAD*, (25)12, 2006, pp. 2747-2756.
19. A. Balasinski, J. Cetin and A. B. Kahng, "Intelligent Fill Pattern and Extraction Methodology for SoC," in *Proc. IWSOC*, 2006, pp. 156–159.

Conference Proceedings AMC XXIV © 2009 Materials Research Society

Post-Direct-CMP dielectric surface copper contamination: quantitative analysis and impact on dielectric breakdown behaviour

Nancy Heylen[1], Yunlong Li[1], Kristof Kellens[1], Youssef Travaly[1], Guy Vereecke[1], Henny Volders[1], Zsolt Tokei[1], Janko Versluijs[1], Jens Rip[1], Els Van Besien[1], Laureen Carbonell[1], Gerald Beyer[1], Paul Fischer[2], Larry Zhao[3], Gaetano Santoro[4], Andrew Cockburn[4], and Olivier Nguyen[4]

[1]IMEC, Kapeldreef 75, Leuven, B-3001, Belgium

[2]Intel Corporation, 5200 N.E. Elam Young Parkway, Hillsboro, OR, 97124-6497

[3]Intel Corporation, Kapeldreef 75, Leuven, B-3001, Belgium

[4]Applied Materials Belgium, Kapeldreef 75, Leuven, B-3001, Belgium

ABSTRACT

In this study we were able to look into the inter-metal dielectric surface copper contamination after a bulk copper and copper barrier film have been removed by chemical mechanical polishing. By using a novel test vehicle – i.e. by switching from blanket to so-called partially patterned wafers - we were able to measure the post-CMP dielectric surface residual copper originating from the M1 Cu-low k damascene structures present on the wafer. We investigated the impact of polishing conditions, cleaning conditions and low-k material properties on the copper contamination level and for the first time established a link between the copper contamination level and reliability of the interconnect structures.

INTRODUCTION

With advancements in ultra large-scale integration (ULSI), integrated microelectronic device dimensions are being scaled down and the effective dielectric constant (k_{eff}) has to be lowered to achieve higher-speed operation and lower power consumption. To meet this demand, high porosity ultra low-k (ULK) materials have been successfully introduced as an interlayer dielectric (ILD) in several integration flows for 45 and/or 32 nm technology nodes. Beyond the 32 nm technology node, dielectric protection layers (DPL) may be left out of the integration schemes with Direct-Chemical Mechanical Polishing (D-CMP) of the ULK material. This makes the implementation of the porous ULK films more difficult because of the impact of the D-CMP step on the k_{eff} value and the interconnect structure reliability. Being able to isolate key contributing factors to process-induced physical and chemical alteration of the ULK material surface and bulk is therefore of paramount importance as damage caused by the D-CMP process step starts to play an important role. We previously reported a D-CMP induced degradation mechanism for porous ULK materials, together with several key contributing factors[1-2]. One of these contributing factors to the ULK damage is surface metallic residues. This study further investigates metallic contamination of porous ULK films integrated with a D-CMP approach, and uses a novel method for measuring residual copper on patterned wafers and correlating the measured copper contamination level to basic reliability data.

EXPERIMENTAL DETAILS

Several relevant material properties of the dielectric films used are listed in table I. To look at the impact of material properties we selected films with significantly different k value, porosity, Young's modulus and hardness. Partially patterned wafers, as shown in figure 1, were used as test vehicle. With this test vehicle we were able to characterize the post-CMP dielectric surface copper contamination by TXRF measurements on the non-printed "checkerboard" structures, and measure reliability of adjacent M1 interconnect structures.

Table I: Relevant material properties of dielectric films used in this work. ULK-1: porous PECVD SiCOH film with ultra low-k 2.5; ULK-2: highly porous film with k of 2.2, achieved by using a different porogen precursor (as compared to ULK-1); ULK-3: experimental film with also k 2.2, achieved by increasing strongly the porosity for ULK-1, but at the same time significantly lower YM and H; LK-4: non-porous PECVD SiCOH film with low k 3.0.

Inter-metal dielectric film	k value	Porosity (%)	YM - Young's modulus (Gpa)	H- Hardness (Gpa)
ULK-1	2.5	25	9	1.25
ULK-1, Exp: UV Cure time +/- 50% Porogen precursor +/- 10%		25+/-2	9 +/- 1	1.25 +/-0.15
ULK-2	2.2	36	ULK-2 < ULK-1	ULK-2 < ULK-1
ULK-3 **= experimental film**	2.2	38	ULK3 << ULK-2 (close to 6)	ULK3 << ULK-2 (close to 0.7)
LK-4	3.0	6	12.5	2

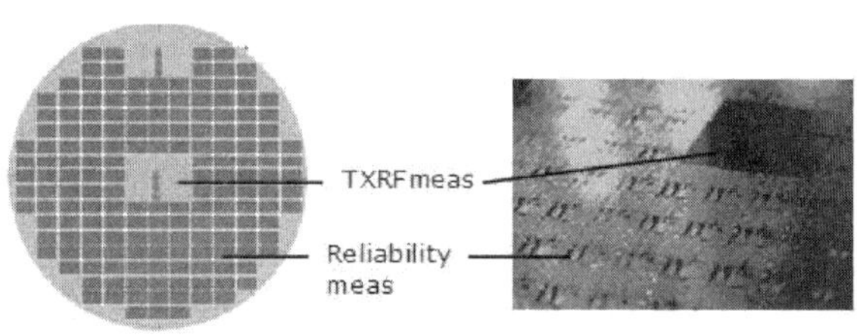

Figure 1: Test vehicle (partially patterned wafer) used to establish a link between ULK surface copper contamination and interconnect reliability. TXRF performed on non-printed checkerboard structures; reliability measurements performed on M1 Cu/ULK interconnect structures.

The M1 Cu CMP process was performed on the 300 mm AMAT Reflexion_LK^{TM} rotary polisher with integrated Desica cleaner, consisting of a megasonic module, two brush modules and a vapor dryer. We evaluated the post-CMP Cu contamination using commercial IMEC's POR Cu and Cu barrier slurries. For a first series of tests, mainly looking at the impact of polishing and cleaning conditions, we used our 65 nm POR Cu barrier slurry which is non-selective to the dielectric film, with < 20% of amorphous silica and a pH of 9.5-10.5, combined with an acidic cleaning solution for megasonic and brushes. To look at the impact of dielectric film properties we used our 32 nm POR selective Cu barrier slurry, with < 5% of colloidal silica and a pH of 2.5-3.0, in combination with the same cleaning solutions. To achieve relatively high levels of Cu residues (as compared to Cu levels post-CMP), single damascene structures were also intentionally contaminated either with an aqueous Cu sulphate solution on a spinning tool or with a controlled PVD Cu flash on an AMAT Endura tool after CMP. Reliability measurements were performed after full passivation for all wafers.

RESULTS AND DISCUSSION

Impact of CMP process conditions on post-CMP Cu contamination

Initially we were polishing blanket metal/ULK systems, for which the metal could be a Cu barrier or a metal hard mask layer. We used TXRF analysis to measure post-CMP surface metallic residues and determined that the level of contamination correlates with the down force used at CMP and the remaining ULK thickness (see figure 2 and table II): a lower down force (DF lowered from 3 to 1 PSI in figure 2) and a lower remaining ULK thickness (between 15 and 75 nm ULK removed in figure 2) result in a lower metal level.

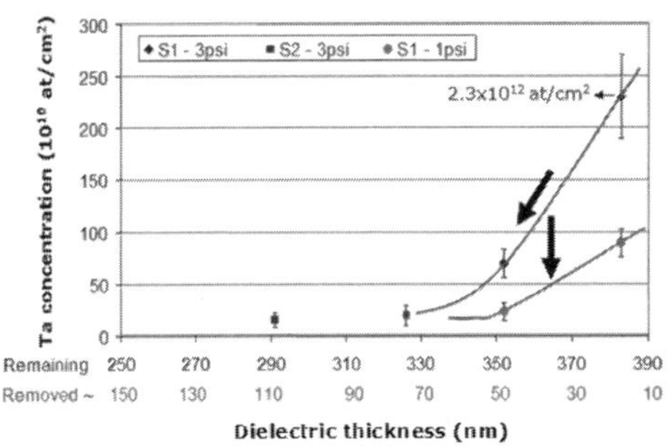

Polished blanket films	Ta measured (10^{10} atoms/cm^2)	Ti measured (10^{10} atoms/cm^2)
ULK-1 + TaN / metal cleared	95.8	0.66
ULK-1 + TaN / ULK 30 nm overpolish	0.34	6.64
ULK-1 + TiN / metal cleared	0.75	16.61
ULK-1 + TiN / ULK 30 nm overpolish	1.40	2.96

Figure 2 (from ref.1): Measured Ta residues on polished TaN/ULK film surface vs. remaining ULK film thickness and down force (1 or 3 PSI). Data obtained on 200 mm polisher with slurries S1 and S2, for organosilicate glass film, not listed in table I.
Table II: Measured Ta or Ti residues on polished TaN/ULK or TiN/ULK film surface after "just clearing" or overpolishing ULK. Data obtained on 300 mm polisher, for ULK-1, listed in table I.

For most of the tests on partially patterned wafers, we used a "standard" porous ULK-1 film, with a Young's modulus (YM) of about 9 Gpa and hardness (H) of 1.25 Gpa (table I). Processing these wafers with IMEC's POR non-selective Cu barrier slurry and acidic cleaning chemicals, initially resulted in post-CMP Cu levels close to 5E+11 atoms/cm^2. Angle Resolved (AR)-TXRF measurements pointed out that the Cu residues on the ULK surface after polishing and cleaning represent a film (and not a particle) type of contamination (see figure3), i.e. adsorbed Cu at this level of contamination. The data in figure 4 also suggested though that the post-CMP Cu level is very much impacted by the polishing and cleaning conditions. With respect to the impact of the polishing conditions - see figure 4, section A and C - we were able to lower the Cu level by extending the process time on the Cu or barrier polish platen, i.e. by overpolishing on the ULK film at the end of the process sequence. These data are consistent with the data shown in figure 2 and table II for blanket samples. With respect to the impact of the cleaning conditions - see figure 4, section B - the Cu contamination was found to be almost one order of magnitude lower (Cu level close to 5E+10 at/cm^2; i.e. close to the detection limit of the TXRF measurement) when skipping the brush cleaning step completely or when switching to new, clean brushes in combination with a replenish of the megasonic cleaning chemicals. These data suggested that while running Cu contaminated wafers through the cleaner, the brushes got loaded with Cu residues. They also showed the importance of implementing an efficient "brush metal cleaning procedure". Running polished oxide dummies through the brushes seems to be an efficient cleaning procedure, i.e. helps in lowering the dielectric surface Cu contamination for wafers that run through the cleaner afterwards.

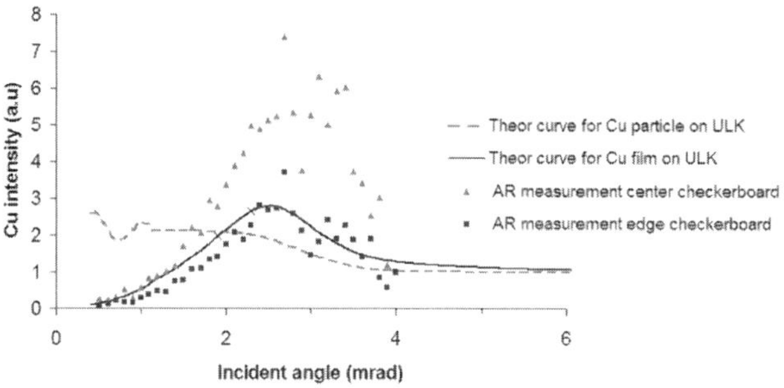

Figure 3: Cu level vs TXRF measurement angle (AR-TXRF), pointing out that the Cu residues on the ULK surface after CMP represent a film (not particle) type of contamination.

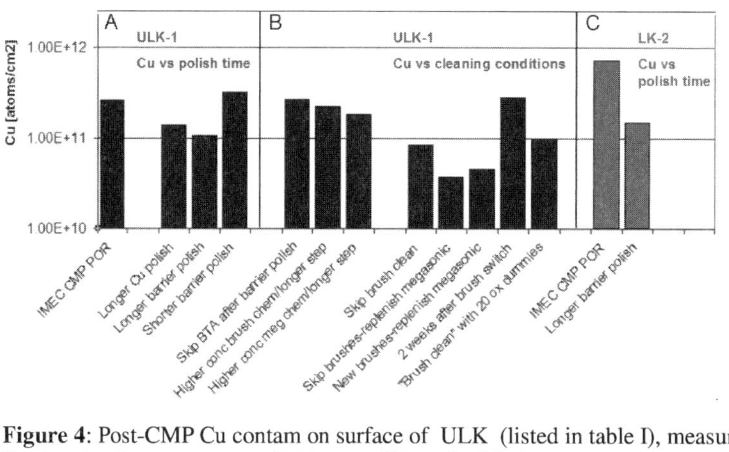

Figure 4: Post-CMP Cu contam on surface of ULK (listed in table I), measured by TXRF. Section A: Cu contam vs. polishing conditions for ULK-1; section B: Cu contam vs. cleaning conditions for ULK-1; section C: Cu contam vs. polishing conditions for LK-2. All wafers are polished with a non-selective Cu barrier slurry and cleaned with acidic chemicals.

Impact of inter-metal dielectric material properties on post-CMP Cu contamination

To look at the impact of the dielectric film properties, we did an experiment in which we changed the porosity, YM and H of ULK-1 by changing the cure time to +/- 50% and the porogen precursor ratio to +/- 10%. The porosity of these experimental films varies between 23-27%, YM between 8-10 Gpa and H between 1.10-1.40 Gpa. We also tested ULK-2, which has a lower k value and higher porosity, and an experimental film ULK-3 that has the same ultra low k value and high porosity as ULK-2, but for which we pushed YM and H to much lower values than for all the other films. All wafers were polished with a selective Cu barrier slurry to avoid an overpolish on the ULK films. The measured post-CMP Cu contamination level was similar for all films tested (no significant difference), except for the experimental film ULK-3. For this "mechanically weaker" film, the Cu level was found to be at least one order of magnitude higher than for the other films which have a significantly higher YM and H (see figure 5).

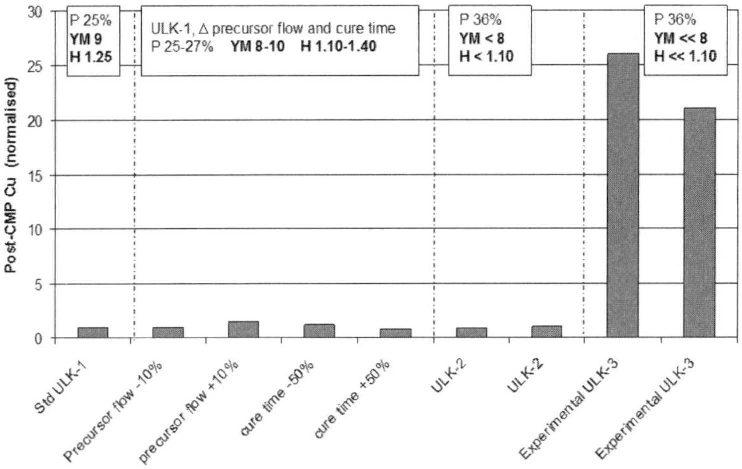

Figure 5: Post-CMP Cu contamination on surface of ULK films, listed in table I; measured by TXRF. All data obtained with a selective Cu barrier slurry and acidic cleaning chemicals.

Impact of dielectric surface Cu contamination on interconnect reliability

Our TXRF data showed that the Cu contamination was found to be almost one order of magnitude lower when switching to new, clean brushes (see figure 4, section B). With respect to reliability, our work suggested that for different cleaning conditions tested on LK-2, i.e. for post-CMP Cu levels up to maximum 1E+12 at/cm^2, there is no issue in terms of degradation of the interconnect reliability (see figure 6).

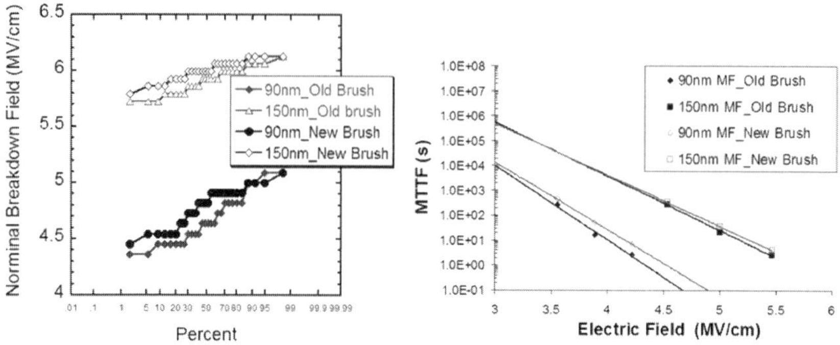

Figure 6: Breakdown field and MTTF data for new cleaner brushes (i.e. post-CMP Cu level 1E+10-1E+11 at/cm^2) vs older cleaner brushes (i.e. post-CMP Cu level 1E+11-1E+12 at/cm^2). MTTF for both CMP conditions similar (in spec). Data obtained for Cu/LK-2 @100°C, for 90/90 nm or 150/150 nm LW/LS MF structures.

To achieve relatively high levels of Cu residues (as compared to the Cu level of < 1E+12 at/cm^2 on a post-CMP reference), single damascene structures were also intentionally contaminated with an aqueous Cu sulphate solution on a spinning tool after CMP. The Cu contamination level on these spin coated wafers appeared to be increasing from about 8E+12 at/cm^2 in the center checkerboard to about 1.2E+15 at/cm^2 at the edge. The data in figure 7 showed there is a clear impact of these higher Cu levels (i.e. > ~1E+13 at/cm^2) on leakage behaviour.

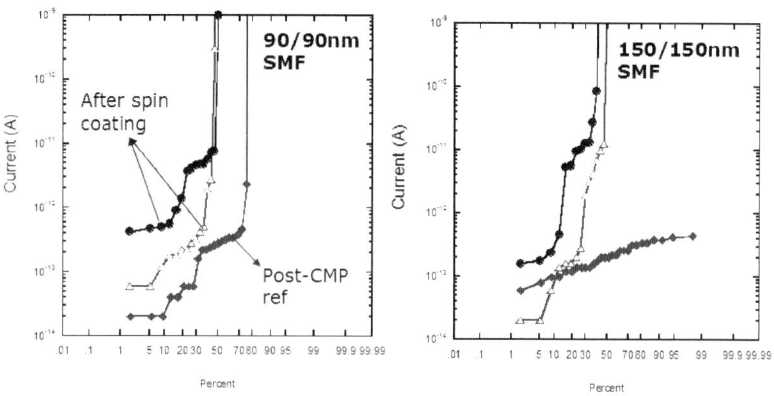

Figure 7: Leakage current data for relatively low Cu level (i.e. post-CMP ref: Cu< 1E+12 at/cm^2) vs high Cu levels (i.e. post-CMP and Cu spin coat: 8E+12 at/cm^2 in center < Cu < 1.2E+15 at/cm^2 at edge). Data obtained for Cu/LK-2 @1MV/cm, 100°C, for 90/90 or 150/150 nm LW/LS MF structures.

We also obtained breakdown voltage and MTTF data for wafers that were intentionally contaminated to uniform, high Cu levels of 6E+15 at/cm^2 with a controlled post-CMP Cu flash on an Endura Cu seed chamber. For these wafers MTTF drops far below the spec of 10 years (see figures 8 and 9).

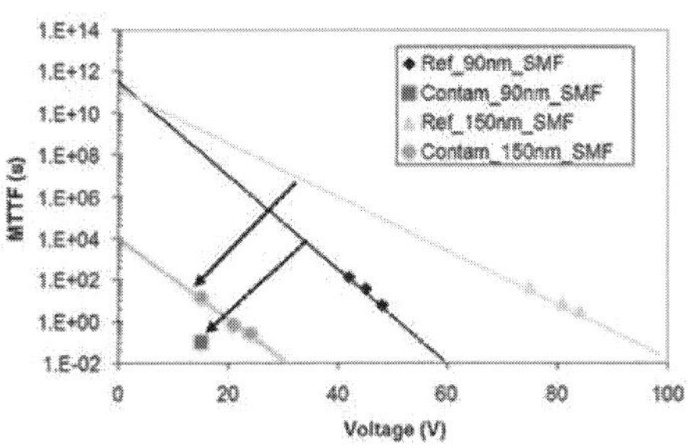

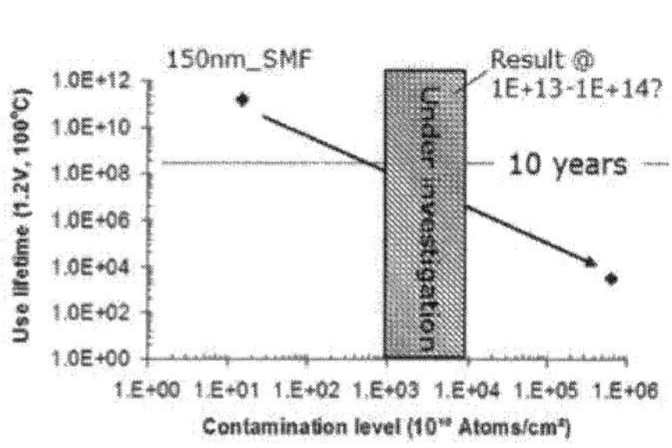

Figure 8 (left): Breakdown voltage and MTTF for relatively low Cu level (i.e. ref. post-CMP: Cu< 3E+11 at/cm^2) vs high Cu level (i.e. Contam. post-CMP and Endura Cu flash: Cu= 6E+15 at/cm^2). Data obtained for Cu/ULK-1 @100°C, for 90/90 or 150/150 nm LW/LS MF structures.
Figure 9 (right): MTTF vs ULK surface Cu contam level. MTTF @ 1.2V, 100°C in spec for Cu <1E+12 at/cm^2; MTTF far below spec for Cu >1E+15 at/cm^2. MTTF for Cu 1E+13-1E+14 at/cm^2 under investigation. Data obtained for Cu/ULK-1, for 150/150 nm LW/LS MF structures.

CONCLUSION

A quantitative analysis of the post-CMP inter-metal dielectric surface copper contamination was performed on partially patterned wafers. For the standard ULK films we measured a Cu level between 5E+10 and 1E+12 at/cm^2, depending on the polishing conditions and cleaner status. The Cu levels were only higher then 1E+12 at/cm^2 when intentionally pushing YM and H of the dielectric film to much lower values. Our work clearly shows a correlation between inter-metal dielectric surface copper contamination levels and dielectric reliability. For the post-CMP Cu contamination levels up to maximum 1E+12 at/cm^2, there is no issue in terms of degradation of interconnect reliability. For controlled, higher Cu levels of at least 1E+13 at/cm^2 we see a clear impact on leakage behaviour and for levels as high as 6E+15 at/cm^2 MTTF drops far below the spec of 10 years. These data can be exploited for definition of contamination specifications that are required for adequate reliability performance.

ACKNOWLEDGMENTS

The authors would like to acknowledge Joaquin Torres (ST Microelectronics) for the fruitful discussions on this topic.

REFERENCES

1. N. Heylen et al., Proc. of AMC/ADMETA 2006
2. Y. Travaly et al, Proc of IITC 2007

Conference Proceedings AMC XXIV © 2009 Materials Research Society

Impact of direct CMP on porous ULK properties and interconnect propagation performance on a wide frequency bandwidth

Michel Gallitre[1,2], Samuel Gall[3], Alexis Farcy[1], Benjamin Blampey[2], Béatrice Icard[3], Vincent Jousseaume[3], Marc Guillermet[3], Philippe Brun[1], Aziz Zenasni[3], Patrick Grosgeorges[3], Vanessa Sbrugnera[2], Jessy Bustos[1], Cédric Bermond[2], Thierry Lacrevaz[2], Bernard Fléchet[2], Maurice Rivoire[3], and Pascal Ancey[1]

[1]STMicroelectronics, 850 rue Jean Monnet, Crolles, 38926, France

[2]Université de Savoie, IMEP-LAHC, Le Bourget du Lac Cedex, 73376, France

[3]CEA-Léti/Minactec, 17 rue des Martyrs, Grenoble Cedex 9, 38054, France

ABSTRACT

Reaching ICs performance specifications is becoming more and more difficult at each new technology node thus requiring new integration schemes. A promising one is based on the removal of the oxide hard mask to reduce the effective k-value but several challenges are associated with direct polishing of porous dielectrics. The effects of direct CMP on ULK materials and interconnect performance were investigated. Measured increases of k-value were correlated to moisture uptake and to the organic contamination on the ULK surface. High-frequency characterizations of multi-level interconnects highlighted the impact of dielectric loss on propagation performance. Thanks to specific processes, it appears nevertheless that their properties can be perfectly restored.

INTRODUCTION

With CMOS integrated circuits dimensions shrinking at each new technology node, porous materials with very low dielectric constants are required to maintain good interconnect performance by decreasing the parasitic capacitance. As from the 45 nm technological node, the alternative of removing the SiO_2 hard mask (HM) as shown on Figure 1 provides a significant decrease in the architecture effective k-value, thus improving propagation. The impact of the removal of a 20 nm thick protective hard mask is quantified thanks to predictive electromagnetic simulations taking into account expected dimensions from the ITRS [1] on Figure 2, highlighting its interest.

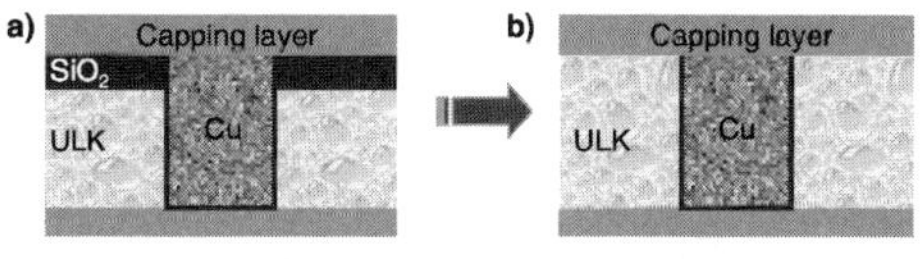

Figure 1. Comparison between stacks a) with hard mask and b) without hard mask.

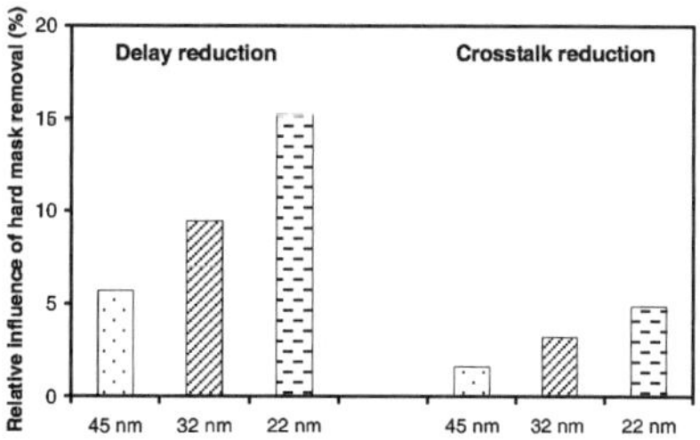

Figure 2. Gain on signal propagation induced by SiO_2 hard mask removal.

As a consequence, direct CMP is being developed to end polishing in the ULK [2]. However many concerns are associated with its effects on the inter-metal insulator [3]. The degradation mechanisms have to be investigated and their impact on interconnect propagation performance quantified. Looking at the increasing frequencies of transmitted signals in ICs, HF characterization and modeling also have to be performed in order to investigate propagation performance. Finally the effects of two kinds of restoration treatments are investigated on SiOCH properties.

EXPERIMENTAL DETAILS

Two porous films were deposited with a porogen approach [4]: SiOCH k = 2.5 with a porosity of 26% and SiOCH k = 2.35 with a porosity of 30%. The average pore radius is 1.2 nm for both materials. Chemical analysis (MIR-IR and ToF-SIMS) on blanket wafers were then performed and allowed an investigation of surface state, chemical uptake and material modification. K-value and contact angles were finally obtained by mercury probe measurements.

It is also of universal importance to study typical transmission lines embedded in SiOCH dielectrics to measure the impact of any degradation during direct CMP on propagation performance. Using the same slurries and cleaning chemicals as previously checked, two and three-level RF test structures embedded in both materials were integrated following the double damascene Trench First Hard Mask architecture [5] as described in Figure 3. These capped and uncapped Metal 2 structures allow the quantification of their different impacts on porous material properties as they correspond to real ICs and worst case scenarios.

The microstrip architecture was chosen in order to concentrate electromagnetic fields in the ULK, thus avoiding interference with Si. The stack is typical of the 45 nm node, combined with designed line widths of respectively 150 nm and 450 nm (to reduce resistance impact compared to other parameters). Two different Metal 1 ground plane topologies described on Figure 7 were furthermore implemented to evaluate potential direct CMP issues also at this level.

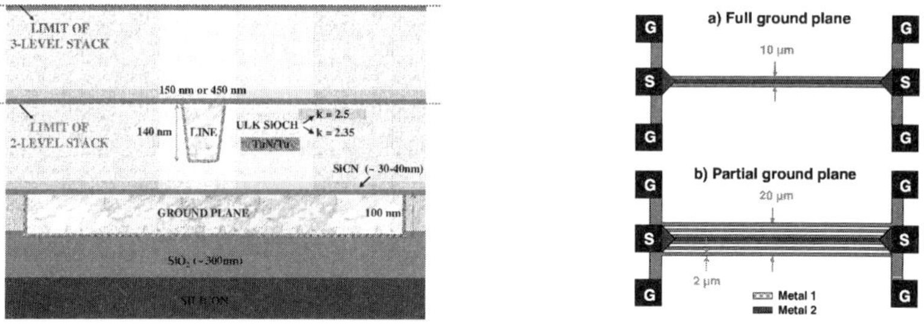

Figure 3. Cross-section and top view of investigated microstrip test structures.

RESULTS AND DISCUSSION

Physical characterizations

First, the organic contamination of porous materials during CMP is underlined on Figure 4 by the increase of the C-H bonds signal observed by MIR-IR after CMP in the [2800-2950]

cm^{-1} range. On the other hand, small changes in OH bonds signal can also be seen in the [3100-3700] cm^{-1} range. Such diffusion from polishing chemicals into the first pores of the insulator is confirmed by the presence of surfactant fragments (C_2H_3 ions) identified by ToF-SIMS (Figure 5) on a total thickness of around 5 nm.

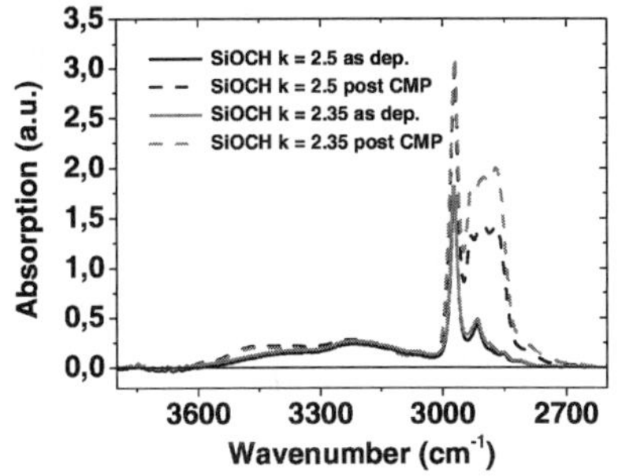

Figure 4. As deposited and post-CMP MIR-IR spectra.

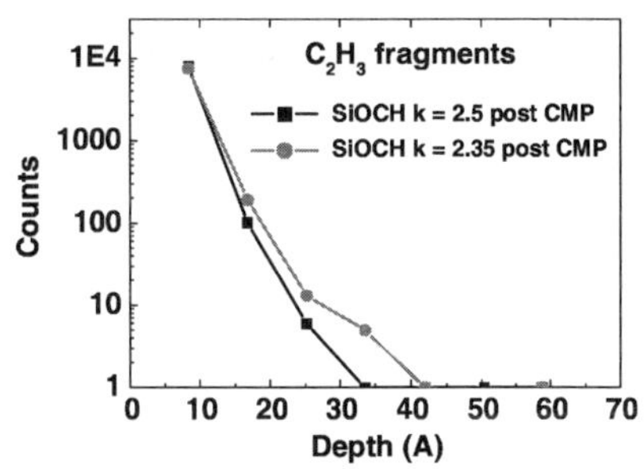

Figure 5. Post-CMP ToF-SIMS analysis of C_2H_3 ions.

The organic signature is associated with a strong k-value increase of the material as its insolating properties are clearly affected on Figure 6. Moreover, the higher organic contamination observed for the most porous material (SiOCH k = 2.35) correlates with a higher k-value increase of nearly 50 %. Measured contact angles also show the hydrophilic nature of the ULK surface after direct CMP.

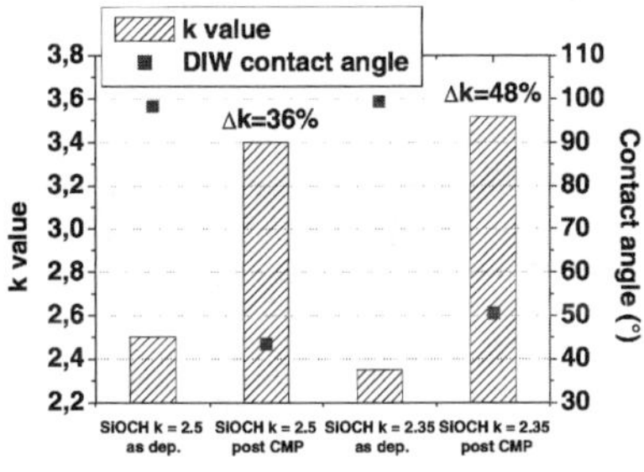

Figure 6. Direct CMP impact on K-value and DIW contact angles for each material.

The proposed hypothesis for these modified properties is that a small organic layer due to the C-H bonds is created on the top of the dielectric surface, making the ULK vulnerable to moisture uptake. An IPA rinse was then performed by spin-coating on some areas of the wafer and highlighted a k-value decrease from 3.4 to 2.8, confirming this idea [6]. It is wise to wonder what impact will be induced on the propagation parameters of specific RF transmission lines.

High-frequency measurements and electromagnetic modelling

After performing S-parameters measurements up to 40 GHz using a Vectorial Network Analyzer, the propagation constant $\gamma = \alpha + j.\beta = ((R + jL\omega)(G + jC\omega))^{1/2}$ was obtained. Results plotted on Figure 7 indicate that the ground plane configuration only holds a small impact on the corresponding propagation parameters whatever the line width. For upcoming technologies, many dishing issues linked to direct CMP on large lines could then be overcome by using this kind of architectures, which will be considered for the following studies.

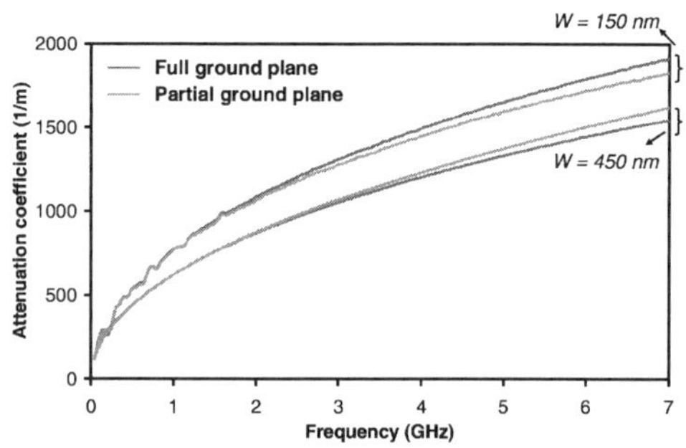

Figure 7. Propagation constant dependence with ground topology.

From previous S-parameters, the elements of the electrical equivalent scheme of these interconnects can also be extracted and are plotted on Figure 8. No skin effect is visible as resistance is nearly constant until 30 GHz which could be expected looking at the small line dimensions, constant capacitance confirms the quasi-TEM mode propagating in these lines and the increase of inductance at lower frequencies can be explained by the larger current loop. On the other hand, very high conductance values are obtained compared to previously measured lines with low-k materials [7].

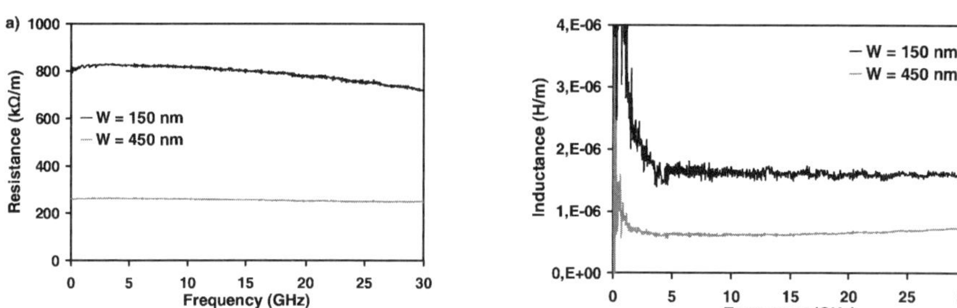

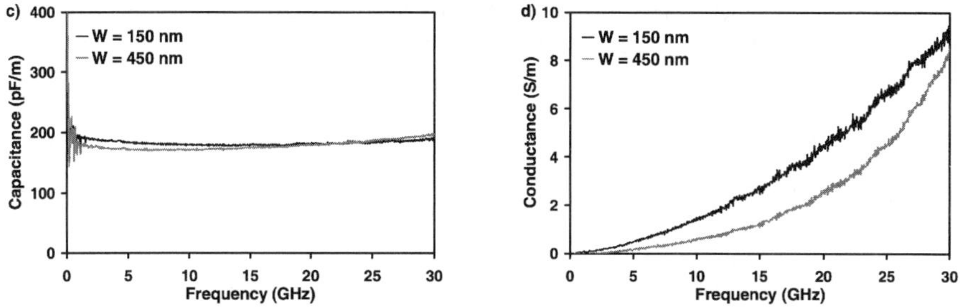

Figure 8. Comparison between RLCG lineic parameters for two line widths.

A specific optimization protocol was furthermore improved to extract porous material properties [8]. This methodology can be applied to any kind of architecture (microstrip or coplanar) and also takes into account neighboring dummies introduced to ensure sufficient Cu density in large dielectric spaces for CMP.

Thanks to this procedure, the increase of the 26 % porous SiOCH's k-value (deposited at 2.5) on the entire frequency range is highlighted on Figure 9. The uncapped 2-level architecture is the most affected with an average k-value of 3.1 but even for the typical ICs on three levels, the resulting 8 % degradation is still significant (k-value of 2.7). It can also be pointed out that the global variation of the ULK k-value versus frequency is limited to 5 % until 20 GHz; no relaxation phenomena thus occurs at these frequencies for porous materials. Significant moisture uptake (OH bonds) finally induces high dielectric loss tangents, with values reaching 10^{-1}, rendering complete RLCG interconnect modeling mandatory as from the 45 nm node if no efficient restoration solution is elaborated. The obtained loss tangent parameters are hence in accordance with the roughly approximated values that can be obtained from the conductance $G = C^{*}\omega^{*}\tan\delta$, validating the proposed methodology.

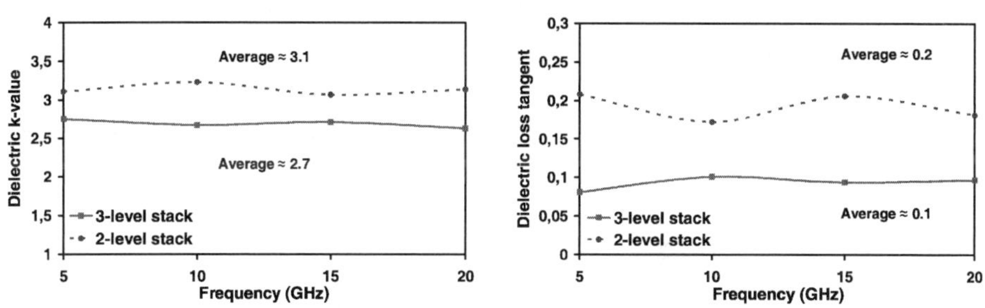

Figure 9. Extracted complex permittivity for material A as a function of frequency for both studied stacks.

The impact of dielectric loss on delay and crosstalk also has to be taken into consideration. Simulation results demonstrate that neglecting the conductance in calculating delay values would result in a 10 % underestimation of delay for the 3-level stack; on the other hand, the crosstalk parasitic level remains nearly unchanged. Time-domain causality issues are also induced by this new problematic, especially if the loss tangent is considered constant on the entire frequency bandwidth which is nearly the case looking at extraction results. As a matter of a fact, the output signal may begin before the input one, which has no physical sense. Therefore new modelling tools are mandatory to avoid these time [9].

As performance issues occur with higher k-value and dielectric loss introduction, restoration treatments are definitely mandatory.

Restoration techniques

First, it is shown on Figure 10 that the organic contamination after CMP can hardly be removed on blanket wafers by a hot plate annealing at 350°C during 40 seconds. A higher anneal time would probably be required. The most efficient technique would finally consist in a NH_3 plasma treatment at the same temperature, resulting in an almost complete removal of CMP-based contamination.

Now moving on to the integrated structures and testing a longer procedure, a low pressure post-integration anneal at 200°C for 2 hours in a controlled N_2/H_2 environment of the previously measured lines is performed and HF characterization occurs a couple of hours later. The static resistance values measured show less than 0.2 % increase after anneal so only capacitance and conductance are modified. Looking at Figure 11 resulting propagation exponent, the corresponding decrease of C and G parameters confirms that this treatment provides a complete restoration of SiOCH k = 2.5.

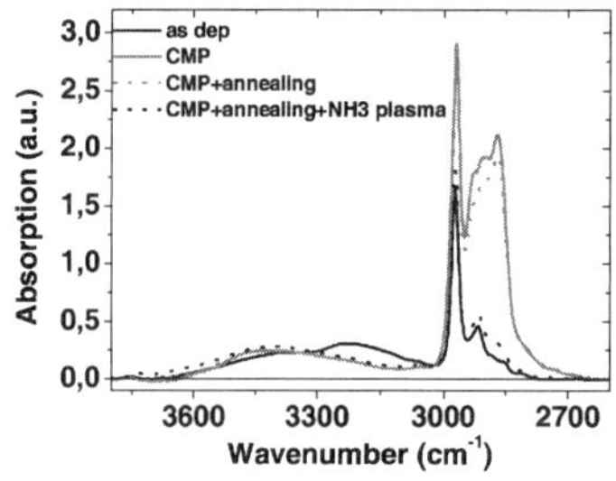

Figure 10. Effects of annealing and NH_3 plasma on SiOCH k=2.35.

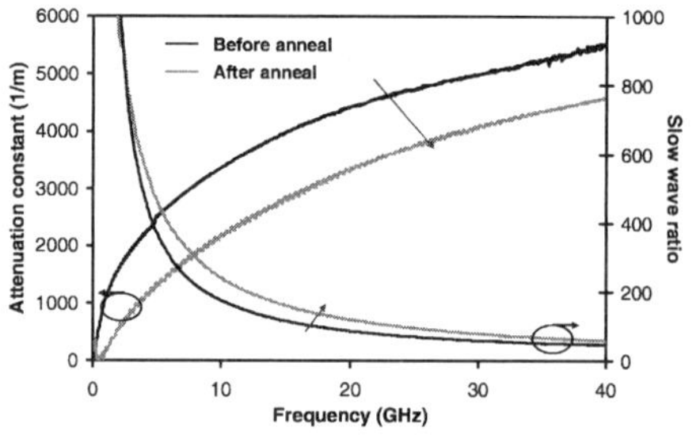

Figure 11. Variation of propagation parameters after final anneal for SiOCH k=2.5.

CONCLUSIONS

Complementary physical and RF characterizations allowed a complete overview of direct CMP influence on interconnects performance. Direct CMP induces significant moisture uptake but also high organic contamination of ULK materials, probably due to a thin layer formation on top of the dielectric surface. More porous dielectrics were also shown as tending to be more sensitive to K-value degradation.

RC models are also no longer sufficient to completely characterize the associated porous materials as dielectric loss now highly contributes to signal propagation and can no longer be neglected. For upcoming nodes, the specific protocol developed in this contribution for ULK complex permittivity extraction will also be applicable to any kind of material or architecture.

The efficiency of restoration techniques was also demonstrated on both blanket and patterned wafers.

ACKNOWLEDGMENTS

These researches are supported by the European Commission's Information Society Technologies Programme (IST) under PULLNANO Project Contract No. IST -026828 and by the MEDEA+ Project 2T103 – FOREMOST (Integration of FOuRty fivE nanometer CMOS Technology).

REFERENCES

[1] International Technology Roadmap for Semiconductors (ITRS), 2006 edition.
[2] S. Kondo, B.U. Yoon, S.G. Lee, S. Tokitoh, K. Misawa, T. Yoshie, N. Ohashi, N. Kobayashi, *Digest of technical papers of the VLSI Symposium 2004*, pp. 68-69.
[3] S. Gall, C. Euvrard, S. Petitdidier, M.Rivoire, F. Durix, *Proceedings of CMP-MIC 2008*, pp. 17-24.
[4] M. Aimadeddine, V. Jousseaume, V. Arnal, L. Favennec, A. Farcy, A. Zenasni, M. Assous, M. Vilmay, S. Jullian, P. Maury, V. Delaye, N. Jourdan, T. Vanypre, P. Brun, G. Imbert, Y. LeFriec, M. Mellier, H. Chaabouni, K. Hamioud, F. Volpi, D. Louis, G. Passemard, J. Torres, *Proceedings of IITC 2007*, pp. 175-177.
[5] R. Fox, O. Hinsinger, E. Richard, E. Sabouret, T. Berger, C. Goldberg, A. Humbert, G. Imbert, P. Brun, E. Ollier, C. Maurice, M. Guillermet, C. Monget, V. Plantier, H. Bono, M. Zaleski, M. Mellier, J.P. Jacquemin, J. Flake, B.G. Sharma, L. Broussous, A. Farcy, V. Arnal, R. Gonella, S. Maubert, V. Girault, P. Vannier, D. Reber, A. Schussler, J. Mueller, W. Besling, *IEDM Technical Digest*, pp. 81-84 (2005).
[6] T-S. Kim, T. Konno, T. Yamanaka, R.H. Dauskardt, *Proceedings of CMP-MIC 2008*, pp. 21-26.
[7] K. Jooyong and D.P. Neikirk, *Electrical Performance of Electronic Packaging*, pp. 93-96 (2003).
[8] T. Lacrevaz, B. Fléchet, A. Farcy, J. Torres, M. Gros-Jean, C. Bermond, O. Cueto, B. Blampey, G. Angenieux, J. Piquet, F. De Crecy, *Microelectronic Engineering 82*, pp. 548-553, (2005).
[9] S. de Rivaz, T. Lacrevaz, M. Gallitre, A. Farcy, B. Blampey, C. Bermond, B. Fléchet, *Proceedings of SPI 2008*, pp. 61-64.

Conference Proceedings AMC XXIV © 2009 Materials Research Society

Porous SiOCH ultra low-k recovery treatments after direct CMP process

Hamed Chaabouni[1,2], Gregory Imbert[1], Laurent Luc Chapelon[1], Maxime Mellier[1], Delphine Fossati[1], Christophe Licitra[3], Thierry Chevolleau[2], Olivier Joubert[2], and Pascal Ancey[1]

[1]STMicroelectronics, 850, rue Jean Monnet, Crolles, 38926, France

[2]LTM/CNRS, 17, rue des martyrs, Grenoble, 38054, France

[3]CEA/LETI - MINATEC, 17, rue des martyrs, Grenoble, 38054, France

Abstract

Direct CMP integration approach has been investigated and was found to damage the porous ultra low k properties leading to weak performance and reliability of interconnects. To overcome these issues, additional restoration step is proposed. This work discusses UV cure restoration treatment effects on porous dielectrics. The treatment reveals a beneficial impact on the dielectric especially by lowering its integrated k-value and improving its reliability allowing a reliable integration.

Introduction

As device dimensions shrink, RC delay in interconnects becomes more significant. In order to minimize this delay increase, porous ultra low-k (ULK) dielectrics are integrated starting from the 45 nm technology node. In a standard integration scheme, a hard mask is used atop the ULK dielectric in order to perform the chemical mechanical polishing (CMP) of copper interconnects. In order to achieve the ITRS requirements, a new integration scheme called direct CMP is used to decrease the effective k on interconnects [1]. However, direct CMP generates damages impacting ULK integrity and performance [2]. Different solutions have been investigated in order to prevent direct CMP induced damages. In particular, the introduction of a supplementary dielectric capping for the ULK protection has already been studied [3] and presented promising results. Nevertheless, this approach induces changes in the integration scheme. In this work, direct CMP damages will be studied through different characterization techniques, and thereafter restoration treatments based on thermal treatments will be investigated.

Experimental

The ULK material under investigation is a porous SiOCH dielectric with a k-value of 2.5, an average pore diameter of 2 nm and a porosity of 26%.

Firstly, electromagnetic simulations are carried out in order to determine potential gains from direct CMP integration scheme in respect of the 45 nm technology node dimensions.

Direct CMP damages and thereafter restoration treatment effects are studied on blanket wafers. The dielectric constant change was evaluated by corona technique [4]. Multi-Internal reflection (MIR) was used to analyze the chemical bonding. Ellipsometric Porosimetry (EP) analysis were performed in the visible range on an ellipsometric porosimeter consisting of a rotating polarizer spectroscopic ellipsometer coupled with a vacuum chamber which is operated at a pressure ranging from 10^{-3} torr to saturation vapor pressure (Ps) of the absorbent. Toluene is used as absorbent in our study (Ps=25 torr at room temperature, refractive index at 633 nm). Secondly, effects of restoration treatments on copper lines are investigated. Using atomic force microscopy (AFM), the nanometer scale surface roughness of copper after restoration treatments is determined. Kelvin Force Microscopy (KFM) is carried out to extract the copper work function variations.

Moreover, direct CMP as well as restoration treatments are studied on patterned wafers to assess their impact on electrical performance. For that purpose, dual damascene process is used to integrate copper/ULK structures with respect to the 45 nm technology node dimensions. Copper lines resistances, line to line capacitance, RC delay and interline leakage current are measured using serpentine/comb structures. Finally, from electromagnetic simulation, the integrated dielectric constant is extracted using TEM cross sections and capacitance measurements.

I. The direct CMP approach

Direct CMP is an integration approach implemented since 45 nm node consisting in the removal of the dielectric hard mask during CMP. This approach is essentially used to perform the standard integration scheme by lowering the effective k-value, improving CMP variability and canceling the undercut issues. Figure 1 is an illustration of the two integrations process flows. Indeed, the hard mask (SiO_2 dielectric cap) has a negative contribution and raises the dielectric effective k-value due to its permittivity of 4.2. Moreover, uniform hard mask retention could not be ensured, due to the CMP rates dependency on the metal pattern density and due to the bad CMP selectivity on the hard mask.

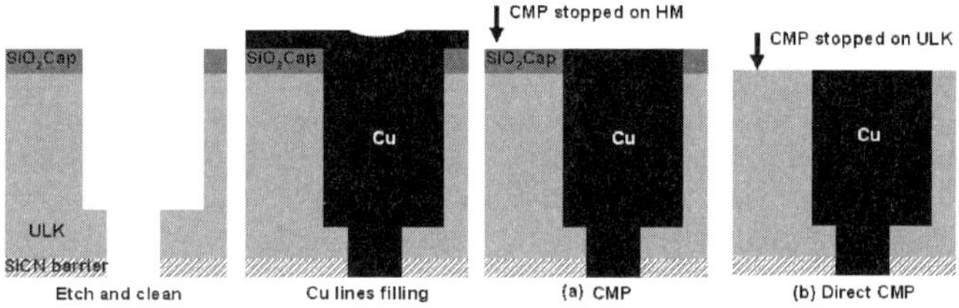

Figure.1: Dual damascene integration scheme (a) standard approach with residual SiO2 cap, and (b) direct CMP approach

Finally, since the hard mask and the ULK material have different densities, an undercut is created during lines etching and leads to metallization issues. Figure 2 shows a TEM cross section imaging of the interconnect structure after the integration through the two different processes. Direct CMP approach ensures straight copper lines formation by removing the undercut zone.

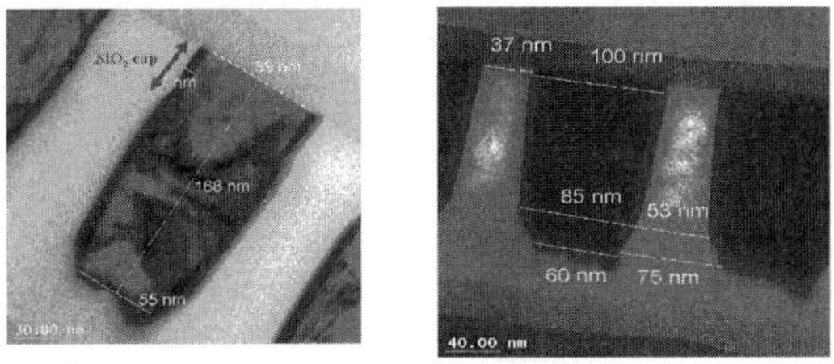

Figure.2: TEM imaging: integration with (a) residual SiO_2 cap, and (b) direct CMP on 45 nm technology node

Figure 3 shows electromagnetic simulation extraction results. It is demonstrated that a 6.3% gain in the effective k-value could be obtained for the 45 nm node with the direct CMP integration approach.

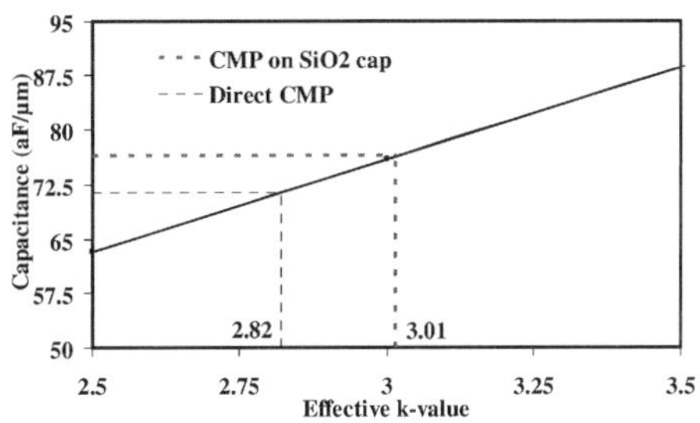

Figure.3: Effective k-value extraction by electromagnetic simulations

Direct CMP approach seems to be mandatory. However, effects on ULK material have to be studied.

II. Direct CMP damages

In the direct CMP approach, the ULK is directly exposed to CMP, and despite the improvement achieved by adjusting slurries species [2], the ULK is damaged. In this part, the damages induced by CMP are characterized. Figure 3 shows permittivity measurements on blanket ULK wafers and demonstrates a drastic increase of the k-value after direct CMP process. This increase is independent of the ULK thickness suggesting a volume degradation of the ULK.

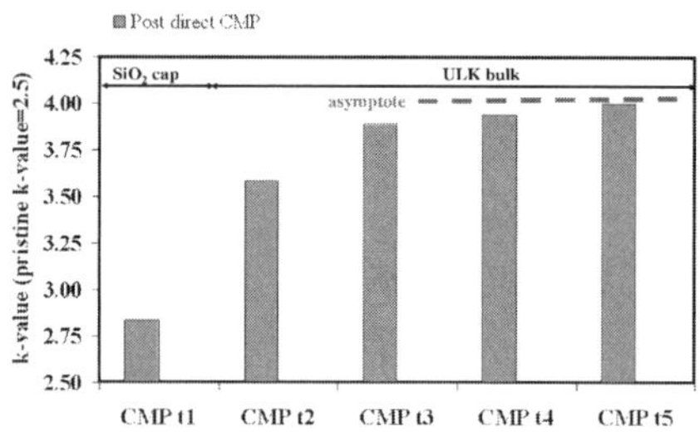

Figure.4: k-value measurements function of direct CMP time and ULK thickness

Figure 5 presents multi-internal reflection (MIR) analysis and show H_2O peak increase. Degradation due to the water intrusion into the ULK is assumed.

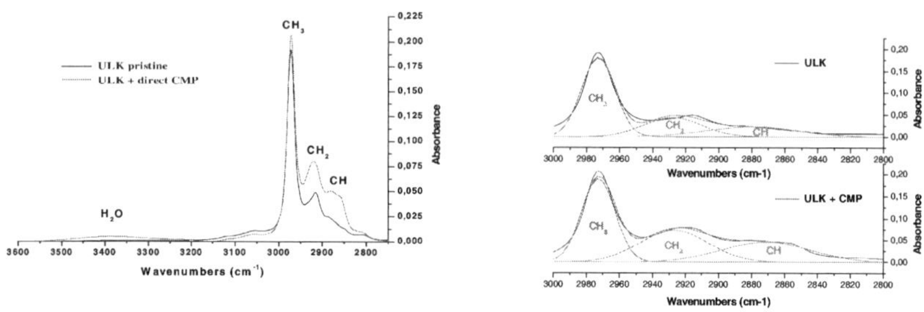

Figure.5: Multi-internal reflection analysis after direct CMP

Besides, figure 5 (right) shows the increase of CH and CH_2 species related to the CMP surfactant presence. Indeed, water absorption is enhanced by surfactants adsorption on the ULK surface and consequently the surface hydrophilisation. The peaks deconvolution and area ratio calculations show that the amount of CH and CH_2 species related to the CMP surfactants adsorption have doubled. Whereas, CH_3 methyl groups amount still not changed (Table1). Obviously, from these results, it can be assumed that the ULK dielectric degradation is mainly due to the water intrusion on the pores rather than to a morphological or a chemical induced damage. Consequently, this degradation seems to be reversible.

Table 1: Extracted MIR data: peaks area calculation

Peak	ULK pristine (peak area)	ULK post CMP (peak area)	Peaks ratio
H_2O	0.2981	1.09032	3.65
CH	1.56	3.10	2
CH_2	1.91	3.79	2
CH_3	4.5	4.77	1.06

Optical measurements evidence an increase of the refraction index value to 1.499 in agreement with the water intrusion fact. Finally, EP analyses show that toluene solvent cannot be absorbed into the pores since they are partially filled by water. Consequently, the empty porosity is estimated at 3% (Table.2).

Table 2: Ellipsometric porosimetry and contact angle analyses

	refraction index at 633 nm	measured porosity	contact angle
ULK pristine	1.32	26%	90°
ULK + CMP	1.499	~3%	49°

Finally, reliability measurements prove that while removing the SiO_2 dielectric cap by direct CMP, the ULK breakdown field is drastically degraded and consequently interconnects reliability is impacted.

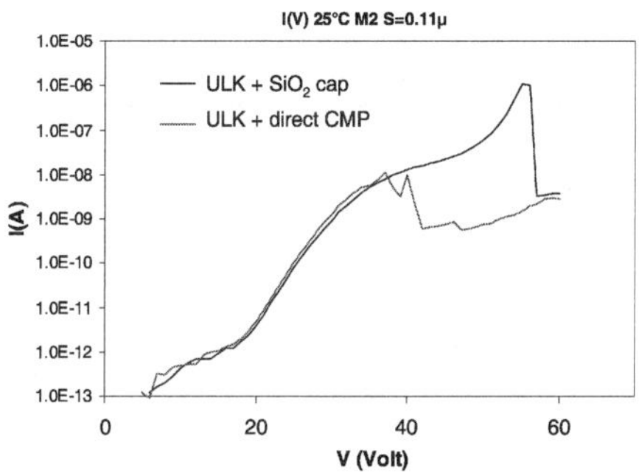

Figure.6: Leakage measurements and breakdown field comparison

All these characterizations show direct CMP damage importance and prove that restoration treatments are needed.

III. Restoration solutions

In this work, treatments that would preserve the standard integration scheme while removing the adsorbed surfactants, desorbing the absorbed water and hydrophobizing the ULK surface are investigated. All these processes are based on thermal treatments.

1. Post CMP anneal

The treatment consists of a post CMP anneal of the ULK at different temperature from 100° to 400°C for 30 minutes.

EP analysis using toluene as solvent combined with contact angle analysis after the different anneal temperatures reveal that adsorbed surfactant removal and water desorption start to be relevant at 300°C. On the one hand, the sudden increase of the measured porosity and the

contact angle and on the other hand the decrease of the refraction index from 1.481 to 1.357 show that at least an anneal temperature of 400°C is requested to remove surfactants from surface and water from ULK pores (Table.3). Moreover, k-value measurements demonstrate that with the 400°C anneal the k-value is decreased and almost totally recovered.
These results validate the assumption about the reversibility of the direct CMP damage.

Table 3: Ellipsometric porosimetry

	refraction index at 633 nm	measured porosity
ULK pristine	1.32	26%
ULK + CMP	1.499	~3%
ULK + CMP + anneal (5 min at 100°C)	1.488	~3%
ULK + CMP + anneal (5 min at 200°C)	1.481	~5%
ULK + CMP + anneal (5 min at 300°C)	1.413	17.5%
ULK + CMP + anneal (5 min at 400°C)	1.357	19.5%

However, at this high temperature, copper lines are roughened (Fig. 6) and consequently electromigration is drastically impacted. For this reason, post CMP anneal at such high temperatures has to be avoided.

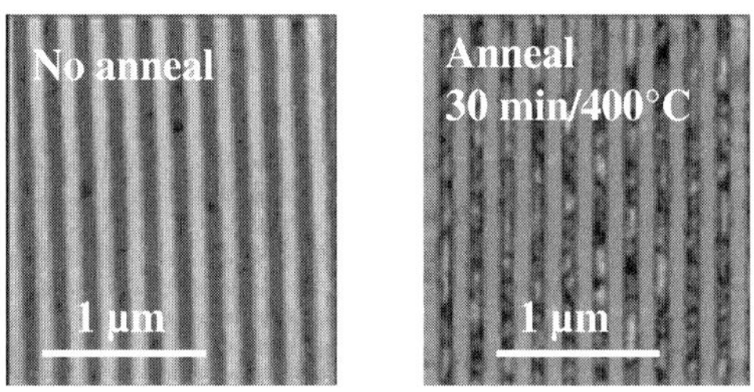

Figure.6: Cu line roughness increase with post CMP anneal (30 min/400°C)

2. Post CMP UV cure treatment

In order to decrease the restoration treatment duration and temperature, different UV cure treatments at lower temperatures (≤ 300°C) are investigated as restoration solutions of the ULK after direct CMP process.

Figure 7 shows that with a brief UV cure treatment (3minutes/300°C) the k-value dropped closely to its initial value.

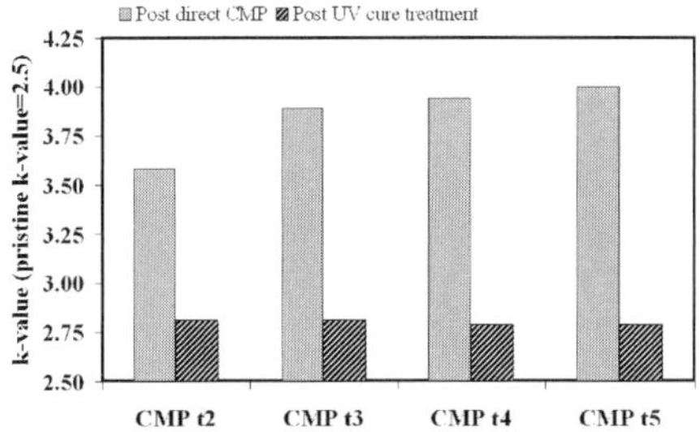

Figure.7: k-value measurements after UV cure restoration

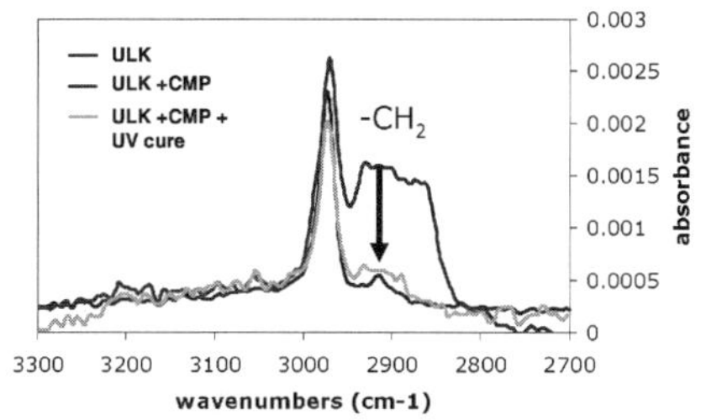

Figure.8: Multi-internal reflection analysis after UV cure restoration

Besides, both MIR analyses and contact angle measurements exhibit respectively a total removal of the adsorbed surfactants and a recovery of the ULK hydrophobicity. In addition, EP analysis shows that the ULK refraction index is lowered from 1.499 to 1.326 and the opened empty porosity is estimated to 20.5%, which correspond to the desorbed water amount during UV cure treatment. Moreover, the EP analysis with water as solvent proves that after the UV cure treatment, the ULK have no more affinity with water demonstrating dielectric stability after restoration.

Regarding the effect on the copper lines, AFM analysis conducted on copper blanket wafers after UV cure treatment show that almost no effect on the surface roughness has been observed.

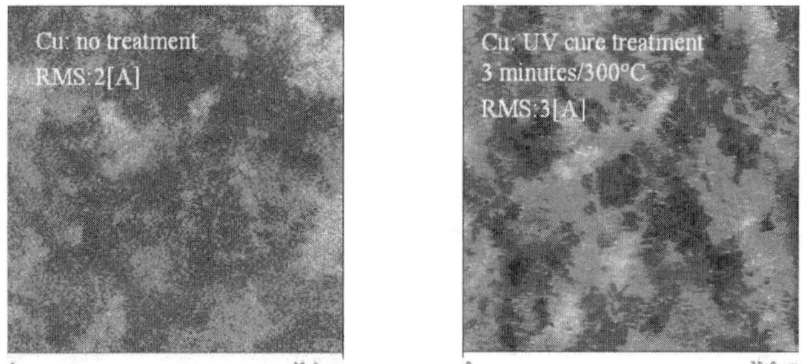

Figure 9: Atomic Force Microscopy analyses on copper surface (a) before (b) after UV cure treatment (3 minutes/300C°)

Moreover, KFM analyses demonstrate that UV cure treatment has a beneficial effect on the copper lines work function. Indeed, the UV cure treatment has increased the minimum energy needed to remove an electron from copper surface (Figure10). This change has doubtlessly a beneficial effect on interconnect leakage current.

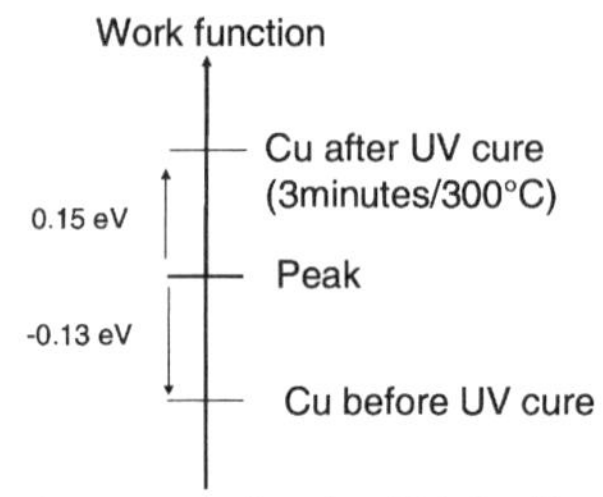

Figure.10: Copper work function extraction by Kelvin Force Microscopy before and after UV cure treatment (3 minutes/30°C)

Finally, electrical performances are assessed through line-to-line capacitance (Figure.11). UV cure treated wafers exhibit a 8% decrease in capacitance. Besides, performing leakage measurements the porous ULK presents higher breakdown fields with the UV cure treatment (Figure.12).

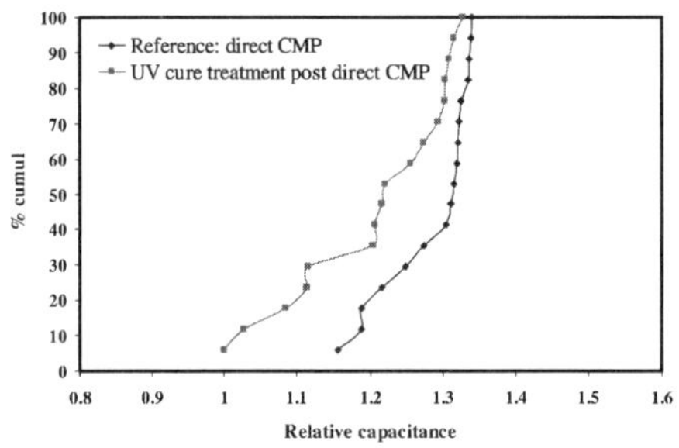

Figure.11: Capacitance measurement before and after UV cure

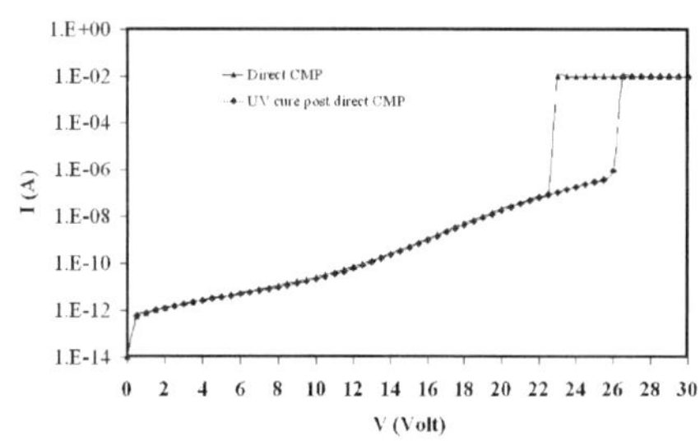

Figure.12: Leakage measurement before and after UV cure

Combining electrical measurements and TEM imaging, electromagnetic simulations are conducted and exhibit a 8% decrease of the extracted k-value and consequently of the effective k-value after integration.

	integrated k-value	Effective k-value
Direct CMP	2.87	3.22
Direct CMP + UV cure treatment	2.60	3

Conclusion

Integration damages due to direct CMP process were investigated. To overcome these issues UV cure treatment has been proposed and its ability to remove adsorbed water and repair the damages of ULK surface has been evidenced. Particularly, electrical results on dual damascene structures evidence a gain on capacitance, interline leakage and integrated dielectric constant. Moreover, effect of the UV cure treatment on copper lines has been investigated and demonstrates that no damaging effect has been noticed making the investigated treatment an interesting option for sub-45 nm nodes

References

[1] Kondo et al., Proc. International Interconnect Technology Conference 2006. pp. 164-166
[2] S.Gall et al., in: Proc. of the Advanced Metallization Conference 2007, Ed. By A.J. McKerrow, Y. Shacham-Diamand, S. Shingubara, Y. Shimogaki, Cof. Proc. AMC XXII, MRS Warrendale PA, 2008. pp. 115-121
[3] LL.Chapelon et al, Proc. Materials for Advanced Metallization 2008. Accepted for publication in Microelectron.Eng. (2008).
[4] D.Fossati et al , Proc. Materials for Microelectronics and Nanoengineering 2006.

Conference Proceedings AMC XXIV © 2009 Materials Research Society

Characterization of Copper Corrosion Phenomena with Scanning Kelvin Probe Force Microscopy

Shohei Shima, Katsuhiko Tokushige, Akira Fukunaga and Manabu Tsujimura
Ebara Corporation, 4-2-1 Honfujisawa, 251-8502, Japan
Email: shima.shohei@ebara.com

ABSTRACT

The surface potentials of Cu, Ta and Ru were evaluated by using atomic force microscope (AFM) with Kelvin probe force microscopy (KFM). Characterization of the phenomena of pattern dependent corrosion in Cu damascene interconnections was carried out with regards to surface potential. The surface potential of Ta was found to be larger than that of Cu, while that of Cu was larger than Ru. On the other hand, electrochemical oxidation reduction potential of these three metals in DIW and in 0.5M H_2SO_4 was highest for Ru and lowest for Ta. These results showed that the metal with higher surface potential measured by Kelvin Force Microscopy is electrochemically less noble. Pattern dependent corrosion phenomena, such as the fast corrosion rate in Cu lines with pads and line edge recess formation in Cu lines with Ta barrier, is successfully explained by the variation of surface potential. We can predict the creation of local corrosion failures in Cu damascene interconnections by using nanometer resolution KFM technique.

INTRODUCTION

Preventing copper corrosion is still one of the main issues in metal chemical mechanical planarization (CMP) process. There are many types of corrosion, such as galvanic[1], photo-assisted [2] and pattern dependent corrosion [3]. Photo-assisted corrosion is easily prevented by eliminating exposure of PN junctions to light [4), 5]. Galvanic corrosion is an intrinsic phenomenon between two different metals having different corrosion potential. In the CMP process, inhibitors and chelate agents are introduced in slurries to form a passivation layer on the copper surface, and the pH of slurries is adjusted to decrease the electrochemical potential difference between the copper and barrier metals. Pattern dependent corrosion is also inherent, and installation of dummy patterns is adopted to decrease corrosion.

In ULSI copper metallization, corrosion failures are very local phenomena and microscopic in size. Potential and current are the basic electrochemical factors that cause corrosion on metals in electrical contact with electrolyte. It is thus necessary to detect the corrosion potential and current with nanometer scale resolution in order to properly understand the corrosion mechanism. In general, most of the electrochemical characterization technique are macroscopic in size and conducted using blanket samples.

Corrosion studies using KFM technique have been carried out by many researchers. However there have been few applications of KFM to copper lines in ULSI [6]. We conducted KFM measurement in air to detect the surface potential of copper damascene interconnections. Differences in surface potential among Cu, Ta and Ru films were clearly detected. Analyzing the line and space of 500 nm CMP test patterns, we were able to propose the models that cause line edge erosion and acceleration of etching in Cu lines with pads in terms of the KFM surface potential.

EXPERIMENT

We used SPA400 AFM equipment manufactured by SII Nanotechnology Inc. KFM potential was evaluated using a Si cantilever coated with Au whose spring constant was 3 N/m. We measured KFM potential in air, not in vacuum. As water layer is adsorbed on the surface of thin films in air, KFM measurements are usually performed in a high vacuum condition to remove the water adsorption layer on the surface of samples in order to obtain stable potential values. However we considered that the surface potential measured in air is similar to the actual condition in a corrosion environment, because absorbed water layer works as an electrolyte in an electrochemical experiment. Silver conductive paste was coated around the observation area to make a stable potential measurement.

For basic evaluation, three metals, Cu, Ta and Ru, were examined whereas Cu damascene structure now using Ta and developing Ru as barrier metals. We also measured patterned Cu damascene lines in which the pattern layout is Sematech CMP test structure 725A888AZ.

Electrochemical testing was also performed to evaluate oxidation reduction potential of the three metals in DIW and in 0.5M H_2SO_4, using a potentiostat with three electrode system.

RESULTS AND DISCUSSION

KFM and electrochemical oxidation reduction potential of Cu, Ta and Ru

First of all, we evaluated the KFM potential on blanket films of copper and two barrier metals, Ta and Ru, to clarify the relationship between KFM potential and electrochemical potential. Figure 1 (A) shows images and profiles of both topographic and KFM potential in Ta/Cu blanket films. Cu was sputtered onto Ta film by masking with a polymide tape to make a Ta/Cu boundary. The step height at the boundary between Ta and Cu was nearly 20 nm. KFM potential change is clearly observed near the boundary and the KFM potential of Ta area is higher than that in Cu area. KFM potential difference between Ta and Cu is 150 mV. On the other hand, KFM potential of Ru in Fig.1(B) is smaller than Cu and exhibit a potential difference of 540 mV compared to Cu. The potential values of Cu are different between the two samples. We think this is due to the difference in sample condition.

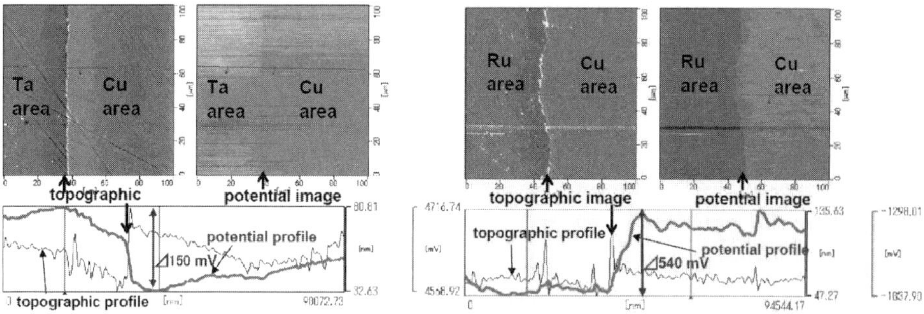

(A) (B)

Figure 1 AFM topographic and KFM potential images with their line profiles, (A) Ta/Cu films,(B) Ru/Cu films.

Figure 2 shows oxidation reduction potential of Cu, Ta and Ru films both in DIW and in 0.5M H_2SO_4. The electrochemical potentials of Ru are highest and those of Ta are lowest in DIW and 0.5M H_2SO_4. These results show that Ru is nobler than Cu and Cu is nobler than Ta. This indicates that Cu is easy corroded when Cu comes into contact with Ru due to galvanic corrosion. On the other hand, Cu is not corroded with Ta barrier metals, because Cu is nobler than Ta. The KFM potential difference data are inserted in Fig.2, and we can see that there is quantitatively close relationship between KFM potential and the standard potential of reduction/oxidation.

Comparing the two kinds of measured potentials shown in Fig.1 and Fig.2, we find there is an inverse relationship between KFM potential and electrochemical potential. A noble metal like Ru exhibits small KFM potential and a less noble metal like Ta possesses a large KFM potential. In KFM technique, images of contact potential difference (CPD) are acquired using the non-contact mode. The principle of KFM is following: An A.C. voltage is applied between the probe and the sample. The resulting surface potential (electrostatic force) is detected by a lock-in amplifier and a feedback adjusts a D.C. voltage on the probe V_{DC} until the CPD is compensated. Potential at the sample surface then equals $-V_{DC}$. The compensation voltage is used to render the images. As shown in Figure 3, brighter regions in CPD (KFM potential) images correspond to relatively lower work function. The metals which show lower work function exhibit lower standard potential. Therefore the KFM technique is markedly useful in evaluating Cu galvanic corrosion failures in the CMP process.

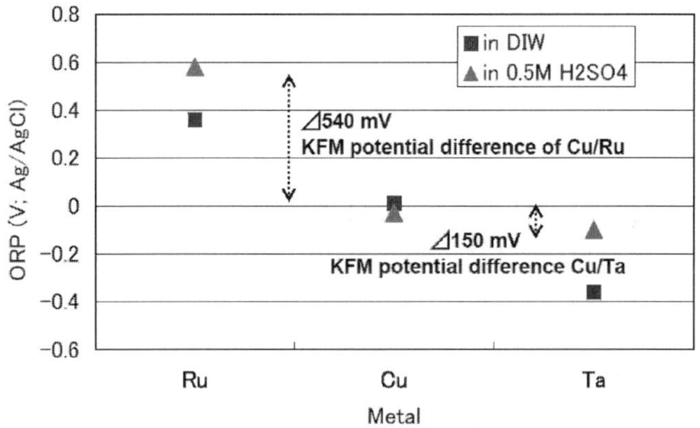

Figure 2 Oxidation reduction potential of Ru, Cu and Ta in DIW and in 0.5M H_2SO_4. KFM potential difference data are inserted to compare with ORP data.

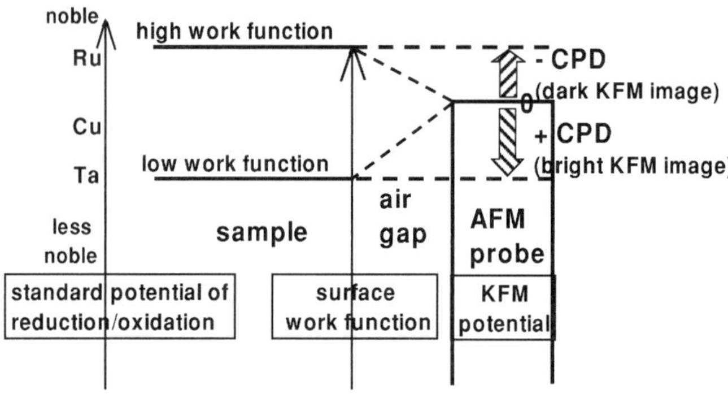

Figure 3 Schematic energy band's diagram of CPD (KFM potential) measurement and the relation among standard potential, work function, and KFM potential image.

KFM potential of damascene copper interconnections

We conducted KFM potential mearsurements of Cu damascene interconnections for Sematech CMP test chips. Figure 4 shows a line and space pattern layout with lines of 0.5 um in width in a Sematech 725A888AZ chip. Several lines with pads (size is 120x120 um) are located in the dense patterned area with dummy lines. One is isolated from the dense area and the others are located at the edge and in the center of the dense area.

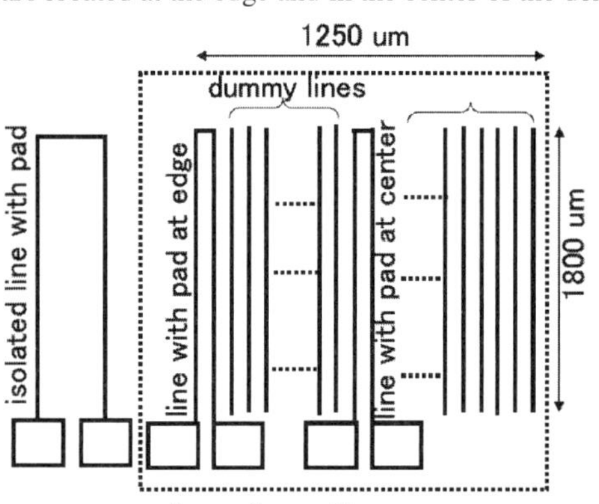

Figure 4 Line pattern layout of Cu damascene interconnections used in KFM evaluation. Sematech 725A888AZ.

Estimation for KFM potential was performed after dipping a chip in 0.5M H_2SO_4 for 1min followed by rinsing in DIW for 1 min and drying. Figure 5 shows the data of topographic and KFM potential images of the isolated line with pads. The Cu line is recessed due to corrosion seen in the topography image. KFM potential on the recessed Cu line is clearly higher than in the insulator area. For a similar Cu line with pads which is located at the center of the dense dummy lines, KFM potential is also higher than in the surrounding dummy lines as shown in Fig.6. This line with pads is more corroded than surrounding dummy lines which have no pads. These results indicate that pads can affect the property of higher KFM potential on lines.

We examined corrosion rate by measuring the etching depth of lines. Figure.7 shows an etching depth profile of the isolated lines with pads after dipping in 0.5M H_2SO_4. Etching depth is larger at line edge (L & R), namely because the corrosion rate is accelerated there. We also compared etching rate between lines with pads and without pads. Figure.8 shows etching depth profile of a line with pads and four dummy lines without pads. The line with pads apparently is corroded more than the dummy lines. Etching rate of all lines is higher at line edge (L & R). We can understand that there are two kinds of corrosion acceleration phenomena from Fig.7 and Fig.8. One is "line edge corrosion" and another is "pads effect".

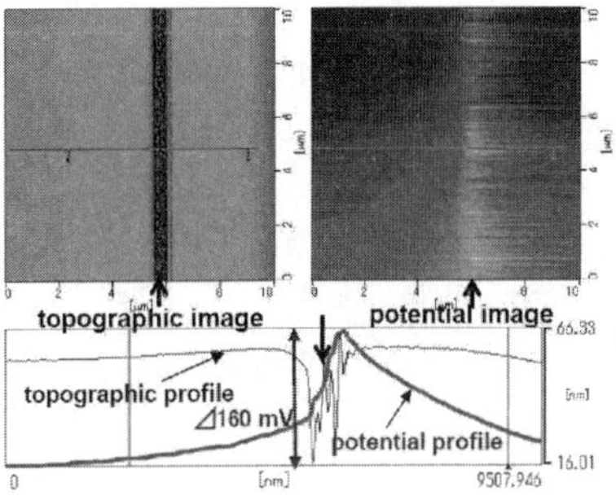

Figure 5 Topographic and KFM potential images of isolated Cu line with pads.

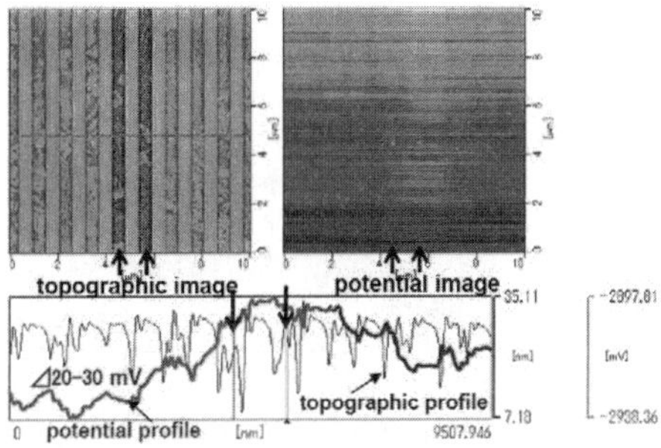

Figure 6 Topographic and KFM potential images of Cu line with pads and dummy lines.

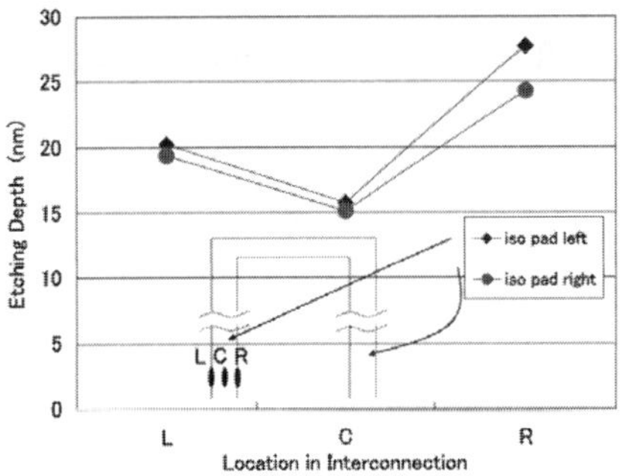

Figure 7 Etching depth profile of isolated Cu line with pads (refer to Fig.4).

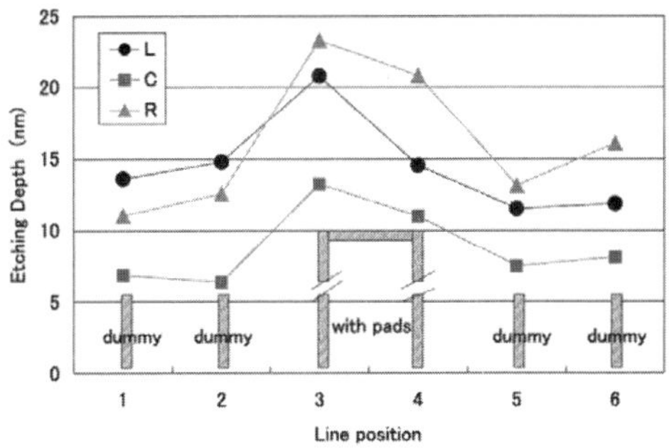

Figure 8 Etching depth profile of Cu line with pads and dummy lines (refer to Fig.5).

We examined KFM potential distribution along the line with pads in order to understand the "pad effect". Figure 9 shows KFM potential difference at different line width areas in an isolated line with pads. KFM potential is highest in the large Cu area and the smallest in the narrowest line, so KFM potential decreases with decrease in area. This result indicates that KFM potential at pad area may be highest and so the pad area shows less noble property. Consequently, lines connected to pads have higher KFM potential (less noble) than dummy lines which have no pads. This phenomenon resembles "area ratio effect in galvanic corrosion".

We now consider another galvanic effect such as "line edge corrosion" using Fig.10 and Fig.11. Figure 10 represents topographic and KFM potential images on Cu pad with small TEOS dummy area. Figure 11 shows a schematic explanation for line edge corrosion and represents KFM potential distribution and etching depth profile in cross-section of a line with Ta barrier metal. Ta films inherently show higher KFM potential as shown in Fig.1 (A). Ta barrier metals in Fig.10 also show higher KFM potential than Cu. Now consider the KFM potential of Cu at the boundary between Cu and Ta. The KFM potential distribution in Cu line with Ta barrier cannot be uniform and Cu may have rather higher KFM potential near Cu/Ta boundary as shown in Fig.11. This higher KFM potential at the periphery of Cu lines causes less noble properties near the Cu/Ta boundary and thus may cause local edge corrosion in Cu lines.

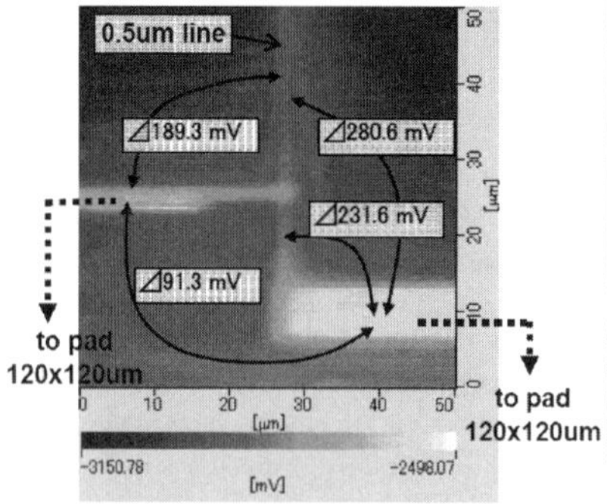

Figure 9 KFM potential distribution image along an isolated Cu line with pads.

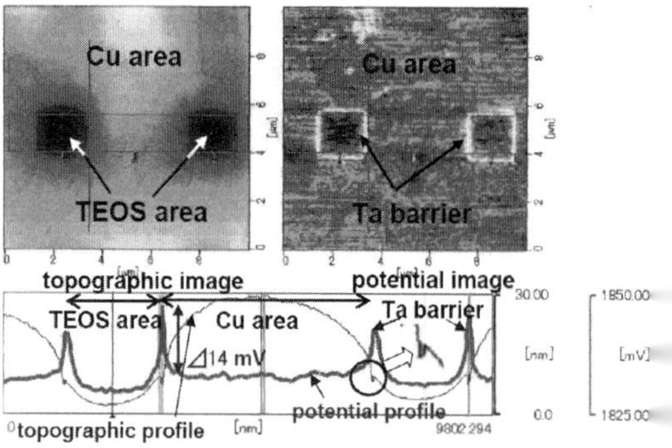

Figure 10 Topographic and KFM potential images of Cu pad with TEOS dummy area. High resolution images of Ta barrier metal.

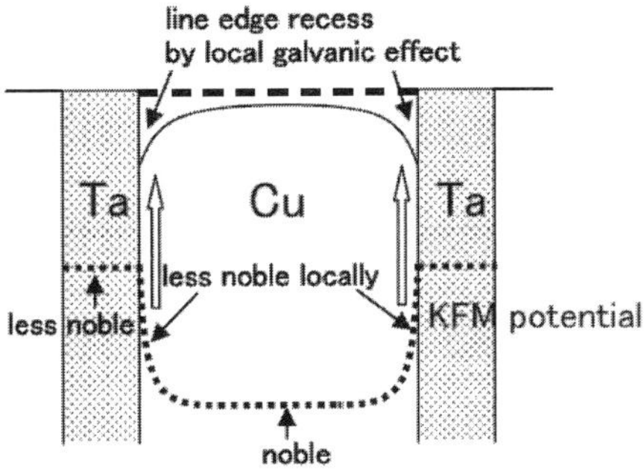

Figure 11 Schematic cross-section of a Cu damascene line with Ta barrier for explaining the model for Cu line edge corrosion with Ta barrier metal.

CONCLUSIONS

KFM potential distribution in Cu damascene lines was successfully measured in air. KFM potential in air was markedly related to oxidation reduction potential owing to surface adsorption layer of water onto metal. Among Cu, Ta and Ru metals, Ru showed smallest KFM potential, and the less noble Ta exhibited highest KFM potential. We were able to estimate oxidation reduction potential by the KFM technique. We measured KFM potential distribution in copper damascene structure. Corrosion rate of Cu lines with pads was larger than that of dummy lines without pads. This was due to the KFM potential difference depending on pattern. KFM potential of Cu lines was proportional to the amount of Cu area; the large Cu area with pad showed higher KFM potential and dummy lines without pad showed smaller KFM potential. As a result, lines connected to pad were easily corroded in electrolyte. A model of the mechanism of line edge corrosion in Cu lines with Ta barrier can be proposed in terms of non-uniformity KFM potential in Cu. The KFM potential at the edge of the Cu line may be higher than in the bulk area due to the higher KFM potential of Ta. We are convinced that KFM is a potential tool for studying corrosion mechanism in Cu lines. Further studies for the high resolution and stable observation of KFM potential are needed.

REFERENCES

1. D. Ernur, S. Kondo, D. Shamiryan, and K. Maex, Microelectr. Eng. 64 (2002) 117.
2. Y. Homma, S. Kondo, N. Sakuma, K. Hinode, J. Noguchi, N. Ohashi, H. Yamaguchi, and N. Owada, J. Electrocem. Soc. 147 (2000) 1193.
3. P. Wrschka, J. Hernandez, G.S. Oehrlein, and J. King, J. Electrocem. Soc. 147 (2000) 706.
4. USA Patent No.6,153,043.
5. USA Patent No.6,251,787.
6. A. Dominget, J. Farkas, and S. Szunerits, Microelectr. Eng. 83 (2006) 2355.

Conference Proceedings AMC XXIV © 2009 Materials Research Society

Barrier CMP Criteria for ULK Integration for 32nm BEOL Technology and Beyond

Jihong Choi[1], Wei-tsu Tseng[2], Hyun-ki Kim[3], Laertis Economikos[2], Qiang Fang[4], Eden Zielinski[1], Edward Engbrecht[2], Craig Child[1], Moosung Chae[5], and Yongsik Moon[1]

[1]Advanced Micro Devices, Hopewell Junction, NY, 12533

[2]IBM Microelectronics, Hopewell Junction, NY, 12533

[3]Samsung Electronics, Hopewell Junction, NY, 12533

[4]Chartered Semiconductor, Hopewell Junction, NY, 12533

[5]Infineon Technologies AG, Hopewell Junction, NY, 12533

ABSTRACT

In this paper, we present various process issues and challenges in the direct ultra low-k (ULK) polishing scheme for 32 nm and beyond and how they affect chemical mechanical planarization (CMP) specifications. We present the relation between removal target, removal rate, process variability, topography, and selectivity for 32 nm ULK integration. We also discuss other important challenges, such as slurry-ULK compatibility and defectivities, and offer examples of results for commercial barrier slurries developed for 32 nm BEOL ULK integration.

INTRODUCTION

Dual damascene process without remaining etch hard masks and etch stops between via and trench has a significant advantage for the effective dielectric constant (k) over bi-layer dielectric scheme where the hard mask remains on ULK after CMP. In this process scheme, barrier CMP process is required to remove certain amount of ULK and Cu below barrier and hard mask. The materials being removed in this step include the diffusion barrier, hard mask, ULK, and Cu. In 32 nm technology with direct-ULK polishing scheme, the requirement for hard mask, ULK, and Cu removal is often greater than the barrier removal requirement. Hence, the primary materials removed in this step are hard mask, ULK, and Cu -- even though this step is called 'barrier CMP'. With its porous microstructure, ULK is vulnerable to CMP both chemically and mechanically, so the barrier slurry for this application must meet special criteria to avoid chemical and mechanical ULK damage during CMP.

Another important criterion for barrier CMP process in this integration scheme is to meet removal target with good uniformity and throughput. CMP removal target is determined with consideration of CMP removal variation to make sure no hard mask remains after CMP regardless of pattern factor, wafer-to-wafer variation, and within-wafer variation. This is a critical specification for the CMP process, since remaining hard mask can lead to a critical time-dependent dielectric breakdown (TDDB) problem. As trench depth becomes shallower for 32 nm and beyond, electroplating thickness can also be reduced to reduce slurry consumption and increase CMP tool throughput. In a typical three-platen CMP process with reduced plating thickness, CMP throughput tends to be determined by process time of Cu clearing step or barrier CMP step. Hence, barrier CMP removal rate is becoming more and more important since it should meet the rigorous removal target while meeting throughput specifications. In addition to these challenges, barrier CMP must also meet increasingly stringent defectivity requirements.

EXPERIMENT

We tested five different commercially available barrier CMP slurries with two different pads and categorized these processes into two different groups: high selectivity between Cu and ULK with high ULK removal rate, and relatively non-selective with comparable Cu and ULK removal rate. Each process was first optimized to meet the resistivity target and removal target, then we evaluated topography, leakage current, throughput, and defectivity for each process. We used 45 nm and 32 nm 300 mm test pattern wafers from 1x BEOL layer to 2x BEOL layers with ULK (k = 2.4) as dielectric for barrier CMP process development. To make general conclusions about barrier CMP processes, we evaluated different Cu bulk and Cu clearing processes, including eCMP Cu bulk process.

DISCUSSION

Topography and Cu puddle formation

One of the most critical defect types caused by post-CMP topography is so called a 'puddle' defect. Figure 1 schematically shows the mechanism of Cu puddle generation. After metal1 (M1) CMP, topography remains on the wafer surface due to the selectivity between Cu, liner, and dielectric (a). M1 topography transfers to M2 layer if the dielectric erosion exceeds a certain limit, allowing Cu to fill the recessed ILD area in the metallization process in M2 (b). In M2 CMP process, Cu in the recessed ILD region remains after CMP; this Cu residual can bridge nearby Cu lines, causing a severe short problem. This type of defect is highly pattern design-dependent, and various models have been developed for CMP pattern dependency [1,2,3]. Design optimization can give more process margins to CMP. However, to avoid this type of defect for a given chip layout, CMP process for the Cu clearing step should have enough Cu removal ability to clear as much Cu as possible in the recessed dielectric area. Also, barrier CMP process should have certain Cu removal ability.

In most cases of CMP process for dual damascene, Cu is recessed after Cu clearing step due to the high Cu removal rate of the platen 2 (Cu clearing CMP) process. This topography is reduced or reversed in platen 3 (or barrier CMP) process by using ILD-fast, Cu-slow process; hence, the final topography is largely dependent on the post-platen 2 topography and removal rate selectivity on platen 3.

To avoid Cu puddle defects in the wide ILD area, high Cu removal rate in platen 2 and platen 3 is desirable. However, too much Cu removal rate in platen 2 can result in high dishing, which requires high selectivity (ILD:Cu) in platen 3 for good final topography. High selectivity in platen 3 means suppressed Cu removal rate and increase in ILD removal rate; this results in high ILD recession in the wide ILD area and, hence, high probability of Cu puddles in the next layer. As a result, there is a tradeoff between platen 2 Cu removal rate and Cu puddle generation, and a CMP process should be optimized to be free of Cu puddles for a given chip layout. Barrier slurry selectivity between Cu and ULK should be optimized with consideration of platen 2 selectivity and Cu clearing ability in platen 3 for puddle prevention. Figure 2 shows examples of the test structures processed with different platen 2-platen 3 combinations.

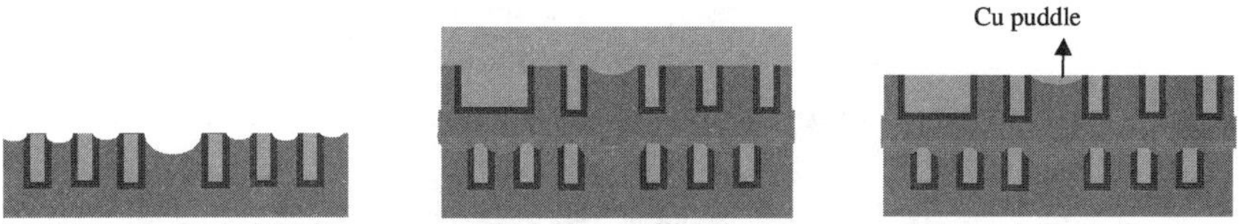

(a) Topography in M1 layer (b) M2 metallization (c) post M2 CMP

Figure 1. Cu puddle generation.

(a) High topo with no Cu puddle
(high Cu rate on P2-high Cu rate on P3)

(b) Cu puddle with medium topo
(low Cu rate on P2-high Cu rate on P3)

(c) Good topo with no puddle
(high Cu rate on P2-low Cu rate on P3)

(d) Cu puddle with good topo
(low Cu rate on P2-low Cu rate on P3)

Figure 2. Examples of Cu puddles with different platen 2-platen 3 slurry combinations.

Barrier CMP removal target

Removal target is determined mainly to achieve good electrical reliability. TDDB, which can be caused by underpolishing, is the greatest concern. CMP should remove hard mask and damaged ULK area from hard mask deposition process. Also, a non-uniform sidewall etch profile at the border of hard mask and ULK can cause liner deposition, Cu fill, or TDDB issues even with good Cu coverage and fill. Hence, barrier CMP step also should remove this non-uniform sidewall profile. With these minimum removal requirements, removal non-uniformity coming from tool-to-tool variation, consumable lifetime dependency, pattern dependency, wafer scale non-uniformity, wafer-to-wafer variation, and lot-to-lot variation should be also considered. Once the statistical confidence level is known for CMP removal non-uniformity from every source, the removal target is determined to ensure no underpolishing-induced reliability issues. Figure 3 schematically shows this relation. The CMP removal target is determined by etch sidewall profile, ULK damage depth, and CMP uniformity across chip and wafer.

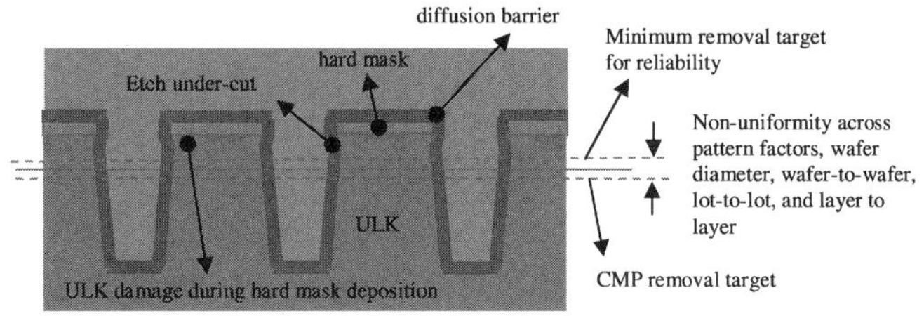

Figure 3. CMP removal target in direct ULK CMP scheme.

ULK compatibility

Another concern for barrier slurries in direct ULK polishing scheme is the interaction between the slurry and ULK film. Chemical attack into porous ULK can change film properties at the surface area. We observed that k-value and reflective index (RI) are differently affected by different barrier slurries [1]. Blanket wafer tests showed significant surface property changes due to barrier CMP process. This property change could be recovered with thermal treatment. In full integration wafer test, effective k-value was not affected by barrier slurry and ULK interaction. However, for different types of ULK films, the possible impact on TDDB by ULK-barrier slurry interaction and surface modification is one of the most fundamental checkpoints for barrier CMP process development.

Removal rate and throughput

As technology nodes shrink, plating over burden thickness is also being reduced since the trench depth becomes shallower. Hence, in the most common three-platen CMP processes, overall manufacturing throughput is being limited by copper clearing step (platen 2) or barrier CMP step (platen 3) since the bulk Cu removal step (platen 1) time can be decreased. In many cases, the barrier CMP process becomes the bottleneck process and the minimum requirement of the ULK removal can significantly limit overall CMP process throughput. With given process variability and uniformity, barrier CMP removal rate becomes a very important factor for overall throughput improvement.

Cu:ULK selectivity and tradeoffs between topography and leakage current

To avoid Cu puddle formation and to deliver enough room for litho focus error, post-CMP topography should be as small as possible. In a process that requires significant polishing, non-selective barrier CMP generates less topography than selective CMP since the saturated ratio between Cu removal rate and ULK removal rate is close to 1. As seen in Figure 4, ULK is more effectively removed by highly selective and high ULK removal rate CMP. However, in high selectivity CMP, ULK erosion in wide ULK area becomes more severe, which can lead to Cu puddle formation or litho focus issues in next layer.

Figure 5 shows leakage current and resistance from selective and non-selective barrier CMP processes. Identical resistance indicates same amount of removal between two processes and same Cu line height. However, leakage current shows difference between selective process and non-selective process as seen in figure 5(a). Near the wafer edge area, where Rs is lower, we observed high leakage current for non-selective CMP. On the other hand, selective CMP showed low leakage current near the edge area.

Figure 6 schematically shows post-CMP surface status between selective and non-selective CMP. With non-selective CMP, better topography can be achieved. However, the damaged ULK area is not effectively removed between lines, and it causes high leakage current that can also lead to TDDB failure. On the other hand, with selective CMP, surface topography becomes worse by Cu protrusion. However, by effective removal of damaged ULK between lines, there is more process margin for good leakage and TDDB. In addition to the consideration of overall removal rate, this tradeoff between topography, leakage, and TDDB reliability should be also considered in barrier CMP process development.

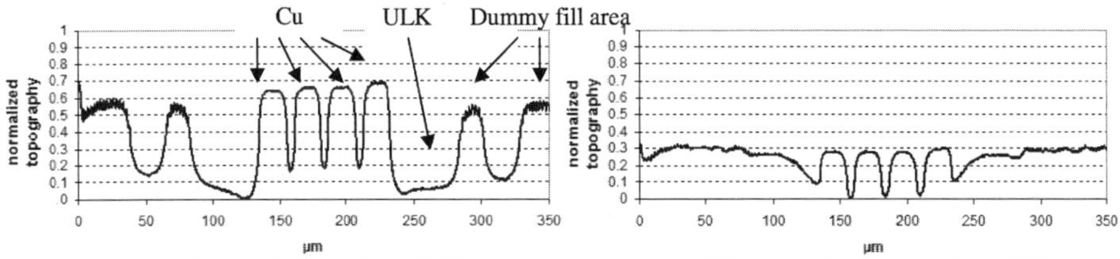

(a) selective barrier CMP (b) non-selective barrier CMP

Figure 4. Post-CMP topography from selective barrier CMP vs. non-selective barrier CMP.

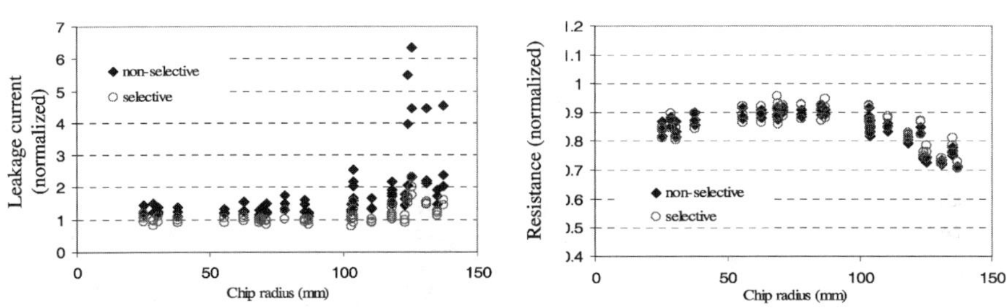

(a) Leakage current vs. chip radius (b) Resistance vs. chip radius

Figure 5. Leakage current and resistance from selective and non-selective barrier CMP.

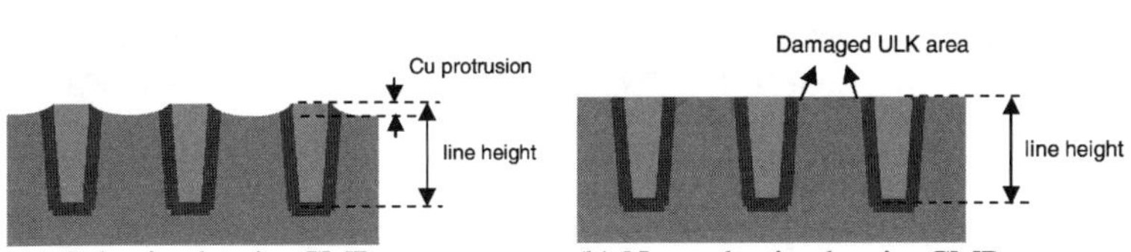

(a) Selective barrier CMP (b) Non-selective barrier CMP

Figure 6. Schematics of post-barrier CMP surface.

Defectivity

Figure 7 shows examples of most common BEOL CMP defect types. Largely, these defects can be categorized as polishing scratch (PS), polish residue (PR), and hollow metal (HM). Polishing scratch includes micro scratches, macro scratches, and mechanical ULK damage. Polishing residue includes metal residue and organic residue.

As CD shrinks, the impact of polishing scratch on the yield becomes more and more important. In 32 nm technology node, ground rule CD in thin wire metal layer is almost in the same order with the diameter of slurry abrasive particles. Hence, not only the big scratches generated by agglomerated particles but also small scratches (so called micro scratches, which have similar width of particle diameter) can be a fatal defect. Preventing agglomerated particles through advanced filtering system and chemical prevention of particle agglomeration can be only the first order effort for scratch reduction. For micro scratch reduction, control of particle size, shape, concentration, and slurry chemical modulation is becoming more and more important in addition to agglomeration prevention.

Polishing residue can be metal (Cu or barrier) residue or organic residue. With a large removal target, metal residues are mostly induced by topography, which are called puddle. The ability of Cu (or barrier) clearing in the recessed area is a very important factor for barrier slurry to give greater margins for topography and overpolishing in the Cu clearing step (platen 2). Besides metal residue, organic residues are also often observed. This type of residue can come from the pad or can be caused by non-optimized post-CMP cleaning. However, it is mostly likely caused by a barrier slurry that has wrong types or concentration of organic materials.

Hollow metal is corrosion type defect that can be induced by corrosion in CMP process or line edge coverage problems in etch, liner deposition, and filling defect in the plating process. Also, queue time between plating and CMP significantly affects this type of defect. Even though hollow metal is largely dependent on other processes and queue time, corrosion inhibition is still a key criterion for barrier slurry. In addition to the direct corrosion effect from barrier slurry, possible interaction between Cu clearing slurry (platen 2) and barrier slurry should be also considered.

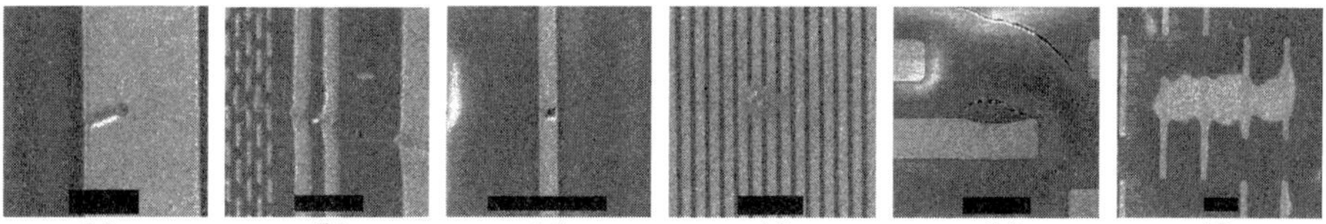

(a) (b) (c) (d) (e) (f)

Figure 7. CMP-related defect types: (a) micro scratch with embedded particle, (b) macro scratch, (c) hollow metal, (d) organic residue, (e) ULK damage (chatter mark), and (f) Cu puddle.

CONCLUSIONS

This paper summarizes key challenges in barrier CMP process development for a direct-ULK polishing BEOL integration scheme for 32 nm and beyond. Tests of various commercially available consumables for ULK test pattern wafers showed tradeoffs between topography and leakage current. They also showed that the process margins for barrier CMP selectivity, removal rates, and topography are largely dependent on CMP uniformity and statistical confidence levels on the amount of local material removal at critical macro areas.

ACKNOWLEDGMENTS

This work was supported by the independent SOI technology development and bulk CMOS projects at the IBM Microelectronics Div. Semiconductor Research and Development Center, Hopewell Junction, NY 12533.

REFERENCES

1. O.G. Chekina, L.M. Keer, and H. Liang, J.Electochem.Soc., vol. 145, no. 6, pp. 2100-2106, June 1998.
2. T.H. Smith, S.J. Fang, D.S. Boning, G.B. Shinn, and J.A.Stefani, 1999 CMP-MIC Conference, pp. 97-104, Santa Clara, CA, Feb.1999.
3. J.Choi, ISBN-10: 3639001664, Chap.4, VDM Verlag, May 29, 2008
4. W. Tseng, D. Kioussis, S. Manikonda, H. Kim, J. Choi, F. Zhao, L. Economikos, N. Klymko, M. Chace, S. Molis, M. Chae, E. Engbrecht, E. Zielinski, C. Truong, D. Watts, 213th Meeting of The Electrochmical Society, Phoenix, AZ, May 18-23, 2008

Topography Study on 45nm CuCMP Low-k Process and its Impact on Defectivity and Resistance

Feng Zhao, Benfu Lin, Hong Yu, Lup San Leong, Xian Bin Wang, Mei Sheng ZHOU, Haigou Huang, Hong Wu, Fan Zhang, Jun Chen, and Bei Chao Zhang

Technology Development, Chartered Semiconductor Manufacturing Ltd., 60 Woodlands Industrial Park D, Street 2, Singapore, 738406, Singapore

ABSTRACTS

Topography study was conducted under different polishing condition, and it's impact on defectivity and in line test performance was investigated. The topography study showed that final dishing and erosion performance is mostly driven by barrier polishing condition. With existing high selectivity barrier slurry, local Cu protruded in narrow dense arrays, and slightly higher down-force can suppress the Cu protrusion. Low selectivity barrier slurry shows less topography, and affords less polishing scratch. Electrical data shows barrier polishing condition has significant impact on resistance rather than Cu clearing over-polishing step.

INTRODUCTION

The progressively decreasing feature size of the circuit components has significantly increased the need for chemical mechanical planarization (CMP) of Copper to build up multiple layer copper interconnects.[1] It's critical for Cu CMP to demonstrate satisfying performance in terms of defectivity, topography, and within-wafer Rs uniformity control. The Cu CMP process involves the removing of Cu and the barrier over the field regions and polish directly to low-k film to define the features. Furthermore, the stringent requirement of the decreasing minimum depth of focus (DOF) in lithography process also poses additional challenge to Cu CMP in maintaining low topography accumulated over multi-levels.[2]

During CMP, materials of different densities polish at different rates. Typically, metal features of all sizes are depressed with respect to the surrounding dielectric. Generally, topography results from overpolishing. What is being attempted is to determine the difference between the metal lines' height compared with the surrounding dielectric material between the lines. That dielectric is harder than the metal, resulting in the metal being polished away more rapidly than the dielectric. Residues are another problem for topography control of advanced interconnects, related to the polish of the current layer or overpolish of the previous layer. Metal residues cause shorts and device failure. If there is residue on top of the metal grating, the current used to measure copper thickness will give incorrect information.

In this paper, topography study was conducted under different polishing condition, and it's impact on defectivity and in line test performance was investigated as well.

EXPERIMENT

The Cu CMP studies were conducted on an Applied Materials Reflexion LK CMP tool. 300 mm patterned Cu wafers were polished using standard production polishing recipes. Platen 1 is used primarily for bulk removal of the Cu above the patterned and etched features. Then, the wafers are transferred to platen 2, and the residual Cu and barrier metal are removed simultaneously. Finally, the wafers are subjected to a buffing step on platen 3 to smoothen the surface, as well as to remove any remaining barrier metal.

To prevent slurry residues and particles from contributing to the defect count during defect inspection, the wafers were cleaned in an integrated post-CMP cleaner immediately after polishing. All defects on each wafer were reviewed and classified using a scanning electron microscope (SEM), to build up a defect pareto chart for each wafer. On top of that, SEM images of all scratches were captured.

For local Cu topography characterizations of different polishing condition, AFM was utilized by scanning structure of certain pattern density.

DISCUSSION

Generally speaking, prior to CMP, Cu protruded in narrow dense arrays, but recessed in big Cu pad. During bulk Cu removal process, wafer planarized with certain Cu remaining, followed by Cu clearing step. Barrier polishing (BP) removed oxide and polished directly into low k film and corrected topography to certain amount, which depends on the selectivity of barrier slurry.[3]

On wide CMP macro (100x100um metal pad), Cu and barrier polishing condition has significant impact on dishing rather than erosion. (Figure 1) More dishing was observed with the increasing Cu clearing over-polishing time (OP +5Δ, **1Δ=10%**), but less dishing with the increasing of barrier polishing time. Furthermore, it shows inversely relationship between oxide removals and dishing, indicating that barrier polishing has strong topography correction with existing slurry.

As for dielectric erosion (0.07um 50%), it doesn't show much dependent from Cu clearing OP time, but significantly decided by barrier polishing time. Less erosion was observed with longer barrier polishing, vice versa.

Low selectivity barrier slurry shows 10% higher dishing with similar oxide removal, indicating the not-so-good capability for topography correction.

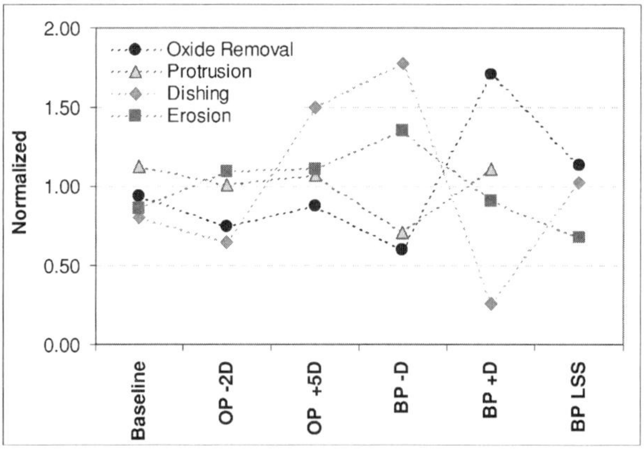

Figure 1. Topography data vs CMP oxide removal. (Dishing Pad: 100x100um, Erosion: 0.07um/50%, Cu Protrusion: 1.2um/90%)

Local topography partition checks by AFM (Figure 2) on narrow dense arrays (1.2um/90%) show Cu dishing about 10nm after Cu clearing step. However, local Cu protruded out as 10nm after barrier polishing for all splits as indicated in Figure 1. It means local Cu topography on dense arrays has been corrected also due to high selectivity of Cu vs oxide, which has been observed on wide metal macros.

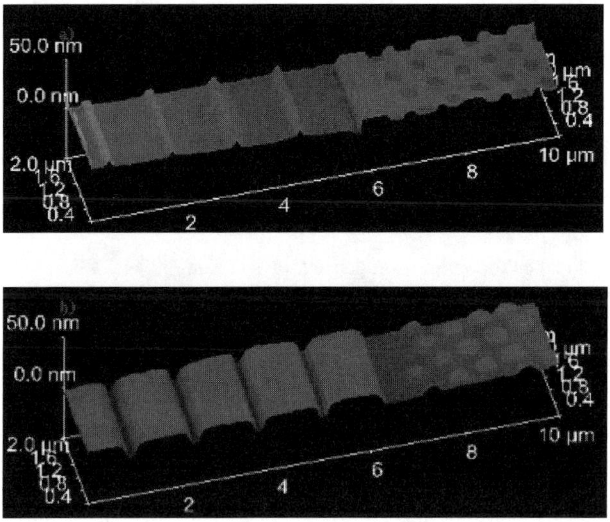

Figure 2. shows the AFM topography profile. a).Post Cu Clearing; b). Post Barrier Polishing.

Barrier polishing time shows significant impact on local Cu protrusion, while over-polishing shows minor impact (Figure 3). Furthermore, slightly less Cu protrusion was observed with high down-force (HDF) during barrier polishing. The strong capability of topography correction during barrier polishing process is due to high selectivity of oxide *vs* Cu with existing barrier slurry.

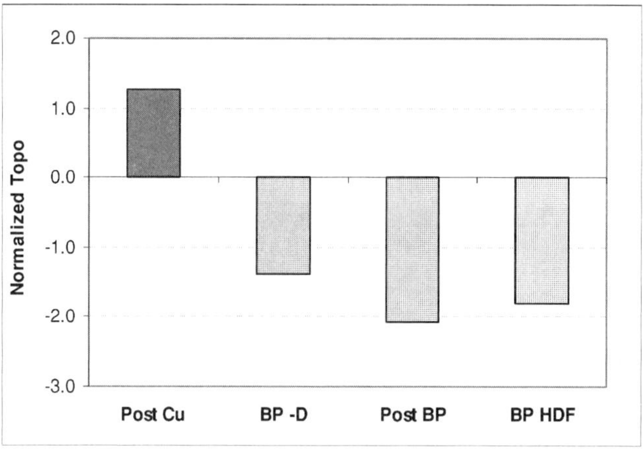

Figure 3. Partition check on local cu topography on dense arrays (1.2um/90%)

Defectivity Performance

Figure 4 shows the polishing scratch defectivity data. High down-force (BP-HDF) and low-selectivity (BP-LSS) approach affords superior micro polishing scratch performance, which seems related with less Cu protrusion. Actually, it was reported that low selectivity barrier slurry affords less polishing scratch and good topography.[4]

Furthermore, dramatic increase of polishing scratch was observed with longer barrier polishing (BP +Δ). It indicates that polishing scratch is possibly coming from barrier slurry, and longer polishing time enhances the possibility for the slurry abrasives to attack the wafer surface. However, interestingly, the polishing scratch is not reduced with short of polishing time (BP -Δ). So it means that polishing scratch is probably saturated then, and can't be further reduced.

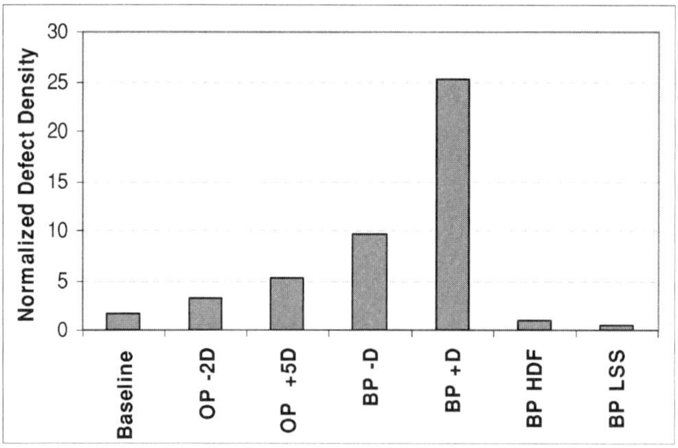

Figure 4. Post CMP Polishing scratch Pareto (HDF: High Down Force, LS: Low Selectivity Slurry)

Figure 5 shows the AES analysis spectra and SEM images. Area 1 is on the polishing scratch, Area 2 is on Cu and Area 3 is on dielectric. C, O, Si and Cu atoms were observed on all areaa. However, high intensity of Si was found on area 1 than area 2. It indicates that main composition of polishing scratch is silica, which is believed coming from P3 barrier slurry, as the main component of P1/P2 Cu slurry is aluminum oxide.

a)

b)

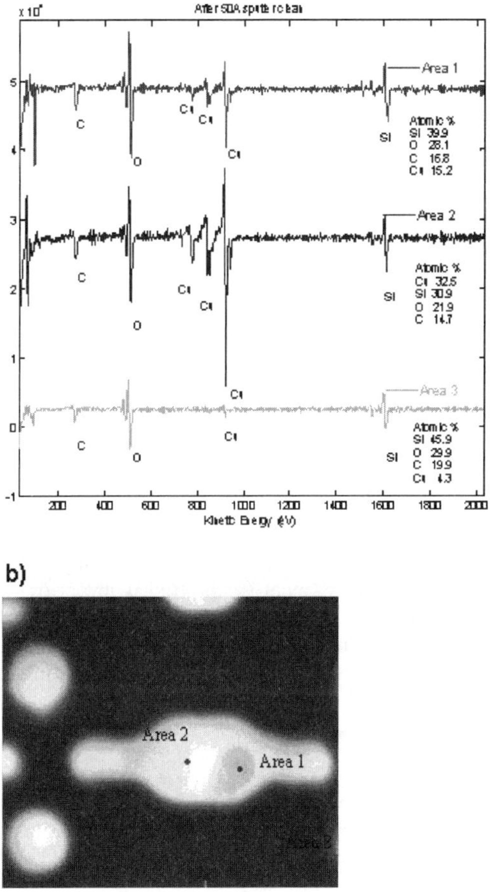

Figure 5. AES analysis result. a). Auger spectra; b). SEM image of polishing scratch (Area 1 on scratch, Area 2 on Cu and Area 3 on Area 3 is on dielectric)

Resistance Performance

Electrical resistance (Figure 6) is mostly driven by barrier polishing condition, rather than Cu clearing over-polishing step. Increasing over-polishing (OP +5Δ) or reducing (OP -2Δ) time has minor impact on resistance, and the delta is within 5%. However, increasing or decreasing barrier-polishing time with 1Δ (10%) will change the Rs delta to 15%. Furthermore, 2X-4X more response on resistance than thickness was observed.

Considering the topography and oxide removal data, it indicates that the resistance is mostly driven by oxide removal. With more oxide removal, some Cu removed, hence increase the resistance. However, topography is reduced with more oxide removal due to high selectivity of barrier slurry, indicated in figure 1.

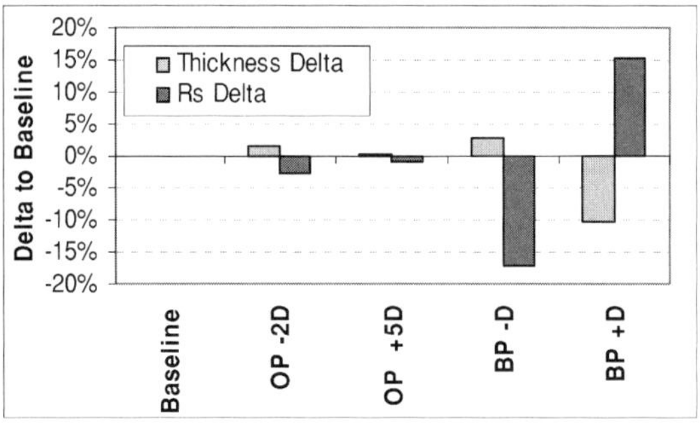

Figure 6. The impact of polishing condition on thickness and resistance.

CONCLUSIONS

The topography study showed that final dishing and erosion performance is mostly driven by barrier polishing condition. With existing barrier slurry, local Cu protruded in narrow dense arrays, and slightly higher down-force can suppress the Cu protrusion, which is good for polishing scratch reduction. Low selectivity barrier slurry shows less topography, and affords less polishing scratch. Electrical data shows barrier polishing condition has significant impact on resistance rather than Cu clearing over-polishing step.

ACKNOWLEDGMENTS

The authors would like to thank the QRA-FA group for their help in Auger analysis, and especially thank Rao Ramesh for the sample preparation and useful discussion. The acknowledgement is extended to YAO Vianna, and Chai KuokChin for defects scan and analysis.

REFERENCES

1 Parshuram B. Zantye, et al., Materials Science and Engineering R 45, 89 (2004).
2 Pike, Michael et al., ISSMC, 75 (2004)
3 L. Economikos, X. Wang, et al., IEEE-IITC Conference, IITC, 233 (2004).
4 F. Zhao et al., ICPT, 155 (2007)

Conference Proceedings AMC XXIV © 2009 Materials Research Society

Porous ULK Dielectric Integrity during CMP and post-CMP Cleans Processing

Laura M Matz[1], Mary K Haas[1], Mark L O'Neill[1], Rebecca A Sawayda[2], and Ann M Meyers[2]
[1]Electronics Division, Air Products, 7201Hamilton Blvd R4203, Allentown, PA, 18195
[2]DuPont Air Products NanoMaterials L.L.C, 2441 W. Erie Dr. Tempe, Tempe, AZ, 85282

ABSTRACT

With the introduction of porous ultra low k (ULK) films into back end of line (BEOL) interconnect structures, chemical compatibility of integrated process chemistries needs to be considered. In this study, CMP slurry optimization was demonstrated to provide minimal impact to ULK films with a wide range of porosities and k values (2.5-2.0). The initial slurry formulation had significant impact to the dielectric constant due to the penetration and retention of a chemical species into the ULK porous network. Through modification of the slurry components, the k impact was minimized to 0.1 k shift. Initial verification of the slurry compatibility was demonstrated on a ULK film with a k value of 2.5. Further studies with ULK films with k values of 2.2 and 2.0 showed similar compatibility. Once the relationship between slurry chemistry and chemical retention was understood, several potential slurry formulations were identified. The relative polish rates of various critical materials (Cu, barrier, dielectric) were tested with several of the formulations, leading to electrical testing with interconnect structures incorporating both ULK 2.5 and 2.2 films and demonstration of excellent electrical properties with very good distributions.

INTRODUCTION

Incorporation of ULK dielectrics at advanced technology nodes (45nm and beyond) has been successfully demonstrated by several device manufacturers [1]. Reducing the dielectric constant of dielectric materials can be accomplished by introducing porosity into an organosilicate glass network [2]. The presence of porosity makes the dielectric material more susceptible to damage during integration processing than previous dense dielectrics due to the open pores. Dielectric damage or modification can occur through several mechanisms include oxidation and densification from exposure to plasma processes [3]. In addition, existence of pores in the ULK makes the material more susceptible to penetration into the film. If not removed prior to subsequent integration processes, the effective dielectric constant of the ULK can be increased [4], reducing the potential performance gain anticipated with ULK implementation.

For BEOL copper interconnect integration, a key process is the chemical mechanical polish (CMP) step utilized to remove the excess copper and barrier metals from the damascene metal fill process. Traditionally, a dielectric hardmask was used on top of the interconnect stack as a strong dielectric for direct polish. However, device scaling requirements for lower capacitance have necessitated shrinking or removal of the hardmask and direct polish into the (interlevel dielectric) ILD layers. This study addresses the fundamental interplay between the electrical performance of ULK films and the diffusion and retention of slurry components in these materials.

EXPERIMENT

For the studies presented in this work, a PDEMS™ porous ILD process with diethoxymethylsilane (DEMS™ ILD precursor) and α-terpinene (ATRP) as the organic porogen precursor was employed to provide porous OSG films[2]. Deposition was performed on a 200mm P-5000 Applied Materials PECVD DXZ chamber. Mercury probe was utilized for all film measurements where dielectric constant, electrical breakdown field and leakage are presented. Modulus and hardness were measured by nanoindentation and bonding properties of the ULK films were analyzed with a Nicolet transmission FTIR tool.

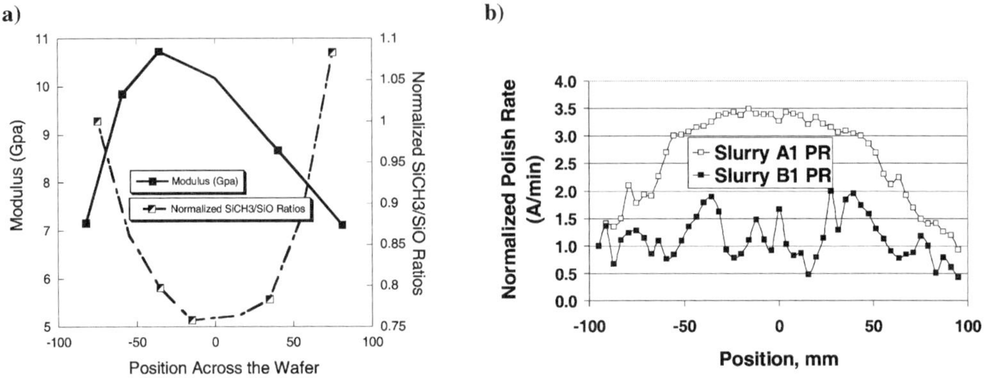

Figure1: Characterization of ULK film properties on CMP polish.
(a) Varied ULK composition test wafer showing distribution in modulus and FTIR ratios with non-uniform UV cure. (b) Normalized direct polish rate for barrier slurries A1 and B1 across varied-composition ULK 2.5 test wafer.

CMP polish experiments were performed on an Applied Materials Mirra-Mesa polisher. An acidic pCMP clean chemistry was used after the CMP polish step. Electrical test structures comprised of M1 Cu lines with PDEMS™ inter-level dielectric (ILD) and a dense organo-silicate glass hardmask were evaluated after M1 CMP processing.

DISCUSSION

Low k Dielectric 2.5 Characterization

While the PDEMS™ dielectric deposition process has been widely adopted at 45nm to deposit porous organosilicate glass films by chemical vapor deposoition, the specific composition of the organosilicate structure of the low k film can be tuned for specific integration schemes [4]. In the FTIR spectrum of an ULK film, two key regions to assess are the Si-CH_3 stretch at 1259 cm^{-1}, and Si-O backbone around 1100cm^{-1}. As shown in Figure 1a, there is a direct correlation between the $SiCH_3$/SiO peak area ratios and the modulus of the film where higher carbon films have lower modulus. It has been reported that advantages to a higher carbon film may result in less integrated damage so the desired film properties are dictated by overall device integration requirements [4]. ULK film properties can be modified through PDEMS chemistry modifications as well as tuning the post deposition UV cure for desired film properties. To generate slurry compatibility across a range of ULK film compositions, a non-uniform UV cure process across an 8" wafer was established. The modulus to $SiCH_3$/SiO ratio properties for the non-uniform UV cure process with a ULK 2.5 wafer are presented in Figure 1a. The highest modulus and lowest carbon occur at the center of the wafer. Initial testing of the polish rate uniformity for slurry A1 showed that the polish rates were highly dependent on the film properties and surprisingly, the highest modulus ULK (center) showed the fastest polish rates as demonstrated in Figure 1b. Once slurry A1 was reformulated for better compatibility (slurry B1), the polish rates were more uniform across the wafer, demonstrating the slurry's capability to polish a wide range of ULK compositions, either occurring from slight variations in UV cure conditions (center/edge effects) or through modifications to ULK deposition conditions.

Diffusion into Low K Dielectric Films

It is anticipated that during exposure to various chemistries, either from the direct barrier polish slurry or during the post CMP clean process, penetration of the chemistry will occur through the bulk of the film. Assuming that all chemistries will penetrate through the porous network, it is necessary to drive the equilibria between the ULK pores and the solution chemistry (both specific components and bulk solution) so that the solution components are not retained in the ULK pores after processing. For a ULK 2.5 film, the porosity is 29% with an average pore size of around 1nm. To measure diffusion of probe molecules, the refractive index change of the ULK film is monitored over exposure time to a given probe molecule. In Table I, the diffusion coefficients are reported for 3 probe molecules with varied chemical composition. It is observed that the 3 molecules show significantly different diffusion rates into the ULK 2.5 film where probe molecule A has 2 orders of magnitude greater diffusion coefficient than probe molecule C. These results demonstrate the ability to modulate the penetration and retention of molecular species into ULK films through molecular optimization, addressing the potential for formulation optimization.

Probe Molecule	Diffusion Coefficient (10^{-12} cm^2/s)
A	2.4
B	0.35
C	0.05

Table I: Diffusion coefficients of 3 probe molecules into a ULK 2.5 film.

Split	K value (+/- 5%)	CHx/SiO (+/- 4%)	$SiCH_3/SiO$ (+/- 4%)
ULK 2.5 Control	2.50	1.00	1.00
pCMP Clean only	2.53	0.90	1.04
Slurry A1 + IPA Dry	3.43	6.83	1.07
Slurry A1 + pCMP Clean	3.32	7.15	1.06

Table II: ULK k value and normalized FTIR peak ratios after slurry and clean exposure.

CMP Slurry Compatibility

Initially, slurry A1 and an acidic post CMP clean chemistry was investigated for the dielectric constant impact to an ULK 2.5 film. As presented in Table II, the dielectric constant increased from 2.5 to 3.43 and an increase in the FTIR peak areas in the hydrocarbon region (2700-3000cm^{-1}) was observed. Implementation of a CMP cleaning process after CMP polish showed minimal removal of the hydrocarbon, suggesting that either the retained species was irreversibly bound to the porous network or not soluble in the post CMP clean chemistry or the IPA dry. Given that this is transmission FTIR, the strong signal indicates that constituents from the slurry were adsorbed into the bulk of the dielectric.

Fundamental studies of the slurry chemistry were performed to assess the impact of systematic changes to slurry chemistry. By varying the slurry chemistry, the k increase varies from <0.1 up to 1.0. The relationship between k value increase and higher leakage is shown in Figure 2. The higher leakage formulations also correlated with lower measured electrical breakdown voltages, indicating that retention of molecules in the pores can potentially impact ULK reliability as well as capacitance.

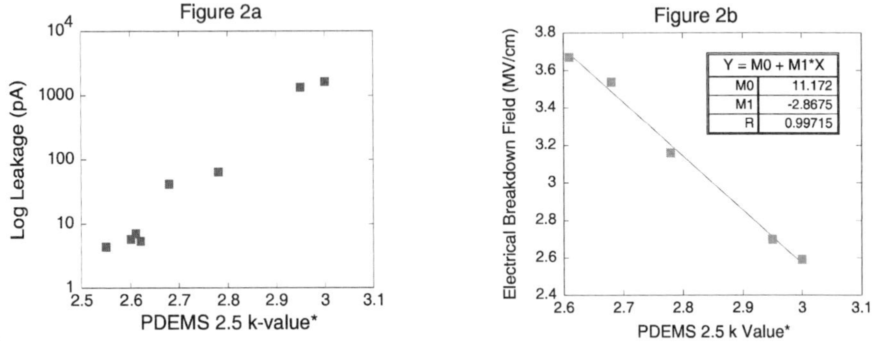

Figure 2: ULK 2.5 k value, leakage and electrical breakdown field results for systematic variation in CMP barrier slurry composition. Figure 2a and 2b show direct correlations between an increased ULK k value and increase in ULK leakage and decrease in electrical breakdown field, respectively. * - PDEMS k value measured after pCMP barrier slurry polish, pCMP cleans , followed by IPA dry

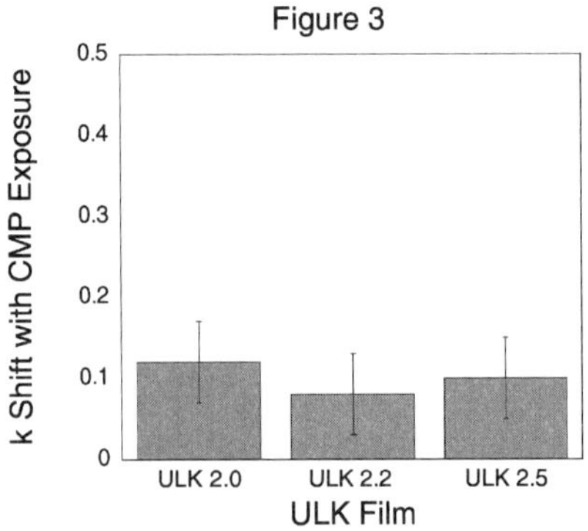

Figure 3: k shift for ULK 2.0, 2.2, and 2.5 films with after CMP polish with CMP6890 followed by pCMP clean and IPA dry. Error for k shift measurements is +/- 0.05.

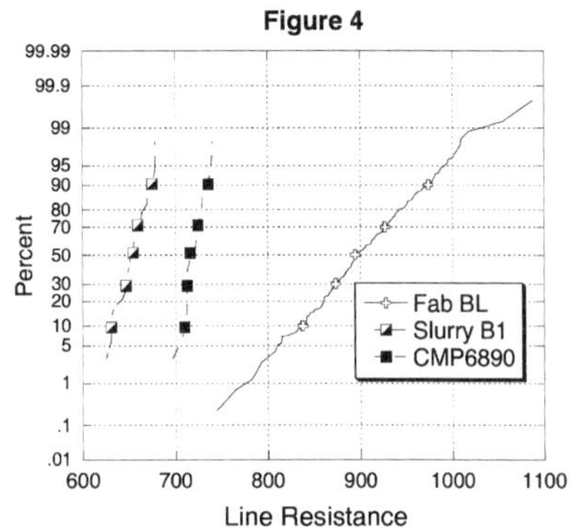

Figure 4: Single damascene Rs distributions for slurry B1 and CMP6890 with the pCMP clean. Nominal Rs values differ due to differences in polish times.

Several barrier slurry formulations were identified with the systematic understanding gained above and an optimal slurry formulation, CMP6890, was then chosen based on desired polish rate selectivities for the single level metal test vehicle. In Figure 3, the dielectric constant shift for CMP6890 combined with a standard pCMP clean chemistry is demonstrated for three ULK films with dielectric constants of 2.5, 2.2 and 2.0. With higher porosity, the expectation would be that any contaminants would show higher retention due to the larger pore volume, however the k shift for all three ULK films is ~0.1, indicating no significant difference as the dielectric constant is reduced and porosity is increased. Tuning of the slurry chemistry enables highly compatible formulations for a wide range of dielectric constants (2.5 to 2.0) and porosities (29% to 49%).

Single level metal integration with ULK 2.5 and 2.2 films was evaluated once the optimized slurry MP6890 were identified along with slurry B1. The M1 resistance profiles with both slurry formulations ›llowed by the acidic pCMP clean and an IPA dry are shown in Figure 4. Relative to the fab baseline rocess, it is noted that the distributions for both slurry formulations appear very tight, suggesting that the nemistry modifications made for ULK compatibility and extendibility were not at the expense of polish erformance. Further testing with CMP6890 has also been demonstrated with ULK 2.2 films as well as dditional pCMP cleans chemistries. By incorporating ULK compatibility as part of the initial screening, CMP cleans and slurry chemistries can be tuned for minimal impact while maintaining their required ntegration performance.

'ONCLUSIONS

Previous reports have demonstrated potential for surface modification through silanol formation when 'LK films were exposed to CMP slurries. In this study, we focused on controlling the retention of slurry omponents in the ULK films. This was achieved through an understanding of the ULK film properties, both lm composition and porosity, and the chemical compatibility of the integrated solution chemistries. ptimized slurry formulations were demonstrated with several ULK films ranging from k values of 2.5 to .0. The slurry considerations include tuning the slurry components for insensitivity to variations in ULK lm composition, either due to differences in film deposition or UV cure processing. Fundamental testing ith varied slurry components indicated a dependence on molecular composition of the slurry component. 'he higher degree of retention into the ULK films correlated with higher dielectric constant, higher leakage nd lower electrical breakdown, indicating the possibility for a potential reliability concern as well as apacitance. Minimizing retention of the slurry components into the ULK enabled an optimized slurry to be lentified that was compatible with very high porosity films with a k=2.0. When the compatible slurry was sted for single metal integration performance, the resistance distributions across a 200mm wafer showed nproved performance over the fab baseline. This study highlights that there is a need to focus on chemistry or ULK compatibility, that porous ULK integrated solution compatibility is achievable without degradation) integrated device performance, and that this this focus needs to be an integral part of a slurry development rogram..

.CKNOWLEDGMENTS

We would like to acknowledge the significant contribution of the Air Products Thin Films 'echnology team and Global Aanalytical Sciences group in Allentown, PA. Patterned wafers and electrical sting were performed with DK Nanotechnology. The authors would also like to acknowledge Air Products or permission to publish this study.

:EFERENCES

[1] N.C.M. Fuller et al., "Analysis of Plasma Induced Modification of ULK and eULK Materials: Dual amascene Processing Challenges for 45nm and Beyond BEOL Technologies," IITC, June 2006, p.24.

2] R.N.Vrtis, et al.,, U.S.Patent # 6846515, 25 January 2005.

3] K. Xu et al, "Modifications of porous low-k by plasma treatments and wet cleans," ECS Transactions, 1(2), pp. 409-416, 2007.

4] S.M.Gates et al, "A Porous SiCOH Dielectric with k=2.4 for High Performance BEOL Interconnects," 'onf. Proceedings AMC, 2007, p. 351-7.

3D Integration and Packaging

Conference Proceedings AMC XXIV © 2009 Materials Research Society

New Connection Method Using Wet Etching Technology

Nobuaki Miyakawa

Director, Honda Research Institute Japan Co., Ltd., 8-1 Honcho, Wako, 351-0188, Japan

ABSTRACT

3D stacking technology uses the connection method between the back-side bump of the upper wafer with TSV and the surface bump of the lower wafer. This bump-to-bump connection method needs much bonding load for better electrical connectivity and requires small resistance value dispersion. To reduction of the contact resistance value between TSV and surface bump, we have proposed and verified direct connection method between TSV and surface bump by cropping out TSV from back-side of the upper wafer using wet etching technology. We have tried the prototype of 3 layer-stacked devices by using the direct connection method with wet etching technology and it showed sophisticated functionality by testing. The yield of functional devices in the stacked wafer reached more than 60 percent.

INTRODUCTION

3D stacking technology is very promising as a future technology due to many benefits. These include high performance, low power consumption, smaller size, and the enabling of heterogeneous technology integration. Such benefits will be useful in future VLSI devices in which there are multiple connection paths between each chip. There are some technologies for 3D stacking technology, such as wafer-to-wafer, chip-to-wafer and chip-to-chip, and trial examples were reported from several universities and companies [1-10]. However, the stacking technology based on the wafer-to-wafer method includes some process problems, such as wafer thinning method for stacking, isolation method and materials, alignment accuracy and bonding force between stacked wafers etc., though this method is useful for manufacture. For stacking devices, many research organizations have been developing this technology using TSV. However, presently, stacking device technology is not yet well on its way to industrial manufacture due to underdevelopment in this area and there are only a few cases of leading products and businesses based on it. It is very important to produce real results with large-scale wafers, yield data and actual data effected through the practical use of 3D stacking technology.

CONNECTION METHOD BETWEEN EACH LAYER

The conventional connection method is bump-to-bump which needs additional processes for the back-side metal formation, that is, both fabrication processes of an isolation layer to avoid an electrical short circuit between the substrate and the back-side metal, and back-side metal for the stacking between the upper and lower chips. However, the surface of the bump which is connected between each bump is not flat and the bonding force between each layer becomes larger for good connection when numbers of TSV increase. The number of factors of total resistance is five and the resistance value dispersion between bumps of each layer increases for the point to point connection. On the other hand, in Figure1 the right side method has some benefits as the number of factors for total resistance is three, and the resistance value dispersion between each layer is small and stable for face-to-face connection. The total connection process is shortened, because this method does not need an additional isolation layer or the back-side bump of the upper layer.

	Conventional Stacking Technology	Proposed Stacking Technology
Structure	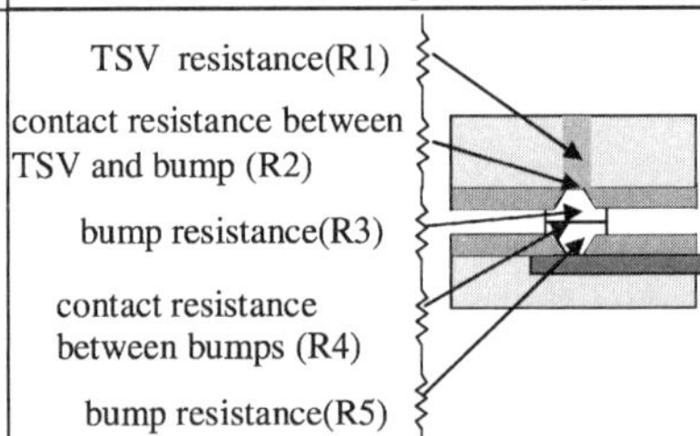	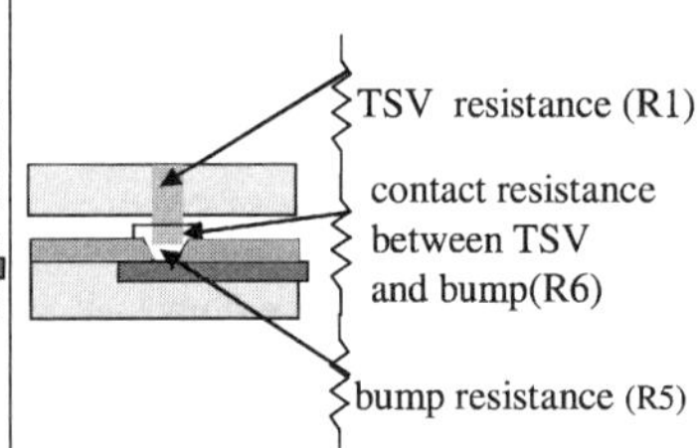
Total resistance value	(R1) + (R2) + (R3) + (R4) + (R5)	(R1) + (R6) + (R5)

Figure 1. Comparison of structure and factors of resistance value between Conventional Stacking method and Proposed method [11]

PROPOSED STACKING PROCESS

The proposed stacking process uses wafer-to-wafer method and comprises upper and lower wafer fabrication, and stacking process. The TSV processes of this method has two steps before/after the MOS transistor fabrication process, because the fabrication process of TSV depends on the temperature. After that multi-wiring, then the wafer is mounted on a support glass for wafer thinning. The two steps in the wafer thinning process are back grinding and wet etching. This wet etching process is an additional wafer thinning process. The lower wafer is also fabricated by these processes. After some surface micro-bumps of the lower wafer are formed, the wafer bow due to the multi-wiring process is compensated for. Following the preparation of the wafers for stacking, the wafer stacking process is started from the alignment and bonding between each wafer. Finally, an adhesive is injected between the stacked wafers.

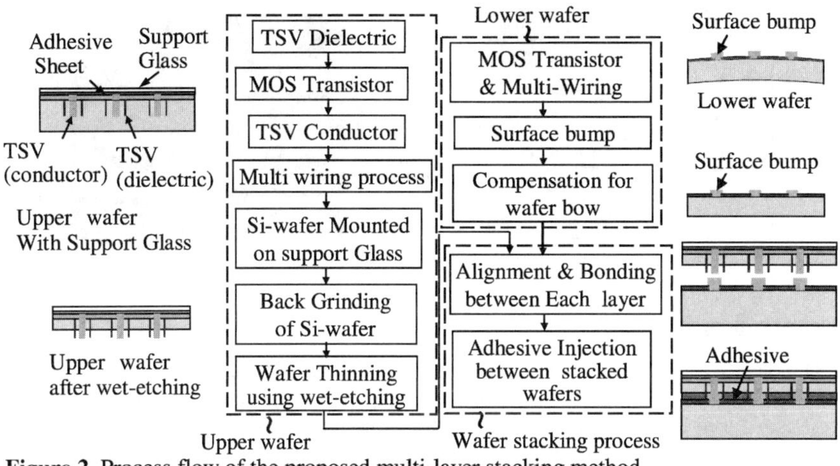

Figure 2. Process flow of the proposed multi-layer stacking method

EXPERIMENTAL DETAILS

Structure of connection portion

This technology adds a wet etching to the conventional grinding and polishing of an upper wafer, and there, the TSV is stuck into the Si substrate. It is a direct stacking method wherein the TSV of the upper wafer is directly connected with a surface-bump of the lower wafer. This structure is also unique from the point of processing temperature, as it affects the formation of MOS transistors and the stacking process based on this technology. The trench hole for TSV structure is opened by Bosch deep-trench etching, the depth and width of both dielectric and conductors is about 35 μ m and 1.5 μ m, respectively. The plane area of conductors is 1.5 μ m x 5.6 μ m and two conductors are used, which are enclosed by the dielectric. The dielectric portion is fabricated by SiO2/poly-si and the conductor portion is fabricated by burying tungsten. The total TSV size is 13.4 μ m x 13.4 μ m and it distributes a 50 μ m pitch at its densest part. The size of a surface bump is 20 μ m x 20 μ m and it is distributed with TSV facing each other. The gap between the upper and lower wafers is about 7 μ m. The injection method of the adhesive is from the edge of the wafer to the center of the wafer.

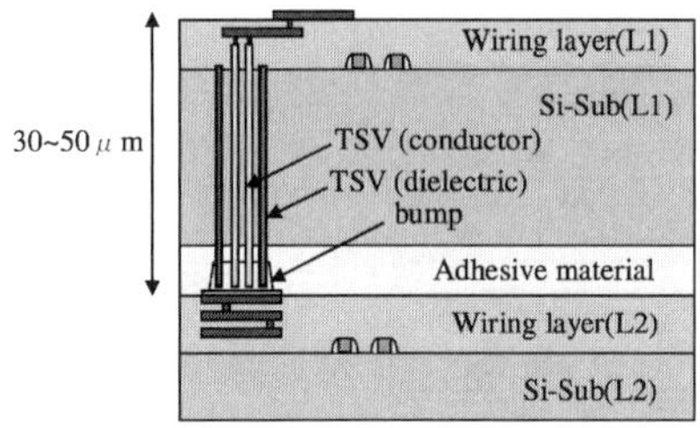

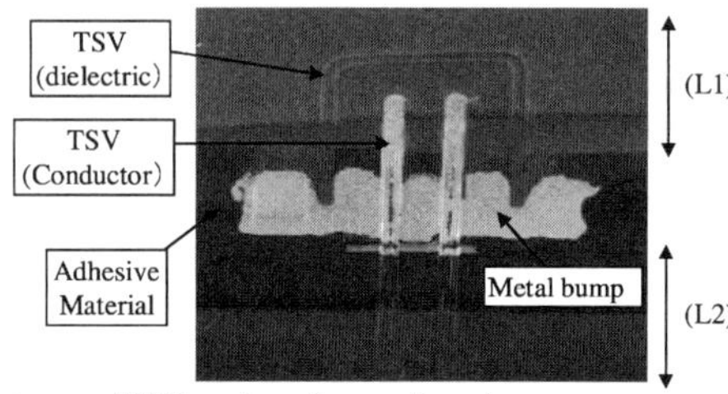

Figure 3. Structure of connection portion between TSV and surface micro-bump

Electrical connectivity

The electrical connectivity and frequency distribution between TSV and micro-bump are important characteristics for the stacking technology, because the yield loss of the stacking process is reduced by improving these factors and the total yield of stacked wafers depends on almost each wafer yield. The interconnection resistance value is checked in the whole wafers using TEG chips for 8-inch wafer. The data is a stand-alone case, a 30-chain case and a 300-chain case after 2-layer stacking, respectively. Almost all values of the stand-alone case are less than 0.7 ohm and the frequency distribution curve is very sharp. The frequency distribution data is made by 6 stacking wafers. The frequency distribution curves of the 30- and 300-chain cases shoe little dispersion, though those they include the parasitic resistance values for the connection between each chain. This data also comes from 6 stacking wafers.

Stand-alone

	1	2	3	4	5	6	7	8	9	10	11
7				0.38	0.38	0.35	0.35	0.40			
6			0.38	0.49	0.38	0.44	0.48	0.33	0.54	0.44	
5		0.37	0.37	0.37	0.35	0.36	0.34	0.35	0.37	0.43	0.47
4	0.43	0.43	0.39	0.45	0.35	0.46	0.36	0.39	0.39	0.52	0.43
3		0.44	0.37	0.40	0.40	0.38	0.38	0.35	0.38	0.46	0.46
2			0.40	0.36	0.39	0.38	0.38	0.45	0.42	0.51	
1				0.42	0.47	0.41	0.49	0.61	0.64		

Notch

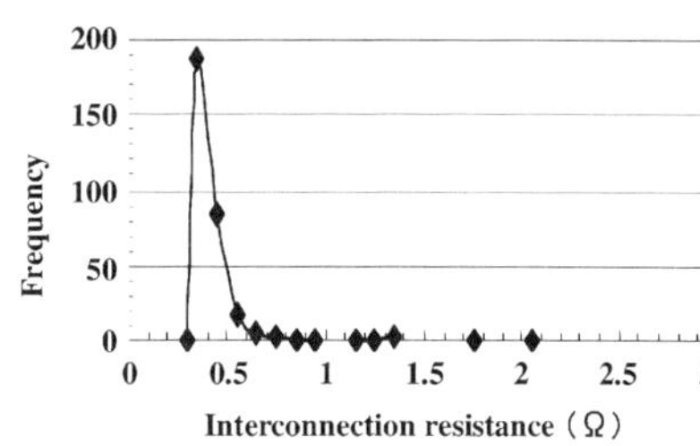

(a) Electrical connectivity between 1st and 2nd layer (b) Frequency Distribution

Figure 4. Electrical Connectivity of stand alone connection between 1st and 2nd layers

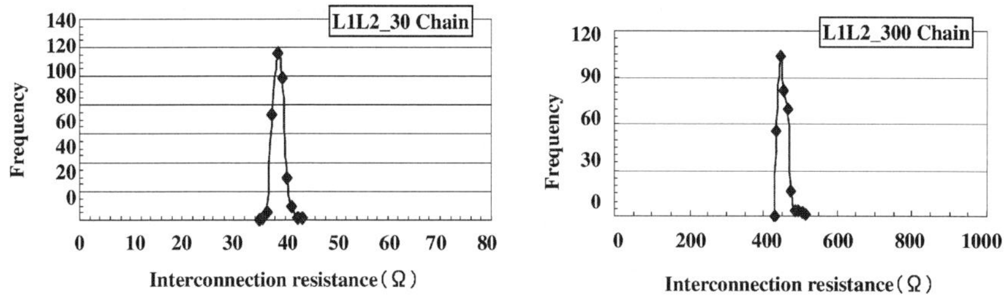

(a) Frequency distribution of 30 chains (b) Frequency distribution of 300 chains

Figure 5. Electrical connectivity of 30 and 300 chain case between 2 layers

The data of 3-layer stacking case from 1st layer to 3rd layer was also measured using 6 stacking wafers. The frequency distribution was less than 6 ohm and the distribution curve is sharp. If too much force is added to the connection portion for bonding, a TSV will snap. However the TSV in this photograph will not snap.

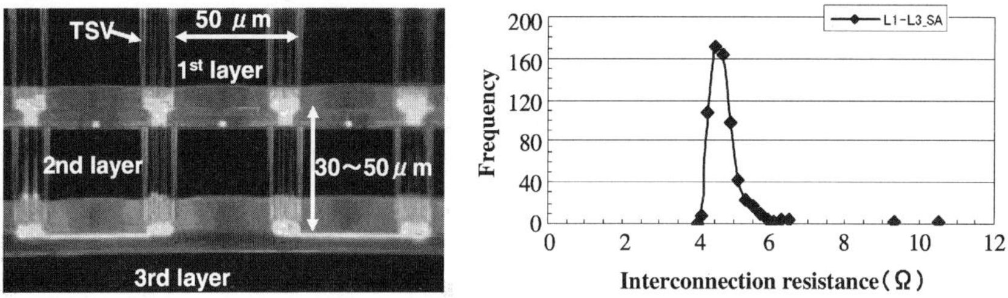

Figure 6. Connection portion between 3 layers and Frequency distribution

Yield of electrical connectivity

The relationship between the number of TSV and relevant yield has been calculated based on the results of the connectivity measurement. The equation for the yield calculation was the poisson distribution.

$$Y=\exp(-D_{TSV} * N_{TSV}) \quad (1)$$

Where D_{TSV} is Defect Density, N_{TSV} is Number of TSV [piece/chip].

This connection method was confirmed to be greater and much more effective after comparing yields of two cases previously researched, and ours,.

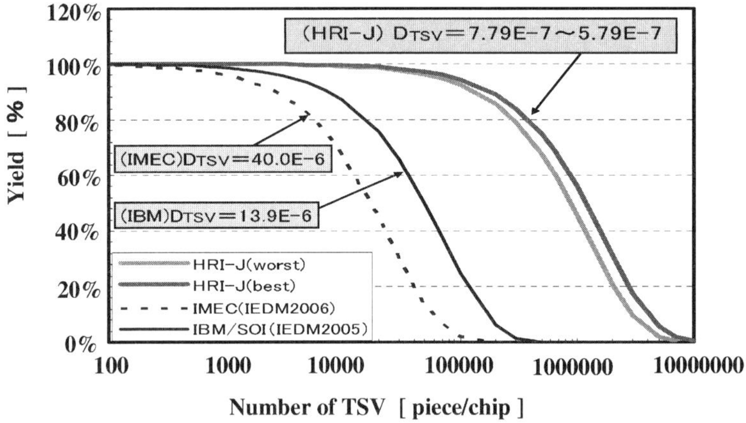

Figure 7. Connection Yield was calculated by the equation of the poisson distribution

The summary of our stacking method is as follows: (1) Stacking technology is wafer-to-wafer method, (2) all back-side processes are removed except wafer thinning, (3) wet etching is used for wafer thinning, (4) total stacking process is simplified and shortened, and (5) TSV and micro-bump are directly connected. This technology is applicable as well for a larger-scale wafer of more than 8 inches.

TRIAL MANUFACTURE

Structure and main function

The prototype of 3-layer stacking devices using our technology has been tried. The purpose of the trial manufacture has verified the effect of our stacking technology and from the results of this trial manufacture we have in fact gotten the characteristics of 3D stacking technology.
Each wafer is structured by a microprocessor (L1), custom circuits including analog circuits (L2), and 64M bit SDRAM (L3), respectively. And wafer size, chip size and usage process is 8 inches, 8.44mm x 4.69mm, and 0.18 μ m CMOS process, respectively. Especially significant, the number of TSV between 1st Layer chip and 2nd Layer chip is 1,056, and total TSV number between 1st wafer and 2nd wafer is 655,584. Stacking method is wafer-to-wafer using Face to Back stacked type.

Experimental results of trial manufacture

The evaluation results of the probe test are shown in Figure 7 after 3-wafer stacking in five cases. The yields gotten from the results comprising all functional tests are more than 60% and these results show the electrical connectivity between each wafer is superior to former experiments. Such values suggest a high probability for industrial fabrication in the near future, although some further investigation remains. The data regarding the relationship between operational frequency and power consumption is shown in Figure 9 with that of an MCM device having the same function. In comparing both devices, the trial device demonstrates two times the operational frequency but less than one-third power consumption of an MCM device under the same test conditions.

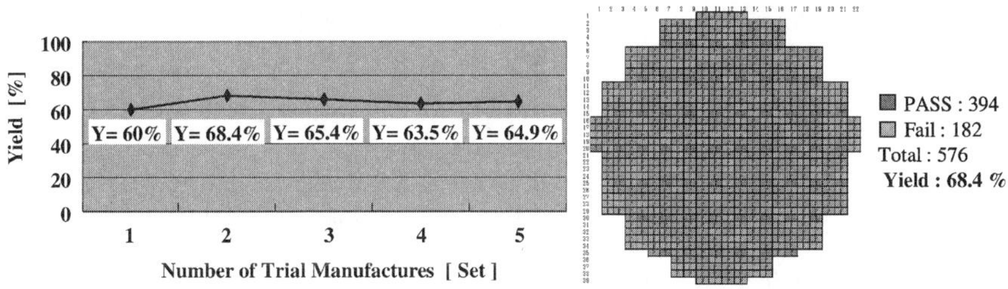

Figure 8. Probe test results of 3-layer stacked devices in five cases and an example of test results

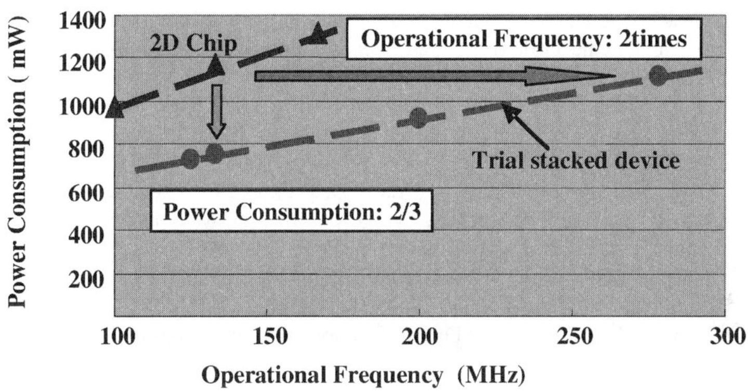

Figure 9. Relationship between operational frequency and power consumption of trial manufacture

CONCLUSIONS

This multi-layer stacking technology proposes a direct connection method between each layer by using wet etching for which the data between each layer shows almost all 100% connection. The proposed method verifies the usefulness of this stacking technology for trial manufacture. The stacking process technology can be used with wafers of more than 8 inches.

ACKNOWLEDGEMENTS

The authors wish to thank Messrs. Takanori Maebashi, Natsuo Nakamura, Shigeto Nakayama, Yutaka Sacho, Nozomi Horino, Eiri Hashimoto and Shinjiro Toyoda for their help. Thanks also to Hitachi, Ltd. Micro Device Division for providing wafer fabrication using 0.18 μm CMOS technology based on 8-inch wafers and evaluating test wafers and devices. And the authors would like to thank Soliton Systems K.K. for the stacked data conversion, Renesas Technology Corp. (formerly SuperH, Inc), Hitachi Information & Communication Engineering, Ltd. and Hitachi ULSI Systems Co., Ltd. for the design of SH4.

REFERENCES

1. M. Koyanagi, H. Kurino, K. W. Lee, K. Sakuma, N. Miyakawa, and H. Itani, "Future System-on-Silicon LSI Chips," IEEE Micro, vol. 18, no.4, 17-22 (1998).
2. Len Schaper, "Copper Electroplated Through Silicon Vias for 3-DIC Interconnection," Advanced Metallization Conference, MINI-Workshop,Oct.18 (2004).
3. SEMATIC Proceedings of "3D Technology, Modeling, and Process Symposium", April 13 (2004).
4. RTI Proceedings of "3D Architectures for Semiconductor Integration and Packaging" Conference 2004 -2007
5. A. Iwata, M. Sasaki, T. Kikkawa, S. Kameda, H. Ando, K. Kimoto, D. Arizono, H. Sunami, "A 3D Integration Scheme Utilizing Wireless Interconnections for Implementing Hyper Brains," ISSCC Dig. Tech. Papers, 262-263 (2005).
6. T. Fukushima Y. Yamada, H. Kikuchi and M. Koyanagi, "New Three-Dimensional Integration Technology Using Self-Assembly Technique," IEEE Int'l. Electron Devices Meeting Conference Digest Technology Papers, 359-362 (2005).
7. A. W. Topol, D. C. La Tulipe, L. Shi, S. M. Alam, D. J. Frank, S.E. Steen, J. Vichiconti, D. Posillico, M. Cobb, S. Medd, J. Patel, S. Goma, D. DiMilia, M.T. Robson, E.Duch, M. Farinelli, C. Wang, R.A. Conti, D. M. Canaperi, L. Deligianni, A. Kumar, K. T. Kwietniak, C. D'Emic, J. Ott, A. M. Young, K. W. Guarini, and M. Leong, "Enabling SOI-Based Assembly Technology for Three-Dimensional (3D) Integrated Circuits (ICs)," IEEE Int'l. Electron Devices Meeting Conference Digest Technology Papers, 363-366 (2005).
8. P. R. Morrow, C.-M. Park, S. Ramanathan, M. J. Kobrinsky and M. Harmes, "Three-Dimensional Wafer Stacking Via Cu-Cu Bonding Integrated With 65-nm Strained-Si/Low-k," IEEE ELECTRON DEVICES, VOL.27, No.5, May (2006).
9. Y. Xie, G. Loh, B. Black and K. Bernstein, "Design Space Exploration for 3D Architectures, ACM Journal on Emerging Technologies in Computing Systems, Vol.2, 65-103 (2006).
10. ISSCC Forum on the " Design of 3D-Chipstacks" Feb.11 (2007).
11. N. Miyakawa, T. Maebashi, N. Nakamura, S. Nakayama, E. Hashimoto and S. Toyoda, "New Multi-Layer Stacking Technology and Trial Manufacture," 3D Architectures for Semiconductor Integration and Packaging, Session 8, 2007.

Conference Proceedings AMC XXIV © 2009 Materials Research Society

3D System Integration Technology and 3D Systems

Takafumi Fukushima, Tetsu Tanaka, and Mitsumasa Koyanagi
Bioengineering and Robotics, Tohoku University, 6-6-01, Aza-Aoba, Aramaki, Aoba-ku, Sendai, 980-8579, Japan

ABSTRACT

We introduce 3D integration technologies using wafer-on-wafer bonding and chip-on-wafer bonding methods. The latter chip-to-wafer 3D integration based on a self-assembly process is called "super chip integration". In the super chip integration technology, a large number of LSI chips can be simultaneously, quickly, and precisely aligned and bonded onto LSI wafers. By using the key technologies including TSV (through-Si via) formation and self-assembly, we successfully fabricated prototype 3D LSI chips.

INTRODUCTION

In recent years, 3D integration technologies have been vigorously studied for the development of vertically stacked IC. In particular, TSV formation consisting of deep silicon etching, insulator deposition, metal filling, and the subsequent planarization of CMP (chemical mechanical polishing) is a representative process for 3D IC fabrication. The 3D integration technologies can provide extremely short wiring connecting between LSI chips, resulting in high LSI performance with the scaling down the device sizes. We have pioneered new 3D integration technologies based on wafer-on-wafer bonding methods in the last 20 years. [1]-[9] The wafer-on-wafer 3D integration has a great advantage in high production throughput. On the other hand, in the last few years, we have newly proposed chip-on-wafer 3D integration using an alignment/bonding technique of self-assembly, which can realize high-throughput batch assembly of a large number of known good dies (KGDs) and overcome yield problems in the wafer-on-wafer 3D integration due to the use of KGDs. In contrast, traditional pick-and-place chip assembly requires a great investment of time to precisely align and bond the many KGDs. We call this new chip-to-wafer 3D integration Super Chip Integration. [10], [11] This paper mainly introduces our key 3D integration technologies including TSV formation, adhesive injection, wafer bonding, microbump formation, and self-assembly and also describes the overview of the Super Chip Integration.

KEY TECHNOLOGIES FOR 3D INTEGRATION

In the past 10 years, we have developed three types of TSVs: the first one is formed with P-doped poly-Si as conductive materials by CVD on a thermal oxidation layer. This TSV process has been published in 1998, where deep trenches with a width of 4 μm and a depth of more than 60 μm can be successfully filled with P-doped poly-Si. [3] The second TSVs are composed of Poly-Si and tungsten by CVD, which is presented in 2001, where very narrow trenches less than 1 μm in width have been already employed for our 3D integration. [12] The first and second TSV processes are generally called Via First Process before FEOL (front end of line). The third TSVs are obtained by CVD of tungsten alone without Poly-Si. The tungsten as conductive materials shows lower resistivity than that of Poly-Si. In addition, low-temperature process at below 400 °C can be applied for TSV formation, which enables us to form huge number of TSVs in KGDs having MOS structures and metal wiring after BEOL (back end of line). The third process is classified into Via Last Process categories. An example of tungsten formation is shown in Fig. 1, in which a thin tungsten layer is conformably formed in high-aspect-ratio TSVs through a thick passivation SiO_2 layer on the top of an LSI chip.

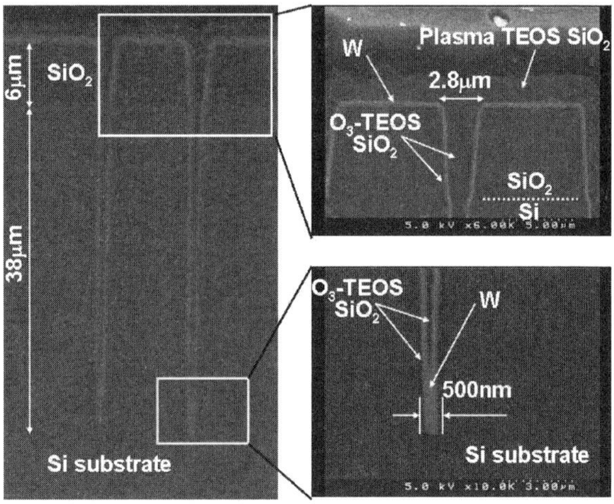

Fig.1 SEM cross-sectional view of conformal W-TSV

As shown in Fig. 2, the following describes a Via First Process we have previously developed. After TSV formation, microbumps are formed by a special lift-off technique, and then, face-to-face (or face-to-back) aligned and bonded to an additional LSI wafer. After that, adhesives are injected into the resulting narrow gap between wafers by using an evacuation technique. [3] The adhesive injection method is resulted from production technologies of liquid crystal display panels and is similar to underfilling in semiconductor packaging technologies. Then, the bonded thick LSI wafer is thinned from the backside until the bottoms of the TSVs are exposed. By repeating these processes, 3D staked LSI wafers can be fabricated as shown in the cross-sectional SEM image in Fig. 3. Figure 4 shows representative prototype 3D LSIs such as 3D image sensor chip [6], 3D shared memory [7], 3D artificial retina chip [8], and 3D microprocessor chip. [9] These chips have three-layer stacked structure and each layer thickness is approximately 30 μm. A number of P-doped poly-Si TSVs with a diameter of 2 μm and In/Au microbumps with a size of 5 by 5 μm and a height of 3 μm can electrically connect between upper and lower layers.

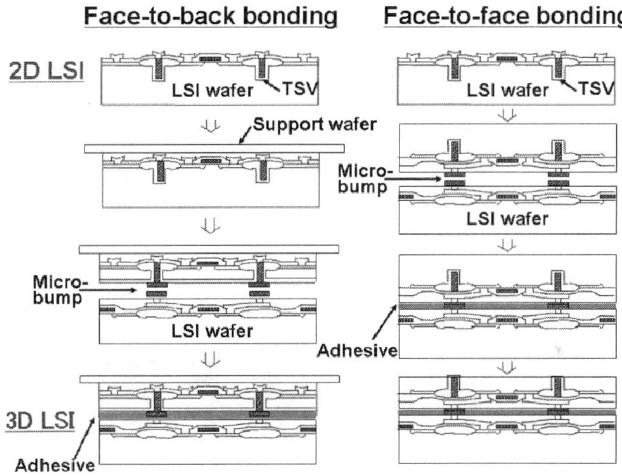

Fig.2 Wafer-on-wafer 3D integration processes

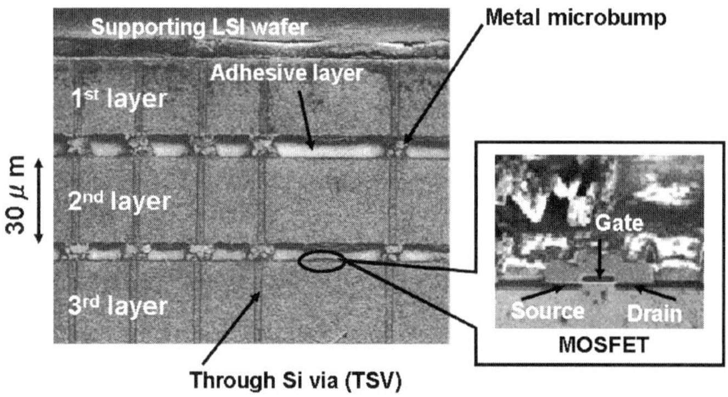

Fig.3 SEM cross-sectional view of 3D microprocessor chip fabricated by wafer-on-wafer bonding

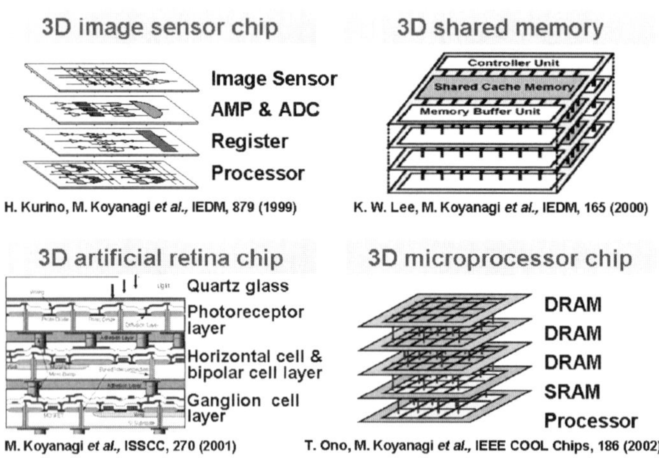

Fig.4 Prototype 3D LSI chips fabricated in Tohoku University

SUPER CHIP INTEGRATION BASED ON SELF-ASSEMBLY

We have proposed for the first time a new 3D integration based on a chip self-assembly technique to solve the yield and throughput problems in conventional wafer-on-wafer and chip-on-wafer 3D integration technology. [10], [11] As shown in Fig. 5, KGDs are selected form completed LSI wafers with/without TSVs in the new 3D integration using self-assembly in which liquid surface tension as a driving force can enable KGDs to be precisely aligned in batch and tightly bonded onto hydrophilic bonding areas formed on the supporting LSI wafer. In addition, various kinds and sizes of KGDs can be integrated using the new 3D integration technology called Super Chip Integration. The conceptual viewgraph of the Super Chip is shown in Fig. 6. Various kinds of thin chips with different sizes such as MEMS chip, sensor chip, CMOS RF-IC, MMIC, power IC, control IC, analog LSI, and logic LSI can be vertically stacked. Thus, Super Chip Integration can provide new potential application using advanced LSIs with small form factor, high packing density, high performance, low power consumption, and new functionality.

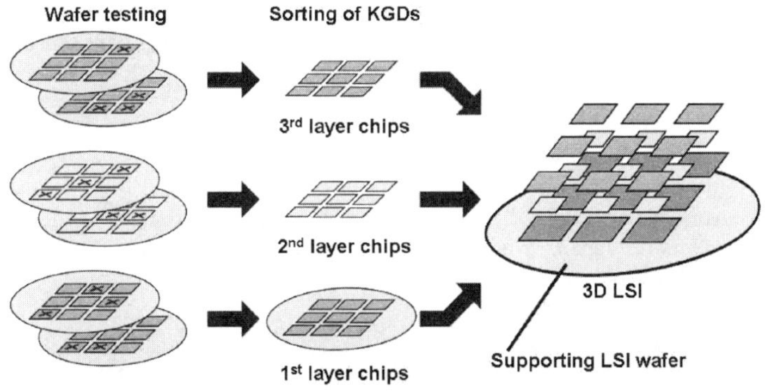

Fig.5 KGD stacking in batch by *Super Chip Integration* using self-assembly

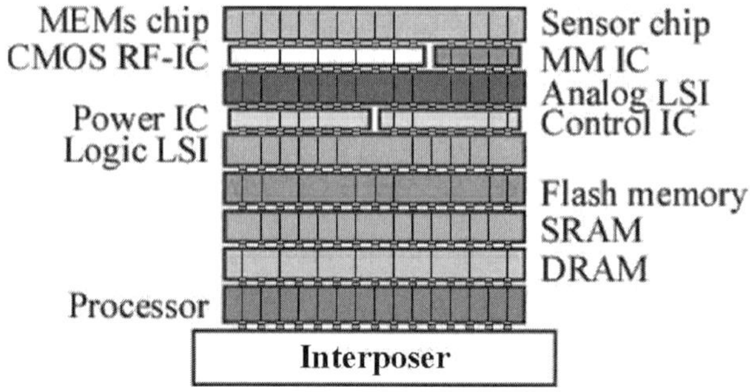

Fig.6 Structure of 3D *Super Chip*

CONCLUSIONS

We have developed 3D integration technologies based on wafer-on-wafer bonding and chip-on-wafer bonding techniques, fabricated prototype 3D LSI chips, and confirmed their basic operation. In addition, new super-chip integration technology using chip self-assembly has been proposed and its potential use is described.

ACKNOWLEDGMENTS

This work was performed in the Micro/nano-machining research and education center at Tohoku University, Japan. We would like to thank the staff of the Micro/nano-machining research and education center (MNC).

REFERENCES

1. Koyanagi, "Roadblocks in Achieving Three-Dimensional LSI", Proc. 8th Symposium on Future Electron Devices, pp. 50-60, 1989.

2. Takata, T. Nakano, S. Yokoyama, S. Horiuchi, H. Itani, H. Tsukamoto and M. Koyanagi, "A Novel Fabrication Technology for Optically Interconnected Three-Dimensional LSI by Wafer Aligning and Bonding Technique," Proc. Int. Semiconductor Device Research Symposium, pp. 327-330, 1991.

3. T. Matsumoto, M. Satoh, K. Sakuma, H. Kurino, N. Miyakawa, H. Itani and M. Koyanagi, "New Three Dimensional Wafer Bonding Technology Using the Adhesive Injection Method," Jpn. J. Appl. Phys., 37, pp. 1217-1221, 1998.

4. M. Koyanagi, H. Kurino, K. W. Lee, K. Sakuma, N. Miyakawa, H. Itani, "Future System-on-Silicon LSI Chips," IEEE MICRO, 18, pp. 17-22, 1998.

5. M. Koyanagi, T. Nakamura, Y. Yamada, H. Kikuchi, T. Fukushima, T. Tanaka, and H. Kurino, "Three-Dimensional Integration Technology Based on Wafer Bonding with Vertical Buried Interconnections," IEEE Trans. Electron Devices, 35, pp. 2799-2808, 2006.

6. H. Kurino, K. W. Lee, T. Nakamura, K. Sakuma, H. Hashimoto, K. T. Park, N. Miyakawa, H. Shimazutsu, K. Y. Kim, K. Inamura and M. Koyanagi, "Intelligent Image Sensor Chip with Three Dimensional Structure," IEDM Tech. Dig., pp. 879-882, 1999.

7. K. W. Lee, T. Nakamura, T. Ono, Y. Yamada, T Mizukusa, H. Hashimoto, K. T. Park, H. Kurino and M. Koyanagi, "Three-Dimensional Shared Memory Fabricated Using Wafer Stacking Technology," IEDM Tech. Dig., pp. 165-168, 2000.

8. M. Koyanagi, Y. Nakagawa, K-W. Lee, T. Nakamura, Y. Yamada, K. Inamura, K-T. Park, H. Kurino, "Neuromorphic Vision Chip Fabricated Using Three-Dimensional Integration Technology," Proc. IEEE Int, Solid State Circuits Conf. (ISSCC), pp. 270-271, 2001.

9. T. Ono, T. Mizukusa, T. Nakamura, Y. Yamada, Y. Igarashi, T. Morooka, H. Kurino and M. Koyanagi, "Three-Dimensional Processor System Fabricated by Wafer Stacking Technology," Proc, Int. Symp. on Low-Power and High-Speed Chips (COOL Chips V), pp. 186-193, 2002.

10. T. Fukushima, Y. Yamada, H. Kikuchi, and M. Koyanagi, "New Three-Dimensional Integration Technology Using Self-Assembly Technique," IEDM Tech. Dig., pp.359-362, 2005.

11. T. Fukushima, Y. Yamada, H. Kikuchi, T. Konno, J. Liang, A. M. Ali, K. Sasaki, K. Inamura, T. Tanaka, and M. Koyanagi, "New Three-Dimensional Integration Technology Based on Reconfigured Wafer-on-Wafer Bonding Technique, IEDM Tech. Dig., pp. 985-988, 2007.

12. Y. Igarashi, T. Morooka, Y. Yamada, T, Nakamura, K. W. Lee, K. T. Park, H. Kurino, and M. Koyanagi, "Filling of Tungsten into Deep Trench Using Time-Modulation CVD Method", Int. Conf. on Solid State Devices and Materials, pp. 34-35, 2001.

Conference Proceedings AMC XXIV © 2009 Materials Research Society

Integration and RF characterization of high density Through Silicon Vias for 3D chip stacking

Lionel Cadix[1], Patrick Leduc[2], Alexis Farcy[1], Miriam Assous[2], Cédric Bermond[3], Léa Di Cioccio[2], Barbara Charlet[2], Anne Roule[2], Laurent-Luc Chapelon[1], Maxime Rousseau[1,2], Christine Fuchs[2], Bernard Flechet[3], Nicolas Sillon[2], and Pascal Ancey[1]

[1]STMicroelectronics, 850 rue Jean Monnet, Crolles Cedex, 38926, France, Metropolitan

[2]CEA-Léti, 17 rue des Martyrs, Grenoble Cedex 9, 38054, France, Metropolitan

[3]Université de Savoie, IMEP-LAHC, Le Bourget du Lac Cedex, 73376, France, Metropolitan

ABSTRACT

3D chip stacking is proposed to overcome conventional 2D-IC issues, using TSVs to ensure vertical signal transmission. Specific integration flows and TSV electrical model are mandatory to achieve 3D products. In this paper, a full 3D process flow was detailed and used to embed specific test structures dedicated to high-frequency characterizations. As a result, 2 and 3µm wide and 15µm deep voidless copper-filled TSVs were successfully integrated. Moreover, a π-shaped electrical equivalent model was extracted, demonstrating the tremendous benefits of TSV compared to standard wire bonding.

INTRODUCTION

Manufacturers introduce more and more remarkable electronic devices into the market allowing the user to surf the Internet, read e-mails, exchange files and watch TV on smart phones or PDAs. These multitask devices require high performance ICs including powerful processors, huge memory capabilities while keeping low consumption. Therefore, innovative solutions have to be developed to overcome conventional 2D IC performance issues: high RC interconnect delay, thermal heating and power consumption. In this way, 3D stacking and high aspect ratio vertical interconnects, so-called Through Silicon Via (TSV), have recently attracted much attention to improve performance of ICs at lower cost. Beyond all the benefits of 3D-IC technology, i.e. high integration density, fast signal transmission, low manufacturing cost and reduced power consumption, 3D integration opens new perspectives in term of heterogeneous technologies stacking enabling more sophisticated than ever system on chip integration [1]. Many different architectures have been proposed [2-3] but whatever the 3D approach (i.e. via first, via last...), TSVs keep a key role and have to transmit a wide range of potentially high frequency signals (digital, power, analog, RF...). According to this signal diversity, it is mandatory to extract TSV characteristics, for instance its parasitic resistance, inductance, capacitance and conductance on a wide frequency range, highlighting the requirement for frequency dependent electrical models. In this paper, a technological solution to process the TSV is presented, as well as static and frequency dependent electrical results. First, the fabrication process of TSV is detailed, according to a via last approach with face to face wafer stacking. Next, a methodology based on high frequency characterizations is investigated to extract equivalent circuit based on [S] parameters, leading to the modelling of high density TSV for 3D-IC on a wide range of frequencies.

INTEGRATION OF HIGH DENSITY THROUGH SILICON VIAS

Wafer stacks including 2-3 µm wide high density TSVs (~10^4 TSV/mm²) were performed according to a via last approach [4]. A top wafer was face to face bonded on a second one using SiO_2 bonding, and thinned down to 15 µm prior to via processing (Fig. 1). The complete process flow for 3D chip stacking is divided into three main steps: wafer bonding, substrate thinning, and TSV integration.

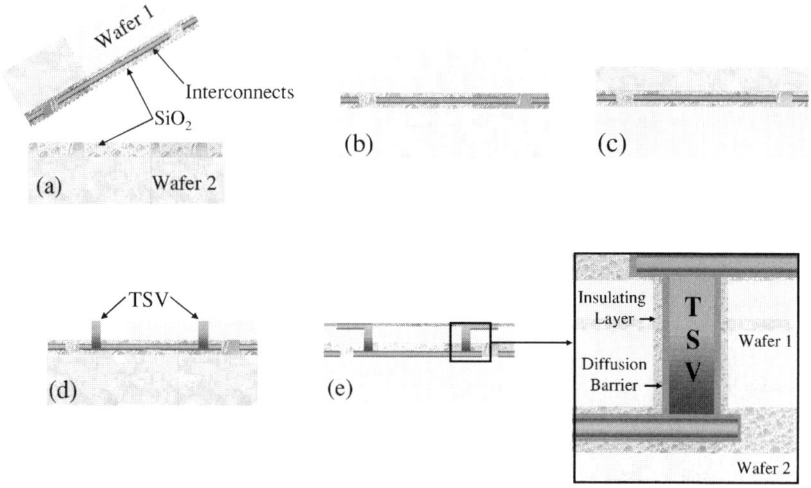

Figure 1: Face to face via last process flow integration (a) before bonding and after bonding (b), Si thinning (c), TSV patterning and filling (d), and top interconnects integration (e)

SiO_2 Bonding

The aim of this technique is to achieve a glue free assembling without strain at room temperature. Before bonding, SiO_2 surface is prepared with Chemical & Mechanical Polishing (CMP) and cleaning. Then, wafers are face to face bonded using direct oxide bonding technique [5-6]. Perfect bonding quality is obtained at a low temperature compatible with copper back-end of line on wafers having relatively high incoming surface topography. Finally, the stacking is stabilized by an annealing step at 400°C under N_2 atmosphere.

Wafer thinning

After bonding, top wafer backside is thinned down to 15 µm using grinding and CMP [9]. A specific edge outlining step is introduced to remove damages at wafer edge within a 3 mm ring. Figure 2 shows resulting two strata-3D stack with 15 µm thin Si. The total thickness variation (TTV) after thinning is about 1µm. Surface roughness is 0.5 nm RMS and the bow is reduced to 20 µm after a stress release step (CMP).

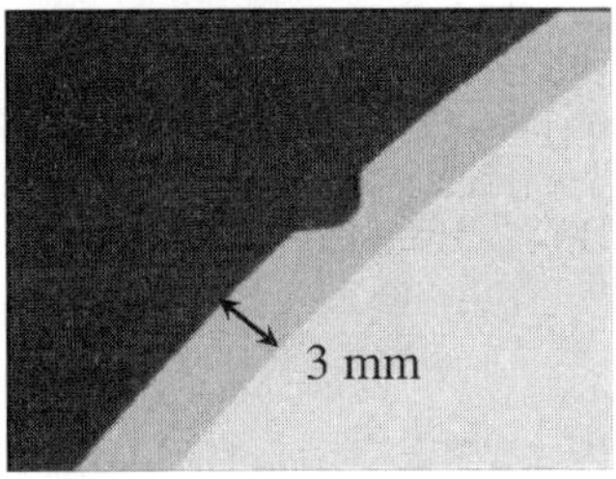

Figure 2: SEM edge view of a wafer stack featuring a top wafer thinned down to 15 µm

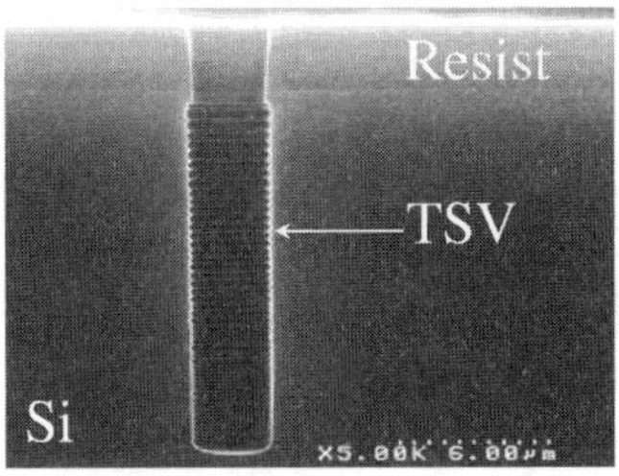

Figure 3: SEM cross section of a 3 µm wide / 15 µm deep TSV after DRIE etching

TSV patterning and filling

2 µm & 3 µm wide and 15 µm deep copper TSVs are integrated to achieve high density 3D integration once bonding and thinning are complete [4]. First, lithography with thick resist (3.2 µm) is performed before deep RIE etching for silicon etching. Good etching profile is required for metallization, enabling integrity and good step coverage. Low undercut (≤ 30 nm) and scalloping (≤ 50 nm) are observed after cleaning and resist stripping compatible with further filling process steps (Fig.3). After CVD deposition of a 200 nm thick SiO_2 layer on the sidewall of the TSVs for isolation, a metallization is performed by successive depositions of MOCVD TiN of 20 nm as a diffusion barrier, 50 nm PVD Cu seed featuring good adhesion with sidewall oxide, 150 nm conformal CVD Cu seed and ECD Cu filling. Specific ECD chemistries and process parameters were selected to obtain voidless filling. Figure 4 shows a FIB SEM cross section of a 15µm deep and 3µm wide TSV after copper filling. To connect TSVs, a top interconnect level is added (ReDistribution Layer or RDL) using conventional damascene process. SiO_2 bonding interface after the complete integration is free of defects.

Figure 4: FIB-SEM cross section of a 3µm wide and 15µm deep TVS after filling

Test devices

To accurately characterize the resistance of single TSV, Kelvin structures were designed and processed. An array of twenty five 3 µm-wide and 5 µm-space TSV connects 80x80 µm² copper pads to top metal level as shown in Figure 5a. Besides, specific RF structures were integrated to evaluate TSV behaviour with frequency. These coplanar structures include Ground Signal Ground pads, access lines and dual via chains. Figure 5b shows a schematic view of the RF devices.

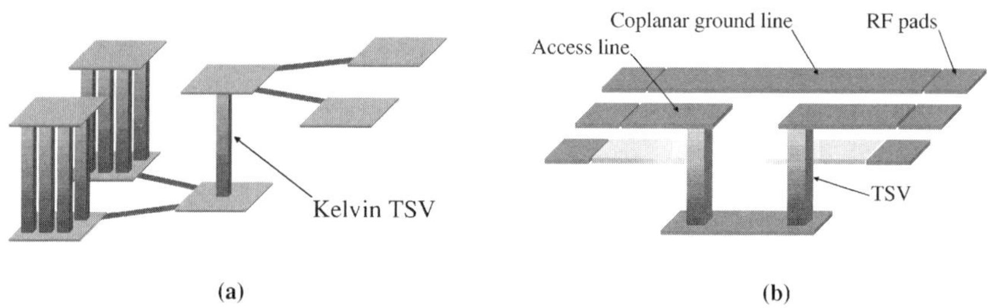

Figure 5: Schematic view of Kelvin TSV (a) and microwave structures (b)

ELECTRICAL CHARACTERIZATION OF TSV

Microwave characterization methodology

RF structures were tested with a Vectorial Network Analyzer (VNA) and 100 µm pitch Ground-Signal-Ground (GSG) probes. The VNA provides frequency sweep from 40 MHz up to 40 GHz, and measures transmitted and reflected waves to build scattering matrix [S]. Both calibration and deembedding operations enable to adjust reference planes to accurately measure the properties of TSVs. In order to remove parasitic effects from wires and probes, a SOLT type calibration enabling to adjust reference planes close to the tip of the microwave probes is carried out. It consists in measuring four standard devices (Short, Open, Line and Thru) to refer the tip of the probe at standard 50 Ω impedance by subtracting parasitic elements of cable and probes. Then, a deembedding step is required to remove pad and access line and adjust reference planes as close as possible to the TSV in order to extract intrinsic TSV [S] matrix. This deembedding is performed by computation and lead to determine intrinsic TSV [S] parameters. The first step consists in calculating [ABCD] matrices from [S] matrices, easier to manipulate. Then [ABCD] matrices of RF pad, access line and full structure, respectively named $[A]_{pad}$, $[A]_{line}$ and $[A]_{device}$, are extracted from deembedding measurements. Finally, TSV matrix is built, removing RF pad and access line effects:

$$[A_{TSV}] = [A_{line}]^{-1} \times [A_{pad}]^{-1} \times [A_{device}] \times [A_{pad}]^{-1} \times [A_{line}]^{-1} \qquad (1)$$

A last calculation step allows to switch from [ABCD] to the TSV chain [S] matrix. Figure 6 shows a summary of the calibration and deembedding effect.

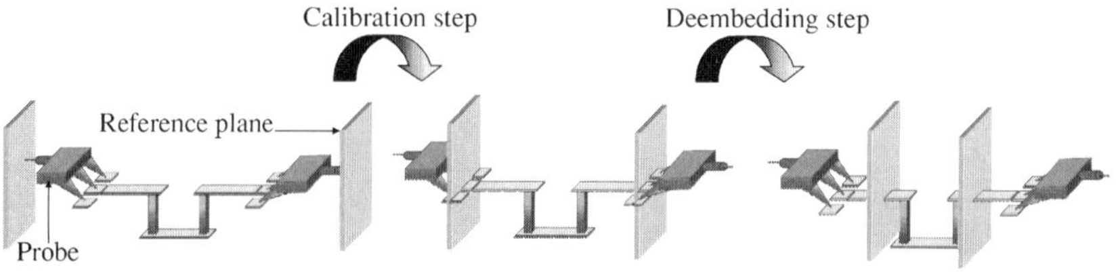

Figure 6: Calibration & deembedding effect

TSV electrical model

After calibration and deembedding step, a model based on extracted RLCG parameters is built from TSV chain A matrix. Figure 7 exhibits the π-equivalent circuit proposed to model TSV: the resistance R and inductance L are part of serial impedance of TSV. The capacitance C and conductance G simulate parasitic effects between TSV and low resistivity Si substrate generally used in most ICs. Note that each element of this model is frequency dependent and fully compatible with SPICE-based simulators.

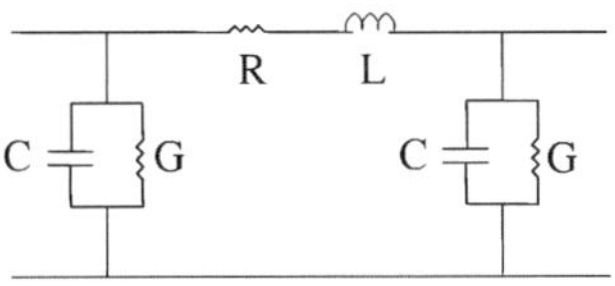

Figure 7: π-equivalent RLCG model of TSV

Microwave characterization

Figures 8 and 9 show extracted resistance, inductance, capacitance and conductance per TSV versus frequency.

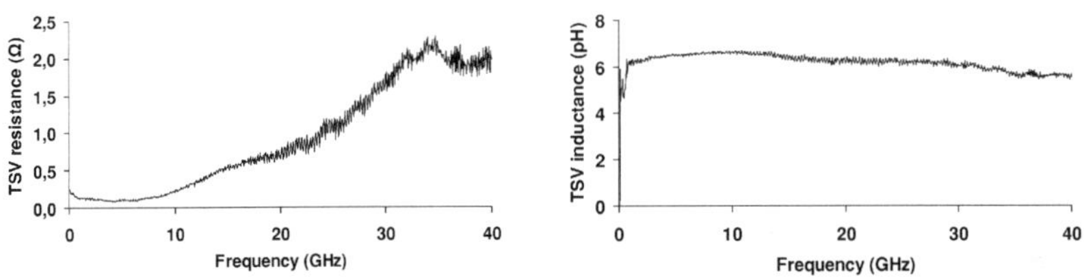

(a) (b)

Figure 8: Evolution of RL parameters with frequency: (a) Resistance, (b) Inductance

According to figure 8a, the resistance increases with frequency after a stationary phase at 100 mΩ (from 40 MHz to 10 GHz), which meets specifications required for usual products. Then resistance increases up to 2.5 Ω at 40 GHz. This growth begins at 10 GHz and is a consequence of the skin effect. At high frequency, electric currents are rejected at the conductor periphery, in a small metal band called skin depth, leading to resistance increase. Figure 8b shows that extracted inductance is quasi constant at ~ 6 pH, with a slight decrease at high frequencies.

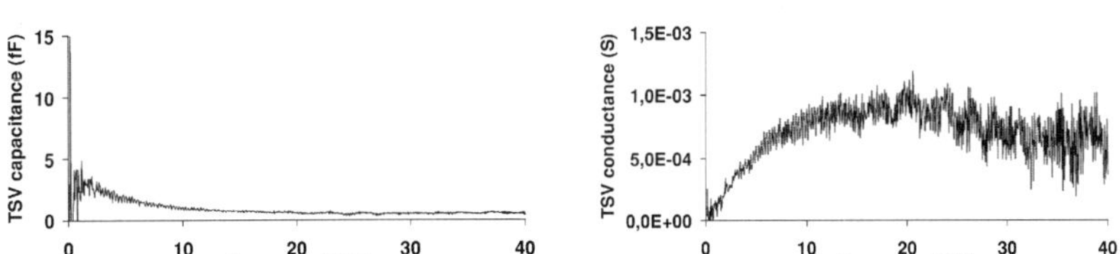

(a) (b)

Figure 9: Evolution of CG parameters with frequency: (a) Capacitance, (b) Conductance

Besides, according to figure 9a, capacitance strongly depends on frequency and drops from 15 fF to ~1 fF at 15 GHz. This lumped capacitance is measured between TSV and top ground planes. Contrary to the most cases in conventional 2D interconnects, conductance is non-null, and increases with frequency (Fig.9b). Actually, this frequency dependence of C and G becomes from the proximity of low resistive Si substrate and its lossy behaviour. Finally, as shown in figures 8 and 9, RLCG are frequency dependent, highlighting that a frequency dependent equivalent circuit model is required.

Static characterization & analytical calculation

The resistance of single TSV is measured on Kelvin structures. For 3 µm TSV, resistance is between 105 and 150 mΩ (Fig.10). These values are in good agreement with analytical calculation performed using (2) for 3 µm wide / 15 µm deep TSVs (Fig.10, Fig.11) and with RF measurements at low frequencies.
Equation (2) deals with all the materials included in a TSV and features the evolution of TSV resistance as a function of metal and diffusion barrier resistivities (ρ_{metal} ρ_{bar}), silicon, dielectric and diffusion barrier thicknessees (t_{Si}, t_{diel} and t_{bar}), and TSV radius (r_{via}).

$$R = \frac{\rho_{metal}.(t_{Si} - t_{bar})}{\pi.(r_{via} - t_{diel} - t_{bar})^2} + \frac{\rho_{bar}.t_{bar}}{\pi.(r_{via} - t_{diel} - t_{bar})^2} \quad (2)$$

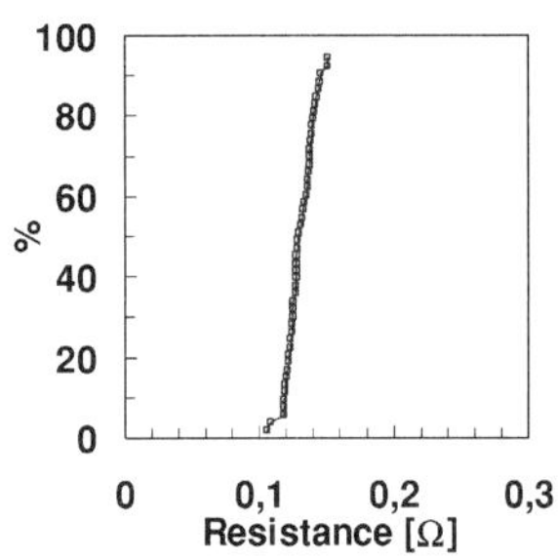

Figure 10: 15 µm deep Kelvin TSV resistance

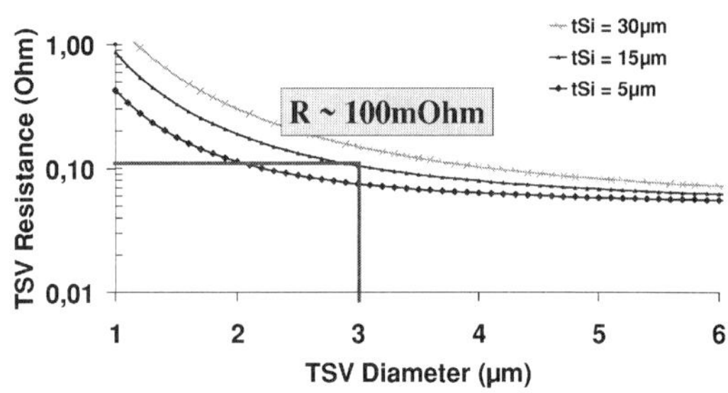

Figure 11: Analytical calculations of TSV resistance

According to figures 8a, 10 and 11, both measured and simulated static values of resistance are in good agreement, which validates TSV model. As a consequence, analytical calculation can be used for predictive simulations, as shown in figure 12, and process tuning. This figure features TSV Aspect Ratio (AR, dashed curves) and iso-response resistance curves. It is possible to give recommendations concerning TSV diameter or silicon thickness based on designer recommendations on resistance.

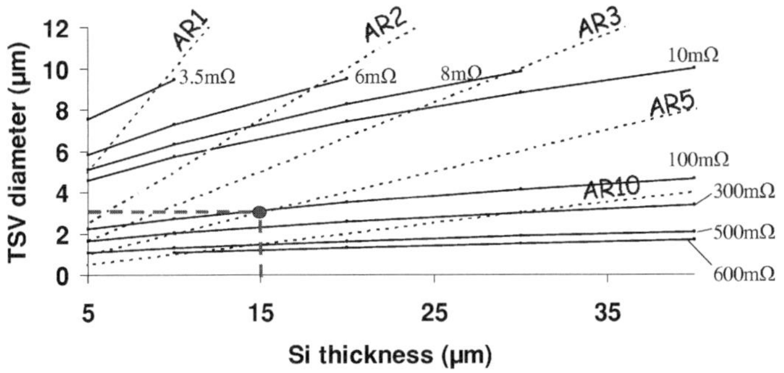

Figure 12: Iso-response resistance curve as functions of TSV diameter and height

Moreover, it is interesting to compare extracted values with properties of typical wirebonds usually involved among System-in-Package. Such wires have a capacitance around 2 pF and inductance of 3 nH. TSV parasitic elements being a thousand times lower, 3D integration should be a solution enabling high bandwidth and high speed communications between stacked chips.

CONCLUSION

In this paper, key technologies to enable 3D chip stacks and an equivalent circuit model of the TSV are presented. This frequency dependent model allows to extract resistance, inductance, capacitance and conductance of a TSV as functions of frequency. TSV static resistance was found to be close to 100 mΩ. Inductance and capacitance are a thousand times lower than conventional wirebonding: all these considerations make TSV a very attractive candidate for high performance products.

REFERENCES

[1] W.R. Davis, J. Wilson, S. Mick, J. Xu, H. Hua, C. Mineo, A.M. Sule, M. Steer, Design & Test of Computers, **22**, 6, pp 498-510 (2005).

[2] K. Sakuma, P.S. Andry, C.K. Tsang, K. Sueoka, Y. Oyama, C. Patel, B. Dang, S.L. Wright, B.C. Webb, E. Sprogis, R. Polastre, R. Horton, J.U Knickerbocker, ECTC, **58**, pp 18-23 (2008).

[3] K. Fukushima, T. Yamada, Y. Kikuchi, H. Tanaka, T. Koyanagi, M. Tohoku Univ., Sendai; J.U. ECTC, **57**, pp 836-841 (2007).

[4] P. Leduc, L. Di Cioccio, B Charlet, M. Rousseau, M. Assous, D. Bouchu, A. Roule, M. Zussy, P. Gueguen, A. Roman, O. Rozeau, M. Heitzmann, J.-P. Nieto, L. Vandroux, P.-H. Haumesser, R. Quenouillere, A. Toffoli, P. Sixt,; S. Maitrejean,; L. Clavelier, N. Sillon, VLSI Tech. Systems and App., 2008, pp 76 – 78 (2008).

[5] B. Aspar, C. Lagahe-Blanchard, H. Moriceau, ECS, **27** (2001).

[6] L. Di Cioccio, P. Gueguen, F. Grossi, P. Leduc, B. Charlet , M. Assous, A Mathewson, J. Brun, D. Henry, P. Batude, P Coudrain, N Sillon, L Clavelier, G Poupon, M Scannell, 4th Int. Conf. and Exhibition on Device Packaging, IMAPS, Scottsdale (2008).

[7] W.P. Maszara, J. Appl. Phys., 64 (**10**), pp.4943-50 (1988).

[8] R.Chatterjee, M. Fayolle*, P. Leduc*, S. Pozder, B. Jones, E. Acosta, B. Charlet*, T. Enot*, M. Heitzmann*, M. Zussy*, A. Roman*, O. Louveau**, S. Maitrejean*, D. Louis*, N. Kernevez*, N. Sillon*, Gérard Passemard**, V. Pol, V. Mathew, S. Garcia, T. Sparks, and Z. Huang, IITC Conf., pp.81-83 (2007).

[9] F. Romagna, B. Charlet, MRS Spring Meeting (2008).

Conference Proceedings AMC XXIV © 2009 Materials Research Society

Through Wafer Via Integration in CMOS and BiCMOS Technologies

Anthony Stamper[1], Paul Andry[2], Mete Erturk[1], Robert Groves[3], Alvin Joseph[1], Peter Lindgren[1], Paul McLaughlin[3], Rosemary Previti-Kelly[1], Edmund Sprogis[1], Kenneth Stein[3], Cornelia Tsang[2], Ping-Chuan Wang[3], and Jim Dunn[1]

[1]IBM, Essex Junction, VT, 05452

[2]IBM, Yorktown Heights, NY, 10598

[3]IBM, Hopewell Junction, NY, 12533

ABSTRACT

This paper discusses the development and manufacturing qualification of IBM's grounded through silicon via (TSV) technology. TSV processing, with the 150μm deep vias grounded to the substrate and contacted from the backside of the chip, has been installed into both AlCu [1] and Cu BEOL technologies. The use of TSV's reduced the ground lead inductance from 160pH to 30pH in wirebond packages and has enabled the use of SiGe power amplifiers operating at 5.8 GHz. This paper discusses the key integration challenges associated with grounded TSV's including deep silicon etch, wafer edge etch profile, TSV lithography/RIE, TSV CVD TiN/tungsten metallization, wafer bending/warpage, and reliability.

INTRODUCTION

TSV's are seeing increasing use in 3D-IC applications, both for chip stacking and for improving chip package IO density or performance. Most TSV applications use either copper or tungsten for the primary conductor and require that the TSV is electrically insulated from the substrate. This paper focuses on grounded TSV integration, where the TSV's are electrically grounded to the silicon substrate. 3D-IC technologies can broadly be classified as using either deep (i.e. >100μm) or shallow (i.e. <10μm) TSV's. Deep TSV technologies use the high etch rate Bosch process for silicon etching and generally do not require the use of handle wafers during backside thinning and packaging. Shallow TSV technologies require the use of a handle wafer after wafer thinning to avoid breakage and may employ either the Bosch process or a standard HBr- or HCl-based silicon etch process. The Bosch process has 3-10X higher silicon etch rate and 8-40X higher resist selectivity compared to standard HBr- or HCl-based silicon etch processes but induces a scalloped undercut etch profile due to the alternating C_4F_8 polymer deposition and SF_6 silicon etch steps, which complicates the metallization processes.

Packaging lead inductance is a major design issue, particularly for RF analog chips such as 5.8GHz WLAN power amplifiers (PA's). In general, the emitter impedance of bipolar junction transistors or the source impedance of MOSFET's degrade their electrical performance [2]. The heterojunction bipolar transistor (HBT) emitter ground leads used in SiGe HBT RF designs are normally contacted to the package either using multiple wire bonds or flip chip solder bumps. Wire bond package ground leads have high inductance, on the order of 160 pH, which results in unacceptable PA insertion loss. Although flip chip solder bumps have low inductance, they increase packaging complexity and are expensive. Grounded TSV's, patterned and etched from the wafer backside, are commonly employed in GaAs chips. In this paper, we discuss the development and manufacturing qualification of grounded TSV's in 200mm wafer silicon-based SiGe RF technologies. In addition to reducing ground lead inductance, moving the ground connection to the chip backside decreases the bond pad area on the chip surface, resulting in up to 20% reduced chip area. For this application, we employ TSV's with bottom width of 3μm grouped in ~ 60μm by 60μm arrays. The ~ 60μm by 60μm TSV arrays have lead inductance at 6GHz of ~ 30pH, dc resistance of ~ 15mΩ, ac resistance at 6GHz of ~ 100mΩ, and have facilitated WLAN 5.8GHz BiCMOS SiGe PA's using low cost wire bond packages [1].

EXPERIMENT

The technologies discussed in this paper use 200mm wafers and standard CMOS and BiCMOS process flows. IBM's TSV process flow is shown in Figure 1 [1, 3-5]. The TSV's are formed in the contact level, above the front-end-of-the-line (FEOL) devices and immediately below the first wiring level. For this grounded TSV application, there are three sequential lithographic patterning and etching steps performed after the FEOL BiCMOS processing is complete: the contact C1 dielectric openings, the TSV dielectric openings, and the deep TSV silicon openings. The TSV's are metallized simultaneously with the C1 contact using a CVD TiN liner and CVD tungsten bulk conductor. Standard tungsten CMP processing, optimized for 2.5μm of tungsten removal, is used. After back-end-of-the-line (BEOL) wafer processing is complete, the wafer backside is ground to 150μm and is blanket metallized prior to dicing.

Wafer breakage, due both to the deep silicon vias and to the heightened wafer bending of backside ground wafers, is a potential problem. Wafer backside grind stresses the wafer and, when wafers are thinned to under ~150μm, a handle wafer or glass carrier is required to provide structural support and avoid wafer breakage. IBM employs 150μm deep grounded TSV's to avoid the added cost and complexity associated with the use of handle wafers.

We investigated various TSV layouts including annular rings [4-5], rectangular 'box' vias, and rectangular bar vias [3]. All data discussed in this paper is for rectangular bar vias. Tungsten was chosen as the TSV conductor over copper because its coefficient of thermal expansion (CTE) is much closer to silicon, which improves thermal cycling reliability (copper, silicon, and tungsten have CTE's of 17, 3.5, and 5.0 ppm/K, respectively), and because tungsten is less reactive than copper during wafer backside grind and metallization. CVD tungsten has excellent TSV gap fill and was easily integrated into the standard tungsten C1 contact module.

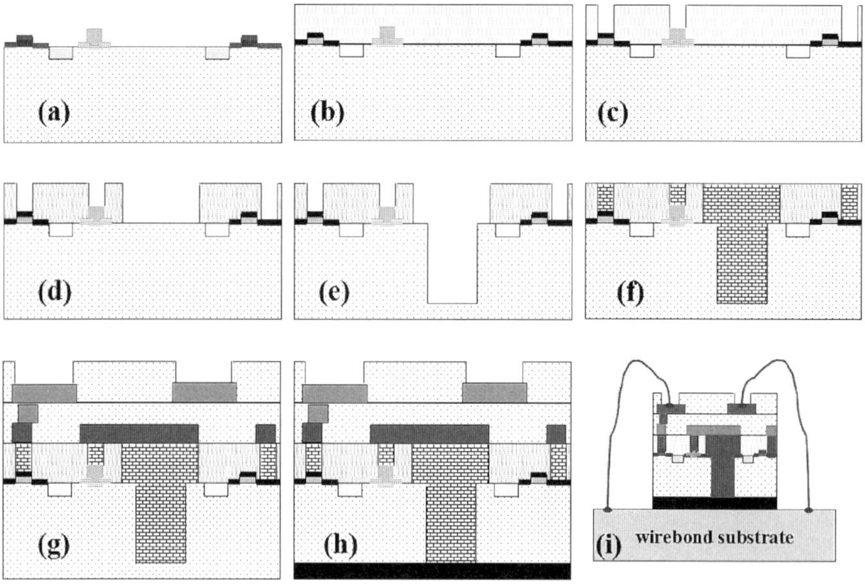

Figure 1: TSV process flow showing (a) FEOL BiCMOS, (b) contact dielectric deposition and planarization, (c) contact C1 patterning and etching, (d) TSV dielectric patterning and etching, (e) TSV silicon patterning and etching, (f) TSV and C1 contact metallization, (g) BEOL wiring, (h) wafer backside grind into the TSV with backside metallization, and (i) wirebond packaging.

DISCUSSION

Deep silicon via etching is a major TSV integration issue. The challenges we faced developing TSV's included RIE undercut, TSV sidewall scalloping induced by the Bosch etch, wafer edge TSV angle variability, and TSV etch rate variability (Figure 2). The RIE undercut issue results in a retrograde TSV profile due to silicon undercut of the C1 dielectric (Figure 2a). Since this retrograde profile severely degrades CVD tungsten gap fill, we etch the TSV using a two mask process, with mask 1 used to etch the C1 dielectric and mask 2 used to etch the silicon, as discussed below (Figure 1). The silicon etch rate is a strong function of TSV width. Figure 2c shows that, for TSV arrays fabricated in adjacent regions, the TSV RIE rate increases significantly as the TSV opening width increases. We resolved this TSV depth variability problem by limiting the TSV layout options to a single width (3µm). Numerous wafer edge TSV RIE problems occurred during development, including non-vertical TSV's (Figures 2b). These problems were solved by performing RIE process design of experiments; and by optimizing the Bosch RIE chamber hardware including replacing the mechanical wafer clamp with an electrostatic wafer chuck.

CVD tungsten integration issues include avoiding WF_6 attack of the silicon, obtaining good gap fill, maximizing deposition rate, and minimizing wafer warpage. WF_6 attack of the silicon is avoided by using a multilayer CVD TiN liner process with appropriate precleans. The tungsten wafer warpage issue is exacerbated by the ~2.5µm tungsten thickness needed to fill the TSV trenches, which have 3µm nominal dimension, but have TSV top openings approaching 5µm due to the two mask RIE process and variation in RIE bias. In general, tungsten gap fill is inversely correlated to both deposition rate and wafer warpage; i.e. gap fill improves as deposition rate decreases and warpage increases (Figures 3 - 5). To achieve acceptable gap fill and deposition

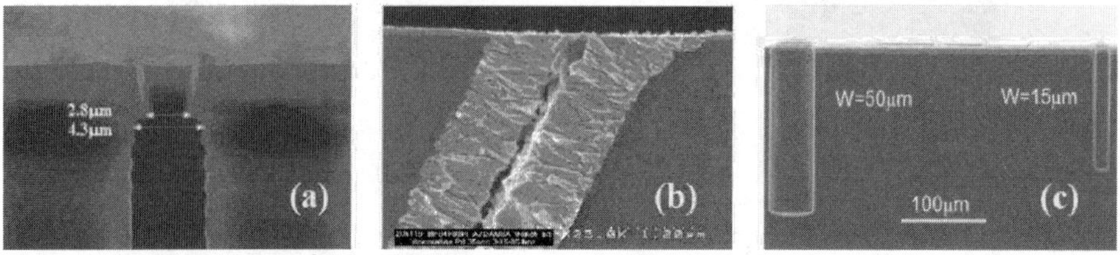

Figure 2: SEM images of TSV RIE problems: (a) TSV C1 dielectric undercut with one mask also showing Bosch RIE process scalloped profile; (b) wafer edge angled profile; and (c) Bosch RIE rate variation vs. TSV width.

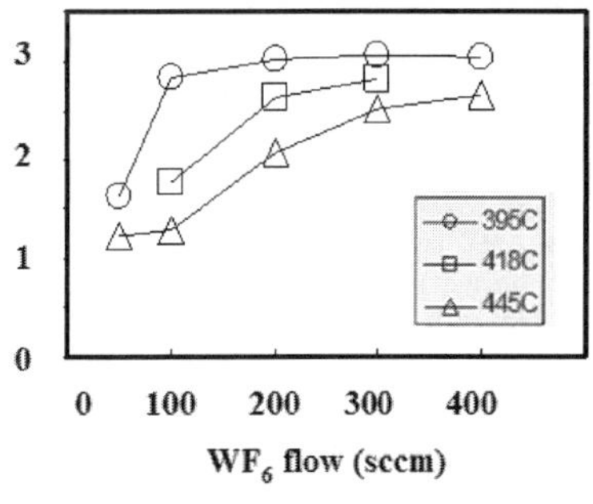

Figure 3: Normalized CVD W stress vs. temperature and WF_6 flow.

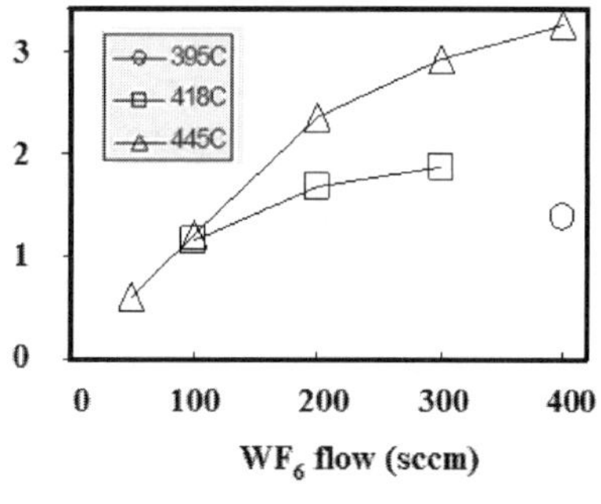

Figure 4: Normalized CVD W deposition rate vs. temperature and WF_6 flow.

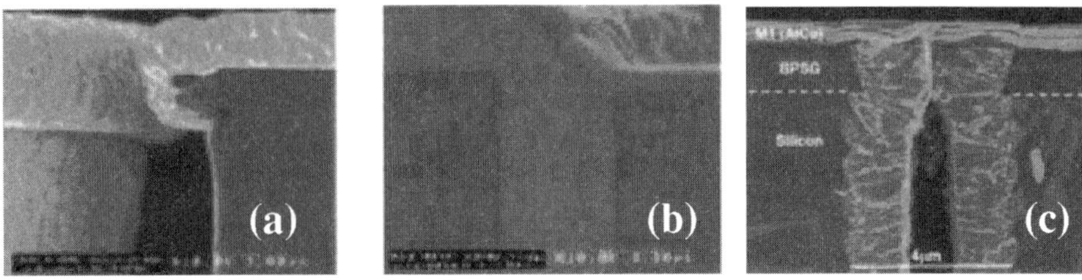

Figure 5: SEM's showing CVD W gap fill for (a) 445C and (b) 395C CVD W; and (c) 395C CVD W deposition with one mask undercut TSV profile.

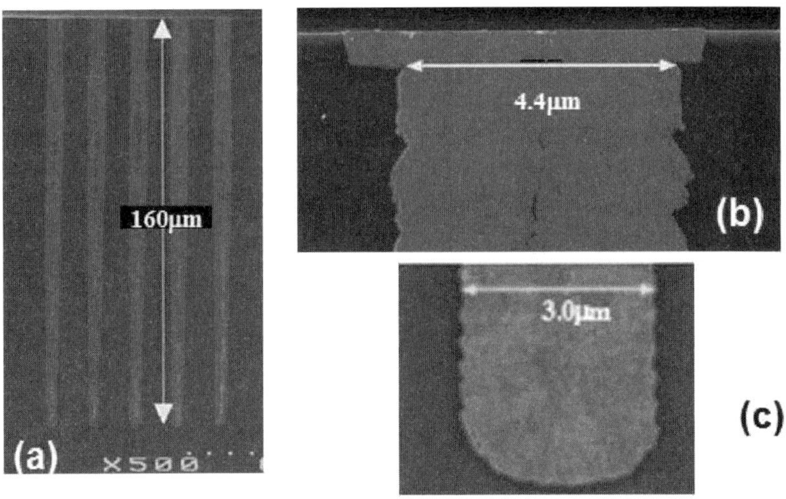

Figure 6: Metallized SEM cross-sections of (a) TSV arrays, (b) TSV top, and (c) TSV bottom.

rate, we use a tungsten atomic layer deposition pulsed nucleation layer, deposit the bulk fill tungsten at 395C, and flow 400sccm of WF_6, which results in a deposition rate of 0.3µm/minute (Figure 5). The high stress 2.5µm thick tungsten film induced a wafer radius of curvature of ~8 meters, which corresponds to 80µm of wafer bending, and required optimizing the tungsten CMP equipment wafer chucking process.

IBM chose a 2-mask TSV integration scheme to eliminate the overhang or undercut profile that occurs with Bosch TSV etching. The SEM cross-section in Figure 6b shows a ~5.6µm TSV opening in the C1 BPSG dielectric opening and ~4.4µm TSV top opening in the silicon substrate. While an undercut TSV etch profile is not a problem if the wafer surface dielectrics are removed prior to metallization or if the TSV is etched on a wafer with no surface dielectric coatings, it degrades the CVD W step coverage in our integration scheme since the wafer surface dielectrics are not removed prior to TSV metallization. As shown in Figure 5c, the one mask BPSG dielectric RIE process step tapers the TSV dielectric opening 85 degrees and the subsequent Bosch silicon RIE process step both scallops and undercuts the silicon. This overhang profile in the one mask TSV structure causes the CVD W to pinch off, resulting in a seam or void in the fill. Our final TSV integration prior to backside grind, which employs the 2-step TSV process (Figure 1) is shown in Figures 6-7. The TSV etch angle is ~ 89.75 degrees, which results in a ~3µm TSV bottom width. The TSV profile shows a slight broadening about 60µm from the wafer surface, which induces a seam (Figure 6a). This seam is embedded several 10's of microns from the bottom and top of the TSV and does not degrade the TSV resistance or yield.

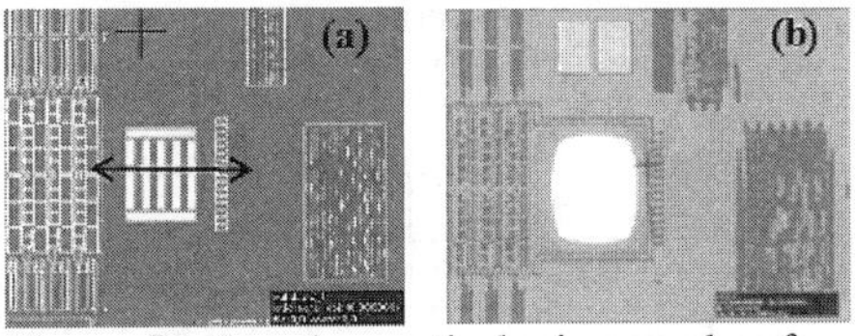

Figure 7: Top view optical micrographs of TSV array (a) post TSV tungsten CMP and (b) post (unoptimized) M1 copper CMP.

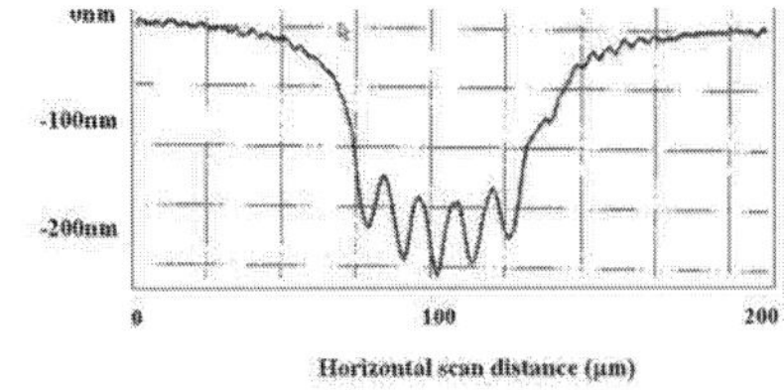

Figure 8: Surface scan across TSV array shown in Figure 7(a) after tungsten CMP showing a ~250nm surface recess.

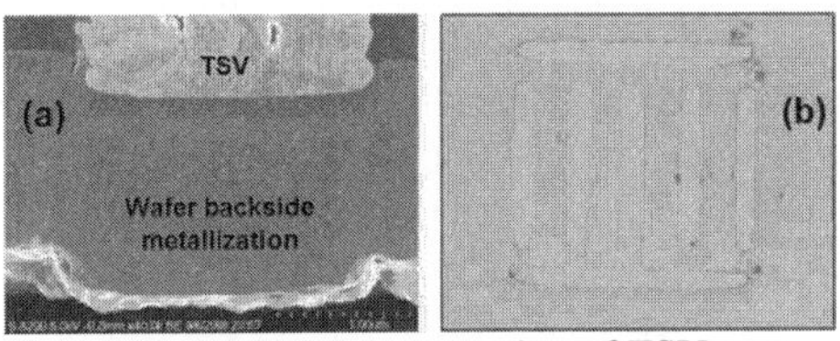

Figure 9: (a) SEM cross-section of TSV post backside grind / metallization; (b) top view image of TSV array post wafer backside grind.

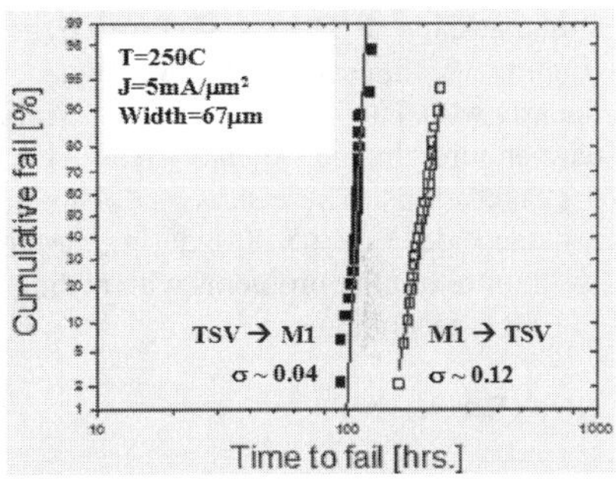

Figure 10: TSV electromigration lifetime data for both electron flow directions (sub-etch AlCu M1).

We expected that the high pattern factor ~5μm wide / ~15μm pitch TSV's would cause dielectric and tungsten dishing during tungsten CMP [6]. The ~ 60μm by 60μm TSV arrays have ~ 30% local exposed tungsten pattern factor which induces ~200-300nm of dishing (Figure 8) during tungsten CMP. The dishing is detectable by optical microscope inspection, which shows halos due to C1 dielectric thinning surrounding the TSV arrays (Figure 7a). Although this TSV dishing is not an issue with sub-etch AlCu M1 wires, it can cause copper puddling with damascene copper M1 wires (Figure 7b). The M1 copper puddling can be eliminated by reducing the dishing of the TSV tungsten CMP process or by increasing the copper M1 over polish.

Images post wafer backside grind are shown in Figure 9. In addition to the previously mentioned benefits of using CVD tungsten vs. electroplated copper for TSV metallization, tungsten also integrates almost seamlessly into wafer back side grind and metallization. We employ standard back side grind processing and, because both of the low global TSV pattern factor (~1%) and the corrosion resistance of tungsten, we have not had significant wafer backside grind integration problems. Wafer back side metallization consists of standard thin refractory and thick low resistivity bulk conductors (Figure 9a) and has been straightforward to integrate, again because we use unreactive tungsten for TSV metallization.

DC resistance variability of the 3μm by 60μm TSV's placed in ~ 60μm by 60μm arrays has been excellent: 15-50mΩ per via array including within- and wafer-to-wafer variability. AC (6GHz) resistance, which averages ~100mΩ per via array, is higher due to skin effects. The low TSV resistance variability is due to a combination of the robust CVD W metallization and the use of parallel TSV's shorted together on the wafer front and back surfaces. Electromigration stressing was performed with both electron flow directions (M1→TSV→chip backside and

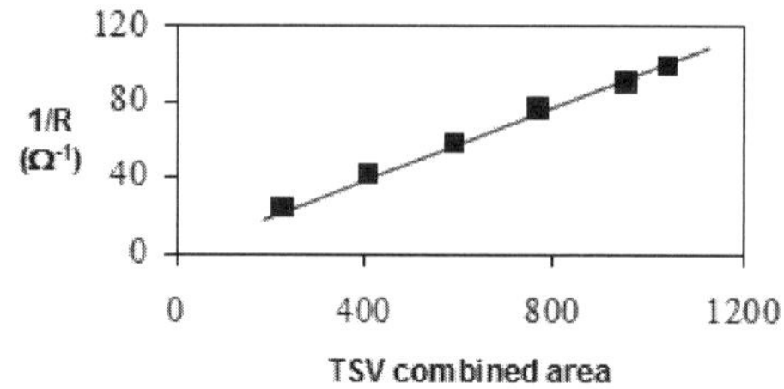

Figure 11: Measured TSV array resistance Vs. combined array TSV area [2].

	DTC		THB		HTS	
TSV Res. (mΩ/array)	T0	500 cyc	T0	500 hrs	T0	500 hrs
Maximum	35	30	31	31	29	24
Mean	30	25	27	26	26	23
Minimum	23	22	23	23	23	22

Table 1: Packaged chip TSV deep thermal temperature storage reliability summary.

chip backside → TSV → M1) with tight fail distributions and all fails occurring in the M1 wires (Figure 10). In general, the low frequency TSV resistance (Figure 11) and inductance [2] scaled linearly with TSV area, as expected. Table 1 summarizes deep thermal cycling (DTC), temperature / humidity / bias stress (THB), and high temperature storage (HTS) data for 50 packaged chips. These stresses did not induce any TSV opens and, in general, the TSV resistance decreased 5-20% during the stresses. Overall, there have been no major resistance, yield, or reliability problems with adding grounded TSV's to 200mm silicon CMOS and SiGe BiCMOS technologies.

CONCLUSIONS

Grounded through silicon vias, processed on the wafer front side, are used in conjunction with blanket wafer backside metallization to reduce ground lead inductance in analog and mixed signal technologies. To minimize packaging complexity and eliminate the need for a handle wafer during backside grind and wafer packaging, IBM chose to qualify 150μm deep grounded TSV's wired out in ~ 60μm by 60μm arrays. The key integration challenges of grounded TSV processing include deep silicon etch, wafer edge etch profile, TSV metallization, wafer bending/warpage, and reliability. IBM's grounded TSV technology was first qualified for manufacturing in 2007 and numerous production designs have been successfully built.

ACKNOWLEDGEMENTS

The authors wish to acknowledge the numerous engineers at IBM who contributed to the RF-CMOS and damascene copper programs. Regarding this paper, we gratefully acknowledge the assistance of Felix Anderson, John Benoit, Steven Boren, William Bosworth, William Cote, Timothy Dalton, Dinh Dang, Daniel Edelstein, Patrick Hughes, Frank Jackson, Alvin Joseph, Jenifer Lary, Brian Lefebvre, Jessica Levy, Stephen Luce, David Meatyard, Stephen St. Onge, Brett Philips, Randall Smith, Daniel Vanslette, and Frank White.

REFERENCES

1. A. Joseph et al., Proc. 2007 IEEE Bipolar/BiCMOS Circuits & Tech. Mtng., 2007, p 198-201.
2. J. H. Wu et al., IEEE Trans. Electron Devices, Vol. 51, number 11, November, 2004.
3. M. Erturk et al., United States patent application 2006/0094163A1, 2007.
4. P. Andry et al., Proc. IEEE 56th Electronic Components and Tech. Conf., 2006, p 837-837.
5. C. Tsang et al.,Mat.Res.Soc.Symp. Proc. V970, Enabling Tech. for 3-D Integ.,2007 p 145-153.
6. See, for example, J.-H. Kim et al., Proc. Adv. Met. Conf., 2006, p 589-593.

Conference Proceedings AMC XXIV © 2009 Materials Research Society

Novel and Production-Worthy Wafer-on-a-Wafer (WOW) Technology Using Self-Aligned TSV (SALT) Interconnects

Nobuhide Maeda[1], Hideki Kitada[1], Koji Fujimoto[2],
Kosuke Suzuki[2], Tomoji Nakamura[3], and Takayuki Ohba[1]
[1] School of Engineering, The University of Tokyo,
2-11-16 Yayoi, Bunkyo-ku, Tokyo 113-8656, Japan
[2] Dai Nippon Printing, 250-1 Wakashiba, Kashiwa, Chiba 277-0871, Japan
[3] Fujitsu Laboratories Ltd, 10-1 Morinosato-Wakamiya, Atsugi, Kanagawa 243-0197, Japan
E-mail: maeda@sogo.t.u-tokyo.ac.jp

Abstract

The world's first self-aligned through-silicon via (SALT) integration process based on wafer scale wafer-on-a-wafer (WOW) stacking with a through-silicon via (TSV) formation after wafer-bonding has been developed. Stacking wafers were thinned down to 20 μm, and TSVs with a diameter of 30 μm was fabricated to connect inter-wafers without electrical failure in chain TSV interconnects. Because the stacking wafers were thinned before TSV formation, there was no metal contamination due to the bump electrodes on the bulk silicon during the grinding process. The TSVs were constructed with a self-aligned technique which is suitable for the multi stack TSVs, while precise alignment and metal-bonding processes are necessary for conventional metal-bump bonding in TSV before the bonding processes. Seven wafers were successfully built up to multi stack TSVs.

Introduction

The process of integrated circuits (ICs) has been driven by the two-dimensional scaling for more than three decades. For the scaling trends toward 32 nm and 22 nm, however, the cost limits, no appropriate solution in lithography, and physical constraints have been arisen. Although the wide bandwidth more than 100 GB/s is required in the next generation of MPUs, the long length and small number of global interconnects restricts the performance of chip sets. Thus, three-dimensional (3D) ICs and several approaches have been proposed [1, 2]. Chip scale 3D-IC is, however, insufficient for the volume production. To overcome these issues, the wafer scale 3D-IC such as wafer-on-a-wafer (WOW) process is anticipated to be low-cost and production-worthy. Via-last and after-bonding processes, in which the TSV formation is carried out after wafer-bonding, were adopted. These processes enabled variations in wafer thickness to be kept thinner, and avoided metal contamination from backside-bumps. Furthermore, since the backside of the stacking wafer is flat and smooth, the bonding process is free from any stress concentration caused by the bump in the TSV before-bonding process. The metal contact between wafers was formed by self-align sputtering of barrier/seed metal after TSV etching. In this paper, the world's first seven-layer WOW-stacked TSV after bonding, the electrical property of the TSV-chain, and the simulation results of stress concentration are discussed.

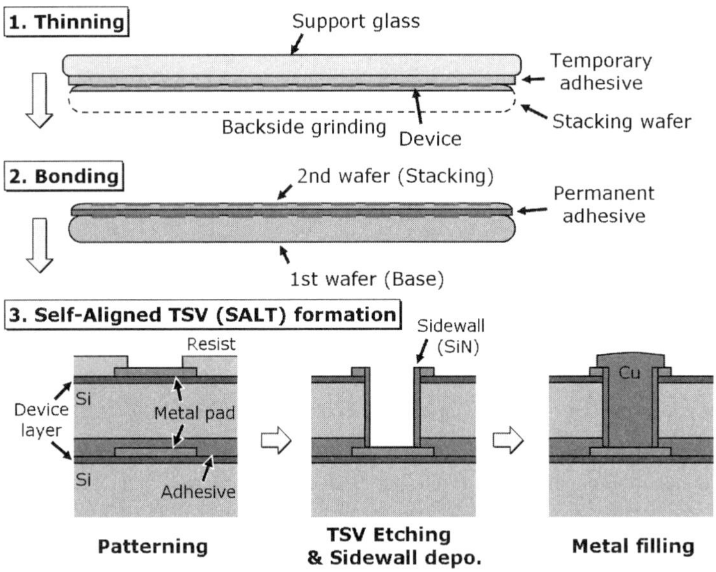

Fig. 1 Process flow of Self-Aligned TSV integration.

Experimental

TSVs and Cu interconnects are formed on an 8-inch wafer and the electrical characteristics are measured by chain structure. The process flow of WOW-basis TSVs after bonding is shown in Fig. 1. The secondary wafer was temporarily bonded with support glass, then ground and polished backside up to 20 μm. This wafer was permanently bonded onto the top of the primary wafer with adhesive. The alignment accuracy of the wafer bonding is about 3 μm. After removing the support glass from the wafer stack, TSV patterning and deep silicon etching were carried out. For electrical isolation, SiN was deposited on the sidewall of the TSVs. Figure 2 shows the cross sectional view of TSV after SiN deposition. No significant scallop shape at the sidewall is observed and thus SiN was formed uniformly. Then an anisotropic etch-back was used to remove SiN formed at Via-bottom. Cu/Ti was sputtered and the plating area was patterned with a dry film resist (DFR). Finally, Cu was plated to fill the TSVs, and the DFR was removed. Repeating those processes, multi stack WOW with TSVs interconnects up to seven-wafer was constructed.

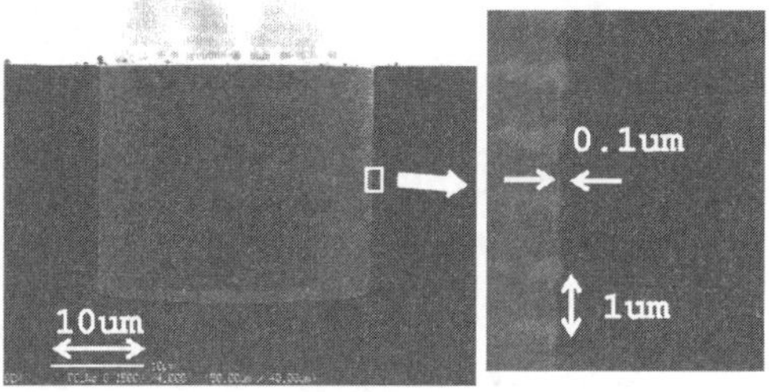

Fig. 2 Cross section of a TSV after sidewall deposition.

Results and discussion

The test structure of the TSVs interconnects using seven-layer WOW is shown in Fig. 3. In this process, as the self-align contact between the TSV and pad are formed after wafer-bonding, there is no bump contact occurring, while precise alignment and delicate bonding is necessary for conventional metal bump bonding of TSVs before the bonding process [3]. This feature is much more advantageous in the finer design rule of TSVs and chips.

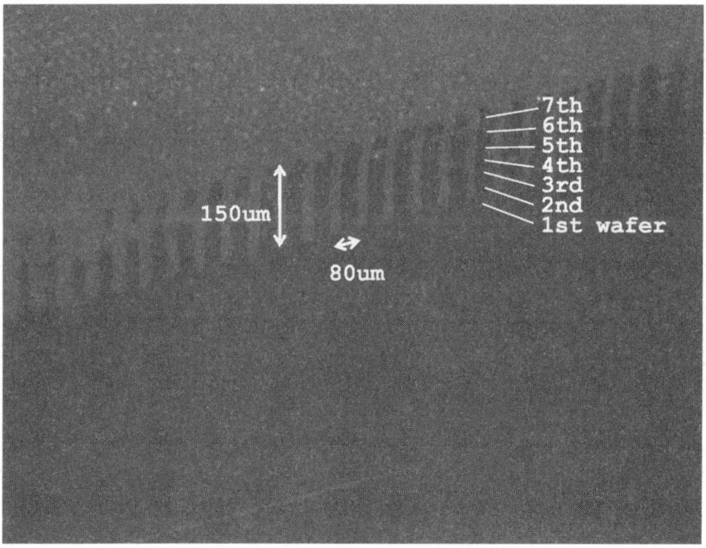

Fig. 3 TSVs after metallization.

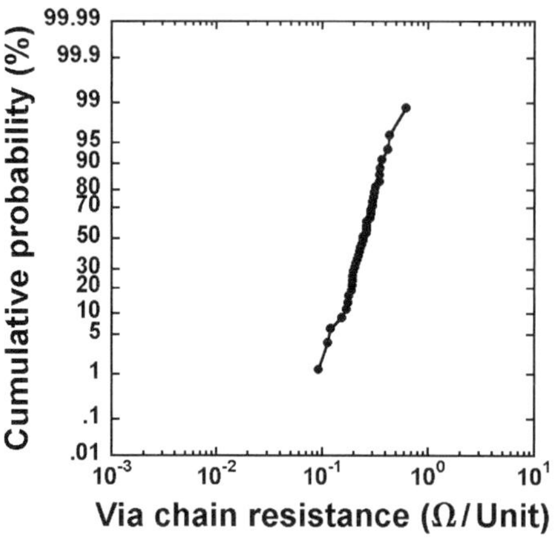

Fig. 4 Via-chain resistance.

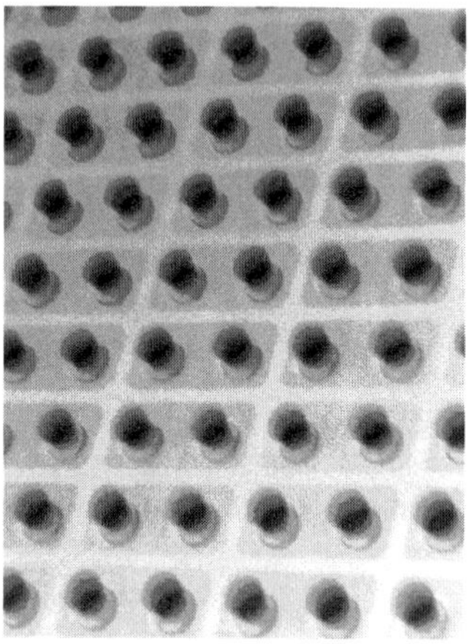

Fig. 5 X-ray CT image of TSV array.

Figure 4 shows the resistance distribution of TSV chain contact at 242 vias in two-layer WOW-stack. No electrical failure and usual distribution were observed at all the chips on the stacking wafer. The medium value of the single-via resistance was as low as 240 mΩ. Because the estimated values for a line resistance and a via resistance are 240 mΩ/unit and 0.6 mΩ/unit respectively, the line resistance is predominant factor for the total resistance. For the stress concentration around single via for various thicknesses of stacking wafer, residual stress is generated in the via as shown in Fig. 6. It was found that the stress of stacking with a thinner wafer is smaller than that of a thicker wafer. Consequently, in case of WOW with thinned wafers, TSV etching is easily performed rather than conventional thick wafer, and wafer bonding is free from stress caused by process temperature in metal bump bonding.

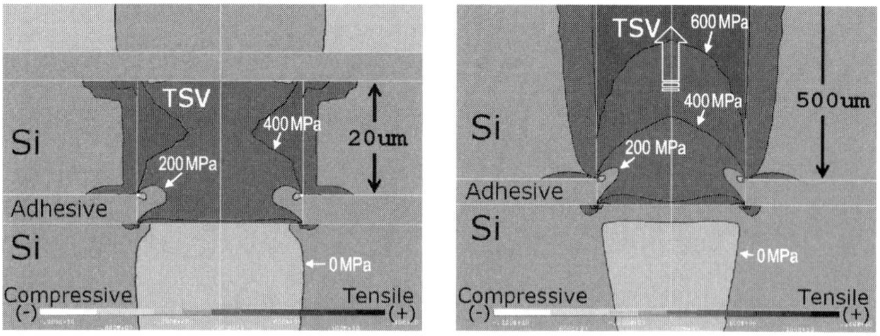

(a) T=20μm (b) T=500μm

Fig. 6 Wafer thickness vs stress concentration.

Conclusion

A new integration of WOW stacking with self-aligned TSV formation after wafer bonding has been developed. Self-aligning TSVs with a diameter of 30 μm are fabricated on a 20 μm-thick wafer. The electrical properties were found to be appropriate. According to the stress simulation, it was clarified that the reduction of stacking-wafer thickness leads to a relaxation of residual stress around the via edge. These results indicate that WOW-basis TSV formation after bonding is the superior method for the forthcoming generation of 3D-ICs, where multiple wafer-stacking, reductions of chip dimensions and smaller TSVs will be applied.

Acknowledgement

We would like to thank DISCO corporation, Nissan Chemical and Sumitomo 3M for thinning the wafers, and providing adhesives and technical support, respectively. This work was carried out in the WOW alliance program.

References

[1] K. Chen, et al, Proc. IEDM2006 (2006) 367.

[2] C. Bower, et al., Proc. ECTC2006 (2006) 399.

[3] R. Labie, et al, Proc. IITC2008 (2008) 19.

Conference Proceedings AMC XXIV © 2009 Materials Research Society

Investigations of Cu Filling in Through-Si Via Holes Using Direct Electroless Plating on CVD-W

[1]F.Inoue, [2]M.Koyanagi, [2]T.Fukushima, [3]K.Yamamoto, [3]S.Tanaka, [4]Z.Wang and [1]S.Shingubara

[1] Department of Mechanical Engineering, Kansai University, 3-3-35 Yamate-Cho, Suita, Osaka, Japan
[2] Department of Bioengineering and Robotics, Tohoku University, 6-6-01 Aza Aoba, Aramaki, Aoba-ku, Sendai, Japan
[3]National Institute of Communication Technology, 588-2 Iwaoka, Nishi-ku, Kobe, Hyogo, Japan
[4] Shaanxi Normal University, 199 South Chang'an Road, Xi'an., PR China

E-mail: sa8m101@edu.kansai-u.ac.jp

ABSTRACT

In order to realize low resistance through-Si via (TSV) electrodes, Cu electroplating is one of the most promising methods. However, with an increase of the aspect ratio of TSV, a formation of conductive seed layer prior to Cu electroplating is becoming more and more difficult. We propose an alternative approach using the electroless plating of Cu, which utilize displacement plating without catalyst. This method is effective for fabricating a low resistance TSV when combined with a barrier layer which is composed of tungsten (W). We found that an addition of Cl ions drastically suppressed the pinch-off at the entrance of the TSV, and it enabled conformal Cu deposition for high aspect ratio TSVs.

INTRODUCTION

In recent studies, tungsten or poly-Si formed by chemical vapor deposition (CVD) have been chosen for filling metals for TSV of three-dimensional integration technology because of their good conformal deposition properties [1,2]. However, tungsten and poly-Si have the disadvantage of large electrical resistivity and are not suitable for high speed operations. For example, the electrical resistivity of CVD-W is 20~30 $\mu\Omega$ cm, while that of the electroless Cu plating is in the range of 2.0~2.5 $\mu\Omega$ cm.

Copper is widely used for metal interconnection in ultra large scale integrated circuit (ULSI) because of its low resistivity and its high reliability against electromigration. Cu electroplating technology has been studied extensively, although it is increasingly more difficult to form a continuous low resistance seed layer on the side walls of via holes. We have developed electroless Cu plating in which Cu-filling in a lower submicron holes in ULSI interconnections is possible [3-6]; Furthermore, we found that the electroless copper plating is able to deposit Cu directly on TaN, WN and W by displacement plating without catalyst. The purpose of this study is to fill Cu in a TSV whose diameter is a few μm. In this study, we investigated the effect of inhibitors, such as sulfopropyl-sulfonate (SPS) and chlorine ions for Cu filling in TSVs. Several of them are known being effective for bottom-up fill of submicron holes.

EXPERIMENT

The depth of TSV hole was 37 μm, and the hole size was 1.2 x 8 μm^2, or 2 x 8 μm^2. Prior to Cu plating, poly-Si (210nm) and CVD-W (120nm) were deposited in the TSV, successively. The electroless plating bath contained copper sulfate pentahydrate($CuSO_4 \cdot 5H_2O$) as a copper ion source, glyoxylic acid monohydorate (CHOCOOH) as a reducing agent, ethylene-diamine-tetra-acetic acid ($C_{10}H_{16}N_2O_8$, EDTA) as a complexing agent, polyethylene glycol (PEG, MW 4000), and tetra methyl- ammonium hydroxide (TMAH) as a pH adjuster. The pH of the plating solution was adjusted 12.5, and the bath temperature was 70 °C. In addition to the above, SPS, and chlorine ions were added. After Cu deposition, the specimens were cut by FIB for cross-sectional SEM and FIB observations. Deposition rate was measured by electro chemical quartz crystal micro balance (QCM). The redox potentials of the barrier layers in the electroless copper plating solution was measured by a two-electrode method using Ag/AgCl as a reference electrode, and the change of the working electrode potential against time was recorded on a personal computer using LabView. After electroless plating, annealing was carried out at 400 °C, 30 min, in vacuum.

RESULTS AND DISCUSSIONS

Displacement plating

Copper is able to deposit on the TaN, WN and W barrier layers without catalyst by immersion to the electroless copper plating solution. In order to understand the mechanism of Cu deposition on the WN barrier layer, the redox potential of the WN, and Cu films were measured in the electroless plating solution at 60°C (Table 1). The redox potential with respect to measurement time is shown in figure 1. The redox potential of WN was lower than that of Cu, indicating that the displacement reaction between WN and Cu ions could proceed. The change in the redox potential of the WN barrier layer is classified into the two stages (Fig. 1). Initially, the potential increases quickly up to that of the Cu electrode. The redox potential then increases in step with that of Cu. This process can be explained as follows. In the first stage, WN is oxidized, which reduced Cu ions and Cu deposit on the WN film. The potential of the WN electrode increases with the amount Cu atoms on the WN surface, reaching the potential of Cu when the entire surface is covered with Cu. In the second stage, the Cu film on the WN electrode become thicker with time by self-catalytic reaction, and the potential of the electrode is maintained at the potential of Cu electrode.

Table 1. Initial potential of barrier films and copper metal in electroless plating solution

Electrode	Temperature of solution (°C)	REDOX potential (mV)
Cu	60	-312
WN	60	-358

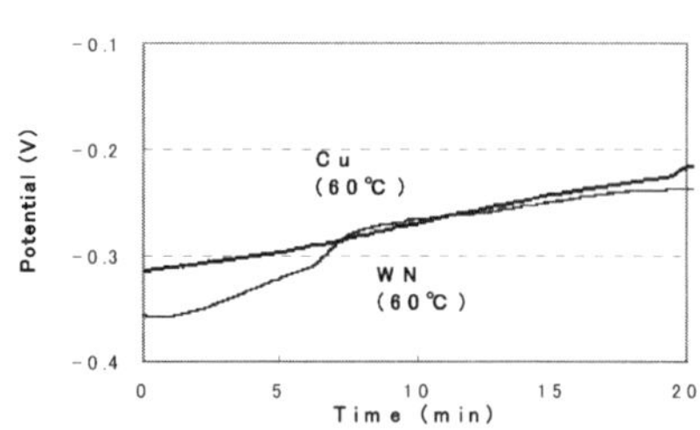

Figure 1. Redox potential of WN barrier films and copper metal in electroless copper plating

TSV filling

For filling submicron via holes, SPS addition is very effective for bottom-up fill of electroless Cu since it acts as a strong inhibitor [3,4]. Diffusion of inhibitor molecules are suppressed at the bottom of fine holes that result in the faster growth of Cu within the hole.

In this study, TSV diameter ranges between 1.2 and 5 μm, and their aspect ratio ranges between 32 and 8. The diameter of TSV is much larger than that of ULSI via holes. We tried to fill TSV by electroless plating only by addition of various additives, such as SPS, PEG, and Cl ions, etc.

At first we tried to fill TSVs with diameters 1.2 μm. Cross-sectional FIB-SIM images of Cu-filled TSV are shown in Fig.2. SPS concentration is 0.1mg/L, plating time is 90 minute. We succeeded in bottom up filling in a submicron via holes by this condition. Cu deposited nearly conformal in TSV. However, Cu thickness of the TSV sidewall gradually decreases toward the bottom. Finally, pinch-off occurs at the entrance of the TSV and a large void was formed at the central axis of the TSV.

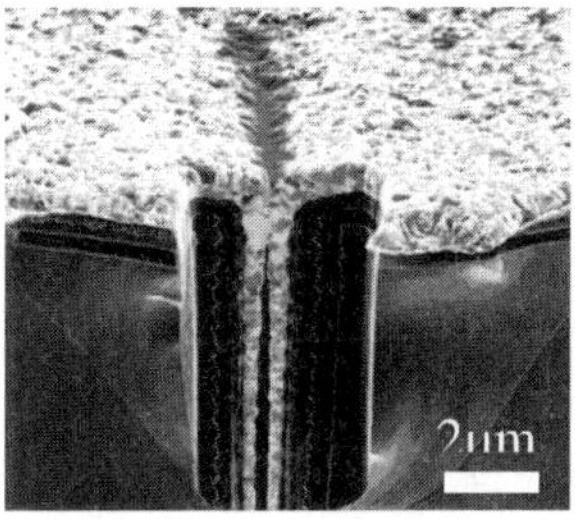

Figure 2. Via structure and cross-sectional FIB-SIM image of electroless Cu plating with SPS 0.1 mg/L, PEG (mw 4000) 500 mg/L, pH 12.5; temperature 70℃; plating time 90 min

Next, we tried to fill the TSV with a diameter of 2.0 μm. Cross-sectional FIB images of TSV after Cu electroless plating are shown in Fig.3. Additives are; (a) PEG only and (b) PEG + SPS. When PEG only is added, the top of TSV is pinched-off. Cu thickness of the surface is much larger than that of the TSV sidewall. When PEG and SPS are added, Cu thickness of the top of TSV is slightly larger than that of the middle of TSV.

Figure 4 shows cross-sectional FIB images when Cl ions are added to SPS and PEG. Cu thickness of the surface is much suppressed, and Cu deposits conformably inside TSV. Cu thickness of the TSV sidewall doesn't change from the bottom to the top. As a consequence, the pinch-off does not occur at the TSV entrance.

Figure 3. Cross-sectional FIB-SIM image of electroless Cu plating. pH12.5; temperature 70℃; plating time 90min (a)PEG 500 ppm, only (b)PEG + SPS 0.05mg/L

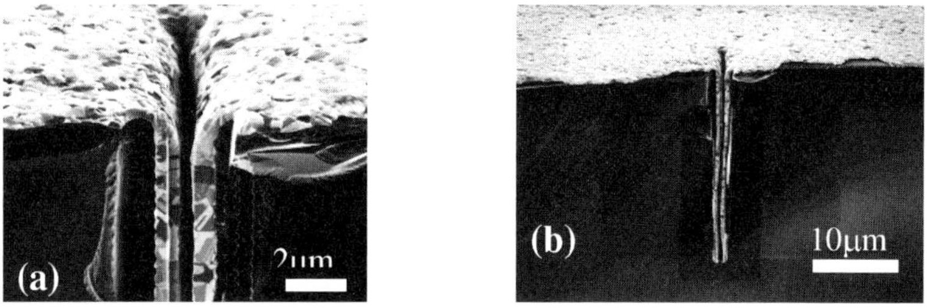

Figure 4. Cross-sectional FIB-SIM image of electroless Cu plating. pH12.5; temperature 70℃; plating time 90min, PEG + SPS 0.1mg/L + Cl 10mg/L, (a)top view (b) cross-sectional view.

In order to understand the effect of Cl ions addition, we observed Cu deposition profile of submicron holes whose diameter and depth were 0.5 and 2 μm. In either case, PEG and SPS are added. Bottom-up tendency occurs and the hole is completely filled as shown in Fig.5-(a) and (b). Furthermore, Cu thickness of the surface was drastically suppressed when Cl ion is added (Fig.5-b). Figure 6 shows time dependence of the surface Cu thickness during electroless plating. Addition of Cl ions to PEG and SPS system decreases Cu deposition rate further at the surface, as compared with PEG, or PEG and SPS added plating baths.

The mechanism of suppression of Cu deposition with addition of Cl ions would be an adsorption of combined molecules with Cl, SPS and PEG. This compound effectively suppresses Cu deposition rate near the entrance of TSV and suppresses pinch-off of TSV. On the other hand, the compound molecule is too large to diffuse into the inside of TSV.

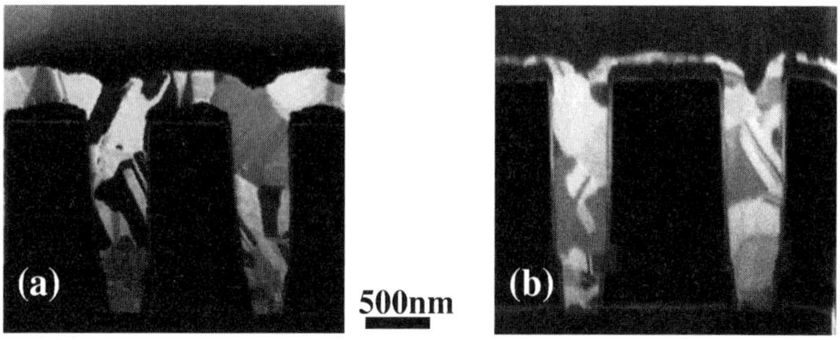

Figure 5. cross-sectional FIB observation of Cu filled submicron via hole.
(a) SPS + PEG (b) SPS + PEG +Cl

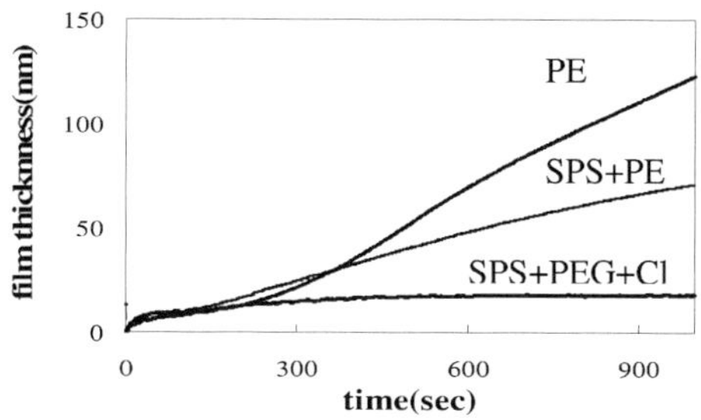

Figure 6. Cu film thickness as a function of time measured by quartz crystal micro balance Additives are: PEG, PEG+SPS, PEG.

CONCLUSION

We studied filling of Cu in TSV with electroless Cu plating by displacement reaction without Pd catalyst. Pinch-off occurred during Cu ELP for high aspect ratio TSVs (AR=20-30, depth=40mm), when SPS and PEG were used. Addition of Cl ions to PEG and SPS system is very effective for suppressing pinch-off, and conformal Cu deposition is possible for TSV which aspect ratio is larger than 10. The present study enables a low resistance TSV with a simple procedure. For future studies, an electroless barrier metals that include W atoms would be important for realizing low cost and low resistance TSV.

ACKNOWLEDGEMENT

We are grateful to Dr. K.Murakami of Osaka University for his kind assistance and suggestions, and Y.Harada, T.Yokoyama and J.Ueda of Kansai University for their experimental assistance.

REFERENCES

1. M. Koyanagi, T. Fukushima, and T. Tanaka
Technical Digest of the International 3D System Integration Conference 31-44 (2008)

2. P. Ramm, A. Sauer, Technical Digest of the International 3D System Integration Conference 355-372 (2008)

3. Z. Wang, O. Yaegashi, H. Sakaue, T. Takahagi, and S. Shingubara,
Journal of the Electrochemical Society, 151 (12) C781-C785 (2004)

4. S. Shingubara, Z. Wang, O. Yaegashi, R. Obata, H. Sakaue, and T. Takahagi ,
Electrochemical and Solid-State Letters, 7 (6) C78-C80 (2004)

5. Z. Wang, T. Ida, H. Sakaue, S. Shingubara, and T. Takahagi,
Electrochemical and Solid-State Letters, 6 (3) C38-C41 (2003)

6. Z. Wang, R. Obata, H. Sakaue, T. Takahagi, S. Shingubara
Electrochimica Acta 51 2442–2446 (2006)

Conference Proceedings AMC XXIV © 2009 Materials Research Society

Chemical Vapor Deposition of Copper from New Precursors

Haizheng Song[a], John A.T. Norman[b], Yukihiro Shimogaki[a]
[a] Dept. of Materials Engineering, The Univ. of Tokyo, 7-3-1 Hongo, Bunkyo-ku, Tokyo 113-8656, Japan.
[b] Air Products and Chemicals, Inc., 1969 Palomar Oaks Way, Carlsbad, California 92011.

ABSTRACT

A new unfluorinated copper precursor $Cu(CH_3C(O)CHC(NCH_2CH(CH_3)OSiMe_2C_2H_3)CH_3)$, (Cu-KI5), is demonstrated to be an effective copper CVD precursor for growing pure copper films onto ruthenium under layers using formic acid (HC(O)OH) as a reducing agent. The effect of various process conditions such as substrate temperature (150-350°C), HC(O)OH partial pressure and Cu precursor partial pressure on the growth rate of copper were investigated. Kinetic data extracted from these experiments enabled 2-D computational simulations for predicting copper deposition into a wide range of trench structures, the results of which indicate excellent conformal deposition. In addition, a fluorinated analogue of Cu-KI5, ie $Cu(CF_3C(O)CHC(NCH_2CH_2OSiMe_2C_2H_3)$, Cu-KI3, was similarly evaluated to provide a comparison of the two new molecules. Both Cu-KI5 and Cu-KI3 are solids at room temperature which melt into thermally stable low viscosity liquids at 65 and 41°C respectively. Both precursors can be heated to elevated source temperatures to yield high precursor partial pressures.

INTRODUCTION

Due to its low electrical resistivity and higher resistance to both EM and stress-induced deformation, copper continues to play a central role in providing low resistance interconnects for the IC manufacturing of logic and memory devices in addition to the burgeoning new field of Through Silicon Vias (TSV) used in 3-D packaging. [1,2].

In considering methods to deposit Cu for seed layers for subsequent copper electroplating or even full copper fill in a trench or via structure, the chemical vapor deposition (CVD) of copper stands out as an attractive technique to provide high quality conformal thin copper films. In the CVD copper technique, a volatile copper-organic complex in the gas phase contacts a target substrate to grow a copper film by a surface chemical reaction. In general, Cu CVD processes have utilized the commercial Cu CVD precursor, Cu(hfac)(tmvs) (hfac=hexafluoroacetylacetonate; tmvs=trimethylvinylsilane) [3] which can deposit pure copper films at ~200°C in a thermal disproportionation reaction. However, this process presents a number of challenges. First, due to the disproportionation reaction, only 50% of the copper present in the precursor can be deposited by CVD. Secondly, the $Cu^{+2}(hfac)_2$ by-product of the disproportionation reaction is a volatile solid which must be managed in the CVD effluent stream. Thirdly, the fluorine present in the (hfac) ligand can lead to contamination at copper /barrier metal interfaces leading to poor adhesion and the expense of incorporating fluorine into a precursor keeps its price relatively high. For all of these reasons, there is a strong interest in developing alternative copper precursors which can deposit copper by a more efficient direct reduction mechanism rather than by disproportionation, are fluorine free and are thermally stable enough to permit high vapor pressures by using elevated source temperatures.

Herein we report a new fluorine free copper precursor, Cu-KI5 ie $Cu(CH_3C(O)CHC(NCH_2CH(CH_3)OSiMe_2C_2H_3)CH_3)$, which meets all of the above requirements and is capable of growing pure copper films by direct reduction using formic acid which is readily available and inexpensive. In addition, we also evaluated Cu-KI3 ie $Cu(CF_3C(O)CHC(NCH_2CH_2OSiMe_2C_2H_3)CF_3)$, which is the fluorinated analogue of Cu-KI5, as a bench mark to compare the performance of the two molecules. The molecular structures of these two compounds are shown in Figure 1. As expected, the Cu-KI3 precursor is more volatile than Cu-KI5, but the latter molecule can easily compensate for this offset by simply increasing its temperature by ~20°C compared to Cu-KI3. Both Cu-KI3 and Cu-KI5 are solids at room temperature but melt into low viscosity liquids at 41°C and 65°C respectively.

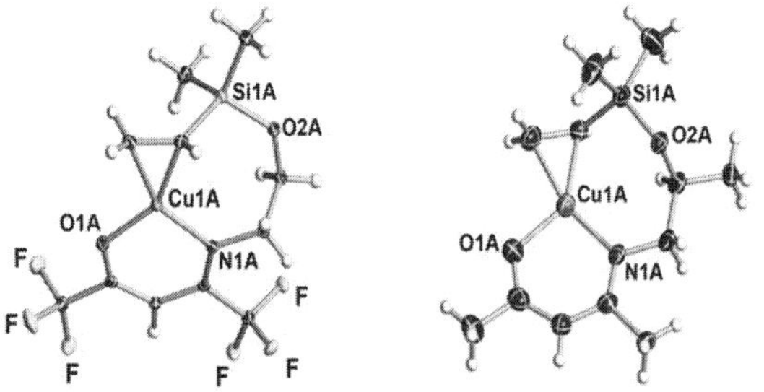

Figure 1. Molecular structures of precursors Cu-KI3 (left) and Cu-KI5 (right).

Two CVD reactor configurations were used in this study. The first one was used to directly compare the performance of the two precursors whereas the second configuration was used to study only Cu-KI5. Since both precursors are thermally stable and do not metallize by a disproportionation reaction, formic acid (HC(O)OH) was used as a reducing agent and all depositions were conducted on ruthenium substrates. Since formic acid is well known to readily form atomic hydrogen upon contact with ruthenium, it is possible that atomic hydrogen is formed during the CVD process and acts to reduce the copper precursors. Alternatively, the formic acid may form an intermediate species with either of the copper precursors which then break down to give copper metal. The effect of process conditions such as substrate temperature and precursor partial pressure was investigated. Based on the kinetic analysis, the CVD process was simulated to provide predictions of step coverage in a trench structure.

EXPERIMENTAL

The CVD copper depositions were conducted using a homemade vertical CVD system which could be configured in two ways. In the first configuration, used to compare the performance of Cu-KI3 and Cu-KI5, the Cu precursor and HC(O)OH vapors are individually flowed through a showerhead after which they mix before contacting the target ruthenium substrate, 5 cm below. This reactor was then modified into a second configuration which was then used solely to evaluate Cu-KI5. The modification comprised of placing a steel plate bored with uniformly and closely spaced 2mm diameter holes in a mesh-like array 1 cm above the substrate to induce a more thorough mixing of the formic acid and Cu-KI5 vapors prior to contacting the substrate. Both of these configurations are shown below in Figure 2.

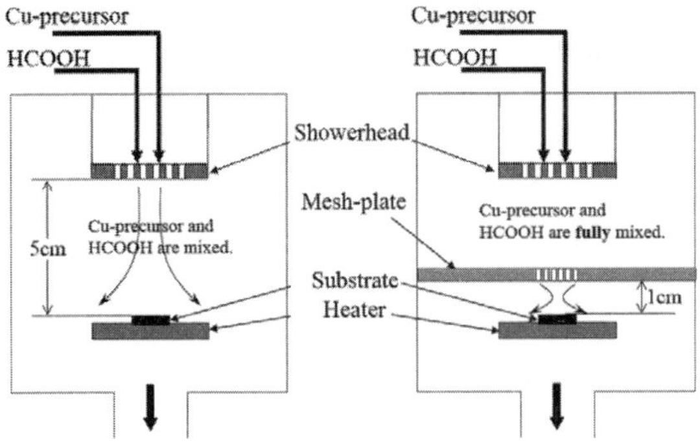

Figure 2. Schematic diagrams of the first configuration (left) and second configuration (right) of the CVD reactor.

In the second reactor configuration a substantially higher copper growth rates for Cu-KI5 were obtained compared to its performance in the original configuration. The CVD process conditions were as follows: The bubbler temperature was 80-100 °C for Cu-KI3 or 100-120 °C for Cu-KI5 typically using 200 sccm of helium carrier gas. Formic acid was delivered from a bubbler heated to 30°C by 100 sccm of helium carrier gas. The transportation lines were about 20°C higher than the bubbler temperature. Substrate temperatures ranged from 150 to 350 °C, reactor pressure was 5 Torr. The partial pressure of the copper precursor in the chamber was varied by adjusting the bubbler pressure or carrier gas flow thereby allowing the growth rate of copper to be determined at a series of precursor partial pressures and substrate temperatures.

Film thickness was measured by surface profiler (Tencor P10) after fabricating a series of copper steps on a processed substrate by traditional photolithography and HNO_3 wet etching the copper film down to the ruthenium underlayer. The resistivity for the as-deposited Cu film was measured by four-point probe method and film purity was evaluated by X-ray photoelectron spectroscopy (XPS; ULVAC-ϕ XPS Model 1600C) and secondary ion mass spectroscopy (SIMS; Physical Electronics PHI 6650). Atomic force microscopy (AFM; Digital Instrument, Nano-Scope III) was used to examine the film morphology.

RESULTS AND DISCUSSION

Cu-KI3 and Cu-KI5 precursors

The structure of ketoimine (KI) type copper molecules was described in an earlier report [4] in detail. In these new molecules, the organometallic copper-olefin bond is a part of a seven membered ring structure which imparts high thermal stability to these precursors and suppresses any disproportionation reaction. In addition, it then becomes possible to eliminate all fluorine from the structure, as in the case of Cu-KI5. The vapor pressures of these two molecules are shown in Figure 3 where, as expected, the fluorinated Cu-KI3 molecule is substantially more volatile than the unfluorinated Cu-KI5. However, both molecules are remarkable stable which allows for high source temperatures which can help Cu-KI5 compensate for its lower VP. In this way, comparable Cu CVD rates can be achieved for Cu-KI5 and Cu-KI3 if their source temperatures are adjusted to give comparable partial pressures in the CVD reactor.

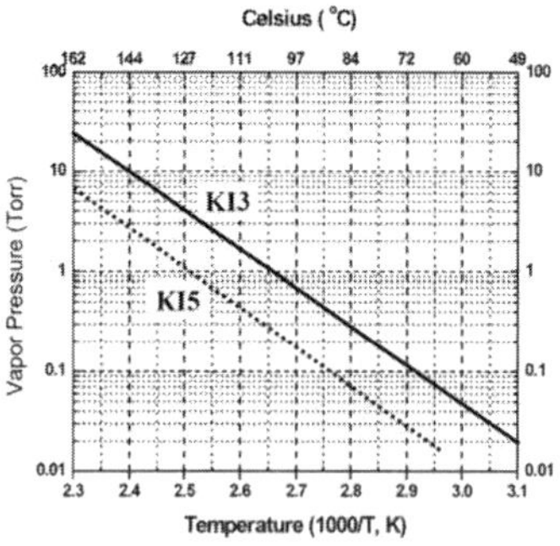

Figure 3. Vapor pressure of Cu-KI3 and Cu-KI5.

Cu-CVD from Cu-KI3 and Cu-KI5 in the first reactor

In the first reactor configuration, using the precursor Cu-KI-3 at a partial pressure ($p_{Cu\text{-}KI3}$) of 0.014 Torr and HC(O)OH at a partial pressure (p_{HCOOH}) of 2.3 Torr, we showed that Cu could be grown on the ruthenium substrates from 150-350°C with copper deposition rates of 1.7-12.2 nm/min with an activation energy of 19.3 kJ/mol which suggested a reaction limited growth regime. Using Cu-KI5 at a partial pressure ($p_{Cu\text{-}KI5}$) of 0.015 Torr and HC(O)OH at p_{HCOOH} of 1.7 Torr over the same substrate temperature range, copper films were grown with lower deposition rates of 2.0-5.2 nm/min and a lower activation energy of 10.6 kJ/mol as shown in Figure 4. Deposition from Cu-KI5 was then further examined in the second reactor.

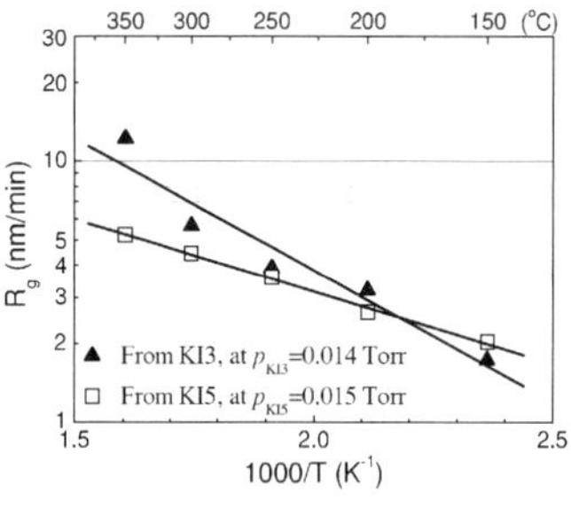

Figure 4. Arrhenius plots of growth rate for Cu-CVD from Cu-KI3 and Cu-KI5 using the first CVD reactor configuration.

The purity of a Cu film deposited from Cu-KI3 at 250°C was measured by XPS. The impurity elements, including F, C, O, N and Si were not detectable by XPS. The minimum resistivity was 2.44 μΩ-cm obtained at 250°C with film thickness of 282 nm. The RMS (root-mean-square) measured by AFM of this film was 17 nm. The immature morphology of Cu film might be responsible for its high resistivity. Similarly, the copper film deposited from Cu-KI5 at 250°C also showed a high purity with the impurity elements of C, O, N and Si not being detectable by XPS.

The effect of p_{HCOOH} on copper film deposition rate is shown in Figure 5. For both Cu-KI3 and Cu-KI5, the deposition rate diminished with reducing p_{HCOOH} although the HC(O)OH/Cu-precursor ratio was much large than unity. This tendency might be due to insufficient mixing of Cu precursor with HC(O)OH in the first reactor configuration. And it seems that Cu-KI5 is more susceptible to formic acid.

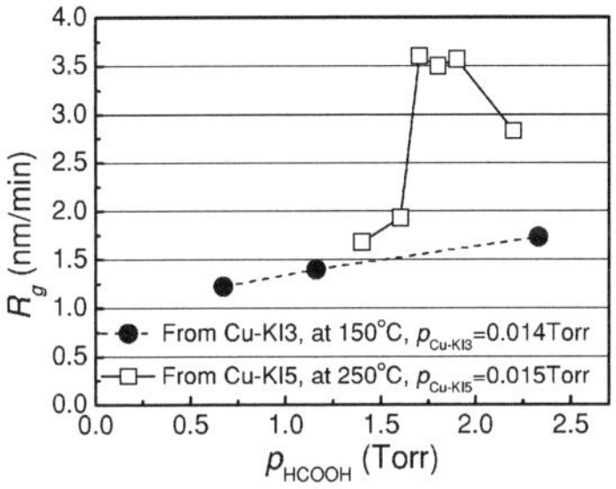

Figure 5. CVD copper deposition rate dependency upon formic acid partial pressure for Cu-KI3 and Cu-KI5 at substrate temperatures of 150 and 250°C respectively using the first reactor configuration.

Cu-CVD from Cu-KI5 in the second reactor

In the second reactor configuration, the Cu-KI5 precursor and HC(O)OH could be fully mixed before reaching the substrate surface resulting in enhanced copper deposition rates compared to the first reactor configuration . The effect of formic acid partial pressure was first examined at a substrate temperature of 250°C and p_{Cu-KI5} around 0.020 Torr. The deposition rate is almost independent of p_{HCOOH} from 0.25-1.8 Torr (Figure 6) and SIMS analysis showed the copper contained no detectable C, H, N, O or Si making it high purity copper.

Figure 7 shows the variations of the deposition rate of the copper films as a function of substrate temperature in the second reactor configuration at p_{Cu-KI5} of 0.021 Torr and p_{HCOOH} of 0.69 Torr. The activation energies were calculated to be about 48.9 kJ/mol under a reaction limited growth regime for temperatures less than 210°C, and about 1.9 kJ/mol under a diffusion limited growth regime above 210°C. The minimum resistivity was 2.36 μΩ-cm obtained at 250°C with film thickness of 480 nm and RMS of 18 nm.

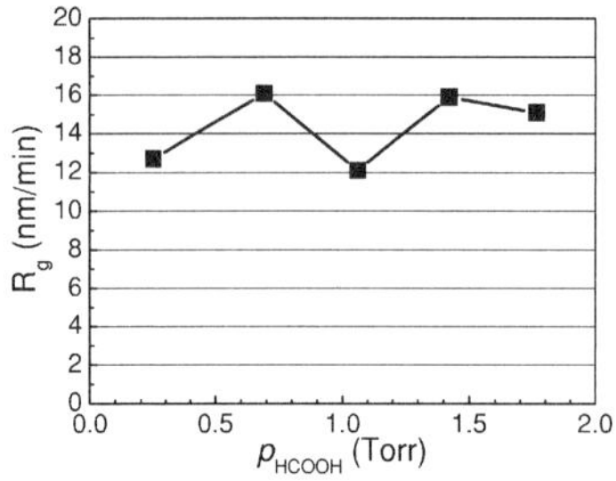

Figure 6. CVD copper deposition rate dependency upon formic acid partial pressure for Cu-KI5 at $p_{Cu\text{-}KI5}$ around 0.020 Torr and substrate temperatures of 250°C using the second reactor configuration.

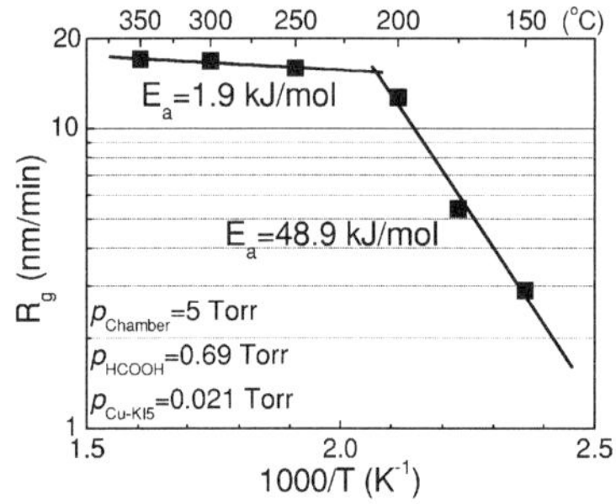

Figure 7. Arrhenius plots of growth rate for Cu-CVD from Cu-KI5 in the second reactor.

The dependency of deposition rate on Cu-KI5 concentration in the CVD reactor was investigated at a fixed HC(O)OH concentration at substrate temperature of 150-250°C (Figure 8a). As expected in the reaction limited regime, the deposition rate at 150 and 200 °C was saturated at high Cu-KI5 concentration. As for the deposition at 250°C, which is the diffusion limited regime at low $p_{Cu\text{-}KI5}$, the deposition rate was also saturated at high Cu-KI5 concentration. It is likely that the high Cu-KI5 concentration and/or the low HC(O)OH/Cu-KI5 concentration ratio result in changing the limited growth regime from diffusion limited to reaction limited. According to the Langmuir-Hinshelwood isotherm,

$$r_g \times \rho_{Cu} = \frac{k_s KC}{1+KC} \tag{1}$$

where ρ_{Cu} is the molar density of copper, the kinetic parameters, surface reaction rate constant (k_s) and adsorption equilibrium coefficient (K), could be calculated (Figure 8b). It is noted that k_s increases with substrate temperature but K decreases with the temperature, which is often reported in catalytic reactions.

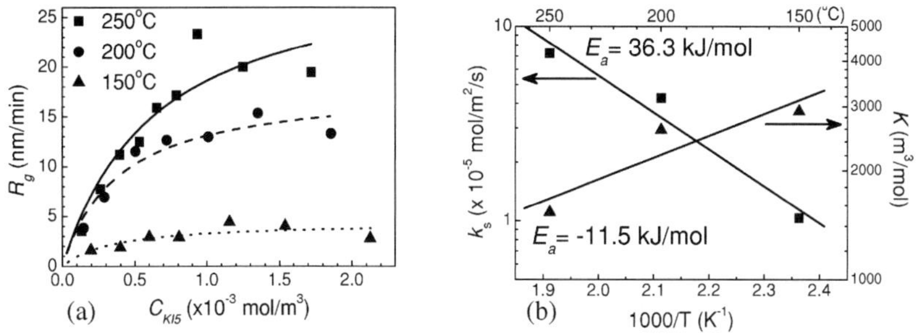

Figure 8. (a) Cu-KI5 concentration dependencies of deposition rate in the second reactor, and (b) Arrhenius plots of k_s and K derived from (a).

At deposition temperature of 200°C and Cu-KI5 concentration of 7.2x10^{-4} mol/m^3, 2-dimensional simulation using k_s and K for step coverage in a trench structure yielded a good step coverage of 0.93 for an aspect ratio of 10 for trench dimensions of 0.1~10 μm wide. This indicates the utility of this precursor for both future ULSI interconnects and TSV applications in 3D packaging schemes.

CONCLUSIONS

The results of this study indicate that both of the new Cu precursors Cu-KI3 and Cu-KI5 can grow pure Cu films by CVD at relatively low temperatures when reacted with formic acid as reductant. Since both molecules are remarkable stable, even higher deposition rates should be possible using direct liquid injection of these precursors as melted liquids if a 'first' reactor configuration is used since in this mode the copper deposition appears to be reaction rate limited. The Cu-KI5 molecule is especially attractive since it is unfluorinated and therefore of lowered cost to manufacture. Even though it is less volatile compared to its fluorinated counterpart, Cu-KI3, its thermal stability permits high source temperatures in compensation without thermal degradation occurring. 2-D computational simulations for predicting copper deposition into a wide range of trench structures indicate excellent conformal deposition from Cu-KI5. For these reasons, Cu-KI5 holds promise as a future Cu CVD precursor for both ULSI interconnect and TSV/3-D structures for either Cu seed layers or full fill.

REFERENCES

1. R. Rosenberg, D. C. Edelstein, C. Hu and K. P. Rodbell, *Annu. Rev. Mater. Sci.* **30**, 229 (2000).
2. R. H. Havemann and J. A. Hutchby, *Proceedings of the IEEE* **89(5)**, 586 (2001).
3. J. A. T. Norman, B. A. Muratore, P. N. Dyer, D. A. Roberts, A.K. Hochberg and L. H. Dubois, *Materials Science and Engineering* **B17**, 87 (1993).
4. J.A.T. Norman, M. Perez, X. Lei and H. Cheng, *ECS Transactions* **3(15)**, 161 (2007).

Low-k Materials

Conference Proceedings AMC XXIV © 2009 Materials Research Society

Porous Low-k Material Etch For 32 nm and Beyond

Catherine Labelle[1], John Arnold[2], Dave Horak[2], Ravi Srivastava[3], Anthony Lisi[4], Hakeem Yusuff[4], Elise Lafosse[5], Joseph Linville[6], Youngjin Choi[4], Nicholas Fuller[7], Kaushik Kumar[4], Ryan Patz[8], Andrew Darlak[8], Kevin Zhou[9], Yifeng Zhou[9], and Jeremiah Pender[9]

[1]Advanced Micro Devices, 2070 Route 52, Building 300, M/S A10, Hopewell Junction, NY, 12533

[2]IBM Research, Albany, NY, 12203

[3]Chartered Semiconductor, Hopewell Junction, NY, 12533

[4]IBM Microelectronics, Hopewell Junction, NY, 12533

[5]Freescale Semiconductor, Hopewell Junction, NY, 12533

[6]Advanced Micro Devices, Hopewell Junction, NY, 12533

[7]IBM Research, Yorktown, NY, 10598

[8]Applied Materials, Albany, NY, 12203

[9]Applied Materials, Sunnyvale, CA, 94085

ABSTRACT

The challenges facing back-end-of-line (BEOL) etch as technology nodes progress are becoming increasingly difficult as the challenges due to shrinking dimensions are compounded by the challenges from new materials integration. Considering a via-first-trench-last (VFTL) patterning scheme, as critical dimensions decrease, key dimension-related etch challenges include CD control, aspect ratio of trench litho film stacks, RIE lag, and LER. The inclusion of low k dielectrics, including porous materials ($k \leq 2.4$), requires the etch to consider the sensitivity of the films to compositional modification (i.e., dielectric damage), polymer interactions with the pores (i.e., rough etch front), and consideration of diffusion effects possible with porous materials. These materials issues have driven the bulk of the novel etch development in recent years. However, as the minimum pitch approaches sub-100nm, new interactions of the materials with this critical dimension need to be considered. The ability of low k materials to be patterned at the aspect ratios required without losing structural integrity will be the key challenge for future technologies' success. This paper will review the key etch challenges as a function of dimensions vs. materials and highlight where their interactions which will drive future work.

INTRODUCTION

Back-end-of-line (BEOL) etch has been challenged in recent years to achieve many things: critical dimensions on the order of ≤ 65nm, increasing via CD shrink to compensate for lithography limits, and most importantly, to date, inclusion of low k or porous low k dielectrics [1]. The issues associated with integrating low k dielectrics have been numerous for etch, most important being the sensitivity of the film to compositional modification during the etch and/or ash [2]. Entire new etch process regimes have been created and explored to successfully etch and integrate these new materials. That said, etch development on this generation of materials is generally maturing, and while more subtle issues are now being discovered, the next wave of etch challenges is coming in the form of interactions of dimensional issues with newer, lower k, materials issues.

Figure 1 shows a typical via-first-trench-last (VFTL) patterning scheme, with key etch issues highlighted at each step. These challenges, as well as some not highlighted in these figures, are categorized as a dimensional or a materials challenge in Table I.

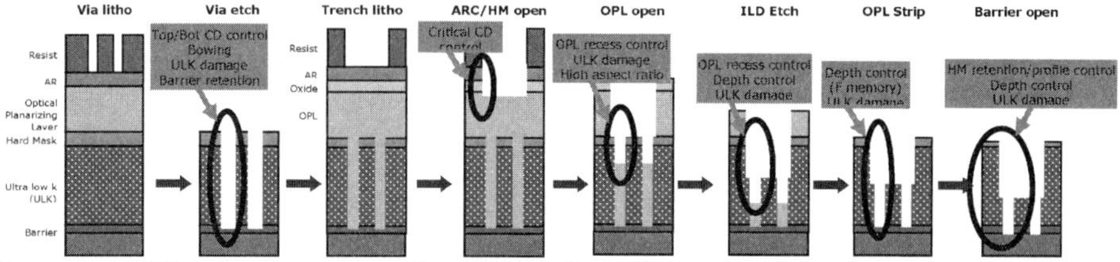

Figure 1. VFTL patterning scheme flow, with etch challenges highlighted at each step.

Table I. BEOL etch challenges categorized by dimension or material.

Dimensions	Materials
CD control	ULK damage (etch/ash) → CD control
CD shrink	CD shrink
Tighter tolerances	Rough etch front
RIE lag	Polymer management
LER	RIE lag
Aspect ratios	Porous materials interactions
Flop over	Flop over
Wiggling	Wiggling
Top profile control	Top profile control
Double patterning	

This paper will review some of these challenges and highlight the key new challenges ahead that result from an interaction of progressively smaller dimensions and progressively lower k materials.

EXPERIMENT

All etch experiments were completed on state-of-the-art 300mm dielectric etch tools at either IBM East Fishkill or Albany Nanotech, both in NY. Dielectric materials were deposited at either IBM East Fishkill or Albany Nanotech. PECVD SiOC-based films with dielectric constants ranging from 3.0 to 2.2 were used [3]. Film stacks generally consisted of a barrier material, the SiOC-based dielectric, and a dielectric hard mask. A trilayer lithography scheme was used to pattern all wafers (as shown in Fig. 1), with different technology critical dimension targets.

DISCUSSION

Dimensions

Table II shows the critical CD parameters for three technology nodes. As expected, both CD targets, as well as the uniformity requirements, decrease with each technology node. State-of-the-art etch tools are needed to achieve these uniformity requirements, and it is already a major challenge to meet 32nm uniformity requirements. It is unclear at present how to achieve the 22nm uniformity requirements, especially considering that current CD-SEM metrology becomes suspect for accuracy around 2nm 3σ.

Table II. CD targets and 3σ uniformity requirements per technology node.

		Technology		
Level	**Parameter**	**45nm**	**32nm**	**22nm**
M1	CD	65nm	50nm	<50nm
	3σ Uniformity	3nm	2.5nm	2nm
Vx	CD	65nm	50nm	<50nm
	3σ Uniformity	3nm	2.5nm	2nm
	Shrink	~20nm	~30nm	~40nm
Mx	CD	65nm	50nm	<50nm
	3σ Uniformity	3nm	2.5nm	2nm

The aspect ratio of the lithography film stack also becomes an issue as critical dimensions decrease. While the dielectric film stack generally scales from one technology node to another, the lithography film stack, in some cases, cannot. For the VFTL scheme in Fig. 1, the OPL aspect ratio can potentially increase from ~4.5 to 7.5 from the 45nm to 22nm node, and flopover of the OPL is a concern at that high an aspect ratio. An example of this OPL flopover issues is shown in Figure 2, where the bending of the OPL shadow-masked the underlying dielectric. Etch chemistry and process regime optimization was able to mitigate this issue.

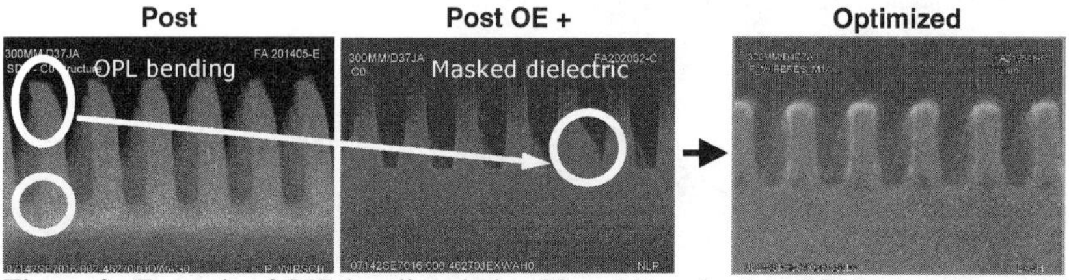

Figure 2. Example of OPL bending during 32nm M1 etch. Post-ME image shows OPL bending, which results in the masked dielectric post-OE+strip. Optimization of the etch avoids the OPL bending, resulting in the good profiles in the far right image.

RIE lag and LER are also key challenges when going from one technology node to the next. At a given technology node, RIE lag generally has a steep slope vs. CD for the smaller CDs, but levels out as CDs become large. Going from node N to N+1 or N+2, the minimum CD decreases and the RIE lag is further down on the steep part of the curve, resulting in increasingly bad RIE lag as the technology nodes progress, assuming the same etch recipe is used for all nodes. Optimization of the etch process is needed to "reset" the RIE lag through-CD curve at each generation in order to meet the technology requirements. Innovative approaches to the etch are needed to succeed with this challenge at each generation. Figure 3 shows an example progression of RIE lag from 45 to 32nm nodes, where optimization of the 32nm process was successful in "resetting" the RIE lag curve. Similar optimization is needed at etch technology node to minimize LER.

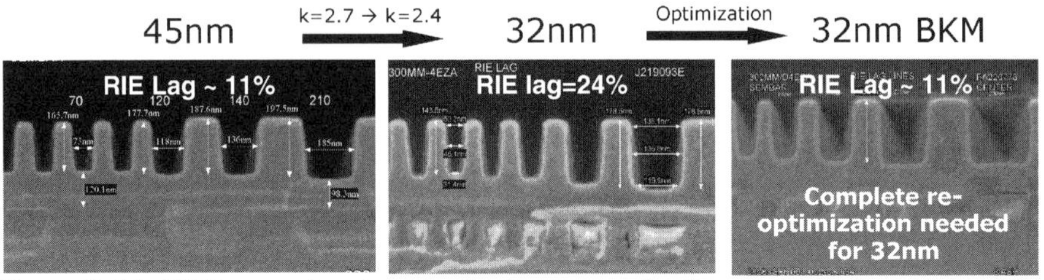

Figure 3. Mx RIE lag as a function of technology node. Smaller minimum dimension at 32nm results in increased RIE lag, but etch optimization able to recover RIE lag to acceptable level.

CD shrink is a focus for via etch, with each technology node requiring additional shrink from etch to compensate for limitations at lithography. However, while shrink of the minimum critical dimension is achievable, the through-CD performance of the shrink is startling. Figure 4 shows the difference in shrink performance for a nominal via vs. a nominal x 2 size via on the same mask. While the nominal via is close to CD target, the 2x via is ~35nm smaller than target, illustrating a very steep through-CD shrink response. Decreasing this through-CD shrink response is an ongoing challenge. Experiments have shown that while the shrink bias may be reduced somewhat, it is at the cost of other profile parameters which whose requirements cannot be relaxed. It is anticipated that the shrink bias will be a key issue for future technology nodes, and may require rethinking of the via designs.

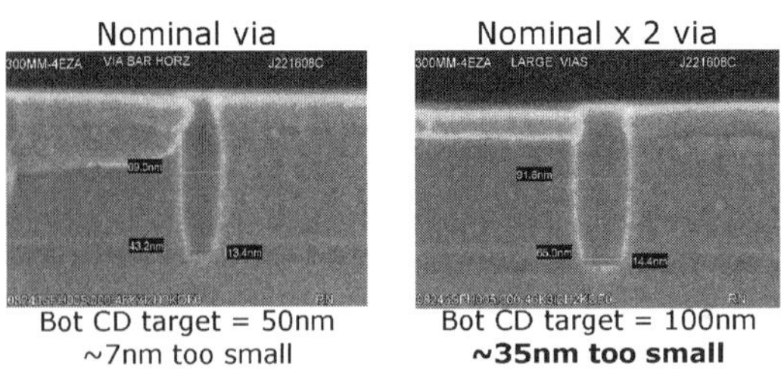

Figure 4. Via CD shrink at two different target CDs on the same mask. Intentional shrink for the nominal via size results in significantly larger shrink for larger-sized vias on the same mask.

The previous results have focused on ≥100nm pitch etch behavior. Recent work has highlighted new challenges for etch for sub-100nm pitch structures. The aspect ratio of the OPL film is a major concern and has shown more severe pattern collapse behavior than observed at previous nodes. Figure 5 demonstrates the challenges for sub-100nm pitch trench structures. OPL collapse or wiggling is predominant and can be triggered even after a clean OPL transfer etch (post-ME image). Optimization of the LER is critical, as LER alone can induce the OPL pattern collapse. Successful pattern definition has been achieved at sub-100nm pitch, but the CD window for no pattern collapse is very small (~8nm in example).

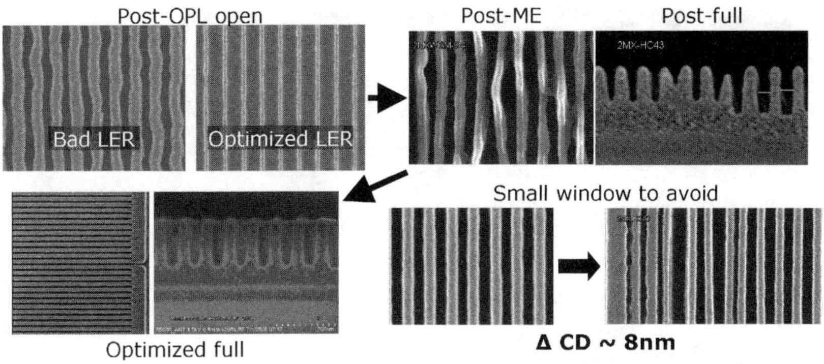

Figure 5. Sub-100nm pitch pattern collapse challenges. LER optimization is the first step to a successful OPL transfer etch, but OPL pattern collapse can still occur during the ME. Process optimization can yield viable sub-100nm pitch structures, but the CD window for no pattern collapse is small.

Materials

As dielectric materials move from low k to porous low k (k=3 → k=2.4 → k=2.2), the materials-related challenges for etch are clear: dielectric damage (i.e., compositional modification, most often demonstrated as C depletion), rough etch front, and porosity effects. Each of these issues could be a paper in of itself. Therefore, only a few highlights will be given here.

Figure 6 shows the progression of ash optimization to go from 30nm/side of hard mask (HM) undercut after DHF removal of the damage layer to ~3.5nm/side. Figure 7 shows the effect of dielectric constant on the dielectric damage behavior. Modifying the dielectric film composition (k=2.2 #2 vs. #1) can influence the degree of compositional modification induced by the plasma etch conditions.

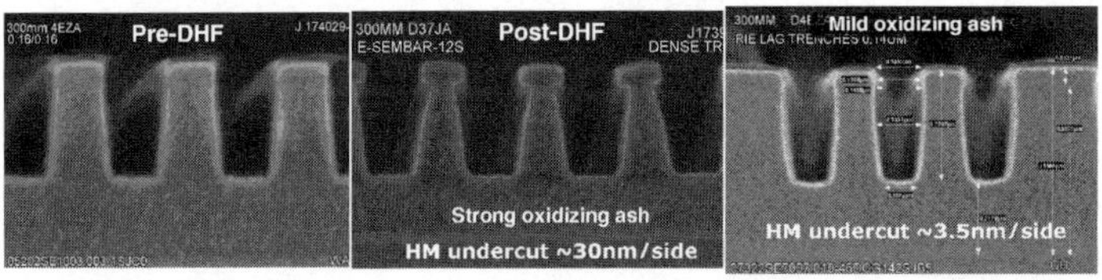

Figure 6. Exposure of a porous dielectric etch to DHF reveals significant dielectric damage. Successive ash optimization was successful in reducing HM undercut to ~3.5nm/side.

Dielectric etch front optimization is needed for porous low k dielectrics, as the porosity provides a micro-masking mechanism if the plasma etch polymers are not managed carefully. In addition, as the dielectric constant is decreased, pitting of the dielectric surface is observed. Modification of the plasma parameters can successfully reduce the pitting and generate a smooth etch front (Figure 8).

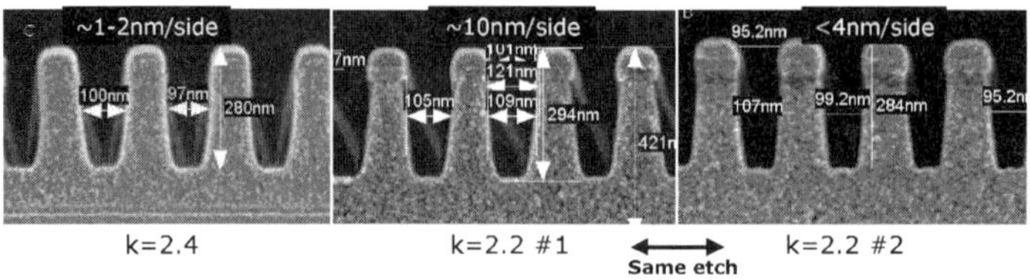

Figure 7. Hard mask undercut as a function of dielectric material.

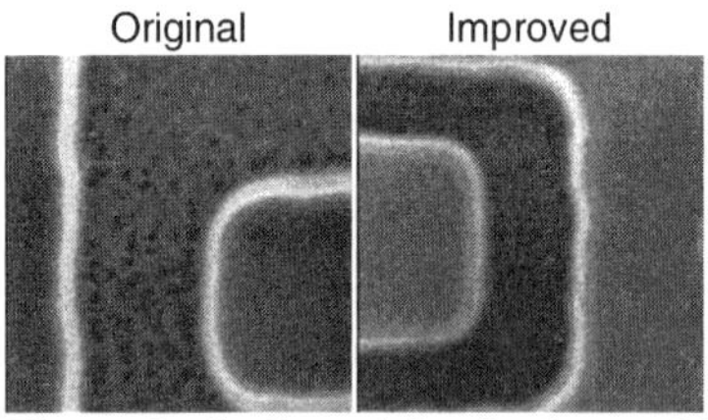

Figure 8. Surface pitting is observed for k < 2.4 dielectrics. Etch optimization can minimize the pitting.

<u>Dimensions + Materials</u>

The interaction of dimensional and material-related challenges is the driving force behind many of the current BEOL etch challenges. Figure 9 demonstrates this succinctly. If the same dielectric damage performance was targeted at different CDs, then the % of the CD taken up by damage would be unacceptable high. Keeping the % CD constant demonstrates that sub-4nm/side HM undercut performance is necessary for ≤ 140nm CDs. Different etch and/or ash solutions may be needed at each CD to achieve the target damage performance.

Critical CD (nm)	Damage (nm/side)	% of CD
40	5	25%
50	5	20%
65	5	15%
100	5	10%
140	5	7%

Damage held constant

→

Critical CD (nm)	Damage (nm/side)	% of CD
40	1.0	5%
50	1.3	5%
65	1.6	5%
100	2.5	5%
140	3.5	5%

% CD change held constant

Figure 9. Fraction of critical CD controlled by the dielectric damage.

Via-to-line spacing has become increasing critical through the technology nodes. Older technologies may have had some leeway with the top via and trench CDs while still achieving good reliability results. Current and future nodes do not have this luxury as spacing is minimal and dielectric damage heavily influences the final space CD (Figure 10).

A new phenomenon observed at sub-100nm pitch is dielectric flopover, often induced by a wet chemical exposure. Dielectric flopover can also occur without WETs exposure if the aspect ratio is too high. If the CD is pushed too high, then OPL flopover happens and the dielectric is masked. Figure 11 shows the different regimes for flopover for a sub-100nm pitch structure. The structural integrity of sub-100nm pitch structures is the number one concern for these technologies and is driving etch, films, and integration work to resolve.

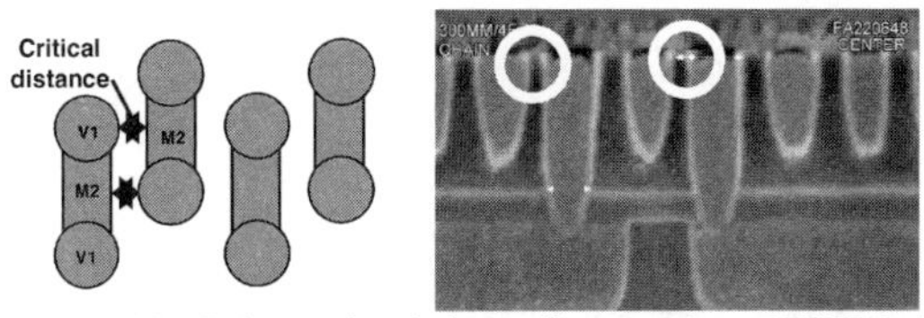

Figure 10. Schematic of via-to-line spacing. XSEM shows critical CD affected by dielectric damage.

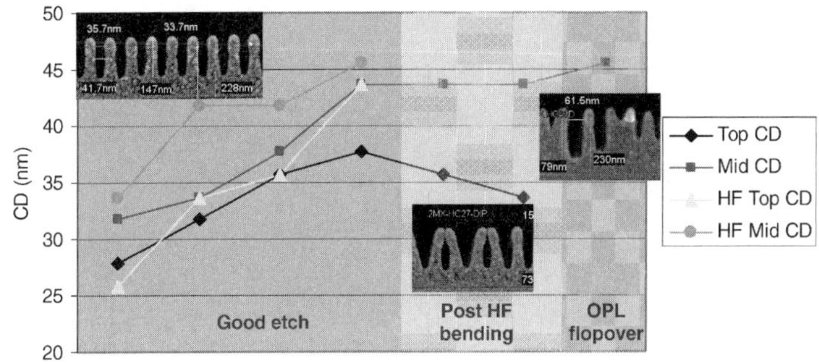

Figure 11. Different structural integrity regimes for sub-100nm pitch structures.

CONCLUSIONS

BEOL etch is increasingly challenging as technologies progress. Both the required dimensional and materials targets define the etch challenges. The interaction of dimensions with the materials is creating the biggest challenges (i.e., <1nm/side HM undercut, via-to-line spacing control). At sub-100nm pitch, dielectric flopover due to both the tight pitch and the porous low k dielectric is a structural integrity problem that will determine the success or failure of future technology nodes.

ACKNOWLEDGMENTS

This work has been supported by the independent Bulk CMOS and SOI technology development projects at the IBM Microelectronics Div. Semiconductor Research & Development Center, Hopewell Junction, NY 12533. This work was performed by the Research Alliance Teams at various IBM Research and Development Facilities.

REFERENCES

1. S. Sankaran, *et al.*, International Electronic Devices Meeting, 2006.
2. T.J. Dalton, *et al.*, IEEE International Interconnect Technology Conference, 154, 2004.
3. A. Grill *et al.*, IEEE International Interconnect Technology Conference, 2008.

Conference Proceedings AMC XXIV © 2009 Materials Research Society

Integration compatible porous SiCOH dielectrics from 45 to 22 nm

S.M. Gates[1], A. Grill[1], C. Dimitrakopoulos[1], V. Patel[1], S.T. Chen[2], T. Spooner[2], E.T. Ryan[3], S.A. Cohen[1], E. Simonyi[1], E. Liniger[1], Y. Ostrovski[1], and R. Bhatia[4]
[1]IBM TJ Watson Research Center, 1101 Kitchawan Rd. / Rte. 134, Yorktown Hts, NY, 10598
[2]IBM at Albany Nanotech, 255 Fuller Rd., Albany, NY, 12203
[3]Advanced Micro Devices, Inc., at Albany Nanotech, 255 Fuller Rd., Albany, NY, 12203
[4]Applied Materials @ Albany Nanotech, 255 Fuller Road, Albany, NY, 12203

ABSTRACT

We have developed two classes of porous SiCOH (pSiCOH) dielectric for integration in back end of the line (BEOL) interconnects. Here, we compare these materials at the k values 2.4 and 2.2. Class 1 ("V1") pSiCOH has a skeleton of O-Si-O bonds, with nm scale pores and carbon bonded as Si-CH3. Class 2 ("V2") contains added Si-C-Si structures, which are chemically stable methylene groups (-CH2-) bonded between 2 Si atoms. This paper compares some properties of the dielectrics, and integration compatibility of the V1 and V2 materials is examined using damascene Cu integration results. These results indicate that V2 pSiCOH is more resistant than V1 pSiCOH to the processing damage of copper damascene integration.

INTRODUCTION

Interconnect structures based on a porous SiCOH (pSiCOH) dielectric with dielectric constant (k) of 2.4 are entering production in IBM's 45 nm technology [1] and will be utilized at 32 nm. Looking ahead, lower capacitance is essential to offset the resistance increase that occurs with shrinking copper dimensions. Success at the 22 and 15 nm nodes with pSiCOH of lower k values (2.2 or lower) will require materials that are engineered for compatibility with integration processing. Simple extensions of the SiCOH dielectric familiar from the 90 and 65 nm nodes are expected to show excessive damage during integration due to etch, resist strip, chemical mechanical polish (CMP) and other process steps.

Here, we compare two classes of pSiCOH with the goals of extension to lower k, controlling the pore structure, engineering resistance to plasma damage, and improved integration compatibility. Class 1 ("V1") has a skeleton of O-Si-O bonds, with nm scale pores stabilized by Si-CH3 groups [2]. Class 2 ("V2") contains added Si-C-Si structures, which are chemically stable methylene groups (-CH2-) bonded between 2 Si atoms. We demonstrate that addition of Si-C-Si offers advantages in the V2 pSiCOH dielectric concerning integration damage.

EXPERIMENTAL DETAILS

The porous dielectrics were deposited by PECVD (plasma enhanced chemical vapor deposition) and cured in a UV cure module, both 300 mm production tools from Applied Materials . Dielectrics we call "V1" are based on an Si-O bond skeleton and were deposited using diethoxymethylsilane (DEMS) and a hydrocarbon porogen, bicycloheptadiene (BCHD). The V1 pSiCOH was described in an earlier paper [2]. Dielectrics with Si-C-Si structures added to the skeleton were prepared by adding a proprietary precursor (called precursor 2) to the reactor gas flow, and we call these "V2" pSiCOH. Films were prepared at a thickness of 400 nm after UV curing for characterization. The effects of UV curing on these materials are described in detail elsewhere [3].

Nanoindentation was used to measure the elastic modulus, and full wafer curvature was used to measure the internal stress. The dielectric constant and the plots of leakage current (J) vs. applied field of the cured pSiCOH films were measured on metal-insulator-semiconductor (MIS) structures, prepared by evaporation of aluminum through a contact mask to form Al dots. To calculate the dielectric constant, k, the MIS capacitance, the area of the dot, and the pSiCOH film thickness were measured. The latter was measured using a profilometer on a step formed in the film.

After integration in the Albany Nanotech center, electrical measurements were made on the integrated test structures, and then cross sections were prepared by wafer cleaving and focussed ion beam (FIB) preparation. Secondary electron microscope (SEM) images were collected from the FIB specimens.

RESULTS AND DISCUSSION

A. Characterization of the dielectrics

Table I describes 4 pSiCOH dielectrics, listing the mechanical properties and porosity of these materials. At each k value, we observe a lower modulus (E) for the V2 pSiCOH. This is consistent with a reduced concentration of Si-O-Si bonds, some of them being replaced by Si-C-Si bonds . There is no significant difference in internal stress in films of the same k value. Table I lists the volume % porosity measured by N_2 adsorption porosimetry . At the k values 2.2 and 2.4, note that V2 pSiCOH requires a slightly lower volume % porosity compared to V1 to reach these k values. We view this lower porous volume as one advantage of V2 pSiCOH compared to V1. A second advantage is seen in Table I in the evaluation of plasma induced damage (PID). This is discussed below.

Positron annihilation lifetime spectroscopy (PALS) data were measured in the laboratory of Professor D. Gidley (University of Michigan). PALS is used here to evaluate the degree of connectivity of the pores, with results shown in Figure 1. At each implant depth (X axis of Fig. 1), the percentage of the Ps escaping to vacuum is measured and plotted as the Y axis of Fig. 1. The % of Ps escape is a measure of the pore connectivity. A rank order in pore connectivity is clearly seen, with V2 k 2.4 showing the lowest % of Ps escape (open triangles). Then 2 dielectrics are similar: V1 k 2.4 and V2 k 2.2. At the k value of 2.2, the V2 material is similar in pore connectivity to the V1 pSiCOH at the higher k value of 2.4. Finally, the V1 k 2.2 pSiCOH with the largest volume % porosity shows the greatest % of Ps escape (solid dots). This data indicates that V2 pSiCOH is characterized by less connected pores compared to V1, which is consistent with the slightly lower volume % porosity for V2 reported in Table I.

We have studied the electrical properties of the same dielectrics as one criteria for use of these materials for BEOL integration. Figure 2 compares current density (J) vs. applied field (E) curves for V1 and V2 pSiCOH films with k=2.4. In the region of low applied field, the curves are equivalent. At the highest applied fields from 5 to 7 MV/cm the conductivity of V2 pSiCOH is higher than V1, possibly indicating a change in conduction mechanism at these fields.

B. Integration compatibility of the dielectrics

Turning now to integration, unit process steps such as plasma resist strip or etch cause dielectric damage and depletion of C in pSiCOH, called plasma induced damage (PID) [4]. We have studied this effect using the methods described in reference [4], using a highly oxidizing O2/Ar plasma. Our results demonstrate that V2 pSiCOH is more stable to oxidizing plasma damage than V1 pSiCOH, and these results will be presented elsewhere [5].

As a learning vehicle, single damascene copper structures patterned in the V1 and V2 pSiCOH dielectrics at k 2.2 were compared after using identical processing for both dielectrics in etch, resist strip, wet clean and metallization. Figure 3 shows SEM cross sections from specimens prepared by FIB after the identical processing. Lines patterned in V1 are on the top in Figure 3, and lines patterned in V2 are on the bottom. The brightest regions in Figure 3 are the tantalum/tantalum nitride bilayer liner surrounding each Cu structure. The full width at half height was measured and averaged over the 4 Cu lines shown in the Figure. This is the average line width, which is approximately 10% larger in V1 compared to V2 pSiCOH. The unit process steps etch, resist strip, wet clean, and liner deposition combined to produce 10 % wider lines in the V1 pSiCOH. Referring to the top row in Figure 1, the V1 structures show line bottom roughness: the trench bottoms have topography that varies from line to line. The sidewalls in V1 are not straight, but slightly bowed. The trench bottoms and bowed sidewalls are expected to produce greater variation in the measured dimensions in V1, compared to V2. Referring to the bottom row in Figure 3, the lines in V2 pSiCOH have flat bottoms, straighter sidewalls, and less variation in dimensions (width and depth) compared to the lines in V1 pSiCOH (top).

Electrical data were measured on wafers prepared identically to the structures shown in Figure 3. The measurements were made on the 1st level, after integration of a complete 2nd level of copper/ulk structures that served to cap and enclose the 1st level. The leakage current was 7X smaller when measured in V2 pSiCOH, compared to V1 p SiCOH. The capacitance (C) of a comb capacitor structure was approximately 10% lower in the V2 test structures compared to V1. Qualitatively, a smaller range of C values was measured in the V2 structure compared to V1.

The lower leakage observed in structures built with V2 pSiCOH current can be explained by a lower local electric field. The average Cu line width is about 10% smaller in V2. Then the line-to-line spacing is about 10% greater. When using the same applied voltage in the leakage measurements, the applied field is smaller by about 10%, resulting in a reduced leakage current. Note that leakage current increases exponentially with field, as seen in Figure 2. The observation of a smaller range in C is significant. Industry-wide, variability of physical and electrical data is observed as copper dimensions shrink. Variability is steadily increasing. The single damascene test produced a smaller range of C in V2 compared to V1 pSiCOH. We attribute the range in C directly to a smaller variation in average line width noted above.

Shown in Figure 4 is the result of an optimized dual damascene process flow using the k 2.2 V2 pSiCOH. The straight sidewalls of both via and trench are noted. The embedded via contact seen in Figure 4 enables a lower contact resistance and a reliability advantage compared to flat via contacts. Some porous dielectrics show rough trench bottoms after the ion bombardment used to create this embedded via contact. Figure 4 illustrates that V2 pSiCOH withstands ion bombardment, yielding relatively flat trench bottoms. This can be understood by comparing the relatively lower volume % porosity in V2 k 2.2 pSiCOH, compared to V1 (Table I).

SUMMARY

Concerning the fundamental properties of V1 and V2 pSiCOH, we have reported here that V2 pSiCOH requires a lower absolute porosity to reach the k value of 2.2. The same is true for k 2.4 pSiCOH, but the difference in the degree of porosity is very small. PALS data show that the pores in V2 pSiCOH are less connected than in V1, when V1 and V2 are compared at the same k value. Concerning the integration of V1 and V2 pSiCOH in our comparative study at k 2.2, V2 pSiCOH is more resistant to integration damage, as evidenced by smaller dimensions in lines patterned using the same process flow. The present results give us confidence that V2 pSiCOH is well suited for integration at the smaller Cu dimensions required by future CMOS technologies.

ACKNOWLEDGEMENT

This work was performed by the Research Alliance Teams at various IBM Research and Development Facilities

REFERENCES

[1] S. Sankaran et al, Proc. IEDM 2006.
[2] S. Gates et al, Proc. AMC 2006, p. 351.
[3] C. Dimitrakopoulos et al, submitted to the AMC 2008.
[4] E.T. Ryan, et al, submitted to J. Appl. Phys.
[5] To be published.

	k 2.4		k 2.2	
Property	V1	V2	V1	V2
Modulus [Gpa]	4.5	3.8	2.7	2.3
Stress [Mpa]	47	51	38	40
N2 Adsorption porosity (%)	30	29	37	33
PID Delta Thickness (Angstroms)	244	123	285	142

Table I : Properties of the 4 different pSiCOH dielctrics compared here.

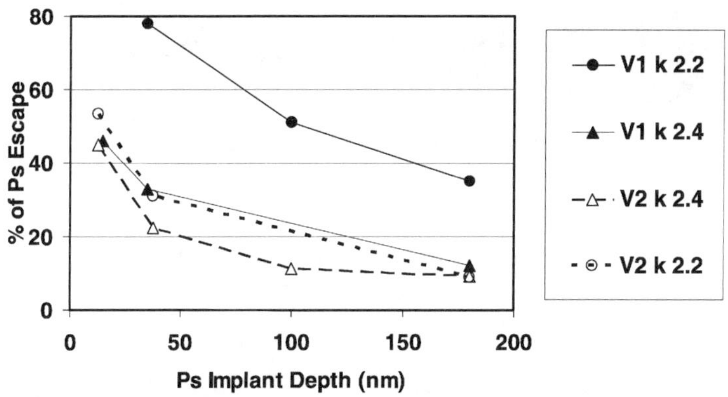

Fig. 1 : Percent of Ps escaping the film vs. implantation depth measured by positron annihilation spectroscopy (PALS).

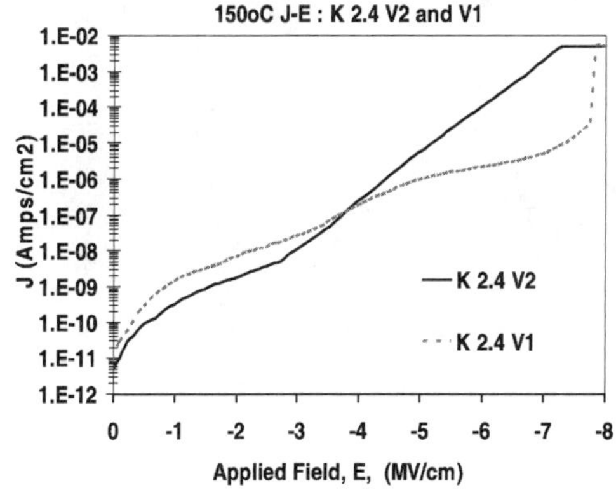

Fig. 2: J-E curves for V1 and V2 pSiCOH k 2.4 measured in MIS test structures.

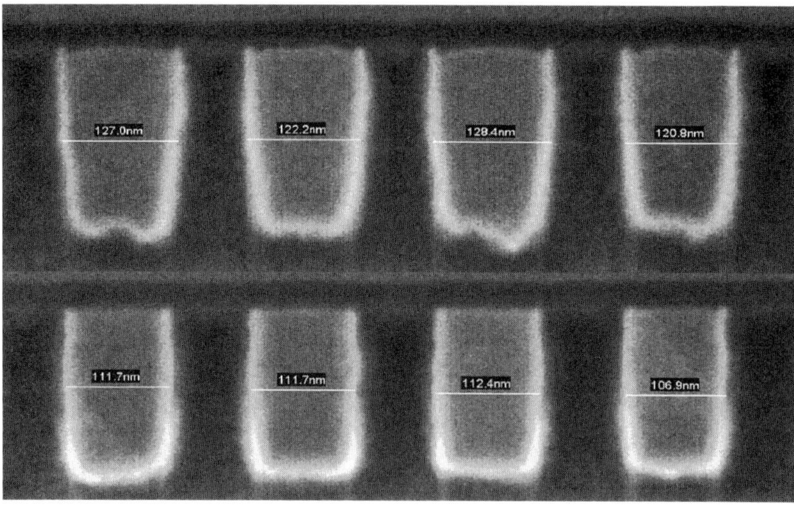

Fig. 3 : SEM of FIB cross sections of single damascene copper lines patterned in V1 pSiCOH (top) and V2 pSiCOH (bottom), both at k 2.2.

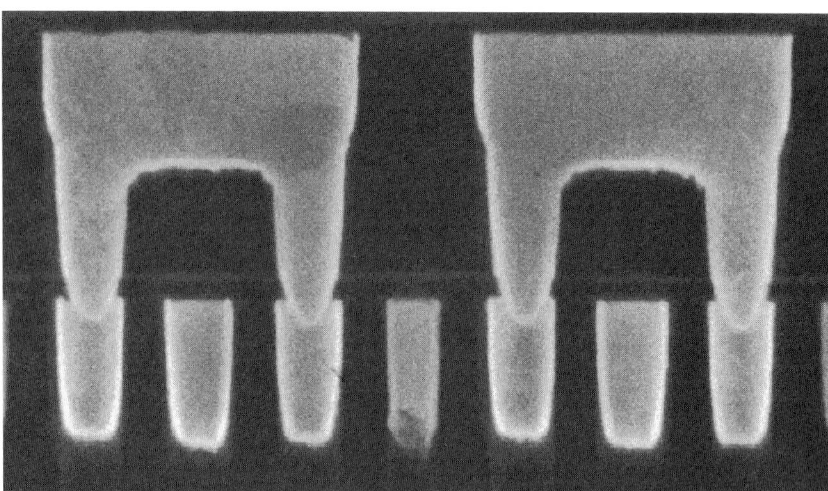

Fig. 4 : SEM of FIB cross sections of optimized dual damascene copper structure patterned in V2 pSiCOH, k 2.2.

Conference Proceedings AMC XXIV © 2009 Materials Research Society

Dielectric constant contributions and structural properties of PECVD porous ULK: Impact of the elaboration process

Olivier Gourhant[1], Vincent Jousseaume[2], Laurent Favennec[1], Aziz Zenasni[2], Patrice Gonon[3], Guillaume Gerbaud[4], and Gilbert Vincent[3]
[1]STMicroelectronics, Crolles, France
[2]CEA-LETI, Grenoble, France
[3]LTM, CNRS-LTM, Grenoble, France
[4]CEA-INAC, Grenoble, France

ABSTRACT

In the trend to decrease design features in microelectronics, porous a-SiOC:H are good candidates to meet the request for lower dielectric constant materials. These dielectrics are elaborated using a two-step process. First, the deposition by PECVD using two types of precursors enables the realization of labile species inclusions in an organo-silicon matrix. Then, a thermal annealing, alone or assisted by UV radiations removes the porogen phase and creates porosity. In this paper, material evolutions induced by process parameters modifications are addressed. It is shown that the deposition temperature and the nature of the post-treatment can influence structural modifications. UV curing is found to enhance SiOSi network reorganization and to break Si-CH_3 bonds, leading to higher film shrinkage. Afterwards, the different contributions of the dielectric constant are evaluated using the temperature dependence of the dipolar contribution. This electrical characterization is performed to enlighten the impact of porosity increase and post-treatment on the different contributions of porous materials permittivity. A continuous decrease of the ionic fraction is observed, which is mainly linked to the SiOSi concentration evolution. Moreover, an increase of the dipolar contribution is sometimes observed, which can be related to the presence of porogen residues.

INTRODUCTION

Porous SiOC:H are currently used as ultra-low K materials in advanced interconnections. Their elaboration consists of the PECVD co-deposition of labile species (i.e. porogen) embedded in an organo-silicon matrix, followed by a post-treatment to remove the porogen phase. This post-treatment is mainly activated by thermal energy but UV radiation is shown to help the process and to enhance mechanical properties of the porous film [1]. Though the integration of such material with K=2.35 has been already demonstrated [2], lower K values are still needed for the 22 nm node and beyond. In general, most studies deal with a global dielectric constant decrease by tuning process parameters. However, very few have investigated their impact on the different contributions of the dielectric constant. In this paper, these dielectric features, as well as material modifications during the process improvement, are explored. In a first part, the enhancement of the porosity and the structural modifications generated by the post-treatment are investigated. Then, the second part will focus on the evolution of the different K-value contributions induced by these process variations.

EXPERIMENT

ULK materials were elaborated in a 300 mm equipment composed of two chambers dedicated, respectively, to the film deposition and to the post-treatment. The deposition of the organic-inorganic thin film (called hybrid material) was carried out within the first chamber by PECVD: an organo-silicon precursor, for the matrix, and an organic precursor, for the labile phase, were simultaneously injected in a capacitive coupled parallel plate reactor using a 13.56 MHz radio frequency excitation. Several deposition temperatures from 200 to 300°C were used. Then, the porogen phase was removed by the post-treatment in the second chamber. Thermal treatment was achieved using a hotplate system at a temperature of 450°C under N_2 atmosphere. UV curing was performed with a hotplate at 400°C placed under a broadband ultra-violet lamp exposition and under He atmosphere.

Fourier Transform Infra-Red (FTIR) analysis was performed with a Biorad QS3300 spectrometer. Solid State Nuclear Magnetic Resonance (NMR) was performed with a Bruker AVANCE 400 to determine the network structure of thin films. Optical properties, such as refractive index at 632.8 nm, and film thicknesses were measured with a Rudolph ellipsometer. The dielectric constant of the porous film was calculated from the capacitance measured in accumulation mode using a MIS structure formed by a mercury drop in contact with the ULK/silicon stack. Capacitance measurements versus temperature (between room temperature and 250°C) and as a function of frequency (between 0.1 to 10^5 Hz) were performed using a probe station connected to a Novocontrol impedancemeter.

DISCUSSION

Several hybrid materials were elaborated by increasing the porogen loading in the plasma gas feed, which is known to induce a direct enhancement of the porogen loading [3]. Figure 1 shows the variations of the dielectric constant, obtained for each post-treatment, versus the porogen ratio in the gas feed. In both cases, two behaviors are observed when the porogen loading increases. The dielectric constant starts to decrease when more porogen phase is included in the hybrid material. Above a porogen loading threshold, the dielectric constant increases, which is linked to a high film shrinkage occurring during the post-treatment. As shown in previous works, the increase of porogen loading enables the porosity enhancement only until a loading limit [3]. Above this threshold, the matrix collapse and the presence of porogen residues induce a porosity decrease. The resulting U-shape evolution of the dielectric constant is observed whatever the value of the deposition temperature between 225 and 300°C. Figure 1 also shows that lower k values are obtained using a thermal treatment alone. This can be explained by a higher matrix shrinkage induced by UV curing.

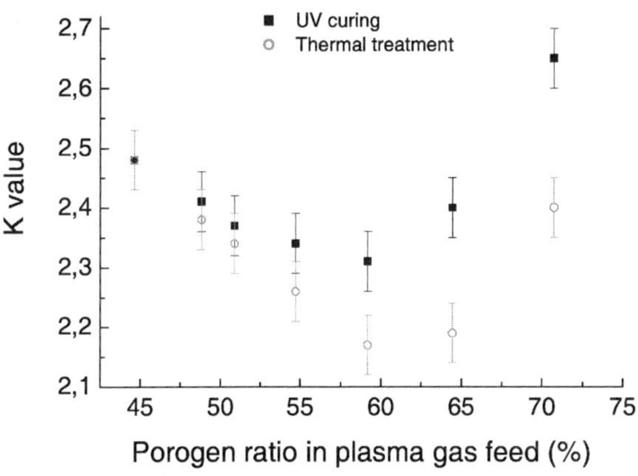

Figure 1: Evolution of the dielectric constant as a function of porogen loading in the plasma gas feed ($T_{deposition}$=300°C).

Figure 2a shows the FTIR spectra of films elaborated at 260°C and treated with UV curing or thermal annealing. These spectra exhibit the same characteristic bonds. A strong band is revealed between 950 and 1200 cm^{-1} which can be mainly assigned to Si-O bond vibration in SiOSi [4]. The large configuration variety of these bonds leads to a broad signature. SiOC bond vibration is also expected in this region, between 1090 and 1120 cm^{-1} [5]. A sharp peak at 1274 cm^{-1} is due to the symmetric deformation of the CH bonds in $SiCH_3$ [4]. As previously shown in the literature, the Si-CH_3 intensity is significantly reduced when the material is UV cured, compared to thermally treated material [1]. The 1050 cm^{-1} contribution of the SiOSi peak is more prominent with UV curing than with thermal treatment which indicates a difference in the SiOSi bonds configuration. This reveals a more important rearrangement of the network by UV curing.

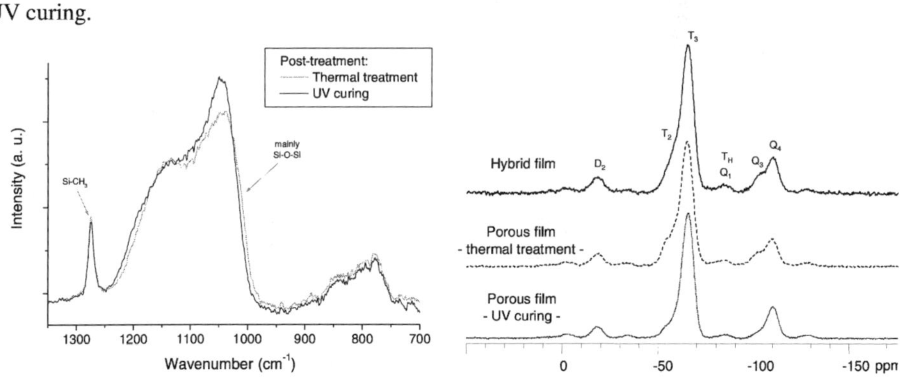

Figure 2: FTIR spectra (a) and NMR ^{29}Si spectra (structure nomenclature defined in [6]) (b) of porous materials issued from the same hybrid film ($T_{deposition}$=260°C).

Figure 2b presents the solid NMR ^{29}Si spectra of the corresponding materials. The main differences between the cured samples are located at the shoulders of the T and Q peaks. The thermal treated film contains T_2 and Q_3 contributions already observed in the hybrid film. These signatures can be mostly related to the presence of silanols in the film. Their strong decrease in the UV cured sample indicates that the silanols condensation has been favored in this case. It can be noticed that the presence of silanols decreases when the deposition temperature increases (not shown). Consequently, a lower impact on the crosslinking is observed for the hybrid material deposited at 300°C. Then, the break of $Si\text{-}CH_3$ bonds and the more important network rearrangement observed with UV curing allow to explain the higher film shrinkage observed with this post-treatment.

In the second part of this work, these materials were electrically characterized to especially enlighten the impact of porosity increase and post-treatment type on the different contributions of the permittivity. Figure 3a presents the evolution of the dielectric constant as a function of frequency for SiO_2 (thermal oxide, K=3.9) and for a porous SiOC:H (K=2.3, UV cured). Spectral analysis shows only a very slight dependence of the dielectric constant with frequency (<5% in the range of the study) and almost no relaxation peak is observed.

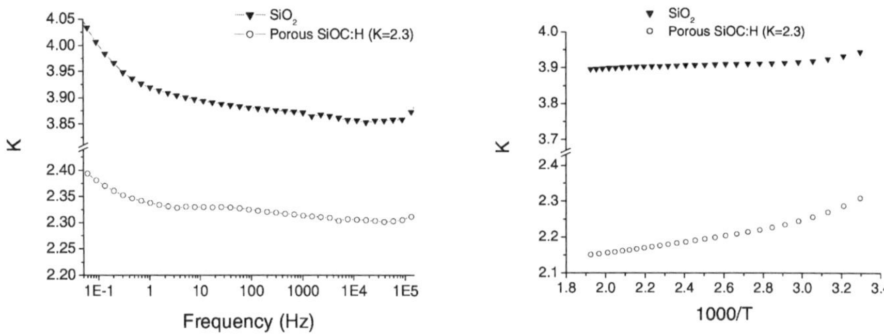

Figure 3: Evolution of the dielectric constant as a function of frequency (measured at room temperature) (a) and as a function of temperature (measured at 100kHz) for SiO_2 (thermal oxide, K=3.9) and for a porous SiOC:H (K=2.3, UV cured).

In order to determine the different contributions of the dielectric constant, namely the electronic, the ionic and the dipolar contributions, the temperature dependence of the dipolar component was used. Figure 3b shows the variation of the dielectric constant (measured at 100 kHz) as a function of temperature for SiO_2 (thermal oxide) and for a porous SiOC:H (K=2.3, UV cured). Almost no dependence is observed for a thermal silicon dioxide, while the existence of a slope proves the presence of dipolar species in the case of porous SiOC:H. Using the Debye equation, the sum of the electronic and the ionic contributions of the dielectric constant was deduced from the intercept of the curve with the Y axis at 1000/T=0 [7]. Then, the different contributions quantification for each material can be calculated, the electronic contribution being deduced from the refractive index in the visible range ($\varepsilon_{elec}=n^2-1$, with n the refractive index).

Figure 4 presents the electronic, ionic and dipolar contributions of the dielectric constant for different elaboration conditions. Among these three contributions, the electronic one is the most important in all cases. The porous film issued from "Hy#3" hybrid material shows similar contributions of the dielectric constant when it is UV cured or thermally treated. Besides, a U-shape behaviour of the electronic contribution as a function of porogen loading is observed, which reflects the porosity evolution as presented in figure 1. In contrast, the ionic contribution of the dielectric constant decreases continuously with the porogen loading. This contribution is mainly correlated to the concentration of SiOSi bonds in the film which shows also a continuous decrease with the porogen loading. Finally, the dipolar contribution remains unchanged for different porogen loading excepted for the highly loaded sample. In this particular sample, the presence of porogen residues can be the source of the higher dipolar contribution.

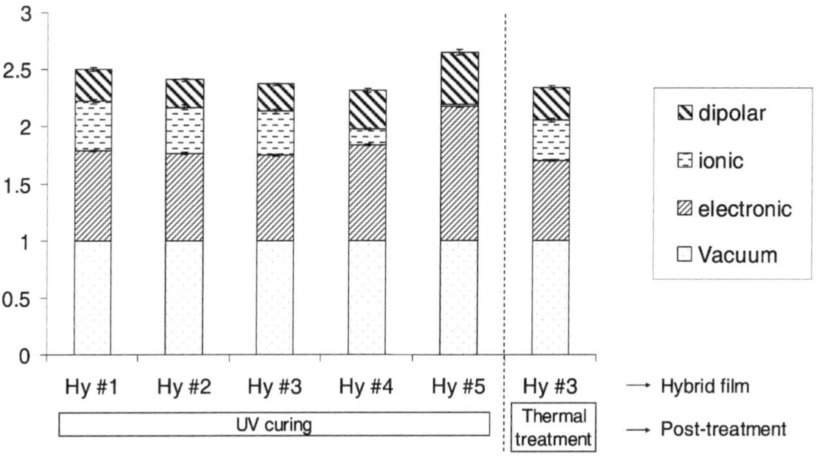

Figure 4: electronic, ionic and dipolar contributions of porous materials dielectric constant for different elaboration conditions (Hy #1 to Hy #5 correspond respectively to hybrid films elaborated with a porogen ratio in the plasma gas feed of 43, 48, 51, 59 and 71% at $T_{deposition}$=300°C).

CONCLUSIONS

In this study, the impact of process parameters on the enhancement of porosity is investigated in order to lower the dielectric constant and materials with different porosity rates were elaborated. It is shown that the deposition temperature and the nature of the post-treatment can influence structural modifications. UV curing is found to enhance SiOSi network reorganisation and to break Si-CH_3 bonds, which leads to higher film shrinkage compared to the case of thermally treated films. Then, the different contributions of the dielectric constant are evaluated using the temperature dependence of the dipolar contribution. A continuous decrease of the ionic fraction is observed, which is mainly linked to the SiOSi concentration evolution. In the meantime, the presence of porogen residues can increase the dipolar contribution of the permittivity.

REFERENCES

1. A. Zenasni et al, J. Appl. Phys. **102** (9) (2007).
2. M. Aimadeddine et al. in *Robust integration of an ULK SiOCH dielectric (k=2.3) for high performance 32nm node BEOL* (International Interconnect Technology Conference, 2007) pp. 175-177.
3. O. Gourhant et al. in *Extendibility of the PECVD Porogen Approach for ULK Materials*, (Mater. Res. Soc. Symp. Proc. **990**, San Francisco, CA, 2007) pp. 45–50.
4. D. R. Anderson, *Analysis of Silicones, chapter 10* (Willey-Interscience, NY, 1974).
5. S. P. Mukherjee et al., Thin Solid Films **14**, 105 (1972).
6. V. Jousseaume et al., J. Electrochem. Soc. **154** (5), G103 (2007).
7. P. Gonon et al, J. Electrochem. Soc., **150** (3), F47 (2003).

Conference Proceedings AMC XXIV © 2009 Materials Research Society

Engineering of Chemical and Physical Properties of Low-k Materials by Different Wavelength of UV Light

Premysl Marsik[1,2], Adam Urbanowicz[1], Patrick Verdonck[1], Karim Ferchichi[1], David DeRoest[3], Lutz Prager[4], and Mikhail R Baklanov[1*]
1AMPS, IMEC, Kapeldreef 75, Leuven, 3001, Belgium
2UFKL, Masaryk University, Kotlarska 2, Brno, 61137, Czech Republic
3ASM Belgium, Kapeldreef 75, Leuven, 3001, Belgiím
4 Leibniz-Institut für Oberflächenmodifizierung, Permoserstrasse 15, 04318 Leipzig, Germany
* corresponding author: baklanov@imec.be

ABSTRACT

SiCOH low-k films were deposited by PECVD with variable flow rates of porogen and matrix precursors and variable substrate temperature and RF power. Two precursor ratios were chosen leading to 1) material with higher target porosity around 33% and target k-value 2.3 and 2) material with lower porosity around 25% and target k-value 2.5. After deposition, the samples were UV-cured on temperature above 400°C in nitrogen ambient. Two types of curing lamps were used for the experiments: 1) lamps "A" emitting photons with energies higher than 6.5 eV (λ<190 nm) and 2) lamps "B" with photon energies below 6.2 eV (λ>200 nm). For all properties evaluated, irradiation at wavelengths below 190 nm resulted in more pronounced changes than at longer wavelengths. Lamps A provide fast decrease of porogen content, conversion of the Si–O–Si bonds from cage to network. However, degradation of Si–CH_3 bonds is significant, as well as formation of H–SiO (Si-H) bonds and amorphous carbon like porogen residue. Lamps B provide more slow porogen removal, do not decrease Si-CH_3 concentration and leave less porogen residues. Application of UV ellipsometry (138 nm -165 nm) for the low- k films, pure matrix material and pure porogen allowed obtaining additional results important for discussions. These findings are well confirmed by quantum-chemical calculations.

INTRODUCTION

Porous low-k dielectric films are introduced as interconnect-dielectric in integrated circuits below 45nm technology node. The actual approach is PECVD co-deposition of SiCOH matrix from selected precursor together with sacrificial organic porogen [1]. The porogen is afterwards removed during annealing [2,3] or UV assisted cure [4]. The resulting material must contain porosity sufficient for lowering the dielectric constant (k) down to target values. The lowered density of the material leads to reduced mechanical properties. The material must be hydrophobic, otherwise the physisorbed water (k~80) in the pores interior deteriorates the k-value. For successful implementation of the porous low-k's into industrial processes, the chemical and plasma damage resistance must be understood and assured.

The UV assisted annealing (UV-cure) is being favored for efficient removal of the porogen in short time [5] and for the changes of the matrix material leading to improved mechanical properties [4,6]. The choice of UV-curing light source has been attracting the interest, because the photons with higher energies cause greater changes to the matrix, improving the mechanical properties, but also can break the Si-CH_3 bond resulting in replacement by Si-H or Si-OH, shrinkage and hydrophilicity [7,8]. The influences of porogen loading and deposition condition and its removal have been studied for annealing cures [3,9]. The purpose of our study is to relate some of possible variables of the deposition and curing to the film compositional and structural properties.

EXPERIMENT

We deposited the low-k material as 200 nm films on Silicon (100) wafers in ASM Eagle 12 system using variable flow rates of porogen and matrix precursors and variable substrate temperature and RF power. Two precursor ratios were chosen leading to 1) material with higher target porosity around 33% and target k-value 2.3, denoted as "CVD1" in this paper and 2) material with lower porosity around 25% and target k-value 2.5 denoted as "CVD2". The deposition conditions were chosen as 1) higher substrate temperature (300°C) and higher RF power (1850 W – 1900 W), further denoted by abbreviation "Hi" and 2) lower temperature (250°C) and lower RF power (1400 W), denoted by abbreviation "Lo".

After deposition, the samples were treated by UV-cure on temperature above 400°C in nitrogen purged ambient. Two curing lamps were used for the experiment: 1) lamp "A" emitting photons with energies higher than 6.5 eV ($\lambda<190$ nm) and 2) lamp "B" with photon energies below 6.2 eV ($\lambda>200$ nm). For both lamps, two curing times were chosen. Shorter time "1" is a half of the longer time "2", which is set as curing time for standard process. Therefore we obtained 16 samples with variable properties. Only already cured samples were studied, however, the thickness was measured before and after UV-cure, allowing calculation of shrinkages of the film.

The optical response of the films has been measured in the range from 2 eV to 9 eV (wavelengths from 620 nm down to 138 nm) using a variable-angle of incidence spectroscopic ellipsometer Sopra GES5 PUV. This tool operates in the rotating analyzer and fixed polarizer configuration with quartz Rochon cubes used as the polarizers, deuterium discharge light source and photomultiplier detector. To achieve a high transmittance for ultraviolet photons, the tool is surrounded by a glove box purged by nitrogen, keeping the oxygen and water concentrations below 1 ppm. For all of the samples, we applied three angles of incidence: 60, 70 and 80 degrees.

The measured spectra of ellipsometric angles Ψ and Δ [10] were fitted by single layer optical model (substrate – layer – ambient) using Marquardt-Levenberg algorithm. The dielectric function ε of the low-k film was modeled by generalized Gauss-Lorentz (G-L) peaks, calculated as rational approximations [11]. The dielectric function ε is recalculated to refractive index n and extinction coefficient k for the reader's convenience.

The porosity measurements were performed on a prototype ellipsometric porosimeter EP10, equipped with a fixed angle of incidence (70 deg) ellipsometer Sentech 801, operating on wavelengths between 350 nm and 850 nm, mounted on a high vacuum chamber with controllable pressure of solvent vapors. The principle of ellipsometric porosimetry is based on pressure dependent adsorption of a fluid (in our case toluene) into the pores. The absorbed toluene modifies the optical properties of the porous sample in the means of effective media and the changes are measured by ellipsometry. The volume ratio of the open pores is then obtained [12].

The infrared absorption of the layers in the range 400-4000 cm^{-1} is measured by a Fourier-transformed infrared (FTIR) spectrophotometer Biorad QS2200 ME with 4 cm^{-1} spectral resolution. The optical response is evaluated as absorbance. The absorbance of the substrate Silicon wafer is subtracted and the resulting spectrum is treated by baseline subtraction and renormalization.

RESULTS AND DISCUSSIONS

Infrared absorbance

We measured the FTIR transmission spectra of the 16 samples and fitted the observed structures using multiple Gaussian peaks. The calculated integrals were normalized to the integral of SiO_2 band. The FTIR spectra show following features [2]: The porogen CH_x absorption band is observed around 2900 cm^{-1}, but the dominant peak at 2973 cm^{-1} is attributed to CH_3 and therefore shows also response from methyl groups bonded in the SiCOH matrix. The $Si\text{-}CH_3$ signature is pronounced well at 1275 cm^{-1} and the SiO_2 band is observed as double peak structure localized at 1130 cm^{-1} (cage) and 1050 cm^{-1} (network). The Si-H bond vibration is detected at 2240 cm^{-1} and 890 cm^{-1}. We observe presence of C=C bonds in all the samples as small band at 1600 cm^{-1}.

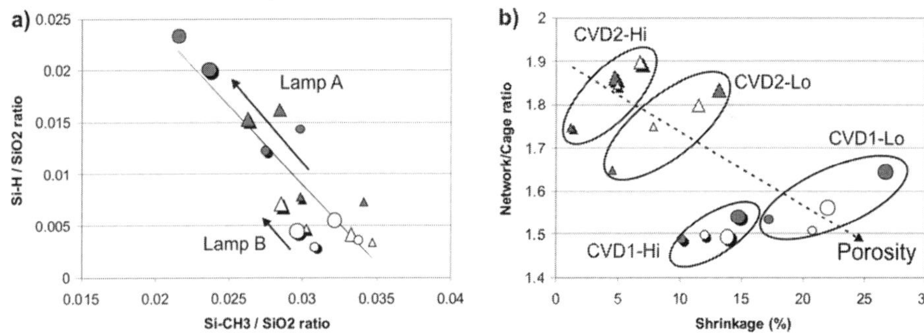

Figure 1 a) Anticorrelation of FTIR peaks related to SiH and $SiCH_3$ bond vibrations, normalized to peak area of SiO_2. The solid line represents the linear regression of the 16 points. **b)** Ratio of FTIR peaks related to SiO_2 network and cage versus the shrinkage of the samples. The dashed arrow follows the direction to higher porosities. - The triangles and circles refer to material CVD1 (higher porosity) resp. CVD2 (lower porosity). The symbols with shadow refer to higher deposition temperature and RF power ("Hi") and the flat symbols refer to lower deposition temperature and RF power ("Lo"). The color of the symbol denotes the curing lamp: Grey symbols for lamp A and white symbols for lamp B. Increasing symbol size denotes increasing curing time.

In figure 1a, we show the integral absorbance of the $Si\text{-}CH_3$ peak at 1275 cm^{-1} related to the integral absorbance of the Si-H peak at 890 cm^{-1}, both relative to the integrated absorption of SiO_2 band between 950 cm^{-1} and 1250 cm^{-1}. There is a strong anticorrelation between the $Si\text{-}CH_3$ peak and the Si-H peak. With increasing curing time, the $Si\text{-}CH_3$ presence is reduced and Si-H bonds are created through a process described in the literature [4,13]. This process is much stronger for the curing lamp A. The energy of the photons of the lamp A (7.2 eV) is sufficient to break the $Si\text{-}CH_3$ bond as expected from quantum-chemical calculations [8]. The experiments applying the UV cure with lamp A to the matrix material itself, without the porogen [14], resulted also in $Si\text{-}CH_3$ reduction, however, no Si-H creation was observed. This suggests that the porogen is the main source of hydrogen in this reaction.

Figure 1b shows ratio of the FTIR peaks assigned to network (around 1050cm^{-1}) and cage structure (around 1130 cm^{-1}) of the SiO_2 skeleton ("networking") compared to shrinkage of the film during the UV-cure. We observe the correlation between the "networking" and the shrinkage with increasing curing time, but this correlation occurs separately in the groups of the same material, deposited in given conditions. In all four groups, lamp A induces more changes than lamp B. Among the groups, this correlation is suppressed in spite of porosity: From material to material the porosity is the leading factor influencing the shrinkage and the "networking": 1) The samples with higher porosities show also higher shrinkage and 2) higher porosity is followed by lower network to cage ratio. It should be noticed that the link between shrinkage and the $SiCH_3$ reduction is also valid only within the groups of given material.

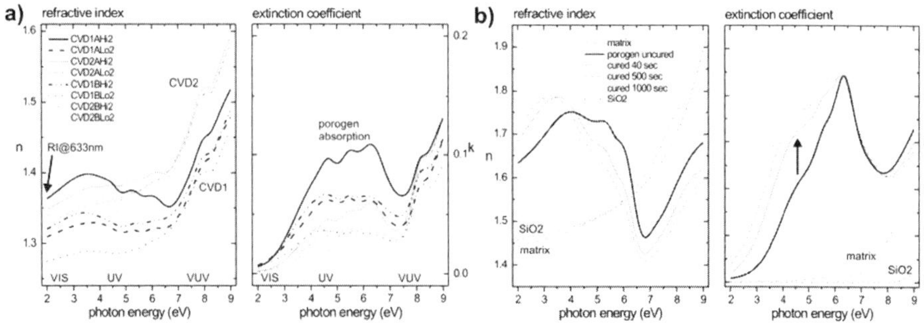

Figure 2 a) The optical functions of selected low-k samples. **b)** Optical functions of porogen films UV-cured for various curing times from 0 sec to 1000 sec and the optical functions of the matrix material. The properties of SiO_2 are plotted for comparison.

UV optical properties

The ultraviolet optical response of the low-k dielectric in the visible and ultraviolet range is determined by presence of porogen and its removal during the cure [15]. We determined the optical properties of the low-k films by spectroscopic ellipsometry and prepared and measured also samples of matrix material alone, without porogen and the organic porogen itself to be able to distinguish the impact of the both constituents. The refractive index and extinction coefficient of the low-k samples are plotted on the figure 2a and the properties of the constituents are plotted on figure 2b together with the properties of SiO_2 for comparison. The global "Cauchy-like" dispersion of the refractive index of the low-k films is following the same trend as the SiCOH matrix and SiO_2; in both cases the dispersion is given by strong absorption of SiO_2 above the measured range, while the porogen contributes to the absorption with the three peak structure between 4 eV and 7 eV in the extinction coefficient. These absorption bands produce the "s"-structure in the refractive index dispersion, shifting up its values in the visible range.

The evolution of the spectra of the low-k during the UV-cure follows the changes observed on the porogen samples and therefore we plotted only the samples with longer curing time in the figure 2a. The deposited porogen layer exposed to the UV light (lamp A in this case) shows fast shrinkage reduction (i.e. removal), but the layer is not removed completely (up to 80%) and the properties are changing [14]. The dominant absorption structure at 6.4 eV is suppressed in spite of growing band around 4.5 eV, followed by the growth of refractive index in the visible ("RI"). In the spectra of the low-k samples, the main difference is observed between the CVD1 and the CVD2 material. The material CVD1 poses higher porosity and so the overall values of refractive index are lower (particularly observable above 7 eV) than the CVD2, however, it contains generally more porogen residues, observable as higher absorption between 4 eV to 7 eV. This absorption is transferred to refractive index as increased values on the lower energies (in the visible range). Therefore, the value of refractive index at 633nm ("RI") carries mixed information about the density of the material and the porogen residues.

As we described in [14], the knowledge of the properties of the constituents and the porosity percentage measured by ellipsometric porosimetry raises the possibility to estimate the volume percentages of the matrix and porogen residue as well, using the three media Bruggeman approximation [16].

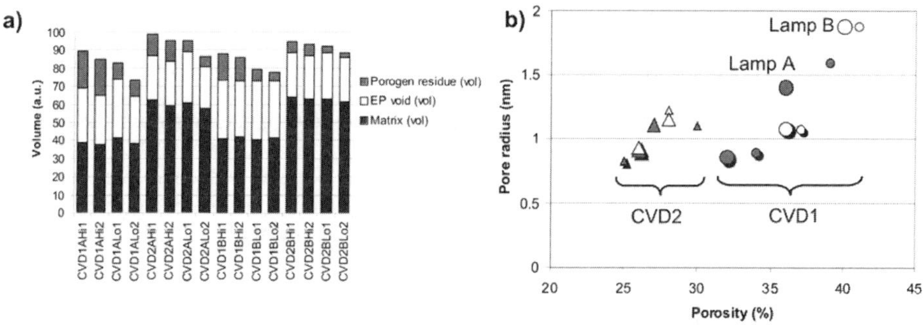

Figure 3 a) Estimated compositions of the samples recalculated to arbitrary volume using the measured shrinkage. **b)** The pore radius and the total open porosity of the samples measured by ellipsometric porosimetry with toluene. For the symbol description see caption of figure 1.

Figure 3a presents the results of estimation of the composition using the porosity measured by EP and the VIS/UV spectra (both refractive index and extinction coefficient). The results were obtained in the form of volume percentages and were recalculated to arbitrary volume quantities applying the measured shrinkage. Therefore, the total heights of the bars in the graph represent the relative thicknesses (i.e. volumes) of the samples and the three parts of the bar represent the individual constituents are in a comparable way. We observe basic difference between the matrix volumes of the material CVD1 (deposited with less matrix precursor and more porogen) and CVD2 (more matrix precursor, less porogen), while the volume of the matrix remains more or less constant during the cure and the shrinkage is occurring through reduction of porogen and void volume. However, there are small differences in the matrix volumes between the samples of the same material. In most of the cases, the matrix volume is reduced with increasing curing time, and the effect is stronger for lamp A, which supports directly the proposed mechanism [13] of shrinkage due to SiCH3 removal. The mean volumes (in arbitrary units) of the matrix for the four groups of samples with given material and curing lamps are CVD1A (38.9±1.7), CVD1B (41.0±0.8), CVD2A (60.0±2.0), CVD2B (62.8±1.2). These data quantitatively show stronger reduction of volume of the matrix for lamp A and greater changes from sample to sample (wider dispersion of values) as well.

It should be noted that the porogen removal (creation of pores) is competing with shrinkage of the samples (collapsing the pore volume). We observed reduction of porosity by 0-

2% for the longer curing time compared to shorter curing time for every given combination of material, deposition and curing lamp. Thus, for given moderate curing times, the shrinkage is the stronger effect and leads to collapse of the pores. However, for the full set of samples, the higher shrinkages are observed on samples with higher porosities, as mentioned before.

Ellipsometric porosimetry

The ellipsometric porosimetry gives the information about porosity percentage as well as the pore size distribution (PSD). For the given samples, we observe narrow PSD with the typical FWHM around 1 nm. In the figure 3b, we plotted the modus of the PSD versus porosity. It is observed that the material with higher porosity (less porogen residue) contains larger pores, That can be seen particularly in the behavior of material CVD1, which is deposited with more porogen precursor and therefore the removal (or residue) of porogen plays more important role.

CONCLUSIONS

All the observed facts might be summarized by following interpretation: The materials are deposited with "intrinsic" porosity around 40% for CVD1 and 60% for CVD2, and most of the space is occupied by porogen. During the UV-cure, the porogen is decomposed and significant part of it is removed. The material shrinks mainly through the pore collapse, while some shrinkage is observed for the matrix itself due to the replacement of $SiCH_3$ by SiH. There is also some rearrangement of SiO_2 backbone towards "network" structure. However, some of the porogen material stays in the pore interior leading to effectively smaller pores. The chemical structure of the residues is different from the as-deposited porogen. There is a transition from the hydrocarbon structure (detected as characteristic bands around 2900 cm^{-1} in the FTIR) towards carbon rich material (with aromatic or amorphous arrangement) giving rise to the absorption band in the ultraviolet (around 4 eV), which should be attributed to π-π^* electronic transitions of sp^2 bonded carbon [17], and also to C=C vibration at 1600 cm^{-1} in the infrared. This phenomenon has been already presented [2] for annealed porogen-based low-k's.

The results of our recent research presented in this paper and in [7,8,15,18], show that chemical and physical properties of low-k films can be controlled by wavelength of UV light. The critical threshold is near 190 nm, which was also proven by quantum-chemical calculations. UV light with $\lambda > 200$ nm allows to remove porogen without significant residue formation and also allow to avoid formation of SiH bonds that have higher polarisability in comparison with Si-CH3 bonds and generally more chemically reactive. However, it was found that chemical resistance of low-k materials to HF solutions and NH_3 plasma almost doesn't depends on presence of SiH groups.

ACKNOWLEDGMENTS

It is our pleasure to thank for the help and valuable contribution S.Eslava, D.Shamiryan, G.Beyer (IMEC), H.Sprey, J.Beynet; K.Matsushita, N.Tsuji, S.Kaneko (ASM) and S.Naumov (IOM).

REFERENCES

1. K. Maex, M. R. Baklanov, D. Shamiryan, F. Iacopi, S. H. Brongersma, and Z. S. Yanovitskaya, Journal of Applied Physics **93** (2003) 8793.
2. S. M. Gates, D. A. Neumayer, M. H. Sherwood, A. Grill, X. Wang, and M. Sankarapandian, Journal of Applied Physics **101** (2007) 094103.
3. A. Grill and V. Patel, Journal of Applied Physics **104** (2008) 024113.
4. A. Zenasni, V. Jousseaume, P. Holliger, L. Favennec, O. Gourhant, P. Maury, and G. Gerbaud, Journal of Applied Physics **102** (2007) 094107.
5. N. Kemeling, K. Matsushita, N. Tsuji, K. Kagami, M. Kato, S. Kaneko, H. Sprey, D. de Roest, and N. Kobayashi, Microelectronic Engineering **84** (2007) 2575.
6. F. Iacopi, Y. Travaly, B. Eyckens, C. Waldfried, T. Abell, E. P. Guyer, D. M. Gage, R. H. Dauskardt, T. Sajavaara, K. Houthoofd, P. Grobet, P. Jacobs, and K. Maex, Journal of Applied Physics **99** (2006) 053511.
7. S. Eslava, F. Iacopi, A. M. Urbanowicz, C. E. A. Kirschhock, K. Maex, J. A. Martens, and M. R. Baklanov, Journal of the Electrochemical Society **155** (2008) G231.
8. L. Prager, P.Marsik, J.W. Gerlach, M.R. Baklanov, S. Naumov, L. Pistol, D. Schneider, L. Wennrich, P. Verdonck, M.R. Baklanov, Microelectronic Engineering **85**(2008) 2094.
9. L. Favennec, V. Jousseaume, G. Gerbaud, A. Zenasni, and G. Passemard, Journal of Applied Physics **102** (2007) 064107.
10. J. Humlicek, in: H. Tompkins, E. Irene (Eds.), Handbook of Ellipsometry, William Andrew Publishing, New York, 2005
11. J. Humlicek, Analysis of spectrum line profiles. analysis of spectrum line profiles, FOLIA Fac. Sci. Nat. UP Brunensis, Physica 36 (1984).
12. M. R. Baklanov,K. P. Mogilnikov, V. G. Polovinkin, F. N. Dultsev, J.Vac.Sci.Technol. **B18** (2000) 1385-1391.
13. J. Ushio, T. Ohno, T. Hamada, S. I. Nakao, K. Yoneda, M. Kato, and N. Kobayashi, Japanese Journal of Applied Physics Part 2-Letters & Express Letters **46** (2007) L405.
14. P. Marsik, article in preparation
15. S. Eslava, G. Eymery, P. Marsik, F. Iacopi, C. E. A. Kirschhock, K. Maex, J. A. Martens, and M. R. Baklanov, Journal of the Electrochemical Society **155** (2008) G115.
16. D. E. Aspnes, Thin Solid Films **89** (1982) 249.
17. M. Gioti and S. Logothetidis, Diamond and Related Materials **12** (2003) 957.
18. A. M. Urbanowicz, B. Meshman, D. Schneider, and M. R. Baklanov. phys. stat. sol. (a) 205, No. 4 (2008) 829.

Novel Oxygen-free Barrier SiC Film (k<3.5) with High Etching Selectivity

J. Nakahira, *Y. Inaishi, S. Nakao, *M. Shinriki, E. Soda, K. Tomioka, S. Chikaki, N. Oda, *S. Hasaka and S. Kondo
Semiconductor Leading Edge Technologies, Inc., 16-1 Onogawa, Tsukuba, Ibaraki, 305-8569, Japan
* Taiyo Nippon Sanso Corporation, 10 Ohkubo, Tsukuba, Ibaraki, 300-2611, Japan
E-mail: nakahira.jyunya@selete.co.jp, Phone: +81-29-849-1480, Fax: +81-29-849-1452

ABSTRACT

A novel Oxygen-free SiC (k<3.5) film was developed for Cu dielectric barrier films. The etching selectivity of the film against low-k film was 40% larger than that of conventional SiCO film, while adhesion strength on Cu film and Cu diffusion rate were kept equal.

INTRODUCTION

For Cu multilevel interconnects of 45-nm-node and higher devices, a decrease in the dielectric Cu barrier constant of SiC films is required, as well as intrinsic low-k films, i.e. porous SiOC films [1][2]. High etching selectivity of low-k film against SiC film is also indispensable for minimizing the thickness of SiC film. Recently, a plasma-enhanced chemical vapor deposition (PE-CVD) SiCO film with a low dielectric constant of 3.7 was developed [3]. However, as the dielectric constant of this film decreases, its composition becomes similar to intrinsic low-k film (SiOC), and the dry-etching selectivity of low-k film against SiCO film tends to decrease.

A novel barrier SiC film, oxygen-less SiC (OL-SiC), was developed using 1-1, divinyl-silacyclopentane (DVScP) precursor to obtain a lower dielectric constant and high etching selectivity of porous low-k film against Cu barrier dielectric film. Because the precursor contained no oxygen and no oxidizer was used in the PE-CVD process, its composition is quite different from conventional SiC film.

EXPERIMENTAL DETAILS

Figure 1 and Table 1 show the chemical structure and the property of the DVScP precursor, respectively. As shown in Figure 2, the vapor pressures of DVScP and tetraethyl orthosilicate (TEOS) are similar. Therefore, DVScP is suitable precursor for PE-CVD.

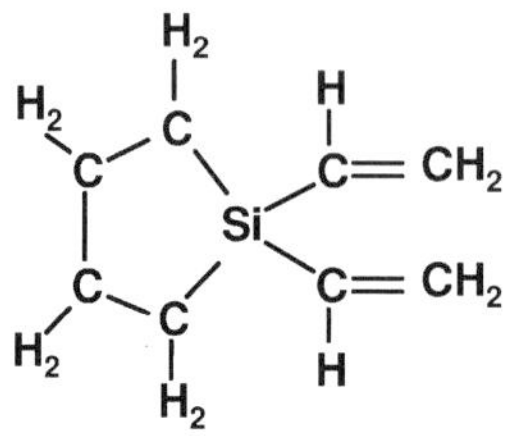

Figure 1. Structure of DVScP.

Table 1. Properties of DVScP.

Boiling Point	65～69℃ (4.4kPa)
Phase @RT, 1aim	Liquid

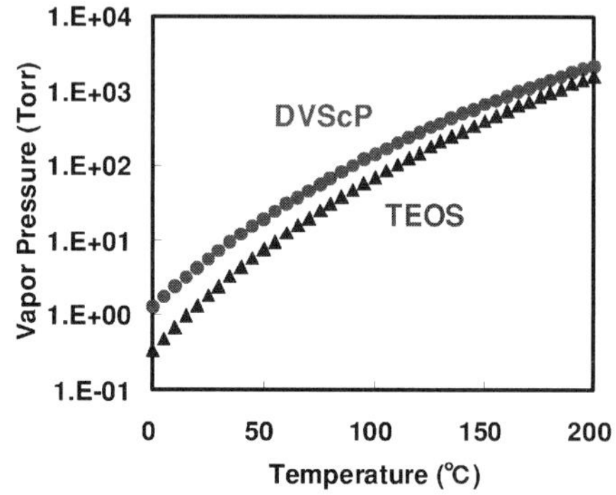

Figure 2. Vapor pressure of DVScP and TEOS.

Figure 3 and Table 2 show the OL-SiC CVD system and deposition conditions, respectively. OL-SiC films were deposited using a conventional PE-CVD system with a vaporizer to supply DVScP gas. As shown in Table 2, no oxidizer gas was used to deposit these films.

Rutherford backscattering spectrometry (RBS), FT-IR spectra, Young's moduli, and dielectric constants were measured using 200-nm-thick blanket films.

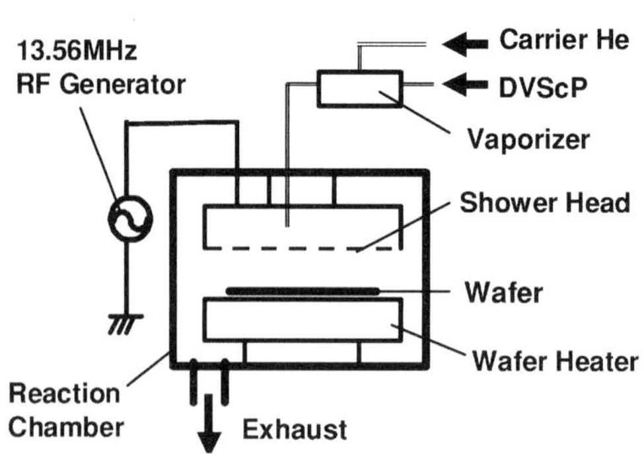

Figure 3. Schematic of OL-SiC CVD System.

Table 2. Deposition conditions of OL-SiC.

Gas	DVScP 10sccm
Pressure	0.4 Torr
Temperature	350°C
RF Power	100~300W

Adhesion strengths of intrinsic SiC films on Cu films were evaluated using a stud-pull test. To improve adhesion strength, glue layers and conventional SiCN (k:5.5) films were used under the SiC films.

OL-SiC, SiCO, and SiCN films, each with a 50-nm thickness, were deposited on chemical-mechanical polished Cu. SiOC (k:2.7) films with 100-nm thickness were deposited on each film, and annealing was performed at 400°C at 1 atm in N_2 ambience for 30 minutes. To evaluate Cu diffusion into the SiC films, Cu concentration depth-profiles were estimated with Cu intensity depth-profiles from a secondary ionization mass spectrometer (SIMS).

The electrical characteristics were evaluated using 200-nm-pitch Cu interconnects. The interconnect structure is shown in Figure 4. The wiring capacitance and line-to-line leakage current were also evaluated.

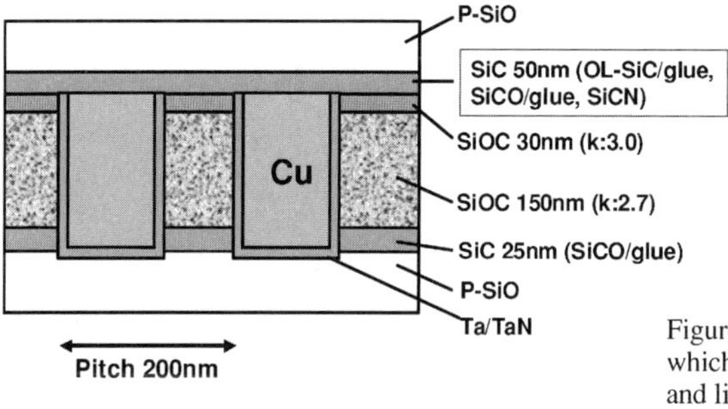

Figure 4. The structure of interconnects which were fabricated to wiring capacitance and line-to-line leakage current.

RESULTS AND DISCUSSION

Basic properties of blanket SiC films were evaluated on a Si substrate. As the FT-IR spectra shows in Figure 5, a large area of a Si-O-Si peak was observed in the spectrum of the SiCO film, and the same peak was not observed in that of the OL-SiC film. On the other hand, other peaks related to carbon, such as C-H, C=C, and Si-$(CH_2)_n$-Si were observed in the spectrum of the OL-SiC film. There are no Si-O-Si bonds contained in OL-SiC. As shown in Table 3, OL-SiC contains 4 times the number of carbon atoms than that of silicon.

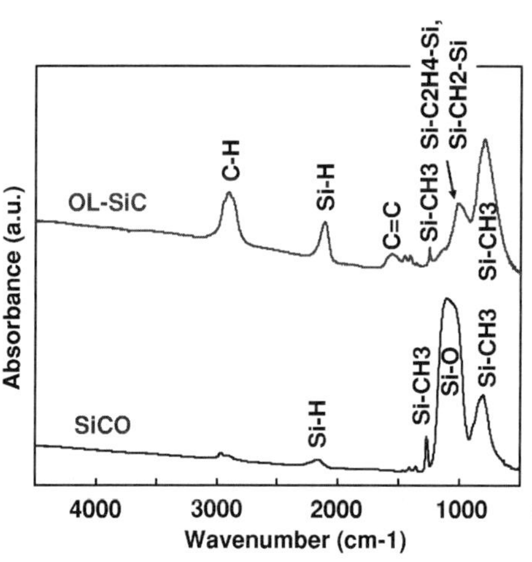

Figure 5. FT-IR Spectra of OL-SiC and SiCO.

Table 3. Compositions of OL-SiC and SiCO by RBS.

	C/Si	O/Si
OL-SiC	3.9	-
SiCO	1.2	1.1

Figure 6 shows the molecular structures of our proposed OL-SiC and SiCO films. In Figure 6(a), the main network of OL-SiC is Si-C_2H_4-Si and another minor network is Si-CH_2-Si. In Figure 6(b), the main network of SiCO is Si-O-Si and another minor network is Si-CH_2-Si.

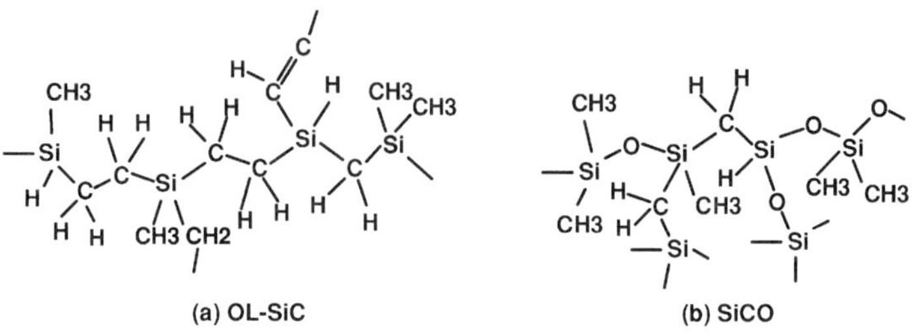

Figure 6. Proposed structures of OL-SiC and SiCO.

Figure 7 shows the dielectric constants and the Young's [moduli of the OL-SiC and SiCO films. At an optimized condition, the dielectric constant of the OL-SiC film was 5% lower than that of the SiCO film with the same mechanical strength. This is attributed to the mechanical strength of the SiC film strengthened by the Si-$(CH_2)_2$-Si networks [4].

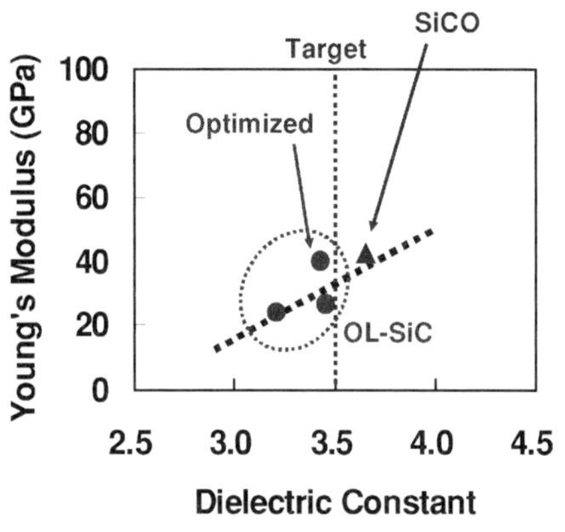

Figure 7. Relationship between Young's modulus and dielectric constant.

Next, the adhesion strength of the SiC films was evaluated. The yield converted from the film-peeling rate is shown in Figure 8. Film peeling occurred when the adhesion strength was less than 50 MPa at the interface between SiC and Cu films. As the Young's modulus of the OL-SiC film increases, the adhesion strength improves. At the optimized deposition condition, the adhesion strength is about the same as that of the SiCO film.

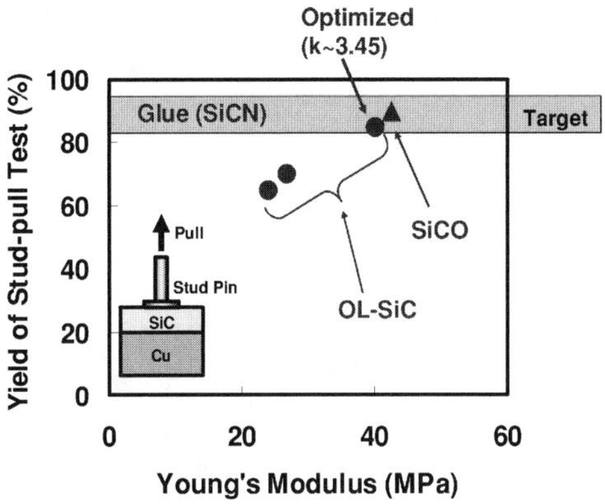

Figure 8. Yield of stud-pull testing as a function of Young's modulus. The peeling rate is converted into the yield.

The Cu diffusion rates during annealing, estimated with Cu intensity depth-profiles of SIMS, are shown in Figure 8. The Cu diffusion rate into the OL-SiC film was similar to that into the SiCO film.

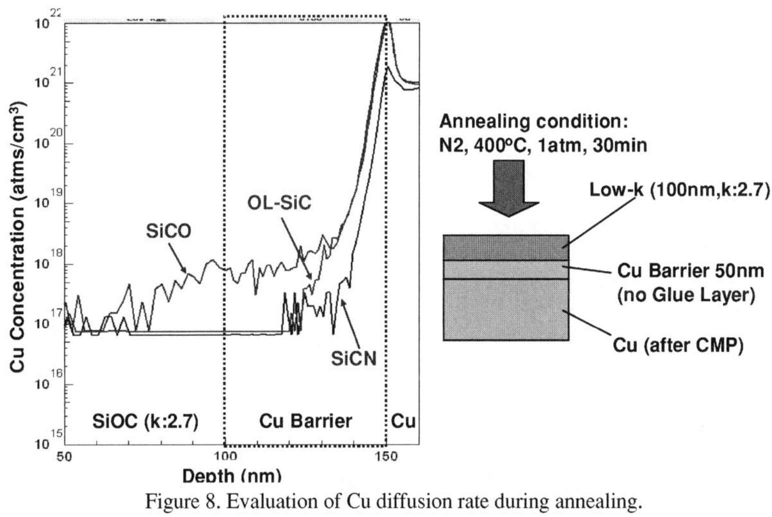

Figure 8. Evaluation of Cu diffusion rate during annealing.

Figure 9 shows the etching rates and the selectivities of various barrier SiC films against low-k (SiOC, k:2.7) films. Mixed etching gas containing CF_4 and Ar was used. The etching rate of the OL-SiC film was approximately 70% of that of the SiCO film. As shown in the transmission electron microscopy (TEM) image of the via bottom, the over-etched depth in the OL-SiC film was obviously suppressed. As the thickness of the SiC film becomes thinner to follow the device scaling, the SiCO film becomes easily over-etched and degrades via reliability. Therefore, this highly selective OL-SiC film is useful for next generation devices.

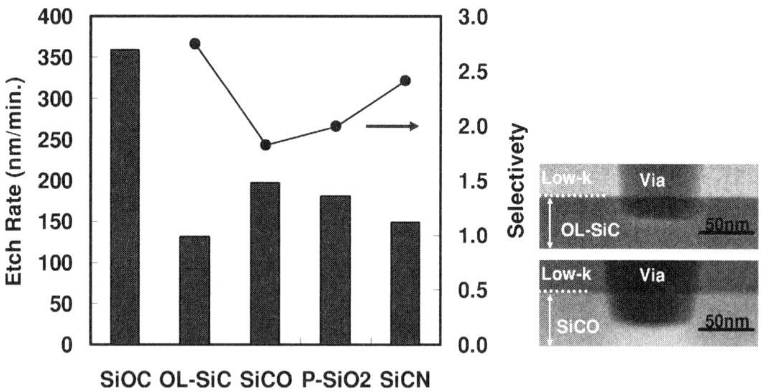

Figure 9. Comparison of the etching rates.

Finally, wiring capacitances were compared between three barrier SiC films, as shown in Figure 10. Cu interconnect sizes were measured using a cross sectional TEM image, and wiring capacitances were simulated using these sizes. As shown in Figure 11, it was confirmed that the Cu barrier SiC film with a low dielectric constant showed a large reduction in wiring capacitances. Although the SiC films include glue layers, their thickness is thin enough to maintain a reduced dielectric constant of the entire SiC film (k<3.5).

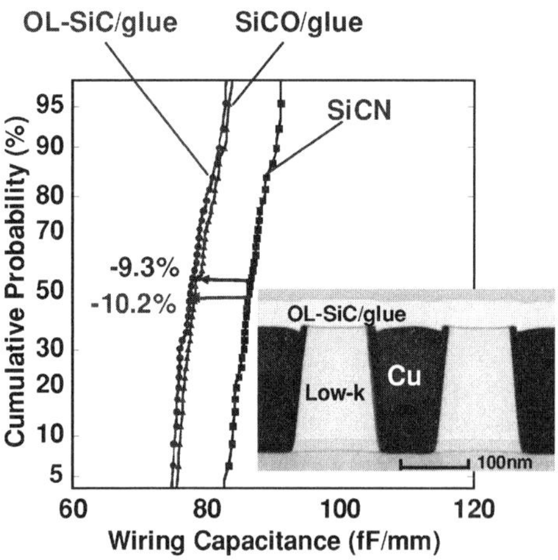

Figure 10. Wiring capacitance reduction by OL-SiC and cross sectional TEM image of Cu interconnects with OL-SiC.

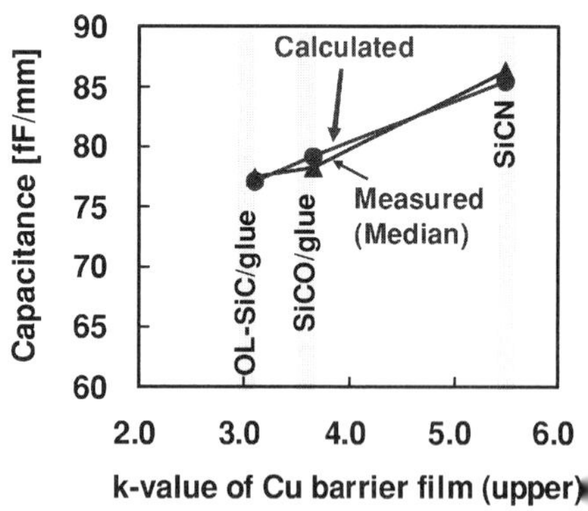

Figure 11. Comparison of measured and calculated wiring capacitances.

Figure 12 shows line-to-line leakage currents in Cu interconnects, 204 mm of line length, with a variety of SiC films. From the viewpoint of brake-down voltage, none of the films exhibited any problems.

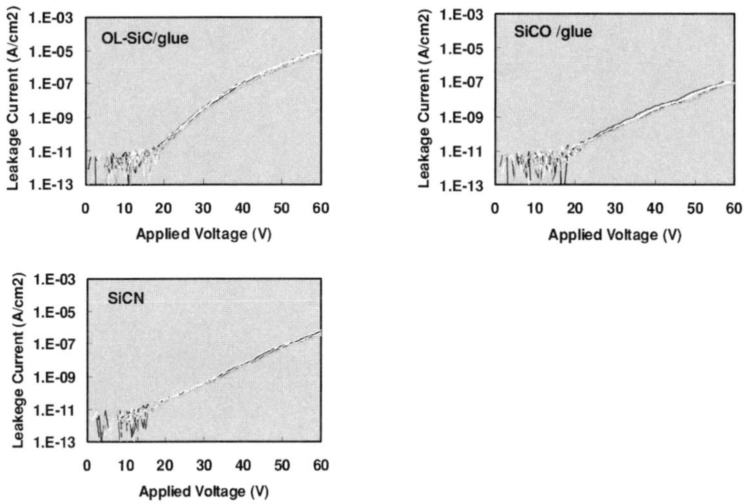

CONCLUSIONS

We developed an OL-SiC (k<3.5) film with a high etching selectivity against low-k films. By increasing Young's modulus, the adhesion strength of OL-SiC on a Cu surface was improved to be equal to that of SiCN film. Therefore, OL-SiC is a promising material as a Cu barrier dielectric layer for 45-nm-node and higher devices.

ACKNOWLEDGEMENTS

The authors would like to thank Y. Seino of the Nanodevice Innovation Research Center for his helpful advice and mechanical strength evaluation of the SiC films.

REFERENCES

[1] M. Bohr, IEEE IEDM. Tech. Dig. (1995) pp.241-244.

[2] M. Ueki, *et al.* IEEE IEDM. Tech. Dig. (2007) pp.973-976.

[3] K. Yoneda, *et al.*, Proc. IITC (2005) pp.220-222.

[4] N. Tajima, *et al.*, Proc. SSDM (2006) pp.152-153.

Conference Proceedings AMC XXIV © 2009 Materials Research Society

Mechanism of plasma damage resistance on porous MSX film with methylene bridge

M. Shimada, K. Watanabe, A. Gawase, H. Miyajima and H. Yano
Process & Manufacturing Engineering Center, Semiconductor Company, Toshiba Corporation
8, Shinsugita-cho, Isogo-ku, Yokohama, Kanagawa 235-8522, Japan
Tel:+81-045-776-4580 Fax:+81-45-776-4100 e-mail:miyoko.shimada@toshiba.co.jp

ABSTRACT

The existence of methylene bridge (Si-CH_2-Si) structure is very effective for improving plasma resistance in porous methyl-siloxane (MSX) film. In this paper, we show its mechanism by investigating porous MSX films with k=2.0. It seems that the sacrifice of the Si-CH_2-Si protected Si-CH_3 bond from vacuum ultraviolet (VUV) lights attacks and therefore k-shift due to decomposition of Si-CH_3 are restrained in the dry etching process.

INTRODUCTION

For a high-end logic device inter layer dielectric (ILD) film in Cu interconnects beyond the 32nm node, various porous low-k films (k=2.0-2.2) have been studied. However, porous low-k films have an inherent disadvantage in the film property. Plasma process induced "damages" have been a severe problem. "Damage" here means deterioration of film properties affected by changes of film structures. Damages induced by plasma processes which decompose Si-CH_3 and C-H bonds in the SiOCH films have been widely known to cause moisture absorption. Consequently, the dielectric constant increases, and it might affect performances and reliability of the interconnects. For this damage problem, we developed porous SOG material based on a new concept[1]. The combination of methylene bridge structure and electron beam (EB) post cure achieved highly secure porous MSX film for a 45 nm node. It showed good balance of plasma resistance and mechanical strength. The existence of methylene bridge in film is very effective for improving plasma resistance. However its mechanism has been unclear.

In this paper, porous MSX films for a 32nm node have been investigated to reveal a role of methylene bridge structure to improve plasma resistance.

EXPERIMENTAL

Two types of porous MSX low-k dielectrics (type A and type B with almost equal k-value 2.0) were prepared [2]. Type A is a conventional MSX film and type B is a new concept MSX film with methylene bridge structure (Figure 1).

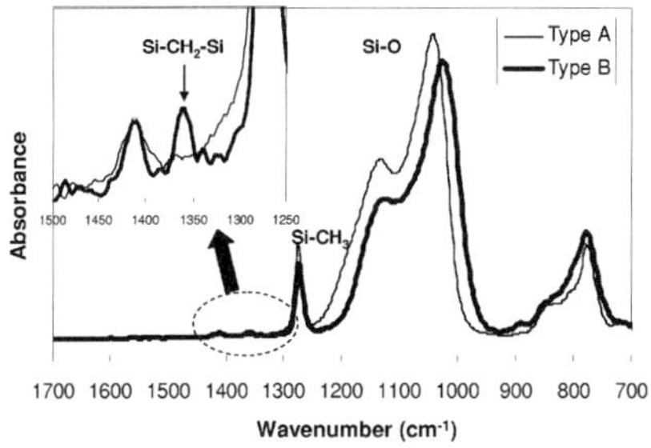

Figure 1
FT-IR spectra of type A and type B film.

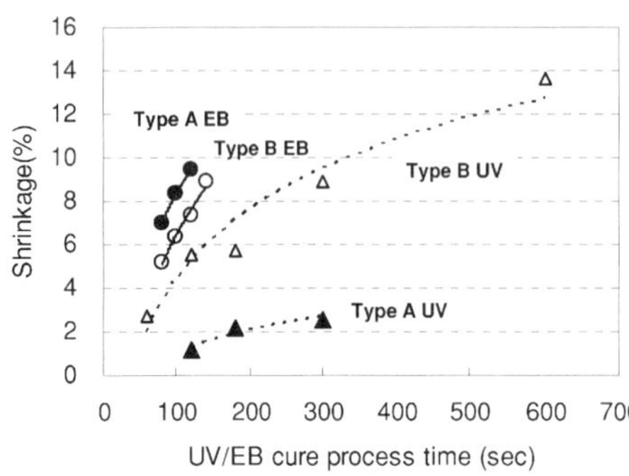

Figure 2 Film shrinkage as function of UV and EB cure time.

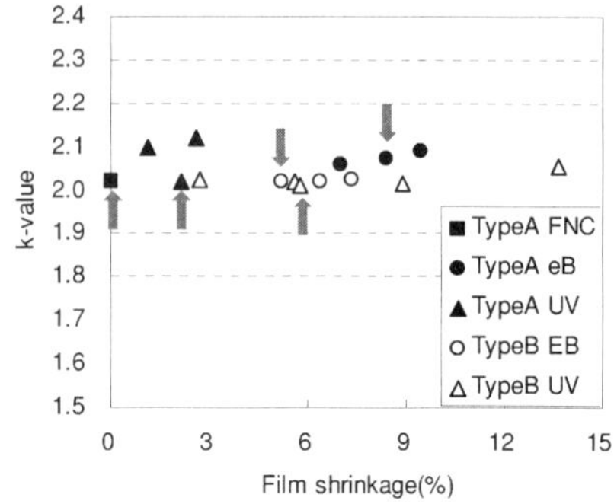

Figure 3 k-value versus film shrinkage.

Porous MSX films were prepared on 300mm wafers with cure treatment after spin coating. A cure method was chosen effectually from thermal method by furnace, EB method [3], ultraviolet light (UV) method [4] because there is a strong relationship between plasma resistance and cure method, initial k-value (before plasma treatment). As shown in Figure 2, film shrinkages of type A and type B exhibit various behavior for EB and UV cure. Figure 3 shows the k-value as a function of film shrinkage. In this study, films with almost the same k-value were chosen (arrow in Fig.3). The cured blanket porous MSX films were exposed to N/H RIE plasma to investigate plasma resistance. The dielectric constant (k-value) was evaluated by Hg probe, and structural properties were determined by Fourier transform infrared (FT-IR) method before and after N/H plasma treatment.

RESULTS AND DISCUSSION

Figure 4 shows k-values and k-shifts in various cure methods for type A and type B films. Here k-shift was defined as increase of k-value after N/H RIE plasma treatment. The smaller the k-shift, the higher the resistance for RIE plasma. It is obvious that the type B films exhibit excellent plasma resistance compared with type A films in all cure methods. It is shown that a methylene bridge is effective for reducing plasma damage in k=2.0 porous MSX film, as in k=2.3 film [1].

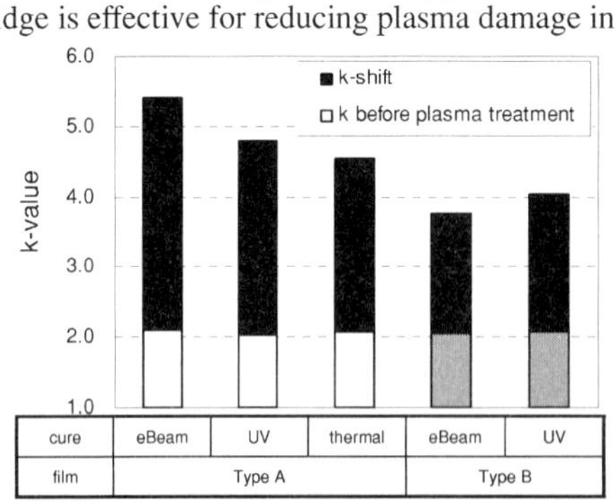

Figure 4
k-values and k-shifts in various cure methods for type A and type B films.

Figure 5 shows k-shifts versus normalized 1270cm^{-1} Si-CH_3 peak intensity in FT-IR spectra. In type B, UV cured films show larger k-shifts than EB cured film, though more Si-CH_3 bonds exist in film before RIE. Moreover, type A film cured by furnace or UV and type B film cured by UV contain almost the same amount of Si-CH_3 bonds, however k-shifts are smaller in type B. From Fig.5, k-shift can not be controlled exactly by the amount of Si-CH_3 bond in cured film. Figure 6 shows k-shifts versus normalized 1360cm^{-1} Si-CH_2-Si peak intensity. It is obvious that k-shift reduces as the Si-CH_2-Si bond increases in cured film. Figure 7 shows Si-CH_3 bond reduction by N/H plasma treatment versus Si-CH_2-Si peak intensity (a) and k-shifts versus Si-CH_3 bond reduction (b). Type A film cured EB especially exhibits large k-shift due to many Si-H bonds by EB cure that is a killer bond (Fig.7(c)). It is shown that Si-CH_3 bond reduction decreases and k-shift is restrained as the Si-CH_2-Si increases in as-cured film. From this result, the existence of Si-CH_2-Si in film suppress the Si-CH_3 reduction due to N/H plasma treatment, therefore k-shift is restrained.

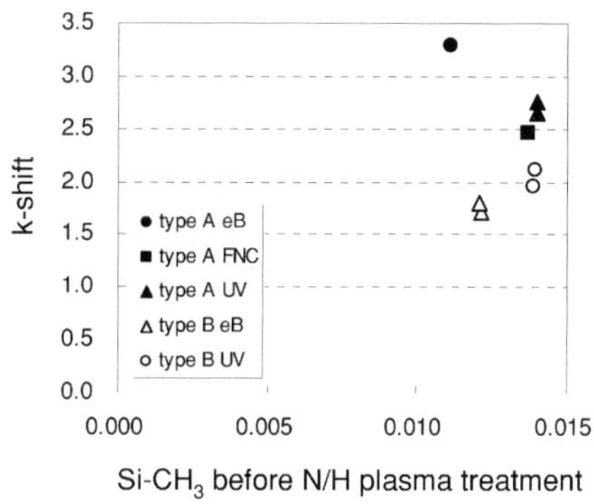

Figure 5 k-shift versus Si-CH_3.

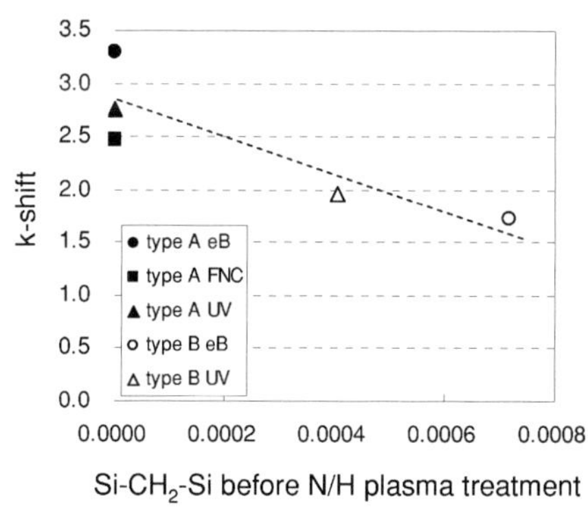

Figure 6 k-shift versus Si-CH_2-Si.

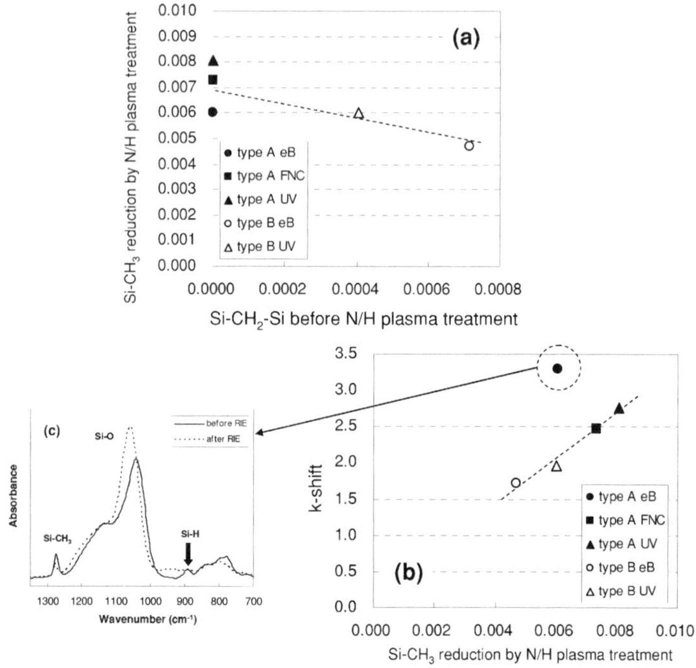

Figure 7 (**a**) Si-CH_3 bond reduction by N/H plasma treatment versus Si-CH_2-Si peak intensity. (**b**) k-shifts versus Si-CH_3 bond reduction . (**c**) FT-IR spectra of type A film cured by EB.

Hori reported that as well as ions and radicals, radiant lighting, especially VUV lights damage low-k film in the dry etching process [5]. The results in Fig.2 suggest the Si-CH_2-Si bond is very sensitive to VUV, so VUV lights are absorbed selectively in the Si-CH_2-Si bond. Figure 8 shows k-shift versus Si-CH_2-Si bond reduction by N/H plasma treatment. It seems that the sacrifice of the Si-CH_2-Si protected Si-CH_3 bond from VUV attacks and therefore k-shift due to decomposition of Si-CH_3, are restrained.

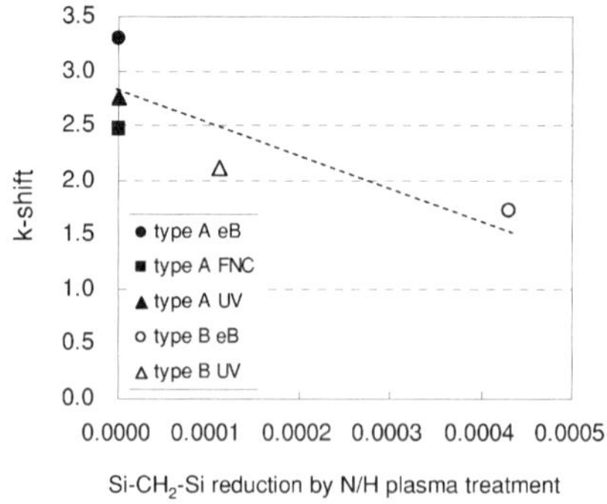

Figure 8
k-shift versus Si-CH_2-Si reduction.

SUMMARY

The methylene bridge structure is effective for reducing plasma damage in k=2.0 porous MSX film for a 32nm node, as in k=2.3 porous MSX film. The Si-CH_2-Si bonds play the role of "barrier" to VUV lights attack for Si-CH_3 bonds. The volume control of Si-CH_2-Si bond in as-cured film is important to improve the plasma damage resistance. The spin-on coating method might be easily controllable for methylene bridge structure.

ACKNOWLEDGEMENTS

The authors wish to thank JSR Corporation for providing its MSX prototypes for evaluation and fruitful discussions.

References

[1] H.Miyajima, AMC2006, p.37
[2] T.Watanabe et al., in Proc. IITC2008, p.208
[3] H.Miyajima et al, in Proc. IEDM204, p.222
[4] K.Fujita et al., in Proc. SSDM2005, p.298
[5] Hori , in Proc. SSDM2007, p.260

Conference Proceedings AMC XXIV © 2009 Materials Research Society

Effect of surface roughness on plasma-induced degradation of porous low-k material

T. Kirimura, *Y. Iba, S. Ozaki, Y. Kobayashi, and Y. Nakata
Fujitsu Laboratories Ltd., *Fujitsu Microelectronics Ltd.
1500, Mizono, Tado-cho, Kuwana-shi, Mie-ken, 511-0192, Japan
kirimura.t@jp.fujitsu.com

ABSTRACT

A hydrophobic property of porous low-k films is significantly degraded during plasma ash treatment in sequential processes of etching and ashing. We found that the ash damage for nano-clustering silica (NCS) is reduced by forming a smooth surface in previous etching process. When an NCS film with a smoother surface was exposed to O_2 plasma for resist ashing, there was less chemical transformation to the film than to an NCS film with a rougher surface; fewer methyl groups were extracted and fewer siloxane groups increased. By roughening the surface with the plasma etching, the NCS exhibited significant chemical transformation as a result of the O_2 ashing. The difference between the chemical transformations in the smooth and in the rough surface results from the difference in density of surface layer densified in the ashing. The dense layer inhibits further penetration of oxygen radicals into the bulk. The results indicate that it is possible to further reduce the plasma-induced damage in sequential etching and ashing steps.

INTRODUCTION

Porous low-k dielectrics are an essential component needed to construct multi-level Cu interconnects of ultra-large system integrations because they can reduce wiring delay that deteriorates the signal propagation performance. Below 45-nm technology nodes, the porous low-k materials should further contribute to the back-end of the line. The outstanding attribute of such materials is related to the chemical constitution of siloxane groups and methyl groups that determines its porous structure and its hydrophobic property. The hydrophobic property is changed to hydrophilic when the porous low-k films are irradiated by plasma used in a sequence of etching and ashing that form trenches and holes. The plasma for etching is excited with fluorocarbon gasses, and the plasma for ashing is based on O_2 gas. In the sequential etching and ashing steps, the O_2 plasma irradiation, in particular, causes significant loss of hydrophobicity, results in a large increase of dielectric constant. There has been a great deal of optimizing the process parameters as a way of minimizing the plasma-induced damage. In most cases reported to date, the parameters have been selected to reduce the damage at the individual etching and ashing steps [1-3], but it is not fully understood that how the etching affects next step of the ashing. A comprehensive analysis in the plasma processes is required to find a way to further reduce the damage. We demonstrate that the film degradation by O_2 plasma treatment depends on etching that has been done before. The effect is strongly related to the level of etch front roughening. The comprehensive reduction in the sequential steps of etching and ashing will improve the performance of low-k/Cu interconnect.

Plasma etching of porous low-k films induces a surface roughening. This feature is attributed to a plasma condition of inhomogeneous polymer deposition and ion irradiation with higher energy [4]. In the reaction, radicals produced in plasma fill the pores of the surface and generate a patchy polymerization that works as micro-mask for etching and results in a rough

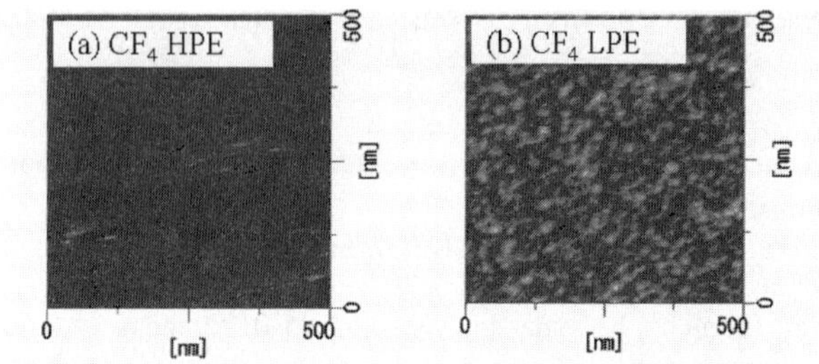

Fig. 1. AFM images of NCS surfaces after partial etchings. Images (a) and (b) were obtained after high-pressure etching (HPE) at 33 Pa and low-pressure etching (LPE) at 7 Pa, respectively.

surface morphology [4,5].

Following the etching, the porous low-k film is exposed to the O_2 plasma to strip resist mask. Oxygen gas is frequently used as a powerful stripping gas because it has higher ashing rate than other gasses such as H_2 and H_2/N_2 [1]. However, the ash treatment transforms the porous low-k film into a SiO_2-like film. The oxygen radicals penetrate into the bulk of the film through the pores, alter the bonding structure of the Si-O-Si in the film, and produce a denser region that gradually expands from the surface to the bulk with the treatment time [7,8].

In our experiment, we have found that for the low-k damage induced in a series of processes of etching and ashing, the porous low-k film with a smooth surface achieved by etching showed a higher barrier of the ash damage than that with a rough surface. In order to describe this phenomenon, we altered the surface roughness by adjusting the etching conditions. The densification reaction and the chemical transformations after the O_2 plasma exposure were then analyzed.

EXPERIMENT

We used the porous low-k material, nano-clustering silica (NCS) that exhibits a low dielectric constant ($k < 2.3$) and a high framework strength (Young's modulus = 10 GPa) [6]. We performed etch and ash treatments in a capacitively coupled plasma reactor operating at the radio frequency (RF) of 13.56 MHz. Blanket NCS film with the thickness of 150 nm was deposited on the Si wafer with spin coating method. The plasma etching with CF_4 gas was conducted in a gas flow rate of 280 sccm (standard cc / min) and an RF power of 100 W at a pressure below 40 Pa. To enhance polymerization during the etching, we used C_4F_6 gas in addition to the CF_4 gas. We generated O_2 plasma for ashing at an oxygen gas flow rate of 500 sccm, an RF power of 100 W, and a pressure of 7 Pa. The film thicknesses before and after etching were measured with a spectroscopic ellipsometer (KLA-Tencor F-5). The surface roughness of the NCS film was characterized by atomic force microscope (AFM) (JEOL JSPM-5400) after the plasma etching. The AFM measurement was performed in an AC non-contact mode at a scale of 500 nm. To measure capacitance and leakage current, an Au electrode was deposited on NCS film coated on a low resistive wafer that was an opposite electrode. To ensure that surface roughness does not affect the capacitance measurement, we compared the result with capacitance of NCS film with silicon carbide capping to obtain a flat surface. As the capacitance of the silicon carbide film is already known, we can estimate the value of only NCS film in the film stack. The depth profile of atomic concentration after etching and ashing was analyzed with secondary ion

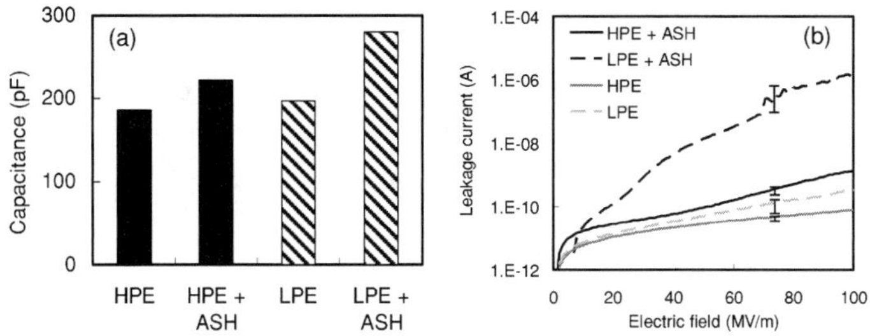

Fig. 2. Film capacitance (a) and leakage current (b) measured after etching and ashing. Etching was done with two conditions, HPE (smooth surface) and LPE (rough surface).

mass spectroscopy (SIMS). The measurement of film density was carried out by X-ray reflection (XRR) method using RIGAKU X-ray analysis tool. The chemical constitution after the etching and the ashing was characterized with Fourier transform infrared (FTIR) spectroscopy in transmission mode.

RESULTS

The surface of NCS film can be altered from rough to smooth depending on the gas pressure during plasma etching based on CF_4 gas. Atomic force microscope images after partial etchings in high and low pressure (33 Pa and 7 Pa) are shown in Figures 1 (a) and (b). In both cases, the etching depth was adjusted to 50 nm, and other process parameters were the same throughout the experiment. The AFM images show that etching under the high pressure forms smooth surface morphology, and in contrast there is considerable surface roughening in the low-pressure condition. The average roughness of the smooth and the rough surface is about 0.4 nm and 1.6 nm, respectively. Following the etchings, the films with the smooth and the rough surface were irradiated by O_2 plasma for 40 s.

To obtain electrical properties of films before and after the ashing, we measured the capacitance and the leakage current (Fig. 2). The capacitance of the film with the rough surface is almost the same as that with the smooth one, but subsequent O_2 plasma exposure provides a higher increase in the capacitance for the rough surface than that for the smooth one (Fig. 2a). A similar difference between the smooth and rough surfaces was observed for leakage current characteristics, shown in Fig. 2 (b). The results indicate that NCS film with the rough surface shows weak tolerance for an O_2 plasma attack.

To investigate the change of film caused by the ash treatment for both the surface morphologies, we measured the film density with the XRR method. For the film with the smooth surface, the XRR spectrum after the ashing was reproduced with double layer model that assumes that density of surface layer is denser than that of bulk layer, as shown in Fig. 3 (a). The density of each layer is listed in Table 1; the surface layer has a density of 2.1 g/cm^3 that is about two times larger than the value of the bulk layer, 1.1 g/cm^3. The surface layer seems to be altered in the SiO_2-like layer because its density is similar to that of silicon oxide film. For the NCS film with the rough surface, the model of densification by the ashing differs from that for the film with the smooth one. In the best fitting result, the XRR data was reproduced not with the double layer model but with a single layer one that has a rougher surface than that after HPE+ASH, as described in Fig. 3(b). As listed in Table 1, the density was turned out to be about 1.3 g/cm^3,

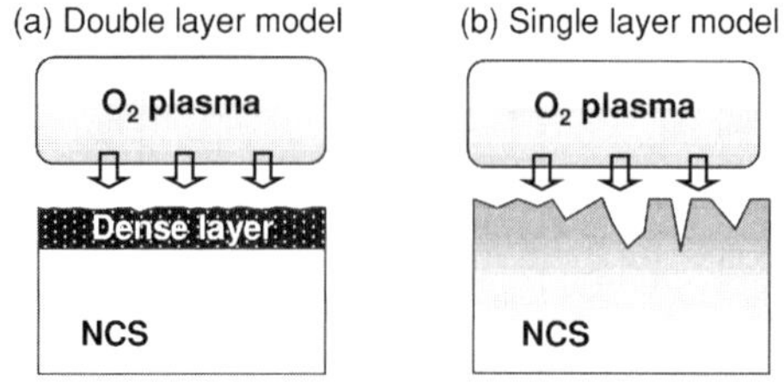

Fig. 3. Fitting models to analyze the XRR result: (a) Double-layer model with smooth surface used to fit film after HPE + ASH (b) Use of single-layer model with rough surface to fit film after LPE + ASH.

Table 1. Density, thickness and surface roughness of NCSs with and without the sequence of etching and ashing, analyzed with models (a) and (b) in Fig. 3. Note difference in densities of pristine film and the films treated in high- and low- pressure etching.

	Density (g/cm^3)	Thickness (nm)	Surface roughness (nm)
Pristine	1.1	150	0.9
HPE + ASH	2.1 (surface)	12 (surface)	1.2
	1.1 (bulk)	77 (bulk)	
LPE + ASH	1.3	89	2.1

which is higher than that of pristine NCS film. These results imply that depth profile of film density after ashing is largely different between the films with the rough and the smooth surface.

To extend the above observation of plasma ash damage for both the surfaces, we measured the depth profiles of atomic concentration before and after the ash treatment with SIMS. The depth profiles of the oxygen, the carbon, and the hydrogen concentration are shown in Figures 4 (a), (b) and (c), respectively. By comparing the profiles before and after the ashing, we see the increase in oxygen concentration and the loss of carbon and hydrogen in the vicinity of a surface. This feature indicates that Si-O-Si species are incorporated and Si-CH_3 species are removed because it agrees with the result obtained by FT-IR spectroscopy [1]. The depth profiles after the ashing have a significant difference between the rough and smooth surface while those before the ashing are strikingly similar between the two (except for the profile of LPE in hydrogen concentration; further experiment is necessary to answer why this profile was largely different from that of HPE). The depth profiles for oxygen show that oxygen atoms deeply

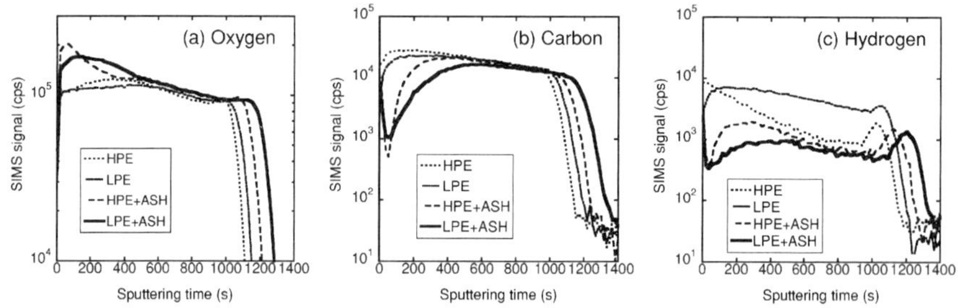

Fig. 4. SIMS profiles of NCS films after etching and ashing (ASH) measured for atomic concentration of (a) oxygen, (b) carbon, and (c) hydrogen. The etching was done with two conditions in which CF_4 high-pressure etching (HPE) and CF_4 low-pressure etching (LPE) are used to obtain smooth and rough surface morphology, respectively.

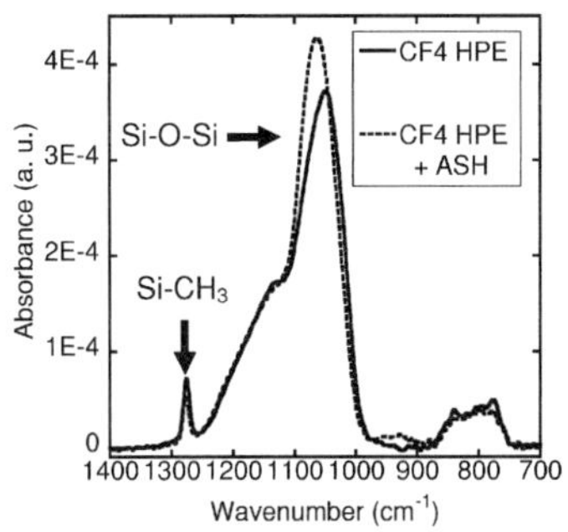

Fig. 5. Absorption spectra of NCS film after high-pressure etching (HPE) and ashing (ASH). Close-up view of peaks for methyl groups (Si-CH_3) and siloxane groups (Si-O-Si).

penetrate the NCS film with the rough surface compared to that with the smooth one. The concentrations of carbon and hydrogen are decreased by the ashing, and the reduction for the rough surface expands to deeper region of bulk than that for the smooth one does. These results demonstrate that the deeper penetration of oxidation provides larger degradation for the rough surface, and the etching condition sensitively changes a degree of subsequent ash damage.

To perform plasma etching of porous low-k film in widespread pressure range, we need to find other etching conditions that can achieve a smooth surface without higher pressure. We have test surface roughness after etchings for extensive parameters. After the test, we found a CF_4/C_4F_6 etch chemistry that can be used to smoothen the surface even when at a low pressure of 7 Pa. In this condition, the gas flow ratio of $CF_4+C_4F_6/C_4F_6$ was tuned to 30. To investigate the O_2-plasma resistance of film etched with the CF_4/C_4F_6 gas chemistry, we measured the ashing time dependence of the plasma-induced damage. To enable the ash damage to be characterized more easily, we used FT-IR spectroscopy that gives data for the amount of the Si-O-Si species and the Si-CH_3 species.

As shown in Fig. 5, the solid line shows an infrared absorption spectrum after etching, and dotted line is that after sequential etching and ashing. These spectra include absorption bands at around 1040 cm^{-1} and 1275 cm^{-1} that can be assigned to the Si-O-Si stretching mode and the Si-CH_3 vibration mode, respectively [10]. The O_2 plasma treatment shifts the absorption bands for the Si-O-Si species to a higher wavenumber. This peak shift is attributed to the change of bonding angle and/or bond length of Si-O-Si [11,12]. The change is also associated with the extraction of Si-CH_3 species [5,13]. In ash treatment, the diminishment of the Si-CH_3 and the incorporation of the Si-O-Si make the NCS hydrophilic, and the chemically damaged film absorbs moisture when exposed into an atmosphere. Although there is wide absorption band for hydroxyl groups between 3000 and 4000 cm^{-1} that show a degree of moisture uptake of the film, the poor signal to noise ratio in the FTIR spectra meant that there was no striking increase in the moisture below the ashing time of 100 s.

The ashing time developments of the Si-CH_3 degradation (Fig. 6a) and the Si-O-Si incorporation (Fig. 6b) are shown for three etching conditions: CF_4/C_4F_6 in 7 Pa (low pressure), CF_4 in 7 Pa (low pressure), and CF_4 in 33 Pa (high pressure). The NCS etched with CF_4 gas in 7 Pa loses the methyl groups at a higher rate than that etched in 33 Pa does, and a similar trend appears in for the Si-O-Si incorporation. Following the etching of CF_4/C_4F_6 in 7 Pa, the ashing

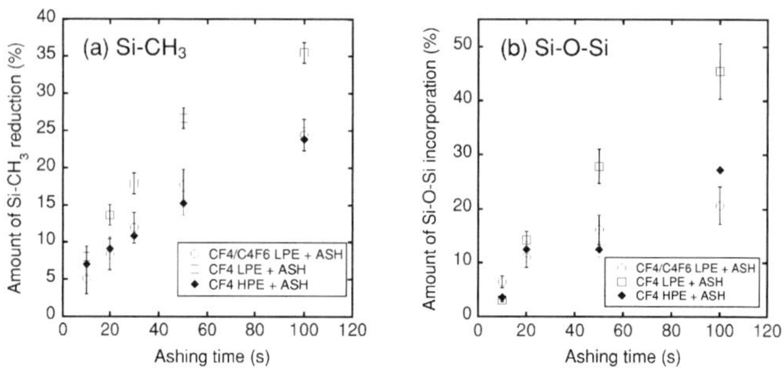

Fig. 6. Ashing-time dependences of Si-CH_3 extraction (a) and Si-O-Si incorporation (b). Note the different effects in three different etching conditions on O_2 plasma-induced damage: CF_4 HPE, CF_4 LPE, and CF_4/C_4F_6 LPE.

time dependence traces on that of CF_4 in 33 Pa, and the O_2-plasma resistance is higher than that of CF_4 in 7 Pa. This indicates that the smooth surface morphology lowers the chemical degradation. The result is significant because the etching of NCS will become available in wide pressure conditions.

DISCUSSION

In etching of porous low-k film, lower-pressures condition roughens the surface. This reaction is related to an interaction of surface with radicals and ions in plasma. When the pressure during fluorocarbon-gasses based etching is lowered, fewer polymers from C_xF_y radicals are deposited on the surface and the kinetic energy of ions becomes high. As seen in the AFM image, in Fig. 1 (b), the etching done in 7 Pa seems to provide deposition of polymers with sub-micron size on the NCS surface. The ion irradiation to wafer with the high energy induces further roughening of the surface while masking the polymers. In contrast, the etching in 33 Pa probably provides a uniform polymer deposition, and the ion bombardment with less energy compared to that in 7 Pa while keeping a flat surface morphology. The surface roughening agrees with previous result for C_2F_6/Ar etching chemistry of porous low-k films [4]. In that research, they observed that an increase of ion energy enlarges the surface roughness of porous low-k films. Using a mixture of CF_4 gas and deposition gas like C_4F_6 should be an effective way to reduce the roughness in the low-pressure etching. As shown in Yin et al's research, additional deposition gas modified the etching chemistry that fills up the pores of low-k films, and produced a flat surface [5]. The C_4F_6/CF_4 etching chemistry shown above also achieves a similar smoothing effect in surface.

We found that how NCS film is densified by O_2-plasma irradiation depends on the surface morphology. As indicated in the XRR result, when the surface is smooth, the density is partially increased. A similar densification at the surface layer is observed in SiOCH film with transmission electron microscope [7,8]. For the rough surface, in contrast, the densification does not follow such local-region formation. The corresponding XRR data show a different result that a whole film density is increased. From these results, we deduce that the dense layer formed in the vicinity of smooth surface resists further diffusion of oxygen radicals into the porous film, and, as a result, the amount of chemical damage is reduced as shown in the Si-CH_3 extraction and Si-O-Si incorporation in Fig. 6. Polymers in surface after etching would show little suppression of the ash damage for the smooth surface because etching based on CF_4 gas does not provide thick-polymer deposition enough to shut out the oxygen radicals. From our experiment of depth profile of carbon with X-ray photoemission spectroscopy, the thickness of polymers remaining at the smooth surface was estimated to be below 5 nm. As the polymers are removed by O_2 plasma exposure, if the deposition works as enough barrier of oxygen attack, the difference in ash damage between films with rough and smooth surface should be the largest at an early point in the ashing. However, the ashing time dependence of FTIR signals in Fig. 6 suggest that the spectra do not show such traces.

SUMMARY

We showed that O_2 plasma-induced damage is suppressed by adjusting the surface morphology during etching. The etching that achieves smooth surface morphology significantly lowers the ash damage to the porous low-k film. As the etching is done with CF_4 gas at a higher pressure or with a mixture of CF_4 gas and deposition gas, a smooth surface is obtainable. The surface morphology is determined by the deposition rate of polymers and the energy of ions during the plasma treatment. The ash treatment results in the NCS film with a smooth surface being partially transformed to a dense SiO_2-like material that reduces the pathways of penetration of oxygen radicals. Finding etching condition that smoothen a surface of post-etching is important in order to reduce the amount of low-k damage in subsequent ashing processes.

ACKNOWLEDGMENTS

We would like to thank Dr. M. Sasaki and Dr. T. Suzuki for measurements and for their helpful discussion.

REFERENCES

1. K. Yonekura, S. Sakamori, K. Goto, M. Matsuura, N. Fujiwara, and M. Yoneda, J. Vac. Sci. Technol. B, **22**, 548 (2004).
2. J. Bao, H. Shi, J. Liu, H. Huang, P. S. Ho, M. D. Goodner, M. Moinpour, and G. M. Kloster, J. Vac. Sci. Technol. B 26, 219(2008).
3. Y. Iba, T. Kirimura, M. Sasaki, Y. Kobayashi, Y. Nakata, and M. Nakaishi, Jpn. J. Appl. Phys. 47, 6923 (2008).
4. Y. Yin et al., J. Vac. Sci. Technol. B 24, 2360 (2006).
5. N. Posseme, T. Chevolleau, O. Joubert, L. Vallier, and N. Rochat, J. Vac. Sci. Technol. B 22, 2772 (2004).
6. I. Sugiura et. al., Microelectronic Engineering 82, 380 (2005).
7. D. Shamiryan et al., J. Vac. Sci. Technol. B 20, 1923 (2002).
8. A. M. Hoyas, J. Schuhmacher, C. M. Whelan, M. R. Baklanov, L. Carbonell, J. P. Celis, and K. Maex, J. Vac. Sci. Technol. B 23, 1551 (2005).
9. D. Eon, M. Darnon, T. Chevolleau, T. David, L. Vallier, and O. Joubert, J. Vac. Sci. Technol. B 25, 715 (2007).
10. The assignment of the peaks is shown in refs. [7] and [9].
11. Y. H. Kim, M. S. Hwang, H. J. Kim, J. Y. Kim, and Y. Lee, J. Appl. Phys. 90, 3367 (2001).
12. J. Y. Kim, M. S. Hwang, Y.-H. Kim, H. J. Kim, and Y. Lee, J. Appl. Phys. **90**, 2469 (2001).
13. N. Posseme, T. Chevolleau, O. Joubert, L. Vallier, and P. Mangiagalli, J. Vac. Sci. Technol. B **21**, 2432 (2003).

Conference Proceedings AMC XXIV © 2009 Materials Research Society

Recombination Probabilities of O and H Atoms on the Surface of Nanoporous Low Dielectric Constant SiOCH films

O.V. Braginsky[1], A.S. Kovalev[1], D.V. Lopaev[1], E.M. Malykhin[1], Yu.A. Mankelevich[1], T.V. Rakhimova[1], A. T. Rakhimov[1], A.N. Vasilieva[1], S.M. Zyryanov[1] and M.R.Baklanov[2]
[1]Skobeltsyn Institute of Nuclear Physics of Moscow State University, 119991 Moscow, Russia
[2]IMEC, Leuven, Belgium, baklanov@imec.be

INTRODUCTION

One of the most difficult challenges during the integration of porous low-k films in ULSI devices is their degradation during plasma treatments. Lost of the carbon containing hydrophobic groups occurs during strip and cleaning processes when exposed to process plasma. Although these effects are discussed in many papers [1], the damage mechanism has not been completely understood because of simultaneous influence of different factors such as ion bombardment, UV radiation etc. Recently [2], it was stated that the depth of damage depends on the diffusion length of active radicals into the pores. The diffusion length depends on porosity, pore size, tortuosity and recombination probability of radicals on pore wall. Understanding of the recombination mechanisms might play a key role in the reduction of plasma damage.

In this work, we studied surface recombination of H and O atoms in «pure» conditions, when ion and UV radiations are excluded. In these experiments low-k sample were placed in plasma downstream. Atoms recombinations were measured by plasma induced fluorescence technique based on the actinometry method.

EXPERIMENT

The experimental set-up is presented in fig. 1. Oxygen and hydrogen atoms were produced in flow of pure O_2 or H_2 at the pressure of 10 Torr after passing the area of 13.56 MHz capacitive discharge with external electrodes. The rf power inputted into the discharge plasma varied from 50 up to 250 W. The samples (SiOCH films – CVD1, CVD2, CVD3, of the 180 nm and 500 nm thickness deposited on a 300 mm silicon substrate of the 400 micron thickness, see Table I) were positioned in discharge afterglow on a boat-support made of teflon on the distance of ~9 cm from the end of electrodes. The gas flow velocity under the samples was about 14.8 km/s.

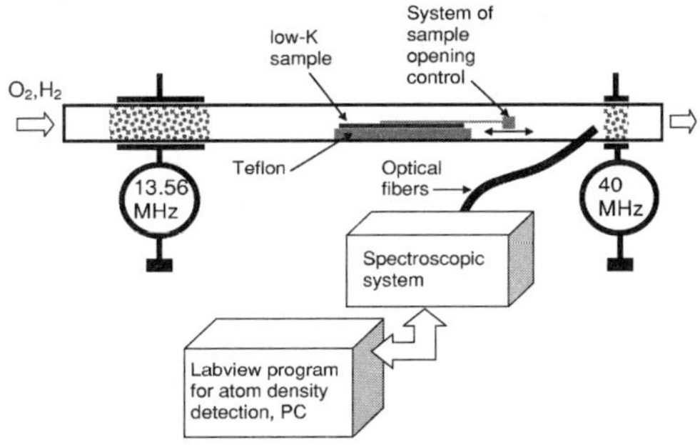

Figure 1. Experimental set-up

Table I. The studied nanoporous low-k materials.

Property	CVD1	CVD2	CVD3
Porosity (%)	24	24	33
Pore radius (nm)	0.8	0.8	0.9-1.0

To determine the density of O and H atoms in discharge afterglow we used the actinometry method on Ar atoms. For this purpose the gas flow after the 13.56 MHz discharge afterglow was «highlighted» by another low-power (~1-3 W) 40 MHz discharge to excite emission of atoms generated by the main discharge. The emission lines corresponding to the transitions in O and H atoms: $O(^5P) \rightarrow O(^5S)$ (777 nm), H (3P) $\rightarrow$ H (2S) (656 nm) as well as Ar atom: $Ar(2p_1) \rightarrow Ar(1s_2)$ (750 nm) were selected, and the ratio between their intensities and the Ar line intensity was detected (fig. 1). The signals corresponding to O and H atoms and produced by 13.56 MHz discharge was as a difference of actinometric signals in two cases when the main discharge was switched on and off. The motion of the auxiliary «highlighting» 40 MHz discharge along the tube allowed detecting an atom concentration change along 13.56 MHz discharge afterglow. Thus, the curves of O and H atom concentration changes were measured in the case, when no low-K samples are in the teflon boat-holder. As example, the curves of the oxygen and hydrogen dissociation degree in 13.56 MHz discharge afterglow are presented in figs. 2a and 2b.

Thus, the loss (recombination) probabilities of oxygen and hydrogen atoms on teflon, i.e. the «empty» boat surface, were obtained from the slops of the given curves in the boat area. The given experiments were carried out at two pressures of 3 and 10 Torr and the gas flow velocities of 3.2 and 7.4 m/s inside the tube. The O and H loss probabilities calculated from the atom density fall in the different conditions coincided with a good accuracy: $\gamma_O^{teflon} = (3\pm0.5)\cdot10^{-3}$ and $\gamma_H^{teflon} = (2\pm0.5)\cdot10^{-3}$. The high enough probabilities of atom recombination on the teflon surface are apparently caused by the fact that it was processed for fixing the samples on it.

Figure 2. Change of the oxygen (a) and hydrogen (b) dissociation degree in 13.56 MHz discharge afterglow. Position of the teflon boat is shown by the shaded rectangular.

The following approach was used to measure the loss probabilities on the low-K sample surfaces not exposing it to any particles except for the atoms. Atoms were detected in far afterglow behind the teflon sample holder. For it, the position of 40 MHz discharge was hard fixed (in the given experiment it was 34 cm from the end of electrodes of the main 13.56 MHz discharge). Further the sample holder was covered with a thin teflon «cover» that could move along the holder and thereby open the given sample surface. Depending on the size of the open sample area where the atom recombination occurs, the density of atoms in the gas flow behind the holder changes. From this change the atom loss probability on the studied material compared to some other material, the loss probability on which is known (in this case it is Teflon), can be extracted. So, the loss probability on low-K materials was determined from the following ratio:

$$\gamma_{O(H)} = \gamma_{O(H)}^{teflon} + \delta\gamma_{sample}, \qquad \delta\gamma_{sample} = A_{O(H)} \cdot \frac{4\upsilon_{flow}}{\upsilon_T} \cdot \frac{S_{flow}}{W_{sample}} \tag{1}$$

where $\gamma_{O(H)}^{teflon}$ is O or H atom recombination probability on teflon, υ_{flow} is a gas flow velocity above the holder, υ_T is a thermal velocity of O or H atoms relatively, S_{flow} is a gas flow cross-section above the sample holder, W_{sample} is a sample width, and A is a slop of the logarithm dependence of the ratio of actinometric signals for the teflon holder with the sample and without it on the sample length: $A_{O(H)} = \partial \ln[(I_{O(H)}/I_{Ar})_{teflon}/(I_{O(H)}/I_{O(H)})_{sample}]/\partial x$.

Since the gas flow velocity can be presented as $\upsilon_{flow} = R_{gas}/S_{flow}$, where R_{gas} – gas flow in cm^3/s at the given pressure, finally the difference in the atom recombination probabilities on studied and known (teflon) samples can be expressed as follows:

$$\delta\gamma_{sample} = A_{O(H)} \cdot \frac{4R_{gas}}{\upsilon_T W_{sample}} \tag{2}$$

As example, the dependences of $\ln\left[(I_{O(H)}/I_{Ar})_{teflon}/(I_{O(H)}/I_{Ar})_{sample}\right]$ on the CVD1 sample length (in fact the open space) for oxygen and hydrogen atoms are presented in figure 3a and 3b respectively. The zero line corresponds to Teflon.

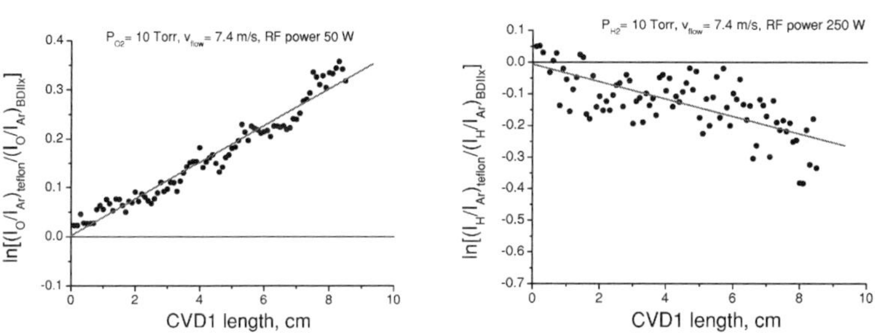

Figure 3. $\ln\frac{(I_{O(H)}/I_{Ar})_{teflon}}{(I_{O(H)}/I_{Ar})_{sample}}$ dependences on the CVD1 sample length (the open space in fact) for oxygen (a) and hydrogen (b) atoms. The zero line corresponds to teflon.

Measurements of the loss probabilities of both O and H atoms for all the low-K materials were carried out during short enough time, no more than ~30 s. The received thus and calculated by using (2) recombination probabilities of O and H atoms for three types of nanoporous low-K materials are presented in Table II. For each material all the experiments were repeated few times (3-5 times), the measurement statistics was presented as an error in Table II.

Table II. The recombination probabilities of oxygen (γ_O) and hydrogen (γ_H) atoms on the low-K surface.

	CVD1	CVD2	CVD3
γ_O	$(5.7\pm0.3)\cdot10^{-3}$	$(6\pm0.3)\cdot10^{-3}$	$(6.5\pm0.4)\cdot10^{-3}$
γ_H	$(1.4\pm0.4)\cdot10^{-3}$	$(1.6\pm0.6)\cdot10^{-3}$	$(3.2\pm0.8)\cdot10^{-3}$

FTIR MEASUREMENTS

FTIR spectra of some samples were measured before and after exposing them in O_2 and H_2 13.56 MHz discharge afterglow. The main conclusion that can be done on these measurements is that actually noticeable changes in the spectra are observed in the case of oxygen. As examples, the FTIR spectra of CVD pristine samples and the samples after the measurement procedure are shown in figures 4-6 correspondently. The reaction time of O and H radicals with low-k films were 45 sec for all data in figures 4-6.

CVD1 sample

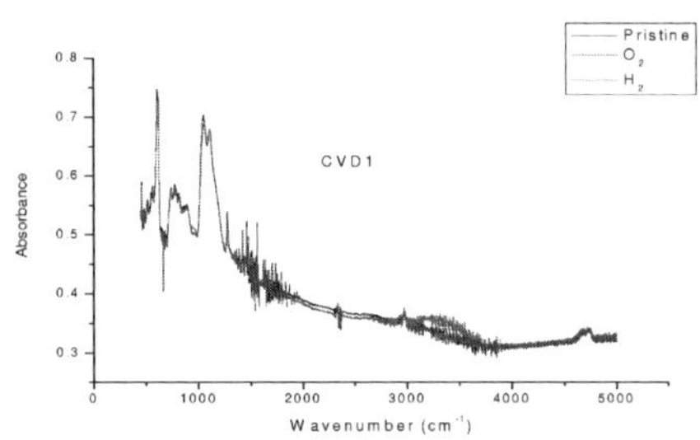

Figure 4. FTIR spectra of CVD1 sample before and after exposure in O_2 and H_2 plasma.

CVD1 sample

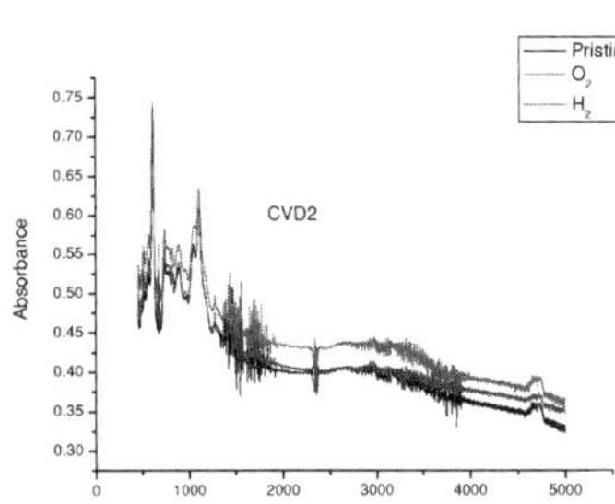

Figure 5 . FTIR spectra of CVD2 sample before and after O_2 and H_2 plasma afterglow.

CVD3 sample

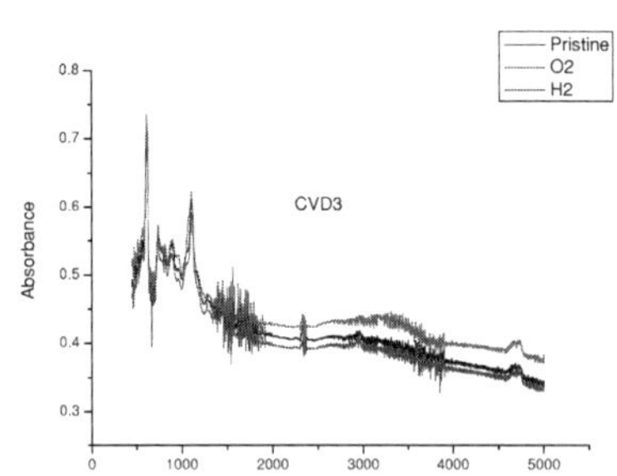

Figure 6. FTIR spectra of CVD3 sample before and after O_2 and H_2 plasma afterglow.

Table III. Relative change of absorption peak at 1275 cm^{-1}

1275 -1260	CVD1	CVD2	CVD3
Pristine	60	10	16
O_2	50	2	13
H_2	59	11	16

The most pronounced changes in FTIR spectra after the treatment in O_2 plasma afterglow is the appearance of hydroxyl bonds and adsorbed moisture that indicates the modification of low-k surface owing to the interaction with oxygen atoms. The low-K surface modification due to H atoms is noticeably slighter. In Table III a relative change of absorption peak at 1275 cm^{-1} (Si-CH_3) for studied films is shown. When exposure time was increased to 20 min, significant damage of the low-K samples by oxygen radicals was observed. In contrast to this, there is no noticeable damage in H_2 plasma afterglow for the same exposition.

MODELLING OF GAS-SURFACE REACTIONS OF O AND H ATOMS RECOMBINATION IN NANOPOROUS FILMS.

The modeling part includes 2 different tasks. First (1), the recombination coefficients were calculated using the kinetic data typical for pure SiO_2. Then (2), probability of O and H radicals consumption in reaction with Si-CH_3 groups were estimated. Comparison of these values with the experimental data allowed to make conclusions about the most important factors that define the depth of plasma damage.

1. Several approaches for study of gas-surface reactions (radical recombination, etching, adsorption, desorption) have been developed and used for different mediums including porous ones [3], silica surfaces without and with an added powder to increase the roughness factor [4]. An analytical model for atom diffusion and Eley-Rideal heterogeneous recombination on pore surfaces was applied in [3] for porous medium of 5 cm in thickness and, from our estimations, with pores of micron sizes. Such diffusion models have limited applicability, from our point of view, for low-k materials of hundreds nanometers in thickness with active near surface region of tens nanometers and pores sizes of about 1-2 nm. In our experiments, typical atomic concentrations is about $\sim 10^{15}$ cm^{-3} and thus average volume per one atom is $100\times100\times100$ nm and it seems to be not fruitful to consider a gas-phase diffusion approach and atomic concentration gradients on much smaller length of tens nanometers.

We developed numerical model of gas-surface reactions on surfaces with different roughness (depending on porosity), which could be covered by a H_2O layer. For a given flux of radicals (H or O), the kinetics of filled surface sites (physisorbed atoms PH or PO for the experiments in H and O, respectively), chemisorbed atoms (SH or SO), and free sites (P* and S*) is calculated numerically. Typical reaction mechanism (reaction rates R [cm^{-2}/s] and coefficients $k=A\cdot T^{B}\cdot exp(-E/T)$) involves following processes: atomic adsorption (reactions 1,2) and desorption (3,4), surface diffusion of physisorbed atoms to the S* sites (5), surface diffusion of physisorbed atoms to the SH sites and its recombination (Langmuir–Hinshelwood heterogeneous recombination (6)) and is presented in Table IV for hydrogen (X=H) and oxygen (X=O).

Exact values of the coefficients k, activation energies E and the total number of the surface sites S_{total}(H)=SH + S* and P_{total}(H)=PH + P* (S_{total}(O) and P_{total}(O) for oxygen) are not available for the surfaces under study. Instead, we have used physically reasonable values (obtained for pure silica), e.g. $S_{total} \sim 2\times10^{12}$ cm^{-2}, $P_{total} \sim 2\times10^{15}$ cm^{-2}, the adsorption reaction coefficients $k_1=k_2\sim10^{-11}$ cm^3/s, frequency factor $\sim10^{13}$ s^{-1} , activation energy $E \sim 2000$ K for desorption of physisorbed atoms (reactions 3) and $E \sim 18000$ K for chemisorbed atoms (reaction 4), and fitted surface diffusion coefficients for processes 5 and 6. More complex mechanism with additional quasi-chemisorbed atoms QH (QO) and sites Q* (activation energy $E \sim 4000$ K for desorption and $Q_{total}\sim10^{14}$ cm^{-2}) shows that dominant recombination process (recombination of physisorbed and chemisorbed atoms in reaction 6) is not changed in our conditions. Such approaches allow us to provide coherent picture of basic atomic loss/production processes in accordance with our experimental measurements. The numerical solution of the balance equations for our conditions (gas pressure 10 Torr, surface temperature 300K) and typical hydrogen and oxygen dissociation degrees ($[H]/[H_2] \sim 0.0003$ and $[O]/[O_2] \sim 0.005$) provides the following surface concentrations [cm^{-2}] for hydrogen and oxygen (Table V).

Table IV. Surface and gas-surface reactions for hydrogen and oxygen.

	Reaction	*A*	*B*	*E*[K]	*R* , X=H	*R*, X=O
1	P* + X → PX	6.0×10^{-13} [cm^3/s]	0.5	0.	2.23×10^{18}	3.49×10^{19}
2	S* + X → SX	6.0×10^{-13} [cm^3/s]	0.5	0.	1.04×10^{14}	1.63×10^{15}
3	PX → P* + X	10^{13} [1/s]	0	2000	2.21×10^{18}	3.46×10^{19}
4	SX → S* + X	10^{13} [1/s]	0	18000	1.65×10^{-1}	1.65×10^{-1}
5	PX + S* → P* + SX	0.02 [cm^2/s]	0	1000	1.15×10^{16}	1.80×10^{17}
6	PX + SX → P* + S* + X_2	0.02 [cm^2/s]	0	1900	1.16×10^{16}	1.82×10^{17}

Table V. Calculated surface concentrations in hydrogen and oxygen gas.

Hydrogen	Oxygen
S* = 9.31×10^{10}	S* = 9.31×10^{10}
SH = 1.91×10^{12}	SO = 1.91×10^{12}
P* = 1.93×10^{15}	P* = 1.99×10^{15}
PH = 1.73×10^{8}	PO = 2.72×10^{9}

As it seen from Table IV, there are balances of adsorption/desorption rates *R*. The rates of atomic loss in recombination reaction 6 from Table IV correspond to atomic loss probability $\gamma_H\sim0.0019$ and $\gamma_O\sim0.007$ for H and O atoms, respectively ($R_6=\gamma_X\cdot v_T\cdot n_X/4$, X=H,O). These low probabilities mean that more than 99% impinging atoms are reflected from the surface. In deference from non-porous medium and one (one-step) reflection, the atom impinging the porous medium could undergo several ($N\geq1$) reflections inside the porous medium before returning to the gas volume. We do not know average number N of such reflections and the depth of penetration and the surface area of pores where these reflections occur. The upper estimate of N is the average number N_{rec} of collisions needed for recombination $N_{rec}= 1/\gamma_O$. If we assume that the typical distance between the two reflections of O atom is about pore diameter *d* (*d*~2 nm in our case) we can estimate the penetration depth $L_{pen}\sim d\, N_{rec}^{0.5}\sim 24$ nm from random walk theory. This depth is in reasonable accord with the depth where CH3 group etching is observed experimentally for our low-k medium.

2. To estimate the loss probability of O and H radicals in their reaction with $SiCH_3$ groups, we considered possible gas-phase analogous reactions of the process $O + CH_3Si \rightarrow CH_3O + Si$ at T=300 K:

$O + CH_3OH \rightarrow OH + CH_2OH$, $k = 5.5e\text{-}15\ cm^3/s$ [5]

$O + CH_3OH \rightarrow OH + CH_3O$, $k = 7.7e\text{-}17\ cm^3/s$ [5]

The reactions with H atoms go on another channel (no CH_3 etching)

$H + CH_3OH \rightarrow CH_2OH + H_2$ $k = 1.27e\text{-}15\ cm^3/s$ [5]

From our FTIR data (reduction of the CH_3 peak I_{CH3} at 1275 cm^{-1} during exposure time t) we have estimated probability of CH_3 etching by O atoms:

$$I_{CH3}(t)/I_{CH3}(0) = \exp(-\gamma_{etch} \cdot v_T \cdot n_O \cdot t/(4 \cdot P_{total}))$$

For CVD1 sample this probability was $\gamma_{etchCVD1} \sim 3 \cdot 10^{-8}$, and for CVD2 sample $\gamma_{etchCVD2} \sim 3 \cdot 10^{-7}$.

One can see that the measured values of O and H radicals loss have the same order of magnitude with the values calculated for the recombination reactions. At the same time, probability of reaction of these radicals with Si-CH_3 groups is smaller or several orders of magnitudes.

CONCLUSIONS

Interaction of H and O atoms with different low-k films is studied. The loss probabilities of O and H atoms are measured in the far plasma afterglow by plasma-induced actinometry method. Modification of low-k films is studied by FTIR measurement. Model of O and H recombination probabilities on low-k is presented. The estimate of the CH_3 group etching probability by O atoms is done. As a result, it is concluded that the main mechanism of the O and H loss is the surface recombination. It is shown that reaction of these atoms with the carbon containing groups has a secondary order contribution. Therefore, the surface recombination defines profile of the radicals concentration in low-k films and, therefore, the depth of plasma damage.

Moreover, it has been shown that hydrogen radicals do not break Si-CH_3 bonds; therefore, the plasma damage can be avoided in case of the hydrogen plasma based resist strip in appropriate conditions. The last conclusion is in agreement with recent experimental observations [6].

ACKNOWLEDGMENTS

This study is carried out in the frame of MSU-IMEC Agreement and was supported by Key Science School (SS-133.2008.2).

REFERENCES

1. K. Maex et al. *J. Appl. Phys.* **93**, 8793 (2003).
2. M.R.Baklanov et al. *Proc. of 2006 8th ICSICT, Part 1*, p. 291, (2006).
3. J. Marschall, J.E. Boulter. *J. Phys. D.* **39**, 3849 (2006).
4. Y.C. Kim, M. Boudart, *Langmuir* **7**, 2999 (1991)
5. G.P. Smith, D.M. Golden et al, *GRI-Mech 3.0,* http://www.me.berkeley.edu/gri_mech/
6. N.Posseme et al. *J.Vac.Sci.Technol.* **B 25**, 1928 (2007).

Conference Proceedings AMC XXIV © 2009 Materials Research Society

Realization of High Stable Porous SiOCH Films (k=2.3-2.4, Er=8.0 GPa) using DcPDMOS by O_2 Addition and UV Cure

Yonghua Xu[1], Manabu Shinriki[2], Yoshiaki Inaishi[2], and Satoshi Hasaka[2]
[1] Tri Chemical Laboratories Inc., 8154-217 Uenohara, Uenohara, Yamanashi 409-0012, Japan
[2] Taiyo Nippon Sanso Corporation, 10 Ookubo, Tsukuba, Ibaraki 300-2611, Japan
E-mail: yonghua.xu@trichemical.com

ABSTRACT

Dicyclopentyldimethoxysilane (DcPDMOS), having large functional CH_x group, was used as a single precursor to deposit low-k dielectric films. As-deposited DcPDMOS films contain high CH_x component. Films with k=2.47, Er=5.4 GPa were prepared after UV cure treatment, attributed to the removal of CH_x species and the increase of Si-O network structure. PALS analysis indicates that the pore size is smaller than 1.5nm in diameter and the pores are almost closed. Furthermore, O_2 was introduced into the deposition chamber during deposition to improve the properties of DcPDMOS films. k=2.37 and Er=8.0 GPa were realized by O_2 introduction, corresponding to the reduction of Si-H content and the increase of Si-O network structure in films. DcPDMOS films deposited with O_2 addition and after UV cure treatment also show high stability in air, due to its low contents of $-CH_2-$, H-SiO-Si/H-SiO_2-Si, cage Si-O, and Si suboxide, which are considered as the major causes for the unstability of SiOCH films.

INTRODUCTION

Lower k value is required for interlayer dielectric materials according to ULSI device scale shrinking. It is shown that introduction of porosity or free volume into dielectric materials by co-deposition with porogen and UV cure is a valid way to achieve lower k value. However, high porosity materials with large and interconnective pores have several integration issues such as low mechanical strength and moisture absorption. It has been investigated that controlling pore size, pore distribution, and the compositions of materials can improve stability, mechanical strength, and resistance to process damages [1]. In this study, we deposited SiOCH films using new developed precursor with high CH_x content without porogen. Porous films with small pore size were formed after UV cure. Moreover, O_2 was introduced during deposition to further improve the properties of films. The characteristics of these films were evaluated, and the effects of O_2 were discussed.

EXPERIMENT

SiOCH films were deposited by PE-CVD using dicyclopentyldimethoxysilane (DcPDMOS), having larger functional CH_x group than current low-k precursors, such as dimethyldimethoxysilane (DMDMOS). O_2 was introduced into the deposition chamber with the carrier gas He at some experiments. Then films were treated by UV cure at 350 ℃ to remove CH_x component. The UV lamp used was a broadband light. Film thickness was measured by spectroscopic ellipsometer and checked by X-ray reflector (XRR), C-V curve was measured with Hg-probe (SSM-495). Then k value was determined from the C-V results and the film thickness. Young's modulus was measured by Nanoindentation (MTS-Nano Indenter XP). Film density was measured by XRR too. Pore diameter and pore-connectivity were measured by Positron Annihilation Lifetime Spectroscopy (PALS). Furthermore, the chemical compositions were analyzed by FT-IR.

RESULTS and DISCUSSION

Figure 1 shows that initial DcPDMOS films deposited at 275 ℃ have high content of CH_x components. This means that organic compositions contained in the precursor remain in DcPDMOS films deposited by PE-CVD. Figure 1 also shows that absorbance of C-H_x bonds was greatly reduced after UV cure treatment, indicating that CH_x was mostly removed from as-deposited films. On the other hand, peak at 1275 cm^{-1} of Si-CH_3 vibration has little change. This corresponds to the facts that the absorbance of -CH_3 species at 2975 cm^{-1} and at 2918 cm^{-1} can be observed after UV cure, while the absorption of -CH_2- at 2951 cm^{-1} and at 2867 cm^{-1} disappeared completely. The k value changed from 2.8 for as-deposited films to 2.47 after UV cure, due to the formation of pores. PALS analysis indicates that the pore diameter is about 1.42 nm. The density of the porous films is 1.16 g/cm^3. At the same time, the absorbance of Si-O bonds has greatly changed after UV cure too. The absorbance at 1144 cm^{-1} and at 1010 cm^{-1} became small while that at 1060 cm^{-1} changed to large. It means that Si-O was transferred from cage structure and Si suboxide structure to network structure. The elastic modulus was increased from 3.9 GPa for as-deposited films to 5.4 GPa after UV cure treatment.

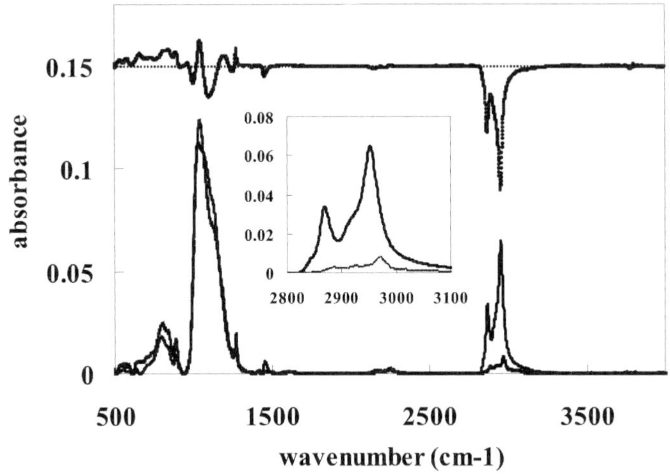

Fig. 1 FT-IR spectra for a DcPDMOS film (500 nm) deposited at 275 ℃, before (—) and after UV cure (----). (The upper shows the difference of IR spectra between after and before UV cure.)

The pore-connectivity of DcPDMOS films was evaluated by PALS system. Figure 2 is the energy spectra of γ-ray emitted by ortho-positronium (o-Ps) annihilation for an UV cured DcPDMOS film before and after SiO_2 (20nm) capping. The emission of 3γ (E<511 keV) correlates with self-annihilation of o-Ps in vacuum. In the case of highly connected open pores, the o-Ps can be annihilated outside the film surface, leading to high rate of 3γ annihilation. Therefore, the ratio of the 3γ annihilation intensities for uncapped sample and capped sample can be used to evaluate pore-connectivity properties [2]. Figure 2 shows that the intensity of 3γ annihilation for uncapped sample is almost the same as that for capped sample. So the pore connectivity of UV cure DcPDMOS film is nearly zero, that is, the pores are closed in film.

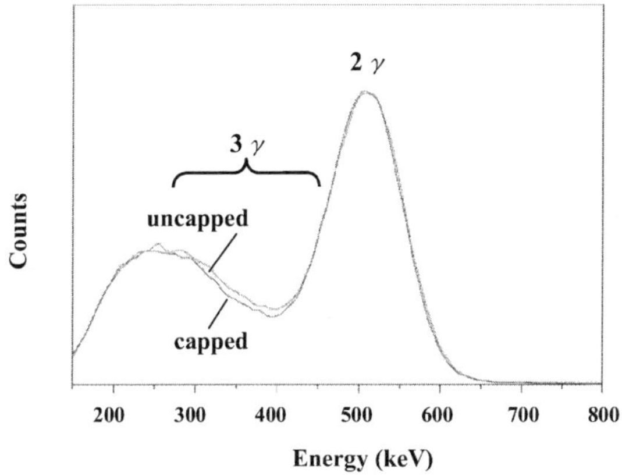

Figure 2 Energy spectra of γ-ray emitted by ortho-positronium (o-Ps) annihilation for an UV cure DcPDMOS film before and after SiO_2 (20nm) capping.

We have figured out that Si-H component is one of the most important influencing factors making k increasing in film by quantum chemistry calculation of polarizabilities of siloxane molecules [3]. Therefore, we tried to reduce Si-H composition by introducing O_2 during deposition. Figure 3 shows that Si-H component got small and Si-O component increased with O_2 introduction.

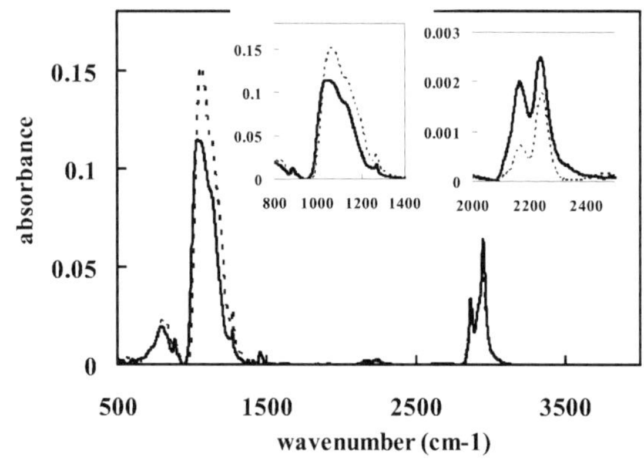

Fig. 3 FT-IR spectra for as-deposited DcPDMOS films (500 nm) without O_2 (——) and with 6 sccm O_2 introduction (----). (Deposition temperature: 275 ℃)

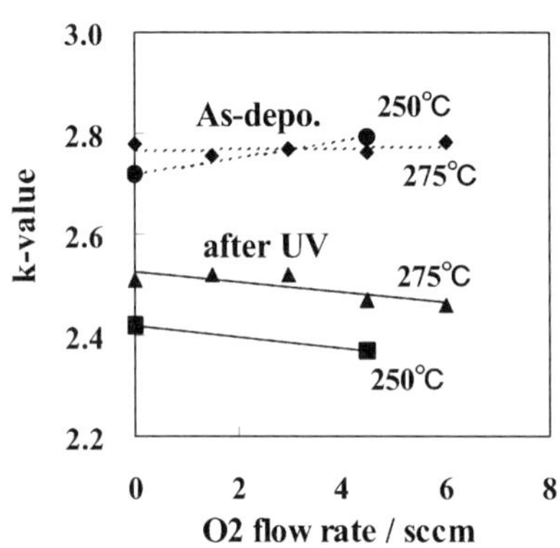

Fig. 4(a) k changes of DcPDMOS films deposited at different temperatures with O_2 flow rate, before (——) and after UV cure (----).

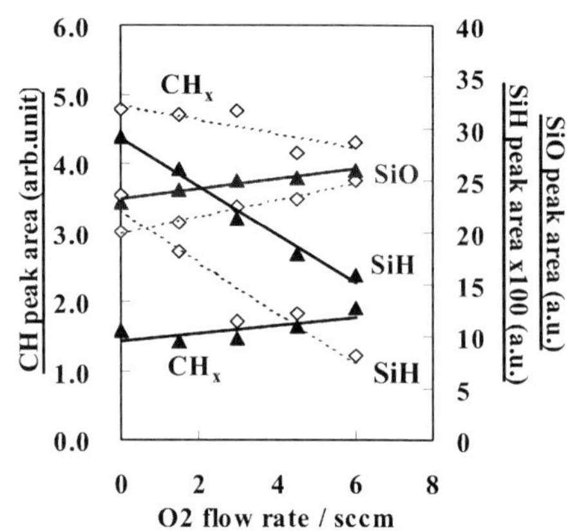

Fig. 4(b) Changes of FT-IR peak areas for Si-H, CH_x, and Si-O for DcPDMOS films (500 nm) deposited at 275 ℃, before (——) and after UV cure (----).

Figure 4 (a) and (b) show the O_2 additional effects on k value and FT-IR spectra areas of $C\text{-}H_x$, Si-H, and Si-O, respectively. The k values of UV cured films decreased with the increase of O_2 flow rate. For the UV cured DcPDMOS films, Si-O component and $C\text{-}H_x$ component got a little high, and only Si-H component was reduced with the increase of O_2 flow rate. Thus, it is considered that the reason for the k reduction is the decrease of Si-H composition. On the other hand, Si-O content can be decreased by deposition at lower temperature although it increases with O_2 addition. A k value of 2.37 for films deposited at 250℃ with 4.5 sccm O_2 was realized. Its Si-H composition was about 30% lower than that in without O_2 introduced films. O_2 additional films also have high elastic modulus of 8.0 GPa, due to its high Si-O network structure and low terminated (≡Si-H) structures. PALS analysis indicates that the pores are closed and its size is about 1.48 nm in diameter.

We found that for as-deposited SiOCH films, which were deposited at low temperature (250℃) and with O_2 addition, the k value had greatly changed from 2.8 to 3.05 after exposed to air for 26 days. We analyzed the composition changes of the films after exposed to air by FT-IR. As shown in figure 5, peaks at 2951-2868 cm^{-1} (-CH_2-), at 2164 cm^{-1} (H-SiO-Si, H-SiO_2-Si), at 1140 cm^{-1} (cage Si-O), and at 1020-1030 cm^{-1} (Si suboxide) became weak, while two new peaks at 3431 cm^{-1} (H bond OH) and at 1721 cm^{-1} (C=O) were formed. These mean that the components of -CH_2-, H-SiO-Si/H-SiO_2-Si, cage Si-O, and Si suboxide are the major causes for the unstability of SiOCH films.

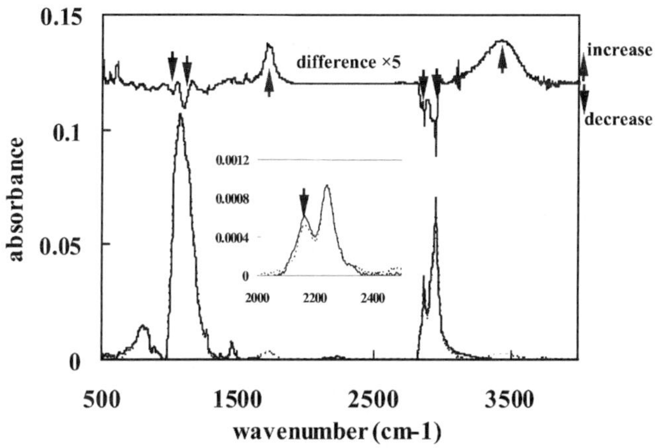

Fig. 5 FT-IR spectra for an as-deposited O_2 introduced DcPDMOS film (500nm) (——) and that exposed to air for 26 days (----). (The upper curve is the difference between them.)

However, as shown in Fig. 6, k value of O_2 introduced films after UV cure has increased only about 1% after exposed to air for 22 days, and was lower than those without O_2 addition. Because sample keeping in N_2 shows no change, we consider that the k increase related to the absorption of gases and reaction, such as H_2O and O_2. Just as mentioned above, H-SiO-Si/H-SiO_2-Si and Si suboxide became small by O_2 introduction, while after UV cure -CH_2- species completely disappeared, and Si-O cage structure and Si suboxide structure changed to Si-O network structure. These mean that the components causing unstability have been reduced. And the high stability of O_2 introduced SiOCH films after UV cure was realized.

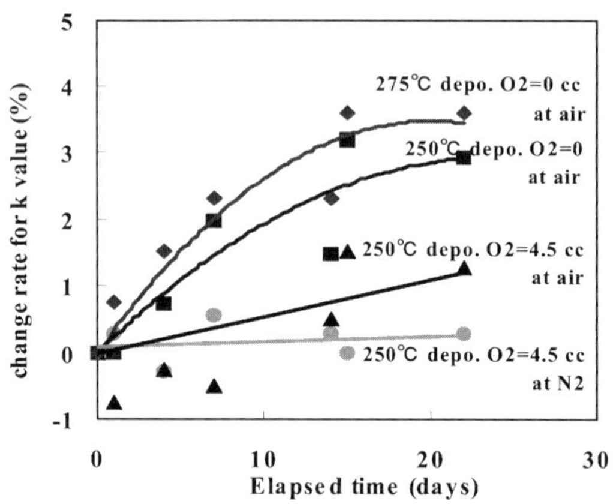

Fig. 6 Changes of k values for three types of porous SiOCH films after exposed to air or N_2.

CONCLUSIONS

SiOCH films containing high CH_x component were prepared without porogen by PE-CVD using DcPDMOS. It is found that the effects of UV cure on DcPDMOS films is the same as that observed on co-deposition films of organosilane and hydrocarbon porogen, but the pore size is smaller than that of most porogen-based SiOCH films. Moreover, introducing O_2 into the deposition chamber reduced Si-H contents for both as-deposited films and UV-cured films. The introduction of O_2 further improved the properties of porous SiOCH films. Properties of k=2.37 and Er=8.0 GPa were performed, and the k value had almost no change after exposed to air for 22 days. Our developed DcPDMOS film might be usable for 45 nm node beyond.

ACKNOWLEDGMENTS

The authors would like to thank Dr. N. Tajima of NIMS for his useful discussion and theoretical support, and thank Dr. T. Ohdaira of AIST for PALS measurements.

REFERENCES

1) F. Ito, K. Hijioka, T. Takeuchi, and Y. Hayashi, *Proc. AMC 2004*, 381 (2004).
2) F. Ito, T. Takeuchi, H. Yamamoto, Y. Hayashi, T. Ohdaira, and R. Suzuki, *AMC 2007*, P.V.3 (2007).
3) N. Tajima, M. Shinriki, Y. Xu, and T. Ohno, *SSDM 2008*, P-2-4 (2008).

Conference Proceedings AMC XXIV © 2009 Materials Research Society

Evaluation of an advanced organic ultra-low-k material

Abdel Karim Ferchichi[1], Youssef Travaly[1], Laureen Carbonell[1], Kris Vanstreels[1], Gerald Bayer[1], Mikhail Baklanov[1], Sumitoshi Asakuma[2], and Michio Nakajima[2]
[1]IMEC, Kapeldreed 75B, Leuven, 3000, Belgium
[2]Sumitomo Bakelite Co., Ltd., Yokohama, Kanagawa, 245-0052, Japan

ABSTRACT:

In this study, we report a new advanced organic ultra-low-k material having particular properties and promising future applications. We evaluate several properties of the material: k-value, porosity, pore size, density, hydrophobicity, refractive index, thermal stability, mechanical properties and direct CMP and N2/H2 striping resistance.

INTRODUCTION:

Over the last decade, various methods have been introduced to produce low dielectric constant layers with porosity (P) ranging from below 5% to 40%. These methods differ in many aspects ranging for the way the porosity is created, the nature of the post deposition treatment and finally, the deposition method in itself. Mapping the most relevant properties of low-k films to IC applications reveals 4 main regions. *Region I* (high k, low P, high mechanical properties) is relevant to memory applications. Dielectric solutions provided so far meet specifications. *Region II* (high k, high P) is irrelevant. *Region III* (low k, high P, large pore size and poor mechanical properties) encompasses dielectric films for logic applications. Finding IC solutions in that region is extremely challenging mainly because of the increasing importance of integration-related issues such as: plasma damage, CMP damage and difficulty of sealing pores. Currently available materials are carbon doped oxides prepared by both CVD and SOG. They are hybrid in nature and contain inorganic and organic parts. The latest is removed with a post deposition treatment. A complete removal of labile species is extremely challenging. *Region IV* (low k, low P, and micro-pores) is the ideal solution for logic applications, but requires the introduction of organic polymers. However, early generations of organic low-k materials were rejected by ULSI technology for several different reasons amongst which the difficulty to integrate [1,2]. Other reasons are for example an insufficiently low k-value, not sufficient coefficient of thermal expansion (CTE), a non-uniform and complicated pore shape, poor mechanical properties …etc.

In this paper, we have studied several properties of the material developed by Sumitomo Bakelite Co., Ltd. The k-value obtained using mercury probe, Ellipsometry Porosimetry (EP) [3,4] was used to determine the porosity, pore size and refractive index, density of the film was obtained by X-ray Reflectometry analysis (XRR), hydrophobicity was evaluated using Water contact angle (WCA) and Water based ellipsometric porosimetry [5], degradation temperature of the film using Thermal Desorption Stability (TDS) and mechanical properties using nanoindentation (NI). All these properties were also evaluated after N2/H2 exposure and Chemical Mechanical Polishing (CMP) to assess the film resistance to N2/H2 strip and direct CMP within a Metal Hard Mask based integration schema.

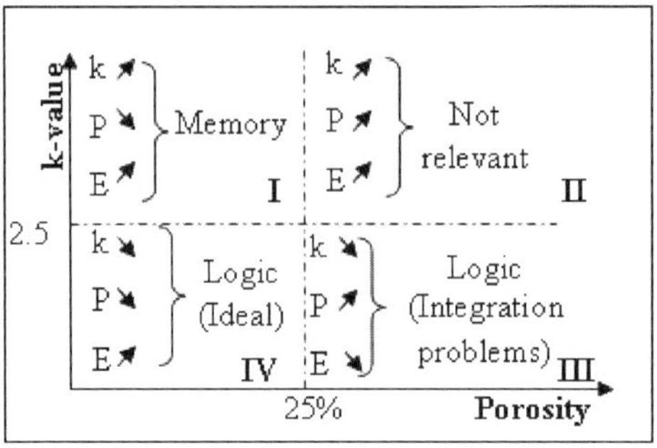

Figure 1. The k-value vs the open porosity. Main regions classified by relevant properties of low-k films.

RESULTS AND DISCUSSION:

Using EP with Toluene as solvent we measured a film porosity and the pore size distribution [3,4]. The polymer film has porosity of ~30% with a micro-porosity characterized by pores with 0.7 nm radius (Figure 2).

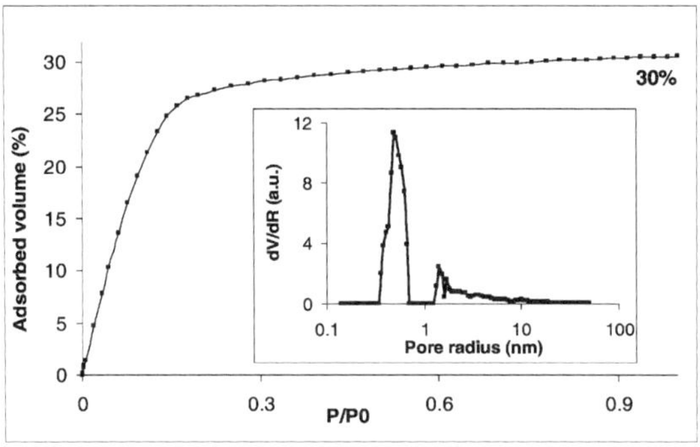

Figure 2. Toluene isotherm of the polymer, the inset shows the pore size distribution.

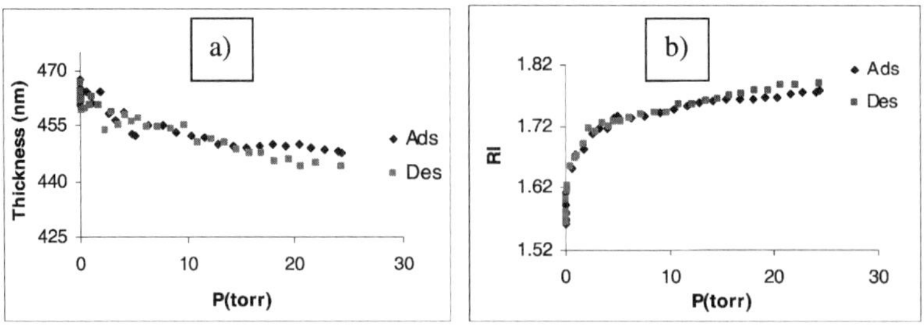

Figure 3. (a) Change of thickness during the toluene adsorption-desorption. (b) Refractive index during the toluene adsorption-desorption

The evolution of the thickness and the refractive index as a function of the pressure demonstrates that this film has properties exceptional for organic low-k, which is the absence of swelling in organic solvents. The refractive index of the film is about 1.55 (Figure 3).
The results of X-ray Reflectometry analysis demonstrate that the bulk of the film appears homogenous with a density of between 1.2 and 1.3 g/cm^3. This value is comparable with those of dense to medium porosity CVD materials. For the thermal stability of the polymer film, the temperature was increased till 600 °C. Thermal desorption (TDS) experiments have allowed detection of different molecules desorption during the temperature increasing. The results demonstrate the degradation of the polymer low-k material starting from 400°C, this allows the resistance of the low-k film to the integration steps.
The film has a k-value of 2.2. Achieving a comparable k-value with CVD films requires an open porosity of ~37-40% (Figure 4). We can observe that the low k-value of this polymer material without high porosity gives the material exceptional properties needed for the ideal application.

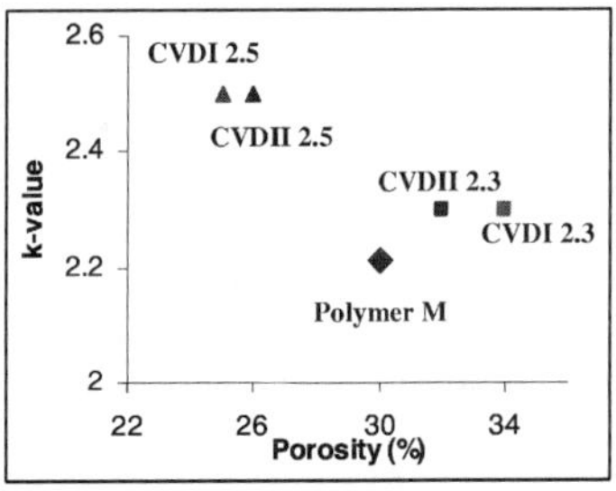

Figure 4. k-value as a function of open porosity

A 1000 nm thick low-k films were prepared for the study of mechanical properties (Elastic modulus and Hardness) by Nanoindentation. The results demonstrate that this organic low-k material has good mechanical properties (E= 5.01 GPa and H= 0.46 GPa) and if to compare with CVD 2.3 materials, it has a smaller k-value but a larger elastic modulus (Figure 5).

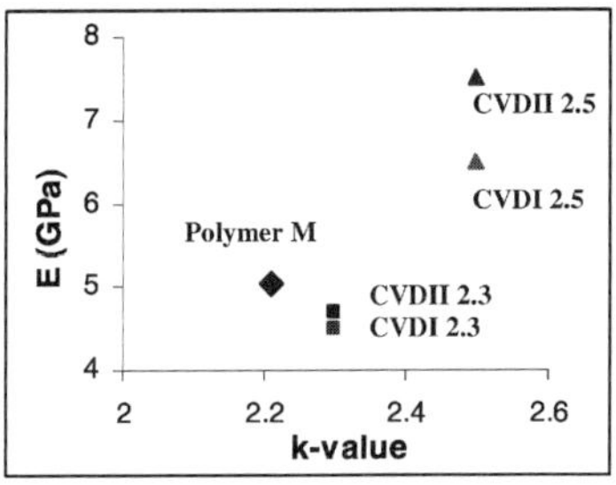

Figure 5. Elastic modulus as a function of k-value.

N2/H2 striping resistance

The characterization of the polymer low-k after the N2/H2 exposure is necessary to study the resistance of the film during the strip. For this study, we have chosen to damage the film with N2/H2 plasma during two different times.

Table I. k-value, E Modulus, porosity and RI of the films after N2/H2 strip.

Material	k-value	E (GPa)	Porosity	RI
30 s of N2/H2 exposure	2.29	5.07	27%	1.51
66 s of N2/H2 exposure	2.13	5.07	23%	1.41

The results obtained after the N2/H2 exposure prove the good resistance of the film to the N2/H2 strip. After 30 s of the plasma exposure, we see a small increase of the k-value, which correlates with a small decrease of open porosity to 27%. The decrease of open porosity is probably related to partial densification of the top part of the film. For the film exposed to N2/H2 plasma during 66 s, even the further decrease of the open porosity, the k-value measured is smaller than the initial ones. One can see that the plasma exposure decreases the refractive index (Table 1). This effect suggests that the decrease of open porosity happens with small simultaneous increase of the total porosity. These results promise a good reaction of the film after N2/H2 strip and we can also support this conclusion with good mechanical properties after 30 and 66 sec of plasma exposure.
Finally, we used the TDS results to analyze the chemical modification on the film and the stability of composition after N2/H2 exposure. The results show appearance of some bonds after plasma exposure but these bonds do not modify the structure of the film and we observe the desorption of the newly formed molecules at low temperature and the degradation of the bulk of the film starting 400 °C.

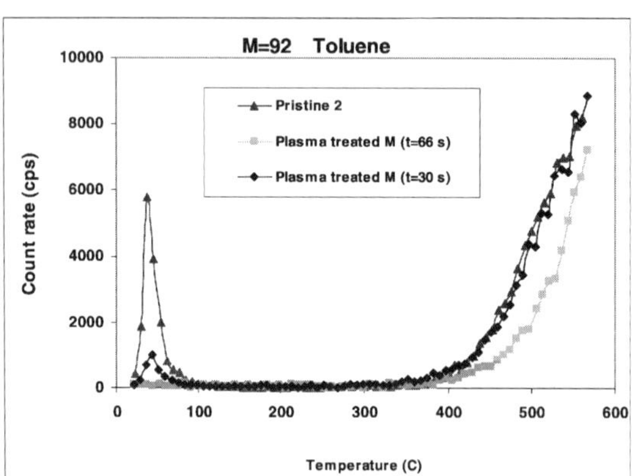

Figure 6. Desorption of the toluene during the increase of temperature before and after damage.

Resistance to direct CMP

Chemical Mechanical Polishing (CMP) introduces some increase of the k-value, which correlates with a small decrease of open porosity, On the other hand, the unchanged mechanical properties and the absence of hydrophilization after CMP suggest there is no chemical degradation.

Table II. The k-value, E Modulus, porosity and contact angle of the films before and after CMP.

Material	k-value	E (GPa)	Porosity	WCA (degree)
Before CMP	2.21	5.01	30%	94
After CMP	2.29	5.08	24%	80

This characterization of the film after CMP shows no big change of the material properties and the resistance comparing to CVD SiCOH based dielectric film, which becomes highly hydrophilic after CMP. For the thermal stability of the low-k film after CMP, TDS results show some change of the structure after CMP with presence of new molecules and chemical bonds (Figure 7). Some of these bonds are broken at low temperature (surface change) but others are strong and they resist high heat treatment. Except for these modifications as a result of the reaction with the chemical polishing composition, degradation of the film was observed, like the film before CMP, starting from 400°C.

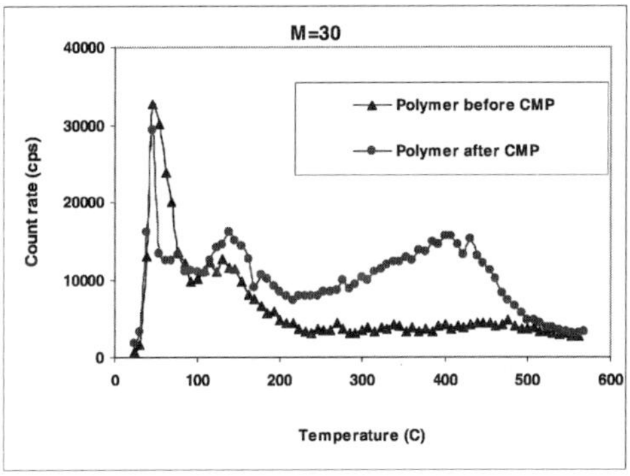

Figure 7. TDS result, appearance of new bonds after CMP

CONCLUSIONS

Different characteristics of the polymer low-k material are reviewed and compared with other CVD material. It appears from our study that this polymer has good structural and chemical properties, resistant to the direct CMP and N2/H2 based strip plasma. Therefore, this polymer is a promising material for future generations of interconnects. Additional studies, such as electric characterization and more in-depth investigations on resistance to plasma damage exposure, are in progress to complete this evaluation. This second part of the work aims to confirm our previous findings regarding the potential use of this polymer low-k material for integration.

REFERENCES

1. G. Maier, *Prog. Polym. Sci.* 26, p. 65 (2001).
2. L. Trabzon, O. O. Awadelkarim, Microelectronic Engineering, V65(4), p. 463 (2003).
3. M.R. Baklanov, K.P. Mogilnikov, V.G. Polovinkin, F.N. Dultsev, J Vac. Sci. technol. B18(3), p. 1385 (2000).
4. M.R. Baklanov, K.P. Mogilnikov, Microelectron. Eng. 64 (114), p. 335 (2002)
5. M.R.Baklanov, K.P.Mogilnikov, Q.T.Le. Microelectron. Eng., 83 (11-12), 2287 (2006).

Conference Proceedings AMC XXIV © 2009 Materials Research Society

Improved Low-k Dielectric Properties Using He/H2 Plasma for Resist Removal

Adam Michal Urbanowicz[1,2], Denis Shamiryan[1], Premysl Marsik[3], Youssef Travaly[1], Alain Jonas[4], Patrick Verdonck[1], Kris Vanstreels[1], Abdelkarim Ferchichi[1], David De Roest[5], Hessel Sprey[5], Kiyohiro Matsushita[6], Shinya Kaneko[6], Naoto Tsuji[6], Shijian Luo[7], Orlando Escorcia[7], Ivan Berry[7], Carlo Waldfried[7], Stefan De Gendt[1,2], and Mikhail Baklanov[1]

[1]IMEC, Kapeldreef 75, Leuven, B-3001, Belgium

[2]Dept. Chem., Katholieke Univ Leuven, Leuven, Belgium

[3]UFKL, Masaryk University, Brno, 601 77, Czech Republic

[4]LPCP, Université Catholique de Louvain, Louvain-la-Neuve, B-1348, Belgium

[5]ASM Belgium, Leuven, Belgium

[6]ASM Japan, Tokyo, Japan

[7]Axcelis Technologies, Beverly, MA, 01915

ABSTRACT

Aurora® ELK low-*k* films, deposited using a PECVD porogen-based process, were treated with 35 s of He/H_2 downstream-plasma (DSP) at varied temperatures of 25 ºC, 200 ºC, 300 ºC and 350 ºC. The plasma modifications were investigated using various physical-chemical methods. Results showed that extended He/H_2 DSP plasma exposure at elevated temperature may lead to *k*-value reduction due to porosity increase without hydrophilisation of the modified layer. Improvement in dielectric constant is accompanied by a small reduction of mechanical strength. The porosity increase was related to removal of porogen residues formed during the ultra-violet (UV) curing. The porogen residue removal signature was reflected in the optical properties in the UV range and the C-depth profile of the investigated low-*k* material.

INTRODUCTION

Prevention of plasma-induced low-*k* damage is a key challenge when integrating porous low-*k* dielectrics. Plasma exposures occur during dielectric etch, and photo-resist mask removal [1]. For the resist removal in the Cu/low-*k* integration process there are two commonly known approaches: (i) a low temperature, low pressure, anisotropic plasma, where the photoresist is removed by an ion-assisted physical process, using oxidizing or reducing plasma chemistries at low temperatures and (ii) an isotropic downstream plasma using a He/H_2 gas mixture where the resist is removed at high temperatures by a thermally activated chemical process [2, 3]. It is known that the reactive species of the low-pressure plasma penetrate into the porous network of the carbon-containing low-*k* films resulting in substantial carbon depletion and subsequent hydrophilisation [4, 5]. The He/H_2 strip approach however, shows no carbon depletion and no hydrophilisation, and is the focus of this study. Different substrate temperatures have been investigated for the He/H_2 process on advanced porous low-*k* materials and show that extended He/H_2 DSP (downstream-plasma) exposure may lead to *k*-value reduction due to porosity increase without hydrophilisation of the modified layer.

EXPERIMENT

Aurora® ELK (180 nm) porous low-*k* films with a *k*-value 2.3 and an open porosity of 32% were used [6]. The deposition process uses a PECVD porogen-based approach in which porogen is removed using a thermally assisted UV cure process. The mean pore radius was determined to be 0.9 nm. The He/H_2 20:1 DSP treatments of blanket low-*k* films were performed in a RadiantStrip320LK asher commercially available from Axcelis Technologies [3]. Four temperatures were used for the study: RT, 200 ºC, 300 ºC and 350 ºC. 7s exposure at 300 ºC wafer temperature was found to remove 150 nm photo resist on blanket resist wafers. Throughout the study, low-*k* films were exposed to the He/H_2 DSP for 5x resist removal time, which corresponds to 35s at 300 ºC. This strip is compared with the conventional low-temperature-anisotropic strip using 10s of O_2 plasma exposure.

RESULTS AND DISCUSSION

Hydrophobic properties

Hydrophobic properties were evaluated using three metrology techniques: (1) Water contact angle (WCA) to determine surface hydrophobic properties, (2) Water-based Ellipsometric Porosimetry (WEP) [7] to determine hydrophobic properties of the pore network, and (3) Fourier Transform Infrared Spectroscopy (FTIR), to determine the absorbance of H_2O and –OH groups and to possibly detect carbon depletion.

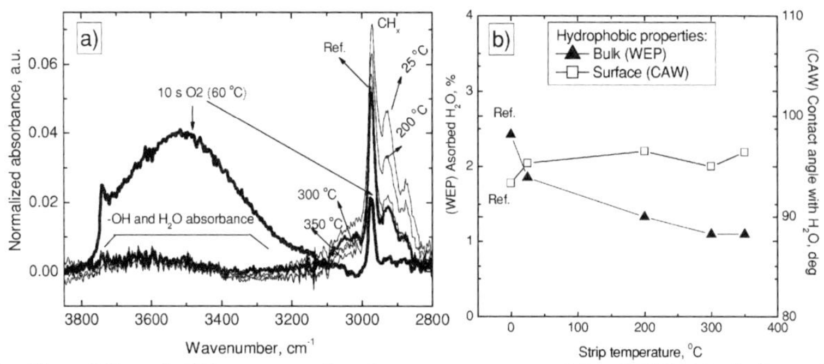

Figure 1. The -OH and CH_x groups absorbance as measured by FTIR (a), bulk and surface hydrophobic properties as measured by WEP and CAW, respectively (b), of as deposited and plasma treated films.

All three methods confirm that the He/H_2 DSP exposure results in no hydrophilisation of the porous low-*k* films. The water contact angle for He/H_2 DSP treated films was determined to be around 90° (equivalent to the pristine films) and the amount of absorbed water is less than 2%, which is even smaller than that of the pristine film, as measured by WEP (Figure 1b). The absorbance of H_2O and –OH groups as well as Si-CH_3 (C-H originated from Si-CH_3) groups are the same or higher as for the pristine film (Figure 1a). The higher CH_x groups absorbance at 25 ºC and 200 ºC might be related to hydrogenation of amorphous carbon [8] like porogen residue in hydrogen plasma and/or residual photoresist deposition from processing chamber walls. An FTIR spectrum of the same film treated with the low pressure anisotropic O_2 plasma is added for comparison and shows significant increase in H_2O and –OH groups.

Structural modification

Time of Flight Secondary Mass Spectroscopy (TOF-SIMS) (Figure 2a) and SEM analysis (Figure 2b) shows that the Aurora® ELK films treated at higher temperatures exhibit a double-layer structure. Moreover TOFSIMS reveals a lower C-content in the top layer. However, the C-depletion is one order of magnitude less than in the case of the anisotropic O_2 plasma exposure. It is presumed that this depletion is due to residual porogen removal. This postulate is supported by FTIR results that show no Si-CH_3 group loss (Figure 2c), and by Spectroscopic Ellipsometry (SE) measurements showing the disappearance of absorption peaks that are attributed to porogen signatures [9] upon the H_2/He DSP treatment. TOF-SIMS also indicates that the time required to sputter the film after the He/H_2 DSP treatments at 300 ºC and 350 ºC is reduced. This could in principle be due to a decrease in thickness or the result of reduced film density.

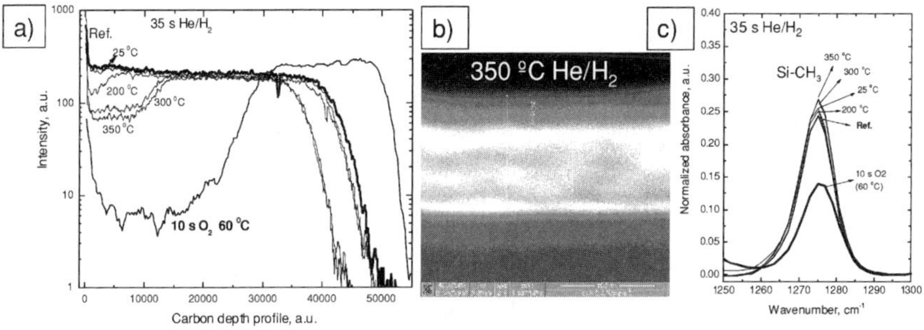

Figure 2. TOFSIMS depth profile (a) SEM of He/H_2 plasma treated film (reference film does not a have double-layer structure) (b), and the Si-CH_3 groups absorbance as measured by FTIR (c)

Three metrology methods, X-Ray Reflectometry (XRR), Ellipsometric Porosimetry (EP) and SE, were applied to probe if the top layer is thinner or less dense, and all three methods indicate a lower density of the top layer. XRR determined a lower critical angle (θ_c) for the He/H_2 DSP treated films at 300 ºC and 350 ºC, signifying lower density of the top layer (Figure 3b). A lower density of the top layer should translate into a higher total open porosity of the film. This was indeed confirmed by EP, which concluded higher porosity (Figure 3a) and bigger mean pore radius of 1.3 nm of the films treated at 300 ºC and 350 ºC. Increased porosity and pore size translate into higher toluene diffusion rates inside the low-*k* films (Figure 3c). The toluene diffusion was measured as described in literature [10]. The sharp increase in toluene diffusivity is related to higher porosity and pore interconnectivity of the top modified layer.

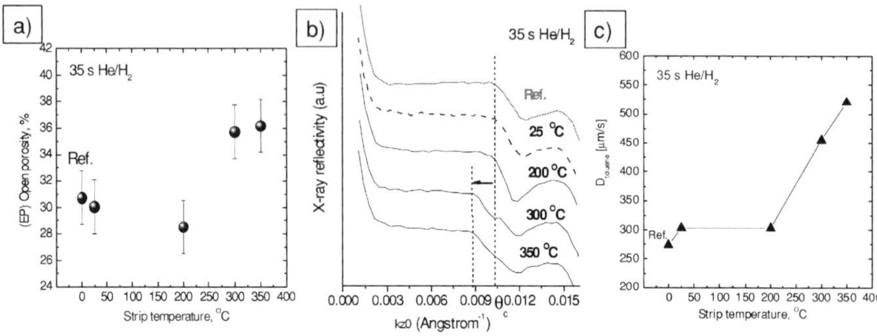

Figure 3 The open porosity as measured by EP (a), the X-ray reflectivity as measured by XRR (b), and the toluene diffusion coefficient (c)

The structural modifications and density changes are also reflected in the optical properties of the film. The modifications can be divided into two groups (Figure 4a). The fist group consists of temperatures up to 200 ºC when porogen removal is very limited and changes might be due to hydrogenation of amorphous carbon [8] like porogen residue and/or organics diffusion from the reactor chamber walls. The second group is for the temperatures of 300 ºC and 350 ºC when the depth of modification is higher. Both the refractive index and the extinction coefficient of the top layer are lower than those of the bottom layer (which was presumed to be unchanged in the as deposited film) indicating porogen removal (Figure 4b).

We presume that the porogen removal mechanism at higher temperatures is similar to the photoresist removal mechanism in He/H_2 DSP plasma proposed by Berry *et al.* [2]. Heating the low-*k* in H_2-based plasma atmosphere supports the evaporation of porogen residue. H* radicals form more volatile carbon hydrides and promote cracking of high mass porogen residue, generating smaller polymer chains of lower vapor pressure. Additionally, depth of penetration of H* radicals into low-*k* pores should also be considered due to H* loss as a result of recombination on low-*k* pore walls.

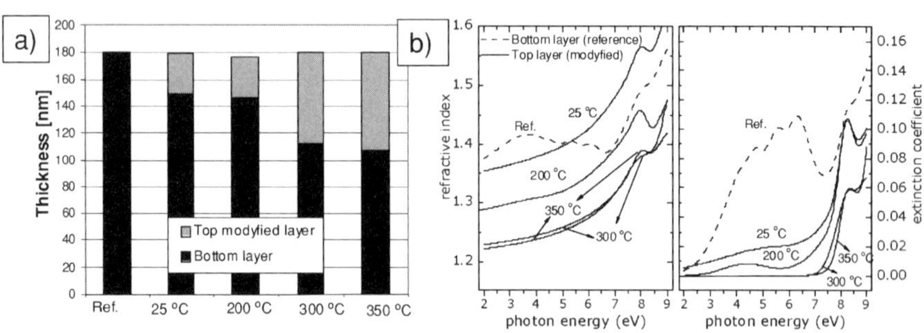

Figure 4. The thickness (a), and optical properties (b) of the top and bottom layer as measured by SE.

Mechanical properties were examined using Nano Indentation (NI). Since the film thicknesses were relatively low, only a relative study was possible. NI showed a small reduction of the Young's modulus for the films treated at higher temperatures (Figure 5a), which can be attributed to the reduced density of the top layer. The electrical measurements using a Hg-probe indicate a small reduction of the effective *k*-value as a result of the lower dielectric constant of the top modified layer (Figure 5b). The results, however, have measurement uncertainty due to the potential error in determining the thickness of the top modified layer. Similar electrical results were observed by Matshushita *et al.* [11], who reported that the *k*-value of a spin-on low-*k* film was reduced after He/H_2 DSP plasma exposure. However, the mechanism of such low-*k* modification was not discussed.

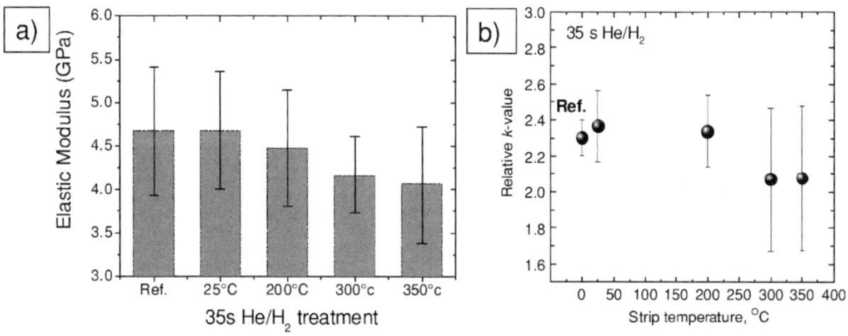

Figure 5. The mechanical properties as measured by NI (a), *k*-value as measured by Hg-probe (b)

CONCLUSIONS

He/H_2 ash DSP treatments impose no damage to the subjected porous low-*k* films as they remain hydrophobic, in contrast to the conventional low temperature anisotropic O_2 ash plasma. He/H_2 DSP exposure at temperatures higher than 300 ºC result in a modified top layer, with reduced density and increased porosity. These modifications are believed to be a result of residual porogen removal and therefore not detrimental, but rather beneficial. The decrease in material density leads to an improvement in dielectric constant with relatively small reduction of the mechanical strength.

REFERENCES

1. K. Maex, M. R . Baklanov, D. Shamiryan, F. Iacopi, S. H. Brogersma, Z. S. Yanovitskaya, J.Appl.Phys., **93** (11), 8793 (2003)

2. I.L. Berry, Q. Han, C. Waldfried, O. Escoria, A. Becknell, SEMICON, (2004)

3. Q. Han, B. White, I. L. Berry, C. Waldfried, and O. Escorcia, Solid State Phenomena, Vols 103-104 (2005), p. 341-344, (2005)

4. M. R. Baklanov, A. M. Urbanowicz, G. Mannaert and S. Vanhaelemeersch, Proceedings of 2006 8th International Conference on Solid-State and Integrated Circuit Technology, Part 1, p. 291-294, Shanghai (2006).

5. A.Urbanowicz, A. Humbert, G. Mannaert, Z. Tokei, and M. Baklanov, Solid State Phenomena, Vol. 134 pp 317-320, (2008)

6. P. Verdonck , D. De Roest , S. Kaneko , R. Caluwaerts , N. Tsuji , K. Matsushita , N. Kemeling , Y. Travaly , H. Sprey , M. Schaekers , G. Beyer., Surface & Coatings Technology Vol. 201, p. 9264–9268 (2007)
7. M.R.Baklanov, K.P.Mogilnikov, and Q.T.Le. , Microelectron. Eng., **83**, 2287 (2006)
8. M.Matsuura, K. Goto, N. Miura, S. Hashii, K. Asai, Mater.Res.Soc.Symp.Proc. Vol 914, 0914-F01-06 (2006)
9. P. Marsik, P. Verdonck, D. Schneider, D. De Roest, S. Kaneko, and M. Baklanov, PSSc, Vol5(5), p. 1253-1256, (2008)
10. P.Mogilnikov,M.R.Baklanov, D. Shamiryan and M. Petkov, Jap. J. of Appl. Phys, Vol. 43, No.1, p. 247, (2004)
11. A. Matsushita, N. Ohashi, K. Inukai, H.J. Shin, S. Sone, K. Sudou, K. Misawa, I. Matsumoto and N. Kobayashi, Proc. IEEE IITC, p. 147, (2003)

Conference Proceedings AMC XXIV © 2009 Materials Research Society

UV assisted curing of plasma damaged porous ultralow-k material after a k-restore process: influence of UV-wavelength and curing ambient

Nicole Ahner[1], Stefan Schulz[2], Matthias Schaller[3], and Lutz Prager[4]
[1]Center for Microtechnologies, Chemnitz University of Technology, Reichenhainer Straße 70, Chemnitz, 09111, Germany
[2]Fraunhofer ENAS, Chemnitz, 09111, Germany
[3]AMC Saxony LLC & Co. KG, Dresden, 01109, Germany
[4]IOM Leipzig, Leipzig, 04318, Germany

ABSTRACT

Low dielectric constant materials (low-k) are widely used for isolation within the interconnect system to lower RC delay and crosstalk. During plasma processes like ashing, etching and cleaning these films are facing heavy damage effects, which increase their k-value. With k-restoration by silylation silanol groups formed at dangling bonds are substituted by $Si(CH_3)_3$ and Carbon content of the damaged surface near area can be increased. The k-restore process can be combined with an UV-assisted thermal curing step to enhance repair effects. In this study we evaluated two repair chemistries and show by FTIR analysis that besides finding a working restore-solution choosing the right UV wavelength for curing is essential. High energy radiation turned out to be destructive to the material and reverses a successful repair achieved by the restore-liquid by breaking Si-C bonds and generating H-Si-O. Additional the modification of the material's structure by using varied process times and curing ambients has been studied.

INTRODUCTION

With decreasing feature sizes, RC-delay and crosstalk are limiting the performance of integrated circuits. Besides lowering the resistance using Cu instead of Al, the integration of low dielectric constant materials (low-k) for isolation is one possibility to overcome those issues [1]. At low-k materials the main focus is on CVD SiCOH materials, where covalently bonded $-CH_3$ groups are incorporated into the SiO_2 lattice. Additionally porosity is introduced by adding an organic porogen during deposition, which is decomposed by a curing step (e.g. UV-curing) to produce a porous ultra low-k (ULK) material. The skeleton of the material consists of a Si-O-Si backbone with $-CH_3$ termination groups, which stop further networking to achieve a loose structure with a lower density than silica [2]. The mechanical weakness of these materials is one major issue and can be diminished by strengthening the network using UV radiation for curing. During this process Si-C bonds are broken and the $-CH_3$ termination groups are changed to -H, which bond as Si-H. The Si-H bonds generate less surrounded free volume and the polymeric chains get closer, which increases the mechanical strength of the material [3].

After deposition the porous dielectric is patterned using CF-containing chemistries in a plasma process, the resist is stripped and etch residues at feature bottom and sidewalls are removed. All those processes can be a source for degradation of ULK dielectric, especially the use of oxygen plasma for resist removal is critical. Radicals (e.g. O_2) react with the carbon in the ULK dielectric, which leads to Carbon depletion in the near-surface areas of the material and a lot of dangling bonds are generated. Those free bonds are easily occupied by hydroxyl groups and silanols are formed. This leads to moisture adsorption and the former hydrophobic surface of the ULK dielectric becomes hydrophilic. Moisture adsorption, densification and carbon depletion lead to increasing k-values and leakage currents [4].

To overcome these issues alternative mild plasma processes are under investigation, which produce less plasma damage and do not use oxygen anymore [5]. Additionally one can try to repair the damaged areas of the ULK dielectric by removing the hydroxyl groups bonded to the silicon atoms and increasing the carbon content. One possible reaction to achieve this is the silylation of the silanol groups, where OH is substituted by $Si(CH_3)_3$ using chemicals like HMDS or TMCS in gas phase or by spin coating [6], [7]. The k-restore process can be combined with a curing step (e.g. thermal, UV-assisted) to enhance the repair effect, accelerate the reaction and improve the properties of the ULK. For UV curing mostly broadband lamps are used, but also light sources for wavelengths below 200nm are considered [8], [9].

In this work two different spin-on k-restore chemistries in combination with an UV-assisted thermal curing have been evaluated. To study UV wavelength dependency of the restore-effect two narrow band UV-sources have been used in different curing ambients. The influence of the curing process on the structure of the ULK dielectric has been determined by FTIR analysis.

EXPERIMENT

A porous CVD SiCOH low-k material of 450nm film thickness deposited on 300mm Si-substrates has been provided by AMD. The dielectric was treated by a 60s ashing process to generate plasma damage and low-k degradation effects. Two different silylating restore chemicals have been provided by a chemical supplier. The liquids were spun on at 100rpm followed by a reaction time of 2min, spin-off at 1000rpm and hotplate drying at 80°C for 1min. For the spin-on the SUSS Microtech RC8 gyrset system was used. During reaction and spin-off the spin-on chamber has been closed to reduce solvent evaporation. The samples have been cured at 300°C using two different ambients (vacuum 10^{-4}mbar and pure nitrogen at 700mbar) for 1 and 5min each and using 2 different UV-lamps (lamp 1: 172nm, XERADEX 20 L40/102; lamp 2: 185nm, UVI 80 U S 15/395). The changes within the material's structure have been determined by FTIR analysis.

RESULTS

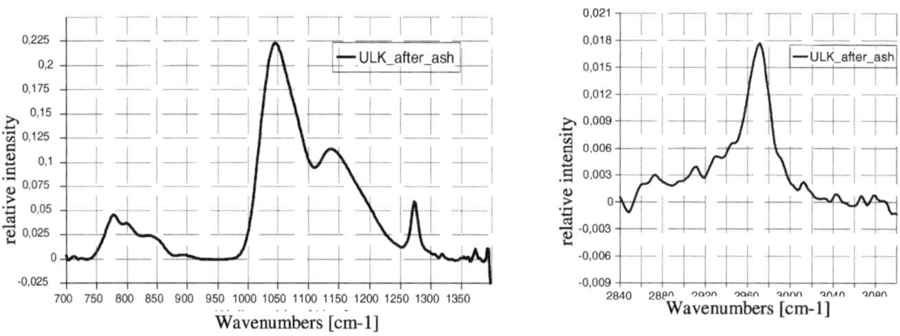

Fig. 1: FTIR-spectrum of the ULK after ash, fingerprint-area (left), functional CH_3 group (right)

The spectrum of the ULK dielectric after ashing is shown in Fig. 1 with the Si-C peaks at 750-870cm^{-1}, the H-Si-O peak at around 900cm^{-1}, the Si-O-Si-backbone (network-peak at 1045cm^{-1} and cage-peak at 1135cm^{-1}), Si-CH_3 at 1275cm^{-1} and the functional -CH_3 group at 2975cm^{-1} [10]. The area of the Si-C and H-Si-O bonds turned out to be the most interesting part of the spectra, so analysis will focus on that. Fig. 2 compares the effects of liquid A and B using lamp 1 and 2 for 5min in vacuum. There is no change visible for liquid A and lamp 2 compared to the reference ULK after ashing. But using lamp 1 decreases Si-C bonds and increases H-Si-O

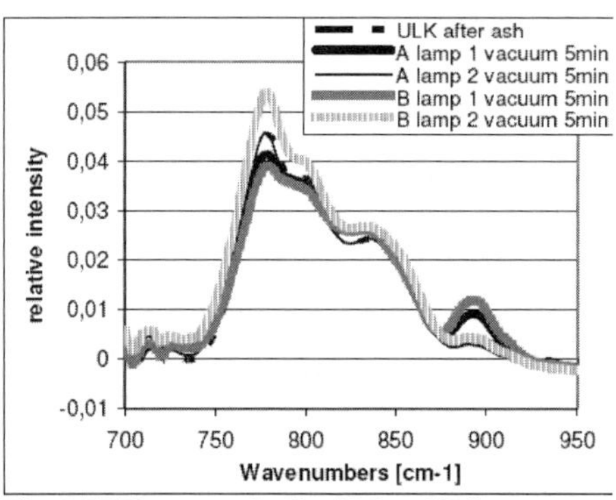

Fig. 2: FTIR-spectra for liquid A and B in vacuum using lamp 1 and 2 for 5min

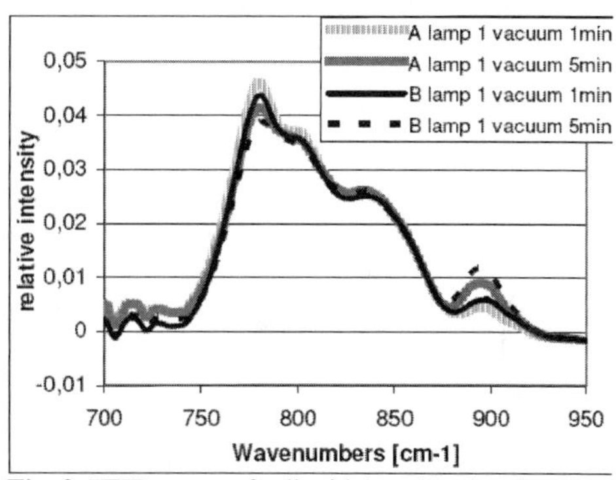

Fig. 3: FTIR-spectra for liquid A and B showing the process time dependencies

bonds. This effect also appears using liquid B and lamp 1. The combination of liquid B and lamp 2 shows an increase in Si-C bonds and no formation of H-Si-O bonds. The higher Si-C content is a hint for a repairing effect of liquid B, liquid A seams to have no effect. The positive effects of liquid B are reversed by a treatment at 172nm UV wavelength. Fig. 3 shows the process time dependencies for lamp 1. The generation of H-Si-O depends on the dose of UV radiation and is clearly visible after 5min of treatment. The strength of the effects of lamp 1 also depends on the curing ambient. Fig. 4 shows that the amount of H-Si-O generated in vacuum after 5min radiation is reached after only 1min in N_2 atmosphere. After 5min treatment in N_2 the content of H-Si-O bonds has increased rapidly and Si-C is much weaker.

Fig. 5 compares the process time dependencies using liquid B and lamp 1 and lamp 2. As already seen, liquid B increases the Si-C content of the low-k material. The increase of the Si-C peak is overlapped by the destructive effects of lamp 1. The amount of Si-C bonds also decreases slightly after 5min treatment using lamp 2, so short UV curing (e.g. 1min) should be preferred. Fig. 6 shows the changes within the Si-O-Si-backbone of the ULK dielectric for liquid B and lamp 1 and 2 for different process times. There is only small structural change visible using lamp 2 and a weakening of network and cage peak using lamp 2 for 5min compared to 1min of treatment. Comparing both lamps the peaks of network, cage and $Si\text{-}CH_3$ decreased using lamp1.

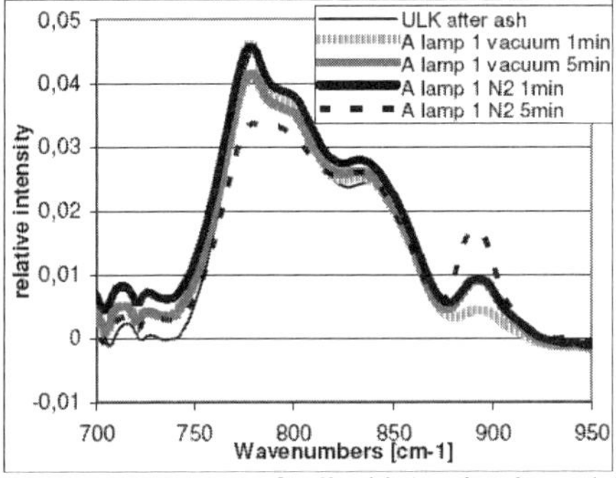

Fig. 4: FTIR-spectra for liquid A using lamp 1 in vacuum and N_2

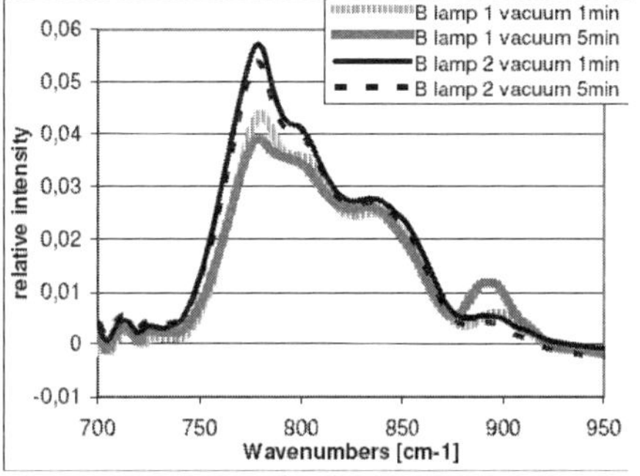

Fig. 5: FTIR-spectra for liquid B using lamp 1 and 2 showing process time dependencies

In Fig. 7 the functional $-CH_3$ group is shown. UV treatment using Lamp 1 decreases the $-CH_3$ content. This effect is also dose depending. For UV curing using lamp 2 the $-CH_3$ content is higher, but also slightly decreases depending on the radiation dose after 5min.

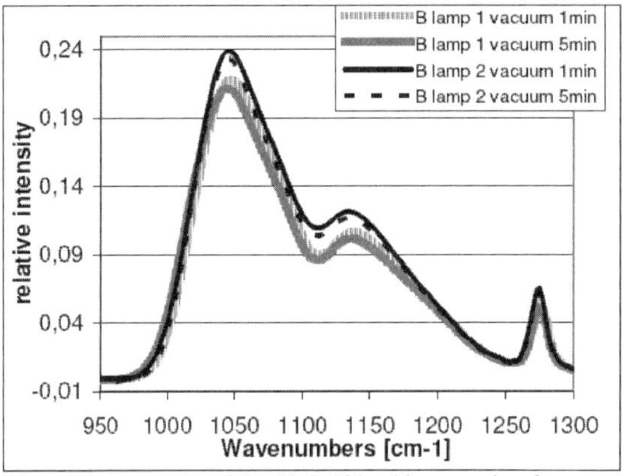

Fig. 6: FTIR-spectra of the Si-O-Si backbone region for liquid B using lamp 1 and 2 for different process times

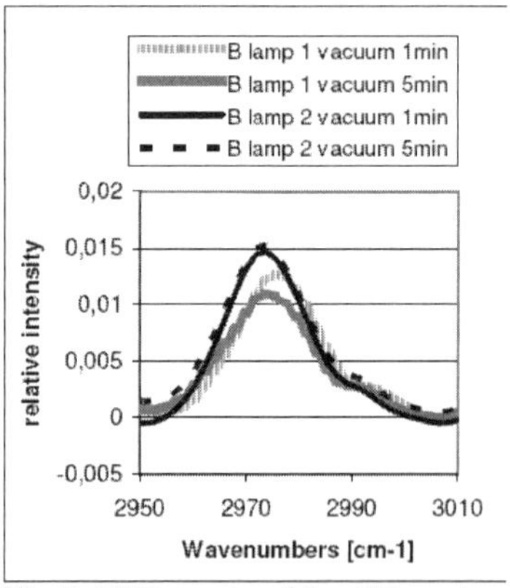

Fig. 7: FTIR-spectra of the $-CH_3$-region for liquid B, lamp 1 and 2 (1 and 5min)

DISCUSSION

As visible in the FTIR spectra liquid A shows no measurable repairing effects after UV treatment, but liquid B increases the content of Si-C bonds in the material. But it is important to choose the appropriate UV-wavelengths for curing. Lamp 1 with 172nm has a destroying character and reverses the positive effects achieved by using liquid B. Si-C links are broken due to the high photon energy, the hydrocarbon part of the methyl group is released and there is a conversion of $-CH_3$ termination groups to SiH. Possible reactions occurring include the formation of O_3-SiH and O_3-Si• + CH_2(H) bonds, but this is not completely understood yet and verified [11], [12]. The amount of Si-H generation and Si-C destruction is depending on the UV-dose and on the curing ambient. The use of lamp 1 (172nm) also effected the structure of the Si-O-Si-backbone of the ULK dielectric and decreased the amount of CH_3 functional groups. Using N_2 atmosphere the Si-C reduction and Si-H increase are much stronger than in vacuum. The breaking of Si-C bonds also effects the decrease in Si-CH_3 visible at $1275cm^{-1}$ and in $-CH_3$ at $2975cm^{-1}$. The analysis showed that besides choosing a working chemistry to repair plasma damage in ULK dielectrics, the curing parameters, especially the used UV wavelength, the process ambient and process duration are essential for successful k-restore.

CONCLUSIONS

The k-restore capabilities of two chemistries in combination with varied UV-curing wavelengths and curing ambients has been studied by FTIR. Liquid B seems to be a promising solution due to its ability to increase the Si-C content of the ULK dielectric. Liquid A did not show a measurable effect. Choosing a suitable UV-wavelength for curing turned out to be essential. The lamp with 172nm showed a destructive character, decreasing Si-C and supporting H-Si-O generation. The positive effects achieved using liquid B have been reversed using this UV wavelength. Lamp 2 with 185nm wavelength only slightly decreased Si-C and did not lead to H-Si-O generation. The destructive effects of lamp 1 were much stronger using a N_2-atmosphere instead of vacuum. So the most promising k-restore process studied here is liquid B in combination with UV-curing at 185nm (lamp 2) for about 1min in vacuum.

ACKNOWLEDGMENTS

The project described in this publication has been funded in line with the technology funding for regional development (ERDF) of the European Union and by funds of the Free State of Saxony. All graphics have been generated by Fraunhofer ENAS.

REFERENCES

1. The International Technology Roadmap for Semiconductor, http://public.itrs.net/
2. K. Maex, M.R. Baklanov, J. Appl. Phys. 93(11), 2003, pp. 8793-8841
3. C.H. Huang, H.L. Huang, Jpn. J. Appl. Phys. 47, 2008, pp. 1532-1535
4. S. Uchida, S. Takashima, J. Appl. Phys. 103, 2008, pp. 073303-1 - 5
5. E. Soda, S. Kondo, J. Vac. Sci. Technol. A, 26(4), 2008, pp. 875-880
6. H. Chaabouni, L.L. Chapelon, Microelectronic Engineering 84, 2007, pp. 2595-2599
7. T. Ono, K. Kinoshita, Jpn. J. Appl. Phys. 45(8A), 2006, pp. 6231-6235

8. S. Eslava, G. Eymery, J. Electrochem. Soc. 155(5), 2008, pp. G115-G120
9. T.-S. Kim, N. Tsuji, J. Appl. Phys. 103, 2008, pp. 064108-1 - 8
10. A. Grill, D.A. Neumayer, J. Appl. Phys. 94(10), 2003, pp. 6697-6707
11. A. Zenasni, U. Jousseaume, J. Appl. Phys. 102, 2007, pp. 094107-1 - 8
12. S.-I. Nakao, J. Ushio: IEEE, 2006, pp. 66-68

Conference Proceedings AMC XXIV © 2009 Materials Research Society

Carbon Dioxide and Oxygen Plasma Induced Damage to Porous OSG Low-k Dielectrics

Hualiang Shi[1], Huai Huang[1], Junjing Bao[1], Ryan Scott Smith[1], Yangming Sun[1], Paul S Ho[1], Andrew Li[2], Jeremiah T Pender[2], Michael D Armacost[2], and David Kyser[2]

[1]Microelectronics Research Center, the University of Texas at Austin, 10100 Burnet Rd. Bldg. 160, Austin, TX, 78758-4445

[2]Applied Materials, 974 E Arques Ave, Sunnyvale, CA, 94085

ABSTRACT

A mechanistic study of CO_2 plasma damage to organosilicate glass (OSG) low-k dielectrics was conducted using both capacitive-coupled reactive ion etching (RIE) plasma and inductive-coupled plasma (ICP) sources with varying plasma density and energy. The plasma damage was characterized using atomic force microscopy (AFM), capacitance-voltage (C-V) measurement, Fourier transform infrared spectroscopy (FTIR), spectroscopic ellipsometer (SE), X-ray photoelectron spectrometer (XPS) and XPS depth profiling. The extent of CO_2 plasma damage was found to depend on plasma density, plasma energy, plasma species, and low-k dielectric chemistry.

With increasing ICP source power or decreasing chamber pressure, the effect of plasma damage such as methyl depletion and moisture uptake became more severe. Under an identical plasma condition, pure ICP plasma induced more damages than RIE plasma. The film shrinkage rate and the depth of carbon depletion layer strongly depended on the low-k dielectric chemistry. In particular, the high carbon doping in OSG, k~2.2, improved the material resistance to plasma damage. Additionally, compared with O_2 plasma, CO_2 plasma induced less plasma damage. By taking into account the kinetics of radical diffusion, reaction, and recombination, the concept of plasma altered layer was proposed and the analysis was employed to quantify the extent of CO_2 plasma damage in OSG low-k dielectrics.

INTRODUCTION

Ultra low-k (ULK) dielectrics with porosity are being introduced to reduce the RC delay for Cu interconnects beyond the 45nm technology node [1]. During plasma processes such as etching and ashing, ULK dielectrics tend to degrade, causing the effective dielectric constant and the leakage current to increase. Such plasma damage limits the further scaling of interconnect structure and has stimulated significant interests to study the plasma damage mechanisms [2-12]. Previous studies revealed that the formation of plasma damage in ULK dielectrics was a complicated phenomenon, which involved not only physical effects arising from ion bombardment, but also chemical effects from chemical reactions between plasma species and ULK dielectrics [10]. In this paper, damage induced by CO_2 plasma in organosilicate glass (OSG) low-k dielectrics was investigated to explore the potential of replacement of O_2 with CO_2 for plasma ashing. Both capacitive-coupled reactive ion etching (RIE) plasma and inductive-coupled plasma (ICP) sources were used here.

Several analytical techniques were employed to evaluate the plasma damage, including atomic force microscopy (AFM), capacitance-voltage (C-V) measurement, Fourier transform infrared spectroscopy (FTIR), spectroscopic ellipsometer (SE), X-ray photoelectron spectrometer (XPS) and XPS depth profiling. Additionally, Le et al. reported that the plasma modified layer could be removed by using dilute HF solution (1%) due to the high etching selectivity of the plasma damage layer over undamaged underlying layer [13]. In this paper, dilute HF (DHF) dipping experiment combined with AFM and SE measurements was performed to investigate the plasma damage such as surface roughness and the depth of the damaged layer. We found that the extent of CO_2 plasma damage depended on plasma density, plasma energy, plasma species, and low-k dielectric chemistry.

On the basis of the kinetics of radical diffusion, reaction, and recombination, Baklanov et al. used Thiele analysis to study the penetration depth of O_2 plasma radical species into porous low-k dielectrics [11]. Following this approach, we monitored the evolution of carbon concentration during plasma treatment in OSG low-k dielectrics and developed a plasma altered layer model to quantify the plasma damage of porous OSG low-k dielectrics. We found that the model provided a good kinetic framework to account for the experimental observation of plasma damage in low-k dielectrics.

EXPERIMENT

Four different OSG films with thickness of ~120 nm were deposited at Applied Materials. These included k~2.2 OSG with high carbon doping, k~2.2 OSG with normal carbon doping, k~2.5 OSG, and k~3.0 OSG. Plasma treatments were performed in a commercial plasma etcher operating at 13.56MHz where the plasma generator was set up to yield ICP plasma with high density but low energy or RIE plasma with low density but high energy. During the experiments, the substrate temperature was kept constant at around -10°C. The flow rate was kept at 20sccm.

The film thickness and refractive index were measured by a J. A. Woollam M-2000 spectroscopic ellipsometer. Bonding configurations were investigated by a Nicolet Magna 560 FTIR Spectrometer. Chemical composition was evaluated by an Axis Ultra DLD XPS Spectrometer from Kratos Analytical Company. Dielectric constant at 1 MHz was obtained by a mercury probe based on a metal-insulator-semiconductor (MIS) structure. The surface roughness was determined by an AFM from Digital Instrument. Diluted HF (1%) to remove the plasma damaged layer was obtained from Air Products.

RESULTS AND DISCUSSION

The results of pure ICP CO_2 plasma induced damage to OSG films were summarized first. Then, the plasma parameters affecting the plasma damage such as plasma density, plasma energy, plasma species, and low-k chemistry were investigated. Finally, the plasma altered layer model was briefly introduced.

Plasma Damage Induced By CO_2 Plasma

As shown in Fig. 1, the XPS C1s peak of OSG film after ICP CO_2 plasma treatment can be deconvoluted into four peaks: C-H/C-C (~284.7eV), C-O-H/C-O-C (~286.4eV), C=O (~287.7eV), and O-C=O (~289.2eV) [14]. To further confirm the bonding configuration changes, differential FTIR spectra were obtained by subtracting the spectrum of pristine OSG from those after ICP CO_2 plasma treatment. As indicated by Fig. 2, ICP CO_2 plasma reduced Si-$(CH_3)_x$ (1280~1250, 870~760cm^{-1}), Si-O-Si suboxide (~1023cm^{-1}), and CH_x (~2960, 2925, 2900,

2865cm^{-1}) bonds, and increased –OH/H_2O (~3100~3800, 935cm^{-1}), C=O (~1720cm^{-1}), and Si-O-Si network (~1063cm^{-1}) structures [15]. These results are consistent with the XPS results.

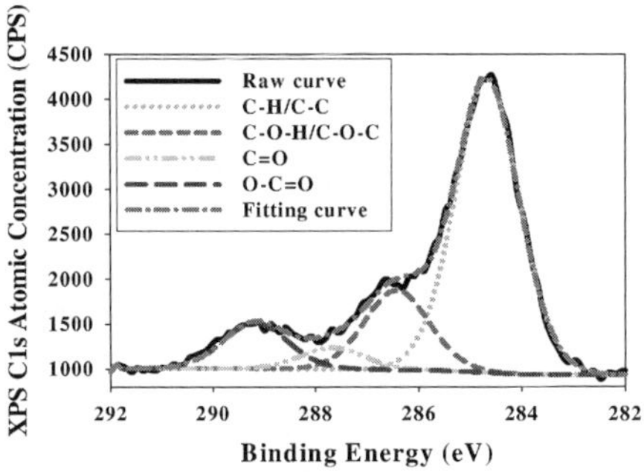

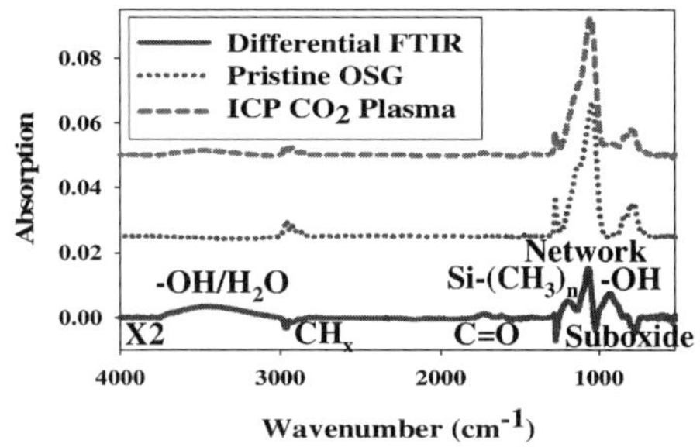

Fig. 1 XPS spectrum for OSG treated by ICP CO_2 plasma

Fig. 2 FTIR spectra for OSG treated by ICP CO_2 plasma

In Fig. 3, methyl depletion (Si-$(CH_3)_x$/Si-O) and moisture uptake (–OH/Si-O) were determined from FTIR measurements and used to evaluate the plasma damage. With increasing plasma treatment time, the Si-$(CH_3)_x$/Si-O ratio decreased linearly while the –OH/Si-O ratio increased linearly. As shown in Fig. 4, these bonding configuration changes occurred concurrently with film shrinkage. Once the methyl group was depleted completely, the film thickness and –OH/Si-O ratio became saturated. Thus, plasma treatment caused methyl depletion (or pore collapse) accompanied by film shrinkage and moisture uptake. Additionally, plasma damage such as film shrinkage, methyl depletion, and moisture uptake were found to depend on plasma density, plasma energy, and plasma species. These effects will be discussed in details later.

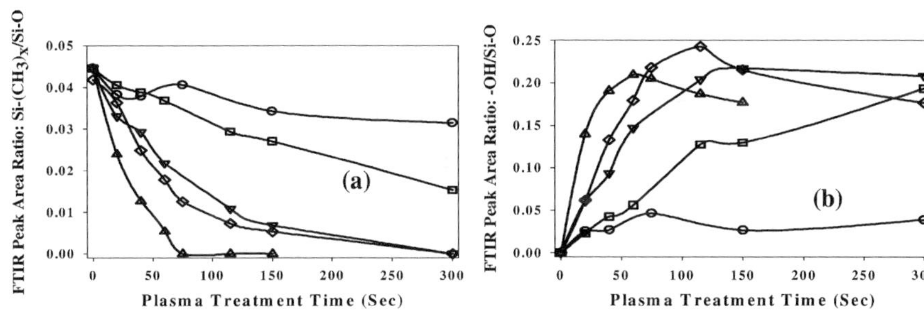

Fig. 3 FTIR peak area ratio (a) Si-CH_3/Si-O and (b) –OH/Si-O for OSG treated by plasma: Δ ICP CO_2, 300W, 5mt, ∇ ICP CO_2, 150W, 5mt, □ ICP CO_2, 150W, 150mt, ◊ ICP O_2, 150W, 5mt, oRIE CO_2, 150W, 5mt

To investigate the origin of the dielectric loss induced by CO_2 plasma, the polarization components were extracted in Fig. 5 by using a combination of analytical techniques including C-V, SE, and FTIR with the Kramers-Kronig dispersion relation, as described in a previous report [16]. After 40-second CO_2 ICP plasma treatment, the dielectric constant obtained from C-V measurement at 1MHz increased significantly from 2.20 to 3.42, while the sum of the electronic dielectric constant and the ionic dielectric constant remained almost unchanged at about 2.10. This indicates that the major dielectric loss was originated from the dipolar contribution such as moisture uptake.

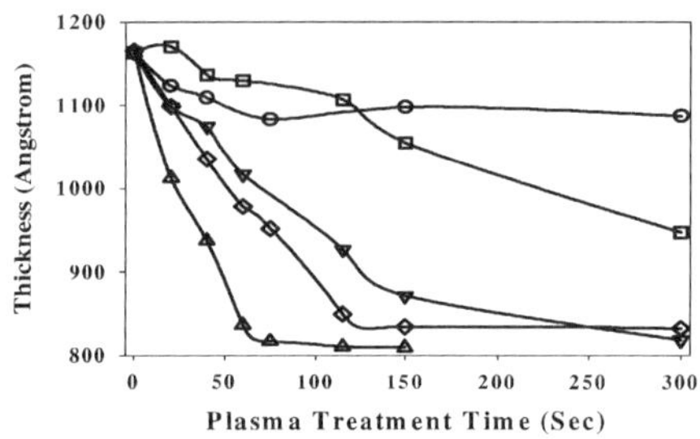

Fig. 4 Thickness for OSG treated by plasma: Δ ICP CO_2,300W,5mt,∇ ICP CO_2, 150W, 5mt □ ICP CO_2,150W,150mt,◊ICP O_2, 150W, 5mt oRIE CO_2, 150W, 5mt

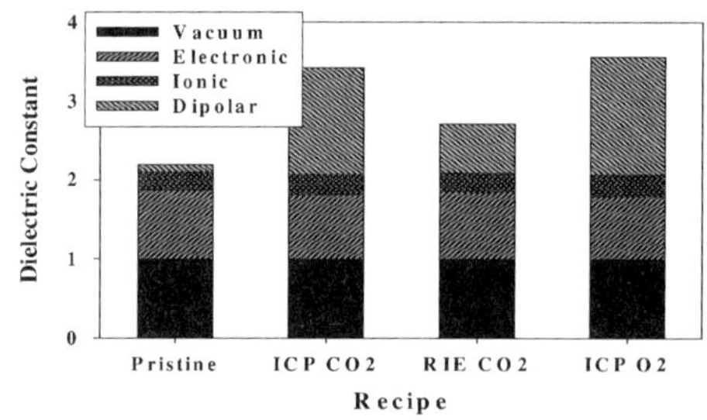

Fig. 5 Dielectric constant components for OSG treated by different plasmas

Plasma damaged OSG films were immersed in a DHF solution with different dipping times to investigate the development of surface roughness and thickness reduction. As shown in Fig. 6, the surface roughness from CO_2 plasma treatment prior to DHF dipping was about the same as the pristine OSG. But after DHF dipping, the surface became much rougher compared to the pristine OSG, probably due to the plasma-induced formation of SiO_2-like structure near the surface. The results indicate that, although DHF cleaning after plasma treatment helped remove the plasma damaged layer, it induced a rougher surface, which presents a difficult challenge to the subsequent barrier deposition.

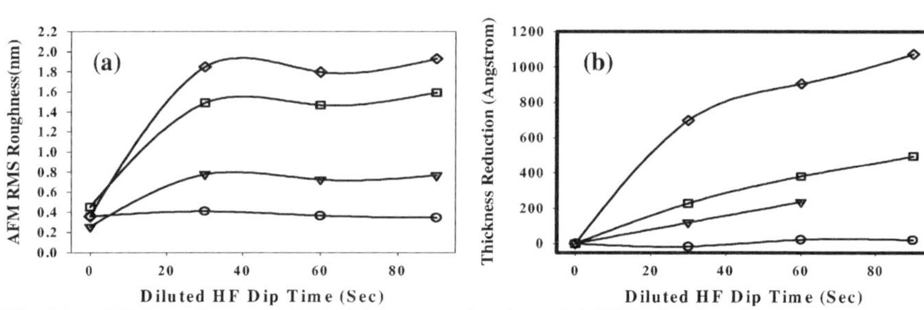

Fig.6 (a) AFM roughness and (b) thickness reduction with DHF dipping for OSG: o pristine, ∇ RIE CO_2, 50W, 5mt, □ ICP CO_2, 50W, 5mt, ◊ ICP CO_2, 150W,5mt

Effect of Plasma Density, Plasma Energy, Plasma Species, and Low-k Chemistry

With increasing power of ICP source, plasma damage became more severe, as indicated by the smaller FTIR Si-CH_3/Si-O ratio in Fig. 3 (a), larger FTIR –OH/Si-O ratio in Fig. 3 (b), more film loss in Fig. 4, severer AFM RMS roughness in Fig. 6 (a) and higher DHF etching rate in Fig. 6 (b). These phenomena can be explained by the fact that plasma density for bond breaking and sputter desorption of low k constituents was increased when the OSG film was subject to a higher ICP source power [17-18].

Interestingly, with the increase of chamber pressure, the plasma damage was reduced, as manifested by higher FTIR Si-$(CH_3)_x$/Si-O ratio in Fig. 3 (a), lower FTIR -OH/Si-O ratio in Fig. 3 (b) and less film loss in Fig. 4. These might be attributed to the lower plasma density or a shorter mean free path resulting from the plasma operation at a higher chamber pressure.

Under an identical plasma condition, pure ICP plasma induced more damages than RIE plasma, as evidenced by the lower FTIR Si-CH_3/Si-O ratio as shown in Fig. 3a, a higher FTIR –OH/Si-O ratio (Fig. 3b), more film loss (Fig. 4), larger AFM RMS roughness (Fig. 6a), higher DHF etching rate (Fig. 6b), and thicker carbon depletion layer in Fig. 7a. Moreover, ICP plasma induced a higher dielectric constant, 3.42 at 1MHz, than 2.72 from RIE plasma (Fig. 5). There are two possible mechanisms differentiating RIE from ICP treatment. First, compared with the pure ICP plasma, RIE plasma has a much lower plasma density. Second, RIE plasma has a much higher ion energy. Due to physical bombardment, surface layer after RIE plasma treatment was more densified or pore size was more reduced, blocking the oxygen radical diffusion [19].

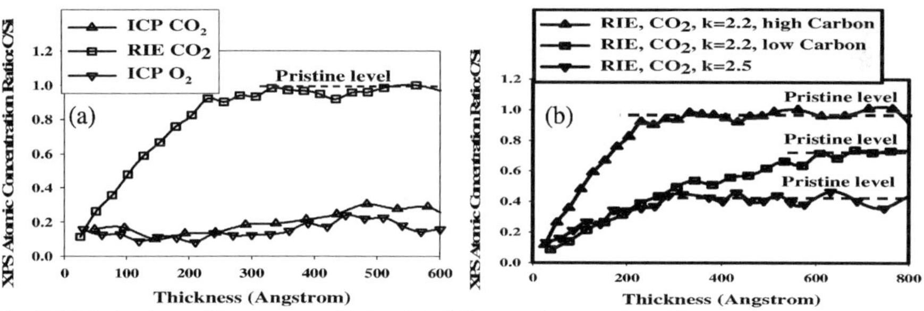

Fig. 7 XPS depth profile (a) same OSG with different plasmas (b) different OSG's treated by same RIE CO_2 plasma

As shown in Fig. 7b and Fig. 8, the thickness of the carbon depletion layer and the film loss rate were strongly dependent on the low-k structure. With the reduction of the k value, there were thicker carbon depletion layer and higher film loss rate [6, 8]. This is probably due to the fact that the porosity and pore size usually increase with a decrease of the k value. Additionally, the high carbon doping in the k~2.2 OSG film improved the material resistance to plasma damage [12]. The high carbon doping level increased the reaction between oxygen radicals and carbon species in the OSG surface, inducing more pore collapse or more densification on the film surface, reducing pore size, and blocking the oxygen radical diffusion.

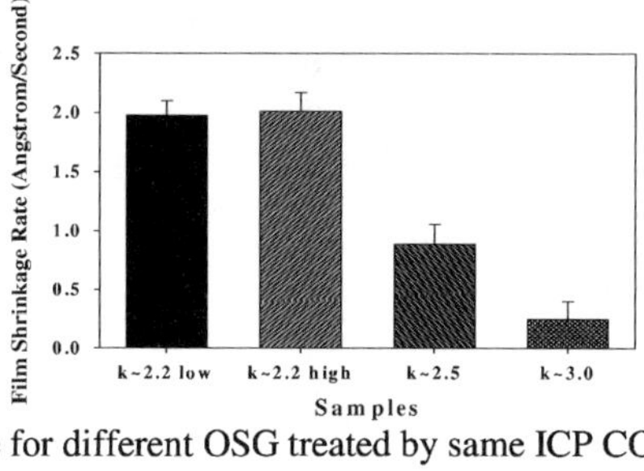

Fig. 8. Film shrinkage rate for different OSG treated by same ICP CO_2 plasma

Compared with O_2 plasma, CO_2 plasma induced less damage, as shown by higher FTIR Si-$(CH_3)_x$/Si-O ratio in Fig. 3 (a), lower FTIR -OH/Si-O ratio in Fig. 3 (b), less film loss in Fig. 4, and thinner carbon depletion layer in Fig. 7 (a). Additionally, CO_2 plasma resulted in a smaller dielectric constant, 3.42 at 1MHz, than 3.57 from O_2 plasma (Fig. 5). These differences might arise due to the lower oxygen radical concentration from CO_2 plasma than O_2 plasma. Two possible mechanisms were proposed for it. First, the bond energy for C=O bond was 799 kJ/mol, much bigger than 495 kJ/mol for O=O bond [20]. So the dissociation of CO_2 plasma was more difficult than O_2 plasma. Second, according to the optical emission spectroscopy, CO_2 plasma species might include CO, C_2, C, O, etc. [21-22], some of which could recombine with oxygen radical.

Plasma Altered Layer Model

By adopting the kinetics of radical diffusion, reaction, and recombination, Baklanov et al. investigated the penetration depth of oxygen plasma radical species inside porous low-k dielectrics by Thiele analysis [11]. Using a similar approach and considering the evolution of carbon concentration in OSG low-k dielectrics, we developed a plasma altered layer model to quantify the plasma damage on porous OSG. Due to the limited scope of this paper, we report here only the final solutions. The detailed derivation will be presented elsewhere [23]. In this model, the depth of carbon depletion layer (L) is $L = \sqrt{\frac{Dr_p}{2(\frac{\psi_0}{\sqrt{2}}\gamma_{reac} + \gamma_{recom})} In(\frac{\phi_0 \gamma_{reac} t_0}{In\sqrt{2}})}$,where r_p is the pore radius, D is Knudsen diffusivity and proportional to r_p, ϕ_0 is the oxygen radical density at vacuum, ψ_0 is the carbon concentration in pristine OSG, γ_{reac} is the reactivity between oxygen radical and carbon species, γ_{recom} is the surface recombination rate of oxygen radical on pore wall, and t_0 is time. With the decrease of oxygen radical density at vacuum (ϕ_0), Knudsen diffusivity (D), and pore size (r_p), or with the increase of carbon concentration in pristine OSG (ψ_0) and surface recombination rate (γ_{recom}), the depth of carbon depletion layer is decreased. The prediction given by this model was consistent with the experimental results.

CONCLUSIONS

The extent of plasma damage depended on the plasma density, plasma energy, plasma species, and low-k dielectric chemistry. For ICP treatment, plasma damage was found to increase with increasing source power or decreasing chamber pressure. Under identical plasma conditions, pure ICP plasma induced more damage than RIE plasma. Compared with O_2 plasma, CO_2 plasma induced less damage. With a decrease of k value, the depth of carbon depletion layer and film loss increased. Additionally, the high carbon doping level improved the material resistance to plasma damage. The plasma altered layer model which was developed can be combined with analytical results for quantitative evaluation of plasma damage. Such an approach can provide a quantitative basis for optimization of the plasma process and low k dielectric materials.

ACKNOWLEDGMENTS

The authors would like to thank G. F. Doyle of Molecular Imprints at Austin, Texas, for the use of the plasma etching chamber and thank Dr. M. R. Baklanov and Dr. J. Im for valuable discussions. This work was supported in part by Applied Materials Corp. and performed at the Microelectronics Research Center and at Center for Nano & Molecular Science & Technology (CNM) at UT Austin supported by NSF.

REFERENCES

1. K. Maex, M. R. Baklanov, D. Shamiryan, F. Lacopi, S. H. Brongersma, and Z. S. Yanovitskaya, J. Appl. Phys., 93, 8793 (2003).
2. J. Sun, D. W. Gidley, Y. Hu, W. E. Frieze, and E. T. Ryan, Appl. Phys. Lett., 81, 1447 (2002).
3. H. Lee, C. L. Soles, E. K. Lin, W. Wu, and Y. Liu, Appl. Phys. Lett., 91, 172908 (2007).
4. H. Cui, R. J. Carter, D. L. Moore, H. Peng, D. W. Gidley, P. A. Burke, J. Appl. Phys., 97, 113302 (2005).
5. M. A. Worsley, S. F. Bent, N. C. M. Fuller, T. L. Tai, J. Doyle, M. Rothwell, and T. Dalton, J. Appl. Phys., 101, 013305 (2007).
6. D. Shamiryan, M. R. Baklanov, S. Vanhaelemeersch, and K. Maex, J. Vac. Sci. Technol. B, 20 (5), 1923 (2002).
7. X. Hua, M. Kuo, G. S. Oehrlein, P. Lazzeri, E. Iacob, M. Anderle, C. K. Inoki, T. S. Kuan, P. Jiang, and W. Wu, J. Vac. Sci. Technol. B, 24, 1238 (2006).
8. N. Posseme, T. Chevolleau, T. David, M. Darnon, O. Louveau, and O. Joubert, J. Vac. Sci. Technol. B, 25 (6), 1928 (2007).
9. S. Xu, C. Qin, L. Diao, D. Gilbert, L. Hou, A. Wiesnoski, E. Busch, R. McGowan, B. White, and F. Weber, J. Vac. Sci. Technol. B, 25, 156 (2007).
10. J. Bao, H. Shi, J. Liu, H. Huang, P. S. Ho, M. D. Goodner, M. Moinpour, G. M. Kloster, J. Vac. Sci. Technol. B, 26 (1), 219 (2008).
11. M. R. Baklanov and A. M. Urbanowicz, N2.6, 2008 MRS Spring Meeting.
12. E. T. Ryan, A. Grill, S. Gates, S. Molis, P. Flaitz, J. Arnold, M. Sankarapandian, S. Cohen, Y. Ostrovski, K. Babich and V. Patel, N2.7, 2008 MRS Spring Meeting.
13. Q. T. Le, M. R. Baklanov, E. Kesters, A. Azioune, H. Struyf, W. Boullart, J. J. Pireaux, and S. Vanhaelemeersch, Electrochemical and Solid-State Letters, 8(7) F21 (2005).
14. Y. Uchida, T. Fukuda, and H. Yanazawa, Surf. Interface Anal. 36, 677 (2004).
15. A. Grill and D. A. Neumayer, J. Appl. Phys., 94, 6697 (2003).
16. H. Shi, J. Bao, J. Liu, H. Huang, R. S. Smith, Q. Zhao, P. S. Ho, M. D. Goodner, M. Moinpour, and G. M. Kloster, the Advanced Metallization Conference 2007,VB.3.
17. H. Nagai, M. Hiramatsu, M. Hori, and T. Goto, Rev. Sci. Instrum., Vol. 74, 3453, 2003.
18. R. J. Shul, C. G. Willison, M. M. Bridges, J. Han, J. W. Lee, S. J. Pearton, C. R. Abernathy, J. D. Mackenzie, and S. M. Donovan, Solid-state Electronics Vol. 42, No.12, 2269 (1998).
19. S. Hwang, G. Lee, J. Min, S. Moon, Surface and Coatings Technology, 174-175, 835 (2003).
20. S. S. Zumdahl, Chemistry, 5th edition, Houghton Mifflin, 2000.
21. P. Yang, R. M. Barnes, J. Vechiarelli, and P. C. Uden, Applied Spectroscopy, Volume 44, No. 3, 531, (1990).
22. P. G. Reyes, E. F. Mendez, D. Osorio-Gonzalez, F. Castillo, and H. Martinez, Phys. Stat. Sol. (c) 5, No. 4, 907, (2008).
23. H. Shi, H. Huang, J. Bao, Y. Sun, J. Im, P. S. Ho, A. Li, J. T. Pender, M. Armacost, and D. Kyser, Appl. Phys. Lett., in press.

Flowable CVD low k (k=2.8) C-doped oxide for inter-metallic dielectrics (IMD) application with 30nm level line/spacing

Sang H. Ahn, Bo Young Lee, Yong Soon Choi, Eunkee Hong, Eun Kyung Baek, Hyeon-Deok Lee, and Joo-Tae Moon

Process Development Team/Memory Division/Semiconductor Business, Samsung Electronics Co., LTD., San #16 Banwol-Dong, Hwasung-City, Gyeonggi-Do, 445-701, Korea, Republic of

ABSTRACT

Low k C-doped oxide (k<3.0) was successfully developed in a flowable chemical vapor deposition (FCVD) reactor for W inter-metallic dielectrics (IMD) application in 30nm device node. The film successfully filled 30nm level trenches with an aspect ratio of up to 9:1. This UV cured C-doped oxide in the trench looked very uniform without any noticeable porosity under the vertical scanning electron microscope. Its dielectric constant k ranged from 2.8 to 3.0 depending on the process conditions with its stability up to one week. UV cure at 350C improved its modulus to 12GPa by 30% compared to thermal cure at 400C. This UV cure improvement was also reflected in wet etch rate ratio (with respect to thermal oxide) in 6:1 BOE: 0.08(UV) vs. 0.95(thermal). Thermal desorption analysis showed no radicals out-gassed up to 600C after 550C UV cure. This UV cured film capped with PECVD oxide was applied to W IMD application at 30nm level without any voids and metal contact holes were dry etched, ashed, and cleaned with little bowing or broadening. W line–to–line capacitance measurement resulted in ~20% reduction with the respect to a reference oxide although its metal contact resistance was twice as high as a reference oxide. This high contact resistance was attributed to post dry etch C-related residue, which could be overcome by dry etch and ashing optimization. FCVD deposition demonstrated gap-fill capability as good as spin-on dielectrics. Subsequent 550C UV cure makes low k C-doped oxide mechanically stronger and more thermally-stable, opening up a large integration window. Therefore, combination of FCVD and UV cure can be a promising solution for advanced gap-fill low k IMD application.

INTRODUCTION

As device size shrinks below 40nm, interconnect resistance and capacitance rapidly increase, making BEOL interconnect a rate-limiting step for a device operation [1]. For a non-volatile memory device, its bit line has one of the smallest pitches among the BEOL interconnect lines so to become the bottleneck for the signal transmission. The combination of Cu and low k dielectrics can certainly improve the device speed most. However, one can still reduce RC delay cost effectively without major integration changes by employing low k dielectric only for W bit line IMD. Typically, low k C-doped oxide (SiOCH) by plasma enhanced chemical vapor deposition (PECVD) has been widely accepted for Cu Damascene processes since 90nm technology node, but it is not suitable for gap-fill applications such as W bit line IMD since its poor gap-fill capability prevents it from filling in the spaces between W lines. Therefore, there are practical needs for a deposition technique for low k dielectrics that can allow the process tunability of a CVD method as well as excellent gap-fill capability of spin-on dielectrics (SOD) technique. Flowable CVD demonstrates both tunability and gap-fill capabilities. A variety of low k spin-on-dielectrics such as HSQ or MSQ type C-doped oxide might as well serve for gap-fill IMD [2, 3]. However, they were very limited in the process tunability and flexibility on the process engineer level. This paper investigates FCVD low k C-doped oxide deposition and UV cure, film properties, and electrical results for a 30nm application.

EXPERIMENT

Flowable low k C-doped oxide (SiOCH) was deposited at T<100C and P<10Torr by flowing together an oxidizing agent and Si & C-containing precursor(s) such as siloxane gas and/or Si alkoxide, which form easily-condensable intermediary species in a gas phase. These intermediary species condense on the substrate, then, flow into a trench or a gap. Once deposited, this film was subsequently cured either thermally or by UV at 350C<T<550C. Thickness was measured by spectroscopic ellipsometer (KLA Spectra Fx) before and after both cures. FTIR (Nicolet) technique was employed to study chemical bond structures. Thermal desorption spectroscopy was used to study film stability after cure. Hg probe (SSM 5100 CV) measured its capacitance to calculate dielectric constant. Mechanical properties were measured by MTS nanoindenter on 1um thick film. The chemical wet etch test was performed in 6:1 BOE with respect to thermal oxide. In order to evaluate gap-fill performance, FCVD film was deposited on W patterns with 30nm level line/space as well as dielectric stacks with a high aspect ratio of up to 9:1. Vertical scanning electron microscope was used for gap-fill analysis. After UV cure, the film underwent O_2-free dry etch, room temperature O_2-ashing and EKC solvent clean prior to metal contact formation. The vertical profile was examined for any damages. In addition, electrical properties such as inter-metallic dielectric capacitance and contact resistance were measured.

DISCUSSION

For advanced device nodes, W IMD must meet the challenging requirements such as narrow and high aspect ratio trench gap-fill, low dielectric constant (k<3), reasonable mechanical strength, thermal stability (>450C), and resistance to etch/ashing damage. It is necessary to understand the relationship among process, structure, and properties of the film in order to meet all the requirements simultaneously. Even though the deposition mechanism of Flowable CVD is not clearly understood, it's possible to identify the key process knobs. According to FTIR spectra on as-dep FCVD film (see Figure 1), a large and broad Si-OH peak was observed from $3000cm^{-1}$ to $3650cm^{-1}$ in addition to CH_x ($2910\sim2965cm^{-1}$), Si-CH_3 ($1275cm^{-1}$), and Si-O-Si ($1060cm^{-1}$, $1100cm^{-1}$). SiO-R_n (R=hydrocarbon) in a siloxane precursor undergoes a hydrolysis reaction in the presence of an oxidizer and becomes an easily- condensable specie such as $SiOR_{n-x}$-$(OH)_x$. $SiOR_{n-x}$-$(OH)_x$ molecules land and condense by van der Waal force on the substrate at T<100C. As n in $SiOR_{n-x}$-$(OH)_x$ increases by higher siloxane/oxidizer flow ratio, the film becomes more fluidic, forming an increasing flow angle which is defined by the film on the bottom and the sidewall of an isolated line pattern. Subsequent thermal or UV cure induces a condensation reaction among $SiOR_{n-x}$-$(OH)_x$ molecules by eliminating H_2O from the film, resulting in Si-O-Si network formation. In fact, Si-OH peaks disappear completely after cure, and Si-O-Si network (1040~60cm^{-1}) to Si-O-Si cage (1100cm^{-1}) bond ratio increases noticeably (see Figure 1)

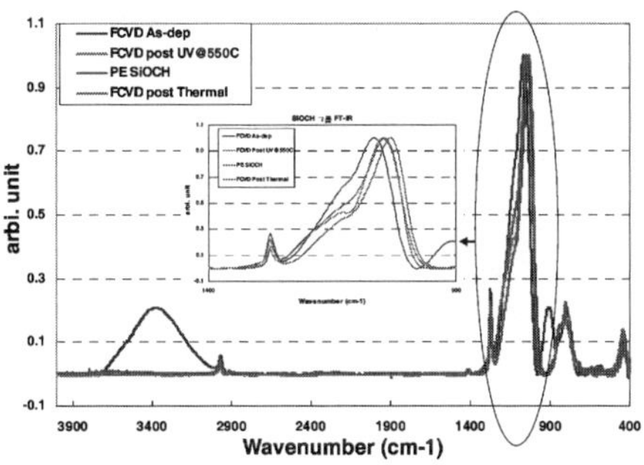

Figure 1. FTIR Spectra for FCVD as-dep, post UV, post thermal, and PE SiOCH (k=3.0) as a reference. Abundant [Si-OH] was observed for as-dep FCVD. The degree of Si-O-Si network increases in the order of as-dep, post thermal, and post UV.

Since Si-CH_3 bond can be correlated with dielectric constant [4] and with the ability of the film to flow, this work puts a major focus on studying dielectric constant and gap-fill capability as a function of [Si-CH_3]. For simplicity, [Si-CH_3] was measured by FTIR and defined to be a peak area ratio of Si-CH_3 around 1275cm^{-1} with respect to Si-O-Si around 1050cm^{-1}. [Si-CH_3] can be tuned in a wide range either by deposition process and/or by post-deposition cure as shown in Figure 2. Dielectric constant of FCVD film increased linearly from 2.6 up to 3.8 as [Si-CH_3] decreases. For a given pattern, it was observed that the number of void-defects decreases with increasing [Si-CH_3]. However, dielectric constant of FCVD film needs to be <3.0 in order to successfully fill 30nm level trenches with aspect ratio of up to 9:1. This post cure C-doped oxide in the trench looked very uniform without any noticeable porosity under the vertical scanning electron microscope (see Figure 3).

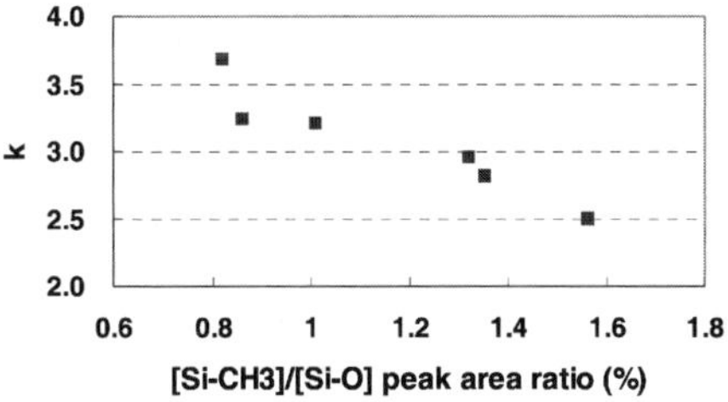

Figure 2. Dielectric constant vs. [Si-CH_3]/[Si-O] peak area ratio(%). Dielectric constant decreases linearly with [Si-CH_3]/[Si-O].

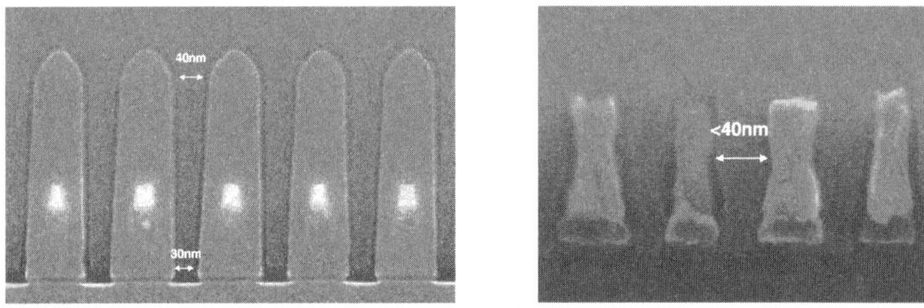

Figure 3. FCVD gap-fill capability: Dielectric stack with aspect ratio > 9:1 (left), and W IMD (right).

Focusing on FCVD film with k<3.0, effects of cure (thermal vs. UV) were studied with respect to film stability, wet etch rate ratio, and modulus. Both thermally- and UV- cured films stay stable without any changes in refractive index and dielectric constant for a week. However, UV cured film showed better thermal stability as thermal desorption spectroscopy showed that the radicals such as $H_2{}^+$ and $CH_3{}^+$ in a UV-cured sample started out-gassing at 50C higher than in a thermally-cured sample (see Figure 4).

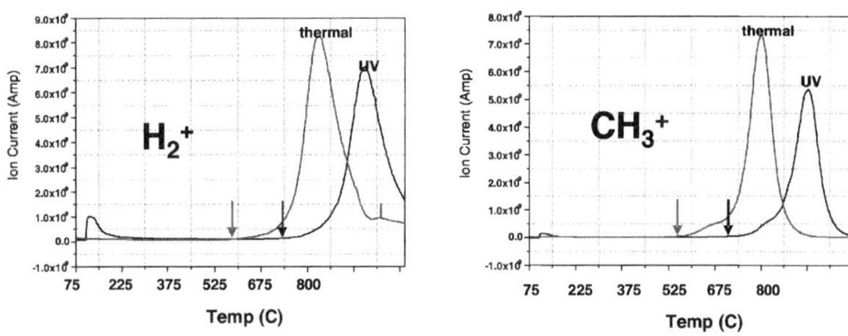

Figure 4. Thermal stability: Thermal cure at 550C vs. UV cure at 550C. $H_2{}^+$ and $CH_3{}^+$ out-gassing started around 550C for thermally-cured film and above 600C for UV-cured film.

UV cure improved its modulus to 12GPa by 30% compared to thermal cure. This is consistent with the fact that higher Si-O-Si network to cage bond ratio was observed for UV-cured film (see Figure 1). UV cure is more effective in forming Si-O-Si network than thermal cure at the same temperature by providing enough energy to rearrange Si-O and $Si\text{-}CH_3$ bonds. This UV cure improvement was also reflected in wet etch rate ratio (with respect to thermal oxide) in 6:1 BOE: 0.08(UV at 350C) vs. 0.95(thermal at 400C). In spite of lower UV cure temperature and comparable $[Si\text{-}CH_3]$, WERR for UV-cured film was 10 times slower due to stronger bond structure. This UV cured film capped with PECVD oxide was applied to W IMD application at 30nm level without any voids. Furthermore, metal contact holes were processed by dry etch, and ash at room temperature, and little bowing or broadening was observed after clean (see Figure 5). The seemingly slight bowing right above W line in Figure 5 can be attributed to a high acceleration voltage for SEM analysis. W line–to–line capacitance measurement resulted in ~20% reduction with the respect to a baseline oxide (see Figure 6), and its metal contact resistance was twice as high as a baseline oxide (see Figure 7). Post dry-etch carbon residues were likely to cause the increase in R_c and requires further improvements in dry etch, ashing, and clean.

Figure 5. Metal contact open on FCVD with PE oxide capping.

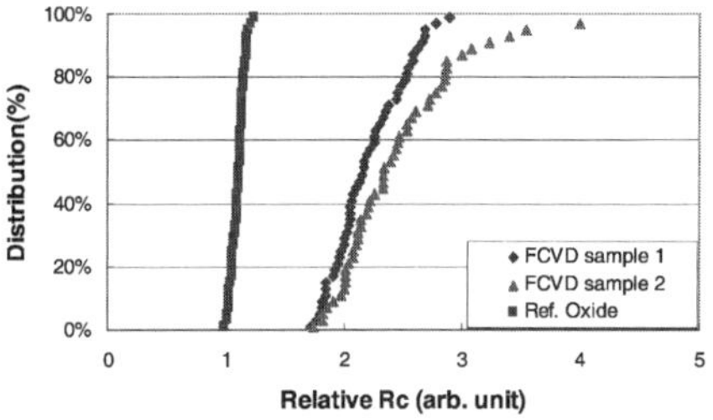

Figure 6. W Metal contact resistance with FCVD.

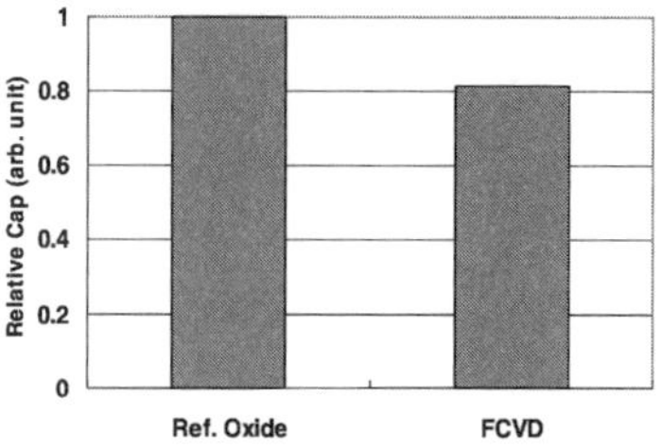

Figure 7. Line-to-line capacitance of FCVD.

CONCLUSIONS

FCVD deposition demonstrated gap-fill capability as good as spin-on dielectrics. Subsequent 550C UV cure makes low k C-doped oxide mechanically stronger and more thermally-stable, opening up a large integration window. FCVD with k~3 was able to reduce inter-metallic capacitance by 20% but contact resistance was twice higher than a reference. Nonetheless, less-polymerizing dry-etch with improved ashing can remove any C-related residues to bring down the contact resistance to a reference level. Therefore, combination of FCVD and UV cure can be a promising solution for advanced gap-fill low k IMD application.

ACKNOWLEDGMENTS

We are thankful for Rs line to accommodate FCVD integration.

REFERENCES

1. Mihail P. Petkov, Technology Readiness Overview March 2003.
2. N. Ahner et el, Microelectronic Engineering 84 2007 (2606-2609).
3. C. H. Diaz, 1999 Symposium on VLSI Technology Digest of Technical Papers p11-12.
4. T.K.S Wang, Thin Solid Films 462-463 (2004) 156-160.

UV cure control of pore interconnectivity for porogen-based SiOC films

Naoto Tsuji, Kiyohiro Matsushita, Shinya Kaneko, Noboru Takamure, and Nobuyoshi Kobayashi
ASM Japan K.K., Nagayama 6-23-1, Tama-shi, Tokyo, Japan
Corresponding Author: Naoto Tsuji: tel.: +81 42 337 6314 fax: +81 42 389 7555
e-mail: n. tsuji@asm.com

ABSTRACT

We have investigated the correlation between k-value, mechanical strength, and interconnectivity of porogen-based SiOC films by using UV curing. With addition of an appropriate amount of oxygen during UV curing, we tried to create a robust porogen-based SiOC film with low pore interconnectivity, comparable to non porogen-based SiOC film. As a result of our investigation, we also have found that the restoration of the open pores proceeded more slowly but steadily, while k-value sharply decreased at the initial stage and slightly increased with curing time.

INTRODUCTION

UV curing is the most promising method for enhancing the mechanical strength of porous SiOC films in the 45nm node and beyond [1, 2]. In addition, UV curing improves the damage resistance of the films since it reduces pore interconnectivity of the non porogen-based SiOC films [3]. A film with low pore interconnectivity suppresses chemical penetration depth through pores, resulting in reduced wet and plasma damage.

In this work, we found that an appropriate amount of oxygen improves the mechanical strength without k-value increase. Moreover, oxygen addition reduces pore interconnectivity more than conventional UV curing. It makes the film robust against wet and plasma damage.

EXPERIMENT

Porogen-based SiOC films (Aurora®ELK; k 2.2~2.7) and non porogen-based SiOC films (Aurora®ULK; k~2.7) were deposited by PE-CVD system. ELK films were followed by UV curing to remove porogen. The dielectric constant (k-value) measurements were performed by Hg-probe, and the elastic modulus and hardness were measured by nanoindentation. The film thickness and refractive index were performed by spectroscopic ellipsometry, and the microscopic properties were determined by Fourier Transform Infrared Spectroscopy (FT-IR). Pore interconnectivity was evaluated by measuring the diffusion coefficient (Dc) of a solvent in the SiOC films [4].

DISCUSSION

Table.1 shows the typical film properties of Aurora®ULK and ELK cured by UV under O_2 free (100% N_2), and O_2 added atmosphere.

Table.1. Film properties of ULK (as-deposited) and ELK (UV cured).

	Film	UV	k	EM (GPa)	H (GPa)	R.I.	Dc (um^2/sec)
Non porogen-based	ULK	as-depo	2.75	8.5	1.4	1.386	0.12
Porogen-based	ELK	w/o O_2	2.47	8.0	1.2	1.359	10.60
	ELK	with O_2	2.48	9.1	1.5	1.406	6.80

The elastic modulus of ELK film was increased by 15% without increase of k-value after being cured by UV under O_2 added atmosphere, as compared to that after curing under O_2 free atmosphere. Dc was reduced by ~35% for O_2 addition.

Figure 1 shows the elastic modulus of ELK film as a function of the Si-CH_3/Si-O ratio from FT-IR spectra. It is about 1GPa higher when ELK was cured by UV under O_2 added atmosphere. This indicates that Si-O network mainly grew, while the amount of Si-CH_3 bonds was sustained after O_2 addition UV curing.

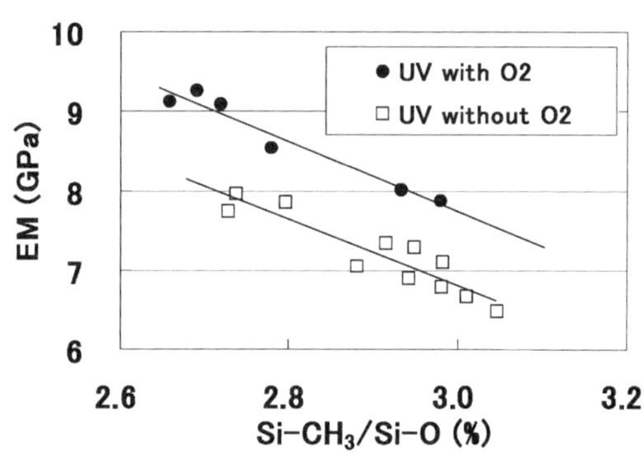

Fig. 1. Elastic modulus of ELK films as a function of Si-CH_3/ Si-O ratio

Figure 2 shows k-value and Dc as a function of UV curing time with O_2 addition. The k-value sharply decreases at the initial stage and slightly increases with curing time. Note that Dc has a peak at around 0.5 (a.u) cure time, and decreases more slowly but steadily afterwards. In the shaded region with a long UV curing time, the values of k and Dc are comparable to those of non-porogen ULK.

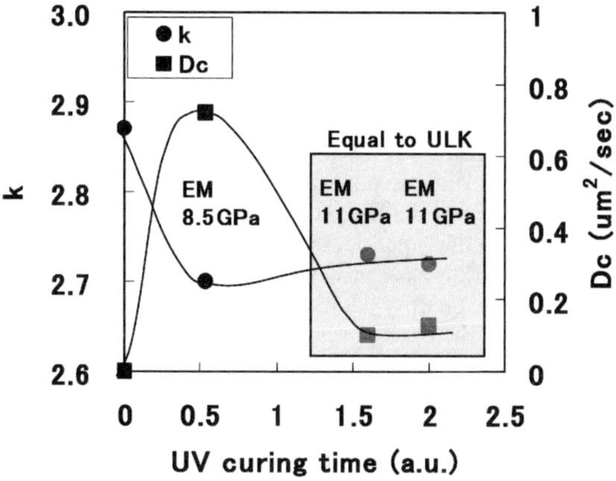

Fig. 2. Dielectric constant (k) and diffusion coefficient (Dc) as a function of UV curing time.

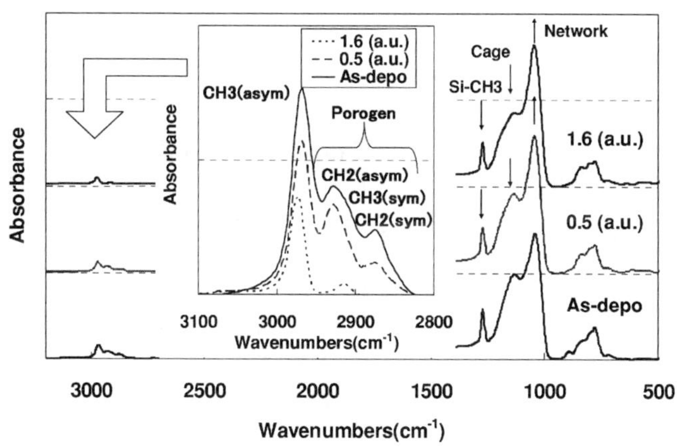

Fig. 3. FT-IR spectra of ELK films (as-deposited and UV cured).

The FT-IR spectra of Aurora®ELK films after deposition and during UV curing are shown in Figure 3. In Fig. 3, the changes in the FT-IR spectra are observed after UV curing in three areas: 1) the CH_x peak area (2850-3000 cm^{-1}), 2) $SiCH_3$ peak area (1285 cm^{-1}) and 3) Si–O–Si network area (1000-1150 cm^{-1}). The FT-IR spectra show that Si–O–Si network structures increase and Si-CH_3 peak decreases with UV curing time, resulting in the decrease of Si-CH3/Si-O ratio. In addition, CH_x bonds (porogen-related groups) rapidly decrease with UV curing time, resulting in the decrease of k-value [5-7].

By comparing Figs. 2 and 3, FT-IR results show that Dc peak corresponds to the initial porogen removal (CH_x bonds), and Dc gradually decreases with the Si-O network growing. During this period, the peak ratio of Si-CH_3/ Si-O and CH_x peaks decrease with UV curing time. It is thought that during UV curing the porogen molecule is disintegrated and diffuses from the film, leaving a porous SiOC network. The reduction of the CH_x peaks is a relatively rapid reaction, and this correlates to the rapid decrease in k-value. The decrease of the Si-CH_3 peak and increase in the SiO network area are associated to the network restructuring and increase in elastic modulus. While k-value sharply decreases at the initial stage and slightly increases with curing time, the restoration of the open pores proceeded more slowly but steadily with the decrease of Dc.

Figure 4 shows k-value and Dc as a function of Porogen/ (Porogen + Si precursor) ratio after 1.6 (a.u) cure time. The results of ELK film in Figures 2 and 3 correspond to those with the ratio of ~ 0.5. By choosing an appropriate amount of porogen, interconnectivity (Dc) and k-value of the films can be controlled continuously.

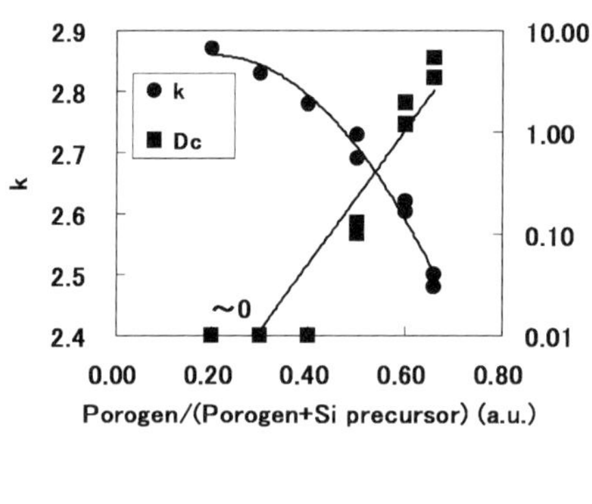

Fig. 4. k-value and Dc as a function of the ratio Porogen/ (Porogen + Si precursor).

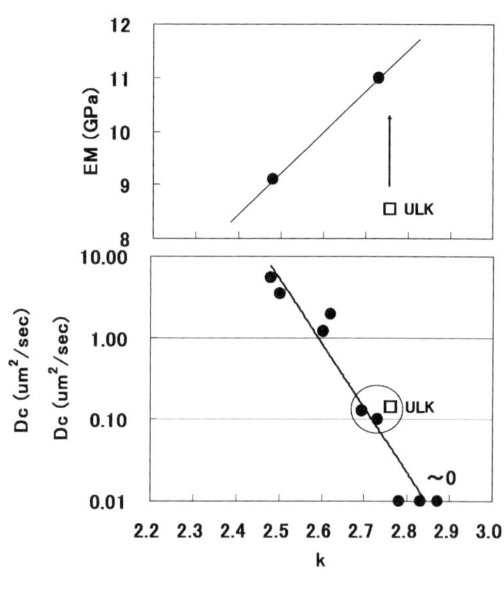

Fig. 5. EM and Dc vs. k-value at various ELK deposition conditions.

Figure 5 shows EM and Dc vs. k-value at various ELK deposition conditions after 1.6 (a.u) cure time. Porogen-based ELK with k-value~ 2.7 has a comparable Dc, however, a much larger EM by O_2-added UV curing, compared to non porogen-based ULK. Furthermore, even ELK with k-value~ 2.5 has a slightly larger EM. This hardening occurs, of course, with UV curing. However, considering the carbon concentration of films, the C/Si ratio of the Si precursor for porogen-based ELK is about 1.3 times higher than the ratio of it for ULK, and the Si-CH_3/Si-O ratio of as-deposited ELK (after curing: k~ 2.5) is 3.5 %, while that of ULK (k~ 2.75) is 2.8 %. Thus an EM enhancement can be due to higher carbon concentration in porogen-based ELK film.

CONCLUSIONS

The correlation between k-value, mechanical strength (EM), and interconnectivity (Dc) was investigated using porogen-based SiOC films after UV curing with addition of an appropriate amount of oxygen. FT-IR results showed that Dc peak corresponded to the initial porogen removal (CH_x bonds), and then Si–O–Si network growth restored open pores and decreased Dc. This restoration of the open pores proceeded slower but more steadily, while k-value sharply decreased at the initial stage and slightly increased with curing time. By choosing a suitable porogen amount, a porogen-based SiOC film with low pore interconnectivity and higher EM has been obtained as compared to non porogen-based SiOC film.

REFERENCES

1. S. I. Nakao, J. Ushio, T. Ohno, T. Hamada, Y. Kamigaki, M. Kato, K. Yoneda, S. Kondo and N. Kobayashi, International Interconnect Technology Conference 2006 (IITC 2006), IEEE, 2006, p. 66-68
2. M. Kato, K. Yoneda, S. Nakano, S. Kondo, N. Kobayashi, N. Matsuki, K. Matsushita, N. Ohara, A. Fukazawa, T. Ikegawa and T. Kimura, Advanced Metallization Conference 2005, p. 271-276
3. K. Matsushita, N. Tsuji, K. Kagami, M. Kato, S. Kaneko and N. Kobayashi, Advanced Metallization Conference 2007, p. 42-43
4. D. Shamiryan, T. Abell, Q.T. Le, and K. Maex, Microelectronic Engineering 70/2-4 (2003) p. 341-345
5. P. Verdonck, D. D. Roest, S. Kaneko, R. Caluwaerts, N. Tsuji, K. Matsushita, N. Kemeling, Y. Travaly, H. Sprey, M. Schaekers and G. Beyer, Surf. Coat. Technol. 201 (2007) p. 9264-9268
6. K. Chan, K. S. Yim, V. Nguyen, S. I. Yi, K. Elsheref, T. Nowak, J. Rocha, A. Demos, D. Witty and H. M'Saad, Advanced Metallization Conference 2007, p. 162-163
7. C. H. Huang, H. L. Huang, C. I. Hung, N. F. Wang, Y. H. Wang and M. P. Houng, Jpn. J. Appl. Phys. 47 (2008) p. 1532-1535

Conference Proceedings AMC XXIV © 2009 Materials Research Society

Spin-on organic low-k material (k= 2.2) with thin Ti liner interconnect

Yukiharu Ono[1], Mihoko Matsutani[1], Takahiro Harada[1], Masahiro Tada[2], and Yoshito Ando[2]

[1]Sumitomo Bakelite Co., Ltd.
Address: 495, Akiba-cho, Totsuka-ku, Yokohama, 245-0052, Japan
Phone: +81-45-812-0914, Fax: +81-45-812-8480, E-mail: onoyukih@sumibe.co.jp
[2]Consortium for Advanced Semiconductor Materials and Related Technologies (CASMAT)
Address: 1-280, Higashi-Koigakubo, Kokubunji, Tokyo, 185-0014, Japan

ABSTRACT

Properties of blanket film and Ti liner interconnect with organic low-k material (k= 2.2) were investigated. The time-dependent dielectric breakdown (TDDB) lifetimes of both the blanket film and Ti liner interconnect exceed 10 years (0.2 MV/cm at 140°C), which means that the k= 2.2 material has a Cu diffusion barrier performance. The RC product of k= 2.2 material is 4% smaller than that of k= 2.65 material. In summary, k= 2.2 material is a strong candidate for the interlayer dielectrics of fine interconnects of the 32-nm generation and beyond.

INTRODUCTION

The RC value of interconnects for LSI operations at higher speeds and/or lower power consumption must be reduced. Cu/low-k interconnects need a barrier metal layer around the Cu line to prevent Cu diffusion. Since barrier metal has a higher resistivity than Cu, it causes increasing Cu line resistance. Therefore, thinning out the barrier metal layer of Cu/low-k interconnects is an advantageous way to reduce RC. However, thinning out the barrier metal layer might lead to Cu diffusion from the Cu line. The use of a low-k material with a Cu diffusion barrier performance is essential for constructing a reliable interconnect with a thin barrier metal layer. We tried using the spin-on type organic polymer material (k= 2.65) [1][2]. In this work, we used a novel spin-on type organic polymer material (k= 2.2) to fabricate single-layer Cu damascene interconnects to evaluate their electrical characteristics and reliability. We used a Ti liner to improve the adhesion performance [3].

Table I shows the properties of k= 2.2 material that we used, which is a non-porogen type material. The dielectric constant k is 2.19 and the pore diameter measured by using PALS is 1.98 nm. The CTE (out-of-plane) is 25 ppm/°C, which is near that of Cu (16.6 ppm/°C). In order to evaluate the Cu diffusion barrier performance of the k= 2.2 material, TDDB measurements were carried out, and the resulting graph is shown in Fig. 1. It can be seen that the k= 2.2 material can guarantee a 10-year lifetime (0.2 MV/cm at 140°C).

Table I. Properties of k= 2.2 material (cure: 400°C/30 min.)

Properties	Unit		Method
Dielectric constant		2.19	Hg probe (@100 kHz)
Dielectric breakdown	MV/cm	3.6	Hg probe
Leak current	A/cm^2	1.7E-10	Hg probe (@1 MV/cm)
		3.8E-9	Hg probe (@2 MV/cm)
Refractive index		1.49	n&k analyzer
Density	g/cm^3	0.89	XRR
*Pore diameter	nm	1.98	PALS
**CTE (out-of-plane)	ppm/°C	25	XRR (30-350°C)
Young's modulus	GPa	4.9 (720nm)	Nanoindentation
Hardness	GPa	0.26 (720nm)	Nanoindentation
Tg	°C	>400	m-DSC
Adhesion		0/100	Tape test (SiCN, Si)
	MPa $m^{1/2}$	0.22	m-ELT (low-k/SiCN)

*Measured by the Advanced Industrial Science and Technology (AIST)
**CTE (out-of-plane): (Cu) 16.6 ppm/°C (k= 2.65 material) 86 ppm/°C

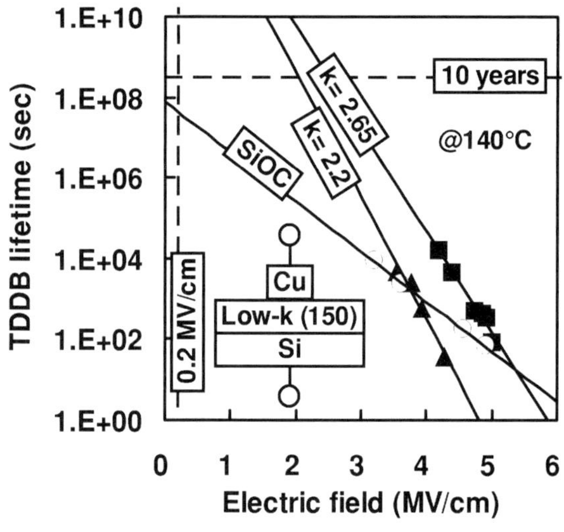

Figure 1. TDDB lifetimes of low-k films

EXPERIMENT

We fabricated single-layer Cu damascene interconnects to evaluate their electrical characteristics. We prepared several types of structures to clarify the differences between their electrical characteristics using a low-k of k= 2.2 material, k= 2.65 material, and SiO. A schematic of the cross-section of a single-layer interconnect is shown in Fig. 2. The barrier dielectric and CMP-cap are SiCN. We evaluated the line resistance of a 20-cm serpentine pattern and the interline capacitance of a 20-cm comb pattern (L/S= 90 nm/90 nm). The TDDB was also measured for a 1-cm comb pattern at 140°C (L/S= 90 nm/90 nm).

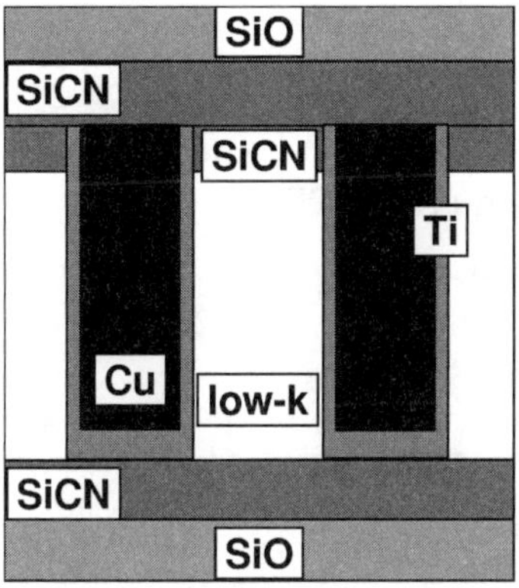

Figure 2. Schematic of M1 layer

DISCUSSION

Line resistance

Fig. 3 shows the 90-nm line resistance distribution of a k= 2.2 material with and without a Ti liner. The line resistance yield with a Ti liner is very high (99%) while that without a Ti liner is lower than 70%. We assume this is due to a low level of adhesion between the low-k and Cu without a Ti liner.

RC reduction

The RC product of the k= 2.2 material is 4% smaller than that of the k= 2.65 material, as shown in Fig. 4. This 4% value is near the 6% that was calculated by using a semiconductor device simulator (Silvaco International, CLEVER).

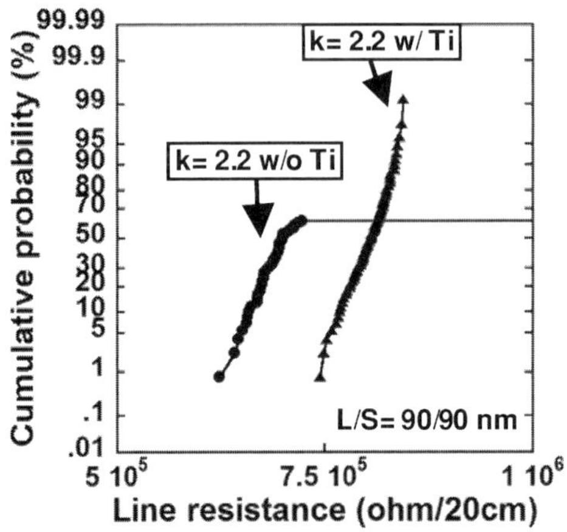

Figure 3. Line resistance of M1 layer with k= 2.2 material

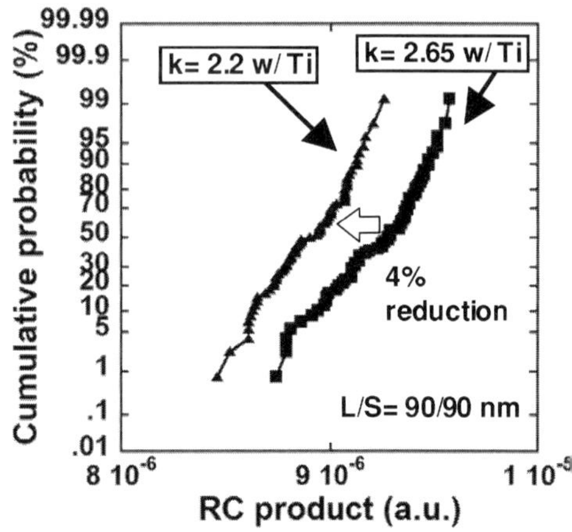

Figure 4. RC product comparison of M1 layer

Cross-section of M1 layer

A FIB-STEM image of the M1 layer with k= 2.2 material is shown in Fig. 5. No Cu penetration into the low-k layer was observed. The thickness of the Ti liner was about 2 nm.

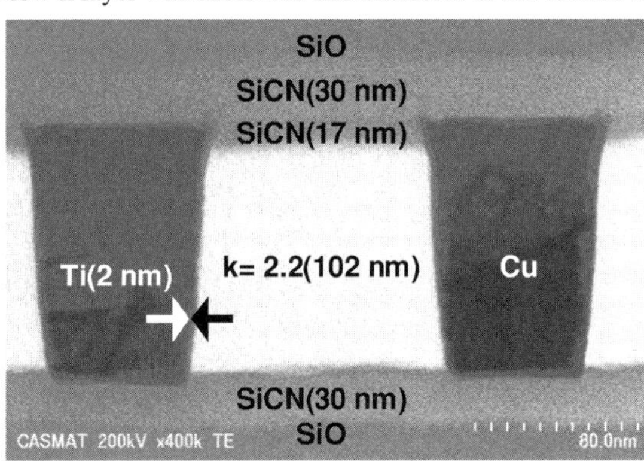

Figure 5. FIB-STEM image of M1 layer with k= 2.2 material (L/S= 90/90 nm)

TDDB measurements

The results of the interline TDDB measurements of M1 layer with k= 2.2 material and SiO are shown in Fig. 6. The straight and dotted lines represent the values for with and without a Ti liner, respectively. The TDDB lifetime of the k= 2.2 material with a Ti liner exceeds 10 years, which is similar to that without a Ti liner, while that of SiO without a Ti liner is much shorter than that of the k= 2.2 material. This means that the k= 2.2 material has a Cu diffusion barrier performance.

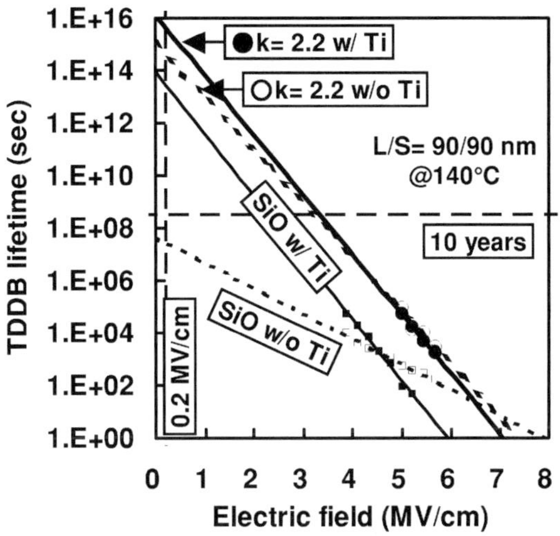

Figure 6. TDDB lifetime of M1 layer with k= 2.2 material and SiO

CONCLUSION

We evaluated the properties of blanket film and Ti liner interconnects with organic low-k material (k= 2.2). The TDDB lifetimes of both the blanket film and Ti liner interconnect exceeded 10 years (0.2 MV/cm at 140°C), which means that the k= 2.2 material has a Cu diffusion barrier performance. The RC product of the k= 2.2 material is 4% smaller than that of the k= 2.65 material. In summary, the k=2.2 material is a strong candidate for the interlayer dielectrics of fine interconnects of the 32-nm generation and beyond.

ACKNOWLEDGEMENT

This research was supported by the New Energy and Industrial Technology Development Organization (NEDO) of Japan.

REFERENCES

1. N. Maeda et al., Proc. AMC2006 (2006) 359
2. N. Maeda et al., Proc. AMC2007 (2007) 521
3. M. Ueki et al., Proc. IEDM (2002) 749

Conference Proceedings AMC XXIV © 2009 Materials Research Society

Ultra Low-k Precursor Solution for 32 nm-node Interconnects

M. Hirakawa, S. Asahina, T. Yamazaki, T. Nakayama and H. Murakami
Tsukuba Institute for Super Materials in Susono Section, ULVAC, Inc.
1220-1 Suyama, Susono, Shizuoka, 410-1231, Japan
E-mail: masaaki_hirakawa@ulvac.com, Phone: +81-55-998-1579, Fax: +81-55-998-1518

ABSTRACT

We have developed spin-on type low-k dielectric named "ULKS". Methylsilicate as the silica source, and the gelation agent were introduced into the precursor solution, and the amount of surfactant and hydrolysis catalyst were also optimized. As the result, the porous silica film with the dielectric constant k=2.0 and the Young's modulus E=10 GPa was obtained. Stability of precursor solution was confirmed over 45 days at room temperature.

Introduction

The 2007 update edition of the International Technology Roadmap for Semiconductors (ITRS) requires interlayer dielectrics with the dielectric constant of 2.1-2.5 and the effective dielectric constant of 2.4-2.8 to be achieved by 2012.

Considering that CVD type low-k materials have limits in reducing the dielectric constant at about 2.4, we have developed spin-on type low-k dielectric named "ULKS". The properties of ULKS are the dielectric constant k=2.2 and Young's modulus E=8 GPa or k=2.1 and E=6 GPa. The ULKS film is a porous silica film made by spin-coating with the precursor solution which contains the siloxane oligomer and the surfactant as a template of pores, and curing the film to remove the solvent and the surfactant.

In this work, aiming at higher performance (k<2.1, E>9 GPa), silica source, gelation agent and hydrolysis catalyst were examined. And stability of ULKS precursor solution was also examined.

Experiment and Results

The developed precursor solution was spin-coated on Si substrates, and the substrates were baked. Successively the hydrophobic treatment was carried out by exposing substrates to hexamethyldisilazane (HMDS) vapor. The dielectric constant and the Young's modulus of the films were measured by mercury probe method and nanoindenter, respectively.

The silica source and the gelation agent in the precursor solution were examined in order to enhance the Young's modulus of the porous silica film, while maintaining low dielectric constant. Figure 1 shows relationship of methylsilicate (MS) concentration and (a) dielectric constant, and (b) Young's modulus. When only MS was added as a silica source in precursor solution, the Young's modulus of the porous silica film E=8.9 GPa was obtained at k=2.1. At 80 % MS concentration, the elastic modulus of the film was largely enhanced to 10 GPa at k=2.2. The dielectric constants exhibit the maximum at 50 % MS concentration and the Young's moduli are increased at higher MS concentrations. The precursor solution at 70-90 % MS concentration can produce films with low dielectric constants and high Young's moduli.

MS is the oligomer of alkoxysilane. The degree of polymerization of the porous silica film using MS solution is higher than using only TEOS. We think the Young's modulus of the porous silica film makes strong.

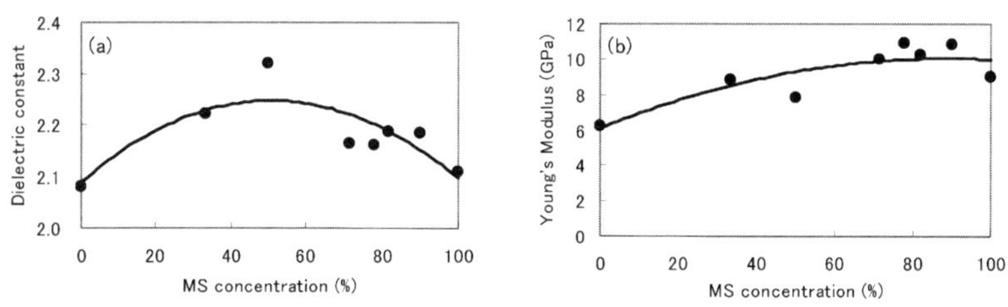

Figure 1 MS concentration in precursor solution dependence of (a) dielectric constant and (b) Young's modulus of the film

Figure 2 shows gelation agent concentration dependence of (a) dielectric constant, and (b) Young's modulus. When the concentration of gelation agent in precursor solution increases, both the dielectric constant and the Young's modulus increase. And the precursor solution which contains the 1.4 times larger amount of surfactant than previously optimized one produces the film with k=2.0 and the Young's modulus 7.6 GPa.

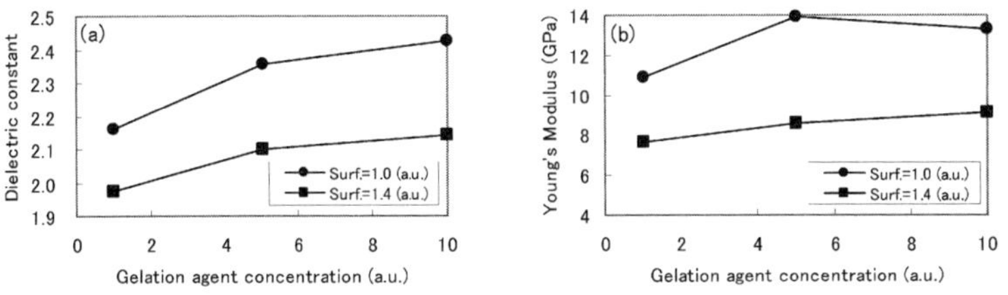

Figure 2 Gelation agent concentration in precursor solution dependence of (a) dielectric constant and (b) Young's modulus of the film.

Another kind of hydrolysis catalyst was introduced into a precursor solution, and the concentration of catalyst was optimized as shown in Figure 3. Using tetraethoxysilane(TEOS) type solution, whose silica is 100 % TEOS, the Young's modulus E=6.2 GPa was obtained at k=1.9. Using MS-TEOS mixture type solution, dielectric constants of the films were decreased at higher catalyst concentrations, and the film with Young's modulus E=10 GPa and k=2.0 was obtained.

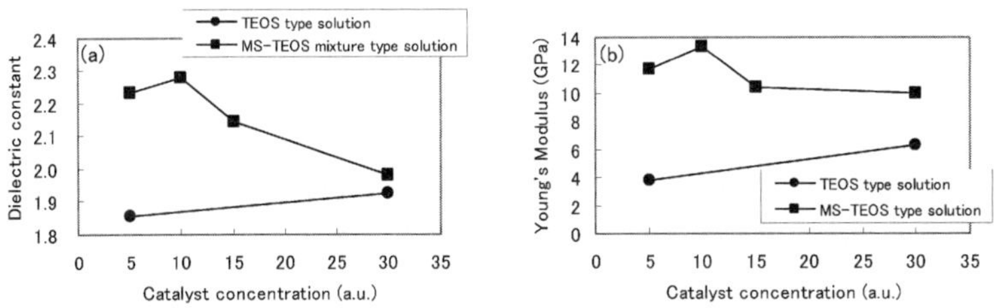

Figure 3 Catalyst concentration in precursor solution dependence of (a) dielectric constant and (b) Young's modulus of the film.

As a result of the optimization of catalyst concentration in precursor solution, improved ULKS films with lower dielectric constant and higher Young's modulus were obtained.

As the next step, stability of optimized ULKS precursor solution was examined. At First , two types of solution were prepared. One is ULKS precursor solution stored at 10 degree C, and the other one is the solution stored at room temperature (r.t.). Then the properties of ULKS films using these solution were examined. Figure 4 shows storage time dependence of dielectric constant.

Between precursor solutions stored at 10 degree C and r.t., no significant difference can be observed in the properties of the films, and the increase of the dielectric constant from initial value is less than 0.1 after 45 days storage. We confirmed that precursor solution can store at r.t..

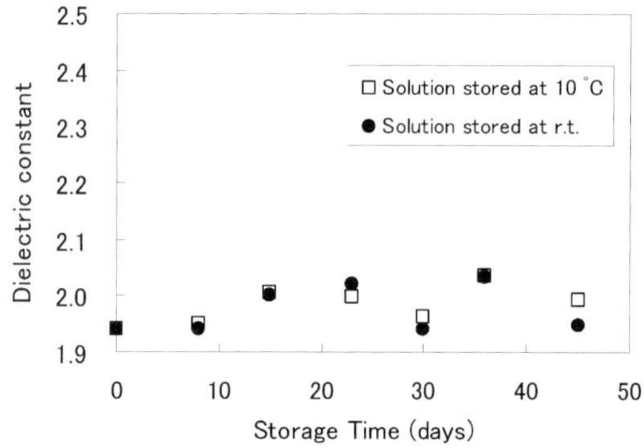

Figure 4 Storage time dependence of dielectric constant of the film.

Summary

Methylsilicate as a silica source, and the gelation agent were introduced into the precursor solution, and the amount of surfactant and hydrolysis catalyst were also optimized. As the result, a porous silica film with the dielectric constant k=2.0 and the Young's modulus E=10 GPa was obtained. Stability of precursor solution was examined over 45 days, and the increase of the dielectric constant from initial value is confirmed to be less than 0.1.

REFERENCES

1. N. Fujii, *et. al.*, Proceedings of SPIE, Vol. 6002, Nanofabrication: Technologies, Devices, and Applications II, 60020N-1, 2005
2. T. Nakayama, et. al., Materials for Advanced Metallization, pp177, 2006

Characterization of pSiCOH for 45nm BEOL interconnects and beyond

J. Widodo[1], T. Ouyang[1] H. Liu[1], W. Lu[1], M. S. Zhou[1], K. H. Leong[2], H. S. Tang[2], A. Jain[2]
[1]Chartered Semiconductor Manufacturing Ltd, 60 Woodlands Industrial Park D, Street 2, Singapore 738406
[2]Applied Materials South East Asia Pte Ltd, 3 Changi Business Park Vista, Singapore 486051
johnnywidodo@charteredsemi.com

ABSTRACT

This paper reports a detailed characterization of porous SiCOH (pSiCOH) film for 45nm BEOL interconnects. Optical, electrical, mechanical, and chemical properties of the bulk film were investigated as a function of deposition temperature. The mechanisms of the curing and the interface behavior of the film with the layer underneath were furthermore discussed.

INTRODUCTION

As the critical dimension of the integrated circuits is scaled down, copper has been introduced to reduce the R components together with low-k or ultra low-k materials to reduce the C components of the RC delay in the BEOL interconnects. Extensive study on various low-k and ultra low-k has been reported over the years[1-2]. In this paper, precursor X and porogen Y have been selected to form porous SiCOH among many other CVD candidates.

EXPERIMENT

By means of PECVD method, pSiCOH films were deposited and cured in Applied Materials™ Producer SE chamber. This paper is divided into 4 sections, mainly discussed about the basic film properties with their stability, the impact of deposition temperature, curing mechanism, and interface behavior characterized with four point bending technique[2] and time of flight secondary ion mass spectroscopy (TOF-SIMS).

DISCUSSION

Film property and aging study

Table 1 summarized the basic properties of pSiCOH film characterized by various techniques. Continuous monitoring of the film optical and chemical properties over 30 days showed that the film is very stable.

Table 1. Film property characterization

Property	Method	Value
Dielectric constant	Hg Probe	2.5 +/-0.05
Refractive index	Spectroscopy ellipsometry	1.35
Stress, MPa	Curvature technique	58
$Si-CH_3/Si-O$ ratio, %	FTIR	0.025
Pore radius, A	Ellipsometry Porosimetry	9.5
Porosity, %	Ellipsometry Porosimetry	23
Hardness, GPa	Nanoindenter	1.1
Modulus, GPa	Nanoindenter	7.1
Breakdown Voltage, MV/cm	Hg Probe	5.5
Leakage current @2MV/cm, A/cm^2	Hg Probe	8.E-08

Impact of deposition temperature on film property

In this study, the deposition temperature was varied from T+/- 8% with 4% increment. Typical FTIR spectra and its peak assignment are described elsewhere[1, 3]. Other important wave numbers to note are $1045cm^{-1}$ and $1133cm^{-1}$ that are assigned to the Si-O Cage and Si-O Network respectively. In figure 1 and 2, it was observed that the Si-H/Si-O and C-H/ Si-O ratio and RI trend down linearly with higher temperature indicating less porogen to be incorporated into the film, which was also signaled by the shrinkage (Shrk) trend. The dielectric constant (k) increases with higher temperature, which is in agreement with the proposed argument mentioned above.

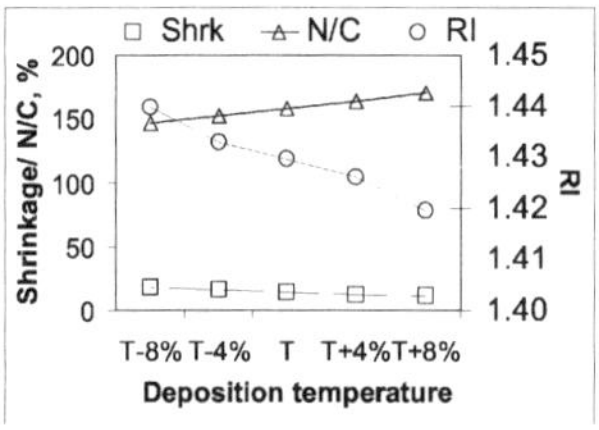

Figure 1. Shrinkage, RI and Si-O Network/Cage ratio as function of deposition temperature.

The trending down of the $Si-CH_3/Si-O$ ratio with higher deposition temperature was a strong indication of less cross-linking as the temperature increased. The proposed argument was strengthened by less increment in the mechanical property compared with the increment in dielectric constant at higher deposition temperature.

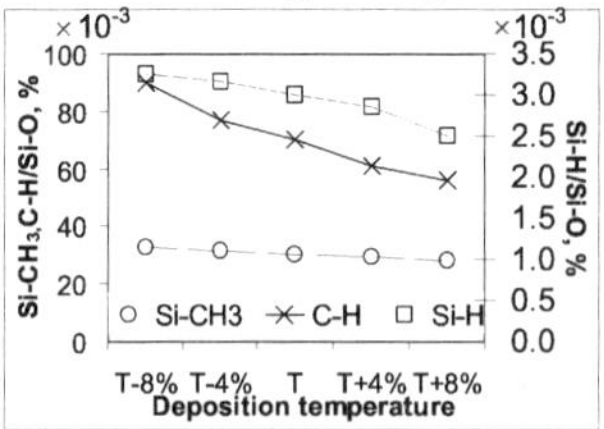

Figure 2. FTIR ratio as function of deposition temperature.

Curing mechanism prediction

In this study, the behaviors of the film property were characterized to predict the curing condition during the main curing mechanisms namely (1) the porogen removal step and (2) the cross-linking step despite the fact that both mechanisms occur simultaneously during curing. A 6000A pSiCOH film was cured from 60s to 540s.

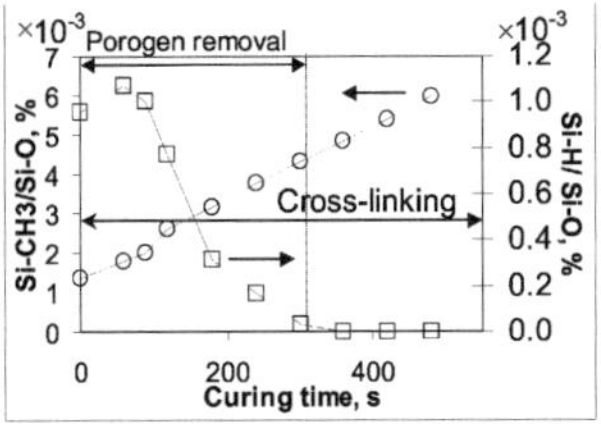

Figure 3. FTIR ratio (precinct-post cure) as function of curing time.

In figure 3, the decomposition of the Si-H bonds was very rapid during the initial 180s suggesting the removal of the porogen to be the main mechanism at this stage. Figure 4 showed that the RI, delta of N/C ratio, and shrinkage trend up linearly with longer curing time while the RI and shrinkage showed slight saturation after 300s. This slight saturation of the RI and shrinkage was accompanied with slight saturation in the delta of Si-H/Si-O ratio (precinct-post cure) indicating the end of porogen removal step. Therefore, porogen removal would be the major curing mechanism from the initial curing until about 180s and ended after 300s of curing time. The continuous increase in the delta of $Si-CH_3/Si-O$ ratio (precinct-post cure, as in Fig.3) showed that the cross-linking occurred throughout the curing. However, cross-linking has only become the main curing mechanism after 300s.

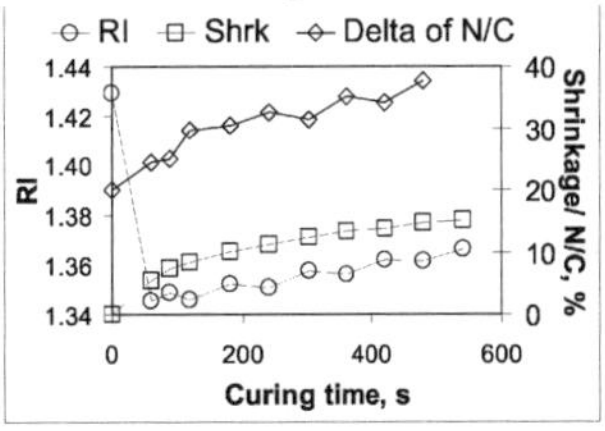

Figure 4. RI, Shrinkage and Delta of Si-O Network/Cage ratio (precinct-post cure) as function of curing time.

Based on the dielectric constant and mechanical property measurement after 300s curing time, it was observed that both dielectric constant and mechanical property increase with longer curing time. These data strengthened the proposed argument of cross-linking as the main mechanism for curing after 300s.

Interface behavior characterization

Optimization of the interfacial adhesion energy is essential to achieve a reliable manufacturing of low-k dielectric[4]. In this study, initial un-optimized film has poor adhesion energy (G_c) at <4J/m^2 towards the layer underneath, which is nitrogen doped silicon carbide film (SiCN). Compositional profile of un-optimized film showed a high intensity of C concentration near the bottom interface between pSiCOH and silicon as shown in figure 5.

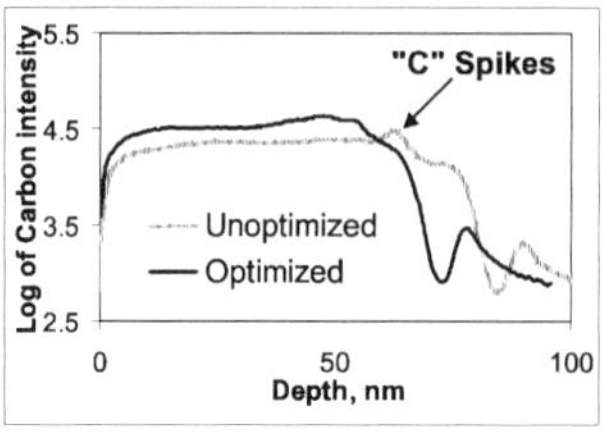

Figure 5. Carbon compositional profile for optimized and un-optimized pSiCOH film.

The key to reduce the C intensity was to have overlapping of the helium carrier for precursors X flow with porogen Y flow in the transition region as shown in figure 6. Compositional profile of optimized film showed carbon free interface as shown in figure 5 and has G_c of 5.5 J/m^2.

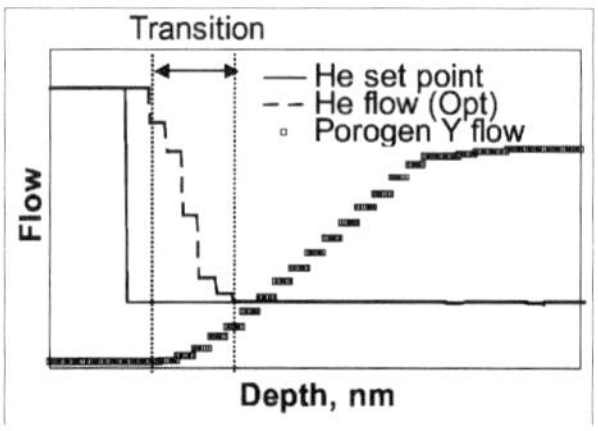

Figure 6. Helium carrier and porogen Y flow profile during transition region for pSiCOH film.

The carbon compositional profile and the adhesion energy as a function of helium (He) carrier of precursors X ramp down speed and as a function of porogen Y ramp up speed were further examined to ensure the robustness of the process. The helium carrier flow for precursors X was varied from He to He+/-17% as shown in figure 7. In addition, the porogen Y was varied from Y to Y-40% as shown in figure 8. Both results are in agreement showing no high carbon intensity at the interface of the film indicating wide process margin. The defect performance was also examined to avoid gas phase nucleation as described elsewhere[4].

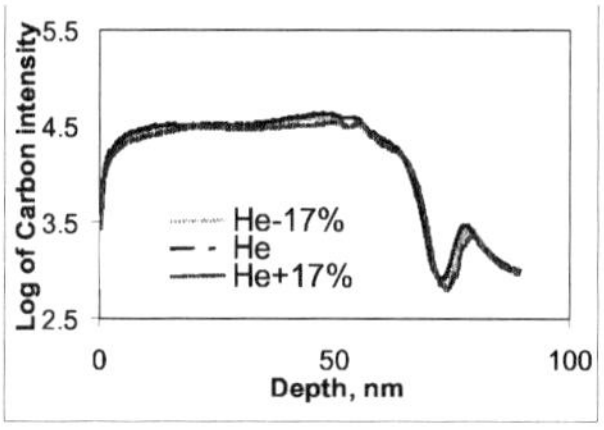

Figure 7. Compositional profile for carbon as function of helium carrier of precursors X ramp down speed

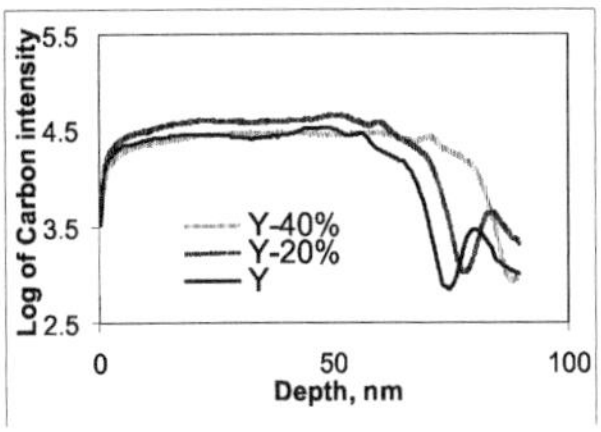

Figure 8. Compositional profile for carbon as function of porogen Y ramp up speed.

The adhesion energy (G_c) was measured for the He and Y variation samples. None of the samples produced G_c lower than $5J/m^2$. In agreement with the previous reported results[4], there is a strong correlation observed between the adhesion energy with the compositional carbon intensity at the interface.

CONCLUSIONS

Detail characterization of pSiCOH has been reported including the interfacial characterization. It is proven that the reported pSiCOH film has a wide process margin for its interfacial property.

ACKNOWLEDGMENTS

The authors would like to thank Applied Materials Inc. and Chartered Failure Analysis Department for their support in this project.

REFERENCES

1. J. Widodo, L. N. Goh, W. Lu, S.G. Mhaisalkar, K.Y. Zeng, L. C. Hsia, J. of Elect. Soc 152(4),G246-G251.
2. M. Damayanti, Z.H. Gan, T. Sritharan, S.G. Mhaisalkar, A. Naman, J. Widodo and H. S. Tan, Thin Solid Films, 504/1-2, 2006, 213-217.
3. J. Widodo, L. N. Goh, W. Lu, S.G. Mhaisalkar, S.Ong, J. L. Sudijono, L. C. Hsia, P. Y. Tan, K.Y. Zeng, Jour. of Vac. Sci.and Tech-B 22-3, 2004, 1030-1036.
4. D. Restaino, S. Molis, A. Grill, V. Patel, M.W. Lane, T. Cheng, A. Demos, S.M. Gates, K. Ida, D. Kioussis, N. Klymko, H. Liu,S. Nguyen, V. Nguyen, J.C. Patel, S. Reiter, H. Shobha, J. Widodo, K.Yim, D. Edelstein, Adv Metalliz Proc. 2007, October 9-11, Albany, NY, USA.

Conference Proceedings AMC XXIV © 2009 Materials Research Society

Hydrogen Dependence of Film Properties of High Tensile Stress UV Cured Silicon Nitride for 45nm Bulk Device

Jingze Tian*, Luona Goh, Jiehui Shu, Wei Lu, Meisheng. Zhou
Technology Development Department, Chartered Semiconductor Manufacturing Ltd,
60 Woodlands Industrial Park D, Street 2, Singapore 738406
* e-mail: tianjz@charteredsemi.com

ABSTRACT

In this paper, we have studied hydrogen dependence of silicon nitride (SiN) film properties during ultraviolet (UV) curing and effect of hydrogen and tensile stress on 45nm bulk device performance. SiN film with high hydrogen and low tensile stress (0.65GPa) grown by PECVD was modified by UV curing to achieve high tensile stress up to 1.6GPa. FTIR spectra were used to demonstrate the evolution of local bonding structure (Si-H, N-H and Si-N bonds) and hydrogen concentration in the UV cured film at different curing stages. The amount of Si-H and N-H components decreases drastically within 2min, and then saturate for longer curing, which governs the film properties of UV cured SiN. We found a close correlation between hydrogen and film properties such as stress, shrinkage, refractive index and wet etch rate (WER). The tensile stress initially increases faster with decreasing hydrogen than film shrinks, but it saturates while the shrinkage continues, which limits maximum achievable stress. Additionally, as hydrogen decreases, refractive index increases and wet etch rate (WER) decreases, thus film becomes densitive. WER shows a linear dependence on the amount of hydrogen in UV cured SiN film. In terms of electronic performance of the device, 45nm bulk device using cured SiN showed significant benefit in the NMOS transistor driver current performance.

INTRODUCTION

Process induced strain in MOSFET channel has become one of the key technologies for enhancing carrier mobility and transistor driver current for the leading edge MOSFET[1-3]. Continuous device shrinkage beyond 45nm requires the use of higher tensile stress (more than 1.5GPa) SiN as contact etch stop layer (CESL) to enhance NMOS carrier mobility. It has been generally accepted that high tensile stress is generated due to hydrogen removal during SiN deposition or subsequent post treatment. However, 45nm device thermal budget requires CESL process temperature not exceeding 400C, in which conventional thermal annealing and plasma treatment yield limited tensile stress of less than 1.2GPa. Recently UV curing on high hydrogen-containing SiN film has become an effective technology [4] to achieve high tensile stress up to 1.6GPa at low substrate temperature (<=400°C). This is because UV photon provides high enough energy to break up Si-H and N-H bonding and remove hydrogen, which results in film shrinkage and significant stress increase. However it has not been well reported on the effect of hydrogen on film properties of UV cured SiN film. In this work, we focus on (1) quantitatively study the correlation of stress, shrinkage, refractive index and wet etch rate (WER) to hydrogen amount of UV cured SiN film; (2) Study the effect of stress and hydrogen on 45nm NMOS carrier mobility.

EXPERIMENTAL DETAILS

SiN deposition and subsequent UV curing are performed by Applied Material integrated Producer SE platform. SiN films were deposited on 300mm single-crystal silicon wafer by decomposition of appropriate mixture of SiH_4, NH_3 and N_2 in a glow discharge plasma in PECVD, in which 13.6MHz radio frequency power was applied to showerhead to ignite and maintain plasma. The deposited SiN films were subjected to UV curing using mercury-based bulb in the range from 1 to 10 min. As-deposited and UV-cured SiN films were characterized by Fourier Transform Infrared Spectroscopy (FTIR) from which local bonding Si-H, N-H and Si-N were quantitatively analyzed. Film thickness, refractive index were measured by KLA-Tencor Fx200 at the wavelength of 633nm. The film shrinkage due to UV curing is estimated by percentage of decrease in thickness. Mechanical stress of SiN film was measured by substrate bow methods with 20MPa accuracy. The wet etch rate (WER) measurement of SiN films was conducted in a solution of 300:1 H_2O: HF.

RESULT AND DISCUSSION

A. Hydrogen dependence of UV cured SiN film properties

Fig.1 shows the FTIR spectra of as-deposited and 10mins UV-cured SiN films. As-deposited

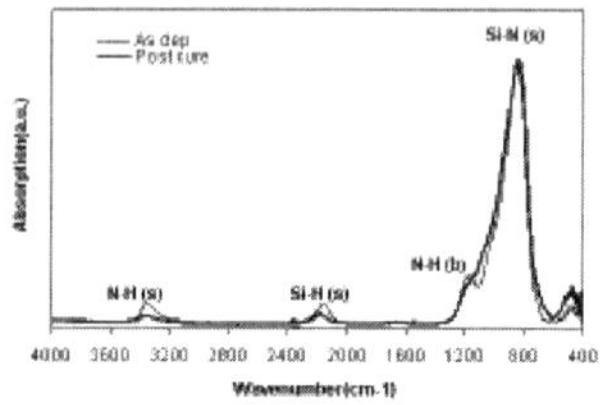

Figure 1. FTIR spectra of as deposited and UV cured SiN films

SiN film exhibits high amount of N-H stretching (at 3340cm^{-1}) and Si-H stretching (at 2158cm^{-1}) and N-H bending (at 1160cm^{-1}), which drastically decreases after 10min UV curing. At the same time, it is observed that the intensity of the Si-N stretching (845cm^{-1}) is slightly increased and its wave-number is shifted to a lower wave-number. Because high energy UV photon significantly breaks the Si-H and N-H bonds, thus causing neighboring hydrogen atoms to recombine and form molecular hydrogen that diffuses out. As a result, the Si- and N- dangling bonds are cross-linked to generate extended strained Si-N bond. Relative hydrogen concentration in the UV-cured SiN film was determined by the integrated absorption in the N-H and Si-H stretching bands as a function of UV curing time, as shown in Fig.2. The amount of Si-N and N-H drastically decreases within 2min, the decrease slows down after 2min. Fig.3 shows wave-number shift of the Si-N and Si-H absorption bands with UV curing time. The main absorption band Si-N peaked at 845cm^{-1} in as-deposited film moves to a low wave-number 835cm^{-1} after 10min curing, which corresponds to near stoichiometric Si_3N_4[4]. Simultaneously Si-H absorption band moves to a higher wave-number upon UV curing, which is probably attributed to Si-H local bonding environmental change[5-6]. As a result of hydrogen removal and strained Si-N formation, SiN film shrinks and the tensile stress largely increases. Fig. 4. shows the dependence of the stress and shrinkage on hydrogen residue in UV cured SiN films. It clearly demonstrates that the large film shrinkage and tensile stress gain is correlated to hydrogen reduction. At the beginning, the stress is seen to change faster than film shrinkage with respect to the H-content in the film. However, shrinkage change rate suppress stress as more hydrogen is

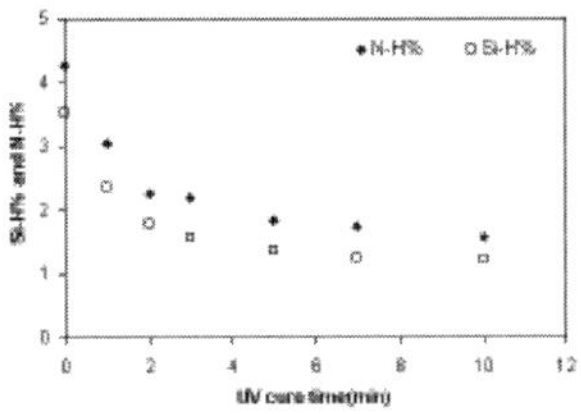

Figure 2. Dependence of N-H% and Si-H% in UV cured SiN on UV curing time

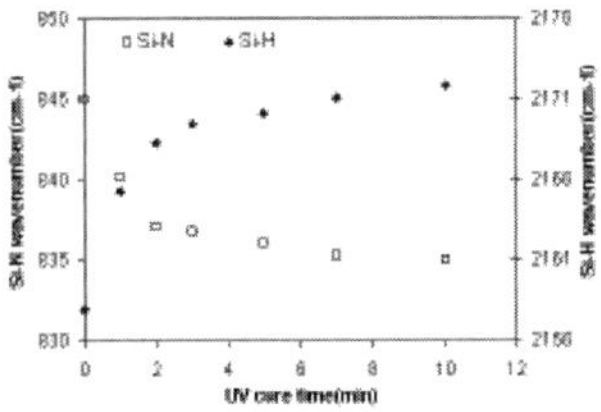

Figure 3. Dependence of Si-N and Si-H wave-number on UV curing time

removed. This explains the phenomenon seen in Fig. 5 that the tensile stress saturates as the film continue to shrink. The stress increases as an quadratic function of the shrinkage, i.e. the stress increases rapidly initially with shrinkage, and saturates stress for more shrinkage. Fig.6 shows refractive index and wet etch rate (WER) of UV cured SiN film as a function of curing time. Increasing curing time leads to the increase of refractive index, which is attributed to hydrogen removal and restructuring of Si-N network. WER significantly decreases within 2 min, and the decrease slows down for longer time. As-deposited SiN film is porous due to presence of large amount of hydrogen. Hence, as hydrogen diminishes upon UV curing, the film become denser, thus resulting in a decrease of WER. WER is closely correlated with hydrogen residue in UV-cured SiN film as shown in Fig.7, from which a linear correlation is observed.

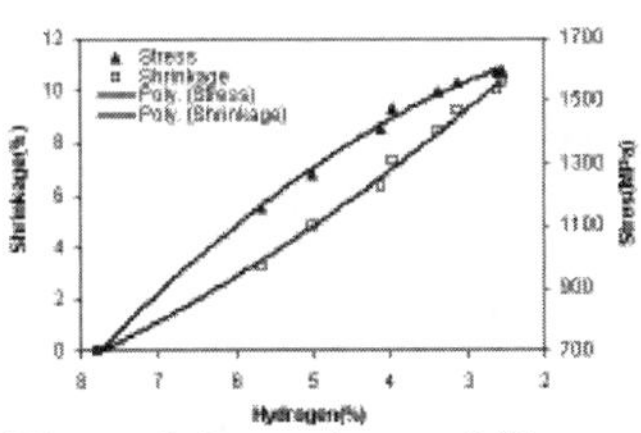

Figure 4. Dependence of film stress and shrinkage on hydrogen

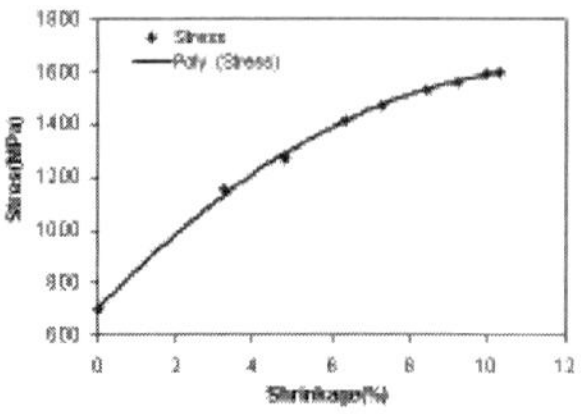

Figure 5. Film stress vs shrinkage due to UV curing

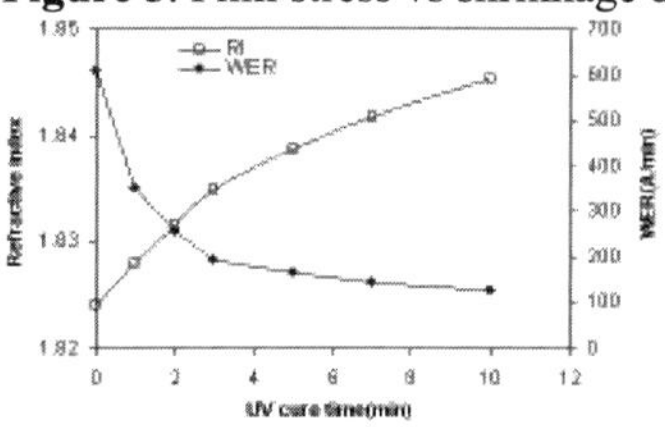

Figure 6. Dependence of refractive index and wet etch rate(WER) on UV curing time

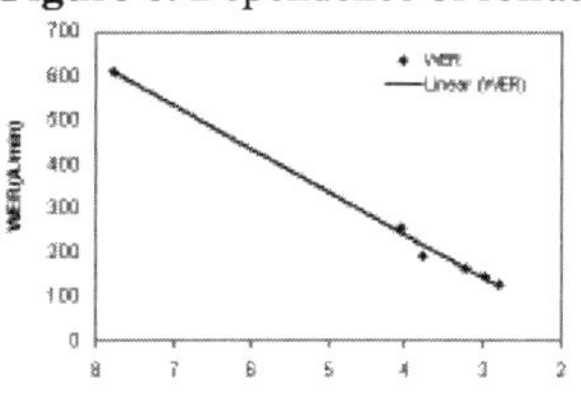

Figure 7. Correlation of WER to hydrogen in UV cured SiN film

As described above, the hydrogen dominates the film properties of UV cured SiN. The hydrogen removal plays an important role in stress enhancement. Hydrogen removal generates micro-voids in SiN film which tends to shrink to minimize surface energy. Film shrinkage combined silicon substrate restraining results in the increase of tensile stress. On the other hand, hydrogen removal results in restructuring of Si-N network and formation of extended strained Si-N bond. In addition, hydrogen removal affects the priority of SiN film, therefore strongly affects the film density and wet etch rate and refractive index.

B. Benefit of 45nm bulk device performance using UV cured SiN

The effect of UV-cured SiN film as contact etch stop layer (CESL) on the 45nm MOSFET performance is investigated. The device fabrication process flow mainly includes shallow trench isolation formation, Well ion implantation, gate electrode formation, S/D formation and salicidation, CESL (PECVD SiN deposition and UV curing) and PMD deposition. CESL split includes 1) single layer SiN with 1 min curing(1x1min), 2) single layer SiN with 7min curing(1x7min), 3) tri-layer dep/cure SiN with 5mins curing(3x5min) and lastly 4) neutral stress SiN with no curing as reference. Fig.8 shows the effect of UV-cured SiN on the transistor drain saturated current(I_d). It is observed that single layer (1x7min) UV-cured SiN increases I_d by about 27%, as compared to neutral stress SiN. Multiple dep/cure SiN (3x5mins) yield the maximum I_d gain due to higher tensile stress resulting from more hydrogen removal, as shown in table1. Fig. 9 shows the NMOS I_{off}/I_{on} performance using 1min vs 7min single layer SiN CESL.

The 7min UV curing is shown to improve NFET I_{off}/I_{on} performance by 4% compared to 1min because of higher tensile stress resulting from more hydrogen removal(see table1). After SiN deposition and UV curing, high tensile stress in UV cured SiN film was transferred to the channel underneath gate oxide. This results in tensile strain in the channel, which enhances NMOS electron mobility, thus increases the drain saturate current (I_d).

Table 1. Hydrogen and stress in different UV cured SiN films

	1x1min	1x7min	3x5min
Hydrogen(%)	5.4	2.9	2.4
Stress(MPa)	1270	1550	1575

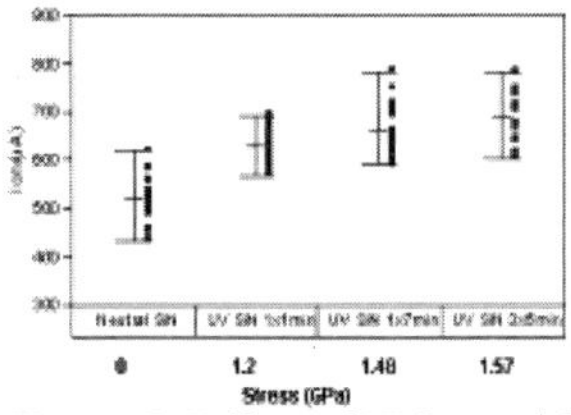

Figure 8. Effect of UV cured SiN on 45nm NMOS drain saturated current(I_d)

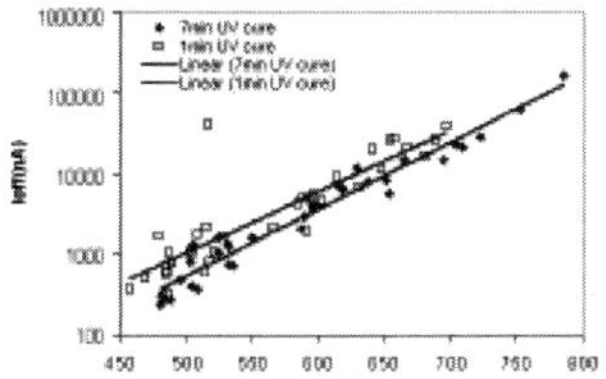

Figure 9. Effect of UV cure time on I_{on}/I_{off} performance of 45nm NMOS

CONCLUSION

We have investigated the film properties dependency on the hydrogen in the UV-cured SiN films. It is well correlated that hydrogen removal due to UV curing results in film shrinkage, tensile stress enhancement, higher refractive index, lower WER and film densification. We have also demonstrated that UV-cured SiN significantly benefits 45nm MOSFET carrier mobility performance.

ACKNOWLEDGMENTS

The authors would like to thank Applied Materials Inc. and Lai Seow Wei and Tan Shyue Seng from Chartered Technology Development Department for their support in this project.

REFERENCES

1 V. Chan, K. Rim, M. Leong , IEEE 2005 Custom Integrated Circuits Conference, (2005), pp.667-674

2 S. Ito, H. Namba, T. Hirata, Microelectronics Reliability, **42**, 201-209(2002).

3 H.S.Yang , IEDM Tech. Dig., (2004), p1075-78

4 Y. Miyangwa, T. Murata, Y. Nishida, Japanese J.of App. phys, **46(4)**,1984-1988 (2007),

5 S. Hasegawa, H.Anbutsu, Y. Kuratal, Philosophical Magazine, **B59(3)**, 365- 375 (1989)

6 G. Lucovsky, J. Yang, S.S.Chao, Physical review, **B 28 (6)**, 3234-3240(1983)

Conference Proceedings AMC XXIV © 2009 Materials Research Society

Characterization of Damage of Low-k Films Between Narrow Cu lines by Micro Beam IR Method

H. Seki[1], N. Tarumi[2], Y. Shimizu[1], Y. Otsuka[1], H. Hashimoto[1], and S. Ogawa[2]
[1]Toray Research Center, Inc.
3-3-7 Sonoyama, Otsu, Shiga 520-8567, Japan
E-mail: Hirofumi_Seki@trc.toray.co.jp
[2]Semiconductor Leading Edge Technologies, Inc.
16-1 Onogawa, Tsukuba, Ibaraki 305-8569, Japan

ABSTRACT

We investigated the process-induced damage to low-k films between narrow Cu lines using the micro-beam IR method. It was found that infrared light can transmit through the narrow spacings between the Cu lines (L/S = 70/70 nm) and the electric-field component of the transmitted light is perpendicular to the Cu lines. In addition, it was observed that the peak intensities of Si-O for Cu-metallized samples were much higher than those for trench-formed samples. We believe that this method is a promising technique for measuring the process-induced damage to narrow interconnects with L/S = 70/70 nm and less.

INTRODUCTION

Porous low-k films have been used as dielectric materials for advanced Cu/low-k interconnects. However, such films are susceptible to damage during their manufacture due to processes such as dry etching, ashing, and barrier metal deposition. The changes in the chemical structure of the films induced by these processes can deteriorate the properties of the films, such as dielectric constants and/or hygroscopic properties. Such "damages" have an adverse effect on the performance and reliability of the Cu/low-k interconnects. Quantitative analysis of low-k films using Fourier transform infrared spectroscopy (FT-IR) shows the porous structure of the low-k films without causing any damage to the films during measurements. We have examined the ashing-induced damages to low-k blanket films in the depth direction by FT-IR [1]. In addition, damages of low-k spaces of 100-nm width between 100-nm width Cu lines have been recently characterized using a micro-beam IR method [2]. In this study, we demonstrated (1) the use of this IR technique in the evaluation of low-k film damage with narrower lines and spaces (L/S = 70/70 nm), (2) the enhancement of the IR signals for samples with L/S structures, and (3) the effect of polarized infrared light on transmittance.

EXPERIMENTAL DETAIL

The structure of the sample used in this experiment is shown in Fig.1. Two different samples with L/S = 70/70 nm were prepared as follows. One sample was first trench etched using a photoresist mask and then subjected to the ashing process. The other sample was a Cu metallized sample followed by chemical mechanical polishing (CMP). The film under investigation was a porous SiOC low-k film (k = 2.6). The samples were investigated by micro-beam IR with a probe size of approximately 100 μm.

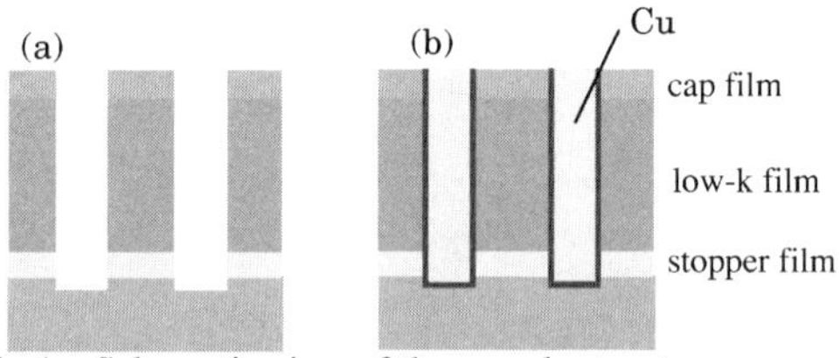

Fig.1 Schematic view of the sample structure
(a) After trench formation (b) After Cu metallization

RESULTS AND DISCUSSION

Effect of polarization on infrared transmittance for the L/S structure.

We investigated the polarization dependence of infrared light for the L/S structures shown in Fig.2. Figure 3 shows the transmittance of infrared light to Si substrate for L/S = 100/100 nm. When the electric-field component of the infrared light was parallel to the copper lines (a), the transmittance was less than 10%. On the other hand, when the electric-field component of the infrared light was perpendicular to the copper lines (b), the transmittance was almost 100%. It is believed that the L/S structure acts as an infrared polarizer. The underlying principle of this phenomenon is discussed in [2]. Therefore, to investigate the damage to low-k films between narrow Cu lines, the measurements should be conducted using infrared light in which the electric-field component is perpendicular to the copper lines, which are separated by an infrared polarizer.

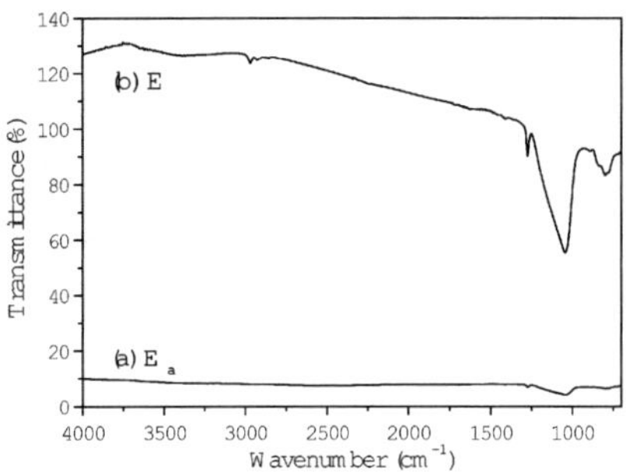

Fig.2 Infrared transmittance to Si substrate for L/S = 100/100 nm

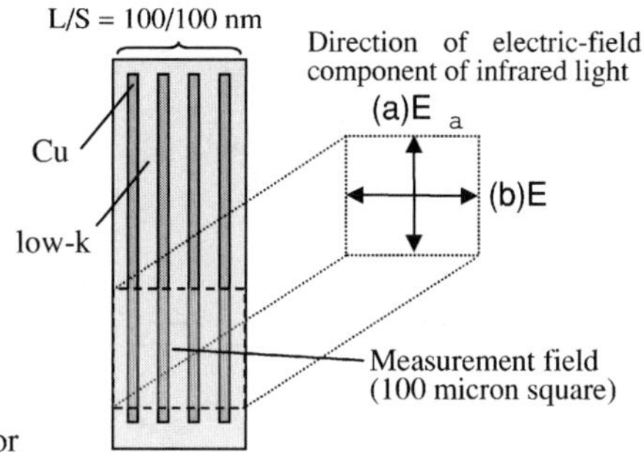

Fig.3 Schematic diagram of the polarized infrared measurement

Enhancement of infrared absorption intensity for the L/S structure.

The IR spectra obtained for the trench-formed sample and the Cu-metallized sample with L/S = 100/100 nm and L/S = 70/70 nm are shown in Fig.4 and Fig.5, respectively. It was found that infrared light could penetrate the narrow spacing sandwiched between the Cu lines, although the wavelength of the infrared light used in the measurements was approximately 10 μm. In addition, it was shown that the intensities of the peak corresponding to Si-O of both the Cu-metallized samples were much higher than those of the trench-formed samples. A similar phenomenon, where the irradiation of infrared light on the film with a Au-, Ag-, or Cu-island structure (Fig.6) enhances the infrared absorption intensity of the molecules adsorbed on these island films, was reported earlier [3,4]. This phenomenon is attributed to the localized electric field generated between the islands by the infrared light; these local fields enhance the infrared absorption intensities due to the transverse collective electron resonance of the valence electrons in the metal islands. It is speculated that the Cu lines play the same role as the Cu islands, and, therefore, the intensity of the infrared absorption for the samples with low-k films between the Cu lines is enhanced.The difference in the enhancement of intensities between the samples with L/S = 100/100 nm and L/S = 70/70 nm is attributed to the difference in the thickness of the films, and the enhancement factor is different for different L/S widths.

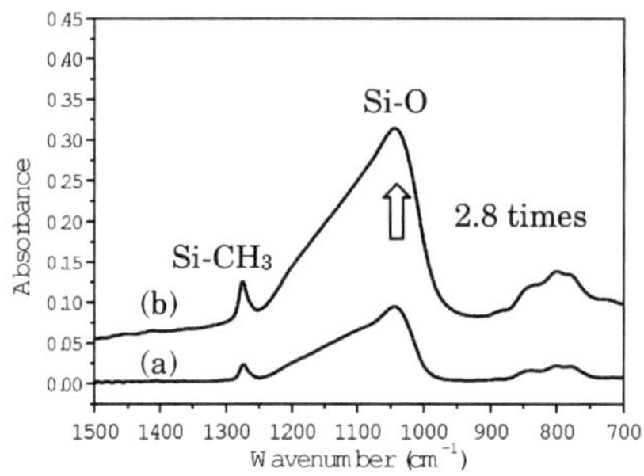

Fig.4 IR spectrum of low-k film between narrow Cu lines (Cu/low-k = 100/100 nm)
(a)After trench formation (b) After Cu metallization

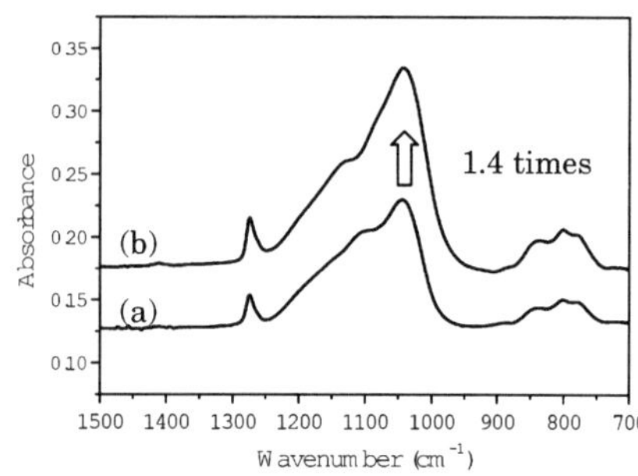

Fig.5 IR spectrum of low-k film between narrow Cu lines (Cu/low-k = 70/70 nm)
(a) After trench formation
(b) After Cu metallization

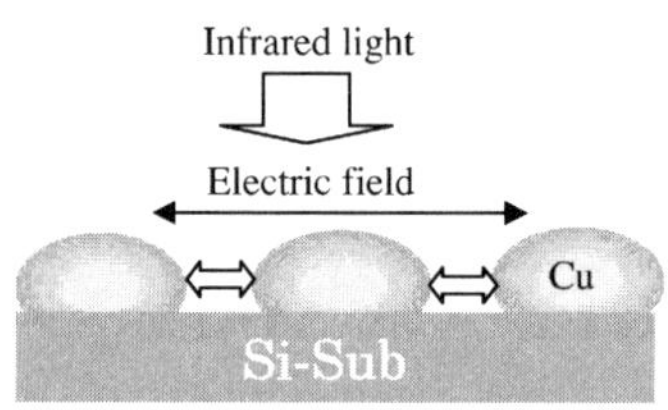

Fig.6 Schematic view of Cu-island film
Electric field strengthens in the gap between Cu islands

Process-induced damage to low-k film with L/S = 70/70 nm.

Figure 7 shows the data obtained from the trench-formed sample and two Cu-metallized samples subjected to different plasma processes with L/S = 70/70 nm. The ratios of the peak intensities of Si-CH_3 and Si-O from the IR data of Fig.7 are shown in Fig.8. From the IR measurements, it was revealed that the fabrication process causes a diminution in the Si-CH_3 concentration.

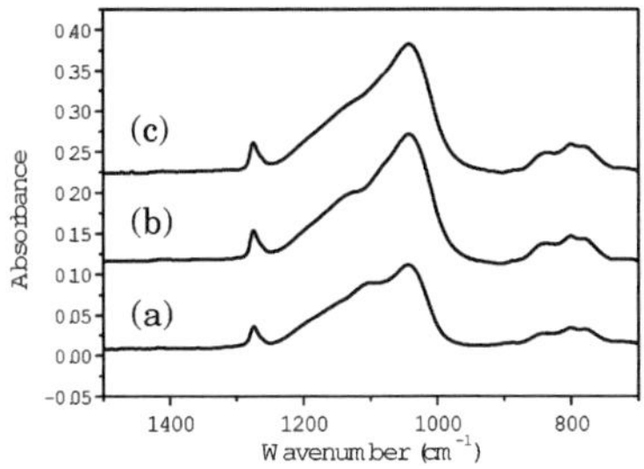

Fig.7 IR spectrum of low-k film between narrow Cu lines (Cu/low-k = 70/70 nm)

(a) trench-formed sample

(b) Cu-metallized sample (plasma process 1)

(c) Cu-metallized sample (plasma process 2)

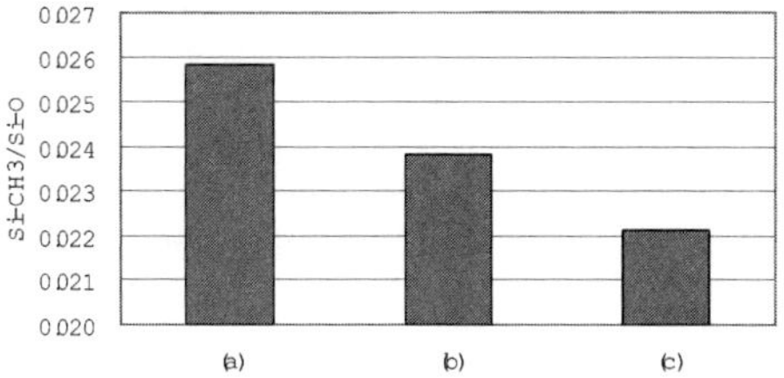

Fig.8 Comparison of the amount of Si-CH_3 in low-k films subjected to different plasma process

CONCLUSIONS

We have demonstrated the damage to low-k films between Cu lines was investigated using a micro-beam IR technique. It was observed that when the electric-field component of infrared light is perpendicular to the copper lines, light can pass though the L/S structure. Moreover, the intensity of infrared absorption of the Cu-metallized samples is much higher than that of the trench-formed sample. It was found that this method can be effectively applied to narrow interconnects with L/S = 70/70 nm or less. We believe that the micro-beam IR technique will be a useful method for the examination of process-induced damages to Cu/low-k interconnects.

REFERENCES

[1] H. Seki et al., in Proc. ADMETA2004, pp. 34.

[2] S. Ogawa et al., in Proc. IITC 2008, pp. 76.

[3] A. Hartstein et al., Phys. Rev. Lett, 45, 201 (1980).

[4] H. Seki et al., Surf. Sci, 506, 23 (2002).

Conference Proceedings AMC XXIV © 2009 Materials Research Society

Proposal of A Simple Method for Evaluation of Wire Bonding Process with Ultra Low-k Materials

Munehiro Hatai and Keisuke Funatsu

Consortium for Advanced Semiconductor Materials and Related Technologies (CASMAT)
1-280, Higashi-Koigakubo, Kokubunji-shi, Tokyo 185-0014, Japan
Phone: +81-42-327-8090, Fax: +81-42-327-8094, E-mail: hataim@casmat.or.jp
Key Words: ultra low-*k*, wire bonding, bump shear

Abstract

A simple evaluation method for interconnect with ultra low-*k* (ULK) materials in wire bonding process is developed. Using this method, three interconnect with ULK materials which have different features are evaluated. As a result of this, it is found that Au bump destruction at bump shearing and Au bump shearing strength increase according as the Young's modulus of ULK increases. This mechanism causing this phenomenon has been discussed.

Introduction

Recently, many ULK materials have been developed in response to requirements for high performance of LSI. However, the strength of ULK generally decreases with the decrease in *k*-value of such material. Therefore, serious problems arise for packaging process in using weak ULK. Concerning the wire bonding process as one of these packaging processes, innovative structures in which interconnects are under the pad are developed to shrink the chip size [1]. However, such structure including metal interconnects is so complicated to be properly analyzed. Therefore, to focus on the direct relationship between the strength of ULK and wire bonding process, a simple layer structure having ULK without metal under pad is adopted. Furthermore, to reduce the turn-around-time, the bump on the pad on the wafer is bonded, and the reliability of the wire bonding process on the pad is evaluated with measuring bump shearing strength.

In this work, destroying in the bump or ULK layer, the relationship between Young's modulus of ULK and the bump shearing strength in the wire bonding process is discussed

Experimental Details

Table 1 indicates the feature list of ULK layer used in this work, and every ULK layer is modified porous methylsilsesquioxane, MSQ. The ULK layer is coated with Spin-on-Dielectric, SOD process and is cured with ultraviolet light.

A simple structure having one ULK layer is used as shown in Fig. 1 on a wafer and the wafer is cut to an appropriate size to be set in wire bonder (UTC-2000 produced by Shinkawa Ltd.) without back-grinding, dicing or mounting on lead frame or interposer. Au bump without loop is prepared on pad with supersonic wave through wire bonder. Table 2 indicates the condition of bump bonding, which is generally used for bonding.

Then, the strength to shear Au bump on pad with shear tester (DAGE 4000 Multi-Purpose Bondtester produced by Dage) is measured as shown in Fig. 2.

Finally, weak ULK layer exists at deeper position under pad than that shown in Fig. 1 (a) since thick layers (ex. passivation) exist above ULK. Therefore, it is evaluated the bump shearing strength of ULK (Sample A) for more actual interconnect structure having thicker SiO layers.

Results and Discussion

(1) Setting of Evaluation Condition

Firstly, the condition of shearing test by one ULK is determined. Figure 3 indicates bump shear strengths measured with different directions on each side. Four shear strength distributions from all sides seem to be divided into two groups, one of which consists of b and d sides, and the other of which consists of a and c sides.

Figure 4 shows the shape of the bump and the direction of the supersonic wave through bump bonding. It is observed that one small hill in front of the bump as indicated with arrow (but there is the other hill behind the bump). It is estimated that the hills on pad are generated as a result of the motion of Au bump caused from supersonic wave application during bonding.

It is found that the hills cause obstruction in shearing bump from the pad in the case that shearing direction is parallel to the supersonic wave (a and c side). The distributions of the shear strengths measured in the direction perpendicular to the supersonic wave in all sides are shown in Fig. 5. All distributions are in good agreement. Thus, appropriate measurement conditions are established.

(2) Evaluation of ULK with Bump Shearing

Table 3 indicates ratios of destructions at bump shearing. All bumps of Sample A and B are destroyed under pads. But about quarter pieces of bumps of sample C are destroyed in Au bump. Since it is recognized to be good that the state of destruction in Au bump at wire bonding process, these results are understood that the shear strength is not enough big.

It is indicated in Fig. 6 cumulative probabilities of shear strengths for all samples are those destroyed under pads. The shear strength increases in the order of Sample A, B, and C. However, it seems that minimum values of all samples are converging.

These results as shown in Table 3 and Fig. 6 are analyzed as follows.

Figure 7 indicates profiles in shearing Au bumps of Sample C. Profiles of three bumps seem to be divided into two slopes changing at about 5 μm and it is considered that each slope corresponds to the different phenomenon. Dispersion of top value in graphs is observed, on the other hand, slopes from 0 point to about 5 μm in graph are almost same. Therefore, it is understood the phenomenon corresponding to the part from about 5 μm in graph causes dispersion of distribution in Fig. 6.

Then, SEM images are obtained to analyze those results. SEM images from Fig. 8 to Fig. 10 indicate those after bump shearing of Sample C. Also it is observed that dark area and light area on the surface after bump shearing in Fig. 8 (in dotted circle). Expanding area as shown with square symbol in Fig. 8, it is seen both areas in detail in Fig. 9. It is considered that the dark area is a remainder after destructed ULK since dark area seems to be a surface after tearing off a layer. Figure 10 indicates an SEM image of cross section after bump shearing and it is considered that light area is SiCN surface and dark area is a part of ULK layer referred to structure in Fig. 1(a). This is the reason why it is considered that the mode of destruction under pad during bump shearing is delamination of ULK from SiCN surface and destruction in ULK.

Therefore, it is well concluded that the slope from 0 point to about 5 μm corresponds to delamination of ULK and the slope from about 5 μm corresponds to ULK destruction in ULK in Fig. 7. This is the reason why destruction in bulk tends to increase dispersion compared to delamination.

Based on the above analysis, it is considered that Sample A and B in Fig. 6 also cause similar phenomenon.

(1) The shear strength of each Sample A, B, and C corresponding to delamination of ULK is almost same since adhesion in Table 1 is almost same.

(2) The order of shear strength of each Sample A, B, and C corresponding to destruction of ULK is the same order since the order of Young's modulus in Table 1 which is closely related to strength of destruction is the order of Sample A, B, and C.

These considerations are consistent that minimum strengths of all samples in Fig. 6 are converging (corresponding to (1)), and the shear strength is in the order of Sample A, B, and C (corresponding to (2)).

It is also analyzed in Table 3 that a part of Sample C which is destroyed in Au bump is related to the highest Young's modulus of three ULK's.

(3) Influence of SiO Thickness

Table 4 indicates that ratio of destruction under pad decreases and ratio of destruction in Au bump increases with the increase in SiO thickness. Almost all measurements of sample having 1000-nm thick SiO layer are destroyed in Au bump and those of sample having 60- or 100-nm thick SiO layer are destroyed under pad.

Figure 11 indicates that even shearing strength destructed under pad increases according as SiO thickness increases. Considering two shear strengths for 100-nm and 500-nm thick SiO layer, it is considered that even though destruction strength of SiO layer is much bigger than that of ULK, only destruction strength of SiO layer does not determine shear strength since shear strength for 500-nm thick SiO layer is not as five times big as that of 100-nm thick SiO layer. Therefore it is concluded that properties of ULK in this structure still influence much on determining shear strength, even if SiO thickness increases.

Conclusion

A simple method of evaluating the reliability of ULK for wire bonding process is developed. Using this method, it is found that how properties of ULK contribute to bump shear strength as well as the reliability.

Reference

1. M. Tagami, H. Ohtake, M. Abe, F. Ito, T. Takeuchi, K. Ohto, T. Usami, M. Suzuki, T. Suzuki, N. Sashida, and Y. Hayashi., "Comprehensive Process Design for Low-Cost Chip Packaging with Circuit-under-pad (CUP) Structure in Porous-SiOCH Film," Proc. IEEE 2005 Int. Interconnect Technol. Conf., pp. 12-14.

Acknowledgements

The authors thank H. Sunami, Y. Kawamoto, and T. Yoshida for encouraging this work. This work was supported by New Energy and Industrial Technology Development Organization (NEDO) of Japan.

Table 1 Features of ULK used this work

Sample	Young's modulus [1] (GPa)	Adhesion[2]	Relative permittivity
A	6.2	1.00	2.40
B	9.5	1.01	2.42
C	13.8	1.01	2.57

1) measured with nanoindentation method
2) measured with nanoscratch method

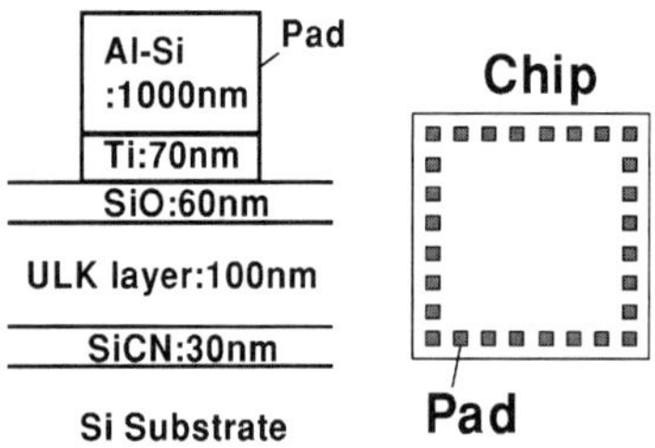

(a) (b)

Fig. 1 (a) Cross section of structure used in this evaluation (b) Pads are located peripherally on a chip. Pad size is 100 μm . Au bump is made on this pad.

Table 2 Condition of Au bump bonding

Frequency of supersonic wave (kHz)	Temperature during bonding ( )	Load of bonding (gf)
120	180	40

Au wire:25 °m

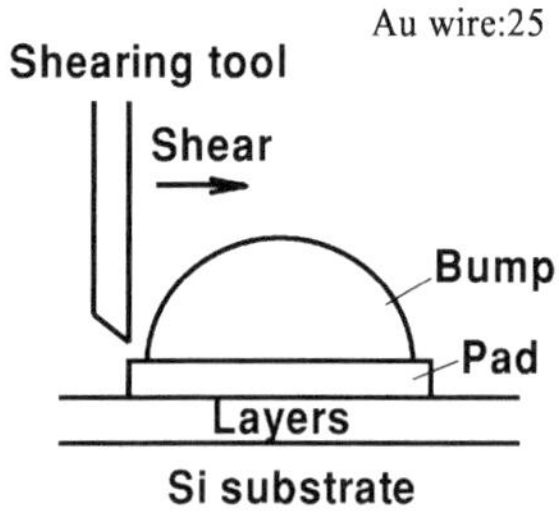

Fig. 2 Method of measuring bump shearing strength The shearing speed was 300μm/sec and shearing position was at the position of the 5μm height above the wafer surface.

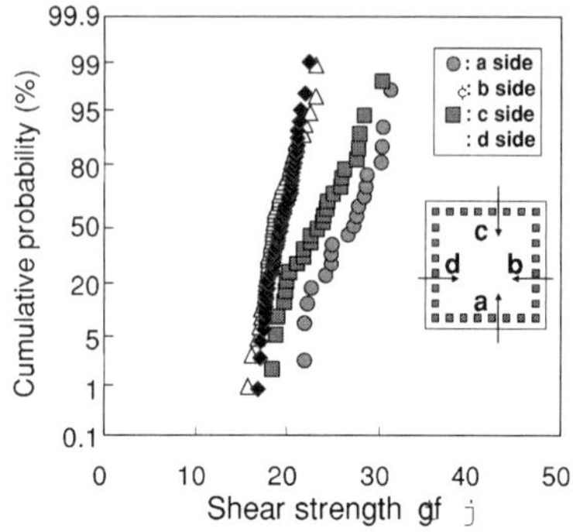

Fig. 3 Cumulative probability of shear strength at each side (Sample A) The direction of shearing is indicated with arrow.

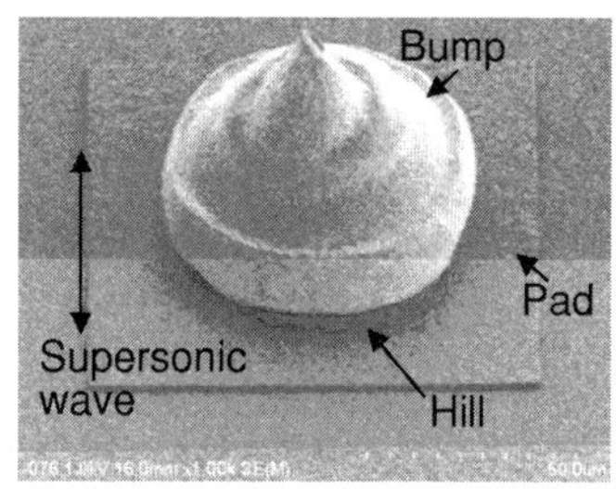

Fig. 4 SEM image of Au bump on pad
There is a hill in the direction of supersonic wave.

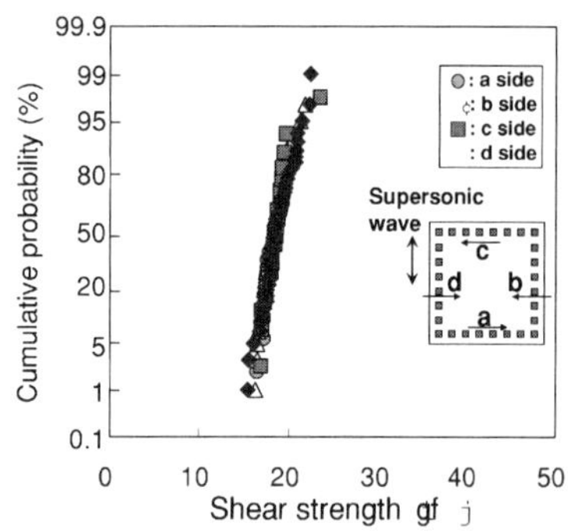

Fig. 5 Cumulative probability of shear strength at each side (Sample A) All the directions of shearing are perpendicular to those of supersonic wave at all sides indicated with arrow.

Table 3 The ratios of these types of destructions

Sample	Ratio of destruction under Pad (%)	Ratio of destruction in Au bump (%)
A	100	0
B	100	0
C	73	27

ƒ The ratio is caluculated with measurement of 144 bumps (1 chip).

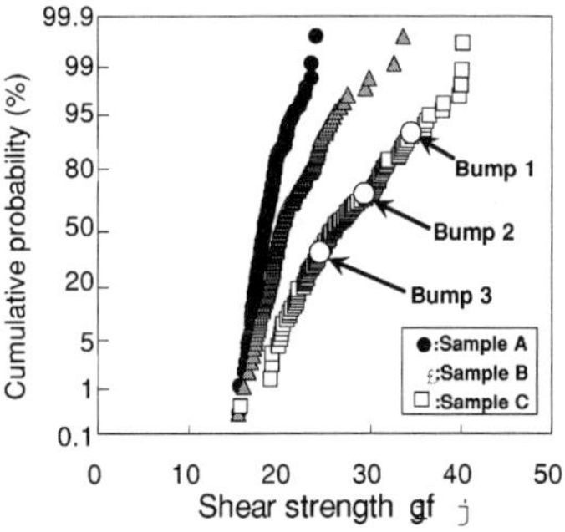

Fig. 6 Cumulative probabilities of shear strength in all samples "Bump" in graph corresponds to Bump in Fig. 7.

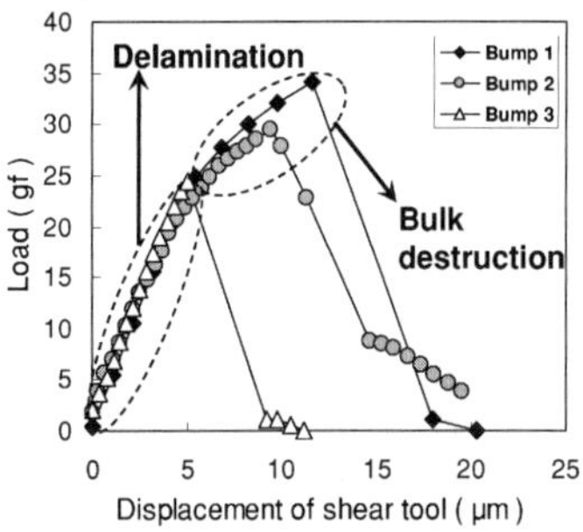

Fig. 7 Profiles in shearing Au bumps from Sample C. The shear strength is defined as maximum load in its profile.

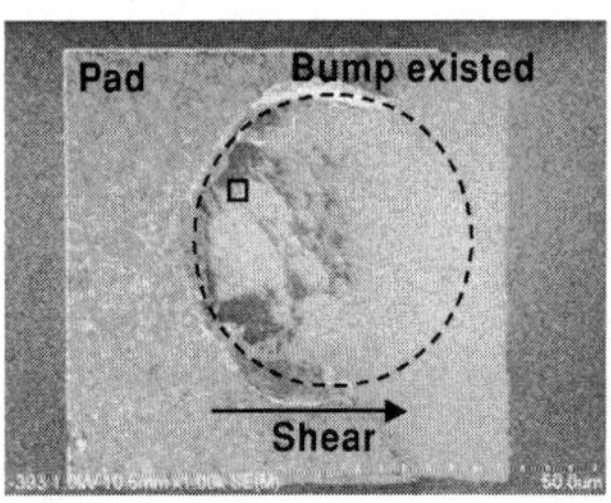

Fig. 8 SEM image after bump shearing of Sample C (:see Fig. 9)

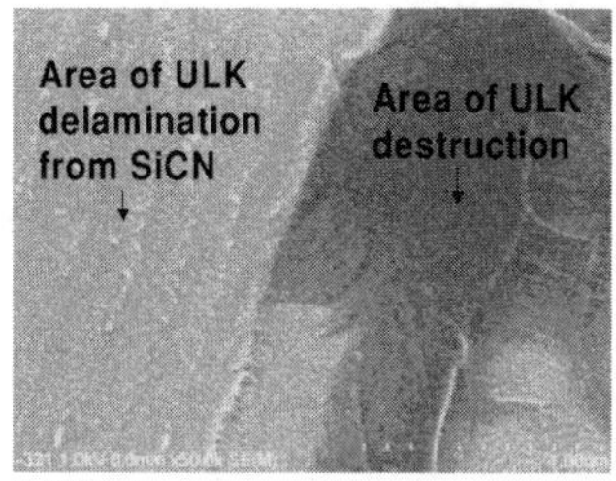

Fig. 9 SEM image expanding area indicated as □ in Fig. 8

Fig. 10 SEM image of cross section after bump shearing of Sample C (by 10° tilt)

Table 4 The ratios of types of destructions for samples having different thicknesses of SiO ULK features are same to those of Sample A

Thickness of SiO (nm)	Ratio of destruction under pad (%)	Ratio of destruction in Au bump (%)
60	100	0
100	100	0
500	95	5
1000	8	92

ƒ The ratio is caluculated with measurement of 144 bumps (1 chip).

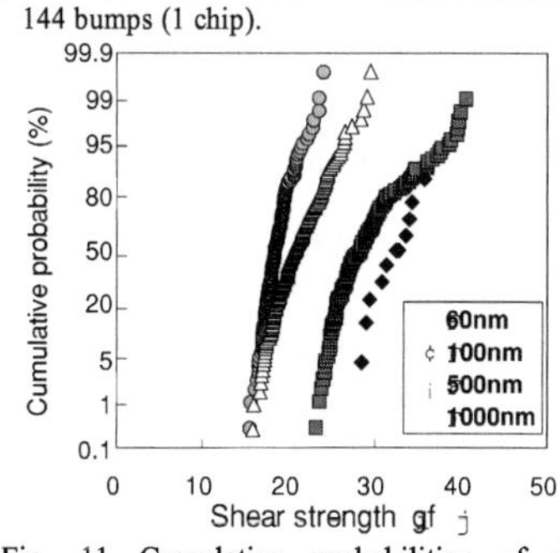

Fig. 11 Cumulative probabilities of shear strength destructed under pad using sample evaluated in table 4

Conference Proceedings AMC XXIV © 2009 Materials Research Society

Thermal stress enhancement and crisis of low-k material around multi-bit interconnections of operating high performance systems

H. Kato, T. Yoshida, M. Nakamura, E. Kondoh and M. Suzuki

University of Yamanashi, Interdisciplinary Graduate School of Medicine and Engineering, 4-3-11 Takeda, Kofu, 400-8511, Japan Tel. +81-55-220-8692
Email: katoh@yamanashi.ac.jp

ABSTRACT

Even now, under the operation environment of high performance integration circuits, the heat generation and the thermal stress are considerably serious, and it is thought that the environment will become more severe in the future. Here, the thermal stress of low-k materials in the operating thermal environment is discussed from the point of view related to the parallel multi-bit interconnections. The maximum thermal stress of the low-k material for the multi-bit line is about four times that of the single line, which is almost equivalent to the ultimate strength of the low-k material.

INTRODUCTION

Interlayer dielectric layers have been altered from their conventional SiO_2 to the low-k material, but the low-k material is fragile and its tolerance against load involved by processes such as CMP etc. is contended [1]. On the other hand, aiming to improve the processing performance of integrated circuits, not only the multi-bit function, but also three-dimensional integration, is discussed. Even now the heat generated in the high performance system is considered serious under ordinary operation [2], and it will become even more severe in the future. We have discussed the thermal stress of global interconnection under the operation environment [3]. Similar attention is necessary for the local interconnection because the thermal stress doesn't strongly depend on the length of the bus line. In this paper, the thermal stress under the operation environment is discussed from the point of view related to the multi-bit interconnection. Here, the multi-bit interconnection means the interconnection technology for data transfer which is realized using parallel signal lines between circuit blocks caused by the increase of command length over than one hundred such as very long instruction word (VLIW) technology. Similar to the crack generation in package due to composite material, the discrepancy in coefficients of thermal expansion involves cracks in the multi-layer structure of LSI.

TREND OF THERMAL ENVIRONMENT

Figure 1 shows an estimated frequency dependence for the temperature increase, ΔT, which is the difference in temperature between the heated junction of the driving gate and the system surface [3]. The data labeled with the filled-in or empty circle/triangle (●/ £ and ○/ ¢) shows both the published and extrapolated supply-voltage/operation-current, respectively. As the frequency increases, the supply voltage V_{DD} is further suppressed to maintain the MOS reliability, but the total operation current I_{op} has the tendency to increase. In addition, the heat generation Q_{op} (=$I_{op}V_{DD}$) turns out to be on the increase. Emergent three-dimensional integration will make the thermal environment more severe.

Although, the energy consumption to drive the multi-bit lines increases as the bit length increases, the main source of the heat is due to the signal processing at the circuits block. The temperature distribution on the substrate surface is assumed to be uniform because of the diffusivity through the substrate body and the power supply lines.

THERMAL STRESS OF LOW-K AROUND MULTI-BIT LINES

With the use of a model that follows the 32nm technology node, the thermal stress is analyzed by finite element method (FEM). In figure 2 a model structure is illustrated, where the lowest layer, L1 is composed of the copper diffusion barrier (Cu DB), the via layer, the trench layer, and the hard mask (HM) layer. The other layers from L2 to L6 are the same as in layer L1, excluding the diffusion barrier layer. The thickness of trench layer with DB and via layer with HM is 90 nm and 80 nm, respectively. The via size is 50x50 nm^2 and the minimum trench width (or the width of metal line) is 57 nm. Here, the barrier metal is ignored because it is very thin and its influence is minimal. The material constants are listed in table 1, where the via layer and the trench layer correspond to the low-k materials. The bus line is set up in the layer L5 with stacked vias, which is a model of intermediate interconnection between adjacent circuit blocks. The thermal stress has tendency to concentrate the fringe of the stacked vias.

The thermal stress does not depend on the length of the bus line and the thermal stress environment of the global and the local interconnection is similar. When the thermal expansion coefficient is α, the length of a material is l, and the temperature rises by ΔT, the elongation Δl is expressed as $\Delta l = l\alpha\Delta T$. When both ends of a material are fixed, the thermal stress is given by $\sigma = E\Delta l/l = E\alpha\Delta T$ with the Young's modulus E, which is independent to the length l. Using FEM simulation, this feature is also confirmed for the interconnection made of the metal line and the low-k material under practical continuum boundary conditions.

Regarding the case of plural parallel interconnections with the stacked vias at the ends, the thermal stress distribution in the three cross sections of (a), (b) and (c) in figure 3 is shown by the

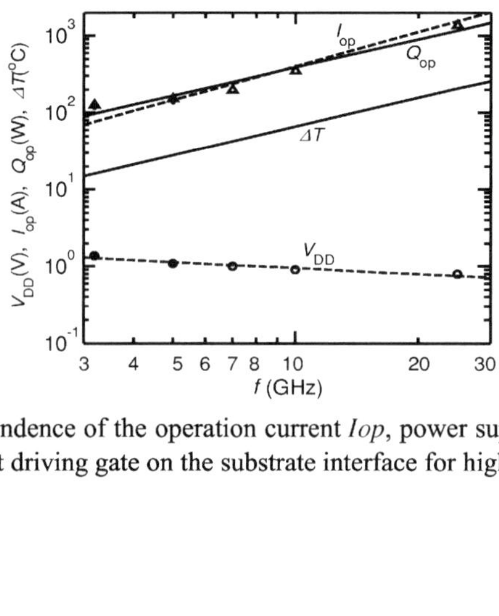

Figure 1. Frequency dependence of the operation current Iop, power supply V_{DD}, power Q_{op}, and temperature increase ΔT at driving gate on the substrate interface for high performance systems.

increase ΔT is 60 °C, which corresponds to the operation frequency of 10 GHz as shown in figure 1. The stress concentration appears at the upper and lower fringe of the stacked vias. The maximum von Mises stress appears at the upper fringe and its value is 55.8 MPa, which is larger than the value 40.1 MPa of an isolated single interconnection. This means that the stress has tendency to increase as the number of bus line increases. This tendency is due to the fact that the stress environment becomes two dimensional for the restricted displacement in the center region because of the influence of adjacent lines. Therefore, it is sufficient to confirm the maximum stress by discussing the stress distribution on the plane (c) in figure 3 or figure 4, which is a cross section along the central bus line. Moreover, the upper limit of the stress for the central line can be confirmed by the two dimensional simulation. This is due to the fact that the aspect ratio of the length against the thickness exceeds 1000 and the displacement of center line is restricted in the two-dimensional plane (c) in figure 3 because of the mirror symmetry.

Figure 5 shows the comparison of the stress distributions in the low-k material that surrounds (a) multi-bit line and (b) the single line. To estimate the upper limit of the von Mises stress for the multi-bit line, figure 5 (a) is calculated by two-dimension simulation with assumption that the displacement perpendicular to the plane is restricted to zero. On the other hand, figure 5 (b) is the result of the three-dimensional analysis for the single line. Because the stress scales of figure 5 (a) and (b) are unified, the area of greater stress in (a) is lost by achroma. Both figures are obtained with a temperature increase of ΔT=60°C. The stress concentrates at the top of the stacked vias and the values of von Mises stress are (a) 155.1 MPa and (b) 40.1 MPa, respectively. The maximum thermal stress of low-k material for the multi-bit line is about four times that of the single line, which is almost equivalent to the ultimate strength of the low-k material [4]. Although the accuracy of the simulated data must be confirmed by careful comparison with the measured data [5], the stress of low-k around stacked vias will be enhanced by the crowded parallel lines.

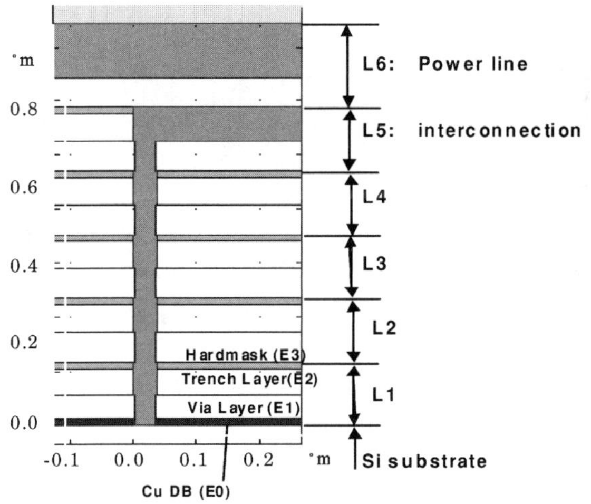

Figure 2. A schematic of the model structure for stress analysis around a stacked via.

Table1. Material constants used in the model.

	Cu D.B.	Via layer	Trench layer	HM layer	copper	Si
Young's modulus (GPa)	100	5.7	1.3	6.3	150	131
Poisson ratio	0.3	0.25	0.25	0.25	0.32	0.27
Coeff. of thermal expansion (10^{-6}/C)	4.15	20	66	20	18.3	3.5

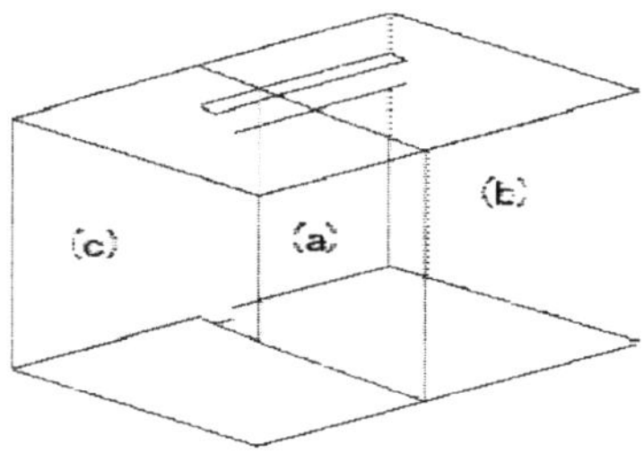

Figure 3. A schematic of stress display sections.

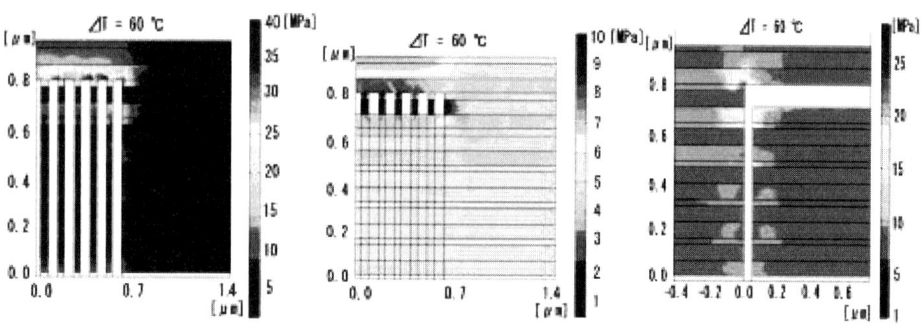

(a) top view (b) rear view (c) side view of center line

Figure 4. A color map of the von Mises stress for 11 interconnections. Because of the mirror symmetry, the half reason is shown and the left line corresponds to the central line in (a) and (b).

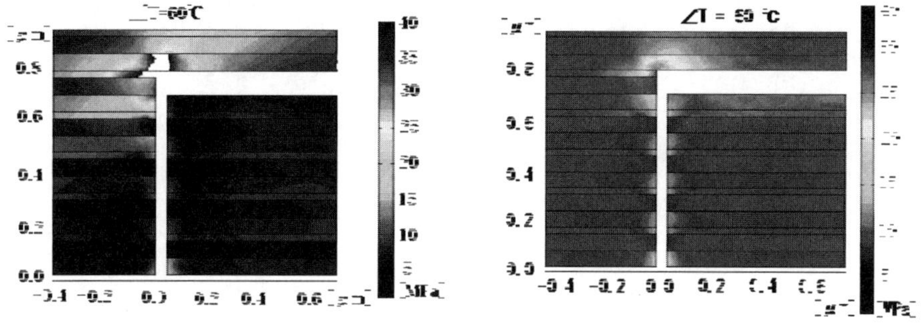

(a) muit-bit line (b) Single line

Figure 5. A color map of the von Mises stress for multi-bit interconnections

CONCLUSIONS

Use of the low-k material in a LSI die involves problem which is similar to that in packages made of composite materials. One of the most serious problems is the crack which occurs in the thermal environment during ordinary operation after shipping. Because the maximum thermal stress of low-k material for the multi-bit line is about four times that of the single line and it is not always possible to detect cracks in the dies using the screening test before shipping, an evaluation on the thermal environment around the multi-bit lines is necessary. This thermal stress is due to the fact that the ultimately compact physical layout in the low-k material is exposed to the heat generation from high-performance devices.

REFERENCES

1. Y. Mochizuki, H. Shibata, M. Tsujimura and H. Hiyama, "Finite element analysis of the stress in vias with Cu/low-k structure", in Proc. of AMC2005, pp.673-678 (MRS, Warrendale 2006)
2. D. Millik et. al., "Advanced package technologies for high-performance systems", Intel technology Journal vol.09, iss.04, pp.259-271(2005)
3. H. Kato, T. Yoshida, T. Tatsuta and E. Kondoh, "Low-k material faces damages induced by mismatch of thermal expansion around the global bus lines when operation frequency is high", abstract of ADMETA pp.82-83 (Tokyo 2006)
4. Y. Xiang X. Chen, T.Y. Tsui, J-I. Jang J.J. Vlassak, "Mechanical properties of porous and fully dense low-k dielectric thin films measured by means of nanoindentaion and plane-strain bulge test technique", J. Mater. Res. vol.21, no.2 pp.386-395 (2006)
5. M. Kodera, S. Kakinuma, Y. Saijo, M. Tsujimura and G. Pezzotti, "Nanometer-scale stress detection of patterned ILD using cathodeluminescence piezo-spectroscopic assessments in a nano-stress microscope", Proc. of AMC 2006, pp.655-660 (MRS, Warrendale 2007)

Reliability and Yield

Conference Proceedings AMC XXIV © 2009 Materials Research Society

Enhancement of the lifetime-model based on the Poole-Frenkel conduction for the SiOCH low-k dielectric used for the sub 65nm technologies

Maxime Vilmay[1,2], David Roy[1], Fabien Volpi[2], and Jean-Marc Chaix[2]
[1]STMicroelectronics, 850 Rue Jean Monnet, Crolles, 38926, France
[2]SIMAP Grenoble INP-CNRS, 1130 Rue de la piscine, Saint Martin d'Hères, 38402, France

ABSTRACT

Due to the size scaling and the introduction of low-k dielectric, the understanding of the mechanisms governing reliability of the inter-metal dielectrics (IMD) used in the copper interconnects is becoming a major concern in the microelectronic industry. This paper presents a reliability study of a dense SiOCH used as IMD in the 65nm technology node. It is demonstrated that the Time To Failure (TTF) of the dielectric is directly linked to the leakage current. This observation enables to develop a lifetime-model based on the Poole-Frenkel (PF) conduction mechanism through the dielectric and leading to an $\exp\left(\sqrt{E}\right)$ TTF dependency. Moreover, this last relationship enables to explain the dependency of the activation energy with the electrical field. An enhancement of the lifetime-model is proposed in agreement with experimental datas.

INTRODUCTION

The introduction of SiOCH low-k dielectrics in copper interconnects associated to the reduction of the critical dimensions in sub 65nm node technologies is a challenge for reliability engineers. Circuit wear-out linked to low-k dielectric breakdown is now becoming a major concern. In the literature, many studies deal with the reliability of the SiOCH low-k and some Time-Dependent Dielectric Breakdown (TDDB) models have been developed to explain the low-k dielectric lifetime. Currently, two main extrapolation lifetime models are used in the industry. The historical and conservative E-model developed by McPherson [5] describes the dielectric lifetime as a function of the external electric field. This model is based on the bond breaking under an external electric field creating defects inside the dielectric. The breakdown occurs when a collapsed defects path happens between the two electrodes. More recently, the $\sqrt{E}$-model [1-4] is going to replace the historical model. This model is based on the square-root dependency of the external electric field of the dielectric lifetime. To explain this model, previous studies suggested that the dielectric lifetime is either due to copper drift [1, 3] or due to conduction mechanism in silicon nitride metal- insulator-metal capacitors [4]. The copper drift hypothesis suggests that the residual copper oxide at the interface between the copper line and the capping is ionized with the electric field. The copper ion can easily drift through the dielectric under this external field. The breakdown occurs when a complete copper path is created between the anode and the cathode. The second hypothesis is based also on the bond breaking but linked to the Poole-Frenkel (PF) or Schottky Emission conduction mechanisms inside the dielectric. These two mechanisms are clearly assisted by the current flowing through the dielectric.

EXPERIMENTAL DETAILS

In this study, copper dual damascene structures were fabricated with the 65nm technology node process with a dense SiOCH low-k dielectrics (ε_r=2.9). The test structures used for our experiments are interdigited combs at metal 1 level with four different spacings between lines (ranging from 90nm to 180nm) with the four associated lengths from 0.6m to 0.9m (Fig.1). Constant Voltage Stress (CVS) and Current-Voltage sweep have been performed to study the dielectric lifetime and the conduction mechanism. These experiments have been carried out between 25°C to 200°C to extract the activation energies of the time to failure and of the conduction mechanism. The time to failure extracted from the CVS tests are all extrapolated at a representative circuit length to compare the lifetime. The well known extrapolation length is described in equation 1:

$$\frac{TTF_1}{TTF_2} = \left(\frac{L_2}{L_1}\right)^{1/\beta} \quad (1)$$

with TTF_1 and TTF_2 the time to failure of the test structure with the respective length L_1 and L_2, and β the Weibull slope. The experiments were performed with a minimum sampling of 15 dies per tests.

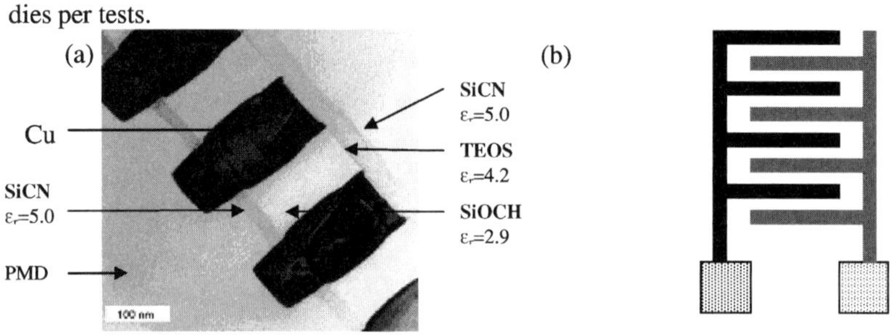

Figure 1. (a) Copper dual damascene stack for the C065 technology node (M1 level).
(b) Interdigited comb test structure.

DISCUSSION

Conduction mechanism

In the range of 2-6MV/cm, the dielectric conduction analysis confirms that the leakage current is consistent with a Poole-Frenkel mechanism (2), in agreement with previous works [1], with extracted parameters (ε_r=3.6, ϕ_b=0.65) physically consistent with the used material. The Poole-Frenkel effect is the lowering of a potential barrier of a trap in the bulk of an insulator when it interacts with an external electric field. The corresponding equation is given here after:

$$J_{PF}(E) = q\mu(T)NE.\exp\left(\frac{-q\left(\phi_{PF} + \beta_{PF}.E^{1/2}\right)}{kT}\right) \quad (2)$$

$$\text{where } \beta_{PF} = \sqrt{\frac{q}{\pi\varepsilon_0\varepsilon_r}} \quad (3),$$

q is the electron charge, $\mu(T)$ the electron mobility, N the number of defects, ϕ_{PF} the barrier height for the PF mechanism, k the Boltzman's constant, ε_0 the vacuum permittivity, ε_r the dielectric permittivity, T the temperature and E the applied electric field.

The figure 2 describes the leakage current dependency as a function of the voltage for various line to line spacing. We can observe, on the figure 3, a perfect merge of the leakage current as a function of the electric field.

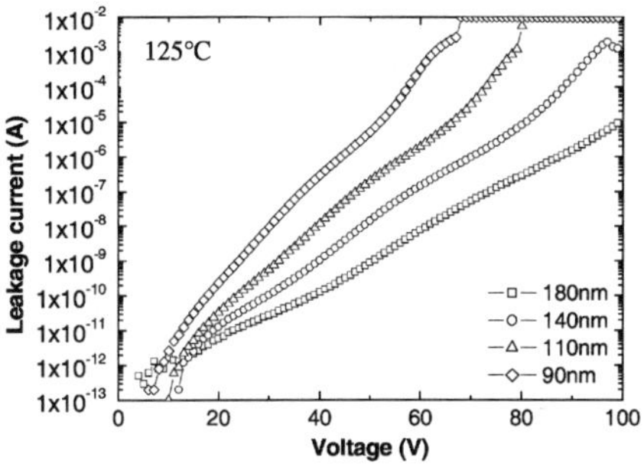

Figure 2. Current-Voltage sweep as a function of the spacing between the copper lines

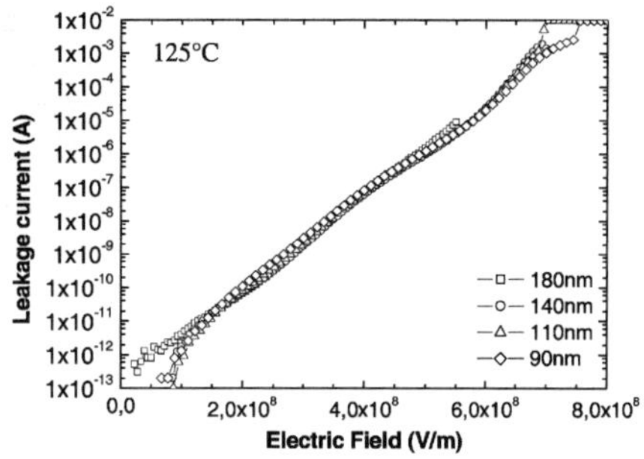

Figure 3. Electric Field dependency of the leakage current as a function of the spacing between the copper lines

Lifetime modeling

Considering the $\sqrt{E}$-model, the two hypotheses, either the copper drift or based on the PF conduction, are linked to the leakage current through the dielectric. The reduction of the spacing between the lines leads to an increase of the leakage and this current must have an impact on the TTF. As reported on the figures 5, we observed on the CVS tests a power law correlation (Eq. 4) between the initial leakage current and the TTF. This dependency is verified whatever the spacing, the voltage and the temperature used.

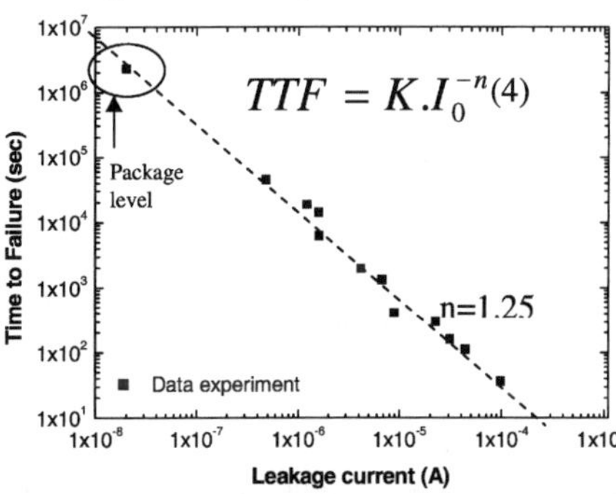

Figure 5. Time to failure dependence as a function of the leakage current through the dielectric.

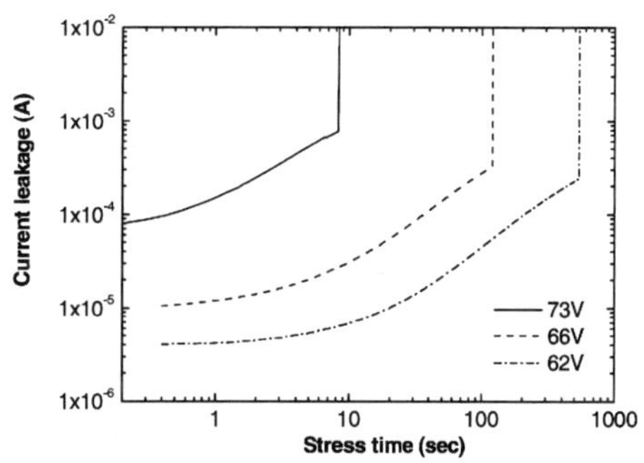

Figure 6. Evolution of the leakage current as a function of the time test and the voltage stress for a spacing of 110nm.

The leakage current during the stress is illustrated in the figure 6. Commonly, two distinct phenomena are reported to describe the temporal evolution of the leakage during the stress. First, the current decreases due to the electron trapping in the dielectric and then in second the current level increases. The hypothesis of the leakage increase is either the defects creation in the bulk of the insulator or due to the time dependency of the copper drift [1]. In our study, only the leakage increase phenomenon is observed. This is due to a low trapping effect in the dielectric. So, the initial leakage is reported as being the conduction current. As observed in the figure 8, the conduction current is coherent with a PF current as describes in the conduction mechanism study (Eq. 5). The time dependency of the current must be totally understood to study the impact of the leakage on the time to failure and will be studied in future works.

Experiments have been carried out for various spacing-temperature-voltage stress conditions and the corresponding n of equation 4 has been extracted. We have observed an experimental variation of extracted n from 0.8 to 2 as describe in the figure 7 with a mean value of 1.4. These fluctuations are not correlated to the spacing, to the temperature and to the voltage range and are then attributed to the experimental error.

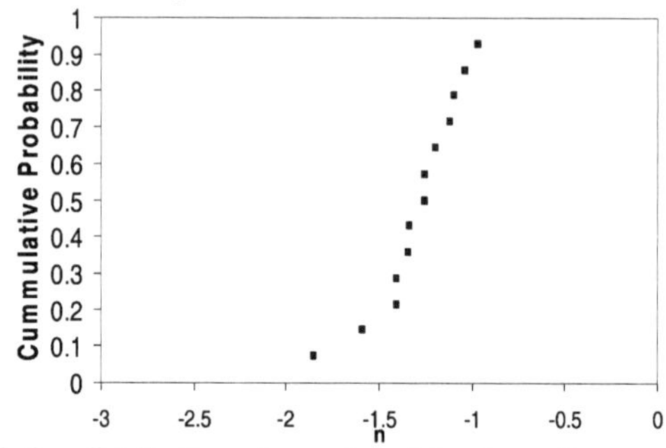

Figure 7. Cumulative distribution of experimental values of exponent *n* in the experiments

As the TTF is linked to leakage current (Eq. 4) and as this leakage is varying with $\exp\left(\sqrt{E}\right)$ (Eq. 5), we can easily express the TTF as a function of the $\exp\left(\sqrt{E}\right)$ (Eq.6). As shown on the figure 9, we observed experimentally this dependence with a large amount of samples. Moreover, the extracted ε_r (2.6) obtained from this model (Eq. 6) is consistent with the used SiOCH dielectric (2.9).

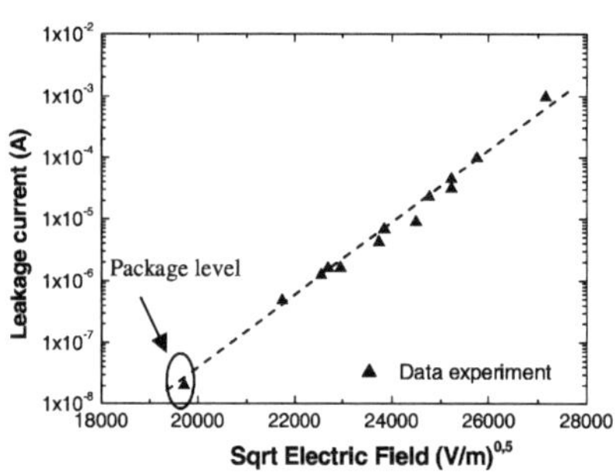

Figure 8. Leakage current dependence as a function of the applied electric field for a spacing of 110nm

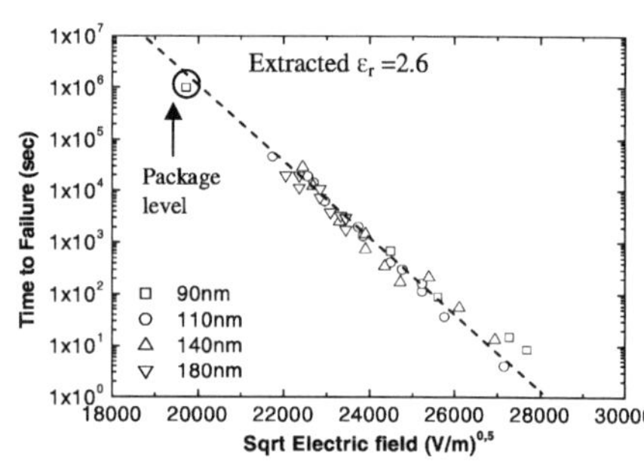

Figure 9. Time to failure dependence as a function of the applied electric field and of the spacing (3 wafers)

$$TTF = K.I_0^{-n} \quad (4)$$

$$I_0 = AE.\exp\left(\frac{-q\Phi + \sqrt{\frac{q^3E}{\pi\varepsilon_0\varepsilon_r}}}{kT}\right) \quad (5)$$

$$TTF = K.(AE)^{-n}.\exp\left[-n\left(\frac{-q\Phi + \sqrt{\frac{q^3E}{\pi\varepsilon_0\varepsilon_r}}}{kT}\right)\right] \quad 6)$$

Moreover, the activation energy of the TTF is often reported to vary as $\exp\left(\sqrt{E}\right)$ [1-4]. We observed in the equations 7, 8 and 9 that the TTF activation energy is linked to the activation energy of the leakage current and proportional to the square-root of the electric field. In fact, these observations were confirmed by the experimental data as illustrated in figures 10, 11 and 12. The activation energies of the TTF and of the leakage current are in a good agreement with the activation energies calculated with the parameters (ε_r and ϕ) extracted from the conduction analysis.

$$Ea(I_0) = -\Phi + \sqrt{\frac{qE}{\pi\varepsilon_0\varepsilon_r}} \quad [\text{eV}] \quad (7)$$

$$Ea(TTF) = -n\left(-\Phi + \sqrt{\frac{qE}{\pi\varepsilon_0\varepsilon_r}}\right) \quad [\text{eV}] \quad (8)$$

$$Ea(TTF) = -n.Ea(I_0) \qquad (9)$$

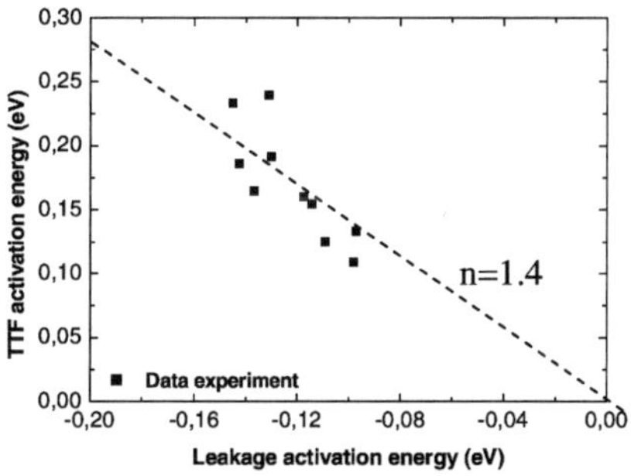

Figure 10. Activation energy of the TTF as a function of activation energy for the leakage current, and comparison with the expected behavior for $n = 1.4$

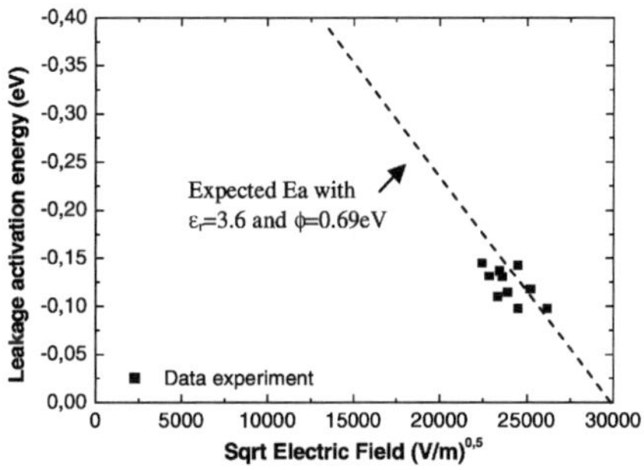

Figure 11. Activation energy of the leakage current as a function of the applied electric field. Comparison with the expected Ea with the parameters extracted from the conduction analysis

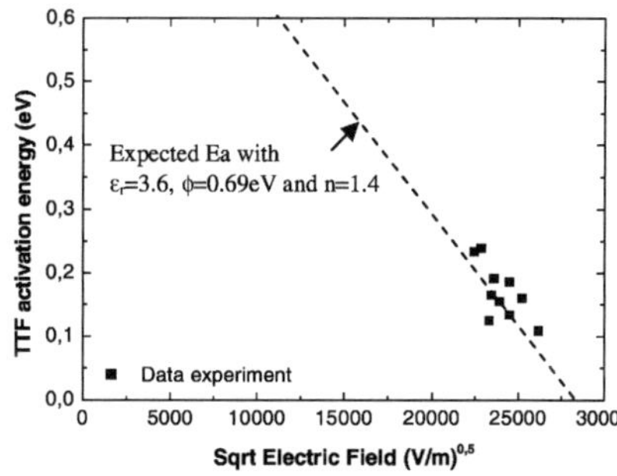

Figure 12. Activation energy of the TTF as a function of the applied electric field. Comparison with the expected Ea with the parameters extracted from the conduction analysis

CONCLUSION

This paper presents the relation between leakage current and time to failure of a dense SiOCH dielectric. An enhancement of the lifetime-model linked to the Poole-Frenkel conduction mechanism is developed and explains the activation energy field dependency. This model should also be developed for porous SiOCH low-k used for advanced technologies.

REFERENCES

1. F.Chen, "*A comprehensive study of low-k SiCOH TDDB phenomena and its reliability lifetime development*", IRPS 2006, pp 46-53

2. N.Suzumura, "*Electric-Field and temperature dependencies of TDDB degradation in Cu/Low-k damascene structures*", IRPS 2008 pp138-143

3. N.Suzumura, "*A new TDDB degradation model based on Cu ion drift in Cu interconnect dielectrics*", IRPS 2006, pp 484-489

4. K.H.Allers, "*Prediction of dielectric reliability from I-V characteristics: Poole-Frenkel conduction mechanism leading to* $\sqrt{E}$ *model for silicon nitride MIM capacitors*", Microelectronic reliability, 2004, pp 411-423

5. J.W.McPherson, "*Complementary model for intrinsic time-dependant dielectric breakdown in SiO2 dielectrics* ", J. Appl. Phys Vol 88 2000 pp5351-5359

Conference Proceedings AMC XXIV © 2009 Materials Research Society

Advanced Bilayer Low-k Dielectric Cap for Reliable High performance Cu-Low k interconnects

Griselda Bonilla[1], Lawrence A Clevenger[2], Chao-Kun Hu[1], Son Nguyen[3], Tien J Cheng[2], Zhiguo Sun[4], Hosadurga Shobha[3], Stephan Cohen[1], Armin Fischer[5], Kaushik Chanda[2], Wu Ping Liu[4], Yiheng Xu[2], Thomas M Shaw[1], Mark S Chace[2], Eric G Liniger[1], Steven E Molis[2], Yun Y Wang[2], Jason Gill[2], Roger A Quon[2], and Glenn Biery[2]

[1]IBM TJ Watson Research Center, 1101 Kitchawan Road, Yorktown Heights, NY, 10598

[2]IBM Semiconductor Research and Development Center, 2070 Route 52, Hopewell Junction, NY, 12533

[3]IBM at Albany Nanotech - Center for Semiconductor Research, 255 Fuller Road, Albany, NY, 12077

[4]Chartered Semiconductor Manufacturing Ltd, 2070 Route 52, Hopewell Junction, NY, 12533

[5]Infineon Technologies A.G., 2070 Route 52, Hopewell Junction, NY, 12533

ABSTRACT

This paper presents the performance of a low k advanced dielectric Cu barrier ($k \leq 4.0$) for 45 nm CMOS devices and beyond while still maintaining equivalent reliability as compared to traditional (k=5.3) SiCNx cap materials. This new PECVD SiCNx/SiCNy bilayer film was developed with varying composition to achieve: 1) a high nitrogen content SiCNx bottom layer with robust mechanical properties, good barrier properties, and excellent interfacial characteristics with Cu. 2) a lower nitrogen content SiCNy top layer with lower dielectric constant, and good plasma RIE selectivity. This paper will discuss the integration and reliability results of this advanced low-k Cu barrier for use in advanced CMOS devices.

INTRODUCTION

The metallization of integrated circuits for high performance sub-45nm CMOS devices requires the use of copper with low-k or ultra low-k dielectrics to reduce RC delay and cross talk between interconnects [1]. This can potentially be achieved by either reducing the dielectric constant of the Inter-Level Dielectric (IMD) films or that of the dielectric diffusion barrier. The latter becomes increasingly important since as the thickness of the low-k dielectric stack decreases with each new generation, the impact of the dielectric barrier on the effective k-value becomes significant [2]. Moreover, a modification to the dielectric barrier film poses fewer challenges to the overall process integration flow and reliability of the BEOL device structures as compared to changing the IMD film, especially since the lower dielectric constant materials must still provide high mechanical, electrical, and fabrication integration robustness.

EXPERIMENT

To study the barrier properties of the advanced dielectric barrier films, the low-k films were deposited on blanket copper for evaluation of oxidation resistance and Cu barrier properties. For their qualification, low-k cap layers were deposited on M1-M3 wiring levels of 45nm CMOS at 0.14 μm pitch. The structures were subsequently evaluated by inline parametric measurements along with electromigration, stress migration, and time-dependent dielectric breakdown (TDDB).

DISCUSSION

RC reduction strategy

An advanced on-chip wiring optimization tool, named AQUAIA [3], was utilized for the extraction of parasitic capacitance. This tool allows the extraction of the per-unit-length C parameters in the presence of highly inhomogeneous dielectric environments, including orthogonal wiring on adjacent layers. The parameters can be outputted by AQUAIA as a function of dimensional parameters or material characteristics.

For a fixed IMD material and Cu line dimensions, a plot of the cap thickness (nm) vs. Ctotal (pF/cm) for a 45nm dual damascene build shows that to obtain a 5% reduction in the total capacitance, the ratio of the k_{cap}/k_{IMD} should be less than 1.5 for a fixed k_{cap} thickness of 30nm (Fig. 1). This corresponds to a k_{cap}=4.0 (k_{IMD}=2.7). On the other hand, by fixing the k_{cap} to 5.3, to achieve a 5% reduction in the Ctotal, a k_{IMD}<2.5 would be required. As previously mentioned, integration of a barrier layer with a k_{eff}=4.0 can be implemented in a timely manner as compared to multiple process changes that would be required to incorporate a k<2.5 porous ULK layer.

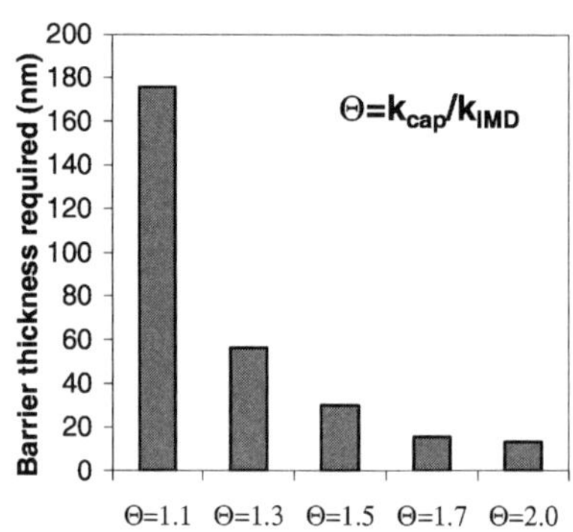

Figure 1. Calculation of the required barrier thickness to achieve a 5% C_{total} reduction from the baseline process.

An additional alternative to modification of the dielectric constant of the diffusion barrier is to reduce the thickness of the base k=5.3 SiCNx layer. For a target 5% C_{total} reduction, the required thickness of the SiCNx layer would be less than 20nm (shown in Figure 1). However, a thinner dielectric barrier would compromise its' role as an etch stop layer and potentially compromise the required diffusion barrier properties. Therefore, this analysis shows that the preferred approach to achieve a reduction in the RC delay is to incorporate the use of a low-k advanced barrier.

Film Development

Apart from the obvious task of preventing the diffusion of mainly Cu, but also other elements like Si, or process gases, there are additional requirements for diffusion barriers in Cu interconnects. In particular, they should: (i) act as an etch stop during the dual damascene processing, and (ii) have good adhesion to Cu and dielectric layers with smooth interfaces. Therefore, defect-free, homogeneous thin barrier films (<50 nm) are needed to maximize barrier properties while not adversely impacting k_{eff}. Additionally, these films must be stable under thermal stressing, form high quality interfaces to the Cu, and yield no detrimental diffusion paths for Cu electromigration.

To meet these requirements, the low-k cap is used in the form a bilayer where a high nitrogen containing layer is deposited adjacent to the Cu, followed by a low nitrogen containing SiCNy low-k layer. Specifically, a 20-50 nm low nitrogen SiCNy layer is deposited by PECVD on a <15nm SiCNx initiation layer as observed in the TEM image shown in Figure 2. TOF-SIMS depth profile exhibits first a carbon-rich layer, followed by a nitrogen-rich nano-laminated layer (Figure 3), as expected. The preclean treatment prior to barrier deposition is also important and has been optimized to improve adhesion of the barrier layer to the Cu and surface.

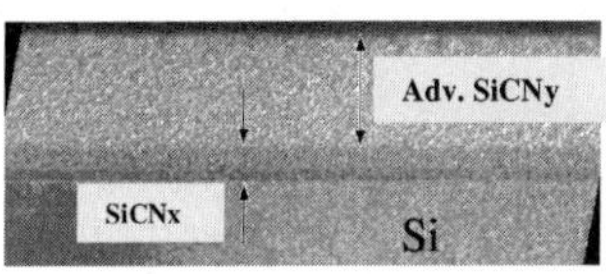

Figure 2. TEM x-section of low-k advanced barriers deposited on bare Si.

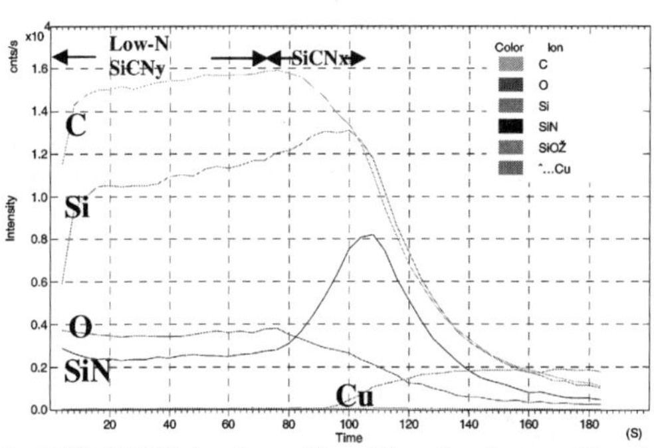

Figure 3. TOF-SIMS depth profile of low-k advanced barriers deposited on a blanket Cu surface.

The electrical and chemical properties of the advanced barrier were compared to a SiCNx baseline film. The effective k-value of the advanced barrier film was demonstrated to be <4.0, compared to a k-value of ~5.3 of the SiCNx baseline film. Breakdown strength of the film was slightly lower than that of the SiCNx single layer (Figure 4), however, the value is still within acceptable limits. Exposing the low-k bilayer atop blanket Cu films to 310°C in air for 24hours shows no Cu oxidation at the SiCNx/Cu interface indicating that the film is resistant to oxidation.

In addition, Cu barrier tests indicate that this film prevents Cu diffusion and is therefore an effective Cu barrier.

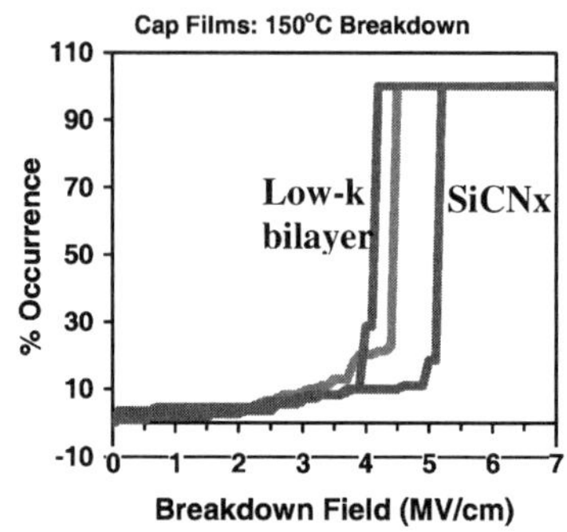

Figure 4. Breakdown field for Low-k bilayer advanced barriers compared to the SiCNx baseline.

Etch selectivity

The etch selectivity was assessed by measuring the barrier loss after the full dual damascene etching sequence of a standard SiCNx layer compared to low-k advanced barriers. A mean thickness of ~2.0 nm was removed for the SiCNx layer compared to ~2.6 nm for the less dense, N-doped topmost layer of the bilayer. While there exists a difference in the etch selectivity of these layers, these differences are much less than expected and can be compensated for by adjusting the thickness of the topmost low-k layer without significantly affecting the effective dielectric constant of the bilayer (<2%). SEM x-sections (Figure 5) of a typical via/line level with an advanced barrier shows well controlled via profiles as well as void-free Cu fill.

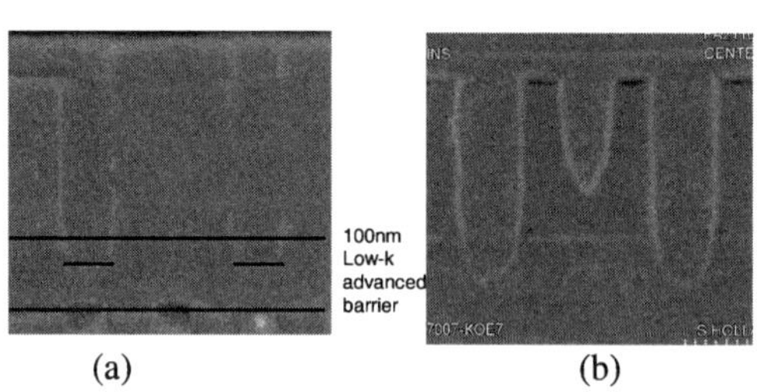

Figure 5. Via etch profiles using standard etch processes and integration flow for the low-k advanced barrier. (a) Etch selectivity is comparable to the SiCNx baseline. (b) Post metallization via chain profile shows no difference with SiCNx baseline process flow.

Electrical Characterization

The same standard integration flow and etch processes were used for both the standard SiCNx and low-k advanced barriers. The contact via resistance measurements from a representative lot are shown in Figure 6a. The low-k advanced barrier film exhibits a lower contact via resistance, attributed to differences in the via opening within the barrier film. Reasonable parametric yields were obtained from a comb-serpentine defect monitor structure as shown in Figure6b.

Figure 7 shows that with the optimized k<4.0 low-k advanced SiCNx dielectric barrier, a C_{total} reduction of ~5% is achieved.

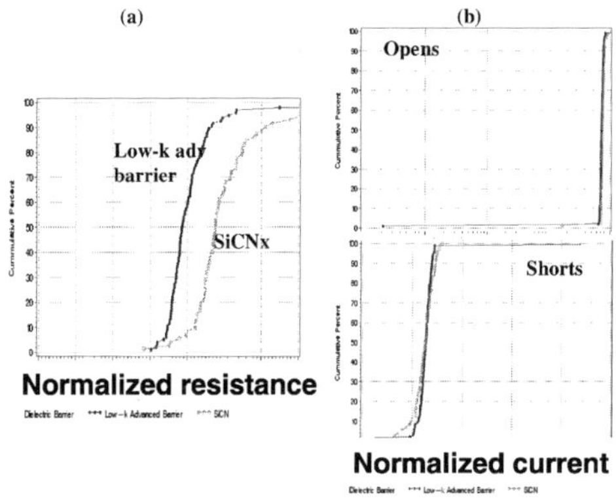

Figure 6. (a) Viachain resistance and (b) defect monitor opens/shorts current yield for builds with $SiCN_x$ and low-k advanced barrier are similar.

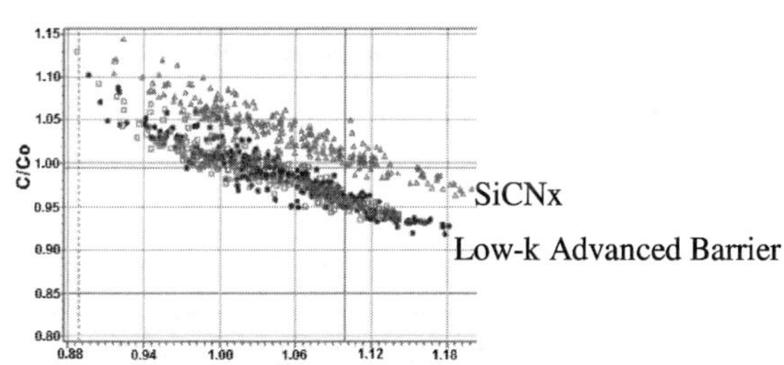

Layer	C reading at target resistance		
	$SiCN_x$	Low K Adv barrier	Delta
M2 total	1.00	0.955	-4.5%
M3 total	1.00	0.955	-4.5%
M4 total	1.00	0.955	-4.5%

Figure 7. A 4-5% capacitance reduction is achieved when the low-k advanced barrier is used in a standard 45nm CMOS build.

Reliability Performance

Electromigration testing was performed under a constant dc current density at sample temperatures from 218 to 351°C. The test structure was either a single or a dual damascene multi-level via-line structure. The electron flow was forced through a single contact via to a 70nm wide line located either at the top or below. Failure was defined as a 10% change in resistance and the failure distribution was assumed to be lognormal. Figures 8(a) and (b) show representative lognormal cumulative time to failure distributions at various sample temperatures and median lifetime as a function of 1/T for wafers with different cap layers. Note that the advanced barrier films show equivalence to the SiCNx baseline film. In addition, the measured activation energy for both barrier layers were equivalent, confirming that the low-k advanced barriers can provide a capacitance reduction with minimal impact to the overall integration process flow.

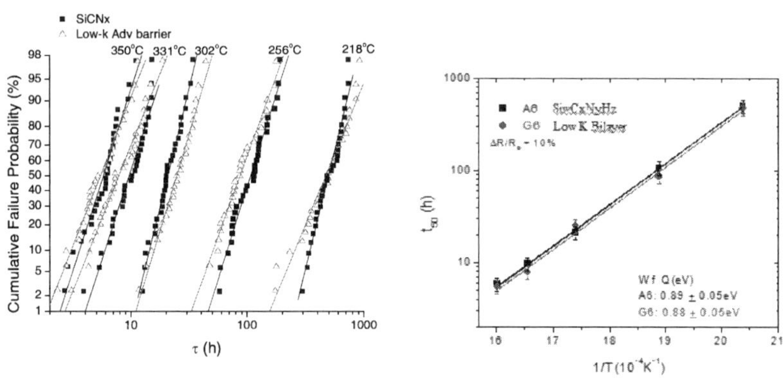

Figure 8. (a) EM lifetime distributions are similar for both the SiCNx baseline and low-k advanced barriers & (b) have the same activation energy.

In addition, builds with low-k advanced barriers also passed stringent stress migration (225°C, 300hrs) and TDDB (5MV/cm, 125°C) reliability qualification testing without any issues.

0.85 Vbdmean	t63(30V)	beta	gamma	tEOL(3.63V/1m)
30.8V	400 sec	0.84	0.806 V^{-1}	> 10,000 hrs

Table 1. TDDB measurements for builds with the low-k advanced barrier film shows no issues.

CONCLUSIONS

Aggressive interconnect dimension scaling and a desirable reduction in RC delay and cross talk has lead to the development and qualification of a new low-k dielectric barrier with a k-value <4.0 for 45nm CMOS dual damascene architecture. The advanced barrier is a bilayer with similar Cu diffusion and oxidation barrier efficiency as standard SiCNx. In addition, up to a 5% C reduction was achieved without degradation of EM, SM, and TDDB reliability performance.

ACKNOWLEDGMENTS

Authors would like to acknowledge the support and original process chemistry from Applied Materials, Inc. [4] that led to the improved low-k advanced barriers reported herein. This work was performed at the IBM Microelectronics, Div. Semiconductor Research & Development Center, Hopewell Junction, NY 12533.

REFERENCES

1. Sankaran et al., IEDM 2006.
2. ITRS Roadmap, 2007.
3. I. M. Elfadel, M. B. Anand et al., "Aquaia: A CAD tool for on-chip interconnect modeling, analysis and optimization", *Dig. Electr. Perf. Electronic Packaging*, Vol. 11, Monterey, CA, Oct. 2002, pp. 337–340.
4. A . Lakshmanan, A Lee, J. Lee and B. H. Kim, US Patent # 7288205 (2007).

Conference Proceedings AMC XXIV © 2009 Materials Research Society

Improvements in Electromigration to Meet the Requirements of Advanced Technology

Roey Shaviv, Hui-Jung Wu, Mandy Sriram, Wen Wu, Anshu Pradhan, Jennifer O'Loughlin, Kaushik Chattopadhyay, Tom Mountsier, and Girish Dixit
Novellus System Inc., 4000 North First Street, San Jose, CA, 95134

ABSTRACT

The reliability of interconnects, and electromigration performance in particular, continues to be a formidable challenge as dimensions shrink from generation to generation. In this paper we demonstrate barrier seed scaling, enabled by HCM IONX PVD technology. We report high activation energy of ~ 1 eV, and J_{max} > 6 MA/cm^2, exceeding the ITRS 2007 requirements for the next several generations by a wide margin. Thinner barrier/seed with increased barrier etch back is shown to increase lifetime. Further increase in electromigration time to fail is achieved by Ge-PSAB, where up to 7 fold lifetime improvement with similar activation energy and J_{max} > 10 MA/cm^2 is reported. Seed alloying with Al is known to improve electromigration lifetime. Here we report on a new, low resistance, seed alloying process enabling high electromigration lifetime. Synergy between these processes enables a significant improvement in electromigration lifetime that provides the reliability margin necessary for continuing technology scaling for future generations.

INTRODUCTION

Meeting the reliability requirements for narrow lines at advanced technology nodes is an increasing technology challenge. The International Technology Roadmap for Semiconductors (ITRS) proclaims that "achieving necessary reliability" is one of the "difficult challenges for ≤ 32 nm".[1] A steep increase in J_{max} between the 65 and 22 nm technology nodes was projected by the ITRS until 2006. The technical difficulty in meeting these high J_{max} levels forced an adjustment to a more modest increase in the ITRS 2007.[1] Meanwhile, as line dimensions shrink, the size of the critical void for electromigration (EM) failure is also reduced, causing a sharp decrease in the mean time to failure.[2] Consequently, an improvement in EM lifetime is required for technology scaling in order to meet the J_{max} requirements as set by the ITRS 2007. Several approaches were proposed in order to achieve this. Examples include selective metal caps,[3] Cu alloying,[4] Cu doping with Al,[5] Ti barrier,[6] self aligned copper silicide,[7] PSAB™,[8,9,10] Cu plating chemistry,[11] and various design modifications.[12]

In a recent communication we addressed the growing challenge of meeting the ever increasing EM requirements in copper interconnects for advanced technologies.[13] We introduced a novel process, Ge-PSAB™,[10,14] that enables up to a 7 fold increase in EM lifetime. Here we present a comprehensive study of the effect of the process space on EM and demonstrate process synergy that leads to significant improvements in EM lifetime. The first part of this paper discusses the effect of barrier and seed scaling on interconnect reliability. Scaling of the barrier/seed films is a key issue for both process integration and reliability. As dimensions shrink film thickness must be scaled in order to allow a sufficient process margin for void-free electroplating of copper. At the same time, the integrity of the barrier film must not be compromised in order to maintain the reliability performance. Recently we reported a significant improvement in time dependent dielectric breakdown (TDDB) with a high density barrier film.[15]

This film opens the process window for thin barriers with high reliability. Here we report statistically significant improvements in EM with a thinner barrier/seed.

The second part of this paper explores synergy among processes that can enable maximizing the benefit from each approach while minimizing the Cu resistance increase penalty associated with Cu doping. Seed alloying with Al was demonstrated by several authors to provide a significant increase in EM lifetime.[16] The effect of alloying on EM lifetime, using a novel, PVD technology has been further explored.[17] An integrated process, combining Ge-PSAB and $CuAl_x$ seed is shown to provide EM enhancement that is larger than that of each process separately. This approach, utilizing process synergy, gives new degrees of freedom for process integration and a path to improved reliability.

EXPERIMENT

Experimental work explores three methodologies for EM improvement. We first optimize the PVD barrier/seed for EM performance. We then investigate $CuAl_x$ PVD seed at varying stoichiometries. Finally we explore the synergy between the previously reported Ge-PSAB and $CuAl_x$ seed for reliability improvement. All experiments are conducted using a two level dual damascene test vehicle, manufactured at the Novellus Customer Integration Center (CIC). A via fed electromigration test structure, with nominal line dimensions of 100 nm by 180 nm by 500 μm, is used for both downstream and upstream EM testing. Via dimensions are 90 nm with nominal aspect ratio of 2:1. Current density range and measurement temperature are 1.5 – 3 MA/cm^2 and 548 – 598 K (275 – 325° C) respectively.

RESULTS AND DISCUSSION

Barrier/Seed Conditions

Hollow Cathode Magnetron (HCM) PVD had been in use for barrier/seed applications in high volume manufacturing for several technology generations. Scaling of line dimensions from generation to generation requires modifications to the PVD process in order to achieve low overhang, high step coverage, and good barrier integrity for thin films. The combination of high density PVD barrier deposition and high density plasma etch back processes (IONX®)[18] leads to the formation of a dense barrier metal that significantly improves reliability.[15] This technology enables continuous sidewall coverage with a high integrity barrier even inside high aspect ratio features, as shown in Figure 1*a*. The incorporation of RF re-sputtered Cu seed[19] enables almost conformal distribution of the seed layer inside the feature. Seed etchback enables extendibility of PVD technology to narrow, high aspect ratio features by providing a low overhang, continuous Cu film, as shown in Figure 1b and 1c, and opening the process window for void free copper fill leading to improved reliability.

The EM performance of the HCM IONX PVD barrier/seed relative to conventional HCM PVD control is shown in Figure 2. A statistically significant increase of 50 % in time to fail is observed in both downstream and upstream testing (confidence level 95 %, $X^2 < 0.02$). Moreover, a significant improvement in the distributions of time to fail is measured. For downstream EM, σ improves from 0.5 for the control to 0.3. For upstream EM, σ improves from 0.6 for the control to 0.2. The activation energy, E_a, for downstream and upstream EM was studied between 275° C and 325° C, as shown, in arbitrary units, in Figure 3. High activation energies of 0.96 and 0.91 eV are measured for downstream and upstream EM respectively.[20] The current acceleration coefficients were measured at 1.6 and 1.1 for downstream and upstream EM respectively, in excellent agreement with theoretical expectations.

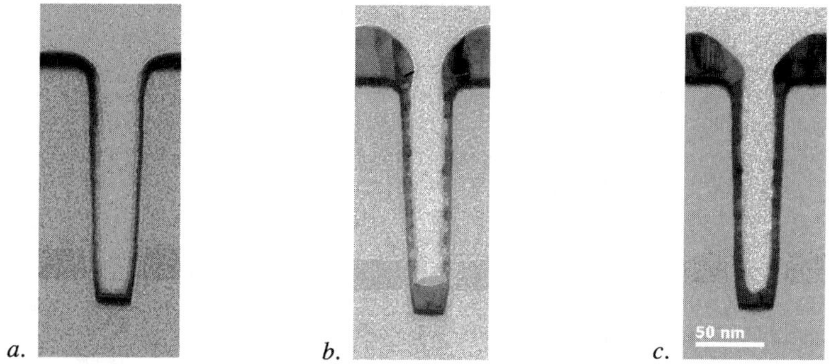

Figure 1. TEM cross sections of the HCM IONX barrier/seed technology in a high aspect ratio 50 nm feature. *a.* High density IONX TAN/Ta barrier, *b.* Same barrier combined with a Cu seed deposition, *c.* Same barrier and seed following the seed etchback step. Final seed coverage is > 5 nm inside the feature.

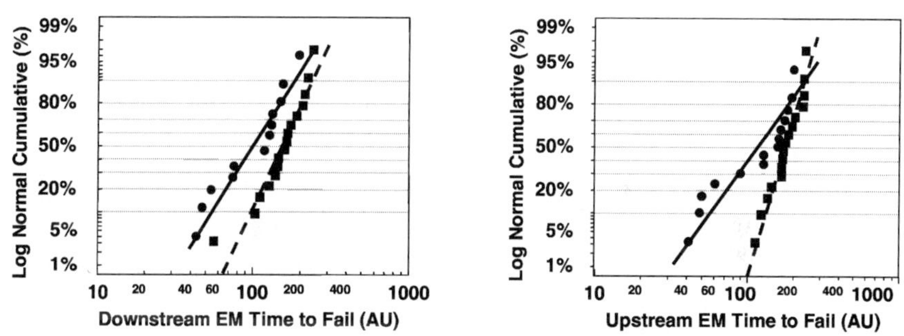

a. *b.*

Figure 2. Effect of barrier/seed on EM: Downstream (*a*) and Upstream (*b*) EM Time to Fail distribution. A statistically significant improvement in lifetime is observed with the IONX high density barrier and resputtered seed process (■) relative to the conventional HCM PVD control (●).

Scaling of the thickness of the HCM IONX PVD barrier/seed is explored in order to enable extendibility of this technology to the 45 and 32 nm nodes. Thus, nominal barrier and seed thickness were reduced by 20 % and 33 % relative to control (65 nm technology), respectively. Similarly, we studied the process window for barrier etchback between 35 % and 125 % etchback, relative to the nominal thickness of the film, using 80 % etchback as control. The EM performance of the thin barrier/seed is presented in Figure 4. Barrier/seed scaling is found to have a small effect on EM performance for both downstream and upstream EM. Thin barrier/seed with 125 % barrier etchback has statistically equivalent downstream EM relative to control (confidence level 95 %, $X^2 = 0.29$). For both conditions $J_{max} > 7$ MA/cm^2. For upstream

EM time to fail is about 50 % longer for thin barrier seed with 125 % barrier etchback, a statistically significant difference (confidence level 95 %, $X^2 < 0.001$). Thin barrier/seed with 35 % barrier etchback shows a significant deterioration in downstream EM performance. Time to fail is lower by about a factor of 3 and σ increases by about a factor of 2 relative to control. J_{max} for downstream EM with these conditions is only 2.1 MA/cm^2. The poorer reliability is attributed to insufficient anchoring resulting in a poor interface between the barrier and the Cu at the bottom of the via. The upstream EM performance with thin barrier/seed and 35 % barrier etchback conditions is statistically equivalent to that of the control (confidence level 95 %, $X^2 = 0.13$). All three upstream EM experiments showed $J_{max} > 6$ MA/cm^2, exceeding the ITRS 2007[1] requirements by a wide margin.

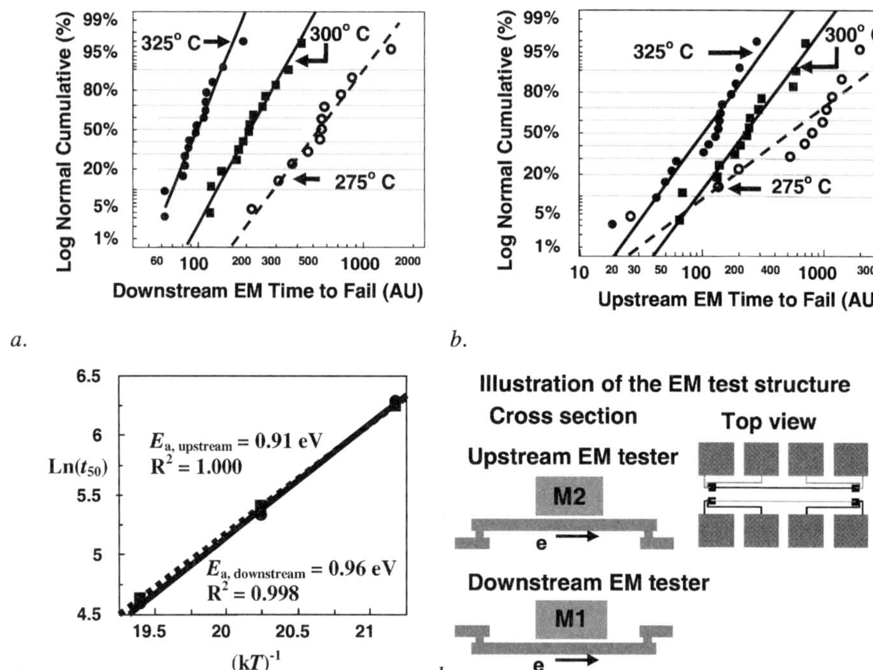

Figure 3. Determination of the electromigration activation energy, E_a, for the HCM IONX PVD barrier/seed processes. Time to fail distribution of downstream (black, 4*a*) and upstream (blue 4*b*) EM. E_a is 0.96 and 0.91 eV for downstream (●) and upstream (■) EM respectively (4*c*). An illustration of the test structure is provided in 4*d*.

The activation energy, E_a, and the current acceleration coefficient, n, for downstream EM with the thin barrier/seed and high etchback were determined as shown in Figure 5. Both parameters, $E_a = 1.00$ eV and $n = 1.4$, are within experimental error relative to the control where $E_a = 0.96$ eV (Figure 4) and $n=1.6$. As mentioned above, J_{max} for the control and the thin barrier/seed with high etchback experiment is > 7 MA/cm^2, exceeding the ITRS 2007[1] requirements by a wide margin.

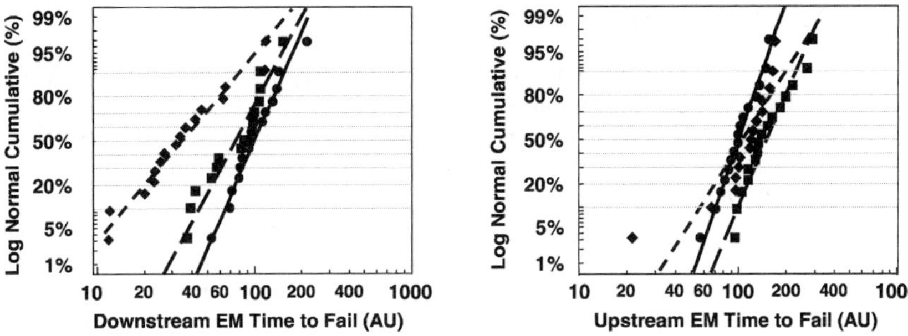

a. *b.*

Figure 4. Time to fail distribution for *a.* downstream and *b.* upstream EM for thin barrier/seed relative to the control (●). Reduced barrier etchback (◆) is detrimental to downstream EM but has an insignificant effect on upstream EM. Increased barrier etchback (■) has an insignificant effect on downstream EM and is beneficial to upstream EM.

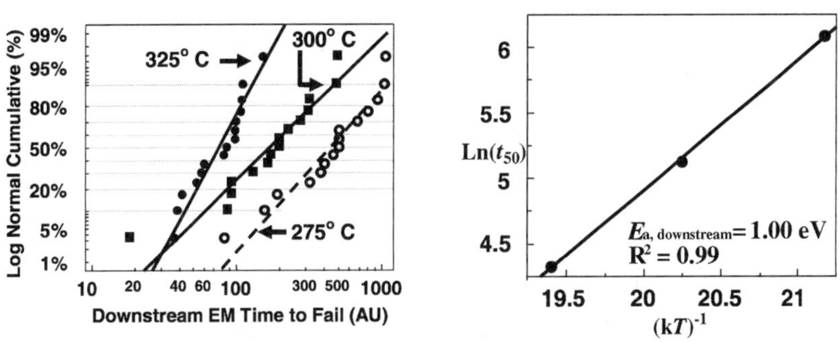

a. *b.*

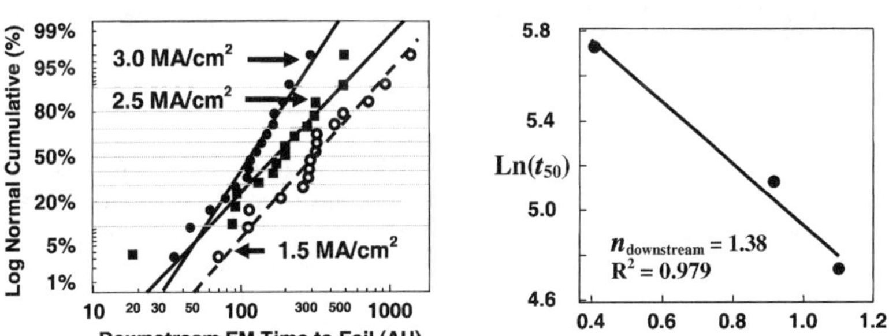

c. *d.*

Figure 5. Time to fail distribution (*a,* and *c*) for determination of the activation energy, E_a, (*b*), and the current acceleration coefficient, *n*, (*d*), for downstream EM with the thin barrier/seed and high etchback.

Process Synergy for Reliability Improvements

While it is apparent from the previous section that J_{max} with the HCM IONX PVD process exceeds the electromigration requirements set by the ITRS 2007,[1] further improvement in time to fail is desired in order to increase process robustness and provide the reliability margin for small line dimensions. In order to achieve this goal we explored the synergy of combining HCM IONX PVD, Ge-PSAB, and a novel $CuAl_x$ alloy seed process.[21]

As discussed above, we recently explored Ge-PSAB and demonstrated a 7 fold increase in EM lifetime.[9] This process involves a patented GeH_4 treatment prior to the deposition of the dielectric diffusion barrier.[10] The activation energy, E_a, for Ge-PSAB is determined here as illustrated in Figure 6. Statistically similar activation energy, of 0.95 eV, is measured for Ge-PSAB and the control (Figure 3).

Additional EM improvement can be achieved with Al doping of Cu.[5] The EM improvement expected with Al alloying is a function of both the Al concentration in the alloy and the deposition process. A novel, patent pending, Al doping process was developed to allow segregation of the Al dopant at the interface between the copper metal and the SiC dielectric diffusion barrier at the top of the interconnect lines, as illustrated in Figure 7*a*. As a result, a significant improvement in the efficiency of EM improvement with Al doping is achieved as illustrated in Figure 7*b* and discussed in detail elsewhere.[21] This enables a significant improvement in EM lifetime with a smaller increase in the resistance of the metal line.

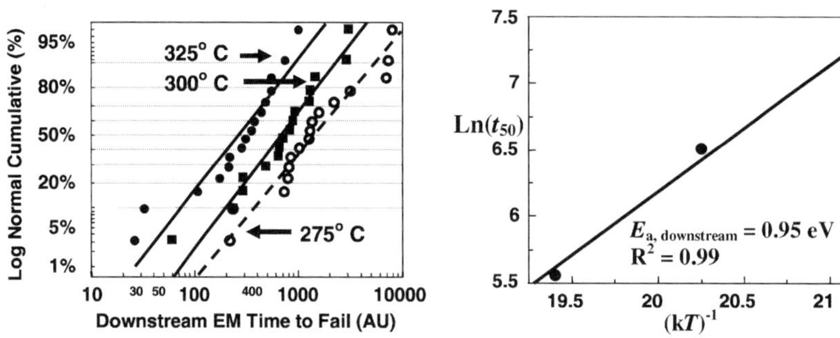

a. *b.*

Figure 6. Time to fail distribution (*a*) and determination of the downstream EM activation energy, E_a, (*b*) for the Ge-PSAB process.

The synergy between Ge-PSAB and Al alloying for EM lifetime improvement is demonstrated in Figure 8 relative to the control. For downstream EM, each process, taken separately, provides about a two fold increase in EM performance. Taken together a four fold increase in EM lifetime is achieved. The effect is even more profound in upstream EM where about a five fold increase in EM lifetime is achieved when each process is taken separately. The synergy between these two processes enables about ten fold increase in the time to fail.

CONCLUSIONS

The reliability of interconnects, and electromigration performance in particular, continues to be a formidable challenge as dimensions shrink from generation to generation. HCM IONX PVD barrier/seed technology enables scaling of film thickness while maintaining high activation energy of ~ 1 eV, thus exceeding the ITRS 2007 J_{max} requirements by a wide margin. Further increase in electromigration time to fail is achieved through the use of either Ge-PSAB or seed alloying with Al. Synergy between these processes is found to enable additional improvement in electromigration lifetime that provides the reliability margin necessary for continuing the technology scaling for future generations.

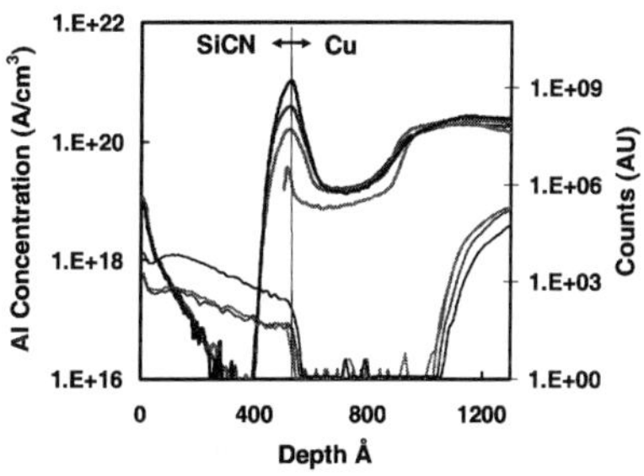

a.

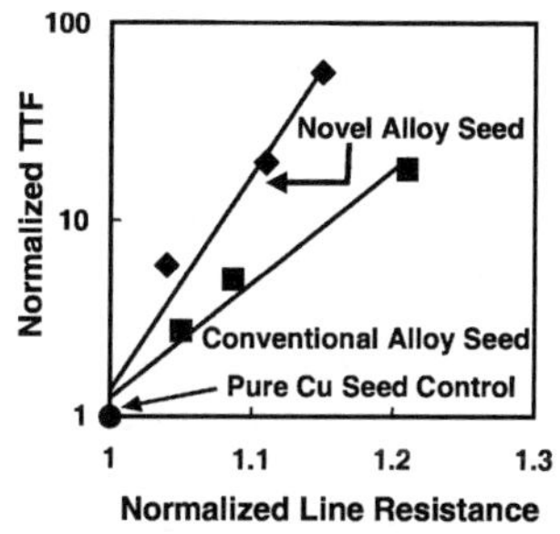

b.

Figure 7. *a.* SIMS depth profile analysis of Al in a (from top) SiCN/plated Cu/alloy Seed/Ta film stacks: Post CMP (red), post SiCN deposition (green), post anneal at 30 min/400° C (blue), post anneal at 30 min/400° C + 36 min/350° C (black) (from top: SiC/CMPed ECD Cu, alloy seed/Ta(N)). *b.* Effect of alloy seed processes on Electromigration.

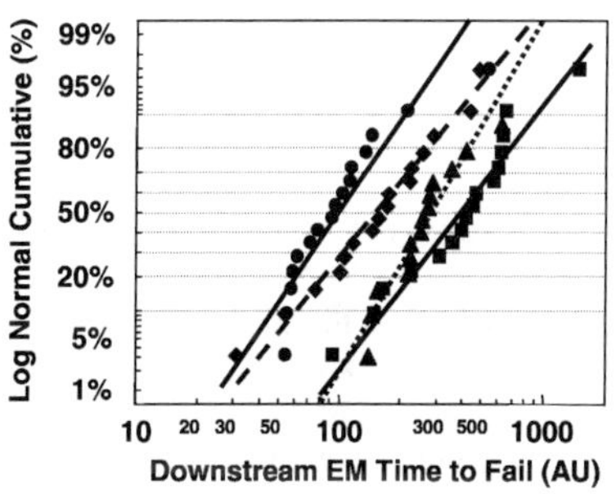

a.

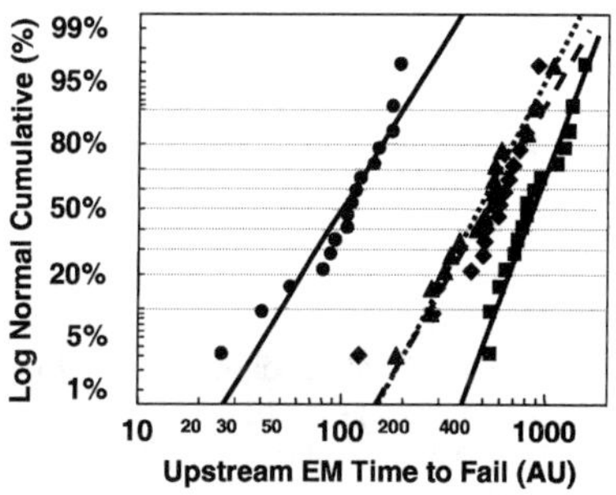

b.

Figure 8. Progressive improvement in, *a.* downstream and *b.* upstream EM, is shown with Ge-PSAB (◆), alloy seed (▲), and Ge-PSAB + Alloy seed (■), relative to control (●) demonstrating process synergy.

REFERENCES

1. ITRS 2007 Edition and ITRS Interconnect 2006 update at ITRS.net. J_{max} is projected to gradually increase to 3.8 MA/cm^2 by year 2019 and the 16 nm technology node.
2. C.-K. Hu, D. Canaperi, S. T. Chen, L.M. Gignac, B. Herbst, S. Kaldor, M. Krishnan, E. Liniger, D. L. Rath, D. Restaino, R. Rosenberg, J. Rubino, S.-C. Seo, A. Simon, S. Smith, W.-T. Tseng, IRPS, p. 222 (2004).

3. P. Moon, V. Dubin, S. Johnston, J. Leu, K. Raol, C. Wu, IEDM 2003, p 841-844.
4. K. Barmak, C. Cabral, Jr. and K. P. Rodbell, J. M. E. Harper, J. Vac. Sci. Technol. B, 24, 6, 2006, p. 2485.
5. S. Yokogawa, Y. Kakuhara, H. Tsuchiya, and K. Kikuta, IRPS 2007, p117 and references therein.
6. W. Wu, H.-J. Wu, G. Dixit, R. Shaviv, M. Gao, T. Mountsier, G. Harm, A. Dulkin, N. Fuchigami, S. K. Kailasam, E. Klawuhn, and R. H. Havemann, IITC 2008.
7. Gosset, L.G., Arnal, V.; Chhun, S.; Casanova, N.; Mellier, M.; Reynard, J.-P.; Federspiel, X.; Guillaumond, J.-F.; Arnaud, L.; Torres, J., AMC 2003, p 321.
8. K. Chattopadhyay, B. van Schravendijk, T. W. Mountsier, G. B. Alers, M. Hornbeck, H. J. Wu, R. Shaviv, G. Harm, D. Vitkavage, E. Apen, Y. Yu, and R. Havemann, IRPS 2006, p 128.
9. H.-J. Wu, J. O'loughlin, R. Shaviv, M. Sriram, K. Chattopadhyay, Y. Yu, T. Mountsier, B. van Schravendijk, S. Varadarajan, G. Dixit and R. Havemann, Semiconductor International, May 2008.
10. Bart van Schravendijk, Thomas W. Mountsier, Mahesh K. Sanganeria, Glenn B. Alers, Roey Shaviv, US patent 7,396,759 (2008).
11. G.B. Alers, X. Lu, J.H. Sukamto, S.K. Kailasam, J. Reid, G. Harm, IITC 2004
12. Luca Benini IEDM 2007.
13. H.-J. Wu, J. O'loughlin, R. Shaviv, M. Sriram, K. Chattopadhyay, Y. Yu, T. Mountsier, B. van Schravendijk, S. Varadarajan, G. Dixit and R. Havemann, Semiconductor International, May 2008.
14. Bart van Schravendijk, Mahesh K. Sanganeria, US patent 6,972,252, 2005.
15. S. B. Law, S. L. Liew, C. S. Seet, E. C. Chua, Y. K. Lim, B. C. Zhang, B. J. Tan, M. K.Ramanan, Roey Shaviv, Hui-Jung Wu, R. L. Graham, Minlin Cheng, and Yee Khiem Teo, AMC 2007.
16. S. Yokogawa, Y. Kakuhara, H. Tsuchiya, and K. Kikuta, IRPS 2007, p117, and references therein.
17. Hui-Jung Wu, Roey Shaviv, Mandy Sriram, Wen Wu, Anshu Pradhan, Jennifer O'Loughlin, Kaushik Chattopadhyay, Tom Mountsier and Girish Dixit, to be published in Semiconductor International October 2008.
18. R. Rozbiki, M. Danek, Method for depositing a diffusion barrier for copper interconnect applications, US patent 6,764,940, 2004.
19. R. Rozbiki, M. Danek, E Klawhun, Method of depositing copper seed on semiconductor substrates, US patent 6,642,146, 2003.
20. Typical activation energy values for electromigration in dual damascene Cu technology range from 0.75 to 0.9 eV. See for example in J. Lloyd , IRPS 2008 tutorial and references therein.
21. Hui-Jung Wu, Roey Shaviv, Mandy Sriram, Wen Wu, Anshu Pradhan, Kie Jin Park, Jennifer O'loughlin, Kaushik Chattopadhyay, Tom Mountsier and Girish Dixit, Semiconductor International, to be published in October 2008.

Conference Proceedings AMC XXIV © 2009 Materials Research Society

Reduction Effect of Line Edge Roughness on TDDB lifetime of Cu/Low-k Interconnects by using CF_3I Etching

E. Soda, T. Suzuki, N. Nakamura, N. Oda, S. Ito, M. Nakamura, S. Kondo, and S. Saito
Semiconductor Leading Edge Technologies, Inc., 16-1 Onogawa, Tsukuba-shi, Ibaraki-ken, 305-8569, Japan

ABSTRACT

Correlation between the line edge roughness (LER) and the time-dependent dielectric breakdown (TDDB) reliability for 100-nm-pitch Cu interconnects was investigated. We controlled LER by using various gases to etch multi-layer photo-resist. CF_3I gas was found to reduce LER better than conventional gases such as CF_4 and CHF_3 because CF_3I had higher etching selectivity of photo-resist against SOG (spin-on glass) film. The LER did not affect measures of electrical performance such as wiring resistance, capacitance, and leakage current, but did affect TDDB lifetime because, according to our simulation, the electric field was strongly enhanced at curvatures in the interconnects. Maximum electric field (E_{max}) was determined to evaluate the effect of LER on TDDB lifetime. Thus, CF_3I etching is promising for creating reliable Cu interconnects with smaller pitches.

INTRODUCTION

Precise control of the critical dimension (CD) and the LER is important in small-pitch Cu interconnects because the effect of LER on TDDB reliability is a critical issue [1]. We previously developed a low-LER and low-damage etching process using CF_3I gas for 200-nm-pitch interconnects on porous SiOC film [2]. The CF_3I etching showed a higher etching selectivity of photo-resist against porous SiOC film compared with conventional perfluorocarbons such as CF_4 gas, because photo-resist damage was suppressed due to the low UV intensity and small amount of F radicals under CF_3I plasma conditions [2].

Although fluorocarbon gases such as C_4F_8, CF_4, C_5F_8 conventionally have been used in porous SiOC etching, their global warming potentials (GWPs) are 8700, 6500 and 90, respectively [3]. Emission of high GWP gases is a serious problem. In contrast to these fluorocarbon gases, the GWP of CF_3I gas is as low as 1 [4]. A simple way of synthesizing CF_3I gas by using a catalyst has been reported [5].

A multi-layer photo-resist has been used for fabricating trench patterns of less than 100-nm pitch to increase the resolution of ArF immersion lithography and the remaining resist thickness during etching. Because etching using fluorocarbon gasses affects the LER, we used CF_3I gas to etch multi-layer photo-resist. After fabricating 100-nm pitch Cu interconnects, we evaluated the effects of their LER on electrical performance, including the TDDB reliability, which had not yet been clarified.

Moreover, two-dimensional electric field distribution in porous SiOC was simulated to examine the correlation between LER and TDDB reliability. We estimated the maximum electric field (E_{max}) distribution from the radius of curvature along the line edge obtained from the experimental results and found the effect of E_{max} on TDDB reliability.

EXPERIMENTAL

One hundred-nm-pitch Cu/low-k (porous SiOC: k=2.6) interconnects were fabricated by 193-nm ArF immersion lithography. Multi-layer photo-resist consisted of ArF photo-resist, anti-reflective coating (ARC), spin-on glass (SOG), and spin-on carbon (SOC). Capacitive coupled plasma (CCP) was used to etch patterned wafers. The amount of LER was varied by using the gases CF_3I, CF_4/CHF_3, and CF_4 when we etched the SOG film. Gases to etch the other film layers were constant, as shown in Fig. 1. The amount of LER (3σ) after SiOC etching was measured by CD-SEM based on 32 measurement points in a trench length of 1 μm.

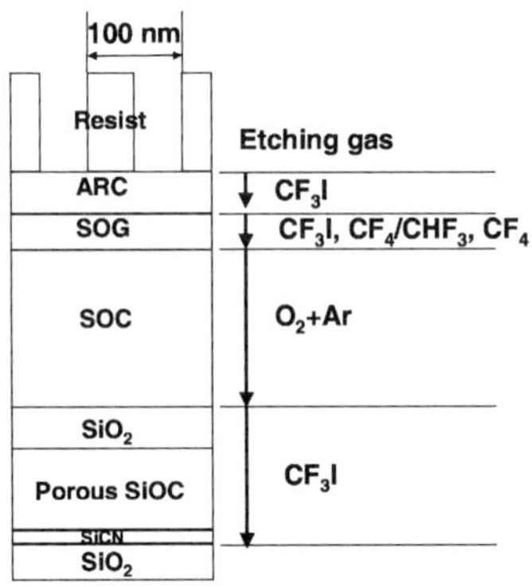

Fig. 1. Film structure and etching gases.

Cu interconnects were fabricated with PVD-TaN/Ta barrier metal film, and then wiring resistance, capacitance, leakage current, and TDDB reliability were evaluated. The maximum electric field in low-k space was estimated by simulation with ENEXSS [6]. Based on E_{max} dependence on the radius of curvature, we estimated the E_{max} distribution and the shortest distance between neighboring E_{max}s obtained from the experimental results. We evaluated the effect of E_{max} on TDDB reliability, which was due to LER.

DISCUSSION

Figure 2 shows the top surface SEM images of 100-nm-pitch trenches after porous SiOC etching and ashing. The LER increased in ascending order with CF_3I, CF_4/CHF_3, and CF_4 used for SOG etching. The mean trench CD was controlled to the same width for all three cases by re-depositing SOG sidewall film during the next SOC etching step.

The amounts of LER are shown in Fig. 3. We attributed the difference in the LER to wiggling of the ArF resist caused by reaction with perfluorocarbon plasma during SOG etching being transferred to roughness in the sidewall. In the case of CF_3I plasma, however, ArF resist was not degraded, because the resist surface has a lower reactivity with CF_3I plasma due to its lower UV intensity and smaller amount of F radicals compared with CF_4 plasma. In the case of CF_4/CHF_3 plasma, a protective layer was deposited on the ArF resist surface by CHF_3 gas. Therefore, these LERs were thought to depend on degradation of the ArF resist surface.

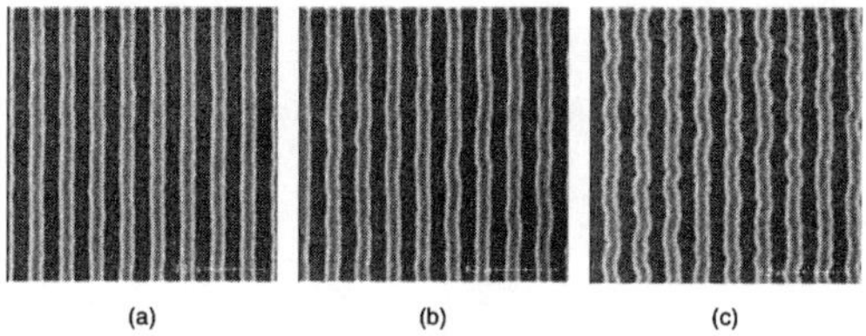

Fig. 2. LER in 100-nm-pitch trenches etched on SOG by (a) CF_3I, (b) CF_4/CHF_3, (c) CF_4.

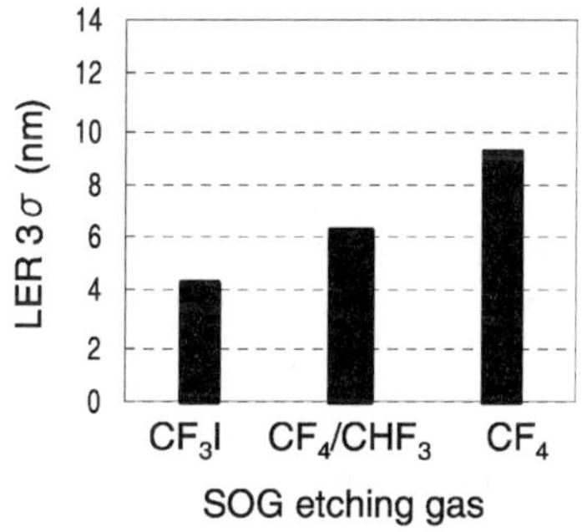

Fig. 3 LER dependence on SOG etching gas at 100-nm pitch.

Next, the electrical performance of 100-nm-pitch interconnects was measured (Fig. 4). Since the line-to-line capacitances were similar among the three cases and did not depend on the LER, the CD values of low-k space were thought to be the same. Although the wiring resistance of the larger LER sample (CF_4) was slightly higher than those of the other samples, the electron scattering effect on the rough surface of the trench sidewall seemed to be observed at 50-nm copper wiring. Since the leakage currents did not depend on the LER either, we concluded that these measures of electrical performance were not affected by LER.

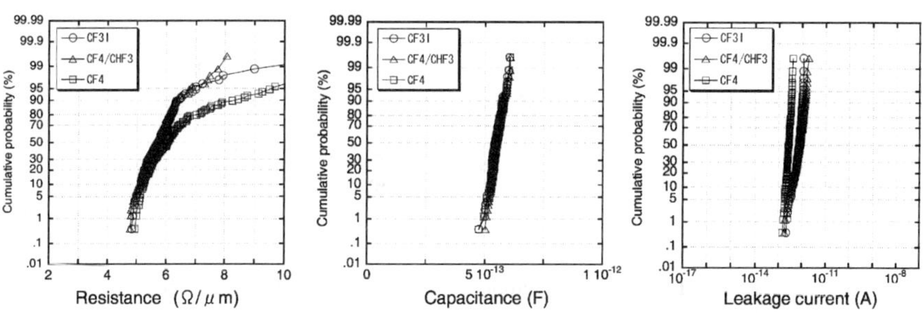

(a) (b) (c)

Fig. 4 Effect of LER on electrical performance in 100-nm-pitch interconnects; (a) resistance, (b) capacitance, and (c) leakage current.

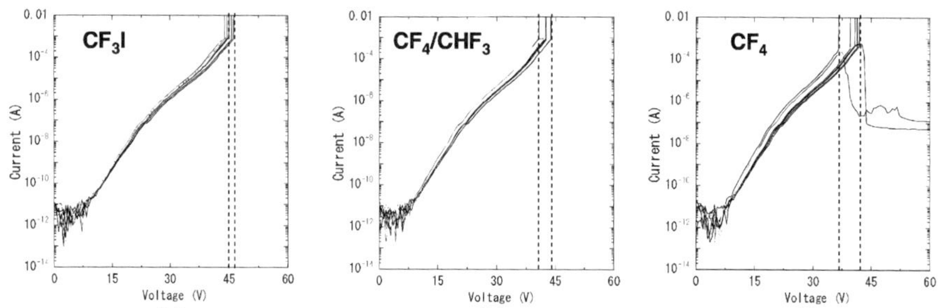

(a) (b) (c)

Fig. 5 Effect of LER on IV characteristics in 100-nm-pitch interconnects; (a) CF_3I, (b) CF_4/CHF_3, and (c) CF_4 as etching gas.

As shown in Fig. 5, however, the breakdown voltage in IV characteristics decreased as LER increased. The breakdown voltage of CF_3I plasma etching was highest because its LER was the smallest. The TDDB lifetimes are shown in Fig. 6. Tested bias voltage was 15 V and the temperature was 150°C. Sixteen chips were measured. The TDDB lifetime showed clear dependence on the LER. The lifetime of 63.2% breakdown was 0.37 hours for CF_4 plasma, 40 hours for CF_4/CHF_3 and 170 hours for CF_3I. Since the degree of leaning of the low-LER sample (CF_3I) was larger than those of the large LER samples, it was thought that the breakdown mode changed from initial defective to time-dependent deterioration. These results show that the TDDB was dependent on etching gas and that CF_3I produced lower LER interconnects that therefore had higher reliability.

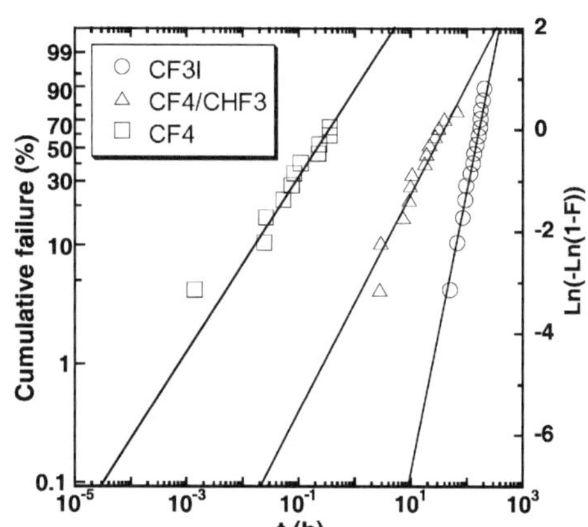

Fig. 6 Effect of LER on TDDB lifetime in 100-nm-pitch interconnects.

An electrical field was simulated to investigate correlation between LER and TDDB reliability. Although a three-dimensional interconnect structure with a dielectric constant of 4.0 was used for the simulation, the electrical field was calculated by two-dimensional structure of top low-k material. The simulated low-k length was 400 nm. Left side Cu voltage of low-k material was 1 V and right side metal voltage was 0 V. Figure 7 shows the electric field simulation based on the actual 45-nm space of 100-nm pitch interconnects. The field intensity increased strongly at bumps on the sidewall. Since E_{max} depended on radius of curvature, we estimated average E_{max} distribution from the experimental results (Fig. 8). As LER increased, the amounts of E_{max} increased. We concluded that these high E_{max} caused TDDB break down.

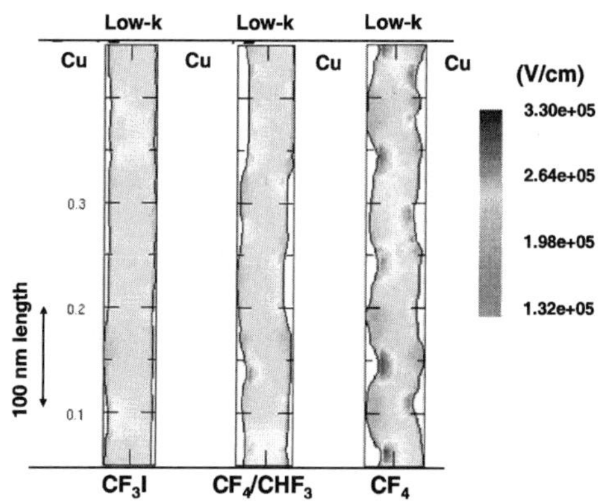

Fig. 7 Electric field simulation in 45-nm low-k space

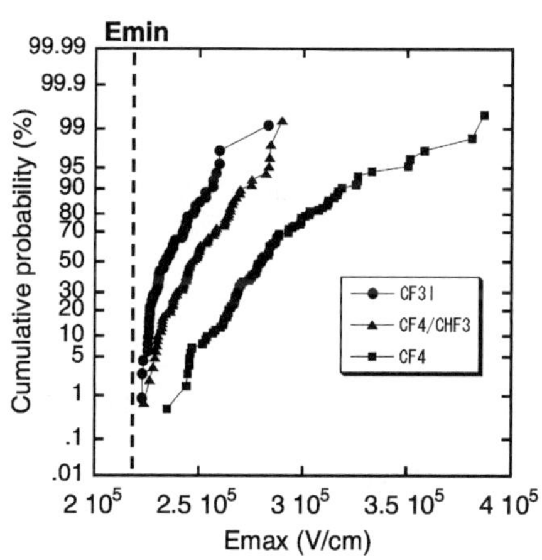

Fig. 8 E_{max} distribution

We simulated the effect of LER on E_{max} depending on low-k space when the LER was assumed to be constant for each etching gas (Fig. 9). The effect of LER on E_{max} increased as low-k space became narrower. Therefore, LER should be suppressed to obtain highly reliable Cu interconnects for 32-nm node technology.

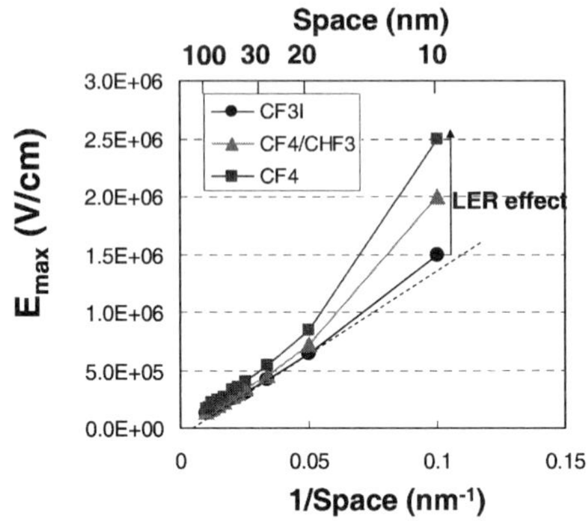

Fig. 9 Effect of LER on maximum electric field depending on low-k space estimated by simulation

CONCLUSION

LER affected TDDB reliability in 100-nm-pitch Cu interconnects. With greater LER, the amounts of E_{max} were greater. Etching with CF_3I gas is promising for fabricating reliable Cu interconnects because it produces low LER with high etching selectivity of photo-resist against SOG film.

ACKNOWLEDGMENTS

The New Energy and Industrial Technology Development Organization, NEDO, supported this work.

REFERENCES

1) F. Chen et al., 46th International Reliability Physics Symposium, pp. 132-137 (2008).
2) E. Soda et al., J. Vac. Sci. Technol. A 26 pp. 875-880 (2008).
3) IPCC Fourth Assessment Report, pp. 33-34 (2007).
4) S. Solomon et al., J. Geophys. Res., 99 pp. 20929-20936 (1994).
5) N. Nagasaki et al., Catal. Today, 88 pp. 121-126 (2004).
6) T. Wada et al., IEICE Trans. Electron., E82-C, p. 839 (1999).

Conference Proceedings AMC XXIV © 2009 Materials Research Society

Scaling Impacts and Challenges on Reliability in Cu/low-κ Interconnects

Shinji Yokogawa
Advanced Device Development Division, NEC Electronics
1120 Shimokuzawa, Sagamihara, Kanagawa 229-1198, Japan
shinji.yokogawa@necel.com

ABSTRACT

In this paper, the scaling impacts on the Cu/low-κ reliability are discussed from the viewpoints of electromigration (EM) and Joule heating. EM lifetime decreases in accordance with decreasing line width at a fixed current density. EM improvement effect of impurity doping has line width dependence and it causes the apparent line width dependence of lifetime. The control of impurity incorporation at the Cu surface is vital for maintaining an efficient tradeoff between resistivity and reliability. The temperature increase due to Joule heating mainly depends on the thermal conductivity of the inter-level dielectric and consumed power on the lines. The multi-level dielectric structures and the layout of lines strongly contribute to the temperature rise due to Joule heating.

INTRODUCTION

Electromigration was considered to be a significant issue with regard to the reliability of interconnects from that sudden rise period. Because Si substrates with large heat capacities are used, the nature of radiation of the interconnects that are covered by SiO2 with high thermal conductivity is outstanding as compared to that of the electric wire usually used. Therefore, this phenomenon allows large current densities and can contribute toward improvement in the speed of integrated circuit functions. However, this significantly high current density induces electromigration, which is a factor that affects the reliability of LSI interconnects.

A decrease in the dimensions of Cu interconnects for a constantly applied current causes an increase in the current density and a decrease in the electromigration (EM) lifetime. Furthermore, the decrease in the cross-sectional area results in an increase in the line resistivity. Because the low-κ dielectrics, generally with a higher degree of porosity, have poorer thermal conductivity characteristics, Joule heating will be enhanced by the resistivity increase. These trends induce lifetime degradation with scaling. Hence, a way to improve EM lifetime and to manage the Joule heating will be needed to satisfy the required current density specification for technology nodes in the future.

In this paper, the scaling impacts on the Cu/low-κ reliability are discussed from the viewpoints of EM and Joule heating. The resistivity and reliability tradeoffs of advanced interconnect technologies were investigated. The electron microprobe techniquc has been used to confirm the impurity distribution. A correlation between the impurity distribution and electromigration phenomena is discussed from the viewpoint of the impurity dependence of the electromigration drift velocity. The scaling impacts on Joule heating was investigated experimentally. The temperature rise due to Joule heating on consumed power on line was investigated about the multi-level structure of dielectrics, the layout of lines, and the resistivity of lines.

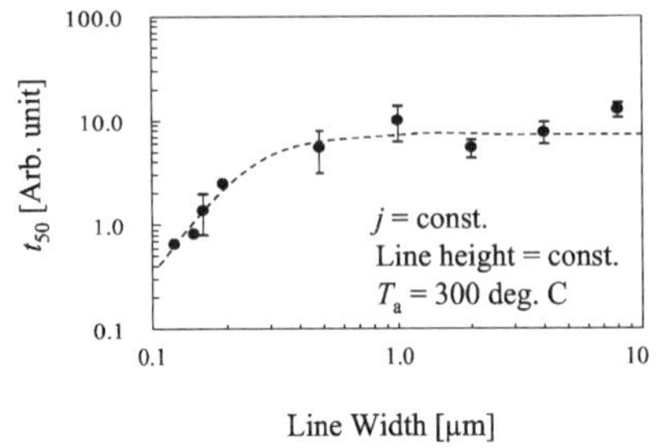

Fig. 1 Lifetime of damascene Cu lines as a function of linewidth at 300 deg C and a current density of 2MA/cm^2. Error bars correspond to 95% confidential interval.

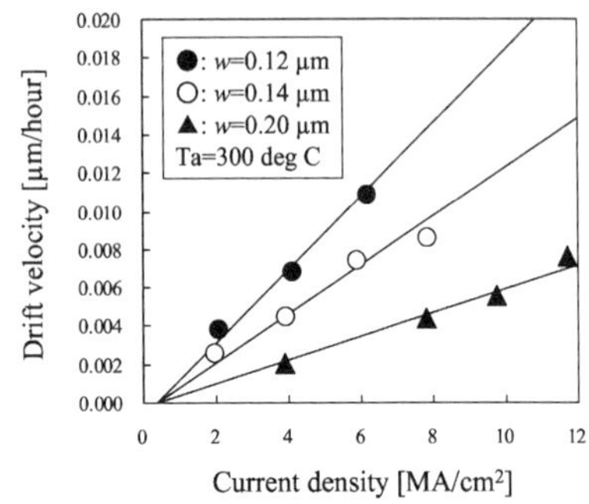

Fig. 2 Drift velocity as function of current density in 90 nm-nodeCu lines.

ELECTROMIGRATION

We investigated the scaling impacts on electromigration-induced mass transport kinetics in narrow damascene Cu interconnects. Lifetime decreases in accordance with decreasing linewidth at a fixed current density (Fig. 1). This is caused by a decrease in the effective incubation time and an increase in the Cu ionic drift velocity [1] (Fig. 2).

A couple of technologies have been proposed to improve these characteristics. We have reported the trade-off characteristics between resistivity and reliability for cobalt–tungsten–phosphorous metal-cap interconnects (CoWP), plasma-enhanced chemical-vapor-deposition self-aligned barrier interconnects (PSAB), and CuAl alloy interconnects (CuAl)[2, 3, 4]. Fig. 3 shows a linewidth dependence of normalized residual resistivity (NRR [4]) for them. For CoWP, NH3 plasma was skipped to investigate an influence of the treatment (CoWP 1) in contrast with the normal condition (CoWP 2). The impurity (Si) concentration of PSAB 1 is adjusted to be smaller than that of PSAB 2. PSAB 3 is not treated with a nitrogen plasma treatment to remove unreacted Si. Hence, the surplus Si atoms will be incorporated in Cu compared to PSAB 1 and 2. The Al concentration in CuAl is adjusted by varying the thickness of the seed CuAl (40, 60, and 90 nm). Observed NRR means increasing ratios of residual resistance due to impurity doping in each linewidth. For PSAB and CuAl, the NRR increases with the linewidth. This result suggests that Cu is doped with impurities and that the impurity concentration is smaller in the narrower line. On the other hand, CoWP indicates a very small NRR, which implies a small impurity concentration in the Cu lines. However, the NRR of the narrow CoWP line is larger than that of the wider line.

Fig. 4 shows the correlation of the NRR and EM improvement ratios for them. The improvement is defined by the ratio of the lifetime of the experimental line to that of the control line (pure Cu). In this study, a cumulative probability lifetime ratio of 0.1% is used. The slopes of the fitted lines represent the EM improvement efficiency. The PSAB and CuAl lines indicate that their EM lifetimes improve with high impurity doping. In other words, the EM improvement in PSAB and CuAl depends on the amount of dopant. In contrast, the CoWP line reveals an exceptional EM improvement efficiency with a small increase in the resistivity.

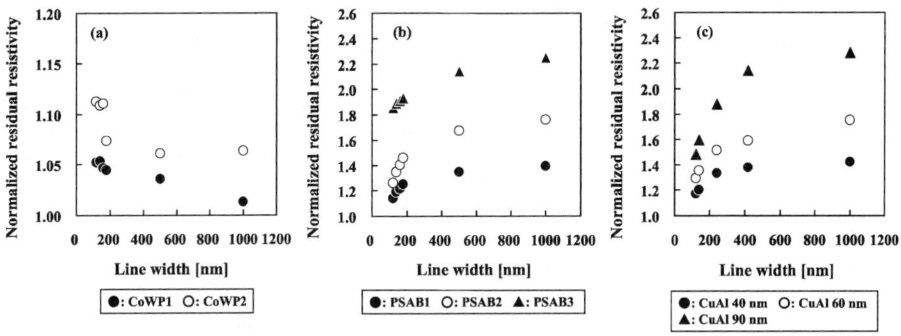

Fig. 3 Normalized residual resistivity (NRR) as a function of line width for CoWP, PSAB, and CuAl. (a) CoWP, (b) PSAB, and (C) CuAl.

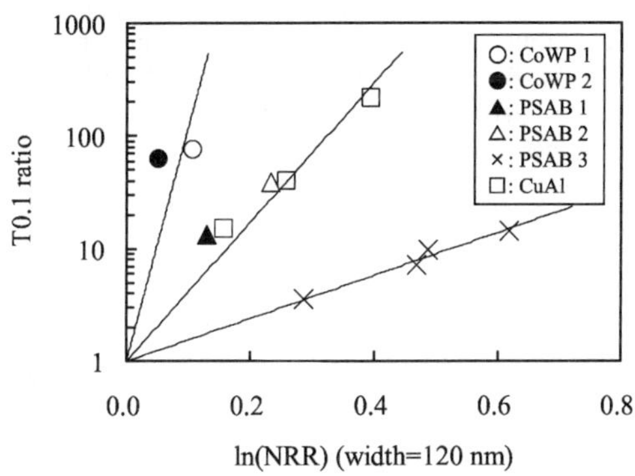

Fig. 4 Comparisons of NRR versus EM improvement ratio for CoWP, PSAB, and CuAl.

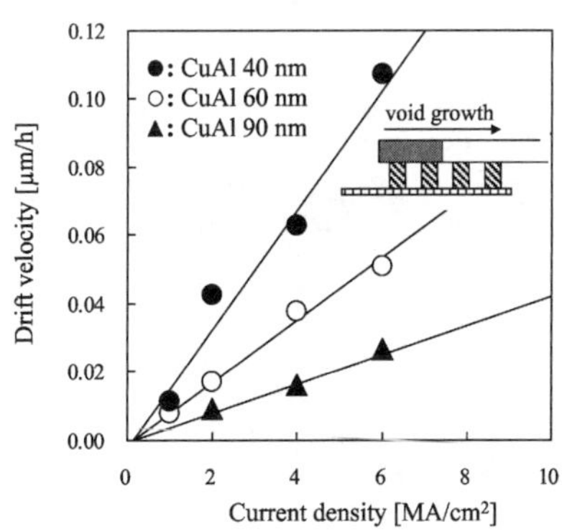

Fig. 5 Current density dependences of drift velocity in 90 nm-node CuAl lines.

The impurity doping of Cu interconnects exhibited a decrease in the drift velocity of the Cu mass transport and an increase in the incubation time of the CuAl alloy interconnects, whereas the critical product of electromigration was found to be independent of the impurity concentration [5]. Fig. 5 shows the current density dependences of the drift velocity for different thicknesses of seed CuAl (different Al concentration). The drift velocities exhibit a linear dependence on the current density for each seed CuAl thickness. Moreover, the threshold current density, j_c, is independent of the thickness, and was found to be 0.15 MA/cm^2. These results suggest that the Al doping causes a change in mobility; however, it does not affect the change in the driving force. Fig. 6 shows that the drift velocity decreases as the NRR (impurity concentration) increases. As shown by Fig. 6(a), the observed drift velocities are independent of the linewidths. And, as shown by Fig. 6(b), the linewidth independence is observed for both the Al doped lines and the Si doped lines. These results suggest that the dominant diffusion is the Cu/SiCN interface diffusion [6], which is independent of the linewidth, and the observed linewidth dependence of lifetime is apparent characteristic that is due to the linewidth dependence of impurity concentration.

Fig. 7 shows the cross-sectional Al-concentration mapping and Al intensity spectrum. Al concentration is observed just at the interface of Cu and the other materials. Particularly, the Al concentration at the top surface of the Cu interconnects (Cu/SiCN interface) seems to be greater than that at the grain boundaries or that in the bulk. A part of the segregated Al contributes to reduce EM-induced Cu diffusion. The control of impurity incorporation at the Cu surface is vital for maintaining an efficient tradeoff between resistivity and reliability.

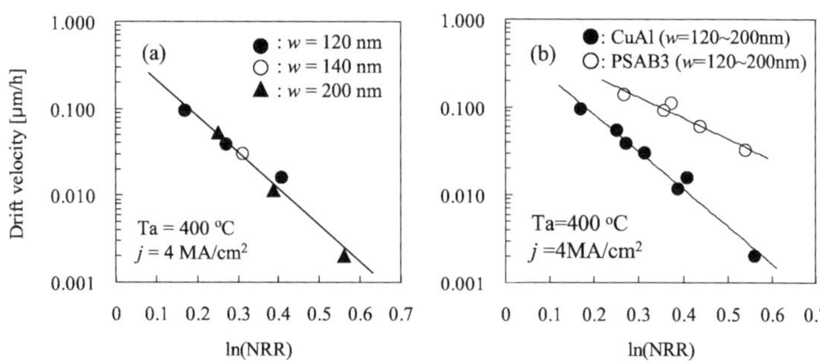

Fig. 6 Cu drift velocity versus NRR. (a)Three types of linewidths in 90 nm-node CuAl interconnects. (b) The comparison of impurity effect between Al and Si.

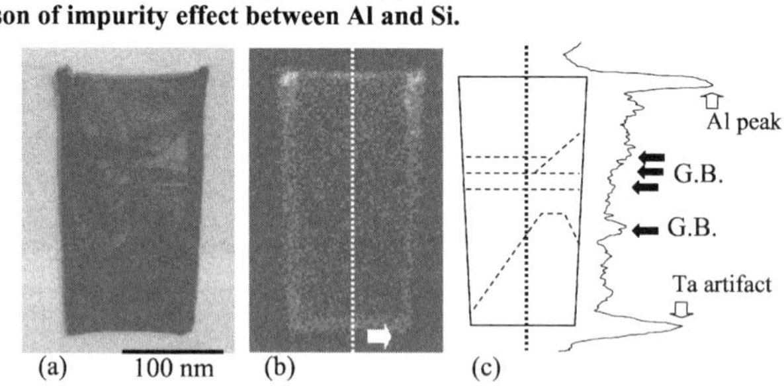

Fig. 7 Cross-sectional mapping and spectrum of Al in CuAl line. (a) Cross-sectional S-TEM observation. (b) EDX mapping. (c) EDX spectrum .

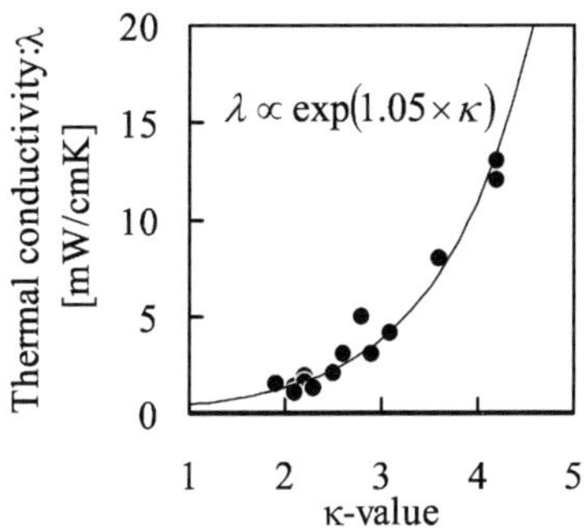

Fig. 8 κ-value vs. thermal conductivity for various dielectrics.

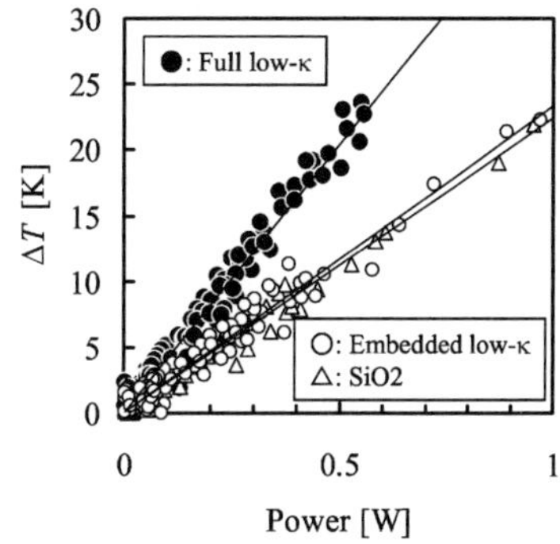

Fig. 9 Low-κ impact on Joule heating in 130 nm-node pure Cu interconnects.

JOULE HEATING

With scaling, the impact of Joule-heating in metal leads will become even more of an issue because the low-κ has poorer thermal conductivity. Fig. 8 indicates κ-value dependence of the thermal conductivity for various dielectrics. It indicates that the thermal conductivity decreases as the κ-value of the dielectrics decreases. It means that a low-κ trend may enhance the Joule heating issues.

As Oda, *et al.* proposed by thermal simulation, the temperature increase of the embedded low-κ structure is much lower than that of the full low-κ structure and is about the same as that of the conventional SiO2 structure [7]. The temperature increase due to Joule heating mainly depends on the thermal conductivity of the interlevel dielectric. Fig. 9 indicates experimental observations. Temperature rise was estimated using the temperature coefficient of resistance of serpentine lines that is electrically floating on Si substrate. It agrees with the previous study by using thermal simulation [7]. These results suggest that the multi-level structure of dielectrics acts key role for the vertical radiation of heated Cu lines. Therefore, as Fig. 10 indicates, the temperature rise is most severe for higher levels of metal.

Advanced technologies accelerate the temperature rise due to resistivity increase. Fig. 11 shows the effects of Al doping on the resistivity, which is categorized into surface, grain boundary, and impurity scatterings as a function of the Al concentrations. This resistivity increase will contribute to a Joule heating enhancement.

Temperature rise due to Joule heating strongly depends on layout of lines. Fig. 12 shows linewidth dependence of the temperature rise. Imaginably, the Joule heating can be normalized by the applied power on the lines. Similarly, the Joule heating can be normalized by the applied power for various impurity concentrations (Fig. 13). Fig. 14 shows line-to-line space dependence of the temperature rise measured by using serpentine lines. The temperature rise increases as the space decreases. It is caused by a mutual heating and a small heat capacity in the area. These results suggest that the scaling accelerates Joule heating, remarkably.

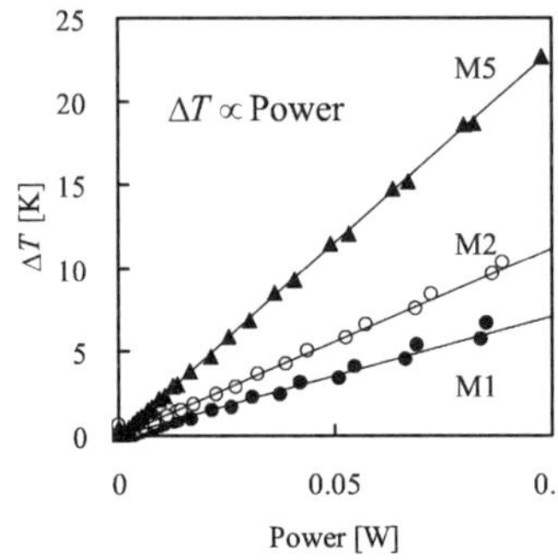

Fig. 10 Power dependence of temperature rise in 55 nm-node multilevel lines.

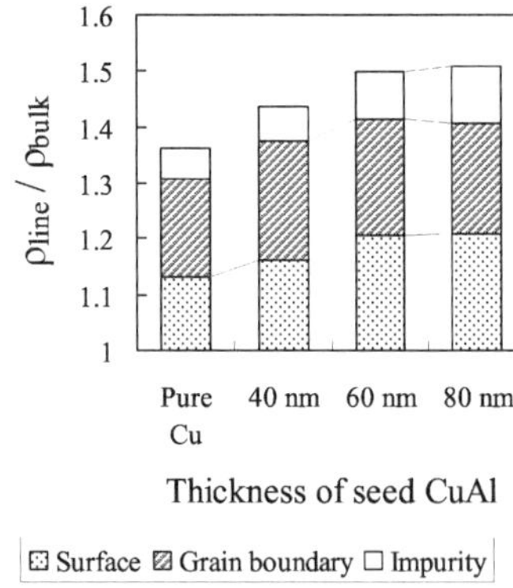

Fig. 11 Al doping effects on resistivity as a function of Al concentration.

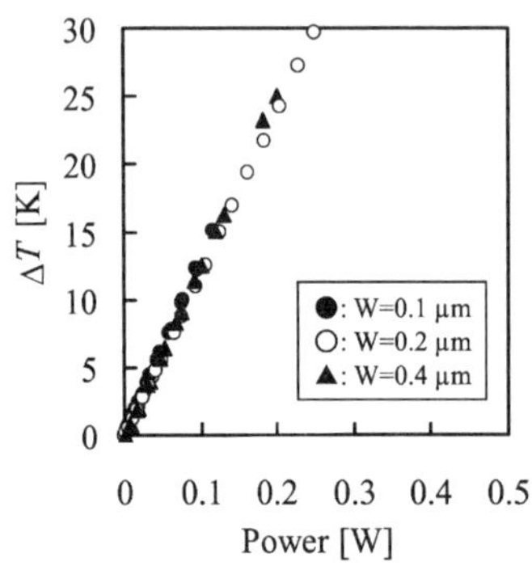

Fig. 12 Line width dependence of temperature rise in 65 nm-node CuAl lines.

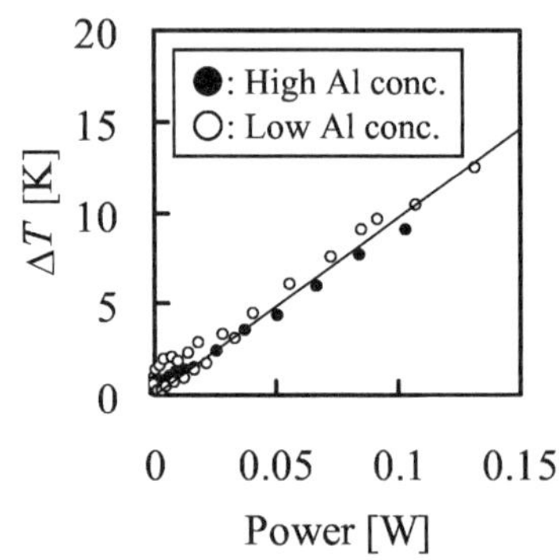

Fig. 13 Al concentration dependence of temperature rise in 90 nm-node CuAl lines.

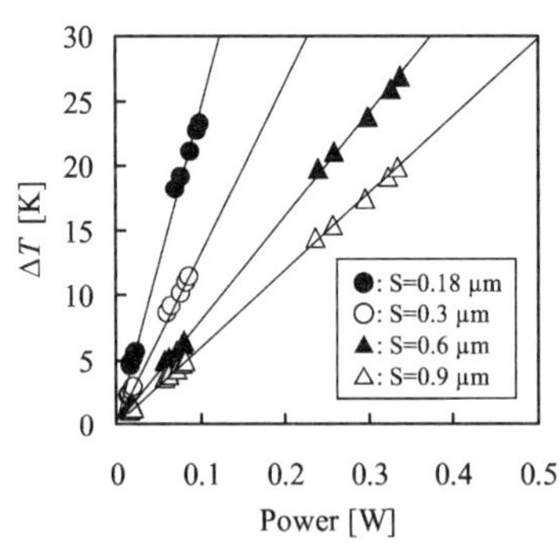

Fig. 14 Space dependence of temperature rise in 130 nm-node Cu interconnects.

It is reported that vias and contact contribute remarkably to a heat radiation to the Si substrate [8]. Therefore, a practicable layout rule for a Joule heating management will be necessary.

CONCLUSIONS

The scaling impacts on EM and Joule heating were discussed. The trade-off characteristics between resistivity and reliability have importance in development of advanced technologies. The control of impurity incorporation at the Cu surface is vital for maintaining an efficient tradeoff between resistivity and reliability. The observed linewidth dependence of lifetime is apparent characteristic that is due to the linewidth dependence of impurity concentration. The interconnect structure and layout strongly contributes to the temperature rise due to Joule heating. The scaling accelerates the Joule heating, remarkably.

ACKNOWLEDGMENTS

The author wishes to thank H. Takizawa and Y. Ajima for their experimental help. The author is grateful to H. Tsuchiya, Y. Kakuhara, T. Takewaki, M. Suzuki, H. Toyoshima, T. Usami, K. Ohto, K. Fujii, Y. Tsuchiya, K. Arita, K. Motoyama, M. Tohara, T. Taiji, T. Kurokawa, M. Sekine, Prof. K. Ueno, for comments and discussions. The authors would like to grateful to T. Kuwata, N. Nakamura, Y. Yamagata, and I. Akiba for their encouragements.

References

1. S. Yokogawa and H. Tsuchiya, *JJAP*, Vol.44, pp.1717-1721 (2005).
2. S. Yokogawa, Ph.D. Thesis, The University of Electro-Communications, 2008 (in Japanese).
3. S. Yokogawa, *et al.*, *Proc. of 2006 IEDM*, pp.85-88 (2006).
4. S. Yokogawa, *et al.*, *IEEE T-ED*, Vol.55, pp.350-357 (2008).
5. S. Yokogawa and H. Tsuchiya, *JAP*, Vol.101, pp.013513-1-6 (2007).
6. S. Yokogawa, *et al.*, *IEEE-TDMR*, Vol.8, pp.216-221 (2008).
7. N. Oda, *et al.*, *JJAP*, Vol.46, pp.954-961 (2007).
8. Ting-Yen Chiang, *et al.*, 2002 Symp. on VLSI Tech. Digest of Tech. Papers, p.38-39.

Conference Proceedings AMC XXIV © 2009 Materials Research Society

Characteristics of Downstream Electromigration of Cu /porous low-*k* Interconnects

Yumi Kakuhara, Shinji Yokogawa, Masayuki Hiroi, and Toshiyuki Takewaki
Advanced Device Development Division, NEC Electronics Corporation
1120, Shimokuzawa, Sagamihara, 229-1198, Japan
phone: +81-42-771-4274, fax: +81-42-771-0957; e-mail: yumi.kakuhara@necel.com

ABSTRACT

Electromigration (EM) of Cu/porous low-*k* interconnects was evaluated. The EM lifetime of the downstream mode was reduced using a porous low-*k* film as a dielectric interlayer in comparison with a SiO_2 dielectric interlayer. The reduction of the EM lifetime was observed only at the low cumulative failure probability, considered as "early failure". It was found that EM lifetime degradation was suppressed by the arrangement of a dummy metal near the metal/via contact. EM lifetime degradation of Cu/porous low-*k* interconnects is likely caused by the mechanical weakness of porous low-*k* film; however, a dummy metal mechanically strengthens porous low-*k* film near the metal/via contact, then EM lifetime degradation is suppressed.

INTRODUCTION

Production of high performance large integrated circuits (LSIs) requires a reduction in resistance-capacitance (RC) delay, so Cu interconnects with a low-*k* dielectric interlayer have been introduced. As the *k* value decreases, the mechanical strength of low-*k* film weakens due to its low modulus. Therefore, as the stiffness of Cu/low-*k* interconnects decreases, certain problems, such as cracking during chemical mechanical polishing (CMP) [1], Cu extrusion [2,3], and stress-induced-voiding (SIV) failure [4,5] have been reported with respect to low-k film application. Moreover, the *k* value has been decreased by introducing porosity into the film, which is called a porous low-*k* film. Then the low-*k* interlayer significantly softens, resulting in more difficulties.

In this study, the effect of porous low-*k* film application for the metal dielectric interlayer on EM lifetime was investigated. It was found that the EM lifetime of the downstream mode was degraded using a porous low-*k* dielectric interlayer in comparison with using a SiO_2 dielectric interlayer, and the arrangement of a dummy metal suppressed this EM lifetime degradation. The cause of EM degradation is likely to be the porous low-*k* mechanical film properties, and the dummy metal fixes porous low-*k* film near the metal/via connection.

EXPERIMENTS

Figure 1 shows a schematic cross-section of the 65-nm-node, three-level Cu dual damascene interconnects used in this study. For the dielectric interlayer, SiO_2 was deposited for the Metal 1 (M1) level and SiOC (k~2.5) of porous low-*k* film was deposited for the Metal 2 (M2) and Metal 3 (M3) levels using chemical vapor deposition (CVD). SiO_2 was deposited between the via and the metal dielectric interlayer as the etching stopper. Ta/TaN and CuAl were successively deposited using physical vapor deposition (PVD) as the barrier metal and the seed layer for electro chemical deposition (ECD) of Cu, respectively. The excess Cu and barrier metal on the dielectric interlayer were removed using CMP to fabricate Cu interconnects. SiCN was deposited using CVD on the Cu interconnects as the cap dielectric layer. We performed package-level electromigration experiments using test structures consisting of M1/via 1/M2 and of M2/via 2/M3, then the EM downstream mode was evaluated. To measure a wide range of failure probabilities of the EM lifetime distribution, two types of test structures were used, as shown in Fig. 2. "TEG 1" tests 20 vias and TEG 2 tests 2 vias [6]. Both the evaluated metal line width and via diameter were 100 nm. The EM experiments were performed at a current density of 2 MA/cm^2 and at a temperature of 350°C. The failure time was defined as the time when the first via failed. The failure criteria were the abrupt resistance change of 0.5% for TEG 1 and 1% for TEG 2. From these criteria, the first via failure was detected. EM lifetimes were plotted with the failure probability of a via [6,7] using the cumulative hazard function. The lower failure probability was measured using TEG 1 and the higher failure probability was measured using TEG 2.

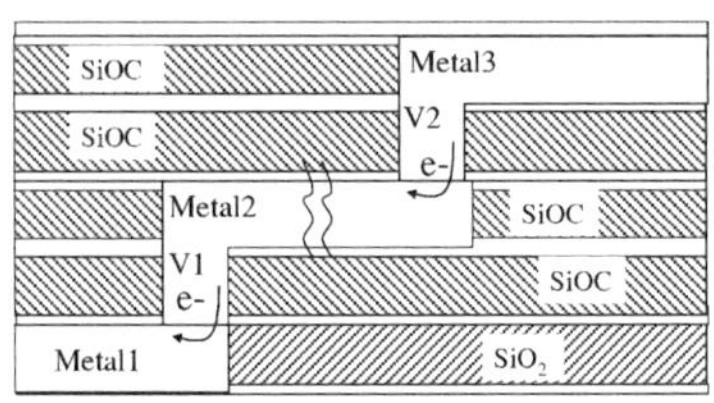

Figure 1. Schematic cross section of three-level Cu interconnects of EM test samples. Downstream from via to metal was evaluated

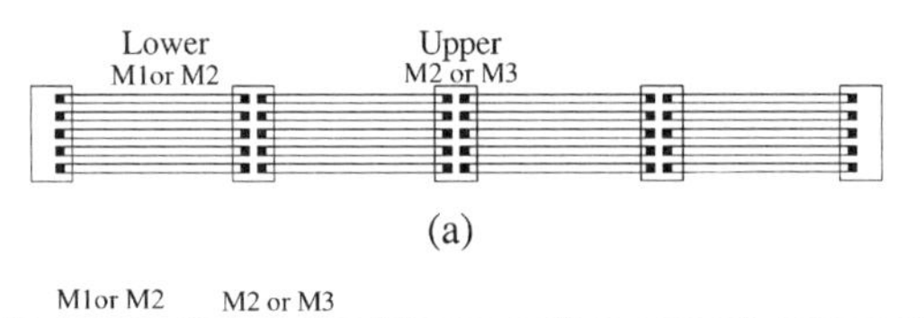

Figure 2. Planar view of EM test structures. (a) TEG 1 has 20 vias tested in downstream mode. (b) TEG 2 has 2 vias tested in downstream mode

RESULTS AND DISCUSSION

Effect of porous low-*k* dielectric interlayer on EM lifetime

Figure 3 shows the EM lifetime of downstream M1/via1 and M2/via2 using TEG 1. The EM lifetime of M2/via2 is shorter than that of M1/via1. In addition, the standard deviation (σ) of the EM lifetime distribution is larger in the case of M2/via2. The notable difference in the EM lifetime is likely due to the difference in the dielectric interlayer, that is, SiO_2 of M1 and SiOC of M2. The EM lifetime of M2/via2 using TEG 2 is also plotted in Fig. 3, where a few samples show extremely short EM lifetimes, considered to be "early failures". The plots of the early failure overlap with the plots of the EM lifetime from TEG 1. These plots from TEG 1 and TEG 2, which cover a wide range of the cumulative failure probabilities, indicate that the EM lifetime of M2/via2 deteriorates only at the lower cumulative failure probability. In addition, bimodal failure is suggested because the slopes of the fit lines (σ) are different between the lower and higher cumulative failure probabilities (1.20 and 0.32), respectively. It is believed that there is a different EM failure mechanism between the lower and higher failure probabilities.

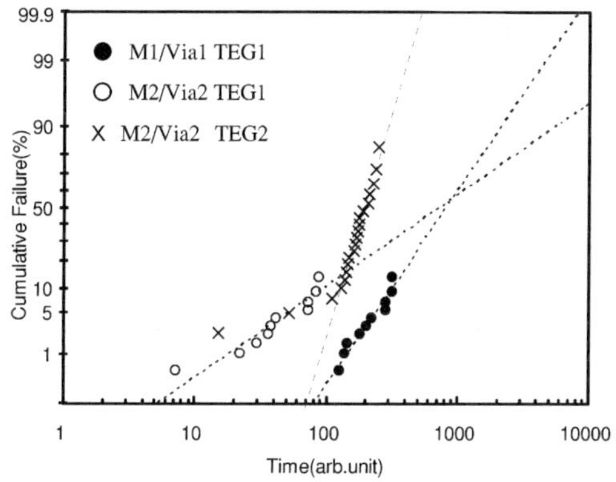

Figure 3. EM lifetime lognormal plots from TEG 1 for M1/via1 and M2/via2, and from TEG 2 for M2/via2. Times to failure of a via are plotted using cumulative hazard function

Effect of dummy metal and via on EM lifetime

To investigate the cause of EM degradation of M2/via2, EM lifetime tests were performed using TEG 1 in which a dummy metal and via were arranged. Figure 4 shows a planar view of the three types of test structures, (a) no dummy metal and via [same as Fig. 2 (a)], (b) with dummy metal, and (c) with dummy metal and via. The EM lifetime plots for each type of test structure are shown in Fig. 5. The EM lifetime is improved by the arrangement of a dummy metal or a dummy metal and via. As the extent of improvement is the same in both cases, improvement resulted from only the dummy metal. Therefore, we confirmed that the cause of EM degradation of M2/via2 is related to the M2 dielectric interlayer.

The EM lifetime and σ of M1/via1 and M2/via2 obtained from these experiments are plotted in Fig. 6. The EM lifetime was defined as the time to cumulative failure probability of 3.3% ($t_{3.3}$). The EM lifetime and σ of M2/via2 using SiOC with a dummy metal and via or with a dummy metal are close to those of M1/via1 using SiO_2. The effect of a dummy metal on EM lifetime improvement indicates that the EM degradation of M2/via2 is likely due to using porous low-*k* film as the M2 dielectric interlayer.

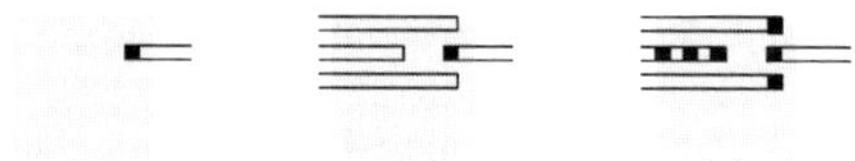

Figure 4. Planar view of TEG 1-type test structure, (a) no dummy, (b) with dummy metal, and (c) with dummy metal and via

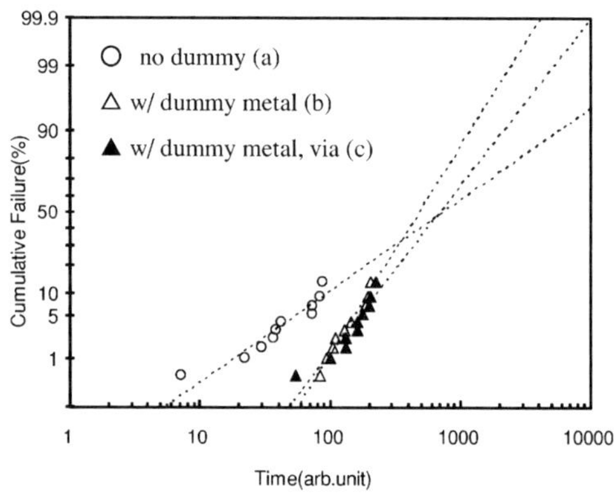

Figure 5. Effect of arranging dummy metal and via on EM lifetime of M2/via2 downstream

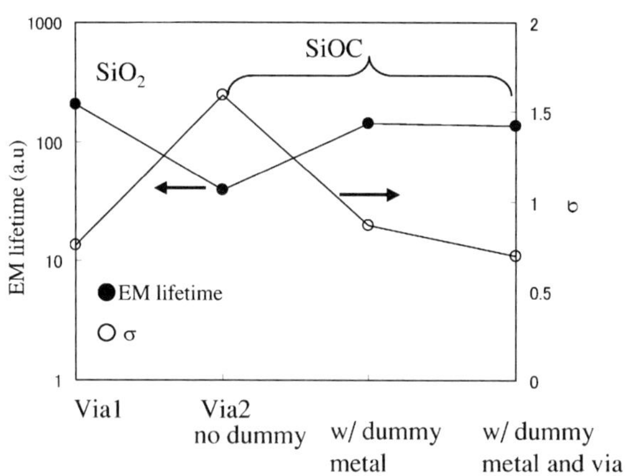

Figure 6. EM lifetime ($t_{3.3}$) and σ of M1/via1 using SiO_2, M2/via2 using SiOC and M2/via2 with dummy metal and via.

We performed failure analyses to investigate the effect of dummy metal and via on EM void formation. The optical beam induced resistance change (OBIRCH) method [8] was carried out to detect failure sites. Figure 7 shows OBIRCH images of the samples after EM testing. EM failure sites were detected in the samples with dummy metal and via and with only a dummy metal with an invisible line image caused by the shutting off of the applied current for OBIRCH measurement at the failure site [Fig. 7(c), (d)]. Meanwhile, in the samples with no dummy metal and via and a short EM lifetime, the failure site could not be detected because all the line images were visible. It is likely that the failure void formed during the EM test was reconnected later. Figure 8 shows SEM images of the EM failure voids with dummy metal and dummy metal and via. A large void was observed under the via in both cases, which suggests that the mechanism of void formation is the same. Figure 9(a) shows an expanded OBIRCH image of the no-dummy sample for detecting a failure site. The via, whose contrast in the OBIRCH image was different from that of the line, was found (indicated by an arrow). The SEM image of the via shows that there is a small slit-like void under the via and it partially connects to M2, as shown in Fig. 9 (b). From this SEM image, the adhesion of M2/via2 was weak is indicated. The formation of the small slit-like void should shorten the EM lifetime due to its small volume.

The effect of the arrangement of a dummy metal on EM lifetime improvement leads to a speculation that a dummy metal fixes SiOC around M2. Consequently, the edge of M2, from which slit-like void generation occurred, is strengthened mechanically, and the adhesion of M2/via2 is retained. Therefore, the mechanical stability restricts the generation of slit-like voids from the M2 edge to under the via (Fig. 10).

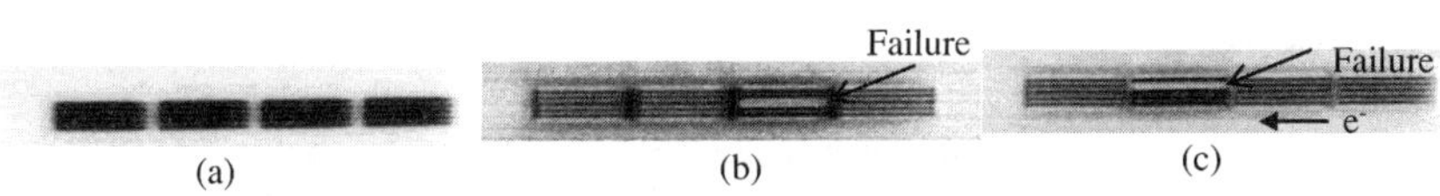

Figure 7. OBIRCH images of EM failed test structures. (a) no dummy, (b) with dummy metal, and (c) with dummy metal and via. Failure sites are indicated by arrows

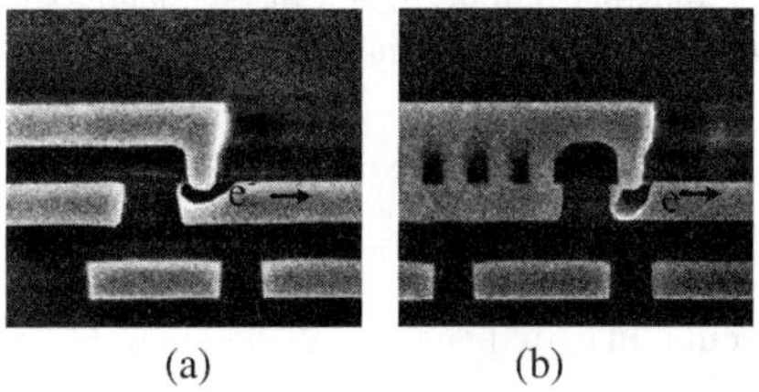

Figure 8. SEM images of EM failure voids of test structures (a) with dummy metal and (b) with dummy metal and via

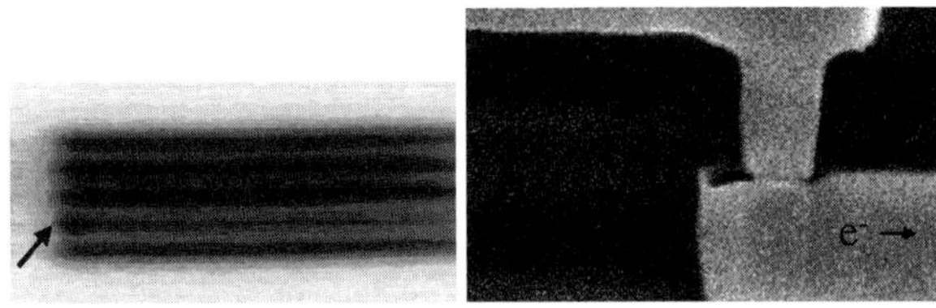

(a) (b)

Figure 9. Expanded OBIRCH image shows failure site by contrast between via and line. (a) OBIRCH image of no-dummy sample after EM test. Failure site is indicated by arrow. (b) Small slit-like void observed under via.

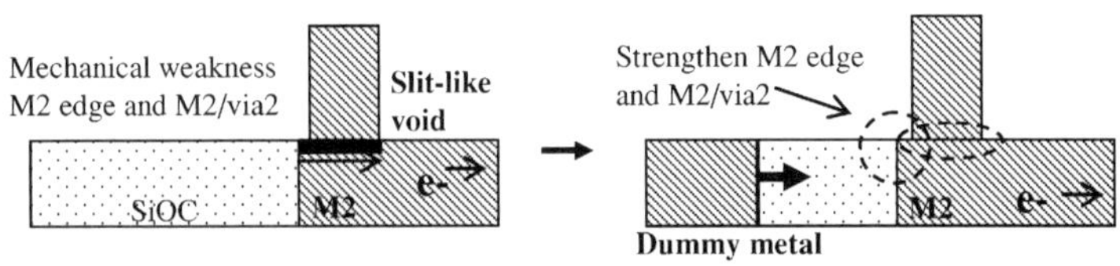

Figure 10. Effect of dummy metal on EM downstream mode. Dummy metal strengthens M2 edge and M2/via2, and restricts formation of slit-like void

CONCLUSIONS

The EM lifetime of the downstream mode was degraded by applying porous low-*k* (SiOC) as the metal dielectric interlayer. EM degradation occurred in the lower failure probability. The failure void was a small slit-like void from the metal edge to the metal/via interface. By arranging the dummy metal near the metal/via contact, EM lifetime degradation was eliminated. This suggests that EM degradation is caused by mechanical instability due to the porous low-*k* interlayer, thus the dummy metal mechanically strengthens the porous low-*k* near the metal/via contact then enhances EM reliability.

ACKNOWLEDGMENTS

The authors would like to thank N. Nakamura, M. Suzuki, and I. Akiba for their comments and encouragement.

REFERENCES

1. M. Kodera, S. Uekusa, H. Nagano, K. Tokushige, S. Shima, A. Fukunaga, Y. Mochizuki, A. Fukuda, H. Hiyama, M. Tsujimura, H. Nagai, and K. Maekawa, J. Electrochem. Soc. 152, G506 (2005).

2. K.-D. Lee, E. T. Ogawa, S. Yoon, X. Lu, and P. S. Ho, Appl. Phys. Lett. 82, 2032 (2003)
3. F. L. Wei, C. L. Gan, T. L. Tan, C. S. Hau-Riege, A. P. Marathe, J. J. Vlassak, and C. V. Thompson, J. Appl. Phys. 104, 023529 (2008).
4. C. J. Zhai, H. W. Yao, A. P. Marathe, P. R. Besser, and R. C.Blish, IEEE Trans. Device. Mater. Reliab. 4, 523 (2004)
5. J. -M. Paik, H. Park, and Y.-C Joo, Microelectron. Eng. 71, 348 (2004)
6. H. Tsuchiya, and S. Yokogawa, Microelectron. Reliab. 46, 1415 (2006)
7. M. Gall, C. Capasso, D. Jawarani, R. Hernandez, H. Kawasaki, and P.S. Ho, J. Appl. Phys. 90, 732 (2001).
8. K. Nikawa, and S. Inoue, Proceedings of 34th Annual International Reliability Physics Symposium, 1996, pp346-354

Conference Proceedings AMC XXIV © 2009 Materials Research Society

Copper Ion Drift as an Early Mode Failure in Bias Temperature Stressing

Hiroshi Miyazaki and Daisuke Kodama
Renesas Technology Corp., 4-1 Mizuhara, Itami-shi, Hyogo, 664-0005, Japan
E-mail: miyazaki.hiroshi@renesas.com

ABSTRACT

Time-dependent dielectric breakdown (TDDB) caused by Cu^+ drift was investigated using metal-insulator-semiconductor (MIS) capacitors. A numerical simulation and an experimental approach were used to estimate the field acceleration parameter of TDDB lifetime. This was estimated to be 0.5 cm/MV for SiOC films, which is distinctly smaller than the intrinsic acceleration parameter which is 4.5 cm/MV. Consequently, the degradation caused by Cu^+ drift becomes dominant at electric fields lower than a certain threshold if Cu^+ ions are injected into SiOC films. Therefore, Cu^+ drift is a mechanism of early mode failure and not of standard TDDB mechanism in the damascene interconnects.

INTRODUCTION

Time-dependent dielectric breakdown (TDDB) is one of the biggest obstacles to reducing interconnect size because at present our ability to control minimum spacing is marginal. McPherson defined dielectric degradation under bias temperature stressing (BTS) as thermochemical Si-O bond breakage [1]. Chen *et al.* attributed it to Cu diffusion [2]. In the study of electric field dependency, the former is known as the "E-model" and the latter is called the "√E-model". Lloyd proposed the "impact damage model" which assumes the cause to be a different mechanism than the two models above and shows a unique electric field dependency in which TDDB lifetime is proportional to $E^{-1}\exp(-\gamma\sqrt{E}+a/E)$, where E is the electric field [3]. People believe that understanding electric field dependency is the key to understanding TDDB mechanisms. However, it is difficult to actually identify the exact relationship between the TDDB lifetime and the electric field because long term acceleration tests are needed for their definition. In conventional models, it is assumed that the TDDB lifetime is controlled by only one mechanism over the whole electric field. Under this assumption, the TDDB lifetime for a low electric field during actual device operation is extrapolated from the results seen in high electric fields during acceleration tests. However, real TDDB behavior must be more complicated than theoretical models. Lloyd indicated the possibility that Cu diffusion model is applied instead of intrinsic TDDB model (impact damage model) below a certain threshold voltage [3]. The TDDB lifetime has no stronger dependency on the electric field at voltage lower than the threshold. As a result, the lifetime might be significantly shorter than the intrinsic one, due to the occurrence of early mode failures. If the barrier ability of materials surrounding the interconnects against Cu diffusion is insufficient, one has to know exactly the threshold voltage at which mechanism switching occurs, the field acceleration parameter for extrapolation and the permitted Cu contamination level for the process etc. In this paper, we will describe the true relationship between Cu diffusion and TDDB with a numerical simulation and an experimental approach.

EXPERIMENTS

Table 1 shows a description of samples. Standard samples were 100-nm-half-pitch damascene Cu interconnects with carbon doped silicon oxide SiOC(k=3.0) as the intra- and interlevel dielectrics. Metal-insulator-semiconductor (MIS) capacitors were used to investigate the relationship between Cu^+ drift and the TDDB lifetime. Three types of dielectric films, SiOC(k=3.0), SiOC(k=2.7) and O_3-TEOS, were examined. These dielectric films were deposited to a 100-nm thickness onto the n-type Si wafers. The Al-capsulated Cu electrodes (1 mm diameter) were directly deposited onto the dielectric film via the stencil using the resistance-heated evaporation. The detailed procedures of TDDB and triangular-voltage-sweep (TVS) measurements are described elsewhere [4].

Table 1 Sample description.

Sample #	Structure
A	Dual Damascene w/ SiOC(k=3.0)
B	MIS: Al / Cu / SiOC(k=3.0) / Si
C	MIS: Al / Cu / SiOC(k=2.7) / Si
D	MIS: Al / Cu / O_3-TEOS / Si

In our numerical simulation of Cu^+ drift, we basically followed the method of Kim *et al.* including their parameters [5]. The one-dimensional differential equation for Cu^+ transport, taking into account barrier height lowering by field assistance is expressed as follows,

$$\frac{\partial C}{\partial t} = D\cosh\left(\frac{q\lambda E}{2kT}\right)\cdot\frac{\partial^2 C}{\partial x^2} + \left(\frac{q\lambda}{2kT}\frac{\partial E}{\partial x} - \frac{2}{\lambda}\right)D\sinh\left(\frac{q\lambda E}{2kT}\right)\cdot\frac{\partial C}{\partial x} - \frac{2}{\lambda}\frac{q\lambda}{2kT}\frac{\partial E}{\partial x}D\cosh\left(\frac{q\lambda E}{2kT}\right)\cdot C, \tag{1}$$

where t is the time, x is the depth in the dielectric film, k is the Boltzmann's constant, T is the temperature. C, D and q are the concentration, the diffusivity, and the electric charge of the Cu^+ ions, respectively. Here, the jump distance of Cu^+ ions, λ ($\approx$0.27 nm) is a key parameter which dictates the electric field dependency of the TDDB lifetime. The non-uniform electric field can be calculated by using Poisson's equation.

$$\frac{\partial^2 \phi}{\partial x^2} = -\frac{qC}{\varepsilon_r \varepsilon_0}, \tag{2}$$

where ϕ is the potential, ε_r is the dielectric constant, and ε_0 is the permittivity of a vacuum. The space charge built up by Cu^+ injection significantly affects the Cu^+ movement.

TDDB MODEL

Figure 1 illustrates our model of TDDB caused by Cu^+ penetration. As each Cu^+ ion is a defect center which lowers the potential barrier, the defect moves with the Cu^+ ion. In the bias temperature stressing, Cu^+ ions are injected from the anode and drift to the cathode. Therefore, the defects also drift from the anode to the cathode. The piled-up defects near the cathode connect in a series and become a conduction filament because the potential overlap between two neighboring defects lowers the potential barrier which isolates them. When the conduction filament connects both end of the potential barrier of the dielectric film near the cathode, dielectric breakdown occurs. This model provides two hypotheses. Firstly, the stress-induced leakage current is unstable because the conduction filament is repeatedly connected and disconnected by the addition and subtraction of Cu^+ ions. This behavior contrasts with the intrinsic breakdown because its defects are fixed on the Si-O network [6]. Secondly, the Cu^+ arrival time to the cathode dictates the time to soft breakdown. If it is correct, the TDDB inherits a field acceleration parameter from the Cu^+ drift.

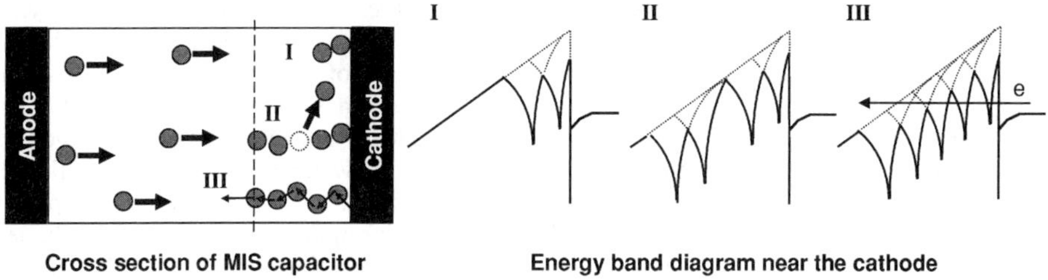

Figure 1. Model of TDDB caused by Cu^+ penetration.

RESULTS AND DISCUSSION

Equations (1) and (2) are numerically solved. Figure 2 shows the Cu^+ profiles in 100-nm-thick SiOC film and the potential distributions when the Cu electrode is positively biased with 20 V and 40 V, respectively. We adopted $6.07\times10^{-9}\exp(-0.71eV/kT)$ cm^2/s as the Cu diffusivity in SiOC [5]. The solid solubility of Cu^+ ions at the Cu/SiOC interface is unknown, however, we gave it as the maximum concentration which forbids a negative electric field within SiOC film. This approximation provides acceptable results compared with Ref. 5.

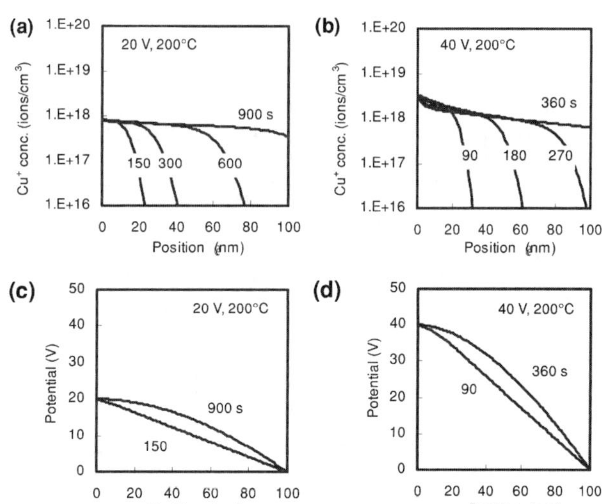

Figure 2. Cu^+ profiles in 100-nm-thick SiOC films for (a) 20 V and (b) 40 V, and potential distributions for (c) 20 V and (d) 40 V, respectively.

The positive space charge, i.e., injected Cu^+ ions, significantly relaxes the local electric field near the anode and suppresses excess Cu^+ injection. Therefore, the Cu^+ concentration in usual TDDB tests remains at most, 10^{18} ions/cm^3 except near the anode. The average distance between two Cu^+ ions is ~10 nm, giving us 10^{18} ions/cm^3, which is too spaced out to be able to create the long conduction filament. Therefore, the breakdown is probabilistic and yields to Poisson distribution, in other words, it can be essentially explained by the percolation theory [6].

Figure 3 shows the leakage current as a function of time during the bias temperature stressing of the various types of dielectric films. In the damascene interconnect (a), the leakage current decreases during the first short period and gradually increases during the second period, resulting in hard breakdown (the dielectric film melts). However, the MIS capacitors (b), (c) and (d) show different behavior. During high voltage stressing, the leakage current abruptly jumps after a gradual decrease during the initial stage, while during low voltage stressing, the leakage current rapidly increases with large fluctuations after a gradual decrease during the initial stage. This behavior agrees with our hypotheses described in the above paragraph. The potential-barrier length from the cathode to the dielectric conduction band on the same energy level is so short during high voltage stressing that a large number of short conduction filaments can break through the potential barrier in a short time. During the lower voltage stressing, a longer conduction filament is needed to break through the longer potential barrier. If a Cu^+ ion is missing from the conduction filament, the electron leak path breaks, resulting in the current decrease. However, when Cu^+ ions are added between two fragments of conduction filament, the leak path is restored. Therefore, these current fluctuations are attributed to the Cu^+ ion movement.

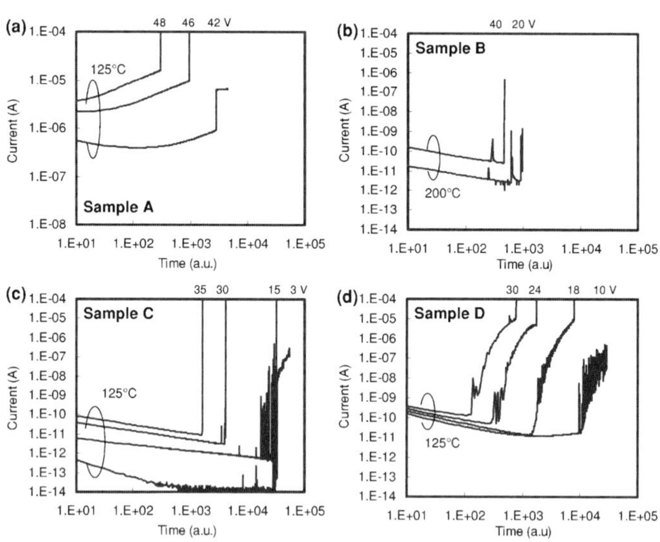

Figure 3 Current-time curves in the bias temperature stressing. (a) is the damascene, (b) is the Cu/SiOC(k=3.0)/Si, (c) is the Cu/SiOC(k=2.7)/Si, (d) is the Cu/O_3-TEOS/Si.

Figure 4 shows the electric field dependencies of the TDDB lifetimes. The damascene interconnect (a) only demonstrates a strong dependency (4.5 cm/MV) due to the intrinsic mechanism. The lifetimes of Cu/SiOC(k=3.0)/Si capacitors (b) are in line with the calculations from the Cu^+ drift model (Cu^+ arrival time to the cathode). This acceleration parameter, 0.5 cm/MV is distinctly smaller than the intrinsic one, 4.5 cm/MV. In examination of Cu/SiOC(k=2.7)/Si capacitors at 125°C (c), the lifetimes plotted on the chart switch to another gradient above 2 MV/cm. The Cu^+ drift is dominant in electric fields lower than 2 MV/cm while the effect of intrinsic defects introduced by the Si-O bond breakage is strong in the higher electric fields. The rapid degradation of Cu/SiOC(k=2.7)/Si at 200°C is due to the thermochemical reaction of Cu, which has less immunity than Cu/SiOC(k=3.0)/Si. From the above, we reasonably speculate that if Cu was really injected through barrier films in damascene structures, the field acceleration parameter would significantly change below a certain threshold. Because the field acceleration parameter is directly related to D and λ in Eq. (1), and not to the amount of Cu^+ ions injected into SiOC films, the shift of the acceleration parameter is inevitable. The Cu/O_3-TEOS/Si capacitors show this peculiar behavior in Fig. 4d. Although the lifetime of Cu/O_3-TEOS/Si is shorter than predicted from the Cu^+ drift model, its dependency yields to the $\sqrt{E}$-model. Another kind of ion simultaneously affects the degradation of O_3-TEOS as described in the latter paragraph.

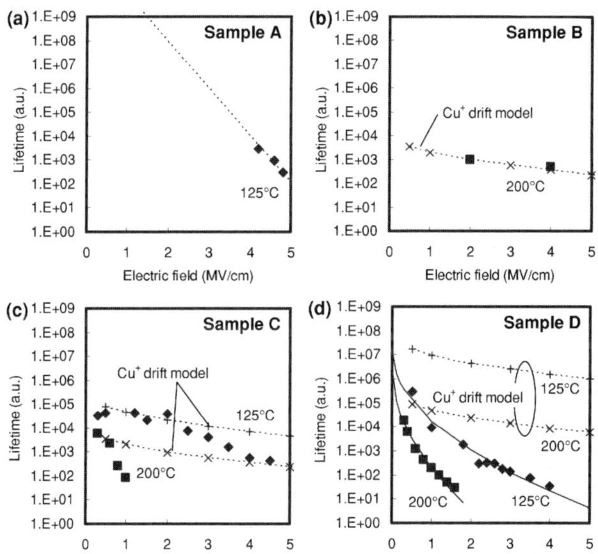

Figure 4 Electric field dependency of TDDB lifetimes. (a) is the damascene, (b) is the Cu/SiOC(k=3.0)/Si, (c) is the Cu/SiOC(k=2.7)/Si, (d) is the Cu/O_3-TEOS/Si.

Figure 5 shows the TVS curves of the above four samples. The damascene interconnect (a) has no mobile ion peak in its TVS curves even after the BTS because the Cu is fully covered with TaN and SiCN barrier. However, the leakage current is point-symmetrically increased by the generation of intrinsic defects during BTS. The Cu/SiOC(k=3.0)/Si capacitor has no mobile ion peak before the BTS but has four peaks after the BTS. First peak can not be due to displacement current caused by Cu^+ drift because there must be a negative current when the voltage is being decreased. This peak is probably caused by a kind of unstable dielectric leakage. The conduction filament might be temporary connected by adding mobile Cu^+ ions. The 2nd and 3rd peaks are negative currents when the voltage is being decreased however we cannot only

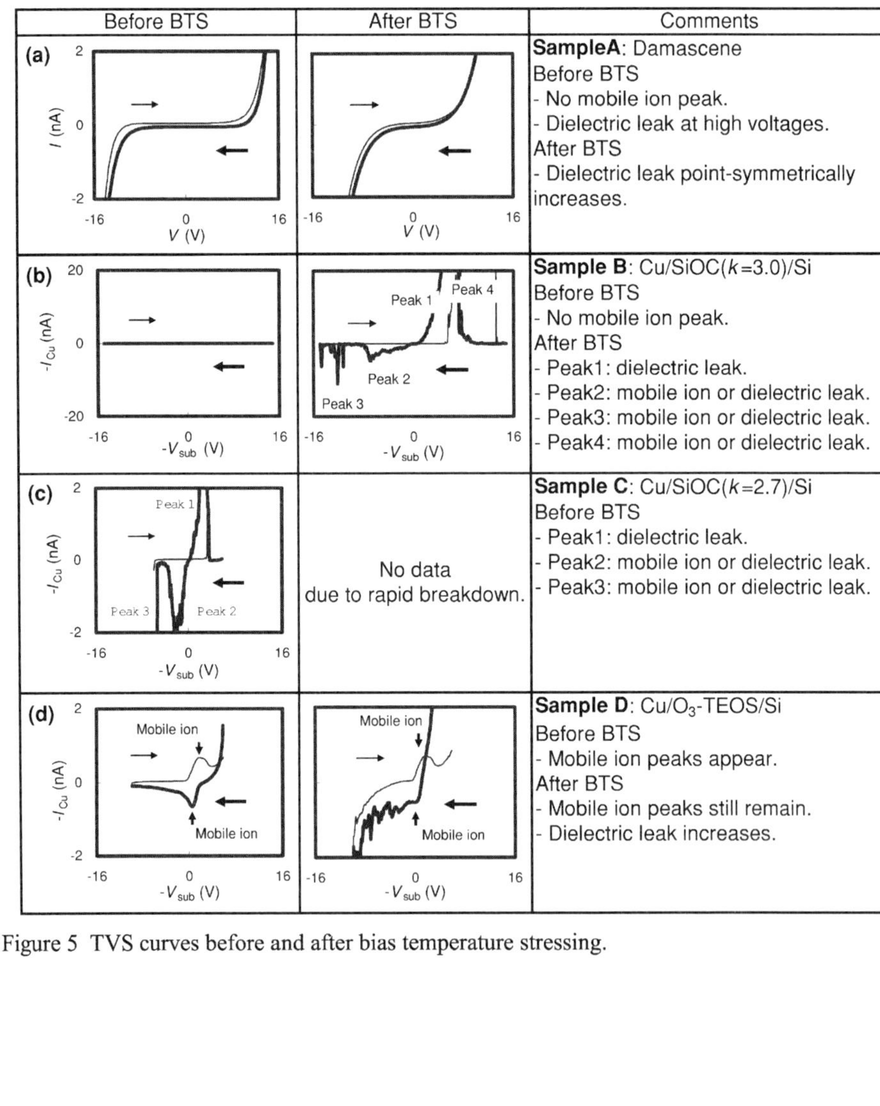

Figure 5 TVS curves before and after bias temperature stressing.

attribute these peaks to mobile Cu^+ ions because the possibility that they are a result of dielectric leakage remains. The Cu/SiOC(k=2.7)/Si capacitor has three peaks even before the BTS because of a thermochemical reaction between Cu and SiOC(k=2.7). The Cu/O_3-TEOS/Si capacitor has typical mobile ion peaks even before BTS. These mobile ion peaks are never due to Cu^+ ions because Cu^+ drift is estimated to be too slow, as shown in Fig. 4d, to generate the displacement current within the short period required for TVS measurements. Moreover, the amount of mobile ions is unchanged after BTS while the fluctuating dielectric leakage is increased by Cu^+ penetration. Therefore, another kind of mobile ion, possibly a proton, must exist. Consequently, TVS measurements unsuccessfully detect mobile Cu^+ ions below 200°C because of their slow movement. However, one should note that Cu^+ ions increase the fluctuating dielectric leakage and provide the distinctly small field acceleration parameter of TDDB lifetimes.

CONCLUSIONS

Without the diffusion barrier, Cu^+ ions rapidly penetrate into the dielectric films and create conduction filaments. The filaments are unstable because of the frequency of Cu movement, therefore large fluctuations are observed in BTS and TVS measurements. From the Cu^+ drift model, the gradual dependence of the TDDB lifetime on the electric field is naturally derived. Therefore, Cu^+ drift causes an early mode failure and cannot be a standard TDDB mechanism.

REFERENCES

1. J. W. McPherson, *J. Appl. Phys.*, vol. 95, no. 12, pp. 8101-8109, 2004.
2. F. Chen, O. Bravo, K. Chanda, P. McLaughlin, T. Sullivan, J. Gill, J. Lloyd, R. Kontra, and J. Aitken, *IEEE-IRPS Proceedings*, pp. 46-53, San Jose, 2006.
3. J. R. Lloyd, C. E. Murray, S. Ponoth, S. Cohen and E. Liniger, *Microelectronics Reliability*, vol. 46, pp.1643-1647, 2006.
4. H. Miyazaki, D. Kodama, N. Suzumura, *IEEE-IRPS Proceedings*, pp. 150-157, Phoenix, 2008.
5. K.-S. Kim, Y.-C. Joo, K.-B. Kim, J.-Y. Kwon, *J. Appl. Phys.*, vol. 100, no. 6, 063517, 2006.
6. K. P. Cheung, *IEEE Transactions on device and materials reliability*, vol. 6, no. 1, pp. 67-74, 2006.

Conference Proceedings AMC XXIV © 2009 Materials Research Society

Photo Misalignment Impacts on the Low-*k* TDDB

H. Tsuchiya, Y. Kakuhara, D. Oshida, M. Iguchi, T. Takewaki, and S. Yokogawa
Advanced Device Development Division, NEC Electronics Corporation
1120 Shimokuzawa, Sagamihara, Kanagawa 229-1198, Japan
hideaki.tsuchiya@necel.com

ABSTRACT

We have investigated the characteristics of the time dependent dielectric breakdown (TDDB) on Cu/low-*k* interconnects with vias. The photo misalignment impacts on lifetime were demonstrated experimentally by using 90 nm and 55 nm technology node. The local field enhancement, caused by the electric field focusing at the edge of vias, could be negligible. So, the lifetime can be simply predicted by actual via-to-line electric field. It was found that the fluctuation of space makes the shape parameter small and it causes pessimistic lifetime prediction. We therefore propose the reduction of space fluctuation influence with the overlay error model. It allows more accurate lifetime prediction.

INTRODUCTION

Time dependent dielectric breakdown (TDDB) on Cu/low-*k* interconnects has been causing a major reliability concern due to the aggressive shrinking of interconnect pitch size without a decreasing of operating voltage [1][2]. Because of its minimal spacing, the structure with misaligned vias is the most critical one. In other words, the misaligned vias determine the TDDB lifetime of the entire LSI. Consequently, it is important to understand clearly about the effect of vias on TDDB lifetime. Previous works have reported about the effects (e.g., [3][4]). However, a detailed analysis of the characteristics has not been carried out.

In this study, we investigate the characteristics of the low-*k* TDDB with vias. To investigate the via misalignment impact on lifetime, we used the test structures with misalignment on the mask pattern. The overlay error model[5] is used to predict a space fluctuation between lines to vias. A prediction of lifetime distributions with via misalignment is discussed.

EXPERIMENT

Fig.1 shows bird's-eye and cross-sectional schematic diagrams of the test structures used in this study. Two-layered (M1/V1/M2) or mono-layered (M2) interdigitated comb capacitors were prepared with 90 nm or 55 nm-node technologies. SiOC (k = 2.9) or SiOCH (k = 2.5) was used as the low-*k* dielectric, and the diffusion barriers used in the interconnect scheme were a SiCN capping layer and TaN/Ta liner. The nominal line-to-line spacing was 140 nm (90 nm-node) or 90 nm (55 nm-node).

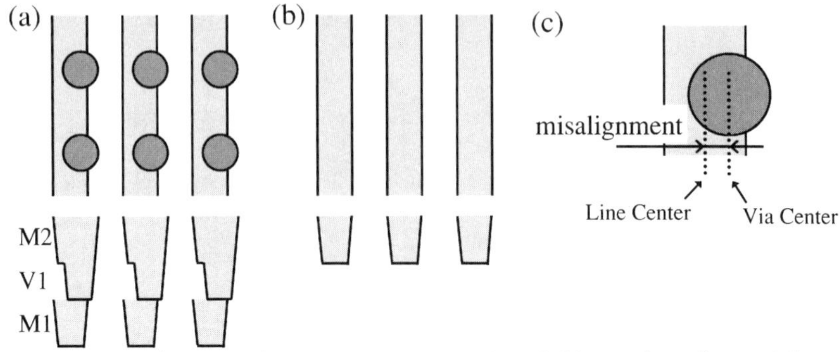

Fig. 1 Schematic diagrams of the test structures. (a) and (b) are via-to-line and line-to-line structure respectively. Upper: bird's-eye view. Lower: cross-sectional view. (c) Definition of misalignment. Misalignment is defined as a distance between the centers of the via and the line.

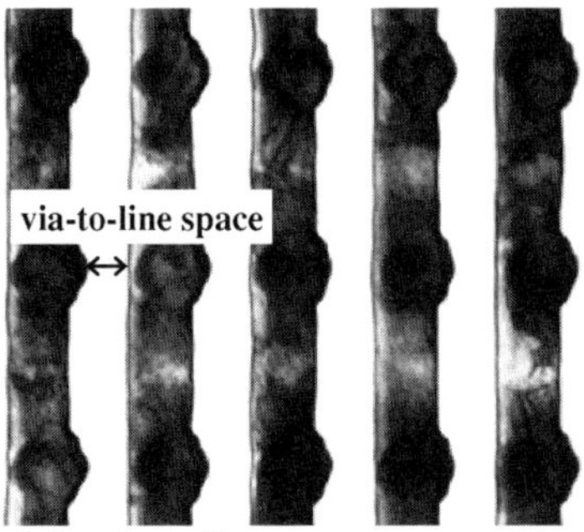

Fig. 2 S-TEM planer view of the via-to-line test structure. The minimum space is defined as average via-to-line space ($s_{\mathrm{v\text{-}l}}$). Effective electric field, E_{eff} is defined as $E_{\mathrm{eff}} = V / s_{\mathrm{v\text{-}l}}$, where V is the stress voltage.

The tests were performed at 150 or 300°C with constant voltage stress. The effective electric field ranges from 1.8 to 5.8 MV/cm which is calculated with actual minimum space obtained by scanning transmission electron microscopy (S-TEM) analysis. Fig. 2 shows a planer view of the via-to-line test structure.

RESULTS and DISCUSSIONS

To investigate the via effects on the lifetime distribution, the via-to-line and the line-to-line structure were tested under the same effective electric field (5.0 MV/cm). The areas (top-down layout view) of two kinds of test structures are the same. Fig. 3 shows Weibull distributions of the results. The via-to-line structure indicates shorter lifetime (scale parameter: η) and smaller Weibull slope (shape parameter:β). To confirm the failure mode in this study, cross-sectional SEM observations were executed. Fig. 4 shows the results. The CMP interface of the metal layer is the failure site for both structures. These results suggest that the degradation of η is caused by some via effect and the smaller β may be caused by the variation of photo misalignments.

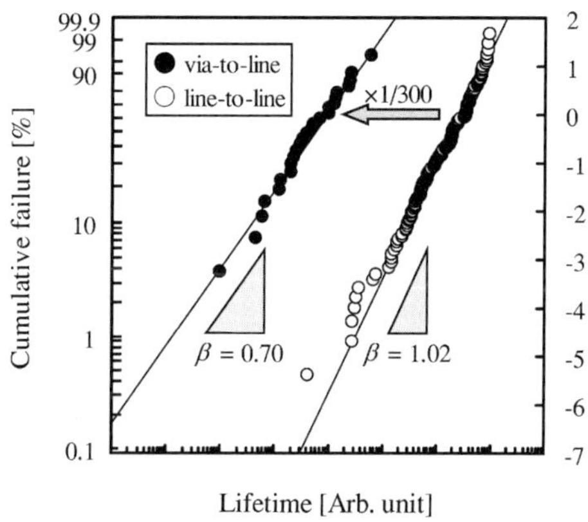

Fig. 3 Weibull distributions of TDDB lifetime. The test was performed with 55 nm-node samples. The data was taken at 300°C, 5.0 MV/cm.

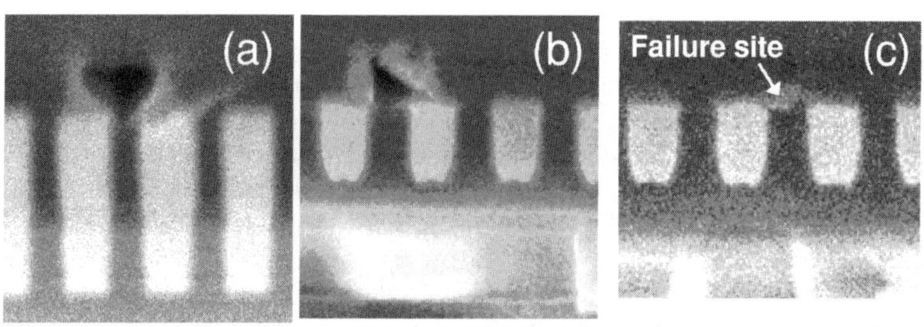

Fig. 4 Cross-sectional inspections of low-k TDDB failure sites. (a) Via-to-line structure. (b) and (c) Line-to-line structure. The nominal line-to-line spacing is 90 nm.

To investigate effects of the via misalignment, four test structures, which demonstrate misalignment on the mask pattern, were used. Misalignments were modified by design and the nominal value ranges from 0 to 40 nm. Fig. 5 shows the misalignment dependence of the lifetime. It indicates that η decreases with misalignment. On the other hand, the β takes a nearly-constant value for the range of the misalignment tested. These four test structures were picked from the same sites, hence the fluctuation of misalignment is supposed to be the same. These results suggest that the via effect which mainly affects the Weibull slope is not the mean value of the misalignment, but the deviation of misalignment.

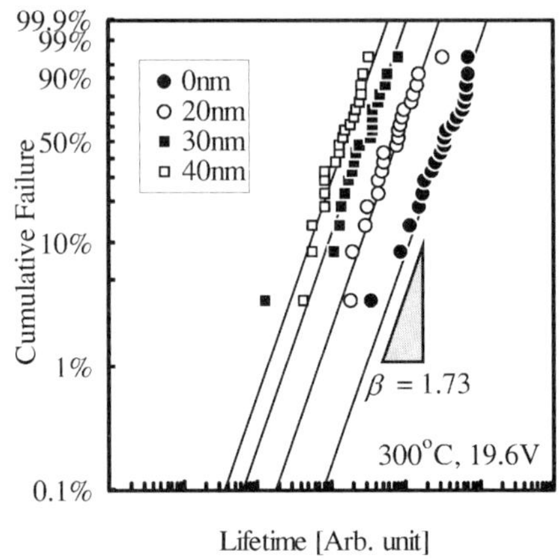

Fig. 5 Misalignment dependence of the TDDB lifetime. The lifetime decreases with misalignment under a constant stress voltage. On the other hand, the Weibull slope takes a nearly-constant value. The test was performed with 90 nm-node by a constant voltage stress.

According to the square root-*E* model [6][7], the lifetime is written as

$$\eta \propto A\exp\left(-\gamma\sqrt{E}\right) \tag{1}$$

where *A* is a constant, γ is the field-acceleration factor and *E* is the electric field. Fig. 6 shows comparison of electric field dependence of lifetime. It indicates that the via-to-line lifetime decreases two orders of magnitude compared with line-to-line lifetime. This degradation can not be explained only by a scaling effect, and hence it is supposed to attribute to the via effect on TDDB. Moreover, Fig. 6 also indicates that the via-to-line lifetime is independent of misalignment. And the γ value is approximately independent of test structure such as the presence of vias and the misalignment. These results suggest that a field enhancement, caused by the electric field focusing at the edge of the via, is negligible. Consequently, the lifetime is simply predicted by actual via-to-line electric field.

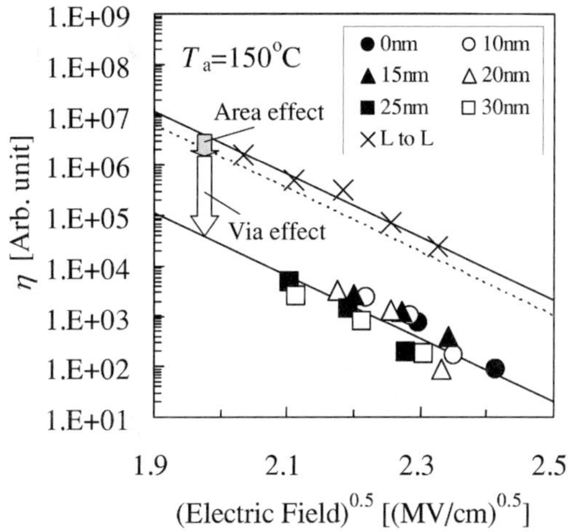

Fig. 6 Comparisons of electric field dependence of lifetime between the via-to-line and the line-to-line structures. The dotted line indicates the area normalized dependence of the line-to-line structure that is estimated by the Poisson area scaling model.

Fig.7 (a) shows the lifetime distribution of via-to-line structure under a constant voltage stress with 0 nm nominal misalignment. These data are taken from the whole area of the wafer. The lifetime distribution is considered to be influenced by fluctuation of space due to misalignment. This is consistent with the observed small β value. We measured the representative misalignment for all of the shots in the wafer. Thus, by using the overlay error model [5], the misalignment along each *x* and *y*-axis (*dx*, *dy*) are able to calculated for any site (*x*, *y*). The misalignments are given by

$$dx = -(\theta_s + \theta_{skew})y + M_x x + \varepsilon_x \qquad (2\text{-}1)$$

$$dy = \theta_s x + M_y y + \varepsilon_y \qquad (2\text{-}2)$$

where θ_s is the swirl error, θ_{skew} is the skew error, M_x is the magnification error along the x-axis, and M_y is the magnification error along the y-axis. In order to confirm the impacts of space fluctuation for lifetime prediction, we de-convoluted the data shown in Fig. 7(a) with actual misalignment of each samples. Fig. 7(b) shows the results whose misalignments were stratified to 5 ranges. The de-convoluted distributions have larger β than in Fig. 7(a). As a result, the space fluctuation makes 1 ppm lifetime (about 100 ppm lifetime in a product level) to be underestimated approximately by a half thousand of magnitude. From the information mentioned above, we found that the lifetime prediction with space fluctuation leads to pessimistic results.

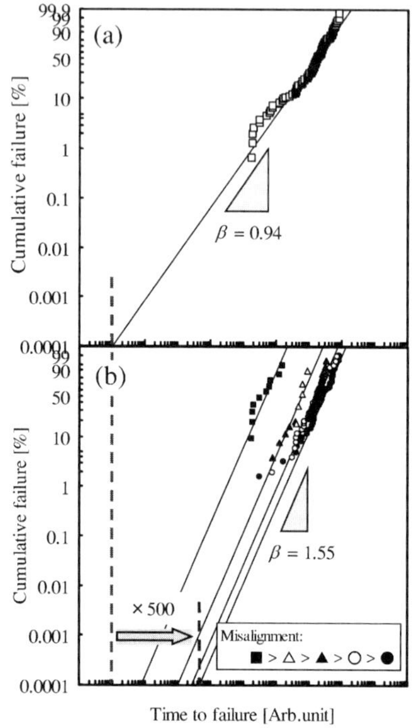

Fig. 7 (a) The lifetime distribution of via-to-line structure taken from the whole area of the wafer. (b) Lifetime distributions of each misalignment. These plots are obtained by de-convolution of the data shown in (a) with actual misalignment. The misalignment is estimated by the overlay error model[5]. The test was performed with 55 nm-node samples. The data was taken at 300°C, 20.0 V.

CONCLUSIONS

We investigated the characteristics of the low-*k* TDDB with vias. The photo misalignment impacts on lifetime were demonstrated experimentally. The local field enhancement, caused by the electric field focusing at the edge of vias, could be negligible. So, the lifetime can be simply predicted by actual via-to-line electric field. The fluctuation of space makes the shape parameter small and it causes pessimistic lifetime prediction. We therefore propose the reduction of space fluctuation influence with the overlay error model, so it allows more accurate lifetime prediction.

ACKNOWLEDGMENTS

The authors thank M. Sakurai and H. Takizawa for technical support and discussions. We also appreciate the encouragement of N. Nakamura, M. Suzuki, and I. Akiba during this work.

REFERENCES

1. S. Yokogawa, *et al.*, *Proc. of 2008 IRPS*, p.144 (2008).
2. D. Oshida, *et al.*, *Proc. of 2008 IITC*, p.222 (2008).
3. H. Miyazaki, *et al.*, *Proc. of 2007 ADMETA*, p.10 (2007).
4. A. H. Fischer, *et al.*, *Proc. of 2008 IRPS*, p.126 (2008).
5. W. H. Arnold, *et al.*, *Proc. of SPIE*, *Optical Laser Microlithography*, vol. 922, p.94 (1988).
6. N. Suzumura, *et al.*, *Proc. of 2006 IRPS*, p.484 (2006).
7. F. Chen, *et al.*, *Proc. of 2006 IRPS*, p.46 (2006).

Conference Proceedings AMC XXIV © 2009 Materials Research Society

Reliability analysis of Cu contacts with various diffusion barriers

Thomas Kauerauf[1], Steven Demuynck[1], Zsolt Tokei[1], Kristof Croes[1], Gerald Beyer[1], and Guido Groeseneken[1,2]
[1]IMEC, Kapeldreef 75, Leuven, 3000, Belgium
[2]K.U.Leuven, Leuven, 3000, Belgium

ABSTRACT

To reduce the transistor contact resistance Cu contacts are a promising alternative to the conventional W contacts. But the use of Cu requires a diffusion barrier, and in this work the reliability of transistors with Cu contacts of different dimensions and various barriers under electrical and thermal stress is studied. For most of the splits high process yield is observed and the t_{BD} measurements at 200°C revealed no lifetime degradation for the thinner barriers. Even after a high temperature anneal up to 10 hours the transistor characteristics remained unchanged, proving the good quality of our process and the feasibility of Cu contacts.

INTRODUCTION

The down-scaling of the contact size leads to a significant increase of the contact resistance [1]. One possibility to reduce the contact resistance is to replace the conventional W-plug contacts by Cu contacts. In literature there are a few reports on this topic available, describing the technological aspects and showing improved device characteristics [1-3]. But Cu diffuses very easily and a diffusion barrier (e.g. TaN/Ta) is essential to avoid yield loss and reduced front-end reliability [4]. The total contact resistance however depends on the thickness of this diffusion barrier [1], and a tradeoff between sufficient barrier reliability and a low contact resistance has to be made.

In this work the reliability and yield of ultra-thin gate dielectrics in the presence of Cu contacts for various barriers and designs is studied. For the first time small 90 nm contacts and the barrier conformality on 300 mm wafers are evaluated. The degradation under high temperature constant voltage stress (CVS) for the different splits is compared as well as the effect of an additional thermal anneal.

EXPERIMENT

The transistors were processed up to metal 1 using 65 nm CMOS technology on 300 mm wafers with SiON gate dielectrics of 1.2 nm EOT and poly-Si gate electrode. The split included three different TaN/Ta barriers (Table 1): a thicker layer as reference (A) and two layers (B, C) where the bottom barrier was partially removed by an Ar resputtering in order to achieve a lower contact resistance [1]. On each wafer transistors and capacitors (with injector) with 150 nm or 90 nm Cu contacts as well as various source/drain (S/D) to poly spacing from 7 μm down to 100 nm were available (Fig. 1).

Split	Barrier	Thickness
A	28 nm TaN/Ta	Thick
B	5 nm TaN/20 nm Ta + short rsp	Medium
C	5 nm TaN/20 nm Ta + long rsp	Thin

Table 1. The splits on contact barrier studied in this work. On splits B and C the barrier was partly resputtered in order to achieve a lower contact resistance.

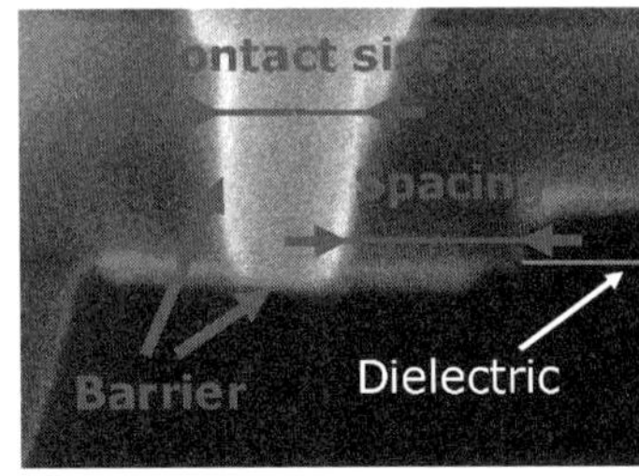

Fig. 1 FIB/SEM image of an S/D contact. The diffusion barrier, contact size (90 or 150 nm) and spacing (100 nm to 7 μm) are indicated.

To evaluate the process yield for the different barriers, contacts and spacings, I_G-V_G measurements in inversion across the entire wafer (167 dies) were performed. The failure criterion chosen was a 10% deviation from the median of all currents at $V_G = 1$ V. A thin barrier at the bottom of the contact is crucial for low contact resistance, but might be harmful for the reliability of the device. To investigate the durability of the barriers, time-to-breakdown (t_{BD}) was measured using CVS at positive gate bias and 200°C in order to enforce Cu diffusion. At least 12 devices were tested for each condition and a soft breakdown criterion of a 100nA current step was used to trigger breakdown. The Weibull distributions are fitted using the Maximum Likelihood Estimation and the voltage acceleration is extrapolated with a power law [5].

To study potential Cu diffusion in the absence of electrical stress for all splits a part of the wafer was exposed to an additional thermal treatment at 400°C. To compare with the initial values the transistor performance as well as the reliability were analyzed after 1 h and 10 h of stress. Moreover, on selected splits FIB/SEM analysis was performed to investigate the barrier uniformity and the Cu filling.

DISCUSSION

The leakage currents revealed high yield and excellent barrier uniformity for all splits across the 300 mm wafers. As an example, in Fig. 2 the data of split C for 150 nm contacts with 7 μm spacing are shown. All failures are located at the edge and were actually caused by a CMP issue. Comparing the splits and devices (Table 2) suggests a slight trend of lower yield with thinner barriers but no significant difference is found. Only the 90 nm contacts of split C suffered from reduced yield due to insufficient contact filling with Cu.

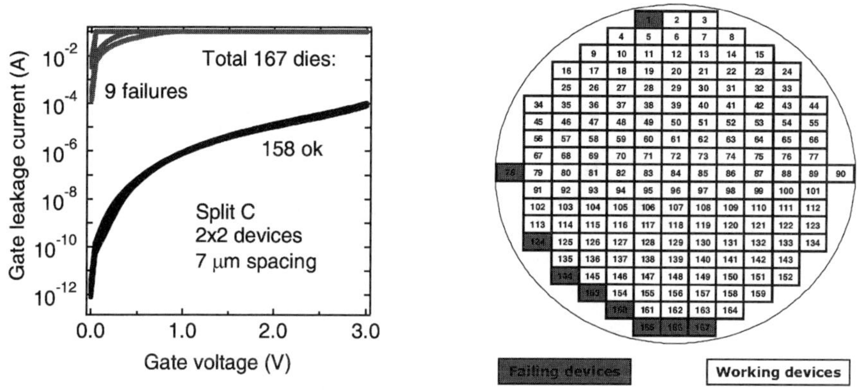

Fig. 2 I_G-V_G characteristics of devices with 150 nm contacts and 7 µm spacing measured across the entire wafer (split C) and the corresponding wafer map. The failing devices are located at the edge of the 300 mm wafer.

Contact [nm]	Spacing [um]	Yield in %		
		A	B	C
150	0.14	98	96	86
	0.235	100	97	95
	7	99	96	95
90	0.1	99	95	58

Table 2. An overview of the yield for the different diffusion barriers, contacts and spacing. Most failures are not Cu related but due to CMP issue.

In Fig. 3 the t_{BD} voltage acceleration for the three splits is shown revealing little difference between the splits. The Weibull slopes β and the exponents of the power law fit are within error-bar identical (Table 3), confirming that the gate dielectric reliability is not affected by the different barriers. Similar results are obtained on 150 nm and 90 nm contacts (Fig. 3), indicating the feasibility to downscale Cu contact barrier technology. In another experiment the t_{BD} was compared for devices of 150 nm contacts with different S/D to gate spacing. Unlike published in [4], for the particular barrier sequences used here, no reduction in lifetime for smaller spacing is detected (Fig. 4), confirming the good stability of these barriers under electrical stress even at high temperature.

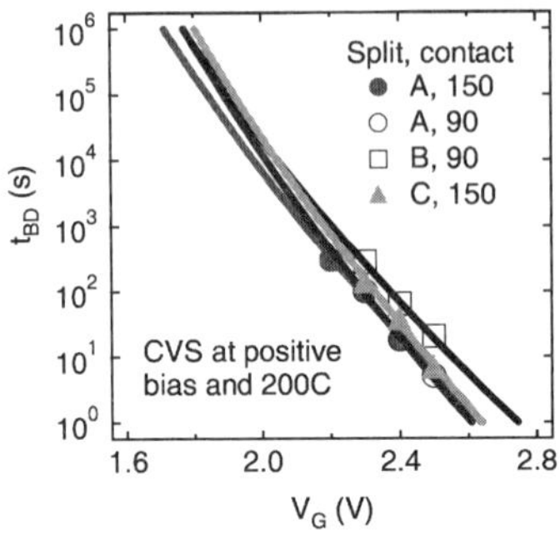

Fig 3. The t_{BD} lifetime measured with CVS at 200°C for the different splits and contacts.

Split	Contact [nm]	β	n
A	150	1.05	32
	90	1.01	35.5
B	90	1.1	31.5
C	150	1.27	36.1

Table 3. The t_{BD} Weibull slopes and voltage acceleration factors for the different splits are within error-bar identical.

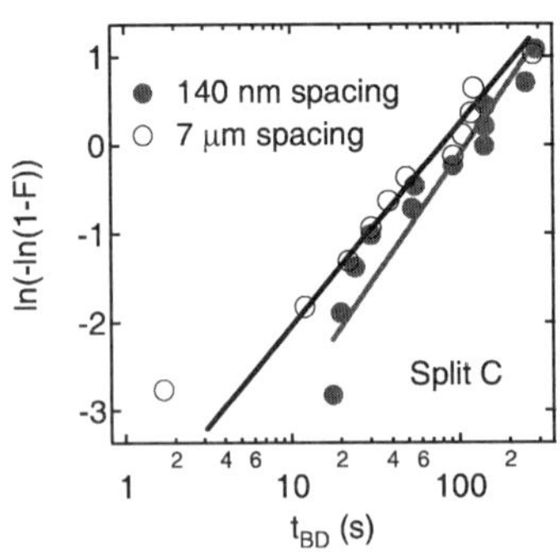

Fig 4. The t_{BD} Weibull distributions of split C for devices with 140 nm (solid) and 7 μm spacing (open) show no Cu related degradation.

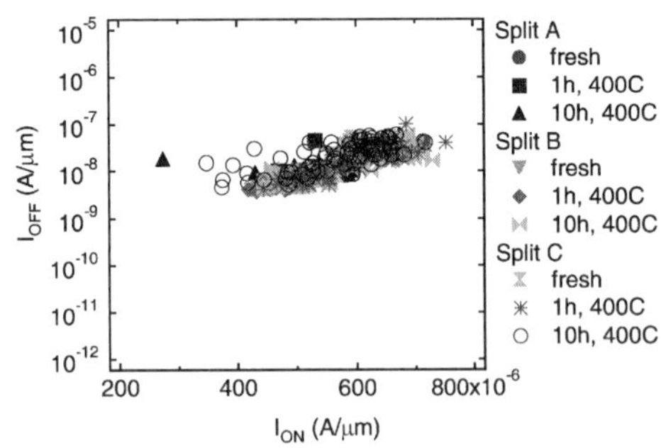

Fig 5. The I_{ON}/I_{OFF} values for the three splits before and after 1 h & 10 h anneal at 400°C. No increase in I_{OFF} is observed.

To qualify Cu contacts for a CMOS process the stability of a multi-level thermal budget has to be guaranteed as well. In a final experiment we therefore exposed parts of the wafers to thermal stress at 400°C and compared key transistor parameters after 1 h and 10 h anneal to the initial values. The obtained results are very promising: all devices survived the stress and the t_{BD} distributions did not show any sign of degradation. The I_{ON}/I_{OFF} ratios for all splits remained identical (Fig. 5), confirming that no Cu diffuses from the S/D contacts into the channel. Only the gate leakage current in inversion of about 15% of the samples had increased for split B after 10 h and split C after 1 h anneal (Fig. 6). The charge-to-breakdown (Q_{BD}) Weibull distributions of split C for transistors and capacitors are shown in Fig. 7. The small differences between the splits are rather caused by die to die variation and the reliability is not affected by the anneal.

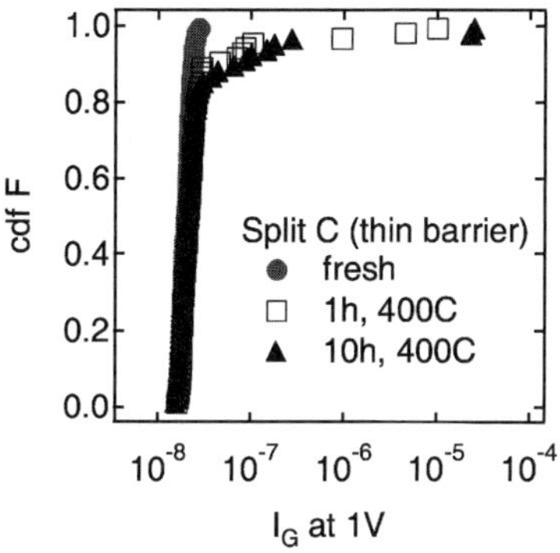

Fig 6. The distribution of the I_G values at 1 V for split C before and after the anneal showing a slight current increase for ~15% of the devices.

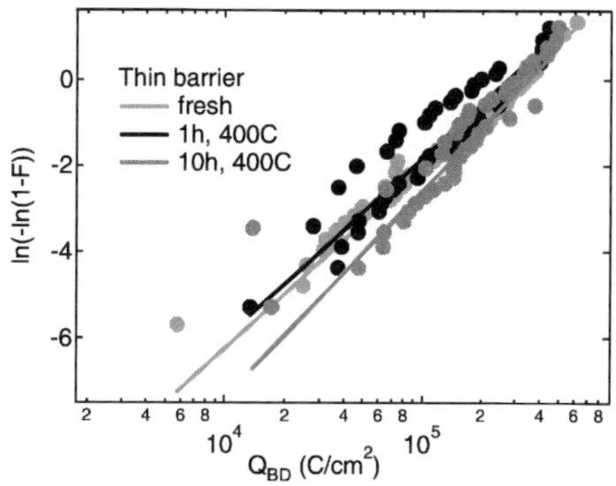

Fig 7. The Q_{BD} Weibull distributions for the thinnest barrier before and after the 400°C anneal for 1 h and 10 h respectively. No significant difference is observed.

CONCLUSIONS

In this work the front-end reliability and yield of Cu contacts for different diffusion barriers, contacts and spacing on 300 mm wafers was studied. High yield for all splits demonstrates good barrier conformality and contact filling across the entire wafer. Electrical stress at 200°C revealed no indication of device failures due to Cu diffusion on 90 nm contacts. Moreover, devices on all splits endure an additional anneal of 400°C for 10 h, showing the compatibility with the thermal budget of multi-level BEOL processing.

ACKNOWLEDGMENTS

The authors would like to thank IMEC's core partners including Intel, Micron, Panasonic, Qimonda, Samsung, TSMC, NXP, Elpida, Infineon and ST Microelectronics.

REFERENCES

[1] S. Demuynck et al., IITC, 2006
[2] R. Islam et al., VLSI 2000
[3] M. Inohara et al., IEDM 2001
[4] G. Van den Bosch et al., IEDM 2006
[5] E.Y. Wu et al., IEDM 2000

Conference Proceedings AMC XXIV © 2009 Materials Research Society

The impact of ash on TDDB of metal-hard-mask-etched Cu/low-k interconnects

M. Pantouvaki, Herbert Struyf, Dirk Hendrickx, Nancy Heylen, Olivier Richard, and Gerald Beyer
IPSI/IMI, IMEC, Kapeldreef 75, Leuven, B-3001, Belgium

ABSTRACT

Plasma damage is a major concern for advanced porous low-k interconnects as it can deteriorate the interconnect performance. Metal hard-mask etch schemes have been adopted to limit the damage caused during resist ashing. In this paper we investigate the effect of ashing conditions used in a metal hard-mask etch scheme on the interconnect reliability. It is found that depending on the ashing conditions, the TDDB lifetime specifications can be met or not.

INTRODUCTION

Porous dielectrics of low dielectric constant (low-k) could potentially reduce the RC delay of advanced technology node interconnects and improve the performance of integrated circuits. However, although the porosity of such materials reduces their dielectric constant, at the same time it increases their sensitivity to plasma damage during integration [1]. Exposure to plasmas destroys surface bonds and depletes carbon, turning these materials hydrophilic. If they are accordingly exposed to moisture, the resulting integrated dielectric constant can be significantly higher than that of pristine films, counteracting the advantage of lower k-value after deposition. At the same time, plasma-induced surface defects can impact reliability, especially close to the polished interface. A detailed study of the integration scheme that takes into consideration patterning, Cu reduction plasmas and chemical mechanical polishing (CMP), for optimum performance of such interconnects is therefore mandatory.

A number of studies have shown that exposure to ashing plasmas in resist-based dielectric etch can have detrimental impact to interconnect reliability, depending on plasma chemistry and conditions [2-4]. In metal-hard-mask based patterning, the ashing of the resist can be performed before the low-k etch, therefore exposing only the top of the dielectric to the plasma and not the sidewalls [5]. However, to ensure that the resist is completely removed, some degree of over-ashing is required. In this paper, the effect of ashing conditions on the porous low-k material reliability is studied in a metal hard-mask (MHM) based patterning scheme. It is shown that optimization of the ashing conditions is necessary also in this case, in order to prevent severe reliability degradation.

EXPERIMENT

The reliability tests were performed in meander-comb structures of 180 nm pitch. The intermetal dielectric material that was used was Black Diamond II (k-value = 2.5, porosity 25%). The patterning scheme is shown in Figure 1. A 30 nm TiN metal hard-mask was used on top of the dielectric. After photolithography, the metal hard-mask was patterned, the resist was ashed and the trenches were etched into the low-k dielectric. Different ashing conditions were tested in an CCP chamber or in a TCP chamber, as shown in Table 1. A diluted HF clean was performed after the low-k etch and in one case also after the MHM patterning. After patterning, a 6 nm

TaN/Ta barrier was deposited, followed by Cu seed. The rest of the trench was filled by electrochemical plating and the excessive Cu was removed by direct CMP. After metallization, the trenches were passivated for protection from the environmental conditions, and they were sintered at 425 ^{0}C in forming gas.

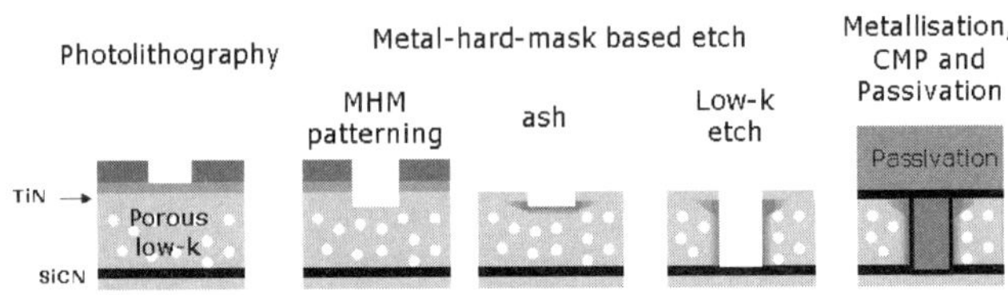

Figure 1. Process sequence.

Table 1.: Ashing conditions

Ash chemistry	Chamber	Dil. HF after strip	Dil. HF after low-k etch	Metal width (nm)
O_2/Cl_2	TCP	-	Yes	87.0
O_2/H_2	TCP	-	Yes	86.0
N_2/H_2	CCP	-	Yes	65
N_2/H_2	CCP	Yes	Yes	70.5
O_2	CCP	-	Yes	74.5
O_2/C_2H_4	CCP	-	Yes	75.5

DISCUSSION

When the ashing conditions are varied, the resist removal rate and the directionality of the ash can also vary. Figure 2 shows the carbon energy-filtered transmission electron microscopy (EFTEM) map after MHM patterning and resist ashing but before low-k etch, for two of the cases examined here: using O_2/Cl_2 and N_2/H_2 ashing chemistries. A dark area in the carbon EFTEM map between the TiN MHM and the SiC(N) indicates carbon depletion in the low-k dielectric. More severe carbon depletion is observed in the case of O_2/Cl_2 chemistry (Figure 2(a)), as compared to the N_2/H_2 chemistry (Figure 2(b)). Although part of this damage will be removed during low-k etch, clean and CMP, a thin damaged layer remaining close to the top of the dielectric spacing could provide additional routes for dielectric breakdown.

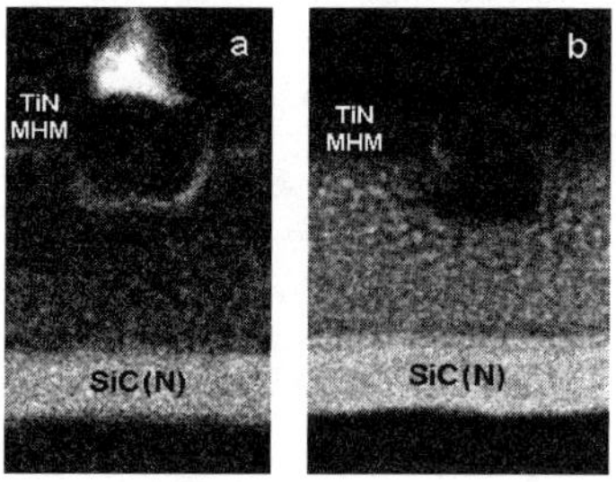

Figure 2. Carbon EFTEM profiles after MHM patterning and ashing, using a) O_2/Cl_2 and b) N_2/H_2 ashing chemistries. A dark area in this map indicates lack of carbon.

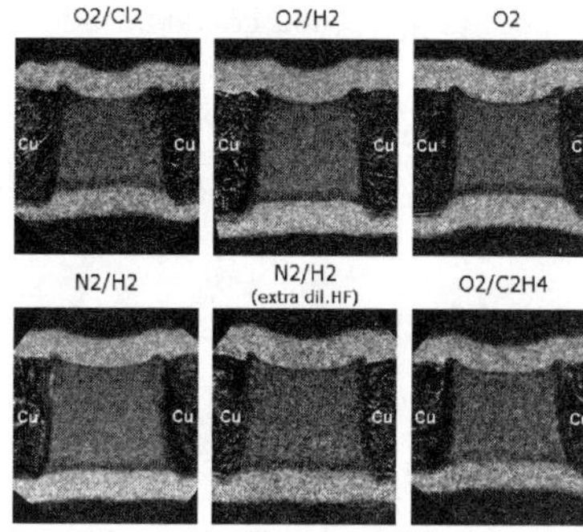

Figure 3. Carbon EFTEM profiles of integrated Cu / porous low-k interconnects patterned using various ashing conditions.

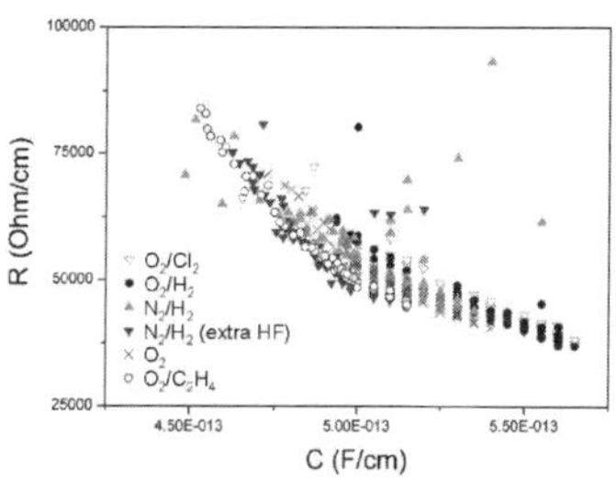

Figure 4. Line resistance versus interline capacitance for different ashing conditions.

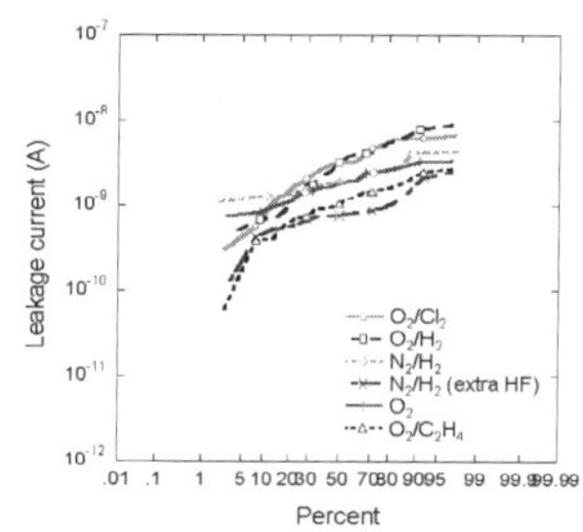

Figure 5. Leakage current distributions measured at 2.5 MV/cm at 100 ^{0}C.

Figure 3 shows the EFTEM maps of the low-k dielectric between the Cu lines after completion of the damascene processing. The O_2/Cl_2 strip chemistry in the TCP chamber results again in more pronounced carbon depletion at the top interface of the low-k with the SiC(N) cap and close to the sidewalls. RC measurements were performed prior to reliability testing and are shown in Figure 4. No major differences were observed from these measurements, as expected in a metal hard-mask patterning scheme. If one compares the results for the same line resistance, a minor decrease in capacitance can be indentified in the cases of O_2/C_2H_4 and N_2/H_2 (extra HF) strip chemistries. Although this trend is small, it comes in agreement with the leakage current measured between the meander and the comb at 2.5 MV/cm and 100^{0}C, which is shown in Figure 5. The condition of 2.5MV/cm at 100^{0}C was selected so that the values are well above the measurement limits of the equipment used, but well below the breakdown voltage in all cases. The leakage current has lower values in the cases of O_2/C_2H_4 and of N_2/H_2 (extra HF) strips. Figure 6 shows characteristic current-voltage curves and the breakdown field distribution of wafers stripped using various conditions. O_2/Cl_2 gives the lowest breakdown field and the largest distribution as compared to the other conditions.

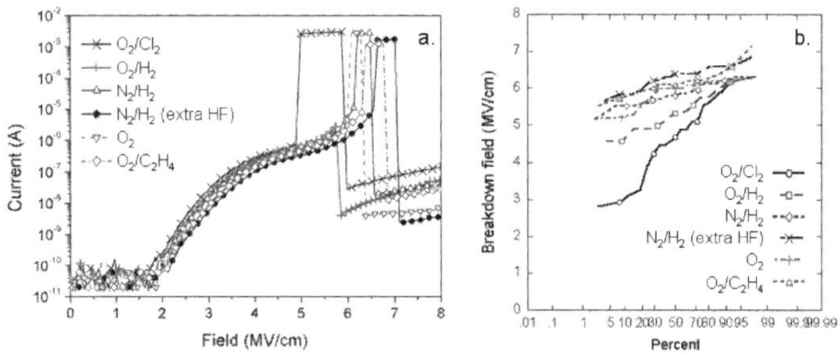

Figure 6. a) Examples of interline current versus voltage curves and b) breakdown field distribution for varying stripping conditions.

Time-dependent dielectric breakdown (TDDB) measurements were performed to estimate the lifetime of the interline dielectric under constant electric field and temperature conditions. The TDDB analysis was done based on the E-model:

$$TTF = A_0 \exp\left(\frac{E_a}{k_B T} - \gamma E\right) \tag{1}$$

where TTF is the time-to-failure, A_0 is a material-dependent constant, E_a is the activation energy, k_B is the Boltzmann constant, T is the temperature, γ is the field acceleration factor and E is the field. Examples of the TDDB curves measured at 4MV/cm and 100^0C are shown in Figure 7. Electron trapping occurs in all cases studied, resulting in an initial decrease of the current with time [4, 6]. However, this effect is less pronounced in the case of O_2/Cl_2, where breakdown happens faster than in other cases. Figure 8 shows the TDDB results for different strips and the parameters extrapolated using lognormal statistics. It was found that the lifetime and the field acceleration factor were higher in the cases of O_2/C_2H_4 or N_2/H_2 strips. Using a O_2/Cl_2 in the TCP chamber for resist ashing results in porous low-k interconnects that do not reach the 10-year lifetime target. The activation energy of the dielectric was determined using the $t_{50\%}$'s from TDDB measurements at different temperatures. The activation energy is ~0.2 eV in the cases of O_2/C_2H_4 and N_2/H_2, in agreement with other publications [4, 6]. The activation energy of the O_2/Cl_2 strip approaches zero, while the distribution of this split is high, implying significant damage of the low-k. From the EFTEM map of Figure 3 the damage can be located to the top (CMP) interface of the low-k, where carbon-depletion is more pronounced. Carbon depletion implies dangling bonds generated during the ashing process. Such broken bonds can be responsible for low-k degradation during reliability testing. Temperature and bias stress cause thermal excitation and polarization of the atoms of the dielectric, and they eventually result in the collapse of the weaker bonds related to defects in the dielectric. If there is more such damage in the interline dielectric, the dielectric breakdown is expected to occur faster.

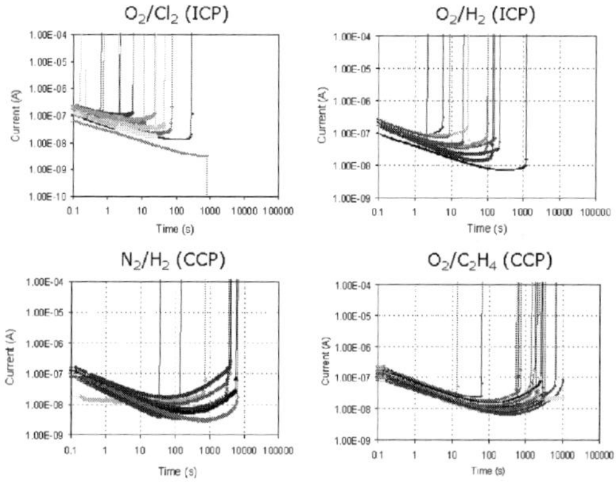

Figure 7. Examples of current versus time curves at 4MV/cm, 100^0C.

CONCLUSIONS

The effect of the stripping conditions on reliability was studied in porous low-k interconnects etched using a metal hard-mask scheme. It was found that different stripping conditions can result in interconnect lifetimes that vary by orders of magnitude, indicating the importance of ash also in metal hard-mask etching.

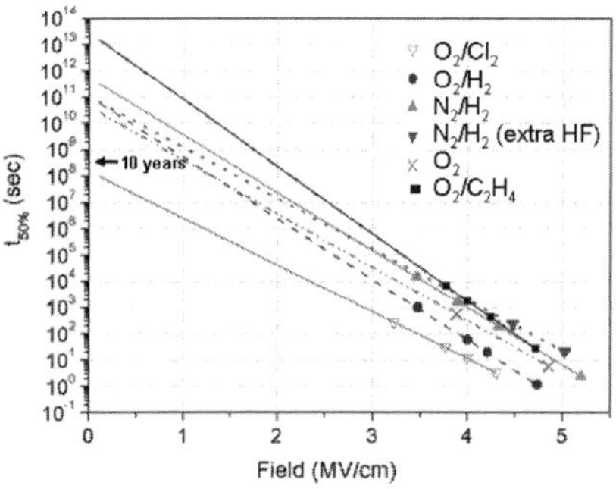

Ashing conditions	t50% (sec) at 0.11MV/cm, 100ºC	σ	Field acceleration factor γ (cm/MV)	Activation energy (eV)
O2/Cl2 (ICP)	$9.7\cdot10^{7}{}^{+1.4e+8}_{-5.5e+7}$	$3.0^{+0.77}_{-0.55}$	$-4.1^{+0.23}_{-0.23}$	$0.024^{+0.3}_{-0.3}$
O2/H2 (ICP)	$5.0^{+8.1}_{-1.9}\cdot10^{10}$	$1.9^{+0.43}_{-0.32}$	$-5.28^{+0.12}_{-0.12}$	
N2/H2 (CCP)	$3.1^{+2.0}_{-1.2}\cdot10^{11}$	$1.5^{+0.44}_{-0.31}$	$-5.02^{+0.11}_{-0.11}$	$0.17^{+0.094}_{-0.094}$
N2/H2 (CCP), extra HF	$7.6^{+9.9}_{-2.1}\cdot10^{12}$	$1.1^{+0.29}_{-0.21}$	$-5.46^{+0.073}_{-0.073}$	$0.23^{+0.13}_{-0.13}$
O2 (CCP)	$2.5^{+1.8}_{-1.1}\cdot10^{10}$	$1.5^{+0.5}_{-0.39}$	$-4.68^{+0.13}_{-0.13}$	
O2/C2H4 (CCP)	$1.5\cdot10^{13}{}^{+5.4e+12}_{-3.9e+12}$	$1.2^{+0.27}_{-0.2}$	$-5.87^{+0.075}_{-0.075}$	$0.18^{+0.13}_{-0.13}$

Figure 8. TTF versus field and TDDB parameters extrapolated using the E-model.

REFERENCES

1. M. R. Baklanov et al, International Conf. On Solid-State and Integrated Circuit Technology, 2006, 4 pp., 2006.
2. J. S. Tsai et al, IEEE IITC 2005, pp. 94-96.
3. M. Aimadeddine et al, Microelectronic Engineering, 82, pp. 341-347, 2005.
4. Y. Li et al, Semiconductor Science and Technology 22(4), pp. 320-325, 2007.
5. H. Struyf et al, IEEE IITC 2005, pp. 30-32.
6. G. Haase et al, IEEE-IRPS 2005 Proceedings, pp. 466-473, 2005.

Conference Proceedings AMC XXIV © 2009 Materials Research Society

First Principle Study of Cu-Sn (111) Surface Diffusion

Satoshi Ebato Junya Inoue Toshihiko Koseki

Department of Materials Engineering, The University of Tokyo
7-3-1 Hongo, Bunkyo-ku, Tokyo, 113-8656 JAPAN
The University of Tokyo, Faculty of Engineering Bldg. 328
TEL +81-3-5841-7110 FAX +81-3-5841-2773 e-mail ebato@metall.t.u-tokyo.ac.jp

ABSTRACT

It is well known that Sn doping is effective in suppressing the electromigration in Cu-based integrated circuits. In this paper, the first principle calculation was introduced to investigate the suppression mechanism. All the calculations were performed on Cu(111) or Cu-Sn(111) surface within local density approximation. The energy calculations suggest that (111) surface of Sn-doped Cu has a p($\sqrt{3}\times\sqrt{3}$)R30°structure. This is compatible with experimental observations of surfaces of Sn-doped Cu. The activation energies for surface diffusion of Cu on Cu (111) surface and Cu-Sn (111) surface with p($\sqrt{3}\times\sqrt{3}$)R30° structure were calculated to be 0.76 eV and 0.99 eV, respectively. These calculated activation energies are compatible with experimental results. It is suggested that the formation of p($\sqrt{3}\times\sqrt{3}$)R30°structure of Cu-Sn suppresses the electromigration of Cu thin-film interconnects.

INTRODUCTION

When an electric current flows through interconnects, the migration of atoms occurs. This phenomenon, called electromigration, is of considerable technological interest because it has been implicated as a major cause of interconnect failure in integrated circuits[1-4]. Lee et al. found that the EM activation energy for pure Cu was 0.73 eV at the conditions of temperature range of 250-450°C and current density of 5×10^5-2.1×10^6 A/cm^2, while the activation energy increased to 0.95, 1.25, and 1.14 eV after alloying Sn element with concentrations of 0.5, 1.0, 2.0 wt.%, respectively[6]. However, it is still not fully understood why Sn is beneficial in resisting EM in Cu line. In this paper the first principle calculation was used to investigate the suppression mechanism of the doped Sn. Surface is known to be the main diffusion pass among the three diffusion passes (bulk, grainboundary, surface) and (111) surface is the most stable surface for

Cu. Accordingly, we studied surface diffusion on Cu(111) and Cu-Sn (111) surface by the first principle calculation.

METHODS

The total energy calculations in this paper were performed using the first principle PAW pseudopotential method within the local–density approximation (LDA) as implemented in the VASP code. The potential structures of Cu-Sn(111) surface are p(2×2) (Cu:Sn=3:1) and p(√3×√3)R30° (Cu:Sn=2:1) (Fig.1). p(√3×√3)R30° structure is observed experimentally on annealed Cu-Sn(111) surface. But the structure of Cu-Sn(111) can be p(2×2) when the surface concentration of Sn is low . Accordingly, stabilities of p(√3×√3)R30° and p(2×2) structures of Cu-Sn (111) surface were investigated by the analysis of the Gibbs energy change of surface reconstruction using long-range order parameters. Second, terrace-ledge-kink (TLK) model (Fig.2) was introduced to model surface diffusion. Surface diffusion was modeled as the diffusion of a Cu adatom (Fig.3). In equilibrium, the activation energy for diffusion (Q) is given by

$$Q=\Delta H_f+\Delta H_m$$

In this equation, ΔH_f is the formation enthalpy of an adatom on a terrace or a ledge, ΔH_m is the activation enthalpy for migration of an adatom. The formation enthalpy of an adatom (ΔH_f) was calculated as the energy difference between bulk state and adatom state of one atom, while the activation enthalpy for migration of an adatom(ΔH_m) was calculated as the energy difference between an adatom on the lowest-energy position and an adatom on the highest-energy position along the migration path.

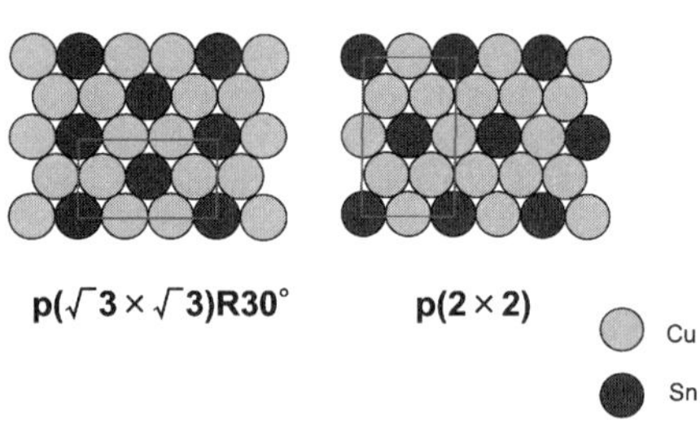

Fig.1 Structures of Cu-Sn(111) surface

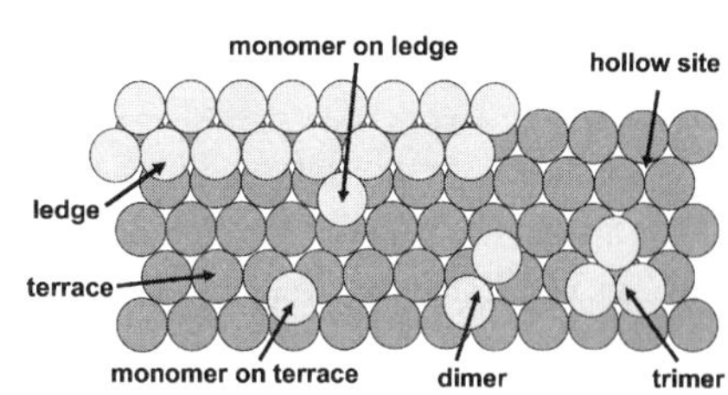

Fig.2 TLK model

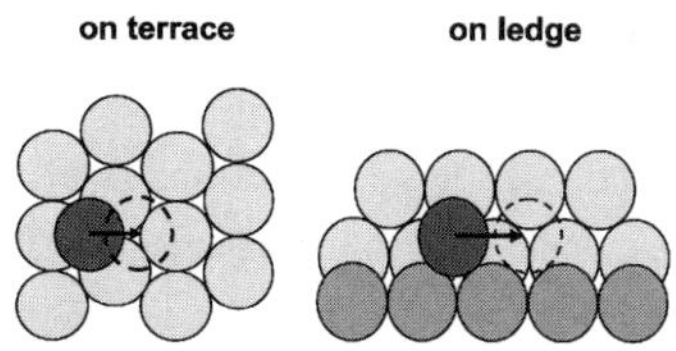

Fig.3 Two types of migration process

RESULTS AND DISCUSSION

Structure of Cu-Sn surface alloy

The analysis of the long-range order parameter change showed that when Sn concentration exceeds a certain value, the long-range order parameter increases drastically to 1 (Fig.4). This result suggests that when Sn concentration exceeds a certain value, the surface reconstruction occurs rapidly. The energy calculations suggested that p(√3×√3)R30° structure is more stable than p(2×2) structure (Fig.5). These results suggest (111) surface of the Sn-doped Cu has p(√3×√3)R30° structure. The results support the experimental observations of p(√3×√3)R30° structure on Cu-Sn(111) surface by G.Contini et al. attained by Auger electron spectroscopy(AES), reflection high-energy electron diffraction (RHEED), and scanning tunneling microscopy (STM)[7].

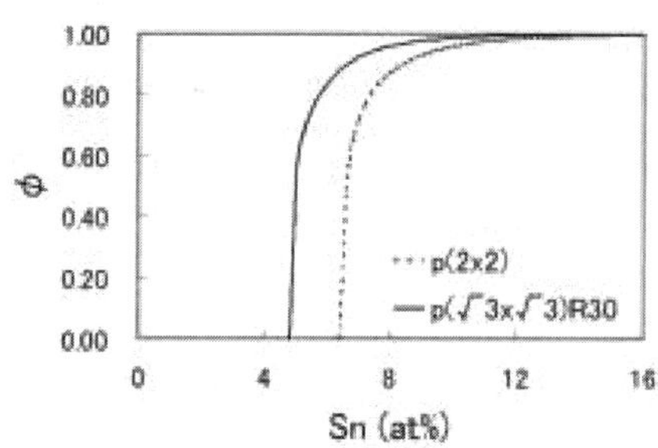

Fig.4 Sn concentration vs long-range order parameter

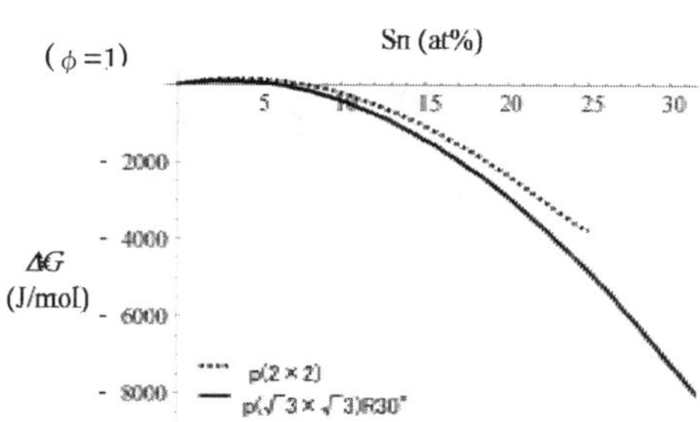

Fig.5 Stabilities of p(2×2) and p(√3×√3)R30°

Activation energies for surface diffusion on Cu(111) and Cu-Sn(111) surface

Calculations of activation energies for surface diffusion of Cu on Cu (111) and Cu-Sn (111) surface were conducted. The calculation results of those on Cu(111) surface suggest that on Cu (111) surface the formation enthalpies of a dimer and a trimer are much higher than the activation energies for diffusion of monomer (Fig.6). Accordingly, it is suggested that the dominant process of diffusion on Cu (111) surface is the diffusion of a monomer on a ledge. The activation energy of diffusion can be estimated to be the smallest value among the activation energies of potential diffusion processes. Therefore, the activation energy for surface diffusion on Cu(111) is estimated to be 0.76 eV. This calculated activation energy is compatible with experimental results[6]. On (111) surface of Cu-Sn, p(√3×√3)R30° structure was observed experimentally and shown to be the most stable structure by our calculations. Therefore, activation energies for surface diffusion on Cu-Sn (111) surface were conducted on p(√3×√3)R30° structure. The ledges of this structure can be classified into 4 types of ledges (ledge A-1, ledgeA-2, ledgeB-1, ledgeB-2) The activation energies for a monomer on these 4 types of ledges were calculated together with the activation energy for a monomer on a terrace (Fig.7). This calculation result suggest that the dominant process of diffusion on Cu-Sn(111) surface is same as that on Cu(111) surface. The activation energy for surface diffusion on Cu-Sn(111) surface is estimated to be 0.99 eV. This result is compatible with experimental results[6].

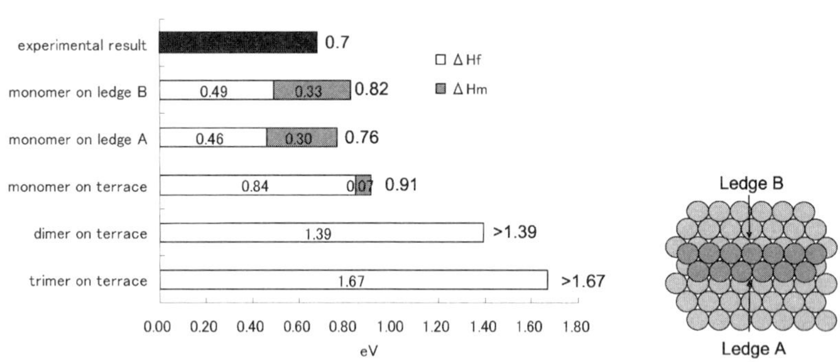

Fig.6 Activation energies for surface diffusion of Cu on Cu (111) surface

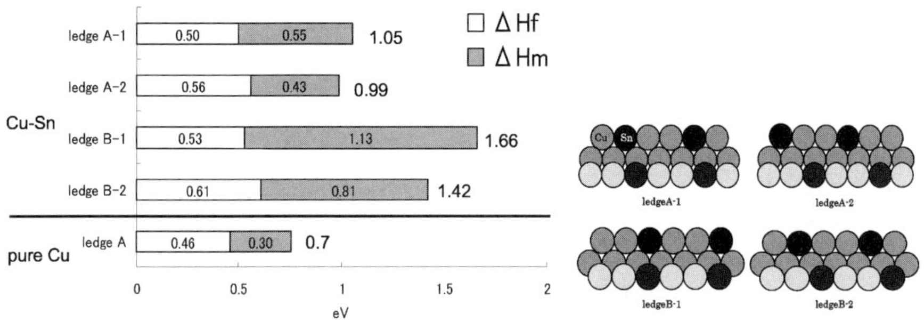

Fig.7 Activation energies for surface diffusion of Cu on Cu-Sn(111) surface (p(√3×√3)R30°structure)

CONCLUSIONS

In this paper, we analyzed the structure of Cu-Sn (111) surface and the behaviors of a Cu adatom on Cu(111) and Cu-Sn (111) surface with the first principle calculation. Among structures of (111) surface of the Cu-Sn alloy, p(√3×√3)R30° structure is the most stable structure according to energy calculations. This result is compatible with experimental observations. The activation energies for surface diffusion of Cu on Cu (111) surface and Cu-Sn (111) surface (p(√3×√3)R30°structure) were estimated to be 0.76 eV and 0.99 eV by the first principle calculation. These calculated activation energies are compatible with experimental results. Consequently, it is suggested that the formation of Cu-Sn surface alloy with p(√3×√3)R30° structure is beneficial in resisting electromigration in Cu interconnects.

REFFERENCES

[1] K.N. Tu, J. Appl. Phys. 94 (2003) 91.

[2] C.K. Hu, L. Gignac, E. Liniger, B. Herbst, D.L. Rath, S.T. Chen, S. Kaldor, A. Simon, and W.T. Tseng, Appl. Phys. Lett. 83 (2003) 869.

[3] C.K. Hu, L.M. Gignac, R. Rosenberg, B. Herbst, S. Smith, J. Rubino, D. Canaperi, S.T. Chen, S.C. Seo, and D. Restaino, Appl. Phys. Lett. 84 (2004) 4986.

[4] E. Zschech, M.A. Meyer, S.G. Mhaisalkar, A.V. Vairagar, A. Krishna-moorthy, H.J. Engelmann, and V. Sukharev, Thin Solid Films 504 (2006) 279.

[5] C.K. Hu et al., J. Electrochem. Soc. 143, (1996) 1001.

[6] K.L. Lee, C.K. Hu, and K.N. Tu, J. Appl. Phys. 78 (1995) 4428.

[7]G.Contini et al. ,Surface Science 405 (1998) 509.

Conference Proceedings AMC XXIV © 2009 Materials Research Society

Quantitative Evaluation of Electromigration Behavior in Cu-Sn Thin Films

Saori Yoshimoto Junya Inoue Toshihiko Koseki
Department of Materials Engineering, The University of Tokyo
7-3-1 Hongo,Bunkyo-ku,Tokyo,Faculty of Engineering Bildg.4–328
e-mail: yoshimoto@metall.t.u-tokyo.ac.jp

ABSTRACT

We investigated the electromigration(EM) and stress induced voiding (SIV) behaviors in thin Cu and Cu-Sn films. The films (100nm-thick) were deposited on Si substrates by magnetron sputtering through a patterned mask. Some of them were further capped by a Ta layer (10nm-thick) in order to suppress vacancy sinks and sources on the free surface of the Cu film. As for the EM tests, the average current density in the narrowest regions of the test structure was maintained to 4.0MA/cm^2, while SIV tests were performed under high temperature of constant value so that constant thermal stresses were applied. The SIV tests showed that Sn addition considerably improves SIV resistance. Moreover, the suppression of void growth by Sn addition is demonstrated through the EM tests.

INTRODUCTION

EM and SIV are serious reliability concerns in today's Cu Dual-Damascene Interconnects because of the downward scaling of feature size in large scale interconnects [1]. A number of investigations have been carried out to quantify the EM behavior in Cu interconnects under high current density exceeding 1.0MA/cm^2[2][3][4]. Most of the studies about EM have been conducted on the multilevel interconnects structure and focused mainly on estimating the MTTF (Mean Time To Failure). Some researches have shown an increased robustness of Cu EM failure by alloying [5][6]. However, EM in the structure can be affected by complex factors such as current density, thermal stress and temperature gradient. These factors have to be separately analyzed to clarify the mechanisms of the improved EM resistance in Cu-alloys,

In this work, we quantitatively evaluated the void formation in Cu and Cu-Sn interconnects during both EM and SIV tests.

EXPERIMENTAL DETAILS

A modified Standard Wafer-level Electromigration Acceleration Test (SWEAT) structure (Fig.1) was used for the investigation of the EM and SIV behaviors [7][8]. Cu or Cu-1.0wt%Sn thin films (100nm-thick) were deposited on a Ta(10nm)/TaN(20nm) /SiO_2(100nm)/Si(725μm) substrate by magnetron sputtering through a patterned mask. Some of the test samples were covered by a Ta layer (10nm-thick) deposited on the Cu film as a capping layer in order to suppress vacancy sinks and sources on the free surface of the Cu film. The operation was made in the same batch without breaking the vacuum. Before the EM and SIV tests, the test samples were annealed at 673K for 30min in Ar-3%H_2 in order to achieve complete grain growth and stress relaxation. Before the EM tests, the samples were kept at 473K so as to initiate SIV voids. In doing so, only the void growth rate can be evaluated in EM tests.

The average current density in the narrow regions of the test structure was maintained to be 4.0MA/cm^2 during the EM experiments. In both EM and SIV tests, the substrate temperature was kept at either 573K or 623K in Ar-3%H_2. After the EM and SIV tests, the void formation in the Cu films was observed with scanning electron microscope and the number and the area fraction of the voids were measured. The thermal stresses induced in the Cu films during SIV tests were investigated by X-ray diffraction.

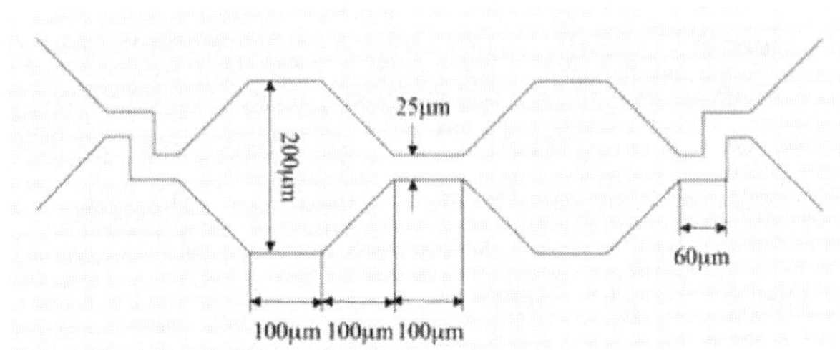

Fig.1 Modified SWEAT structure of EM test pattern

RESULT & DISCUSSION

Microstructure of Cu and Cu-Sn films

The average grain sizes of Cu and Cu-Sn films were evaluated by tunneling electron microscope and the sheet resistivities of the films were measured with a standard four point probes as shown in Table1. Since the average grain sizes of these films were almost equivalent, it is fair to compare the number of voids between the pure Cu and the Cu-Sn films. The segregation behavior of Sn atoms in the Cu-Sn film was investigated by electron spectroscopy for chemical analysis. Fig.2 shows the distribution of Sn atoms in the Cu-Sn film as a function of the etching time. The result indicates that Sn atoms were highly concentrated at the interface of Ta/Cu.

Table1 Average grain sizes and sheet resistivities of Cu and Cu-Sn films

	Average grain size(nm)	Sheet resistivity(μΩcm)
Pure Cu	133	3.45
Cu-Sn	120	3.53

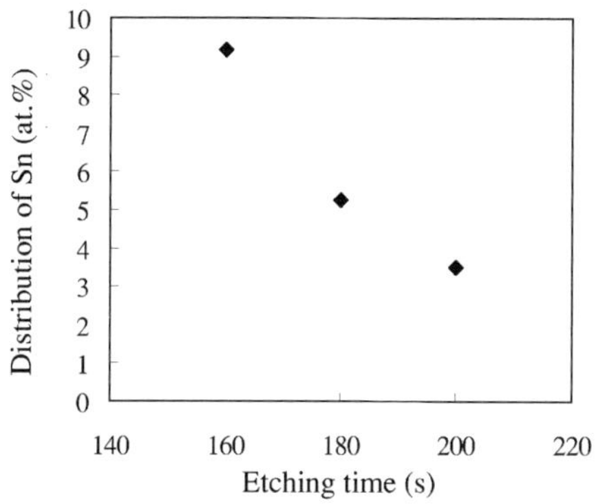

Fig.2 Distribution of Sn in Cu-Sn film

SIV behavior of Cu and Cu-Sn films

Since the thermal stresses in the Cu films were measured to be tensile and about 700MPa at 300K. As a consequence the residual stresses of 573K and 623K were estimated to be about 180MPa and 90MPa, respectively. The surfacic density of voids in the samples during SIV tests is plotted as a function of the annealing time on Fig.3. As it can be seen from the figure, the fraction of voids in pure Cu films is considerably larger than that in Cu-Sn films.

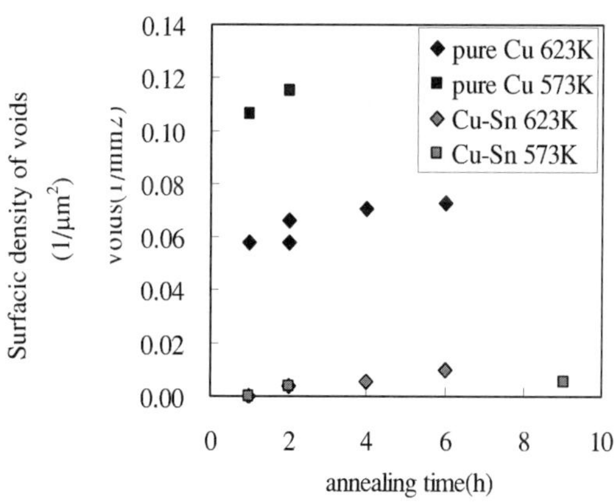

Fig.3 Surfacic density of SIV voids in Cu and Cu-Sn films

EM behavior of Cu film without Ta capping layer

To clarify the mechanism of void growth quantitatively, the total area of voids formed in the sample after the EM tests is plotted as a function of the testing time in Fig.4. The total area of voids shows a linear relationship with the testing time. In order to estimate the activation energy of the void formation, the experiments were stopped after 1 hour and 3 sets of EM tests using 3 different temperatures. From Arrhenius plot, the estimated activation energy of void formation is about 0.5eV, which is almost the same as the activation energy of Cu surface diffusion. The observation of the location of voids leads to a relation between the surfascic area of voids and the current density gradient in the pattern. Moreover, the current density gradient seems to have a stronger relationship with voiding than current density.

EM behavior of Cu and Cu-Sn films with Ta cap layer.

The total area of voids formed in the sample during the EM test at 623K is plotted as a function of the testing time in Fig.5. Since most of the voids were initiated by SIV prior to the EM test, growth rate of voids can be estimated from the area fraction of voids. The comparison of void formation in the Cu and the Cu-Sn films shows that the void growth rate in the Cu-Sn film is much slower than that in the Cu film.

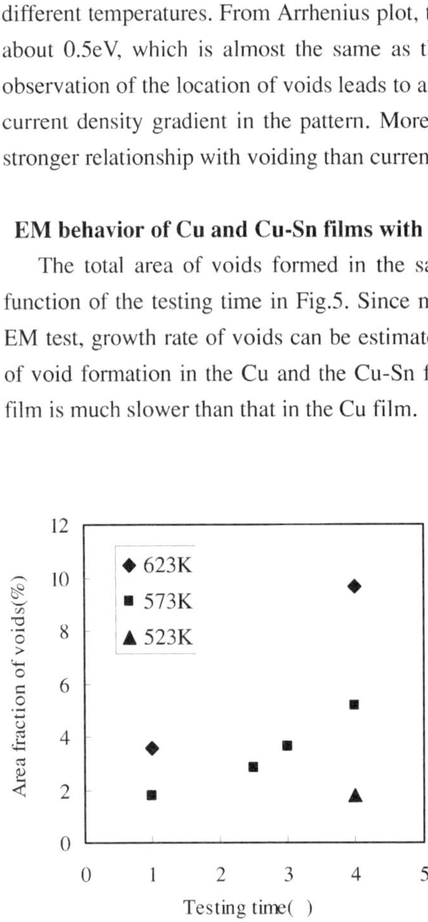

Fig. 4 Total area in Cu film without a Ta capping layer

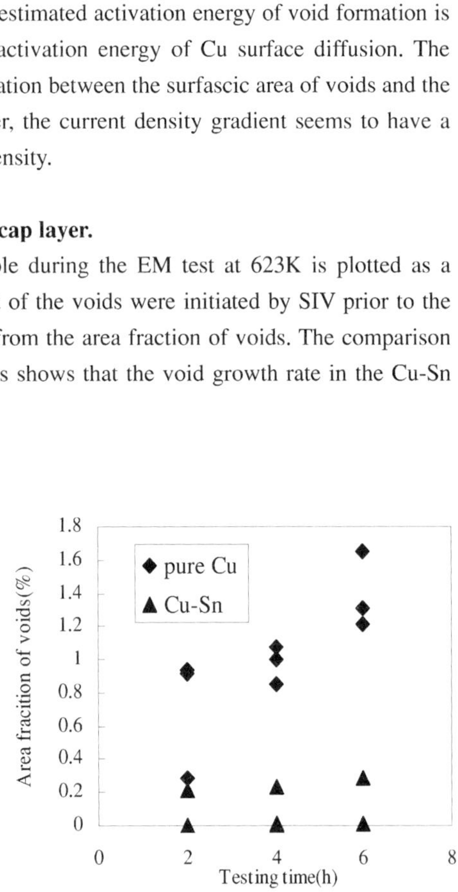

Fig. 5 Total area of voids in Cu and Cu-Sn films with a Ta capping layer at 623K

CONCLUSIONS

Both EM and SIV behaviors in Cu and Cu-Sn films were investigated. The thermal stresses in the films were tensile stresses and measured to be around 700MPa at 300K. The result of the SIV tests shows that the Cu-Sn films have extremely high SIV resistance compared to the Cu films. The result of the EM tests demonstrates that the void growth process, which is considered to be dominated by diffusion at the Ta/Cu interface, is substantially suppressed.

REFERENCES

[1]ITRS: The international Roadmap for Semiconductors 2005 Editions.

[2]L.Arnaud et al.: J.Appl.Phys.93,1(2003)192-204

[3]C.S.Hau-Riege et al.: Applied Physics Letter,78,22(2001)3451-3454

[4]J.R.Lloyd et al.: Microelectronics Reliability,39(1999)1595-1602

[5]K.L.Lee et al.:J.Appl.Phys.78,7(1995)

[6]S.Yokogawa et al.:J.Appl. Phys.101,013513(2007)

[7]M.Kumikawa et al.:Jpn.J.Appl.Phys.31(1992)1147-1149

[8]F.Giroux et al.:Proc.IEEE 1995 Int.Conference on Microelectronic Test Structures,8,(1995)

Conference Proceedings AMC XXIV © 2009 Materials Research Society

Lifetime Prediction of Stress Induced Voiding Failure by Novel Numerical Analysis in Cu Interconnects with an Ultra Low-*k* Dielectric

Shota Nakajima[1], Takenao Nemoto[2] and A.Toshimitsu Yokobori, Jr.[1]
[1]Graduate School of Engineering, Tohoku University
[2]New Industry Creation Hatchery Center (NICHe), Tohoku University
Aoba01, Aramaki, Aoba-ku, Sendai-shi, Miyagi, 980-8579, Japan
Phone and Fax: 022-795-6905, e-mail: nakajima@md.mech.tohoku.ac.jp

Abstract

Newly developed computer aided numerical analysis by using the finite difference analysis (FDA) for vacancy concentration combined with the finite elemental analysis (FEA) for stress distribution was proposed to simulate time sequential change of vacancy concentration by stress induced voiding (SIV) and to predict a lifetime. Comparing of results obtained by computer simulation with those by experimental results, the developed numerical analysis was found to enable to simulate SIV phenomenon. The proposed numerical analysis is also possible to obtain the method of improvement on the SIV life time by optimum stress distribution and vacancy diffusivity.

1. Introduction

As minimizing design rule in the large scale integration (LSI), half-pitch of multilayer interconnects becomes small. Increase in current density as well as residual stress in interconnects likely bring many issues for LSI reliability. SIV has been reported that migration of vacancies and nucleation of voids promoted by residual stress after LSI fabrication resulted in failure[1,2]. A number of theories have been proposed to explain the dependence of SIV failure rate based on the diffusional creep model[3-4]. Moreover, initial voids produced by SIV presumably acts as a starting defect of electromigration (EM) voiding[5]. Therefore, it becomes crucial to analyze phenomenon of migration of vacancies and void growth due to SIV to assure LSI reliability. The computer aided simulation is considered to be a strong tool to clarify the behaviors of vacancies and predict the SIV lifetime. Many researchers have showed the finite elemental analysis to determine stress distribution in interconnects[3,6], however, there are no papers which concerns the simulation for vacancy migration and perdition of lifetime in SIV. Authors have proposed the computer aided numerical analysis for EM voiding by using the FDA combined with the FEA for stress distribution[7]. In this paper, we apply the proposed computer aided FDA to SIV to analyze vacancy migration and to predict SIV lifetime.

Residual stress distribution in Cu/Ta/SiO_2 and Cu/Ta/ULK with damascene structure obtained by the FEA was used to run FDA simulation to estimate vacancy concentration and SIV lifetime. To verify simulation results, Cu/Ta/SiO_2 and Cu/Ta/ULK damascene lines was fabricated and they were stored under high temperature (T=100-200°C) as the accelerated test of SIV.

2. Experimental

Thermal grown oxide was formed on Si. Conventional photolithography and dry etching were applied to form a single damascene structure. Cu/Ta was prepared by sputtering and electrochemical deposition was carried out to form Cu wiring. CMP was used to polish over grown Cu. Conventional annealing process was done at end of LSI fabrication. SIV acceleration test was carried out at 100-200°C for 24-48hrs. Resistance change of damascene lines was measured to evaluate failure time.

3. Numerical Analysis

3.1 Finite Element Analysis (FEA)

A difference in thermal expansion coefficients of wiring material and SiO_2 dielectric produces a residual stress in Cu interconnects after LSI fabrication. A single damascene structure line with Cu/Ta surrounded by SiO_2 or ULK dielectric, as shown in Fig.1, was considered. The three dimensional FEA (3D-FEA) by Marc/Mentat2005 was run to estimate stress distribution. The mesh drawing is shown in Fig.2. The mechanical properties of the materials, such as CTE, Young's modulus, and Poisson's ratio, are listed in

Table 1. Calculations were performed for the case of cooling from the initial stress free temperature, which is assumed to be 325°C, to room temperature (25°C).

3.2 Finite Difference Analysis (FDA)

The basic equation of vacancy diffusion is written as Eq. (1) with the addition of a term of stress diffusion.

$$\frac{\partial C}{\partial t} = D\nabla^2 C - \frac{D\Delta V}{RT}\nabla\cdot\left(C\nabla\sigma_p\right) = D\nabla^2 C - \frac{D\Delta V}{RT}\left(\nabla C\cdot\nabla\sigma_p + C\nabla^2\sigma_p\right) \tag{1}$$

$$D = D_0\exp\left(-\frac{Q}{RT}\right) \tag{2}$$

where C: vacancy concentration, D: diffusion coefficient, σ_p: hydrostatic stress, R: universal gas constant, T: absolute temperature, t: time, ΔV: partial molar volume of vacancy in a metal considered, D_0: diffusion constant, Q: activation energy. To stimulate the realistic vacancy diffusion and concentration, we showed that Eq.(3) should be adopted as the basic equation of vacancy diffusion in which weight coefficients α are added to Eq.(1)[9].

$$\frac{\partial C}{\partial t} = D\left(\frac{\partial^2 C}{\partial x^2} + \frac{\partial^2 C}{\partial y^2}\right) - \alpha\frac{D\Delta V}{RT}\left\{\left(\frac{\partial C}{\partial x}\frac{\partial\sigma_p}{\partial x} + \frac{\partial C}{\partial y}\frac{\partial\sigma_p}{\partial y}\right) + C\left(\frac{\partial^2\sigma_p}{\partial x^2} + \frac{\partial^2\sigma_p}{\partial y^2}\right)\right\} \tag{3}$$

Numerical analysis on hydrogen diffusion based on mass transportation theory was performed by authors[8]. The similar technique has been applied to demonstrate the distribution of vacancy concentration. The FDA is designed based upon the Crank-Nicolson method. The obtained simultaneous equations were solved by the SOR method. Analytical model is set to be 2 dimensional cross section of Cu/Ta line. A total number of grids of FDA is 9928. Values of hydrostatic stress calculated by FEA were interpolated to grids of FDA. The physical properties of the materials are listed in Table 2. This method is able to extend to a dual damascene structure [9]. Even though shape of corner becomes rounded, the modified coordination system to orthogonal from x-y can be applied [10].

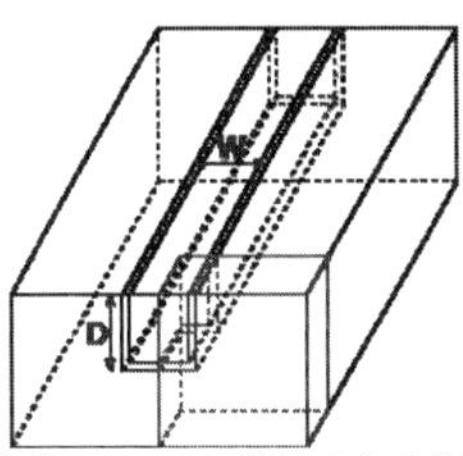

Fig.1 Single Damascene Line Model. Numerical analysis was carried out based up this model.

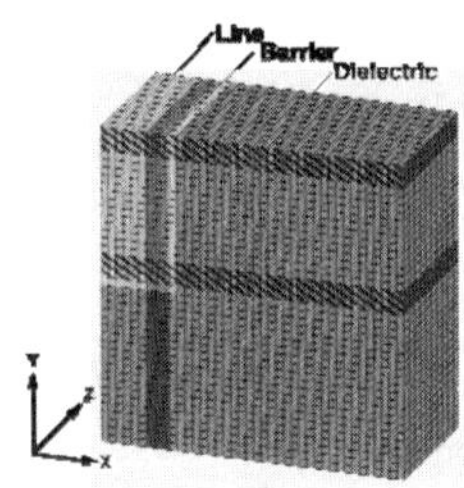

Fig.2 Mesh Drawing of 3D-FEA. Residual stress was estimated by 3D-FEA on this mesh model.

Table1 Mechanical properties of the materials used in the FEA[*11].

Properties	Materials	CTE (ppm/K)	Modulus (GPa)	Poisson' s Ratio
Line*	Cu	17.7	129.8	0.343
Barrier*	Ta	6.5	185.7	0.342
Dielectric 1*	SiO_2	0.51	71.7	0.16
Dielectric 2	Ultra low-κ	40	4	0.29

Table.2 Physical properties of the materials used in the FDA[12-14].

	D_0 [m²/s]	Q [J/mol]	R [J/k mol]	ΔV [m³/mol]
Cu	0.62×10^{-4}	2.0736×10^{5}	8.31	2.0×10^{-6}
Ta	0.12×10^{-4}	4.1296×10^{5}		

4. Results and Discussions

4.1 Experimental results

Results of resistance change after SIV accelerated test were show in Figs.3, 4. Fig.3 shows a relationship between resistance change and storage temperature at 48hrs. The maximum resistance change was found at 150˚C. Fig.4 shows resistance change of Cu line as a function of time at 150˚C. As increasing storage time, resistance change increased and 10% of resistance change, which is defined as a failure, achieved at approximately 24hrs.

4.2 Estimation of thermal stress distribution in a Cu line

The hydrostatic stress distributions of Cu/Ta/SiO_2 and Cu/Ta/ULK at 150˚C are shown in Fig.5(a) and (b), respectively. The large arrow, which points from the 3-D schematic drawing to the FEA stress level profile, identifies the cross-sectional view displayed. Tensile hydrostatic stress was found to appear in the Cu/Ta/SiO_2 and Cu/Ta/ULK line. The maximum tensile stress of Cu/Ta/SiO_2, σ_p=260MPa, was found at the side of line near the bottom corner, while the maximum tensile stress of Cu/Ta/ ULK was only σ_p=154MPa. This result revealed that tensile hydrostatic stress is produced by the difference in thermal expansion coefficients and Young's modulus of SiO_2 and ULK. The maximum hydrostatic stresses estimated from 3D-FEA plotted against temperature are shown in Fig.6. Cu/Ta/SiO_2 shows higher maximum hydrostatic stress than Cu/Ta/ULK when temperature is low than 200˚C.

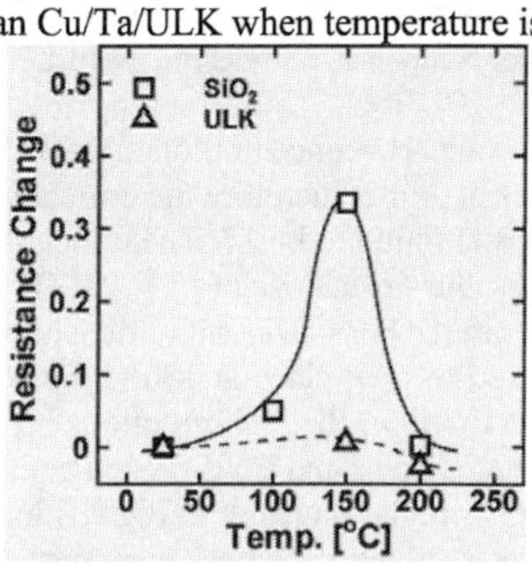

Fig.3 Resistance change of Cu/Ta/SiO_2 and Cu/Ta/ULK at each temperature after 48 hrs.

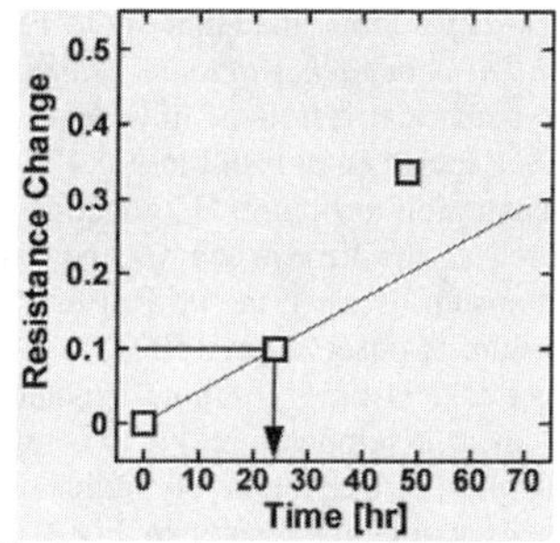

Fig.4 Resistance change of Cu/Ta/SiO_2 at 150˚C

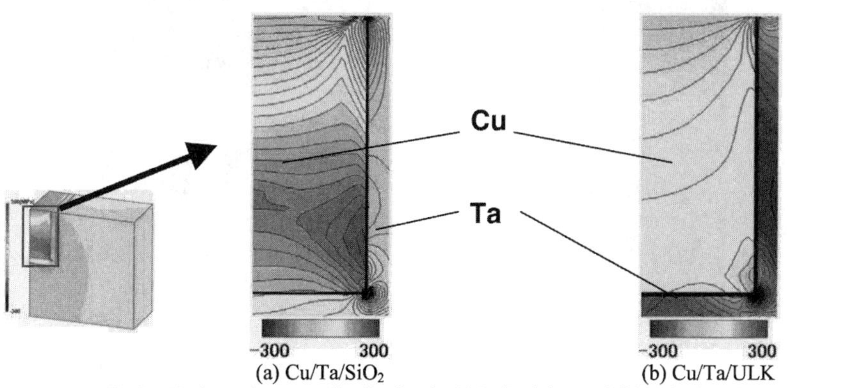

(a) Cu/Ta/SiO_2 (b) Cu/Ta/ULK

Fig.5 Hydrostatic stress distribution in (a)Cu/Ta/SiO_2 and (b)Cu/Ta/ULK at 150˚C.

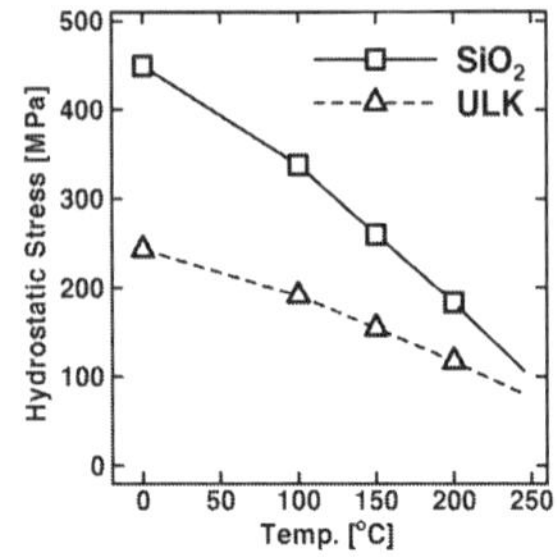

Fig.6 Relationship between hydrostatic stress and temperature. Cu/Ta/SiO_2 always shows higher maximum hydrostatic stress than Cu/Ta/ULK.

4.3 Distribution of vacancies in Cu/Ta line

Figures 7(a),(b) shows distribution of vacancies for 24hrs at 150°C. Vacancies accumulated at the side of Cu line near the bottom corner. The maximum vacancy concentration (C/C_0) is C/C_0=17, where C_0 is vacancy concentration in infinitely distance. The maximum vacancy concentration of Cu/Ta/ULK (C/C_0=12) is lower than that of Cu/Ta/SiO_2 because of lower hydrostatic stress in Cu/Ta/ULK.

4.4 The dependence of temperature on SIV failure

The numerical results of time sequential characteristic of vacancy concentration at 150°C were shown in Fig.8. Experimental result in Fig.4 shows that failure occurs at 24hrs, therefore the critical value of vacancy concentration to failure, C^*_{crit}, can be taken at 24hrs and it was found to be 17. The critical time to failure is defined as the time when vacancy concentration becomes the critical value (C^*_{crit}=17). Figure 9 shows relationship between the critical time to failure and temperature. For Cu/Ta/SiO_2, the shortest critical time to failure is observed at 150°C, then SIV failure is expected to take place at 150°C. For Cu/Ta/ULK, the longer critical time to failure expected to be a longer SIV lifetime. SIV failure rate is reported to behave non-linear to temperature due to complex effects of hydrostatic stress and Cu diffusion coefficient[3-4], and the highest failure rate likely indicates at around 100-200°C. Simulation result in Fig.9 is in good agreement with experimental results in Fig.3 and this also supports that the proposed computer aided numerical analysis in this paper enables to predict SIV life time.

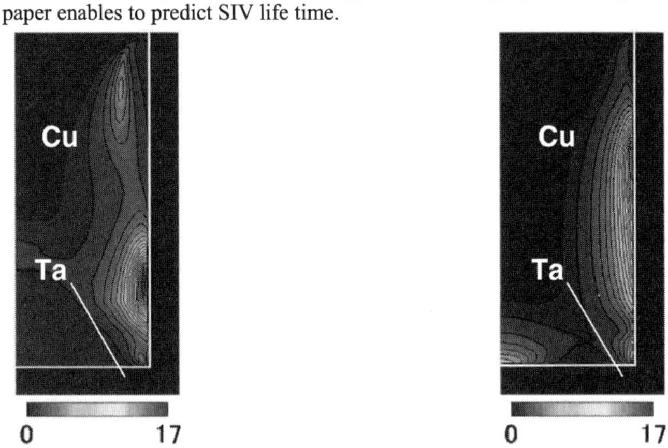

(a) Cu/Ta/SiO_2 (b) Cu/Ta/ULK

Fig.7 Distribution of vacancies of (a)Cu/Ta/SiO_2 and (b)Cu/Ta/ULK

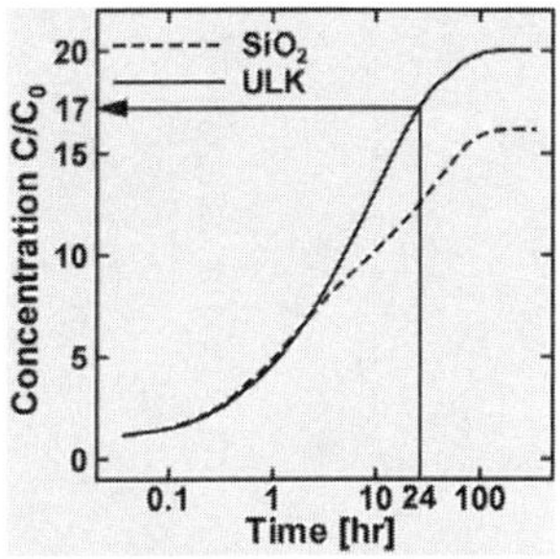

Fig.8 The time sequential characteristic of vacancy concentration at 150˚C

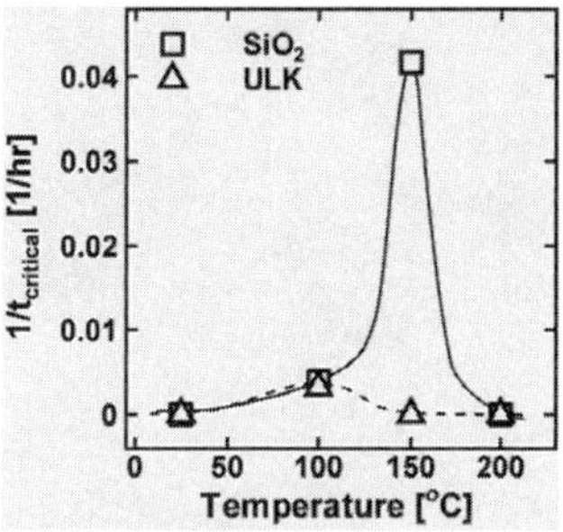

Fig.9 Relationship between the critical time and temperature. The minimum critical time is found at 150˚C.

5. Conclusion

SIV accelerated test used single damascene structure fabricated two kinds of dielectrics was carried out. To verify SIV test results, numerical analysis on the vacancy diffusion and concentration used FEA and FDA was conducted to predict SIV lifetime. The following conclusions were obtained.

1. The maximum resistance change of Cu/Ta/SiO_2 was found at 150˚C.
2. By FEA, Cu/Ta/SiO_2 shows higher maximum hydrostatic stress than Cu/Ta/ULK.
3. By FDA, Vacancy concentration of Cu/Ta/ULK is lower than that of Cu/Ta/SiO_2.
4. The complex effect of hydrostatic stress and diffusion coefficient on vacancy migration was clarified and the numerical analysis in this paper enables to predict SIV life time.

Reference

[1] J.Klema, R.Pyle, and E.Domangue, *1984 IEEE International Reliability Physics Symposium Proceedings 22th Annual*, (1984), pp.1-5.

[2] J.Curry, G.Fitzgibbon, Y.Guan, R.Muollo, G.Nelson, A.Thomas, *1984 IEEE International Reliability Physics Symposium Proceedings 22th Annual*, (1984), pp.6-8.

[3] J.W.McPherson and C.F.Dunn : *Journal of Vacuum Science and Technology B*, 5(1987), 1321-1325

[4] E.T.Ogawa, J.W.McPherson, J.A.Rosal, K.J.Dickerson, T.-CChiu, and L.Y.Tsung et al., *2002 IEEE International Reliability Physics Symposium Proceedings 40th Annual*, (2002), pp.312-321.

[5] T.Nemoto, T.Nogami, *J. Jpn. Soc. Strength and Fracture of Materials*, 40, 3, (2006) pp.67-73.

[6] Y.K.Lim, K.L.Pey, P.S.Lee, N.R.Kamat, J.B.Tan, Thomas Fu and L.C.Hsia, *2007 IEEE International Reliability Physics Symposium Proceedings 45th Annual*, (2007), pp.134-140.

[7] T.Nemoto, A.T.Yokobori,Jr. and T. Murakawa, *2007 IEEE International Reliability Physics Symposium Proceedings 45th Annual*, (2007), pp.634-635.

[8] A.T. Yokobori, Jr, T. Nemoto, K. Satoh and T. Yamada, *Eng. Fract. Mech.* 55, 1, (1996), pp.47-60.

[9] H. Ito, A.T. Yokobori, Jr, *Trans. of the 42th Ann. meet. of the Tohoku Branch in Japanese Soc. for Mech. Engineers, No.2007-1*, (2007), pp.50-51, (in Jappanese)

[10] T. Murakawa, A.T. Yokobori, Jr, T.Nemoto, *J.Japan Inst. Materials*, 70, 12 (2006), pp.987-994, (in Jappanese)

[11] J.M.Paik, H.Park, Y.C.Joo and K.C.Park : *Journal of Applied Physics*, 97(2005), 104513(1-7).

[12] K. Monma, *Kinzokubinran* (Handobook of Metals), Edited by The Japan Institute of Metals, (Maruzen, Tokyo, 1971), pp.167, (in Jappanese)

[13] J.R.Lloyd, *Journal of Physics D, Applied Physics*, 32(1999),R109-R118

[14] H.P.V.Leeuwen, *Engineering Fracture Mechanics*, 9(1977), 951-974

AUTHOR INDEX

SUBJECT INDEX